Stripline and Microstrip Resonators and Bandpass Filters

Alexander Zakharov

Sergii Litvintsev

WILEY

Published by John Wiley & Sons, Inc., Hoboken, New Jersey.

Library of Congress Cataloging-in-Publication Data:

Names: Zakharov, Alexander author | Litvintsev, Sergii author
Title: Stripline and microstrip resonators and bandpass filters / Alexander Zakharov, Sergii Litvintsev.
Description: Hoboken, New Jersey : Wiley, [2026] | Includes bibliographical references and index.
Identifiers: LCCN 2026011277 (print) | LCCN 2026011278 (ebook) | ISBN 9781394374984 cloth | ISBN 9781394375042 adobe pdf | ISBN 9781394375028 epub | ISBN 9781394375066 ebook
Subjects: LCSH: Electric filters, Bandpass | Electric resonators | Wireless communication systems
Classification: LCC TK7872.F5 Z34 2026 (print) | LCC TK7872.F5 (ebook)
LC record available at https://lccn.loc.gov/2026011277
LC ebook record available at https://lccn.loc.gov/2026011278

Cover Design: Wiley
Cover Image: © Stock Image Zone/stock.adobe.com

Contents

Preface

Bandpass filters (BPFs) are essential to wireless communications and radar systems. The performance of these systems largely depends on them. For this reason, BPFs are subject to increased requirements, including increased frequency selectivity, phase linearization, and miniaturization. Most microwave BPFs operating at frequencies above 300 MHz use resonators made of transmission line (TL) segments of various types. Many articles have been devoted to the development of microwave BPFs in recent years, but their comprehensive coverage is not possible due to the limited volume and required knowledge.

The book is focused on resonators and BPFs based on stripline and microstrip TL sections, which, according to the authors, are very promising. Using substrates with different dielectric constants $\varepsilon_r = 2.2 - 100$ enables such filters to operate in a wide range of wavelengths, including meter and millimeter waves. The use of high dielectric constant substrates provides filters with significant miniaturization. The dielectric loss of the materials used, $\tan\delta = (1 - 5) \times 10^{-4}$, is significantly lower than that of dielectrics used in low-temperature co-fired ceramic technology. This provides stripline resonators with a higher unloaded quality factor. Implementing thin (1 mm or less) stripline BPFs is relatively easy, which is problematic for other technologies. The use of thick (2 mm or more) substrates and a low-mounted screen allows the implementation of compact, fully shielded microstrip BPFs with a narrow bandwidth of 1–3%. Thermally stable dielectric substrates provide such BPFs with increased frequency stability.

The book systematizes materials to develop the theory and practice of constructing advanced stripline and microstrip BPFs. Most of the material contained in the book is the result of the authors' research. The most significant results include the following:

New approach to forming transmission zeros in BPF: Transmission zeros (TZs) are introduced into the transfer function to improve the filter's performance. For this purpose, cross-couplings and mixed couplings between resonators are used, but the resonators' role is ignored. Recently, an approach that consists of forming TZs by the resonators themselves has begun to be used.

This approach is noteworthy because it significantly increases the TZ number achieved using cross- and mixed couplings.

Synthesis of nonuniform TL resonators and their optimization: Nonuniform TL resonators, which include stepped-impedance resonators (SIRs) and resonators with stubs, have a significant advantage over uniform TL resonators. This is because the mutual arrangement of their critical frequencies (resonant and antiresonant) can be regulated by changing the shape of the resonators. The synthesis of such resonators, which consists of the transition from predetermined critical frequencies to the function of changing the characteristic impedance along the length, is practically not considered in the literature. In the book, this gap is eliminated. The optimization process of such resonators, expressed by parametric and structural-parametric synthesis methods, is also presented. The synthesis technique of nonuniform TL resonators dramatically simplifies the creation of BPF with several passbands, which has recently received attention.

Sensitivity function of resonance frequencies in analyzing and synthesizing tunable TL resonators: The proposed sensitivity function is the circuit function expressed through its critical frequencies. It allowed us to establish limitations on the tunability of resonators by one or more lumped elements and to obtain exact expressions for determining the unloaded quality factor of tunable resonators. As a result, resonators with improved tunability were synthesized.

BPF with mixed couplings in general for the first time: A general solution to the direct and inverse problems for mixed coupled BPFs is given and described systematically for the first time. An analytical description of triplet and quadruplet BPFs with all mixed couplings is given. A design method of arbitrary order BPFs with mixed couplings between adjacent resonators is established. Compact two-level designs of mixed coupled BPFs are proposed. In mixed coupled BPF, a quadrant arrangement of four TZs on the complex plane is implemented (for delay line filters).

New dual-mode resonators and BPFs: New dual-mode resonators and BPFs are considered. All ends are short-circuited and practically applicable. Analysis of such resonators based on equations for even and odd waves is impossible, and other equations are used. The functional capabilities of dual-mode resonators are expanded using TL SIRs. Meander dual-mode resonators, which are characterized by only one dimension, in contrast to existing two-dimensional TL dual-mode resonators, are considered for the first time.

Invariant of the coupling coefficient in stripline BPF: Authors have shown that coupling coefficient K between stripline resonators depends not on the dielectric constant ε_r of the substrates forming the filter (invariant K) but on their geometric parameters. The value of ε_r only moves the frequency response on the frequency axis. This pattern is helpful since it allows us to extend a filter with suitable parameters, once developed to other frequencies.

Implementing compact narrow-band microstrip BPFs: This possibility appears if a thick substrate, a low-mounted screen, and stepped-impedance resonators are used. The thick substrate provides the resonators with an increased unloaded quality factor, and closely located SIRs provide a small coupling coefficient without increasing the filter's transverse dimensions.

Although this book is explicitly written for stripline and microstrip BPFs, many of its concepts apply to BPFs with other types of TLs.

A distinctive feature of this book is the presentation of many new experimental filter configurations with a description of their design sequences and technical characteristics. It extensively uses electromagnetic and linear computer simulation.

The book is intended for many readers, including R&D engineers, professionals, academics working on microwave filters, students, and postgraduates. In addition to new scientific and technical results, the book contains general information on filter theory. This makes it self-contained and accessible for students.

1

Introduction

Bandpass filters (BPFs) are one of the main elements of wireless communications and radar systems. They perform the function of frequency selection of signals. The rapid development of wireless communications and radar systems places increasing demands on BPF parameters. It includes higher selectivity, more considerable stopband attenuation, lower passband loss, wider electronic frequency tuning range, more linear phase and flat group delay, smaller size, and lower cost. A BPF is a set of resonators interconnected by electromagnetic couplings, and the end resonators are also connected to loads. Microwave BPFs operate at frequencies above 300 MHz, and a significant part contains resonators from transmission line (TL) sections such as coaxial, stripline, microstrip, waveguide, coplanar waveguide, and slotline.

The preference given in the book to stripline and microstrip BPFs is due to several reasons. Stripline BPFs are formed by two dielectric substrates, between which a metal layer with a topological pattern is located. The development of these filters was slowed down for many years by the absence of electromagnetic coupling between the resonators in combline and pseudocombline structures. However, this is not so; there is a coupling between the resonators in the specified structures, which are BPFs. Moreover, the potential capabilities of stripline filters are quite high due to the extensive arsenal of manufactured substrates with various dielectric constants $\varepsilon_r = 2.2–100$.

Compared to monoblock ceramic filters, in which the topological conductive pattern is located on a narrow edge of a ceramic block with some metalized holes, stripline filters have an advantage since their topological pattern occupies a significant area, expanding the possibilities of choosing the best option. The use of thin substrates allows the creation of thin filters, the thickness of which is 1 mm or less, which cannot be realized in monoblock designs.

Microstrip BPFs are notable because their resonators are easily connected to lumped elements, which are small-sized, commercially available capacitors and inductors. The use of thick substrates and a low-mounted shield allows the creation of compact, fully shielded microstrip filters with a bandwidth of 1–3%, which is most in demand in communication systems. The choice of stripline and microstrip BPFs is also because they can be implemented on high dielectric constant substrates, which provide significant miniaturization. It is also essential that these filters can be manufactured in a batch method using single-sided and double-sided photolithography.

Over the past decade, stripline and microstrip resonators and BPFs have undergone significant developments in the theory and practice of their design, resulting in many papers in scientific and technical journals. The main objective of this book is to systematize new results published primarily by the authors. Due to space limitations and the knowledge required, this book does not attempt to cover newly published results on the wide variety of microwave BPFs. An analysis of the simplest meander dual-mode resonators was carried out, which, unlike other two-dimensional dual-mode resonators, are characterized by only one dimension. Although the book presents theories and design methods applicable to stripline and microstrip BPFs, the presentation applies to other types of TL filters.

Chapter 2 contains the basic concepts of the electrical filters theory. A BPF is a set of electromagnetically coupled resonators. At the same time, a BPF can be considered an electrical circuit (two-port network) with special properties. This chapter defines the basic parameters of a two-port network and the relationship between these parameters.

Chapter 3 introduces the basic concepts related to inline BPFs with parallel-type resonators and simple couplings. In the coupling structure of BPF, such resonators are represented by susceptance slope parameters, and coupling elements are represented by admittance inverters. Many BPFs without cross-couplings are considered within this coupling structure, including BPFs with stripline or microstrip designs. The parameters presented in this chapter include the coupling coefficient between resonators and the external quality factor of end resonators, which determines the filter's matching with loads. The classical method of G. Matthaei is used to design such filters. This design method is also used in subsequent chapters of the book. Appropriate additions are made to this method as the BPFs under consideration become more complex.

Chapter 4 discusses the influence of dissipative loss on the BPF frequency responses. The concept of the unloaded quality factor of resonators and a method for measuring it are defined. Physical factors determining the unloaded quality factor of microstrip and stripline resonators and ways to increase this factor are considered.

Chapter 5 is devoted to TL resonators, the main elements of microwave BPFs. Primary attention is paid to nonuniform TL resonators, which include stepped-impedance resonators (SIRs). In these resonators, unlike uniform resonators, it is possible to provide the required mutual arrangement of critical frequencies (resonant and antiresonant), which gives them a significant advantage and allows for achieving various positive effects in BPF. The specified arrangement of critical frequencies is adjusted by changing the resonator shape, in which its characteristic impedance changes along its length according to a specific law. Also, stubs can be connected to the resonator.

The general patterns of the critical frequency arrangement of such resonators and the synthesis problem consisting of the transition from critical frequencies to the function of characteristic impedance are considered. In stripline and microstrip TLs, the boundaries of change in the characteristic impedance (the largest and smallest values) are limited. This leads to an optimization problem caused by the desire to obtain the required arrangement of some critical frequencies with a minimum change in the characteristic impedance. If resonators do not contain stubs, then the parametric synthesis method is used for their optimization. The structural-parametric synthesis method is used to optimize resonators with stubs. The optimization methods considered in this chapter are used in subsequent Chapters 9, 10, 14, 15, and 17.

Chapter 6 is devoted to BPF with cross-couplings. Using cross-couplings in BPF introduces transmission zeros (TZs) into their transfer function, which generally are complex frequencies. Introducing TZs allows us to obtain higher frequency selectivity, more significant attenuation in the rejection band, more linear phase, and flat group delay in the filter. Cross-couplings are the most traditional way of introducing TZs into the BPF transfer function. The conventional coupling matrix is widely used to describe such filters. This chapter shows the significant difference between reflective-type and through-type resonators with two open ends, through which coupling with neighboring resonators is carried out. Replacing a reflective-type resonator with a through-type resonator sometimes leads to a significant change in the BPF properties. In this case, it is necessary to modify it to obtain reliable frequency responses using the coupling matrix.

Chapter 7 discusses BPF with mixed couplings as the sum of magnetic and electric components assigned plus and minus signs, respectively. Unlike simple couplings, which can be positive or negative, mixed couplings can have two signs and take a zero value. Using mixed couplings allows us to obtain TZs in BPF without cross-couplings. BPFs with mixed couplings have one more TZ than filters with simple couplings for a given number of resonators. This difference is most noticeable in filters with a small number of resonators. The main advantage of mixed coupled BPFs is the many options for the TZ location, making implementing a significant variety of frequency responses possible. Thus, a third-order BPF with mixed cross-coupling has two TZs with 10 possible locations on the complex plane. This chapter presents a general solution to the direct and inverse problems of mixed coupled BPFs. A detailed analysis of triplet BPF with mixed cross-coupling, quadruplet BPF with mixed cross-coupling, and triplet BPF with all mixed couplings was also carried out.

Chapter 8 is devoted to describing mixed coupled, high-order inline and cascaded filters. It presents a new method for designing BPFs of arbitrary order with mixed couplings between adjacent resonators. An analytical method for designing cascade connection of triplet and quadruplet BPFs with mixed cross-coupling is also proposed. It is based on the exact solution of the inverse problem for a single triplet and quadruplet BPF. It is complemented by optimizing the coupling coefficients and the external quality factor of the end resonators for composite filters. For the first time in mixed coupled BPF, a quadrant arrangement of four TZs on the complex plane is implemented for delay line filters.

Chapter 9 introduces resonators with unique properties in which one or two antiresonant frequencies are close to the operating resonant frequency. Their use in BPFs with parallel-type resonators increases the TZ number; each antiresonant frequency of the resonator corresponds to a TZ at a real frequency. These resonators can be lumped-distributed or contain only TL segments. A significant part of the presented resonators are synthesized by the parametric synthesis method described in Chapter 5. The properties of these resonators are described, including the permissible convergence degree of the specified critical frequencies for the implemented parameters of the TL segments and the values of commercially available miniature capacitors and inductors.

Chapter 10 discusses filters with simple couplings and multiple TZs generated by resonators. The two approaches to forming BPF TZs when using cross and mixed couplings, discussed in Chapters 6 and 7, are based on a single concept associated with the physical process of electromagnetic interaction between two or more resonators. When using this concept, the role of resonators in TZ formation is ignored. More recently, another idea has begun to be used. It is based on a physical process associated with the propagation of an electromagnetic wave along the resonator. This process differs significantly from electromagnetic interaction between two or more resonators. In this case, the electromagnetic wave propagation can be controlled by nonuniform TL resonators and serial connection of lumped elements to uniform or nonuniform resonators. The specified control leads to different variants of mutual arrangement of critical (resonant and antiresonant) frequencies of resonators.

Such filters have unique properties. They efficiently implement higher-order TZs in equidistant pairs relative to the filter's central frequency. The nominal values of small-sized capacitors and inductors regulate the TZ positions. The unloaded quality of these elements does not affect the filter's insertion loss at the central frequency. With a small number of resonators N in the filter, the number of its TZs can equal $2N$. The design sequence of such filters includes the design method of G. Matthaei and analytical expressions that allow the implementation of the initially specified TZ location.

Chapter 11 is devoted to BPFs with mixed couplings and multiple TZs generated by resonators. Cross-couplings between resonators are not used in these filters. To date, such filters have the most significant number of TZs. If a filter contains N resonators with two antiresonant frequencies close to the operating resonant frequency, then the TZ number can equal $(3N + 1)$. The properties of such filters are considered. In the design of such filters, analytical expressions are used to allow the implementation of a predetermined TZ arrangement and the classical design method of G. Matthaei.

In Chapter 12, stripline BPFs are presented systematically for the first time, the use of which was previously limited by the assumed absence of coupling between resonators in combline and pseudocombline structures. Two dielectric substrates form these filters, between which a metal layer with a topological pattern is located. Due to the extensive arsenal of manufactured substrates with various dielectric constants $\varepsilon_r = 2.2$–100, these filters can operate in a wide frequency range, covering decimeter and millimeter waves while being completely shielded. It is shown that in combline and pseudocombline structures, there is electromagnetic coupling between stripline quarter-wave and half-wave resonators, and these structures are BPFs. The coupling coefficients between stripline resonators can be simple and mixed, adjustable over a wide range, which allows for a significant variety of frequency responses.

It turned out that the coupling coefficient K between stripline resonators does not depend on the dielectric constant ε_r of the substrates forming the filter (invariant K) but depends only on their geometric parameters. The value of ε_r only moves the frequency response on the frequency axis. This pattern is proper since it allows us to extend a once-developed filter with suitable parameters to other frequency ranges. Stripline BPFs containing two substrates with different ε_r are also considered; such a design provides certain advantages. Invariant K is established for this case. The analysis showed that the thickness of the metal layer significantly affects the simple and mixed coupling between stripline resonators, which should be considered at the BPF design stage. Many experimental samples of stripline BPFs are presented with their characteristics, including a ninth-order BPF, a fourth-order BPF measuring 7.4 mm × 4.2 mm × 2 mm at 1.85 GHz, and a thin filter with a thickness of 1 mm. Electromagnetic simulators are used in designing such filters, and using substrates with a high dielectric constant leads to miniature sizes of stripline BPFs. The advantages of the group technology used to produce stripline BPFs are very promising.

Chapter 13 focuses on compact microstrip BPFs and the resonators they use. Compactness is the close arrangement of resonators to each other, in which a significant part of the substrate is metalized. At the same time, sufficiently wide resonators are used, which provides them with an increased value of unloaded quality factor Q_u. Using soft and hard substrates with different ε_r values allows microstrip filters to operate in a wide frequency range. Experimental samples of filters of the third, fourth, and higher orders are presented, including miniature filters on substrates with high permittivity.

Particular attention is paid to compact BPFs on thick substrates and a low-mounted screen, which contain SIRs. Regularities of the mixed coupling coefficient change, and its magnetic and electric components between closely spaced SIRs on thick substrates and a low-mounted screen are presented. These regularities allow the creation of compact, fully shielded, narrow-band microstrip BPFs with various frequency responses. Increasing the substrate thickness increases the unloaded quality factor Q_u of the microstrip resonator, and using SIRs does not increase the transverse BPF dimensions. These regularities and the low-mounted screen make microstrip filters competitive in implementing narrow bandwidths of 1–3%, which are most in demand in many wireless communication systems.

The problem of increasing the stopband of BPFs has been relevant for many years. There are several methods for solving it. Chapter 14 discusses some unpopular and little-known methods allowing wide stopbands exceeding a decade. It is shown that it is theoretically possible to eliminate all parasitic passbands in a BPF from TL segments, resulting in a filter with a single passband.

Chapter 15 introduces TL resonators tuned in frequency by changing capacitances and inductances. The sensitivity function of the resonant frequency to a change in capacitance or inductance is considered. It is a circuit function similar to the input admittance and is expressed through the critical frequencies of the resonator. At the same time, it characterizes the local tunability of the resonator, i.e., its ability to tune frequency. The introduction and use of the sensitivity function, which is a circuit function, made it possible to establish valuable patterns and relationships between the parameters of tunable resonators. In particular, theoretical limitations of the tunability were established for various options for frequency tuning: with one capacitance, with several capacitances, with the simultaneous use of several capacitances and inductances. These limitations are significant since they answer the questions of what can be expected and what should be strived for. They show that frequency tuning with the help of capacitance and inductance is much more effective than tuning with several capacitances. The introduction of this function makes the reason for a quarter-wave resonator's more effective capacitive frequency tuning obvious compared to a half-wave resonator. Using the sensitivity function, exact expressions are obtained to determine the unloaded quality factor of tunable resonators, which considers losses in the resonator and the tuning elements. Based on this function, TL resonators with improved tunability were synthesized. In some synthesized resonators with capacitive tuning, the superposition effect of resonance regions is observed, which is not typical for traditional resonators (e.g., combline), in which frequency intervals separate the resonance regions.

Chapter 16 is concerned with electronically tunable filters. Various filter options are considered. Among them are two narrow-band varactor-tuned filters, whose bandwidth by 3 dB level does not exceed 2.5%. One filter contains only hairpin resonators, and the second filter uses a combination of hairpin and combline resonators, which makes the frequency response more symmetrical. Combining hairpin and combline resonators is possible because their tunability is the same, and the resonators have synchronous tuning. A set of four filters with ferroelectric capacitors is also considered, which provides tuning in the frequency band from 1.78 to 4.33 GHz. Attention is also paid to the interdigital second-order BPF, which is tuned by smoothly changing the capacitance and switching inductance.

In the final Chapter 17, we discuss dual-mode resonators and BPFs based on them. A new property of many of these resonators is considered, consisting of alternating dual-mode and simple oscillations on the frequency axis. New dual-mode resonators are described, all ends of which are short-circuited, and their traditional analysis based on the equations for even and odd modes is impossible. For this, other equations are used. Dual-mode resonators containing stepped-impedance TL segments, which are characterized by extended functionality, are described. Reconfigurable loop-type dual-mode resonators and filters are also considered.

2

Basic Concepts of Electrical Filter Theory

Bandpass filters (BPFs) are designed for frequency selection of signals. They pass the useful signal in a particular frequency band and suppress other (interference) signals occupying other frequency bands. BPF is an electric circuit in the form of a two-port network (Figure 2.1), connected between the input Z_{01} and output Z_{02} loads. This figure shows the voltage and current variables V_i and I_i $(i = 1,\ 2)$ at ports *1* and *2*, respectively. Variables a_i introduce the incident waves, and variables b_i introduce the reflected waves. The quantities Z_{01} and Z_{02} are the terminal impedances, and E_s is the source voltage. The electric circuit in Figure 2.1 has been studied for a long time [1–4]. This section briefly examines the known provisions of the two-port network, which will be used as reference material in the following chapters.

2.1 *ABCD* Parameters

With the notations adopted in Figure 2.1, the *ABCD* parameters of the two-port network are determined by the equations

$$\begin{aligned} V_1 &= AV_2 - BI_2, \\ I_1 &= CV_2 - DI_2, \end{aligned} \tag{2.1}$$

or in the matrix form

$$\begin{bmatrix} V_1 \\ I_1 \end{bmatrix} = \begin{bmatrix} A & B \\ C & D \end{bmatrix} \cdot \begin{bmatrix} V_2 \\ -I_2 \end{bmatrix} \tag{2.2}$$

Parameters A and D are dimensionless; parameter B has the dimension of impedance; and parameter C has the dimension of admittance. If the network is lossless, then A and D are purely real, and B and C are purely imaginary.

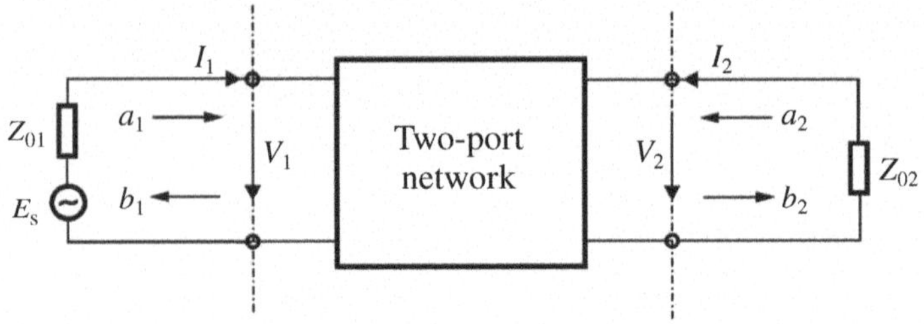

Figure 2.1 BPF as a two-port network.

The matrix in (2.2), comprising the *ABCD* parameters, is called the *ABCD* matrix. It is also known as the chain or transfer matrix.

If the input and output of the two-port network in Figure 2.1 are swapped, its transfer matrix will be replaced by

$$\begin{bmatrix} A_t & B_t \\ C_t & D_t \end{bmatrix} = \begin{bmatrix} D & B \\ C & A \end{bmatrix}. \tag{2.3}$$

If the two-port network is a reciprocal network, then

$$AD - BC = 1. \tag{2.4}$$

If the two-port network is a symmetrical network, then

$$A = D. \tag{2.5}$$

If a two-port network is symmetrical and its left half has $A'B'C'D'$ parameters, then the transfer matrix of the two-port network is written as

$$\begin{bmatrix} A & B \\ C & D \end{bmatrix} = \begin{bmatrix} 1 + 2B'C' & 2A'B' \\ 2C'D' & 1 + 2B'C' \end{bmatrix}. \tag{2.6}$$

The *ABCD* parameters for two cascaded two-port networks (Figure 2.2) are determined by multiplying their matrices:

$$\begin{bmatrix} A & B \\ C & D \end{bmatrix} = \begin{bmatrix} A_a & B_a \\ C_a & D_a \end{bmatrix} \begin{bmatrix} A_b & B_b \\ C_b & D_b \end{bmatrix} = \begin{bmatrix} A_aA_b + B_aC_b & A_aB_b + B_aD_b \\ C_aA_b + D_aC_b & C_aB_b + D_aD_b \end{bmatrix}. \tag{2.7}$$

By repeating the multiplication (2.7), we can determine the transfer matrix for any number of cascaded two-port networks. Table 2.1 contains the *ABCD* parameters of the most common two-port networks.

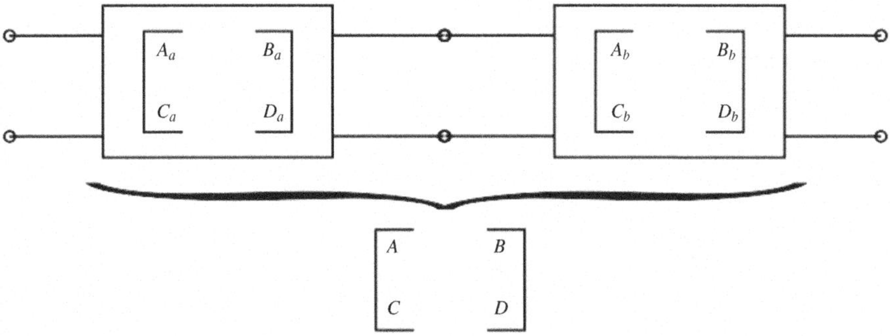

Figure 2.2 Cascade connection of two two-port networks.

Table 2.1 *ABCD* parameters of some two-port networks.

No.	Schematic	*ABCD* parameters
1	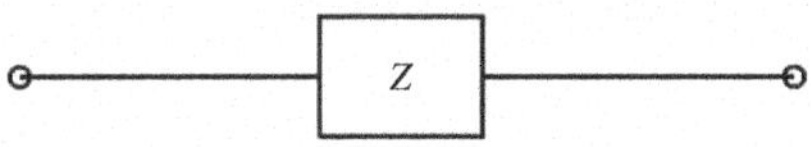	$A = 1, B = Z,$ $C = 0, D = 1$
2	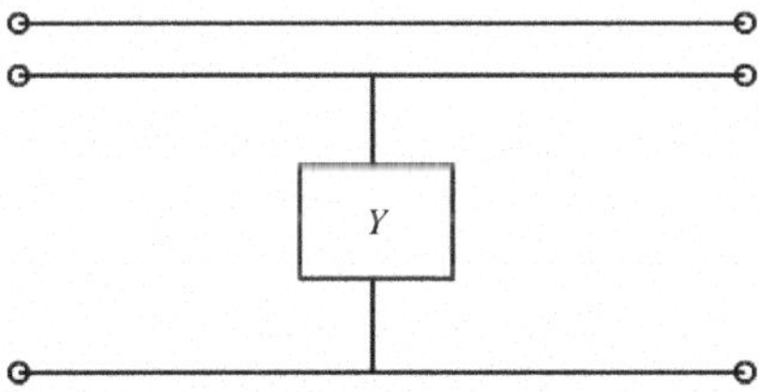	$A = 1, B = 0,$ $C = Y, D = 1$
3	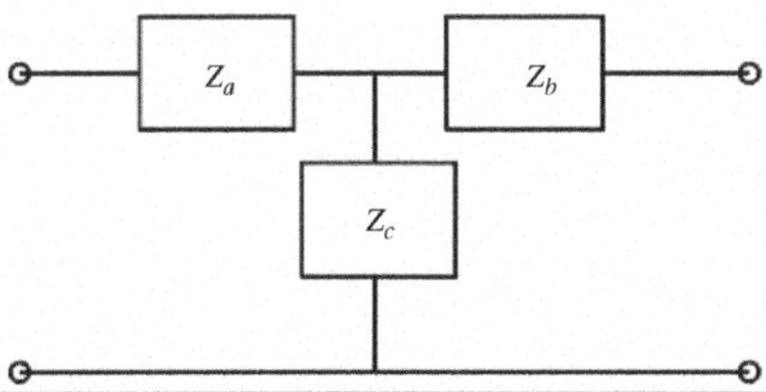	$A = 1 + \frac{Z_a}{Z_c},$ $B = Z_a + Z_b + \frac{Z_a Z_b}{Z_c},$ $C = \frac{1}{Z_c}, D = 1 + \frac{Z_b}{Z_c}$

(*Continued*)

Table 2.1 (Continued)

No.	Schematic	*ABCD* parameters
4	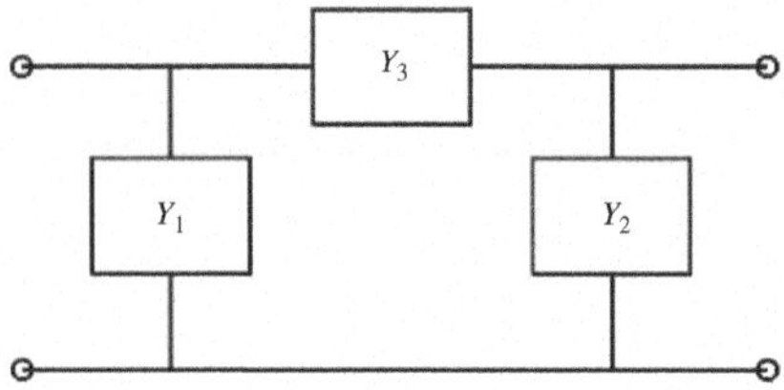	$A = 1 + \frac{Y_2}{Y_3}$, $B = \frac{1}{Y_3}$, $C = Y_1 + Y_2 + \frac{Y_1 Y_2}{Y_3}$, $D = 1 + \frac{Y_1}{Y_3}$
5	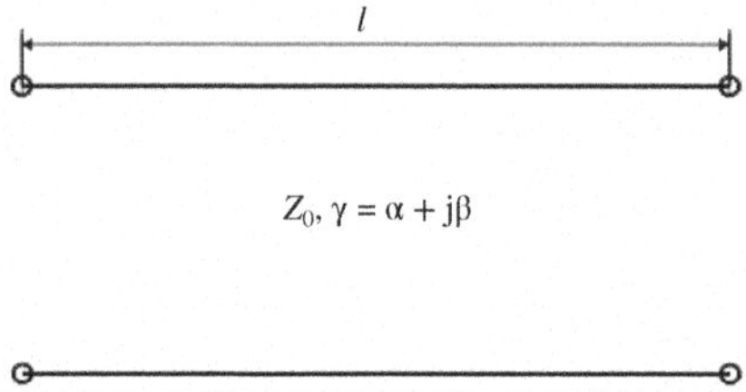	$A = \cosh(\gamma l)$, $B = Z_0 \sinh(\gamma l)$, $C = \frac{\sinh(\gamma l)}{Z_0}$, $D = \cosh(\gamma l)$

2.2 Open-circuit and Short-circuit Parameters

With the notations adopted in Figure 2.1, the open-circuit impedance or Z parameters of the two-port network are determined by the equations

$$\begin{aligned} V_1 &= Z_{11} I_1 + Z_{12} I_2 \\ V_2 &= Z_{21} I_1 + Z_{22} I_2 \end{aligned}. \tag{2.8}$$

The Z_{ij} $(i, j = 1, 2)$ parameters form the so-called open-circuit impedance or Z matrix, which is denoted by $[\mathbf{Z}]$. Parameter Z_{11} is the input impedance on the 1–1′ side when terminals 2–2′ are open. Parameter Z_{22} is the input impedance on the 2–2′ side when terminals 1–1′ are open. Parameter Z_{12} is the ratio V_1/I_2 when terminals 1–1′ are open. The meaning of parameter Z_{21} is explained similarly. If the two-port network is reciprocal, then $Z_{12} = Z_{21}$. If the network is symmetrical, $Z_{12} = Z_{21}$ and $Z_{11} = Z_{22}$. For a network without loss, all Z parameters are purely imaginary.

When analyzing filters, sometimes we need to express open-circuit impedances Z_{11} and Z_{22} through *ABCD* parameters. In this case

$$Z_{11} = A/C, \tag{2.9a}$$

$$Z_{22} = D/C. \tag{2.9b}$$

If two networks with impedance matrices $[\mathbf{Z}']$ and $[\mathbf{Z}'']$ are connected in series, then the $[\mathbf{Z}]$ matrix of the resulting network is written as

$$[\mathbf{Z}] = [\mathbf{Z}'] + [\mathbf{Z}'']. \tag{2.10}$$

With the notations adopted in Figure 2.1, the short-circuit admittance or Y parameters of the two-port network are determined by the equations

$$\begin{aligned} I_1 &= Y_{11}V_1 + Y_{12}V_2, \\ I_2 &= Y_{21}V_1 + Y_{22}V_2. \end{aligned} \tag{2.11}$$

The Y_{ij} $(i, j = 1, 2)$ parameters form the so-called short-circuit admittance or Y matrix, which is denoted by $[\mathbf{Y}]$. Parameter Y_{11} is the input admittance on the side 1–1′ when terminals 2–2′ are short-circuited. Parameter Y_{22} is the input admittance on the side 2–2′ when terminals 1–1′ are short-circuited. Parameter Y_{12} is the ratio I_1/V_2 for the case when terminals 1–1′ are short-circuited, and voltage V_2 is applied to terminals 2–2′. The meaning of parameter Y_{21} is explained similarly. If a two-port network is reciprocal, then $Y_{12} = Y_{21}$. If a network is symmetrical, $Y_{12} = Y_{21}$ and $Y_{11} = Y_{22}$. For a network without loss, all Y parameters are purely imaginary.

If we need to express the short-circuit admittance Y_{11} and Y_{22} regarding $ABCD$ parameters, we use

$$Y_{11} = D/B, \tag{2.12a}$$

$$Y_{22} = A/B. \tag{2.12b}$$

If two networks with admittance matrices $[\mathbf{Y}']$ and $[\mathbf{Y}'']$ are connected in parallel, then the $[\mathbf{Y}]$ matrix of the resulting network is written as

$$[\mathbf{Y}] = [\mathbf{Y}'] + [\mathbf{Y}'']. \tag{2.13}$$

Matrices $[\mathbf{Z}]$ and $[\mathbf{Y}]$ are mutually invertible [5]

$$[\mathbf{Y}] = [\mathbf{Z}]^{-1}. \tag{2.14}$$

2.3 Scattering Parameters

With the notations adopted in Figure 2.1, the scattering or S parameters of the two-port network are determined by the equations

$$\begin{aligned} b_1 &= S_{11}a_1 + S_{12}a_2, \\ b_2 &= S_{21}a_1 + S_{22}a_2, \end{aligned} \tag{2.15}$$

or in matrix form

$$\begin{bmatrix} b_1 \\ b_2 \end{bmatrix} = \begin{bmatrix} S_{11} & S_{12} \\ S_{21} & S_{22} \end{bmatrix} \begin{bmatrix} a_1 \\ a_2 \end{bmatrix}. \tag{2.16}$$

The matrix (2.16) comprised of the S_{ij} $(i,j = 1, 2)$ parameters is called the scattering matrix or S matrix and is denoted by $[S]$.

The parameters S_{12} and S_{21} are called the transmission coefficients, whereas S_{11} and S_{22} are the reflection coefficients. The advantage of the S parameters is that they can be measured directly at microwave frequencies. The S_{ij} $(i,j = 1, 2)$ parameters are complex and can be represented as a modulus $|S_{ij}|$ and a phase φ_{ij}. For lossless two-port network

$$|S_{11}|^2 + |S_{21}|^2 = 1. \tag{2.17}$$

2.4 Relationships Between Two-port Network Parameters

The two-port network parameters considered earlier are the functions of the complex variable $s = \sigma + j\omega$. There is a relationship between these parameters, which is shown in Tables 2.2–2.5 for the case $Z_{01} = Z_{02} = Z_0$ and $Y_0 = 1/Z_0$.

2.5 Immittance Transformation by Two-port Network

A two-port network transforms immittance. With the help of the transfer matrix, the transformation laws acquire the simplest form. Let us denote by Z_L the impedance of the load connected to terminals 2–2′ of a two-port network with

Table 2.2 *ABCD* parameters expression via other parameters.

Parameter	S	Y	Z
A	$\dfrac{(1+S_{11})(1-S_{22})+S_{12}S_{21}}{2S_{21}}$	$\dfrac{-Y_{22}}{Y_{21}}$	$\dfrac{Z_{11}}{Z_{21}}$
B	$Z_0\dfrac{(1+S_{11})(1+S_{22})-S_{12}S_{21}}{2S_{21}}$	$\dfrac{-1}{Y_{21}}$	$\dfrac{Z_{11}Z_{22}-Z_{12}Z_{21}}{Z_{21}}$
C	$\dfrac{1}{Z_0}\dfrac{(1-S_{11})(1-S_{22})-S_{12}S_{21}}{2S_{21}}$	$\dfrac{-(Y_{11}Y_{22}-Y_{12}Y_{21})}{Y_{21}}$	$\dfrac{1}{Z_{21}}$
D	$\dfrac{(1-S_{11})(1+S_{22})+S_{12}S_{21}}{2S_{21}}$	$\dfrac{-Y_{11}}{Y_{21}}$	$\dfrac{Z_{22}}{Z_{21}}$

Table 2.3 Z parameters expression via other parameters.

Parameter	S	$ABCD$	Y
Z_{11}	$Z_0\frac{(1+S_{11})(1-S_{22})+S_{12}S_{21}}{(1-S_{11})(1-S_{22})-S_{12}S_{21}}$	$\frac{A}{C}$	$\frac{Y_{22}}{Y_{11}Y_{22}-Y_{12}Y_{21}}$
Z_{12}	$Z_0\frac{2S_{12}}{(1-S_{11})(1-S_{22})-S_{12}S_{21}}$	$\frac{AD-BC}{C}$	$\frac{-Y_{12}}{Y_{11}Y_{22}-Y_{12}Y_{21}}$
Z_{21}	$Z_0\frac{2S_{21}}{(1-S_{11})(1-S_{22})-S_{12}S_{21}}$	$\frac{1}{C}$	$\frac{-Y_{21}}{Y_{11}Y_{22}-Y_{12}Y_{21}}$
Z_{22}	$Z_0\frac{(1-S_{11})(1+S_{22})+S_{12}S_{21}}{(1-S_{11})(1-S_{22}))-S_{12}S_{21}}$	$\frac{D}{C}$	$\frac{Y_{11}}{Y_{11}Y_{22}-Y_{12}Y_{21}}$

Table 2.4 Y parameters expression via other parameters.

Parameter	S	$ABCD$	Z
Y_{11}	$Y_0\frac{(1-S_{11})(1+S_{22})+S_{12}S_{21}}{(1+S_{11})(1+S_{22})-S_{12}S_{21}}$	$\frac{D}{B}$	$\frac{Z_{22}}{Z_{11}Z_{22}-Z_{12}Z_{21}}$
Y_{12}	$Y_0\frac{-2S_{12}}{(1+S_{11})(1+S_{22})-S_{12}S_{21}}$	$\frac{-(AD-BC)}{B}$	$\frac{-Z_{12}}{Z_{11}Z_{22}-Z_{12}Z_{21}}$
Y_{21}	$Y_0\frac{-2S_{21}}{(1+S_{11})(1+S_{22})-S_{12}S_{21}}$	$\frac{-1}{B}$	$\frac{-Z_{21}}{Z_{11}Z_{22}-Z_{12}Z_{21}}$
Y_{22}	$Y_0\frac{(1+S_{11})(1-S_{22})+S_{12}S_{21}}{(1+S_{11})(1+S_{22})-S_{12}S_{21}}$	$\frac{A}{B}$	$\frac{Z_{11}}{Z_{11}Z_{22}-Z_{12}Z_{21}}$

$ABCD$ parameters. Then, the input impedance Z_{in} at terminals 1–1′ is expressed by the equality

$$Z_{\text{in}} = \frac{AZ_L + B}{CZ_L + D}. \tag{2.18}$$

If admittance Y_L is connected to terminals 2–2′, then input admittance Y_{in} at terminals 1–1′ is expressed by the formula

$$Y_{\text{in}} = \frac{C + Y_L D}{A + Y_L B}. \tag{2.19}$$

If the transmission line segment (see Table 2.1) is used as a two-port network, then the input immittances are expressed by the corresponding equalities

$$Z_{\text{in}} = \frac{Z_L \cosh(\gamma l) + Z_0 \sinh(\gamma l)}{(Z_L/Z_0)\sinh(\gamma l) + \cosh(\gamma l)} = \frac{Z_L + Z_0 \tanh(\gamma l)}{1 + (Z_L/Z_0)\tanh(\gamma l)}, \tag{2.20}$$

Table 2.5 S parameters expression via other parameters.

Parameter	$ABCD$	Y	Z
S_{11}	$\dfrac{A+B/Z_0-CZ_0-D}{A+B/Z_0+CZ_0+D}$	$\dfrac{(Y_0-Y_{11})(Y_0+Y_{22})+Y_{12}Y_{21}}{(Y_0+Y_{11})(Y_0+Y_{22})-Y_{12}Y_{21}}$	$\dfrac{(Z_{11}-Z_0)(Z_{22}+Z_0)-Z_{12}Z_{21}}{(Z_{11}+Z_0)(Z_{22}+Z_0)-Z_{12}Z_{21}}$
S_{12}	$\dfrac{2(AD-BC)}{A+B/Z_0+CZ_0+D}$	$\dfrac{-2Y_{12}Y_0}{(Y_0+Y_{11})(Y_0+Y_{22})-Y_{12}Y_{21}}$	$\dfrac{2Z_{12}Z_0}{(Z_{11}+Z_0)(Z_{22}+Z_0)-Z_{12}Z_{21}}$
S_{21}	$\dfrac{2}{A+B/Z_0+CZ_0+D}$	$\dfrac{-2Y_{21}Y_0}{(Y_0+Y_{11})(Y_0+Y_{22})-Y_{12}Y_{21}}$	$\dfrac{2Z_{21}Z_0}{(Z_{11}+Z_0)(Z_{22}+Z_0)-Z_{12}Z_{21}}$
S_{22}	$\dfrac{-A+B/Z_0-CZ_0+D}{A+B/Z_0+CZ_0+D}$	$\dfrac{(Y_0+Y_{11})(Y_0-Y_{22})+Y_{12}Y_{21}}{(Y_0+Y_{11})(Y_0+Y_{22})-Y_{12}Y_{21}}$	$\dfrac{(Z_{11}+Z_0)(Z_{22}-Z_0)-Z_{12}Z_{21}}{(Z_{11}+Z_0)(Z_{22}+Z_0)-Z_{12}Z_{21}}$

$$Y_{\text{in}} = \frac{Y_0 \sinh(\gamma l) + Y_L \cosh(\gamma l)}{\cosh(\gamma l) + (Y_L/Y_0)\sinh(\gamma l)} = \frac{Y_L + Y_0 \tanh(\gamma l)}{1 + (Y_L/Y_0)\tanh(\gamma l)}, \tag{2.21}$$

where $Y_0 = 1/Z_0$.

2.6 Bandpass Filter Characteristics

The energy transfer from terminals 1–1′ to 2–2′ (see Figure 2.1) is characterized by the insertion loss function

$$L_A = -20 \log |S_{21}| \,[\text{dB}]. \tag{2.22}$$

We use the natural logarithm of the number A to the base 10: $\log A = \log_{10} A$. Since there is a relationship between the parameters S_{ij} and $ABCD$ (see Table 2.5), this function is also expressed by the formula

$$L_A = -10 \log \frac{4}{|A + B/Z_0 + CZ_0 + D|^2} \,[\text{dB}]. \tag{2.23}$$

A minimum insertion loss of constant value is sought in the passband (denoted as BW, bandwidth) of a BPF. This band should be combined with the frequency spectrum of the useful signal. In the BPF design, fractional bandwidth $\text{FBW} = \text{BW}/f_0$ is often used, where f_0 is the center frequency.

To the left and right of the passband, the BPF has rejection bands. These bands contain parasitic or interfering signals, for the suppression of which it is desirable to have sufficiently large insertion loss.

Between the BW and rejection bands, there are transition bands. The frequency selectivity of the filter is characterized by attenuation L_A, which is provided at a detuning value Δ (MHz) from the central frequency f_0:

$$\text{Right-sided selectivity} \quad L_A\,(f_0 + \Delta), \tag{2.24a}$$

$$\text{Left-sided selectivity} \quad L_A\,(f_0 - \Delta), \tag{2.24b}$$

$$\text{Both-sided selectivity} \quad L_A\,(f_0 \pm \Delta). \tag{2.24c}$$

The greater L_A and the smaller Δ in (2.24), the higher the filter selectivity. A reduction in the transition bands accompanies increasing filter selectivity.

With the direction of energy propagation indicated earlier, its reflection from terminals 1–1′ is characterized by the return loss function

$$L_R = 20 \log |S_{11}| \text{ [dB]}. \tag{2.25}$$

The energy reflection from the filter is also characterized by the voltage-standing wave ratio (VSWR), which is determined by the parameter S_{11}:

$$\text{VSWR} = \frac{1 + |S_{11}|}{1 - |S_{11}|}. \tag{2.26}$$

When a signal passes through a filter, it is delayed. An important parameter is the group delay, which is defined as

$$\tau_d = -\frac{d\varphi_{21}}{d\omega} \text{ [s]}. \tag{2.27}$$

For the signal to pass through the BPF without distortion, the values of L_A and τ_d must be constant over the passband.

When transmitting energy through the filter in the opposite direction from terminals 2–2′ to 1–1′ (see Figure 2.1), it is necessary to change the indices in the aforementioned expressions accordingly.

2.7 Chebyshev Low-pass Prototype Filter

Historically, low-pass filters were the first to be studied. It was found that their insertion losses can be described by various mathematical functions: Butterworth, Chebyshev, and Gaussian. The most widely used filters are those with a Chebyshev frequency response, which has equal-wave pulsations. They are called Chebyshev filters. The knowledge gained about low-pass filters was extended to high-pass, bandpass, and rejection filters. In this case, certain frequency transformations were used. The listed issues are considered in detail in Refs. [6–9]. The following is only the minimum information related to Chebyshev low-pass filters, sufficient for further consideration of BPFs.

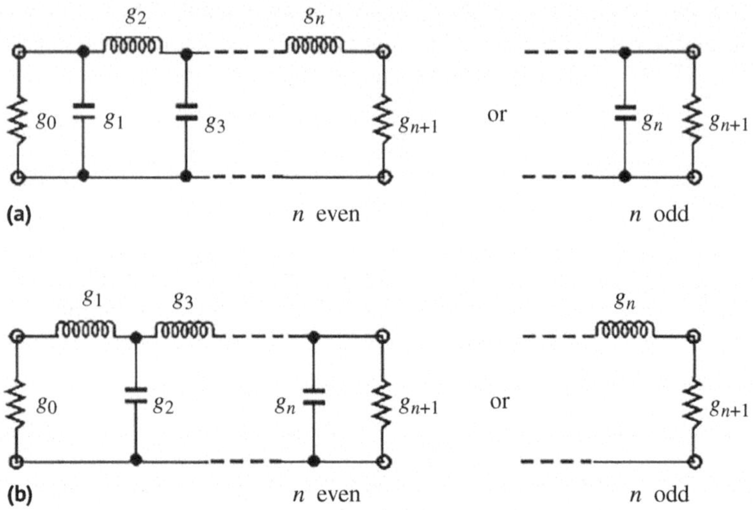

Figure 2.3 Low-pass ladder filters. (a) Element with g_1 parameter is capacitance. (b) Element with g_1 parameter is inductance.

Figure 2.3 shows two dual circuits of n-pole low-pass ladder filters. The parameters g_i $(i = 1, 2, \ldots, n)$ are the values of the reactive elements, where g_0 and g_{n+1} are the source resistance or conductance.

The parameters g_i $(i = 1, 2, \ldots, n)$ of reactive elements are normalized to make the source resistance or conductance g_0 equal to one $(g_0 = 1)$ and the cutoff angular frequency Ω_c to be equal to one $(\Omega_c = 1)$.

The task of synthesizing low-pass filters (see Figure 2.3) is to determine the g_i parameters for a given transfer function. The Chebyshev low-pass filter response with equal-ripple L_{Ar} in passband and cutoff frequency Ω_c is depicted in Figure 2.4. The insertion loss function is written as

$$L_A = 10 \log \left[1 + \varepsilon^2 T_n^2 (\Omega)\right], \tag{2.28}$$

where $T_n(\Omega)$ is the n-order Chebyshev function of the first kind

$$T_n(\Omega) = \begin{cases} \cos\left(n \cos^{-1} \Omega\right), & \Omega \le \Omega_c, \\ \cosh\left(n \cosh^{-1} \Omega\right), & \Omega \ge \Omega_c. \end{cases} \tag{2.29}$$

The ripple constant ε in (2.28) is defined by equality

$$\varepsilon = \left(10^{L_{\mathrm{Ar}}/10} - 1\right)^{1/2}. \tag{2.30}$$

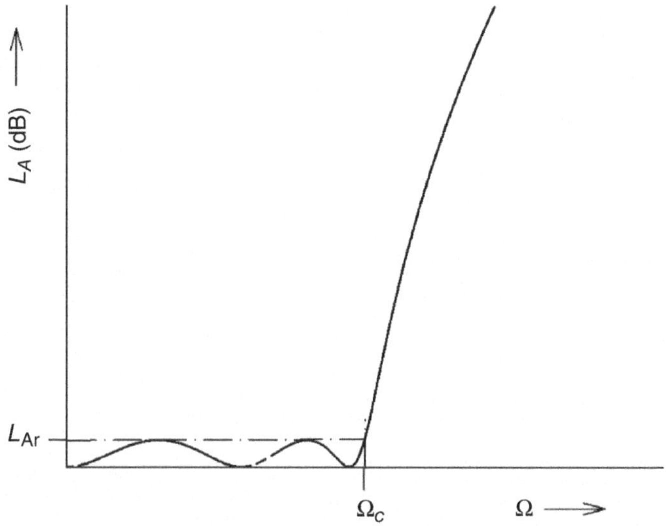

Figure 2.4 Chebyshev (equal-ripple) low-pass response.

For Chebyshev low-pass ladder filters (see Figure 2.3) having an insertion loss function (2.28) with a passband ripple L_{Ar} (dB) and the cutoff frequency $\Omega_c = 1$, the g_i parameters are calculated by the formulas [7]:

$$g_0 = 1, \quad g_1 = \frac{2a_1}{\gamma}, \quad g_i = \frac{4a_{i-1}a_i}{b_{i-1}g_{i-1}}, \quad i = 2, 3, \dots, n,$$

$$g_{i+1} = \begin{cases} 1.0, & n \text{ is odd}, \\ \coth(\beta/4), & n \text{ is even}, \end{cases} \tag{2.31}$$

where

$$\beta = \ln\left(\coth\frac{L_{Ar}}{17.37}\right), \qquad \gamma = \sinh\left(\frac{\beta}{2n}\right),$$

$$a_i = \sin\left(\frac{[2i-1]\,\pi}{2n}\right), \qquad i = 1, 2, \dots, n,$$

$$b_i = \gamma^2 + \sin^2\left(\frac{i\pi}{n}\right), \qquad i = 1, 2, \dots, n.$$

Table 2.6 presents the calculated g_i parameters of low-pass filters with different L_{Ar} values and the number of reactive elements $n = 1$–10.

Table 2.6 Element values for Chebyshev low-pass filter (Figure 2.3) for different n and pulsation level L_{Ar} ($g_0 = 1, \Omega_c = 1$).

L_{Ar}	n	g_1	g_2	g_3	g_4	g_5	g_6	g_7	g_8	g_9	g_{10}	g_{11}
0.01 dB	1	0.0960	1.0000									
	2	0.4488	0.4077	1.1007								
	3	0.6291	0.9702	0.6291	1.0000							
	4	0.7128	1.2003	1.3212	0.6476	1.1007						
	5	0.7563	1.3049	1.5773	1.3049	0.7563	1.0000					
	6	0.7813	1.3600	1.6896	1.5350	1.4970	0.7098	1.1007				
	7	0.7969	1.3924	1.7481	1.6331	1.7481	1.3924	0.7969	1.0000			
	8	0.8072	1.4130	1.7824	1.6833	1.8529	1.6193	1.5554	0.7333	1.1007		
	9	0.8144	1.4270	1.8043	1.7125	1.9057	1.7125	1.8043	1.4270	0.8144	1.0000	
	10	0.8196	1.4369	1.8192	1.7311	1.9362	1.7590	1.9055	1.6527	1.5817	0.7446	1.1007
0.1 dB	1	0.3052	1.0000									
	2	0.8430	0.6220	1.3554								
	3	1.0315	1.1474	1.0315	1.0000							
	4	1.1088	1.3061	1.7703	0.8180	1.3554						
	5	1.1468	1.3712	1.9750	1.3712	1.1468	1.0000					
	6	1.1681	1.4039	2.0562	1.5170	1.9029	0.8618	1.3554				
	7	1.1811	1.4228	2.0966	1.5733	2.0966	1.4228	1.1811	1.0000			
	8	1.1897	1.4346	2.1199	1.6010	2.1699	1.5640	1.9444	0.8778	1.3554		
	9	1.1956	1.4425	2.1345	1.6167	2.2053	1.6167	2.1345	1.4425	1.1956	1.0000	
	10	1.1999	1.4481	2.1444	1.6265	2.2253	1.6418	2.2046	1.5821	1.9628	0.8853	1.3554

(Continued)

Table 2.6 (Continued)

L_{Ar}	n	g_1	g_2	g_3	g_4	g_5	g_6	g_7	g_8	g_9	g_{10}	g_{11}
0.2 dB	1	0.4342	1.0000									
	2	1.0378	0.6745	1.5386								
	3	1.2275	1.1525	1.2275	1.0000							
	4	1.3028	1.2844	1.9761	0.8468	1.5386						
	5	1.3394	1.3370	2.1660	1.3370	1.3394	1.0000					
	6	1.3598	1.3632	2.2394	1.4555	2.0974	0.8838	1.5386				
	7	1.3722	1.3781	2.2756	1.5001	2.2756	1.3781	1.3722	1.0000			
	8	1.3804	1.3875	2.2963	1.5217	2.3413	1.4925	2.1349	0.8972	1.5386		
	9	1.3860	1.3938	2.3093	1.5340	2.3728	1.5340	2.3093	1.3938	1.3860	1.0000	
	10	1.3901	1.3983	2.3181	1.3417	2.3904	1.5536	2.3720	1.5066	2.1514	0.9034	1.5386
0.5 dB	1	0.6986	1.0000									
	2	1.4029	0.7071	1.9841								
	3	1.5963	1.0967	1.5963	1.0000							
	4	1.6703	1.1926	2.3661	0.8419	1.9841						
	5	1.7058	1.2296	2.5408	1.2296	1.7058	1.0000					
	6	1.7254	1.2479	2.6064	1.3137	2.4758	0.8696	1.9841				
	7	1.7372	1.2583	2.6381	1.3444	2.6381	1.2583	1.7372	1.0000			
	8	1.7451	1.2647	2.6564	1.3590	2.6964	1.3389	2.5093	0.8796	1.9841		
	9	1.7504	1.2690	2.6678	1.3673	2.7239	1.3673	2.6678	1.2690	1.7504	1.0000	
	10	1.7543	1.2721	2.6754	1.3725	2.7392	1.3806	2.7231	1.3485	2.5239	0.8842	1.9841

2.8 Relationship Between Transfer and Reflection Functions

The filter insertion loss L_A, which is a transmission function, is determined by the scattering parameter $|S_{21}|$. The return loss L_R and VSWR, which are reflection functions, are determined by the scattering parameter $|S_{11}|$. In the absence of loss, the parameters $|S_{21}|$ and $|S_{11}|$ are related by equality (2.17), which means that there is a one-to-one correspondence between the transmission and reflection functions. This allows the return loss L_R or VSWR to be used in Table 2.6 instead of the ripple level L_{Ar}.

The following formulas allow us to determine the other parameters if one of them is known. Table 2.7 shows the correspondence between the L_{Ar}, L_R, and VSWR values of lossless filter:

$$L_R = 10\log\left(1 - 10^{-L_{\text{Ar}}/10}\right)\ [\text{dB}], \tag{2.32}$$

$$L_{\text{Ar}} = -10\log\left(1 - 10^{L_R/10}\right)\ [\text{dB}], \tag{2.33}$$

$$\text{VSWR} = \frac{10^{-L_R/20} + 1}{10^{-L_R/20} - 1}, \tag{2.34}$$

$$L_R = -20\log\frac{\text{VSWR} + 1}{\text{VSWR} - 1}\ [\text{dB}]. \tag{2.35}$$

Table 2.7 Correspondence between L_{Ar}, L_R, and VSWR values of a lossless filter.

L_{Ar}, dB	0.01	0.1	0.2	0.5
L_R, dB	-26.383	-16.426	-13.467	-9.636
VSWR	1.101	1.355	1.538	1.984

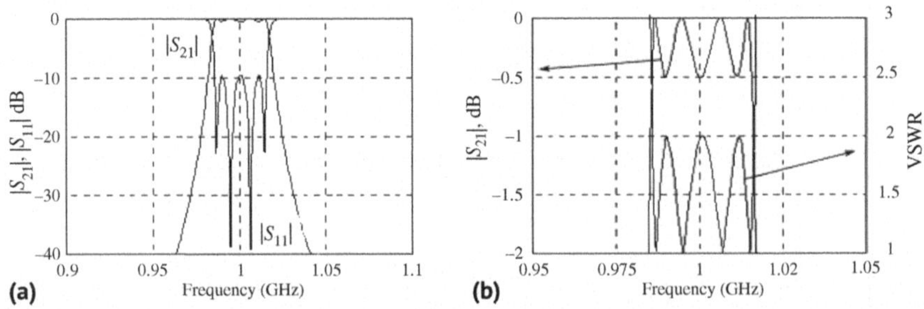

Figure 2.5 Frequency responses of a fourth-order lossless BPF. (a) Illustration L_{Ar} and L_R; (b) Illustration L_{Ar} and VSWR.

For clarity of the existing pattern and with some nonobservance of the sequence of presentation, Figure 2.5 shows the frequency responses of a fourth-order lossless BPF with a central frequency of $f_0 = 1.0$ GHz and FBW = 0.03. Its characteristic values L_{Ar}, L_R, and VSWR correspond to the last column in Table 2.7.

Note that expressions (2.32)–(2.35) are invalid for filters with loss, and the patterns shown in Table 2.7 and Figure 2.5 do not hold.

References

1 S. Darlington, Synthesis of reactance 4-poles which produce prescribed insertion loss characteristics: including special applications to filter design, *J. Math. Phys.*, vol. 18, no. 1–4, pp. 257–353, 1939, https://doi.org/10.1002/sapm1939181257.

2 E. A. Guillemin, *Introductory Circuit Theory*. Wiley, 1953.

3 E. A. Guillemin, *Synthesis of Passive Networks: Theory and Methods Appropriate to the Realization and Approximation Problems*. New York: Wiley, 1957.

4 H. Carlin, The scattering matrix in network theory, *IRE Trans. Circuit Theory*, vol. 3, no. 2, pp. 88–97, 1956, https://doi.org/10.1109/TCT.1956.1086297.

5 S. Seshu and N. Balabanian, *Linear Network Analysis*. Wiley, 1959.

6 G. C. Temes and S. K. Mitra (Eds.), *Modern Filter Theory and Design*. New York: Wiley, 1973.

7 G. L. Matthaei, L. Young, and E. M. T. Jones, *Microwave Filters, Impedance-Matching Networks, and Coupling Structures*. New York: Artech House Books, 1980.

8 J.-S. Hong, *Microstrip Filters for RF/Microwave Applications*, 2nd ed. New Jersey: Wiley, 2011, https://doi.org/10.1002/9780470937297.

9 R. J. Cameron, C. M. Kudsia, and R. R. Mansour, *Microwave Filters for Communication Systems*, 2nd ed. Hoboken, NJ: John Wiley & Sons, Inc., 2018, https://doi.org/10.1002/9781119292371.

3

Bandpass Filters with Parallel-type Resonators and Simple Couplings

In this chapter, we consider an N-order inline bandpass filter (BPF), which has a generalized scheme as shown in Figure 3.1. The BPF contains parallel-type resonators and J inverters. All couplings between resonators and end resonators with loads are simple. Most planar BPFs can be considered within this scheme.

The resonators are represented by the resonant frequency ω_0, at which their input admittance is equal to zero

$$Y_{\text{in}}\left(j\omega_0\right) = jB\left(\omega_0\right) = 0. \tag{3.1}$$

The second essential characteristic of a resonator is the susceptance slope parameter:

$$b = \frac{\omega_0}{2j} \left.\frac{dY_{\text{in}}}{d\omega}\right|_{\omega=\omega_0}. \tag{3.2}$$

Admittance inverters are used as coupling elements between resonators and between end resonators and loads.

Within the generalized circuit (Figure 3.1), various planar BPFs can be considered. Their topologies are shown in Figure 3.2, an explanation of which is given in the subfigure.

3.1 G. Matthaei's Design Method

The admittance inverter, also called the J inverter, was introduced by G. Matthaei [1]. It is a two-port network with a transfer matrix

$$\begin{bmatrix} A & B \\ C & D \end{bmatrix} = \begin{bmatrix} 0 & \pm 1/jJ \\ \mp jJ & 0 \end{bmatrix}. \tag{3.3}$$

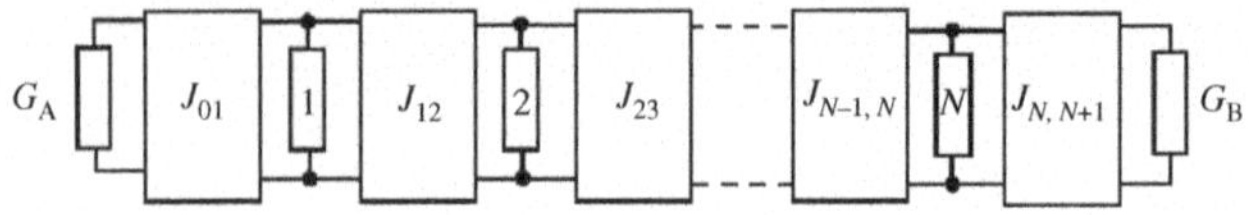

Figure 3.1 Generalized circuit of N-order BPF with parallel-type resonators and admittance inverters.

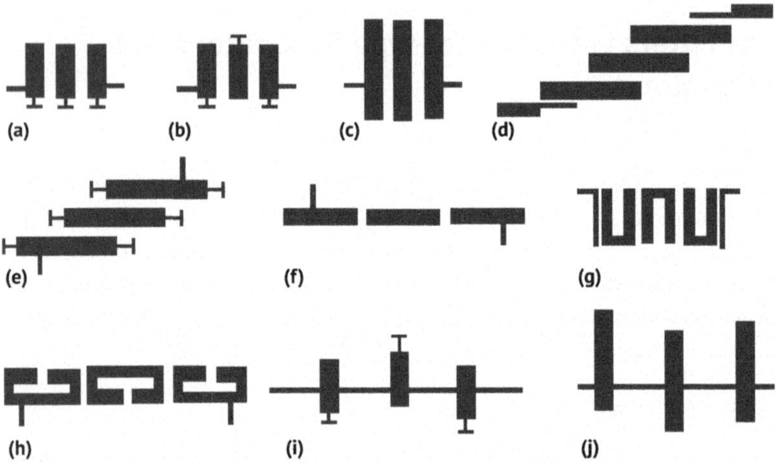

Figure 3.2 Topologies of some planar inline BPFs. (a) Combline. (b) Interdigital. (c) Pseudocombline. (d) With parallel-coupled open-circuited end $\lambda/2$ resonators. (e) With parallel-coupled short-circuited end $\lambda/2$ resonators. (f) With end-coupled $\lambda/2$ resonators. (g) With hairpin resonators. (h) With open-loop resonators. (i) With $\lambda/4$ resonators coupled by $\lambda/4$ segments. (j) With $\lambda/2$ resonators coupled by $\lambda/4$ segments.

The admittance inverter is characterized by parameter J. When connecting two resonators with resonant frequency ω_0 and slope parameters b_1, b_2 by an inverter, we have a coupling coefficient

$$K = J/\sqrt{b_1 b_2}. \tag{3.4}$$

The more significant the parameter J, the greater the coupling between the resonators and between end resonators and loads. Figure 3.3 shows the most common admittance inverters used in BPF design.

Taking into account (3.4), these inverters provide the following coupling coefficients:

$$K = 1/\omega L\sqrt{b_1 b_2} \text{ for Figure 3.3a,} \tag{3.5a}$$

$$K = -\omega C/\sqrt{b_1 b_2} \text{ for Figure 3.3b,} \tag{3.5b}$$

$$K = Z_c^{-1}/\sqrt{b_1 b_2} \text{ for Figure 3.3c.} \tag{3.5c}$$

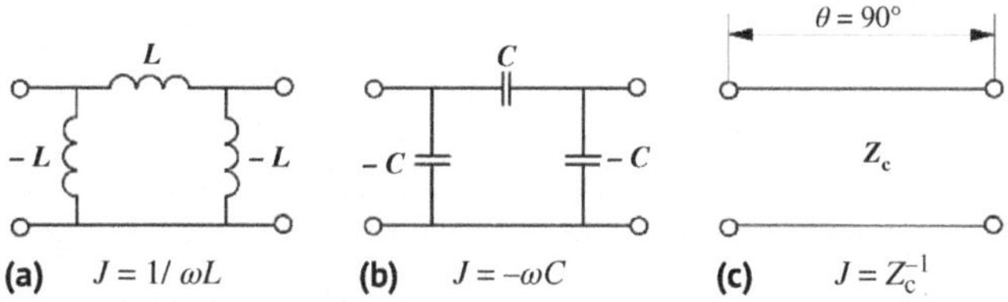

Figure 3.3 Admittance inverters. (a) With inductances. (b) With capacitances. (c) With transmission line segment.

Admittance inverters with inductances (Figure 3.3a) and transmission line segment (Figure 3.3c) provide a positive K, and a J inverter with capacitances (Figure 3.3b) provides a negative K.

The use of J inverters has led to the BPF design method based on classic expressions [2] containing g parameters of low-frequency prototype:

$$J_{i,i+1} = \text{FBW}\sqrt{b_i b_{i+1}/g_i g_{i+1}}, \tag{3.6a}$$

$$J_{01} = \sqrt{G_A b_1 \text{FBW}/g_0 g_1}, \tag{3.6b}$$

$$J_{n,n+1} = \sqrt{G_B b_n \text{FBW}/g_n g_{n+1}}, \tag{3.6c}$$

where FBW is the fractional bandwidth, and b_i is the susceptance slope parameter of the resonators. The inverter $J_{i,i+1}$ (3.6a) represents the coupling between resonators, and the inverters J_{01} (3.6b), $J_{n,n+1}$ (3.6c) represent the coupling between end resonators and loads. As a rule, the conductivities G_A and G_B in (3.6) have the same value, $G_A = G_B = 0.02\ \Omega^{-1}$.

After the J parameters of inverters are determined using (3.6), the transition to the values of L, C, and Z_c indicated in Figure 3.3 is carried out according to obvious expressions:

$$L = \frac{1}{J\omega_0}, \tag{3.7a}$$

$$C = \frac{-J}{\omega_0}, \tag{3.7b}$$

$$Z_c = \frac{1}{J}. \tag{3.7c}$$

Expressions (3.6) can be presented in an alternative form [2]:

$$K_{i,i+1} = \frac{\text{FBW}}{\sqrt{g_i g_{i+1}}} \text{ for } i = 1, \dots, n-1, \tag{3.8a}$$

$$Q_{e1} = \frac{g_0 g_1}{\text{FBW}}, \tag{3.8b}$$

$$Q_{eN} = \frac{g_n g_{n+1}}{\text{FBW}}. \tag{3.8c}$$

Using initial data and classical formulas (3.8), we determine the coupling coefficients between adjacent resonators $K_{i,i+1}$ and the external quality factor of the end resonators Q_{e1}, Q_{eN} of the Chebyshev filter, which corresponds to a given n and ripple level L_{Ar} in the passband. The quantities Q_{e1} and Q_{eN} are the inverse of the coupling coefficients between the end resonators and the loads.

Formulas (3.6) and (3.8) are equivalent and can be used for BPF design. A mixed use is also possible; for example, the coupling coefficients are determined by formula (3.8a), and the couplings of end resonators with loads are determined by formulas (3.6b) and (3.6c).

One of the main advantages of the admittance inverters [1] is that they provide the required values of the coupling coefficients (3.4) without changing the resonant frequencies of the resonators ω_{0i} ($i = 1, 2, \ldots, n$). The values of ω_{0i} are equal to the central frequency of the filter passband ω_0:

$$\omega_{01} = \omega_{02} = \ldots \omega_{0n} = \omega_0. \tag{3.9}$$

Expression (3.9) represents the condition for synchronous tuning of BPFs.

In the considered G. Matthaei's design method, an approximation consists of replacing the resonator conductivity in the vicinity of the resonance frequency ω_0 by a straight-line segment with a steepness parameter b. This method gives satisfactory results with a relative bandwidth FBW = 0.15–0.20. Existing computer programs with a linear circuit simulator and an electromagnetic simulator allow optimization and extension of this method to filters with wider bandwidths. Note that narrowband BPFs with a bandwidth FBW = 0.01–0.05 are most in demand in wireless communication systems.

There are other methods of designing BPFs [3]. However, G. Matthaei's design method is the most popular. Numerous examples of its use for BPFs of various designs are given in Puglia [4, 5]. Note that there is a similar design method by S. Cohn [6, 7], which uses impedance inverters (K inverters) and is extended to BPF with series-type resonators.

The concepts of coupling coefficient and external quality factor Q_e of resonators used in designing BPFs with J inverters require more detailed consideration.

3.2 Coupling Coefficient Between Resonators

When two resonators interact, an exchange of electromagnetic energy occurs between them. The normalized value of the exchange energy between resonators determines the coupling coefficient. It can be calculated using the formula [8]

$$K = \frac{f_o^2 - f_e^2}{f_o^2 + f_e^2}, \tag{3.10}$$

where f_e and f_o are the frequencies of the even and odd modes, respectively (called the coupling frequencies). If $f_o > f_e$, then $K > 0$; if $f_o < f_e$, then $K < 0$. The K values are expressed in units and in percentages.

Figure 3.4a shows a pair of resonators with a characteristic impedance $Z_0 = 10\ \Omega$, which are quarter-wave ($\theta_0 = \pi/2$) at a resonant frequency $f_{01} = 1.0$ GHz. These resonators are coupled to each other by a J inverter with capacitances, and their coupling to the loads is very weak. It is achieved by capacitances of small value $C_{01} = C_{23} = 0.02$ pF. With such a weak coupling to the loads, two peaks are visible on the transfer characteristic (Figure 3.4b, solid curve), corresponding to two coupling frequencies. To identify these frequencies, it is necessary to use the phase characteristic, which is shown by the dotted lines in Figure 3.4b. Both characteristics in this figure were obtained using the linear circuit simulator of Microwave Office software. The phase value φ of S_{21} between the two peaks is negative $\varphi < 0$. This indicates a negative value of the coupling coefficient between resonators [8] and the fulfillment of the condition $f_o < f_e$.

When using quarter-wave resonators, the slope parameter of which is $b = \pi/4Z_0$ [1], expressions (3.5) can be represented in a form that is more preferable for determining the elements of J inverters for a given coupling coefficient

$$L = \frac{2Z_0}{\pi^2 f_0 K}, \tag{3.11a}$$

$$C = \frac{-K}{8 f_0 Z_0}, \tag{3.11b}$$

$$Z_c = \frac{4Z_0}{\pi K}. \tag{3.11c}$$

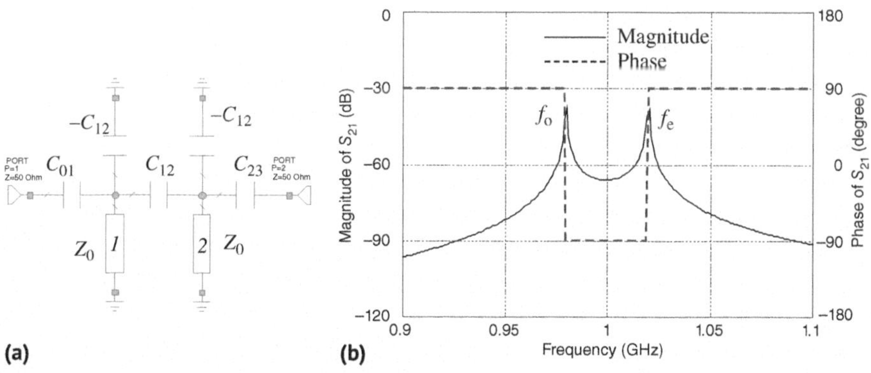

Figure 3.4 Coupling between resonators by a J inverter with capacitances. (a) Schematic. (b) Magnitude and phase of S_{21}.

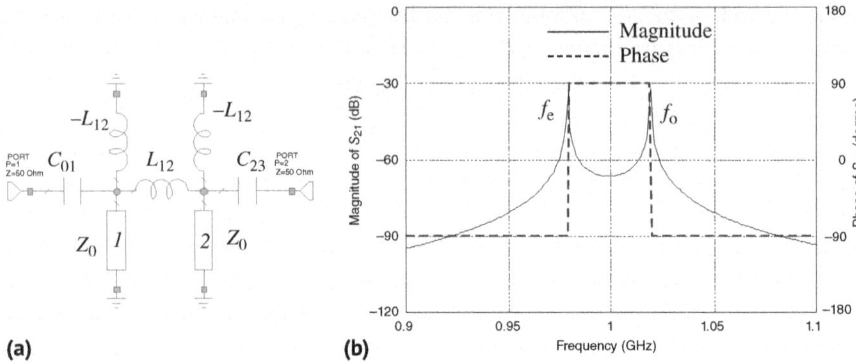

Figure 3.5 Coupling between resonators by a J inverter with inductances. (a) Schematic. (b) Magnitude and phase of S_{21}.

Let us set the coupling coefficient $K = -0.04$, and from (3.11b), we obtain the value of the capacitance of J inverter in the circuit in Figure 3.4a: $C_{12} = 0.5$ pF. This value corresponds to the frequency responses in Figure 3.4b, where $f_o = 0.98$ GHz and $f_e = 1.02$ GHz. Substituting the specified coupling frequencies into (3.10) yields $K = -0.04$.

Figure 3.5a shows the circuit schematic of the previous pair of resonators, which are weakly coupled to the loads by capacitances $C_{01} = C_{23} = 0.02$ pF. The resonators are coupled to each other by a J inverter with inductances. We set the coupling coefficient $K = 0.04$, and from (3.11a), we obtain the value of the inductance of the J inverter in the circuit in Figure 3.5a: $L_{12} = 50.66$ nH. Figure 3.5b shows the simulated frequency responses of the circuit under consideration for the specified values of its elements. The phase value φ of S_{21} between two peaks is positive $\varphi > 0$. This indicates a positive value of the coupling coefficient between the resonators [8] and the fulfillment of the condition for the even and odd modes of coupled oscillations $f_o > f_e$. In this case, $f_e = 0.98$ GHz and $f_o = 1.02$ GHz. Substituting these values into (3.10) gives $K = 0.04$.

In this case, as in the previous one, there is a complete correspondence between the calculated value of K and that obtained by the linear simulator using formula (3.10). We also note that in both cases, the resonant frequency of the resonators did not shift under the influence of the coupling elements. It remained equal to the previous value.

Let us consider the cases when there are no negative elements in the J inverters, and a capacitance or inductance couples the resonators. Figure 3.6a shows the frequency responses of the previous pair of resonators, coupled by a capacitance C_{12}, the value of which is the same as in the circuit in Figure 3.4a: $C_{12} = 0.5$ pF. The frequency responses have shifted to the left by 0.02 GHz. At the same time,

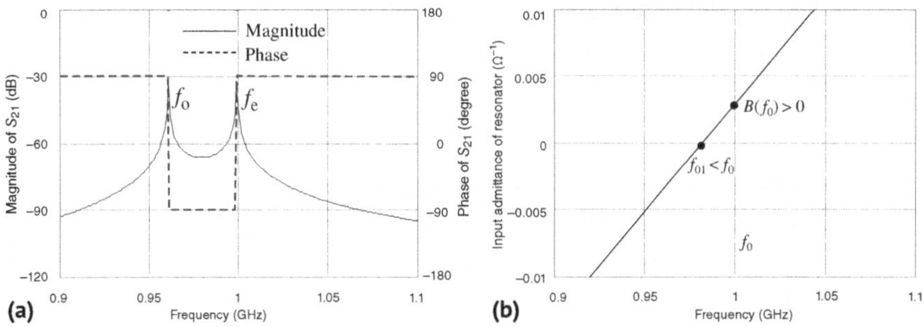

Figure 3.6 Coupling between resonators by capacitance. (a) Magnitude and phase of S_{21}. (b) Input admittance of resonator.

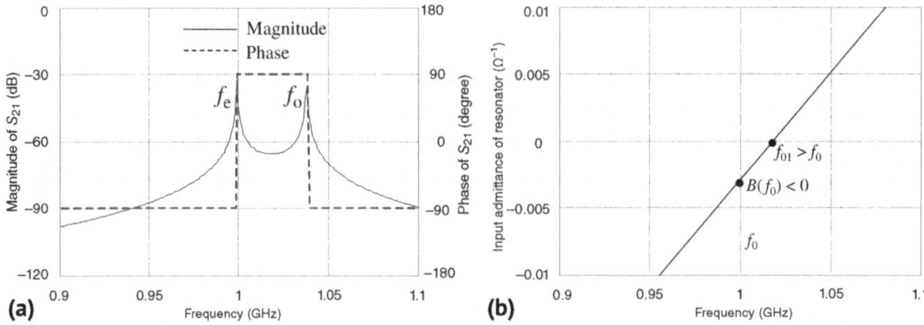

Figure 3.7 Coupling between resonators by inductance. (a) Magnitude and phase of S_{21}. (b) Input admittance of resonator.

the sign of the coupling coefficient and its value have remained the same. This shift effect is due to the coupling capacitance; the resonant frequency of the resonators has decreased by 0.02 GHz. Its new value is $f_{01} = 0.98$ GHz, which is reflected in Figure 3.6b.

Figure 3.7a shows the frequency responses of the same pair of resonators, coupled by inductance L_{12}, the value of which is the same as in the circuit in Figure 3.5a: $L_{12} = 50.66$ nH. The frequency responses have shifted to the right by 0.02 GHz. The sign of the coupling coefficient and its value have remained the same. The shift effect in this case is due to the coupling inductance; the resonant frequency of the resonators has increased by 0.02 GHz. Its new value is $f_{01} = 1.02$ GHz, which is reflected in Figure 3.7b.

In both cases shown in Figures 3.6b and 3.7b, the value of input admittance $B(f)$ at the same central frequency f_0 is determined by the approximate equality

$$B(f_0) \approx -b(f_{01} - f_0), \tag{3.12}$$

where $(f_{01} - f_0)$ represents the magnitude of the resonator detuning caused by the influence of the coupling element. The value of the input admittance (3.12) is used in the coupling matrix on its main diagonal.

3.3 External Q-factor of Resonators

When a resonator is connected to an external load, its external quality factor Q_e is determined by the equality [2]

$$Q_e = b/G_L = bR_L, \tag{3.13}$$

where G_L is the load admittance, R_L is the load resistance, and b is the susceptance slope parameter of the resonator.

Figure 3.8 shows the most common resonators with characteristic impedance Z_0 and electrical length θ. The load is connected conductively at a distance θ' from one end of the resonator and is characterized by a normalized coordinate q. The connection point is indicated by the arrow on the right. The corresponding expressions determine the input admittance of these resonators at the load connection point:

$$B = -Z_0^{-1}\left[\cot\theta' - \tan(\theta - \theta')\right], \quad 0 < \theta' \leq \pi/2, \text{ for Figure 3.8a} \tag{3.14a}$$

$$B = Z_0^{-1}\left[\tan\theta' + \tan(\theta - \theta')\right], \quad 0 \leq \theta' \leq \pi, \text{for Figure 3.8b} \tag{3.14b}$$

$$B = -Z_0^{-1}\left[\cot\theta' + \cot(\theta - \theta')\right], \ 0 < \theta' < \pi, \text{ for Figure 3.8c} \tag{3.14c}$$

Substituting these expressions into (3.2), we determine the parameter b of the resonators under consideration:

$$b = \frac{\pi}{4Z_0 \sin^2\theta'} \text{ for Figure 3.8a} \tag{3.15a}$$

$$b = \frac{\pi}{2Z_0 \cos^2\theta'} \text{ for Figure 3.8b} \tag{3.15b}$$

$$b = \frac{\pi}{2Z_0 \sin^2\theta'} \text{ for Figure 3.8c} \tag{3.15c}$$

For a quarter-wave resonator, the parameter $b \to \infty$ at $\theta' \to 0$. For a half-wave resonator with two open ends, $b \to \infty$ at $\theta' \to \pi/2$. For a half-wave resonator with two short-circuited ends, $b \to \infty$ at $\theta' \to 0$ and $\theta' \to \pi$.

Multiplying equalities (3.15) by R_L, we obtain expressions for the external quality factor Q_e of the resonators under consideration, which is a function of the connection coordinate θ'. When designing filters, it is necessary to determine the coordinate θ', at which the required value of Q_e is ensured. For this, a relatively simple transition from expressions (3.15) to the following ones is used:

$$\theta' = \sin^{-1}\sqrt{\frac{\pi R_L}{4Z_0 Q_e}} \text{ for Figure 3.8a} \tag{3.16a}$$

$$\theta' = \cos^{-1}\sqrt{\frac{\pi R_L}{2Z_0 Q_e}} \text{ for Figure 3.8b} \tag{3.16b}$$

$$\theta' = \sin^{-1}\sqrt{\frac{\pi R_L}{2Z_0 Q_e}} \text{ for Figure 3.8c} \tag{3.16c}$$

Let us determine the coordinate θ' of a quarter-wave resonator with $Z_0 = 10\ \Omega$, when connecting to which a load $R_L = 50\ \Omega$ provides an external quality factor $Q_e = 30$. Substituting the specified values into (3.16a) gives $\theta' = 21.21°$.

Expressions (3.13) and (3.15) allow us to accurately determine by calculation the external quality factor of transmission line resonators (Figure 3.8) with conductive connections of loads to them.

There is a general method for determining the external quality factor of various resonators with double-sided loads using the transfer function $|S_{21}|$ [2], which uses the concept of the resonator's loaded quality factor, which can be implemented by measurement or computer simulation.

The loaded quality factor of a resonator Q_L, loaded on both sides, is determined by the equality [2]

$$\frac{1}{Q_L} = \frac{1}{Q_{e1}} + \frac{1}{Q_u} + \frac{1}{Q_{e2}}, \tag{3.17}$$

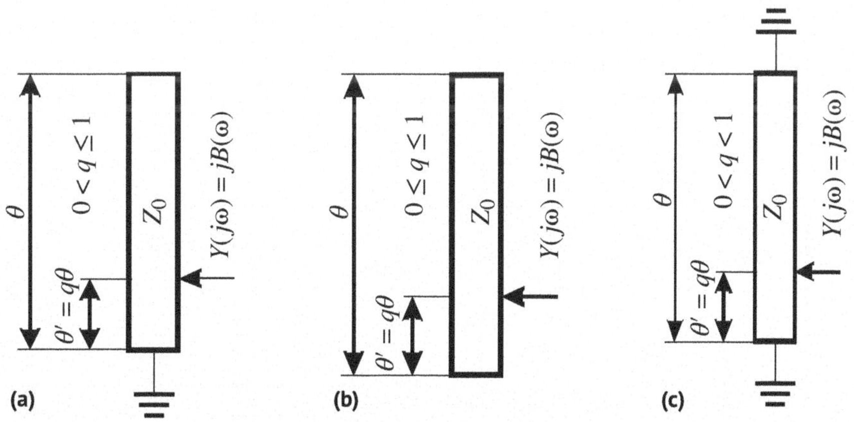

Figure 3.8 Resonators with conductive load connection. (a) $\lambda/4$ resonator. (b) $\lambda/2$ resonator with open-circuited ends. (c) $\lambda/2$ resonator with short-circuited ends.

where Q_{e1} and Q_{e2} are the external quality factors due to the left-hand and right-hand loads, respectively, Q_u is the unloaded quality factor of the resonator. The value of Q_L can be determined from the transfer function

$$Q_L = \frac{f_0}{f_2 - f_1},$$

where f_0 is the resonant frequency of the resonator, and f_2 and f_1 are the cutoff frequencies of the frequency band with a relative attenuation level of 3 dB, measured from the insertion loss at the central frequency L_{A0}.

In simulation, we have the right to set $Q_u = \infty$, which will lead to a simplification of expression (3.17)

$$\frac{1}{Q_L} = \frac{1}{Q_{e1}} + \frac{1}{Q_{e2}}. \tag{3.18}$$

If, during simulation, we provide a very weak coupling with the right-hand load, then from (3.18) it follows that

$$Q_{e1} \approx Q_L. \tag{3.19}$$

Let us apply this method to the previously considered $\lambda/4$ resonator with the coordinate of the conductive connection of the left-side load $\theta' = 21.21°$, providing the value $Q_e = 30$. The measurement circuit is shown in Figure 3.9a. The value of the capacitance C_{12} is chosen equal to 0.01 pF to provide a weak coupling with the right-side load and to be able to use expression (3.19).

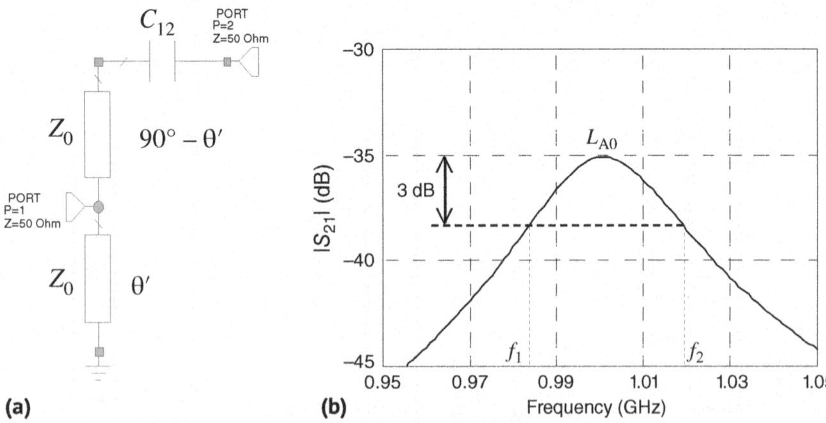

Figure 3.9 Determination of external quality factor of resonator. (a) Schematic: $f_0 = 1$ GHz, $Z_0 = 10\ \Omega$, $\theta' = 21.21°$, $C_{12} = 0.01$ pF. (b) Simulated insertion loss curve.

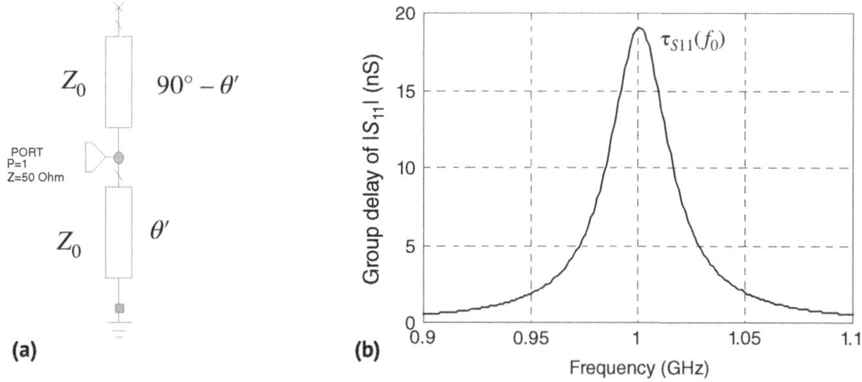

Figure 3.10 Determination of Q_e value by group delay curve. (a) Schematic: $f_0 = 1$ GHz, $Z_0 = 10\ \Omega$, $\theta' = 21.21°$. (b) Simulated group delay curve.

In Figure 3.9b, the simulated insertion loss function is presented. The difference in the boundary frequencies of the specified band was $f_2 - f_1 = 33$ MHz. Considering (3.18) and (3.19), we obtain $Q_e = 30.3$; the difference with the exact value of Q_e was 1%. The considered method was applied to a $\lambda/4$ resonator with a conductive load connection to evaluate its accuracy. However, it can be used with a large number of different resonators. Moreover, the connection of the load to the resonator can also be different; for example, a gap can be used, as shown in Figure 3.2d, g.

There is also another way to calculate Q_e, based on the relationship of this value with the group delay $\tau_{S11}(f_0)$ of S_{11} at the resonance frequency f_0 of the resonator [8]

$$Q_e = \frac{\pi f_0 \tau_{S11}(f_0)}{2}. \tag{3.20}$$

To simulate the function $\tau_{S11}(f)$ of $\lambda/4$ resonator at a certain conductive connection point, the simplest circuit shown in Figure 3.10a is used. We still use the connection point with the coordinate $\theta' = 21.21°$. Figure 3.10b shows the simulated curve $\tau_{S11}(f)$. Its value at the resonant frequency is $\tau_{S11}(f_0) = 19.1$ ns. Substituting this value into (3.20) yields the external quality factor $Q_e = 30$, which coincides with the theoretical value.

3.4 Bandpass Filter Design Examples

Let us consider examples of determining the BPF parameters with J inverters using classical expressions (3.6) and (3.8). We will also evaluate the influence of the values of L_{Ar}, L_R, and VSWR on the frequency selectivity of the BPF. Although the insertion loss L_A and return loss L_R are negative values, they are often assigned the sign "+," which shortens the notation and simplifies the formulations. In further consideration, we will also change the sign in front of these values, keeping in mind that they are negative.

As an example, let us set the following parameters of the fourth-order BPF: center frequency $f_0 = 1$ GHz, bandwidth BW = 50 MHz (FBW = 0.05), passband ripple $L_{Ar} = 0.1$ dB ($L_R = 16.426$ dB, VSWR = 1.355). From Table 2.6, we find the g parameters: $g_0 = 1, g_1 = 1.1088, g_2 = 1.3061, g_3 = 1.7703, g_4 = 0.8180, g_5 = 1.3554$. To implement the specified parameters, we use BPF with J inverters containing capacitances (Figure 3.11). The transverse branches of J inverters are not shown in this figure. The filter uses $\lambda/4$ resonators with characteristic impedance $Z_0 = 10\ \Omega$ and parameter $b = 0.07854\ \Omega^{-1}$. Using expressions (3.6), we determine the J parameters of the inverters: $J_{01} = J_{45} = 0.008416\ \Omega^{-1}$, $J_{12} = J_{34} = 0.003263\ \Omega^{-1}$, $J_{23} = 0.002582\ \Omega^{-1}$. Substituting these parameters into (3.7b) leads to the values of the capacitances used in the J inverters: $C_{01} = C_{45} = 1.339$ pF, $C_{12} = C_{34} = 0.519$ pF, $C_{23} = 0.411$ pF. All parameters of the BPF in Figure 3.11a are defined. Its simulated frequency responses are shown in Figure 3.11b as solid lines.

We use the same initial data for the filter with J inverters containing inductances (Figure 3.12). The transverse branches of the inverters are not shown. In this filter, the end resonators are conductively connected to the loads, which distinguishes it from the previous filter. Using expressions (3.8), we determine the coupling coefficients $K_{12} = K_{34} = 0.0415$, $K_{23} = 0.0329$, and external quality factors $Q_{e1} = Q_{e4} = Q_e = 22.176$. Substituting the coupling coefficients into (3.11a), we determine the inductances of the J inverters: $L_{12} = L_{34} = 48.83$ nH, $L_{23} = 61.59$ nH. To ensure the value $Q_e = 22.176$ of the end resonators, it is necessary to use the coordinate of the conductive connection of the loads $\theta' = 24.88°$. All parameters of this filter are

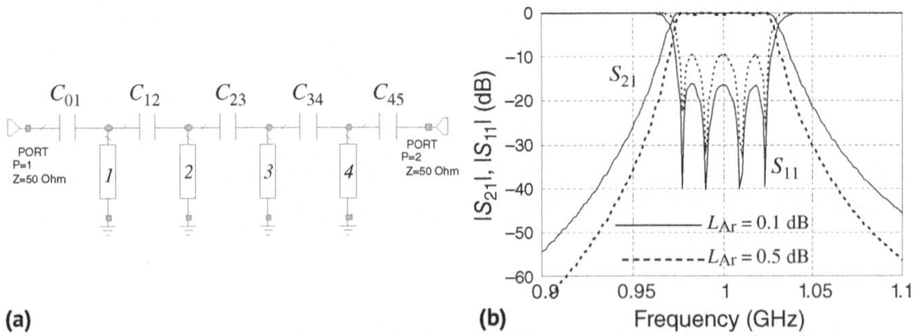

Figure 3.11 Fourth-order BPF with J inverters containing capacitances. (a) Schematic. (b) Simulated frequency responses.

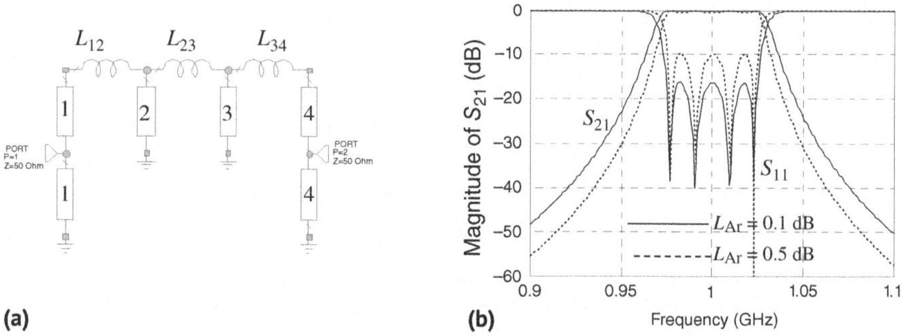

Figure 3.12 Fourth-order BPF with J inverters containing inductances. (a) Schematic. (b) Simulated frequency responses.

defined, and its simulated frequency responses are shown in Figure 3.12b by solid lines.

We increase the pulsation level in the initial data, setting it equal to $L_{Ar} = 0.5$ dB ($L_R = 9.636$ dB, VSWR = 1.984). This pulsation level corresponds to the g parameters (Table 2.6): $g_0 = 1$, $g_1 = 1.6703$, $g_2 = 1.1926$, $g_3 = 2.3661$, $g_4 = 0.8419$, $g_5 = 1.9841$. Repeating the calculations for the filter in Figure 3.11a with these g parameters, we sequentially obtain the J parameters: $J_{01} = J_{45} = 0.006857\ \Omega^{-1}$, $J_{12} = J_{34} = 0.002782\ \Omega^{-1}$, $J_{23} = 0.002338\ \Omega^{-1}$ and inverter capacitances $C_{01} = C_{45} = 1.091$ pF, $C_{12} = C_{34} = 0.443$ pF, $C_{23} = 0.3721$ pF. The frequency responses of the filter in Figure 3.11a at $L_{Ar} = 0.5$ dB are shown by the dotted lines in Figure 3.11b.

Let us recalculate the filter parameters in Figure 3.12a with the pulsation equation $L_{Ar} = 0.5$ dB. For this, we use formulas (3.8), (3.11a), and (3.16a). As a result, we obtain $K_{12} = K_{34} = 0.0354$, $K_{23} = 0.0298$, $Q_{e1} = Q_{e4} = Q_e = 33.406$, and $L_{12} = L_{34} = 48.83$ nH, $L_{23} = 61.59$ nH, $\theta' = 20.05°$. The simulated frequency responses corresponding to this case are shown in Figure 3.12b by dotted lines.

The filters in Figures 3.11a and 3.12a have practically identical frequency responses, even though they differ by J inverters and connection of end resonators to loads. Increasing the pulsation level leads to an increase in the frequency selectivity of Chebyshev filters. Note that the pulsation level $L_{Ar} = 0.5$ dB is quite acceptable, since the corresponding values of the reflection parameters VSWR = 2 and $L_R = 10$ dB are widely used in commercially available filters.

Thus, the analysis established that two groups of formulas (3.6) and (3.8) used in the BPF design are equal and lead to the same results.

References

1 G. L. Matthaei, Direct-coupled bandpass-filters with λo/4 resonators, in *1958 IRE National Convention Record*, New York: IRE, 1958, pp. 98–111.
2 G. L. Matthaei, L. Young, and E. M. T. Jones, *Microwave Filters, Impedance-Matching Networks, and Coupling Structures*. New York: Artech House Books, 1980.
3 G. C. Temes and S. K. Mitra (Eds.), *Modern Filter Theory and Design*. New York: Wiley, 1973.
4 K. V. Puglia, A general design procedure for bandpass filters derived from low pass prototype elements: Part I, *Microw. J.*, pp. 22–38, 2000, https://www.microwavejournal.com/articles/3102-a-general-design-procedure-for-bandpass-filters-derived-from-low-pass-prototype-elements-part-i.
5 K. V. Puglia, A general design procedure for bandpass filters derived from low pass prototype elements: Part II, *Microw. J.*, vol. 43 pp. 114–136, 2001.
6 S. B. Cohn, Direct-coupled-resonator filters, *Proc. IRE*, vol. 45, no. 2, pp. 187–196, 1957, https://doi.org/10.1109/JRPROC.1957.278389.
7 S. B. Cohn, Parallel-coupled transmission-line-resonator filters, *IEEE Trans. Microw. Theory Tech.*, vol. 6, no. 2, pp. 223–231, 1958, https://doi.org/10.1109/TMTT.1958.1124542.
8 J.-S. Hong, *Microstrip Filters for RF/Microwave Applications*, 2nd ed. New Jersey: Wiley, 2011, https://doi.org/10.1002/9780470937297.

4

Influence of Dissipative Loss on Bandpass Filter Frequency Response

Reducing the unloaded quality factor Q_u of a resonator leads to a change in the bandpass filter (BPF) frequency responses. The smaller the Q_u value, the more significant these changes are.

4.1 Effect of Loss at Center Frequency of Passband

The influence of unloaded quality factor Q_u of the resonator on BPF insertion loss in the center of the passband can be determined by the formula of S. Cohn [1]

$$\Delta L_{A0} = \frac{4.343}{\text{FBW}} \sum_{i=1}^{n} \frac{g_i}{Q_{ui}} \text{ [dB]}. \tag{4.1}$$

In expression (4.1), the value ΔL_{A0} represents the attenuation increase at the central frequency. For Chebyshev filters with an odd number of resonators n, the attenuation at the central frequency L_{A0} equals the attenuation increase $L_{A0} = \Delta L_{A0}$. If n is even, the pulsation value L_{Ar} is added to ΔL_{A0}: $L_{A0} = L_{\text{Ar}} + \Delta L_{A0}$. The g_i values are the parameters of the low-frequency prototype.

If all g parameters are equal to 1, $g_1 = g_2 = \cdots = g_{n-1} = g_n = 1$, and the unloaded quality factor of all resonators is equal to each other $Q_{ui} = Q_u$, then expression (4.1) takes the simplest form

$$\Delta L_{A0} = \frac{4.343n}{\text{FBW}Q_u}. \tag{4.2}$$

Formulas (4.1) and (4.2) are approximate. It follows directly from them that a decrease in the bandwidth and the unloaded quality factor Q_u of resonators leads to an increase in insertion losses at the central frequency of the bandwidth.

Table 4.1 shows the effect of the third-order Chebyshev BPF pulsations on the value of ΔL_{A0}. The analysis used expression (4.1) and the values of g parameters from Table 2.6. Figures 3.11 and 3.12 show that increasing the pulsation level in

Table 4.1 Effect of L_{Ar} pulsations on ΔL_{A0} for the third-order Chebyshev BPF with fractional bandwidth (FBW) = 0.03 and $Q_{ui} = 250$.

L_{Ar}, dB	0.01	0.1	0.2	0.5
$\sum g_i$	2.2284	3.2104	3.6075	4.693
ΔL_{A0}, dB	1.29	1.86	2.09	2.72

the Chebyshev BPF leads to some increase in selectivity. Table 4.1 shows that increasing selectivity is accompanied by an increase in insertion loss at the center frequency.

If the FBW value in Table 4.1 is reduced threefold, setting it equal to FBW = 0.01, then the insertion loss will increase threefold. Its value will vary within the range from 3.87 to 8.16 dB. These values may be unacceptable in some cases of practical design of communication equipment, which somewhat limits the use of microstrip resonators for designing narrowband filters.

4.2 Influence of Loss on BPF Performance

In previous years, considerable attention has been paid to studying the influence of dissipative losses on the filter performance within and outside the passband [2–5]. The analytical relationship between the increment of insertion loss at the passband edges and the unloaded quality factor Q_{ui} of resonators is quite complex. We will only note that this increment can be several times greater than the increment of insertion loss (4.1) at the central frequency [6].

Computer simulation allows us to most easily illustrate the influence of Q_u on the filter performance. Figure 4.1 shows the simulated frequency responses of the third-order BPF for different values of Q_u. The parameters of the lossless filter in the simulation are as follows: $f_0 = 1$ GHz, FBW = 0.03, $L_{\mathrm{Ar}} = 0.5$ dB. The solid curves correspond to the unloaded quality factor of the resonators $Q_u = 250$, the dashed ones to the value of $Q_u = 550$, and the dotted ones to $Q_u = \infty$.

Let us note the features of the BPF frequency responses associated with a decrease in the unloaded quality factor of the resonators Q_u:

- The insertion loss increases at the center frequency and the passband ends.
- The attenuation increment at the passband edges is greater than at the center frequency.
- The pulsations are smoothed out and then disappear.
- The filter selectivity is preserved, since at a significant distance from the passband, the insertion loss hardly changes.

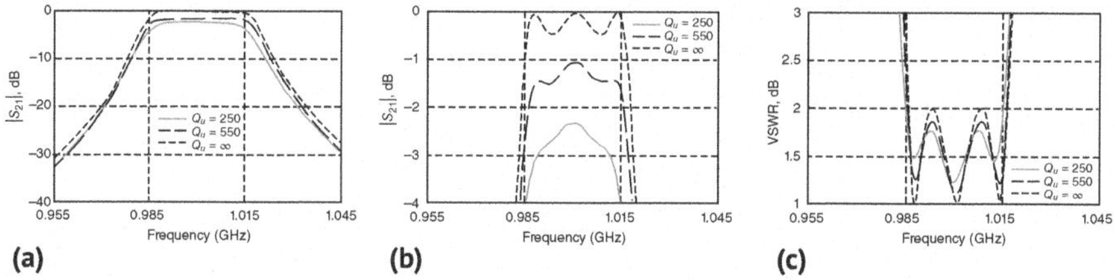

Figure 4.1 Simulated frequency responses of a third-order BPF for different Q_u values ($Q_u = 250, 550, \infty$): (a) Insertion loss. (b) Passband insertion loss. (c) VSWR.

- The reflection characteristics in the passband, which include voltage-standing wave ratio (VSWR) and return loss L_R, are less affected by the unloaded quality factor of resonators, with a decrease in Q_u they improve somewhat.
- By measuring VSWR or L_R of a lossy filter, we can approximately determine the pulsation value L_{Ar} of a lossless filter using (2.33)–(2.35).

Dissipative loss in the coupling elements also affects the BPF frequency responses. However, they contribute less than the dissipative loss of the resonators.

4.3 Unloaded Quality Factor Q_u of Resonator

To determine the unloaded quality factor Q_u of the resonator, the circuit in Figure 4.2 is used, where the resonator is represented by its admittance $Y(\omega)$, and admittance inverters J_{01} and J_{02} ($J_{01} = J_{02} = J_0$) provide weak coupling of the resonator to the loads. Initially, the loaded quality factor of the resonator Q_L, loaded on both sides (3.17), is determined. For this, expression (3.18) is used, which we will present in a slightly different form

$$Q_L = \frac{f_0}{\mathrm{BW}_{3\,\mathrm{dB}}}, \tag{4.3}$$

where $\mathrm{BW}_{3\,\mathrm{dB}}$ is the frequency band with a relative attenuation level of 3 dB, measured from the insertion loss L_{A0} at the center frequency (Figure 3.9b).

After the values of Q_L and L_{A0} are established, it is necessary to determine the unloaded quality factor Q_u of the resonator. The solution to this problem is presented in Sagawa et al. [7]:

$$Q_u = \eta Q_L, \tag{4.4}$$

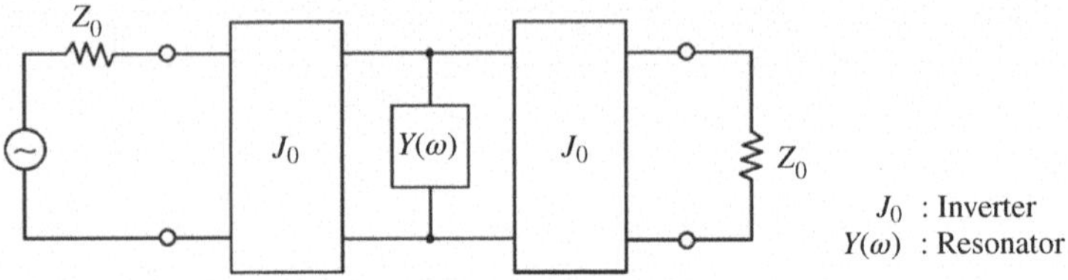

Figure 4.2 An equivalent circuit for determining Q_u and Q_e.

where

$$\eta = \frac{\sqrt{\text{antilog}\,(L_{A0}/10)}}{\sqrt{\text{antilog}\,(L_{A0}/10)} - 1}. \tag{4.5}$$

Since $Q_u > Q_L$, the coefficient (4.5) is always greater than 1 ($\eta > 1$). Thus, if $L_{A0} = 20$ dB, then $\eta = 10/9$, and if $L_{A0} = 40$ dB, then $\eta = 100/99$.

Knowing the values of Q_L and L_{A0} also allows us to determine the resulting external quality factor of the resonator

$$Q_e = \frac{Q_{e1}Q_{e2}}{Q_{e1} + Q_{e2}} = Q_L\sqrt{\text{antilog}\,(L_{A0}/10)}. \tag{4.6}$$

The method considered is suitable for computer simulation and experimental measurements.

In practice, extracting the unloaded quality factor of resonator Q_u from the measuring results of BPF frequency responses is often necessary. Note that this can only be done approximately, based on the formulas of S. Cohn (4.1) or (4.2).

Let all filter resonators have the same unloaded quality factor $Q_{ui} = Q_u$. Let us use expression (4.1), which in this case can be represented as

$$Q_u = \frac{4.343}{\text{FBW}\Delta L_{A0}}\sum\nolimits_{i=1}^{n} g_i. \tag{4.7}$$

The value of ΔL_{A0} in (4.7) is determined by measurement. The FBW of the lossless filter and the g parameters used in calculating Q_u are unknown and should be determined. To determine them, it is sufficient to measure two more quantities: one of the reflection characteristics (L_R or VSWR) and FBWs with a relatively high attenuation level of 20–40 dB. Knowing L_R or VSWR, we can approximately determine the ripple value L_{Ar} using equalities (2.32)–(2.35). These equalities are exact in the loss absence. In the presence of loss, they become approximate. Knowing the ripple value, we can easily determine the sum of g parameters in (4.7), using Table 2.6 for the g parameters of Chebyshev filters.

The measured FBW value, already known pulsation level L_{Ar}, and the transfer function of Chebyshev or Butterworth filters given in Matthaei et al. [6] allow us

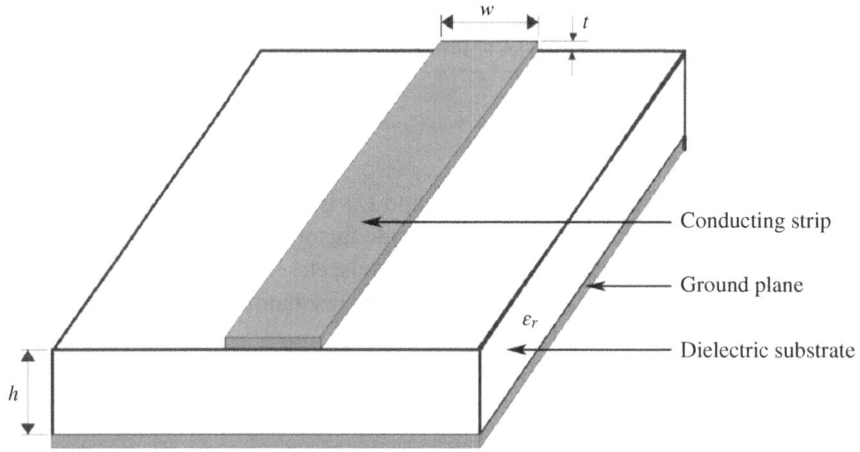

Figure 4.3 Microstrip transmission line.

to determine the FBW value. For example, we will set the following initial data: $\Delta L_{A0} = 2$ dB, $L_R = 14$ dB, the relative bandwidth at attenuation level of 30 dB is FBWs(30 dB) = 0.145. We will substitute the value $L_R = 14$ dB into (2.33) and obtain $L_{\text{Ar}} = 0.176$ dB. We will round this value to $L_{\text{Ar}} = 0.2$ dB, and according to Table 2.6, we will determine the sum of g parameters included in (4.7). As a result, we will obtain $\sum g_i = 3.6075$.

In Matthaei et al. [6], Figure 4.3 shows the insertion loss graphs of the Chebyshev filter with a pulsation value of 0.2 dB. At $n = 3$, attenuation of 30 dB is provided when (FBWs/FBW) – 1 = 2.5. We determine FBW = FBWs/3.5 = 0.03 from the last equality. Substituting the values $\Delta L_{A0} = 2$ dB, $\sum g_i = 3.6075$, and FBW = 0.03 into (4.7) gives $Q_u \approx 261$.

The considered method has often been used [8] to determine the unloaded quality factor Q_u of tunable resonators with ferroelectric capacitors.

4.4 Physical Factors Determining Unloaded Quality Factor Q_u of Microstrip and Stripline Resonators

The quality factor of a planar transmission line Q_t is understood as the unloaded quality factor of the half-wave segment of this line, including losses in conductors and dielectrics

$$Q_t = \frac{8.686\pi}{(\alpha_c + \alpha_d)\,\lambda_g}, \tag{4.8}$$

Table 4.2 Measured unloaded quality factor Q_u of microstrip half-wave resonators on a substrate with $\varepsilon_r = 92$ ($\tan\delta = 0.0003$) in the frequency range $f_0 = 1.5\text{–}2$ GHz.

Width of center strip	$w = 3$ mm	$w = 2$ mm	$w = 1$ mm
Q_u ($h = 1$ mm)	270–300	250–280	170–190
Q_u ($h = 0.5$ mm)	—	170–190	140–160

Let us compare the measured values of the unloaded quality factor Q_u of microstrip resonators with the calculated values of Q_t. One of the half-wave resonators with both open ends ($\lambda/2$) was fabricated on a polished substrate with a thickness of $h = 1$ mm, with $\varepsilon_r = 92$, length $L = 9$ mm, and width $w = 2$ mm. The substrate was metallized with copper by vacuum deposition, and the metallization thickness was 8 μm. Its measured values are $Q_u = 275$ and $f_0 = 1916$ MHz. For this frequency, formulas (4.13) and (4.14) give the calculated value of $Q_t = 301$. The measured value of Q_u is less than the calculated value of Q_t by 8.6%, and parameter χ in (4.11) takes the value $\chi = 0.914$. Table 4.2 systematizes the measured values of Q_u of some half-wave resonators manufactured on a substrate with $\varepsilon_r = 92$ and $\tan\delta = 0.0003$.

The measured Q_u value of a microstrip $\lambda/4$ resonator of width $w = 1$ mm, made on the same substrate, was $Q_u = 171$ at frequency $f_0 = 2000$ MHz. It is 24% lower than the calculated value $Q_t = 223$. In this case, $\chi = 0.767$. The short circuit at the end of the quarter-wave resonator was implemented using the metallized end face of the substrate. The short-circuited end introduces additional losses and reduces the Q_u value of a microstrip resonator. The presented experimental data showed that the unloaded quality factor Q_u of half-wave microstrip resonators is approximately 10% lower than the quality factor of the corresponding transmission line Q_t, obtained by calculation. For quarter-wave resonators, the Q_u values are lower than the Q_t values by 15–30%.

4.4.2 Stripline Resonators

Figure 4.5 shows a symmetric stripline. The notations do not require special explanations. As before, the quality factor Q_t of stripline is expressed by the formula (4.8). The conductor attenuation constant α_c of symmetric stripline with copper conductors of relatively small value ($t/b \leq 0.1$) is represented by the formula [12]

$$\alpha_c = \alpha_c' \frac{\sqrt{f\varepsilon_r}}{b} \text{ [dB/unit length]}. \tag{4.16}$$

To determine the auxiliary value α_c' in (4.16), the graphs [12] are used, which are presented in Figure 4.6. These graphs are calculated without taking into account the surface roughness, which somewhat reduces the value of α_c.

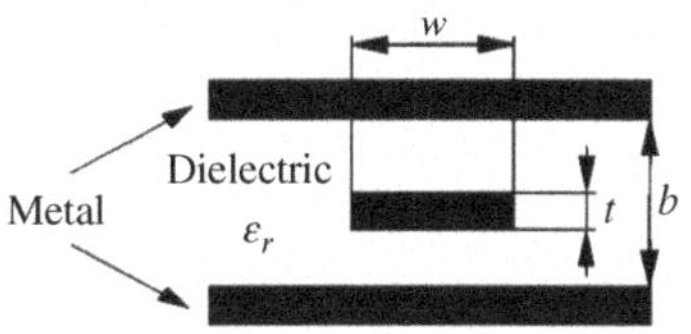

Figure 4.5 Symmetric stripline.

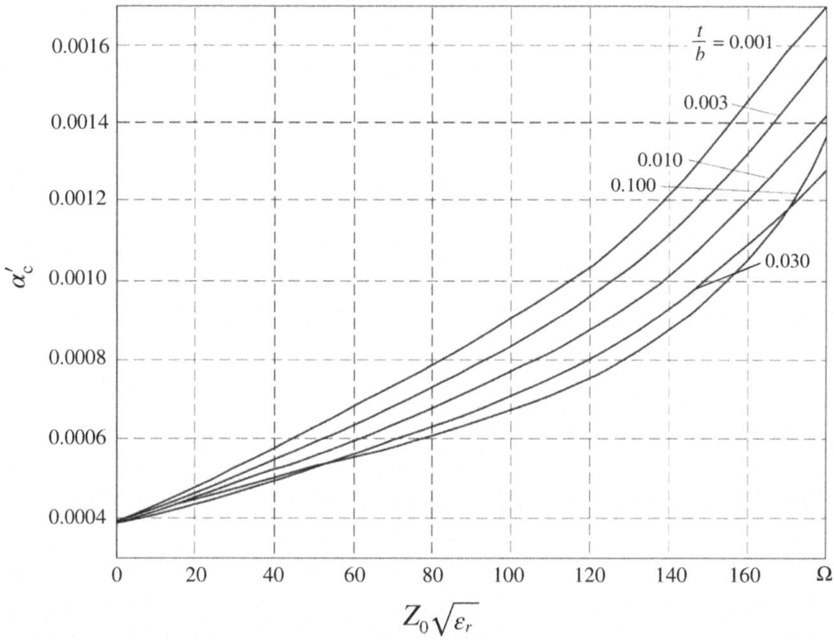

Figure 4.6 Graphs for determining auxiliary quantity α'_c, used in (4.16).

The dielectric attenuation constant α_d of symmetric stripline is conveniently determined using the approximate equality [13]

$$\alpha_d = 91f\ \tan\delta\sqrt{\varepsilon_r}\ [\text{dB/unit length}]. \tag{4.17}$$

Based on expressions (4.16) and (4.17), the dependences of the quality factor Q_t of symmetric alumina stripline on frequency are presented in Figure 4.7. These dependences are built for three thicknesses of stripline: $b = 4, 2, 1$ mm. The thickness of the metal strips t is the same in all cases, $t = 20\ \mu$m. The ratios $t/b = 0.005$, 0.01, and 0.02 correspond to each case. The graphs in Figure 4.7 show that the quality factor of a stripline Q_t increases with increasing thickness b and width

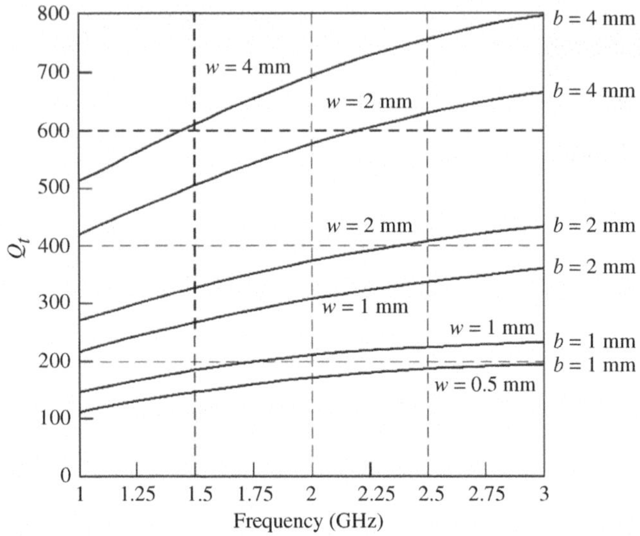

Figure 4.7 Quality factor Q_t of symmetric alumina stripline versus frequency f.

of the central conductor w and frequency. Striplines have higher Q_t values than microstrip lines (Figure 4.4).

One of the advantages of stripline is that they can be used to create thin and fully shielded BPFs that operate over a wide frequency range.

Based on expressions (4.8), (4.16), and (4.17), the Q_t valucs of thin stripline ($b = 1.016$ mm) were calculated for some permittivity values $\varepsilon_r = 9.7, 4.5$, and 2.2. The calculation data are presented in Figure 4.8. The value $\varepsilon_r = 9.7$ refers to alumina with $\tan\delta = 0.0002$. In the calculations, the width of the central strip is equal to $w = 1$ and 1.6 mm. The value of $\varepsilon_r = 4.5$ corresponds to the "soft" dielectric TMM-4 [14] with $\tan\delta = 0.0017$. Another "soft" dielectric RT/Duroid 5880 [15] has the values of $\varepsilon_r = 2.2$ and $\tan\delta = 0.0009$. The width of the central strip with "soft" dielectrics is equal to $w = 1$ and 0.8 mm.

Note that the conductor quality factor Q_c (4.9) of stripline, which considers only the loss in copper conductors (4.16), is practically independent of the ε_r value. For the same width w of different striplines, their Q_c values close to each other. Thus, for all three striplines under consideration, we have $Q_c \approx 475$ at $w = 1$ mm and the frequency of 10 GHz. The values of the dielectric quality factor Q_d (4.9) of stripline, which takes into account only the loss in the dielectric (4.17), differ significantly from each other at this frequency: 440 ($\varepsilon_r = 9.7$), 330 ($\varepsilon_r = 2.2$), and 260 ($\varepsilon_r = 4.5$). The decrease in Q_d is associated with the dielectric loss tangent, which equals 0.0002, 0.0009, and 0.0017, respectively. The contribution of two types of losses to the stripline quality factor Q_t can be characterized by the ratio α_d/α_c,

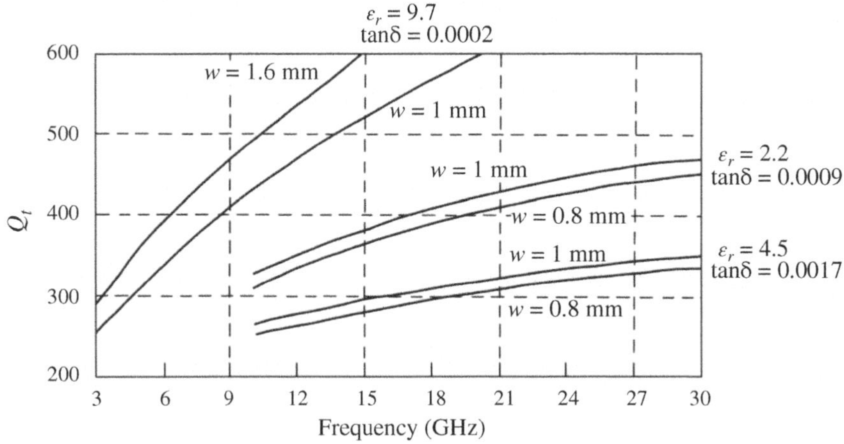

Figure 4.8 Quality factor Q_t of thin stripline ($b = 1.016$ mm) in the frequency range 3–30 GHz.

the values of which are equal to 0.096 ($\tan\delta = 0.0002$), 0.425 ($\tan\delta = 0.0009$), and 0.802 ($\tan\delta = 0.0017$) at the frequency of 10 GHz.

The higher the $\tan\delta$, the more significant the contribution of dielectric loss to Q_t. This contribution increases with increasing frequency. At the frequency of 30 GHz, the ratio α_d/α_c is equal to 0.735 ($\tan\delta = 0.0009$) and 1.39 ($\tan\delta = 0.0017$). In the latter case, the attenuation in the dielectric is already greater than in copper conductors. We should consider the tan value when choosing a dielectric for thin stripline filters. Thus, Rogers produces metallized sheets of RO4003 ($\varepsilon_r = 3.38$) and RO4350 ($\varepsilon_r = 3.48$) dielectrics for $\tan\delta = 0.0027$ and 0.0040, respectively. Due to large dielectric loss at high frequencies, thin stripline resonators based on these materials will have low unloaded quality factor Q_u values and can only be used in wideband filters.

Stripline resonators and BPFs can be used in the millimeter range, at frequencies exceeding 30 GHz [16]. Figure 4.9 shows the frequency dependence of the stripline quality factor Q_t in the millimeter-wave range. The calculations were performed for the RT/Duroid 5880 dielectric with the values $\varepsilon_r = 2.2$ and $\tan\delta = 0.0009$ [15]. The left curve corresponds to the stripline thickness $b = 1.016$ mm and the central conductor width $w = 0.8$ mm. The right curve is obtained for the values $b = 0.508$ mm and $w = 0.4$ mm. The Q_t value increases with increasing frequency. The left curve is limited by the frequency of 63 GHz, at which the value of $Q_t = 700$. The right curve at the frequency of 120 GHz has the value $Q_t = 570$.

The Q_t values in Figure 4.9 show that the stripline quality factor Q_t values are sufficient for implementing thin millimeter-wave BPFs.

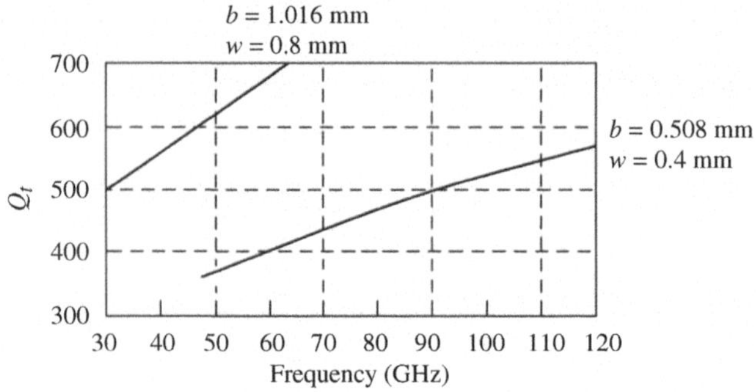

Figure 4.9 Quality factor Q_l of thin stripline ($\varepsilon_r = 2.2$, $\tan\delta = 0.0009$) in the frequency range 30–120 GHz.

References

1 S. Cohn, Dissipation loss in multiple-coupled-resonator filters, *Proc. IRE*, vol. 47, no. 8, pp. 1342–1348, 1959, https://doi.org/10.1109/JRPROC.1959.287201.

2 H. W. Bode, *Network Analysis and Feedback Amplifier Design*. New Jersey: D. Van Nostrand Company, Inc., 1945.

3 J. Taub and B. Bogner, Design of three-resonator dissipative band-pass filters having minimum insertion loss, *Proc. IRE*, vol. 45, no. 5, pp. 681–687, 1957, https://doi.org/10.1109/JRPROC.1957.278417.

4 E. Fubini and E. Guillemin, Minimum insertion loss filters, *Proc. IRE*, vol. 47, no. 1, pp. 37–41, 1959, https://doi.org/10.1109/JRPROC.1959.287105.

5 J. F. Cline and B. M. Schiffman, Tunable passive multicouplers employing minimum-loss filters, *IEEE Trans. Microw. Theory Tech.*, vol. 7, no. 1, pp. 121–127, 1959, https://doi.org/10.1109/TMTT.1959.1124635.

6 G. L. Matthaei, L. Young, and E. M. T. Jones, *Microwave Filters, Impedance-Matching Networks, and Coupling Structures*. New York: Artech House Books, 1980.

7 M. Sagawa, K. Takahashi, and M. Makimoto, Miniaturized hairpin resonator filters and their application to receiver front-end MICs, *IEEE Trans. Microw. Theory Tech.*, vol. 37, no. 12, pp. 1991–1997, 1989, https://doi.org/10.1109/22.44113.

8 A. Zakharov and M. Ilchenko, Unloaded quality factor of transmission line resonators with capacitors, *IEEE Trans. Circuits Syst. I Regul. Pap.*, vol. 67, no. 7, pp. 2204–2215, 2020, https://doi.org/10.1109/TCSI.2020.2971112.

9 R. A. Pucel, D. J. Masse, and C. P. Hartwig, Losses in microstrip, *IEEE Trans. Microw. Theory Tech.*, vol. 16, no. 6, pp. 342–350, 1968, https://doi.org/10.1109/TMTT.1968.1126691.

10 A. Presser, RF properties of microstrip line, *Microwaves*, Vol. 7, no. 3, pp. 53–55, 1968.

11 E. J. Denlinger, Losses of microstrip lines, *IEEE Trans. Microw. Theory Tech.*, vol. 28, no. 6, pp. 513–522, 1980, https://doi.org/10.1109/TMTT.1980.1130112.

12 T. Saad (Ed.), *Microwave Engineers' Handbook*. Dedham, Massachusetts: Artech House, 1966.

13 V. I. Wolman (Ed.), *Handbook on the Calculation and Design of Microwave Stripline Devices* [in Russian]. Moscow: Radio i Svyaz', 1982.

14 Rogers Corporation, TMM® thermoset microwave materials. Data Sheet. http://www.rogerscorp.com.

15 Rogers Corporation, RT/duroid® 5870 /5880 high frequency laminates. Data Sheet. http://www.rogerscorp.com.

16 M. Makimoto and S. Yamashita, *Microwave Resonators and Filters for Wireless Communication*, vol. 4. Berlin, Heidelberg: Springer Berlin Heidelberg, 2001, https://doi.org/10.1007/978-3-662-04325-7.

5

Transmission Line Resonators

Transmission line (TL) resonators have been used for a long time [1]. Their main difference from the lumped *LC* resonator is that they have many resonant and antiresonant frequencies. These frequencies are also called critical [2]. The input immittances are reactance functions of infinite order at negligible dissipative losses. Resonant frequencies are the zeros of the input admittance Y_{in} or the poles of the input impedance Z_{in} of resonator. Antiresonant frequencies are the poles of the input admittance Y_{in} or the zeros of the input impedance Z_{in} of resonator. These immittances satisfy the equality

$$Z_{\mathrm{in}} = \frac{1}{Y_{\mathrm{in}}}. \tag{5.1}$$

The multitude of resonant frequencies of the resonators leads to a multitude of passbands in a bandpass filter (BPF) (Figure 3.1). The antiresonant frequencies of resonators are one of the reasons for the appearance of attenuation poles in a BPF at real frequencies ($s = j\omega$).

5.1 Uniform Resonators and Their Critical Frequencies

Figure 5.1 shows the most common TL resonators. The input immittances of a quarter-wave resonator (Figure 5.1a) are determined by the equalities

$$Y_{\mathrm{in}}\,(j\omega) = -jZ_0^{-1}\cot\theta, \tag{5.2a}$$

$$Z_{\mathrm{in}}\,(j\omega) = jZ_0\tan\theta, \tag{5.2b}$$

where $\theta = \omega l/v$ is the electrical length of the resonator, l is the length of the resonator, v is the propagation speed of the electromagnetic wave, and Z_0 is the characteristic impedance of the TL segment.

Resonant frequencies ω_{0n} ($n = 1, 2, \ldots$) correspond to resonant electrical lengths θ_{0n}, and antiresonant frequencies ω_{pn} correspond to antiresonant electrical

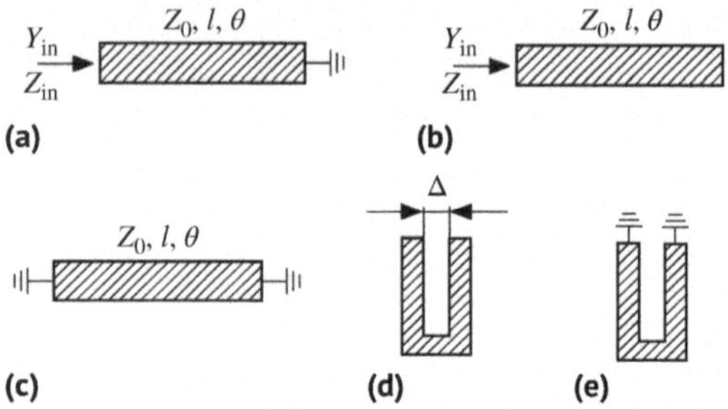

Figure 5.1 TL resonators. (a) Quarter-wave resonator. (b) Half-wave resonator with both open-circuited ends. (c) Half-wave resonator with both short-circuited ends. (d) Hairpin resonator with both open-circuited ends. (e) Hairpin resonator with both short-circuited ends.

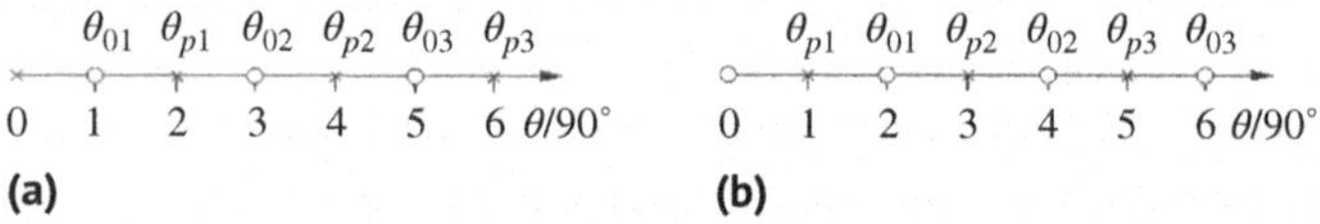

Figure 5.2 Critical electrical lengths of TL resonators. (a) Quarter-wave resonator. (b) Half-wave resonator with both open-circuited ends.

lengths θ_{pn}. From (5.2), it follows that the values θ_{0n} and θ_{pn} are determined by the corresponding equalities

$$\theta_{0n} = (2n-1)\pi/2, \tag{5.2$'$a}$$

$$\theta_{pn} = n\pi. \tag{5.2$'$b}$$

These critical electrical lengths are shown in Figure 5.2a. The resonant frequencies are related to each other as odd numbers 1:3:5... and the antiresonant frequencies are related to each other as even numbers 2:4:6....

From expression (2.9a) and Table 2.1 (line 5), it follows that the input immittances of a half-wave resonator with both open ends (Figure 5.1b) are determined by the equalities

$$Y_{\text{in}}(j\omega) = jZ_0^{-1}\tan\theta, \tag{5.3a}$$

$$Z_{\text{in}}(j\omega) = -jZ_0\cot\theta. \tag{5.3b}$$

Expressions (5.3) indicate that the resonant θ_{0n} and antiresonant θ_{pn} electrical lengths of a half-wave resonator with both open ends are determined by the corresponding equalities

$$\theta_{0n} = n\pi, \tag{5.3$'$a}$$

$$\theta_{pn} = (2n-1)\pi/2. \tag{5.3$'$b}$$

These critical electrical lengths are shown in Figure 5.2b. Their arrangement is mutually opposite to that in Figure 5.2a.

The critical electrical lengths of the quarter-wave and half-wave resonators are located equidistantly. The interval between two adjacent resonant and antiresonant electrical lengths is a constant value equal to $\pi/2$.

An important value characterizing a resonator is the ratio of the fundamental frequency to the closest resonant frequency

$$R = \omega_{02}/\omega_{01}. \tag{5.4}$$

The larger R (5.4), the farther apart the resonant frequencies ω_{02} and ω_{01} are, and the wider the BPF rejection band based on these resonators. The half-wave resonator has $R = 2$, and the quarter-wave resonator has $R = 3$, providing it with the aforementioned advantage.

A half-wave resonator with both short-circuited ends (Figure 5.1c) has the same resonant frequencies as a half-wave resonator with both open-circuited ends. Electrical lengths at these frequencies are determined by equality (5.3a). However, antiresonant electrical lengths will not be determined by equality (5.3b). Their position will depend on the connection coordinate to the resonator.

It is important to note that for all resonators, including those shown in Figure 5.1, the resonant frequencies do not depend on the connection point (coordinate) to the resonator. The position of the antiresonant frequencies on the frequency axis depends on the connection point to the resonator. Moreover, there are special points on resonators, which lead to the disappearance of some resonant frequencies, and antiresonant frequencies take their place. Thus, for a quarter-wave resonator (Figure 5.1a), the second resonant frequency ω_{02} disappears if the connection point is located by a third of the resonator length $l/3$ from the open end. For a half-wave resonator (Figure 5.1b), the same effect of ω_{02} disappearance is observed if the connection point is located by a quarter of the resonator length $l/4$ from the open end. This effect is used in BPF to suppress the parasitic passband closest to the main passband, which expands the filter's stopband.

Hairpin microstrip resonators in Figure 5.1d, e are often considered as half-wave resonators. However, this is correct at a relatively large distance Δ between their sleeves. The analysis in Ref. [3] by electromagnetic simulation showed that a decrease in Δ for the resonator in Figure 5.1d leads to a pairwise convergence of the resonant frequencies $\omega_{01} - \omega_{02}$ and $\omega_{03} - \omega_{04}$. This is accompanied by a decrease in the ratio R (5.4), which can reach $R = 1.6$ at small Δ.

In the microstrip hairpin resonator in Figure 5.1e, everything happens the other way around. As Δ decreases, the resonant frequencies $\omega_{01} - \omega_{02}$ and $\omega_{03} - \omega_{04}$ move away from each other; the ratio R (5.4) increases and can reach $R = 2.4$.

5.2 Nonuniform Transmission Line Resonators

The characteristic impedance $Z_0(x)$, $x \in [0, l]$ of nonuniform TL resonators varies along their length l. This allows us to change the ratio between critical frequencies, which improves some properties of resonators and devices based on them. Nonuniform TL resonators, whose characteristic impedance $Z_0(x)$ varies continuously, are called continuously impedance resonators (CIRs) [4–8]. Using the delay time t_d of these resonators rather than their length l is convenient. Then τ varies in the range $0 \le \tau \le t_d$. The characteristic impedance Z_0 is a function of the coordinate τ, $Z_0 = Z_0(\tau)$.

Traditionally, when considering nonuniform TL resonators, the input impedance Z_{in} is used predominantly, rather than the input admittance Y_{in}. Therefore, the resonant frequencies will be the poles ω_{pn} of Z_{in}, and the antiresonant frequencies will be the zeros ω_{0n} of Z_{in}.

The resonance frequencies of resonators ω_{pn} are the square roots of eigenvalues λ_n of the Sturm–Liouville boundary value problem [9], which is determined by the second-degree differential equation for the voltage of a nonuniform TL

$$-U'' + q(\tau)U = \lambda U, \quad \tau \in [0, t_d], \tag{5.5}$$

and boundary conditions corresponding to the open-circuit or short-circuit conditions at the segment ends. The potential $q(\tau)$ in (5.5) is a real-valued function determined by the characteristic impedance [9]

$$q(\tau) = \left[\frac{Z_0'(\tau)}{2Z_0(\tau)}\right]^2 - \left[\frac{Z_0'(\tau)}{2Z_0(\tau)}\right]' . \tag{5.6}$$

The differential equation for the line current determines the antiresonant frequencies ω_{0n}. The frequencies ω_{pn} and ω_{0n} alternate

$$0 < \omega_{01} < \omega_{p1} < \omega_{02} < \omega_{p2} < \dots . \tag{5.7}$$

Distribution (5.7) corresponds to an open-circuit segment at both ends. If the segment is short-circuited at one end, the frequencies ω_{pn} and ω_{0n} in (5.7) switch their places. The critical frequencies (5.7) must satisfy additional asymptotic conditions (at $n \to \infty$), which depend on the nature of the potential q (5.6).

A smoothly nonuniform TL is a line whose characteristic impedance function $Z_0(\tau)$ is twice differentiable and assumes finite values ($0 < Z_0(\tau) < \infty$) in the interval $\tau \in [0, t_d]$. In such lines, the potential $q(\tau)$ is a continuous, and critical frequencies satisfy the following asymptotic equalities at large n:

$$\omega_{pn} = \left(n - \frac{1}{2}\right)\frac{\pi}{t_d} + \mathrm{O}\left(\frac{1}{n}\right), \quad \omega_{0n} = \frac{n\pi}{t_d} + \mathrm{O}\left(\frac{1}{n}\right), \tag{5.8}$$

for open-circuit segments at one end;

$$\omega_{pn} = \frac{n\pi}{t_d} + O\left(\frac{1}{n}\right), \quad \omega_{0n} = \left(n - \frac{1}{2}\right)\frac{\pi}{t_d} + O\left(\frac{1}{n}\right), \tag{5.9}$$

for open-circuit segments at both ends. Conditions (5.8) and (5.9) indicate that the critical frequencies of CIRs with high serial numbers coincide with similar frequencies of uniform TL resonators.

The inverse problem for CIR was first formulated by Large [8]. It consists of determining the function of the characteristic impedance $Z_0(\tau)$ from the given critical frequencies ω_{pn} and ω_{0n} satisfying the alternation condition (5.7) and the asymptotic condition (5.8) or (5.9). The solution to the formulated problem was obtained by Litvinenko and Soshnikov [9]. Let us denote the critical frequencies of a uniform TL resonator by ω_{pn}° and ω_{0n}°. The critical frequencies distribution of the CIR can be obtained by shifting the initial frequencies of the uniform resonator (Figure 5.3).

If some values of ω_{0k}° and ω_{pl}° of the input impedance of a uniform resonator are replaced with ω_{0k} and ω_{pl}, we obtain the input impedance of CIR:

$$Z(s) = Z_0(0)\, F(s) \coth(st_d) \tag{5.10a}$$

for a half-wave resonator; and

$$Z(s) = Z_0(0)\, F(s) \tanh(st_d) \tag{5.10b}$$

for a quarter-wave resonator.

In these expressions, $s = \sigma + j\omega$, $Z_0(0)$ is the characteristic impedance at $\tau = 0$ and

$$F(s) = \frac{\prod_k \left(s^2 + \omega_{0k}^2\right) \quad \prod_l \left[s^2 + \left(\omega_{pl}^{\circ}\right)^2\right]}{\prod_k \left[s^2 + \left(\omega_{0k}^{\circ}\right)^2\right] \quad \prod_l \left(s^2 + \omega_{pl}^2\right)}. \tag{5.11}$$

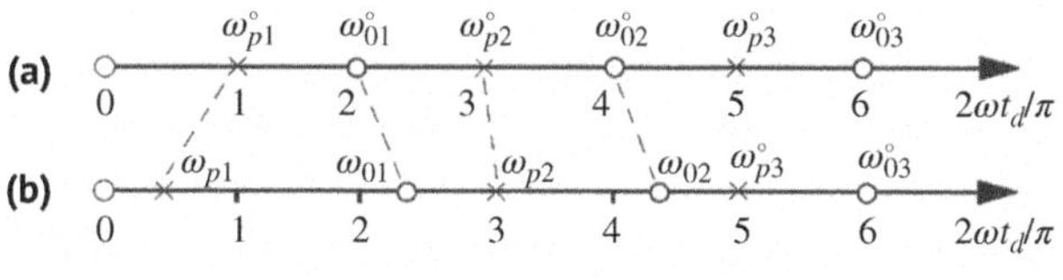

Figure 5.3 Critical frequencies of input impedances of quarter-wave resonators. (a) Uniform. (b) Nonuniform.

The characteristic impedance $Z_0(\tau)$ of CIR is expressed in a closed form [9]:

$$Z_0(\tau) = Z_0(0)\frac{\prod_{i=1}^{K}\mu_i^2\left|c_{ij}\right|^2}{\prod_{j=1}^{K}\beta_j^2\left|d_{ij}\right|^2}, i, j = 1, 2, \ldots, K, \tag{5.12}$$

where μ_i are the frequencies ω_{0k} and ω°_{pl} of the numerator of (5.12), numbers β_j are the frequencies of the denominator of (5.12), K is the total number of changed critical frequencies, $\left|c_{ij}\right|$ and $\left|d_{ij}\right|$ are the determinants of square matrices of order K, the elements c_{ij} and d_{ij} of which are calculated as follows:

$$c_{ij} = \frac{\mu_i\tan(\mu_i\tau)\tan(\beta_i\tau) + \beta_j}{\mu_i^2 - \beta_j^2};\quad d_{ij} = \frac{\beta_j\tan(\mu_i\tau)\tan(\beta_j\tau) + \mu_i}{\mu_i^2 - \beta_j^2}, \tag{5.13}$$

where i is the column number and j is the row number. Expressions (5.12) and (5.13) determining the function $Z_0(\tau)$ are valid for half-wave and quarter-wave CIRs. The transition from ω_{pn} and ω_{0n} to $Z_0(\tau)$ using (5.12) and (5.13) became possible only after mathematicians solved the inverse boundary spectral problem, the elements of which are presented by Kachalov et al. [10].

For example, if two initial critical frequencies ω°_{01} and ω°_{p1} of a regular open-circuited section changed, then the function $Z_0(\tau)$ of the obtained irregular section is written in the expanded form as

$$Z_0(\tau) = Z_0(0)\left(\frac{\omega_{01}\omega^{\circ}_{p1}}{\omega^{\circ}_{01}\omega_{p1}}\right)^2 \times$$

$$\times\frac{\begin{vmatrix}\dfrac{\omega_{01}\tan(\omega_{01}\tau)\tan(\omega^{\circ}_{01}\tau)+\omega^{\circ}_{01}}{\omega_{01}^2-(\omega^{\circ}_{01})^2} & \dfrac{\omega_{01}\tan(\omega_{01}\tau)\tan(\omega_{p1}\tau)+\omega_{p1}}{\omega_{01}^2-\omega_{p1}^2}\\ \dfrac{\omega^{\circ}_{p1}\tan(\omega^{\circ}_{p1}\tau)\tan(\omega^{\circ}_{01}\tau)+\omega^{\circ}_{01}}{(\omega^{\circ}_{p1})^2-(\omega^{\circ}_{01})^2} & \dfrac{\omega^{\circ}_{p1}\tan(\omega^{\circ}_{p1}\tau)\tan(\omega_{p1}\tau)+\omega_{p1}}{(\omega^{\circ}_{p1})^2-\omega_{p1}^2}\end{vmatrix}^2}{\begin{vmatrix}\dfrac{\omega^{\circ}_{p1}\tan(\omega_{01}\tau)\tan(\omega^{\circ}_{01}\tau)+\omega_{01}}{\omega_{01}^2-(\omega^{\circ}_{01})^2} & \dfrac{\omega_{p1}\tan(\omega_{01}\tau)\tan(\omega_{p1}\tau)+\omega_{01}}{\omega_{01}^2-\omega_{p1}^2}\\ \dfrac{\omega^{\circ}_{01}\tan(\omega^{\circ}_{p1}\tau)\tan(\omega^{\circ}_{01}\tau)+\omega^{\circ}_{p1}}{(\omega^{\circ}_{p1})^2-(\omega^{\circ}_{01})^2} & \dfrac{\omega_{p1}\tan(\omega^{\circ}_{p1}\tau)\tan(\omega_{p1}\tau)+\omega^{\circ}_{p1}}{(\omega^{\circ}_{p1})^2-\omega_{p1}^2}\end{vmatrix}^2} \tag{5.14}$$

As another example, consider a quarter-wave CIR with zero ω_{0n} of input impedance offset from ω°_{0n}. Using expressions (5.10b) and (5.11), we write down the input impedance of such a CIR

$$Z(s) = Z_0\frac{s_2 + \omega_{0n}^2}{s_2 + \left(\omega^{\circ}_{0n}\right)^2}\tanh(st_d). \tag{5.15}$$

According to (5.12) and (5.13), we obtain the characteristic impedance function for input impedance

$$Z_0(\tau) = Z_0(0)\left[\frac{1 + (\omega_{0n}/\omega_{0n}^{\circ})\tan(\omega_{0n}^{\circ}\tau)\tan(\omega_{0n}\tau)}{1 + (\omega_{0n}^{\circ}/\omega_{0n})\tan(\omega_{0n}^{\circ}\tau)\tan(\omega_{0n}\tau)}\right]^2. \tag{5.16}$$

The arrangement of the second antiresonant frequency $\omega_{02} = 4.3\omega_{p1}^{\circ}$ of the CIR is shown in Figure 5.4a. Figure 5.4b plots the function (5.16) graph corresponding to this case.

The function $Z_0(\tau)$ of the CIR with the second antiresonance frequency ω_{02} shifted has three extremes. In the general case, with a displacement ω_{0n}, the function $Z_0(\tau)$ has $(2n - 1)$ extremes. Examples of nonuniform TL resonators are considered in Refs. [11–13].

From (5.12) and (5.13), we get the expression that determines the ratio of the characteristic impedances at the ends of different CIRs

$$\frac{Z_0(t_d)}{Z_0(0)} = \left(\frac{\omega_1^{\circ}}{\omega_1}\frac{\omega_3^{\circ}}{\omega_3}\cdots\frac{\omega_{2n-1}^{\circ}}{\omega_{2n-1}}\right)^{\pm 4}, \tag{5.17}$$

where degree is "+4" for short-circuited CIR and "–4" for open-circuited CIR. Expression (5.17) contains critical frequencies, which have only odd numbers with continuous numbering. In the considered examples (Figure 5.4), the critical frequency with even numbers was changed; therefore, $Z_0(t_d)/Z_0(0) = 1$.

Suppose the fundamental resonant frequency ω_{p1} of the quarter-wave resonator is reduced by half, making it equal to $\omega_{p1} = \omega_{p1}^{\circ}/2$. In that case, the characteristic impedances at the ends of the resonator will differ by 16 times $Z_0(t_d)/Z_0(0) = 16$ (5.17).

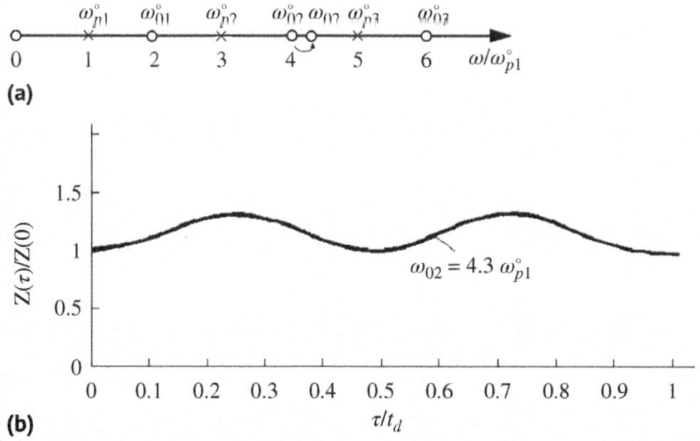

Figure 5.4 Quarter-wave CIR. (a) Critical frequencies. (b) Functions of characteristic impedance.

5.3 Stepped-impedance Resonators

Nonuniform TL resonators, whose characteristic impedance $Z_0(x)$ varies stepwise, are called stepped-impedance resonators (SIRs) [14–19].

The general view of SIR with N equal length segments is shown in Figure 5.5a. Using the delay time t_d of these resonators rather than their length l is convenient. The delay time of one section is $t'_d = t_d/N$, the electric length of the SIR is $\theta = \omega t_d$, and the electric length of one section is $\theta' = \omega t'_d = \theta/N$. For Richards' frequency variable [20]

$$S = j\Omega = j\tan\omega t'_d = j\tan\theta', \tag{5.18}$$

the transfer matrix of SIRs with N segments is written as [21]

$$\begin{bmatrix} A & B \\ C & D \end{bmatrix} = (1 - S^2)^{-N/2} \times \begin{bmatrix} a_0 + a_2S^2 + \ldots + a_NS^N & a_1S + a_3S^3 + \ldots + a_{N-1}S^{N-1} \\ b_1S + b_3S^3 + \ldots + b_{N-1}S^{N-1} & b_0 + b_2S^2 + \ldots + b_NS^N \end{bmatrix}, \tag{5.19}$$

where N is an even number; $a_0 = b_0 = 1$; and the remaining coefficients of polynomials are positive numbers depending on characteristic impedances Z_{0i}, $i = 1, 2, \ldots, N$. If N is an odd number, the highest degrees of polynomials (5.19) change their places.

The input impedance of SIR with N segments is described by a reactance function of variable S (5.18) of degree N. If the SIR is a half-wave resonator (two open ends), then

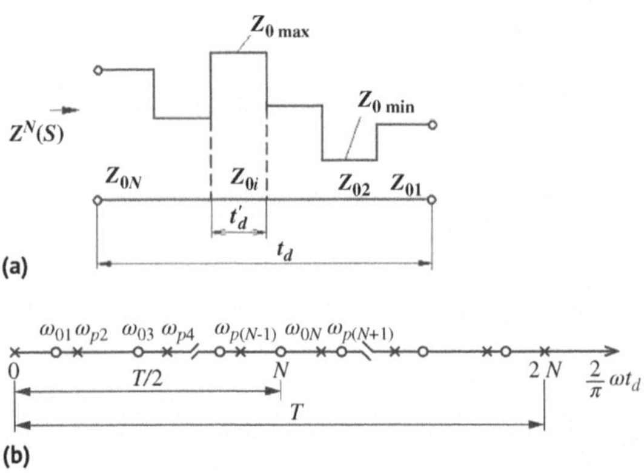

Figure 5.5 SIR of half-wave type with segments of the same length. (a) Schematic. (b) Critical frequencies of input impedance.

$$Z^N(S) = \frac{A}{C} = \frac{a_0 + a_2S^2 + \ldots + a_NS^N}{b_1S + b_3S^3 + \ldots + b_{N-1}S^{N-1}} \tag{5.20a}$$

$$= H\frac{\left(S^2 + \Omega_{01}^2\right)\left(S^2 + \Omega_{02}^2\right)\ldots\left(S^2 + \Omega_{0N/2}^2\right)}{S\left(S^2 + \Omega_{p1}^2\right)\left(S^2 + \Omega_{p2}^2\right)\ldots\left(S^2 + \Omega_{p(N-2)/2}^2\right)}, \tag{5.20b}$$

where H is the impedance level, and Ω_{0j} and Ω_{pj} are the critical frequencies on the Ω-axis. If the SIR is a quarter-wave resonator (open- and short-circuit ends), $Z^N(S) = B/D$. Even and odd polynomials in (5.20a) and (5.20b) change their places.

The critical frequencies of half-wave SIR with N segments are shown in Figure 5.5b. These frequencies alternate with each other. The even polynomial of the variable S is located in the numerator of $Z^N(S)$ (5.20), and the odd polynomial is located in the denominator of $Z^N(S)$. The even and odd polynomials in (5.20) are reversed if the far ends of the SIR are short-circuited.

The critical frequencies of SIR have certain features [22]. These frequencies are periodic. Within each period, they are located symmetrically about the center of the period. The critical frequency ω_N with number N is located in the center of the first period (with continuous numbering of frequencies). At this frequency, the electrical length of the composite segment becomes a quarter-wave

$$\theta' = \omega_N t'_d = \pi/2 \tag{5.21}$$

and the Richards' frequency variable (5.18) goes to infinity

$$S = j\Omega_N = j\infty. \tag{5.22}$$

The critical frequency ω_N can be either the zero of the function $Z^N(S)$ (5.20) or the pole. It depends on the highest degree of polynomials of the numerator and denominator (5.20). If the degree of the numerator polynomial (5.20) is higher than the degree of the denominator polynomial, then the frequency ω_N is the zero of $Z^N(S)$. Otherwise, the frequency ω_N is the pole of $Z^N(S)$. The characteristic impedances of segments Z_{0i} do not affect this critical frequency. These characteristic impedances also do not affect critical frequencies with multiples of $N/2$. The entire spectrum of critical frequencies of SIR is determined by the value of t'_d and by the first $(N-1)$ critical frequencies. The frequencies of the right half of the first period are the mirror image of the frequencies of the left half of the period relative to the center frequency of the period ω_N.

The task of SIR synthesis is determining the delay time t'_d of composite segments and their characteristic impedances Z_{0i} according to the required set of resonant ω_{pi} and antiresonant ω_{0i} frequencies. This task is solved in a certain sequence:

1. The number N of composite SIR segments equals the number of the critical frequencies ω_N with continuous numbering.

2. Condition (5.21) leads to an equality that determines the delay time of each composite segment:

$$t'_d = \frac{\pi}{2\omega_N}. \tag{5.23}$$

3. Using the frequency transformation (5.18) and t'_d value (5.23), we transit from the given critical frequencies ω_{pi} and ω_{0i} to the critical frequencies Ω_{pi} and Ω_{0i} on the Ω-axis

$$\Omega_i = \tan\left(\omega_i t'_d\right). \tag{5.24}$$

4. Knowing the critical frequencies Ω_{pi} and Ω_{0i}, it is easy to write the input impedance $Z^N(S)$ in the form (5.20b).
5. To go from (5.20b) to (5.20a), we need to multiply the numerator and denominator factors or use Vieta's formulas [23]. For the numerator (5.20a), these formulas look like

$$\begin{aligned}
&a_{(N-2)}/a_N = -\left(\Omega_{01}^2 + \Omega_{02}^2 + \cdots + \Omega_{0N/2}^2\right)\\
&a_{(N-4)}/a_N = \Omega_{01}^2\Omega_{02}^2 + \Omega_{01}^2\Omega_{03}^2 + \cdots + \Omega_{0(N/2-1)}^2 + \Omega_{0N/2}^2\\
&\text{------------------------------------}\\
&a_0/a_N = (-1)^{N/2-1}\left(\Omega_{01}^2\Omega_{02}^2\cdots\Omega_{0N/2}^2\right)
\end{aligned} \tag{5.25}$$

6. The determination of characteristic impedances Z_{0i} according to the function $Z^N(S)$ is based on the Richards' theorem [20, 24]. It consists of a successive lowering of the order of the reactance function at each calculation stage. The characteristic impedance of the segment adjoining the input is $Z_{0N} = Z^N(1)$. The transition from function $Z^N(S)$ to $Z^{N-1}(S)$ is performed according to the formula:

$$Z^{N-1}(S) = Z^N(1)\frac{Z^N(S) - SZ^N(1)}{Z^N(1) - SZ^N(S)}. \tag{5.26}$$

Note that the numerator and denominator of (5.26) contain the factor $\left(1 - S^2\right)$, subject to cancellation. Upon determining the $Z^{N-1}(S)$, we find the next characteristic impedance $Z_{0(N-1)} = Z^{N-1}(1)$, and so on.

For example, let us define the Z_{0i} of the quarter-wave SIR, the input impedance of which is described by the third-order reactance function

$$Z^3(S) = \frac{14S + 4S^3}{1 + 8S^2}.$$

We determine the characteristic impedance $Z_{03} = Z^3(1) = 2$. Using (5.26), and canceling the numerator and denominator of the obtained expression by $\left(1 - S^2\right)$, we pass from $Z^3(S)$ to:

$$Z^2(S) = \frac{12S}{1 + 2S^2}.$$

The value of Z_{02} is $Z_{02} = Z^2\ (1) = 4$. Carrying out the same sequence of actions, we pass from $Z^2\ (S)$ to $Z^1\ (1) = 8S$. Finally, we get the value of $Z_{01} = Z^1\ (1) = 8$.

When considering SIR, it is convenient to set $t_d = \pi/2$. In this case, $\omega_N = N$, and the critical frequencies of the uniform resonator are expressed in natural numbers. The solution of the inverse problem is performed in a specific sequence. From the sets of resonant $\{\omega_{pn}\}$ and antiresonant $\{\omega_{0n}\}$ frequencies, the transition is made to the frequencies $\{\Omega_{pn}\}$, $\{\Omega_{0n}\}$ on the Ω-axis (5.24) and then to the input impedance $Z^N\ (S)$ (5.20). Using the expressions $Z_{0N} = Z^N\ (1)$ and (5.26), the characteristic impedances Z_{0i}, $i = 1, 2, \ldots, N$ are determined.

Determining the characteristic impedances Z_{0i} of SIR from the function (5.20a) is carried out according to the Richards' procedure, which is recurrent and requires certain labor costs. These calculations are best done with a computer program that divides polynomials by $(1 - S^2)$. If the specified program is not available, then to calculate Z_{0i}, we can use the formulas [25] that are presented in Table 5.1 for SIRs with both open-circuited ends, and in Table 5.2 for SIRs with open- and short-circuited ends.

Table 5.1 Formulas for determining the characteristic impedances Z_{0i} $(i = 1, 2, \ldots, N)$ of SIR with both open-circuited ends.

$N = 2$	$N = 3$
$Z_{02} = \dfrac{a_0 + a_2}{b_1}$, $Z_{01} = Z_{02}\dfrac{a_0}{a_2}$	$Z_{03} = \dfrac{a_0 + a_2}{b_1 + b_3}$, $Z_{02} = \dfrac{a_0 + Z_{03}b_3}{b_1 - a_0/Z_{03}}$, $Z_{01} = Z_{02}\dfrac{a_0}{a_2}$
$N = 4$	$N = 5$
$Z_{04} = \dfrac{a_0 + a_2 + a_4}{b_1 + b_3}$,	$Z_{05} = \dfrac{a_0 + a_2 + a_4}{b_1 + b_3 + b_5}$,
$Z_{03} = \dfrac{a_0 - a_4 + b_3 Z_{04}}{b_1 + \frac{a_4 - a_0}{Z_{04}}}$,	$Z_{04} = \dfrac{2a_0 + a_2 + (b_5 - b_1)\, Z_{05}}{b_1 - b_5 + \frac{a_4 - a_0}{Z_{05}}}$,
$Z_{02} = \dfrac{a_0 + a_4 Z_{03}/Z_{04}}{b_1 - a_0\,(1/Z_{03} + 1/Z_{04})}$,	$Z_{03} = \dfrac{a_0 + a_4 Z_{04}/Z_{05} - b_5\,(Z_{04} + Z_{05})}{b_1 - a_0\,(1/Z_{04} + 1/Z_{05}) + b_5 Z_{04}/Z_{05}}$,
$Z_{01} = \dfrac{Z_{02}/Z_{04}}{Z_{03}}\dfrac{a_0}{a_4}$	$Z_{02} = \dfrac{a_0 + b_5 Z_{03} Z_{05}/Z_{04}}{b_1 - a_0\,(1/Z_{03} + 1/Z_{04} + 1/Z_{05})}$,
	$Z_{01} = \dfrac{Z_{02}/Z_{04}}{Z_{03} Z_{05}}\dfrac{a_0}{b_5}$

(Continued)

Table 5.1 (Continued)

$N = 6$

$$Z_{06} = \frac{a_0 + a_2 + a_4 + a_6}{b_1 + b_3 + b_5}, \quad Z_{05} = \frac{2a_0 + a_2 - a_6 + (b_5 - b_1) Z_{06}}{b_1 - b_5 + (a_4 + 2a_6 - a_0)/Z_{06}},$$

$$Z_{04} = \frac{a_0 (3 + Z_{05}/Z_{06}) + a_2 + a_6 Z_{05}/Z_{06} - b_1 (Z_{05} + Z_{06})}{b_1 - b_5 Z_{06}/Z_{05} - (a_0 + a_6)(1/Z_{05} + 1/Z_{06})},$$

$$Z_{03} = \frac{a_0 - a_6 (Z_{04}/Z_{05} + Z_{04}/Z_{06} + Z_{05}/Z_{06}) + b_5 Z_{04} Z_{06}/Z_{05}}{b_1 - a_0 (1/Z_{04} + 1/Z_{05} + 1/Z_{06}) + a_6 Z_{05}/(Z_{04} Z_{06})},$$

$$Z_{02} = \frac{a_0 + a_6 Z_{03} Z_{05}/Z_{04} Z_{06}}{b_1 - a_0 (1/Z_{03} + 1/Z_{04} + 1/Z_{05} + 1/Z_{06})}, \quad Z_{01} = \frac{Z_{02} Z_{04} Z_{06}}{Z_{03} Z_{05}} \frac{a_0}{a_6}$$

Table 5.2 Formulas for determining the characteristic impedances Z_{0i} $(i = 1, 2, \ldots, N)$ of SIR with open- and short-circuited ends.

$N = 2$

$$Z_{02} = \frac{a_1}{b_0 + b_2}, \quad Z_{01} = Z_{02} \frac{b_2}{b_0}$$

$N = 3$

$$Z_{03} = \frac{a_1 + a_3}{b_0 + b_2}, \quad Z_{02} = \frac{a_1 - Z_{03} b_0}{b_0 + a_3/Z_{03}},$$

$$Z_{01} = \frac{Z_{02}}{Z_{03}} \frac{a_3}{b_0}$$

$N = 4$

$$Z_{04} = \frac{a_1 + a_3}{b_0 + b_2 + b_4},$$

$$Z_{03} = \frac{a_1 + (b_4 - b_0) Z_{04}}{b_0 - b_4 + a_3/Z_{04}},$$

$$Z_{02} = \frac{a_1 - b_0 (Z_{03} + Z_{04})}{b_0 + b_4 Z_{04}/Z_{03}},$$

$$Z_{01} = \frac{Z_{02} Z_{04}}{Z_{03}} \frac{b_4}{b_0}$$

$N = 5$

$$Z_{05} = \frac{a_1 + a_3 + a_5}{b_0 + b_2 + b_4},$$

$$Z_{04} = \frac{a_1 - a_5 + Z_{05} (b_4 - b_0)}{2b_0 + b_2 + (a_5 - a_1)/Z_{05}},$$

$$Z_{03} = \frac{a_1 - b_0 (Z_{04} + Z_{05}) + a_5 Z_{04}/Z_{05}}{b_0 + b_4 Z_{05}/Z_{04} - a_5 (1/Z_{04} + 1/Z_{05})},$$

$$Z_{02} = \frac{a_1 - b_0 (Z_{03} + Z_{04} + Z_{05})}{b_0 + a_5 Z_{04}/Z_{03} Z_{05}},$$

$$Z_{01} = \frac{Z_{02} Z_{04}}{Z_{03} Z_{05}} \frac{a_5}{b_0}$$

$N = 6$

$$Z_{06} = \frac{a_1 + a_3 + a_5}{b_0 + b_2 + b_4 + b_6}, \quad Z_{05} = \frac{a_1 - a_5 + (b_4 + 2b_6 + b_0) Z_{06}}{2b_0 + b_2 - b_6 + (a_5 - a_1)/Z_{06}},$$

$$Z_{04} = \frac{a_1 + a_5 Z_{05}/Z_{06} - (b_0 + b_6)(Z_{05} + Z_{06})}{b_0 (3 + Z_{06}/Z_{05}) + b_2 + b_6 Z_{06}/Z_{05} - a_1 (1/Z_{05} + 1/Z_{06})},$$

$$Z_{03} = \frac{a_1 - b_0 (Z_{04} + Z_{05} + Z_{06}) + b_6 Z_{04} Z_{06}/Z_{05}}{b_0 - b_6 (Z_{05}/Z_{04} + Z_{06}/Z_{04} + Z_{06}/Z_{05}) + a_5 Z_{05}/Z_{04} Z_{06}},$$

$$Z_{02} = \frac{a_1 - b_0 (Z_{03} + Z_{04} + Z_{05} + Z_{06})}{b_0 + b_6 Z_{04} Z_{06}/Z_{03} Z_{05}}, \quad Z_{01} = \frac{Z_{02} Z_{04} Z_{06}}{Z_{03} Z_{05}} \frac{b_6}{b_0}$$

The given formulas are obtained as follows. According to the equality $Z_{0N} = Z^{(N)}(1)$ established in Ref. [20], the characteristic impedance Z_{0N} of the segment adjoining the input is determined by the expression that contains only the coefficients of the polynomials

$$Z_{0N} = \frac{a_0^{(N)} + a_2^{(N)} + \ldots + a_N^{(N)}}{b_1^{(N)} + b_3^{(N)} + \ldots + b_{N-1}^{(N)}}.$$

Knowing Z_{0N} and using (5.26), we determine the coefficients $a_i^{(N-1)}$, $b_i^{(N-1)}$ of the input impedance of a lower degree $Z^{(N-1)}(S)$. Substituting these coefficients into the upper equality, we determine the characteristic impedance of the next segment $Z_{0(N-1)}$, and so on.

5.4 Synthesis Example of Stepped-impedance Resonators at Given Critical Frequencies

Consider an example of three harmonic suppressions using SIR with a short-circuited end [26]. From the general patterns considered earlier, it follows that to suppress three harmonics, it is sufficient to use SIR with $N = 4$, which has a fourth-order input reactance function:

$$Z^4(S) = \frac{a_1 S + a_3 S^3}{b_0 + b_2 S^2 + b_4 S^4}. \tag{5.27}$$

To ensure the suppression of three harmonics, it is necessary to place the critical frequencies within the first half-period $T/2$: $\omega_{p1} = 1$, $\omega_{p2} = 2.5$, $\omega_{01} = 2$, $\omega_{02} = 3$ (Figure 5.6). The antiresonant frequency ω_{02}, is the fourth (in the case of continuous numbering), located in the center of the period T, and is the center of symmetry. Therefore, the number N of composite segments of SIR is 4. Other critical frequencies ω_{p3}, ω_{03}, ω_{p4}, and ω_{04} are located symmetrically concerning ω_{02}. The lowest resonant frequency ω_{p1} coincides with the frequency of the main signal ω_0. Antiresonant frequencies ω_{01}, ω_{02}, ω_{03} coincide with the first three harmonics of the main signal $2\omega_0$, $3\omega_0$, $4\omega_0$ and will suppress them.

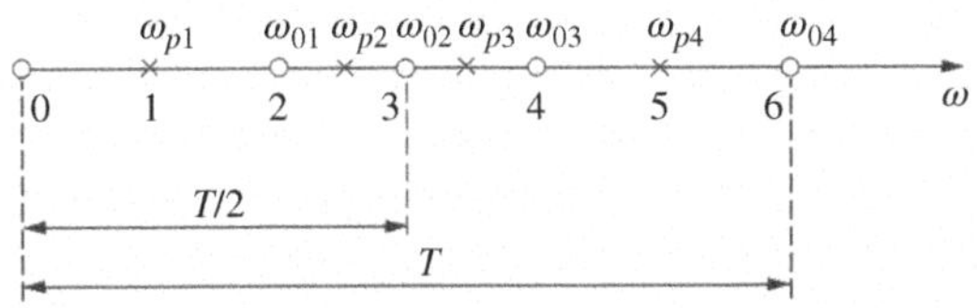

Figure 5.6 Critical frequencies of SIR with $N = 4$, suppressing three harmonics.

The reactance function $Z^4(S)$ (5.27) can be formed in the abovementioned sequence.

1. The antiresonant frequency $\omega_{02} = 3$ is the center of symmetry. It determines the delay time of the composite segment (5.21): $t'_d = \pi/6$.
2. Using the frequency transformation (5.24), we determine the critical frequencies of the first half-period on the Ω-axis: $\Omega_{p1} = 0.57735$, $\Omega_{01} = 1.7320$, $\Omega_{p2} = 3.7320$.
3. We write the function $Z^4(S)$ in the form (5.20b)

$$Z^4(S) = H\frac{S(S^2 + \Omega_{01}^2)}{\left(S^2 + \Omega_{p1}^2\right)\left(S^2 + \Omega_{p2}^2\right)} = H\frac{S(S^2 + 3)}{(S^2 + 0.33333)(S^2 + 13.9278)}. \tag{5.28}$$

4. The H value in (5.28) does not affect the location of the critical frequencies so that it can be arbitrary. Assuming $H = 100$ and carrying out multiplication, we finally pass to the notation form (5.27)

$$Z^4(S) = \frac{300S + 100S^3}{4.64273 + 14.2615S^2 + S^4}; \tag{5.29}$$

5. The determination of the characteristic impedances Z_{0i} of SIR from the function (5.29) can be performed using the Richards' procedure, which is based on expression (5.26). However, these calculations are quite cumbersome. To calculate Z_{0i}, we use Table 5.2 for $N = 4$. In these formulas: $b_0 = 4.64273$, $a_1 = 300$, $b_2 = 14.2615$, $a_3 = 100$, $b_4 = 1$. Calculations using these formulas give:

$$\begin{aligned} Z_{04} &= 20.096\ \Omega, \quad Z_{03} = 26.314\ \Omega, \\ Z_{02} &= 15.635\ \Omega, \quad Z_{01} = 2.5719\ \Omega. \end{aligned} \tag{5.30}$$

Figure 5.7a shows a schematic representation of the obtained solution (5.30) for SIR. The electrical length of each composite segment $\theta' = 30°$ corresponds to the frequency of 1 GHz. Figure 5.7b shows the simulated input impedance function $Z^4(S)$ of the synthesized SIR. The construction was done using the linear circuit simulator of Microwave Office software. The first three antiresonant frequencies ω_{01}, ω_{02}, ω_{03} coincide with harmonic frequencies and suppress them.

Figure 5.8 schematically depicts the topology of the synthesized SIR. The practicability factor $m = Z_{0\max}/Z_{0\min}$ is essential for TL resonators. In the obtained solution (5.30), the value of $Z_{01} = 2.5719\ \Omega$ is very low. This results in a value of $m = 10.231$, which is difficult to implement in planar designs. The value of Z_{01} can

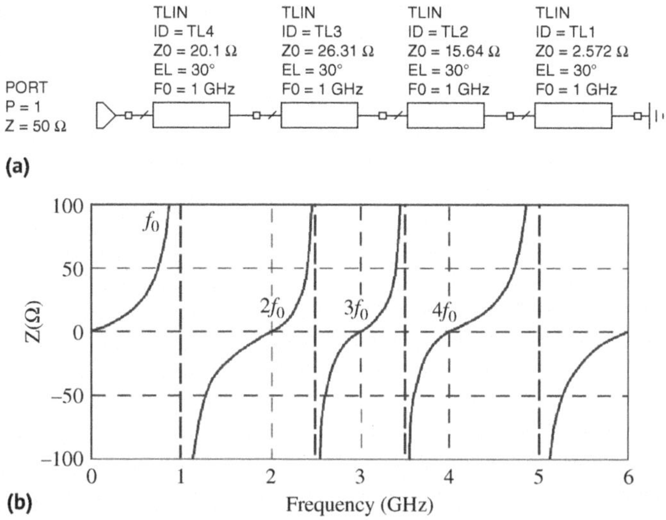

Figure 5.7 Synthesized SIR to suppress three harmonics at frequencies $2f_0$, $3f_0$, $4f_0$. (a) Schematic. (b) Input impedance $Z^4(S)$ (5.29).

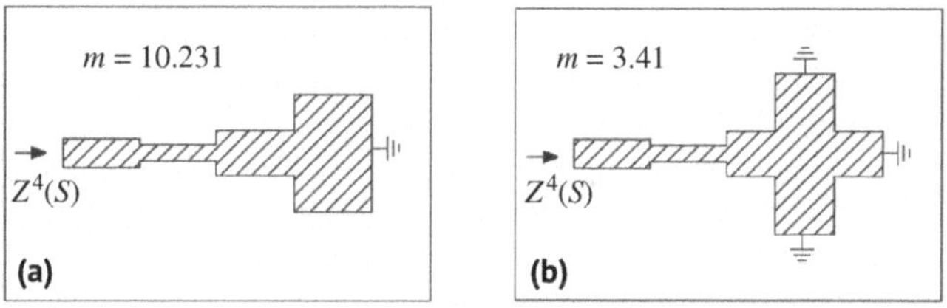

Figure 5.8 Topology of planar synthesized TL resonators that suppress three harmonics. (a) SIR. (b) Resonator with two additional stubs.

be increased three times, up to the value of $Z_{01} = 7.7157\ \Omega$. To do this, it is enough to go from SIR (Figure 5.8a) to a resonator with two additional stubs (Figure 5.8b). In this case, the parameter m will take the value $m = 3.41$, which is quite simple to implement in planar construction.

Figure 5.9a shows the second-order BPF with synthesized resonators, which is characterized by fractional bandwidth FBW = 0.03. The coupling between the resonators is carried out by capacitance. The inductances carry out the coupling between the resonators and loads. The simulated frequency responses of the BPF are displayed in Figure 5.9b, which indicates the presence of attenuation poles on the first three harmonics of the main signal. SIR with N segments can generally continuously suppress $(N - 1)$ signal harmonics.

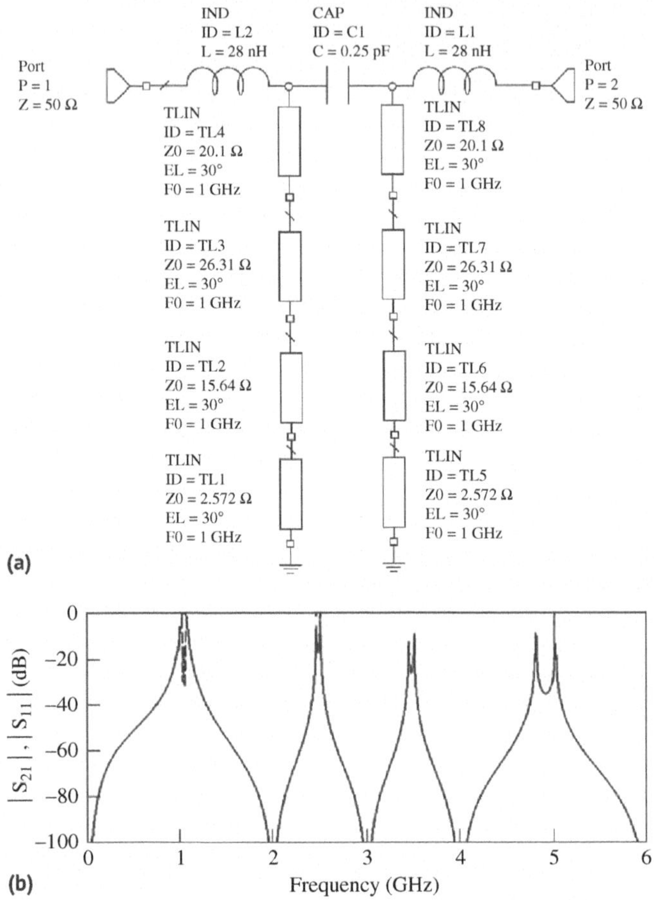

Figure 5.9 Second-order BPF with synthesized SIR. (a) Schematic. (b) Simulated frequency responses.

5.5 Optimization of Nonuniform Transmission Line Resonators

As noted, an important design parameter of nonuniform TL resonators is the $m = Z_{0max}/Z_{0min}$ ratio. This parameter is limited to the value of $m = 10$ for stripline and microstrip TLs.

To confirm this, Table 5.3 presents the characteristic impedances of microstrip lines with $w = 0.1$ mm and $w = 5$ mm on the substrate with different ε_r and $h = 0.5$ mm. This table also contains the ratio of the specified characteristic

Table 5.3 Characteristic impedances of microstrip line with $w = 0.1$ mm and $w = 5$ mm on the substrate with different ε_r and $h = 0.5$ mm.

ε_r	**2.2**	**4.5**	**9.7**	**21**	**38**	**92**
Z_0, Ω ($w = 0.1$ mm)	169.9	127.4	90.5	62.8	47.0	30.4
Z_0, Ω ($w = 5$ mm)	20.44	14.6	10.07	6.88	5.13	3.30
$m = \dfrac{Z_0\ (w = 5\ \text{mm})}{Z_0\ (w = 0.1\ \text{mm})}$	8.31	8.73	8.99	9.13	9.16	9.21

Table 5.4 Characteristic impedances of microstrip line with $w = 0.1$ mm and $w = 5$ mm on the substrate with different ε_r and $h = 1$ mm.

ε_r	**2.2**	**4.5**	**9.7**	**21**	**38**	**92**
Z_0, Ω ($w = 0.1$ mm)	202.6	152.4	108.4	75.3	56.4	36.7
Z_0, Ω ($w = 5$ mm)	35.41	25.6	17.74	12.16	9.11	5.93
$m = \dfrac{Z_0\ (w = 5\ \text{mm})}{Z_0\ (w = 0.1\ \text{mm})}$	5.72	5.95	6.11	6.19	6.19	6.19

impedances. Table 5.4 presents similar values for a thicker substrate $h = 1$ mm. Note that the thinner the substrate, the greater the value of m can be realized.

Since the given ratios between the initial critical frequencies can be realized on the set of characteristic impedance functions $\{Z_{0k}(x)\}$, $k = 1, 2, \ldots$. Then, the limited value of m leads to the following optimization problem. It is required to determine the function $Z_0^*(x)$ among the set of functions $\{Z_{0k}(x)\}$, providing the resonator with the given ratios between its initial critical frequencies has a minimum value of m. The formulated problem involves the optimization of the resonator shape.

The shape optimization of SIRs [14–18] is carried out as follows. SIRs with one or two (symmetrically located) characteristic impedance jumps are considered. The dependence of the ratio of some critical frequencies on the coordinate of the characteristic impedance jump is plotted. The coordinate at which the specified parameter takes an extreme value is fixed. It determines the shape of the "optimal" resonator. This way of solving the problem of SIRs optimizing is somewhat limited. It does not use the theoretical results presented in the classical works of P. I. Richards [20] and H. J. Riblet [21], which was later supplemented in Ref. [27].

The aforementioned expressions (5.11)–(5.13) for CIRs and (5.23)–(5.26) for SIRs allow solving the problem of minimizing the m parameter

$$\min(m) = \min(Z_{0\max}/Z_{0\min}) \tag{5.31}$$

of nonuniform TL resonators by repeatedly solving the inverse problem based on the transition from the critical frequencies ω_{pn} and ω_{0n} to the characteristic impedance function $Z_0(\tau)$. Some of these frequencies are interconnected by conditions determined by the specific problem being solved.

Let us consider the problem of removing the resonance frequencies ω_{p1} and ω_{p2} of the half-wave SIR from each other. In a uniform resonator, we have $\omega_{p2}/\omega_{p1} = 2$. An increase in the ratio ω_{p2}/ω_{p1} causes the BPF rejection band extension. Let us set $\omega_{p2}/\omega_{p1} = 5$, and write down this equality in the form of a coupling condition between ω_{p2} and ω_{p1}

$$\omega_{p2} - 5\omega_{p1} = 0. \tag{5.32}$$

The minimization problem of m (5.31) can be formulated as follows. Under restrictions (5.7) on critical frequencies ω_{pn} and ω_{0n} (independent variables), two of which satisfy the coupling condition (5.32), it is necessary to minimize the objective function

$$m = m\left(\{\omega_{pn}\}, \{\omega_{0n}\}\right), \quad m \geq 1, \tag{5.33}$$

and determine the function $Z_0^*(\tau)$ implementing this minimum.

To solve the formulated problem, we choose an SIR with the number of composite segments $N = 30$. We put $t_d = \pi/2$, then the alternation condition (5.7) takes the form

$$0 < \omega_{01} < \omega_{p1} < \omega_{02} < \omega_{p2} < \dots \omega_{015} < \omega_{p15} = 30. \tag{5.34}$$

The coupling condition (5.32) indicates that the number of independent critical frequencies in sequence (5.34) is 28. The dimension of the objective function minimization problem (5.33) is also 28. Note that the objective function (5.33) is multi-extreme. Various mathematical programming methods can be used to search for the global minimum of this function. The mentioned methods are beyond the scope of this consideration.

The solution to the minimization problem m (5.32)–(5.34) is provided for the critical frequencies $\omega_{01} = 0.664$, $\omega_{p1} = 1$, $\omega_{02} = 3$, $\omega_{p2} = 5$, $\omega_{03} = 5.336$, $\omega_{p3} = 6$. The remaining critical frequencies are located periodically with the period $T = 6$. The minimum value $m_{\min}$ is implemented for optimum function $Z_0^*(\tau)$: $Z_{01} = Z_{02} = \dots = Z_{010} = Z_{0\min}$; $Z_{011} = Z_{012} = \dots = Z_{020} = Z_{0\max}$; $Z_{021} = Z_{022} = \dots = Z_{030} = Z_{0\min}$. The values of the characteristic impedances are normalized by $Z_0^*(0)$. The solid line in Figure 5.10 presents the optimum function. The dotted line shows the arbitrary Z_{0i} of SIR, which contains 30 line segments of the same length.

We can increase the number of segments forming SIR when solving the problem (5.32)–(5.34), but the optimal form of SIR (Figure 5.10) is not changed. The characteristic impedances Z_{0i} take only two values $Z_{0\min}$ and $Z_{0\max}$, and the number of sub-segments is $N = 3$.

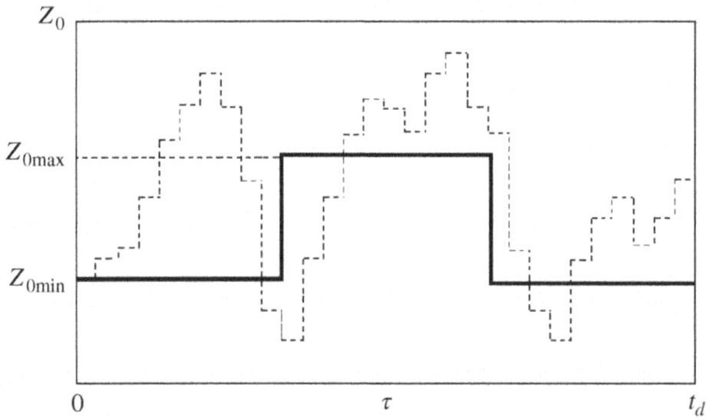

Figure 5.10 Optimal shape of half-wave SIR with 30 sub-sections, providing ratio of resonant frequencies $\omega_{p2}/\omega_{p1} > 2$.

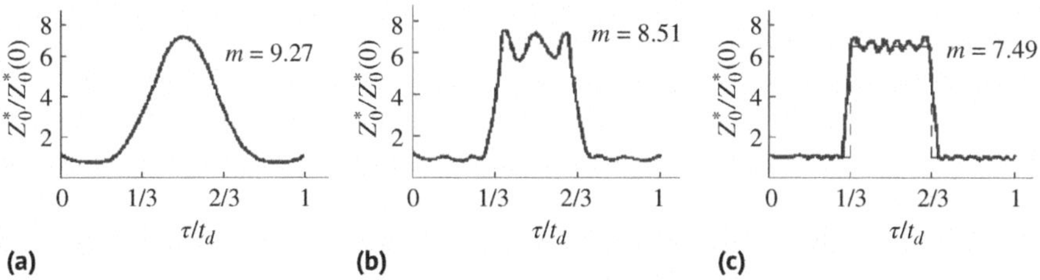

Figure 5.11 Minimization of parameter m of CIR with different numbers K of independent critical frequencies: (a) $K = 5$. (b) $K = 15$. (c) $K = 40$.

Using CIR, let us solve the same problem (5.32)–(5.34). Let us assume that the number of changed critical frequencies ω_{pn} and ω_{0n} determining characteristic impedance $Z_0^*(\tau)$ (5.12) and parameter m (5.33) are equal to 6. Based on the coupling condition (5.32), the dimensionality K of the minimization problem (5.33) is $K = 5$.

Based on the coupling condition (5.32), the dimensionality K of the minimization problem (5.33) is $K = 5$. The minimum value, $m = 9.27$, is attained at the values of variables $\omega_{01} = 0.660$, $\omega_{p1} = 1.004$, $\omega_{02} = 3.416$, $\omega_{p2} = 5.020$, $\omega_{03} = 5.411$, $\omega_{p3} = 6.608$ and is implemented for function $Z_0^*(\tau)$ plotted in Figure 5.11a. The function $Z_0^*(\tau)$ (Figure 5.11b) is obtained as a result of the solution of the same problem, but for the number of independent variables $K = 15$. The minimum value of parameter m is 8.51. At $K = 40$, the minimum value of this parameter is $m = 7.49$. It is implemented in function $Z_0^*(\tau)$ shown in Figure 5.11c by the solid line.

The solutions presented in Figure 5.11 show that an increase in the number of changed critical frequencies ω_{pn} and ω_{0n} of the CIR leads to more accurate reproduction of the optimal solution. Such a solution is shown by the dashed line in Figure 5.11c. It represents a two-level SIR with $m = 6.46$.

SIR has an advantage over CIR. This is explained by the fact that the critical frequencies of SIR are located periodically, which is also typical for very high frequencies. The critical frequencies of CIR with huge serial numbers (5.8) and (5.9) differ from the critical frequencies of SIR. The more critical frequencies $\{\omega_{pn}\}, \{\omega_{0n}\}$ are used in the minimization problem (5.31)–(5.33) of CIR, the greater the convergence between the obtained solution and the optimal SIR. However, complete coincidence cannot be achieved, since this is prevented by asymptotic differences (5.8) and (5.9).

The obtained optimal solution for SIR (Figure 5.10) is characterized by a twofold degeneration: the minimum number of composite segments N (for this problem) and two levels of characteristic impedances $Z_{0\min}$ or $Z_{0\max}$. This shows the desire for optimal resonators to adopt the simplest forms.

The aforementioned two characteristic features for optimal SIR (Figure 5.10) extend to other optimal SIRs, which differ in coupling between critical frequencies (5.32). In practice, resonators are required with different ratios between the critical frequencies, like (5.32), which have a minimum value of m. It is somewhat complex and laborious to minimize the value of m by repeatedly solving the inverse problem for SIR. The method of parametric synthesis is considered as follows. They make it possible to minimize m more simply. These methods are based on the noted characteristics inherent in optimal SIR. Let us give them the mathematical interpretation

$$Z^N(S) = Z^{N\min}(S) \tag{5.35}$$

$$Z_{0i} = Z_{0\min} \text{ or } Z_{0\max} \tag{5.36}$$

Condition (5.35) means that the order of the realized reactance function $Z^N(S)$ should be as small as possible. The condition (5.36) is self-explanatory. We note $Z_{0\min} = Z_0$ and $Z_{0\max} = mZ_0$.

5.6 Parametric Synthesis

The parametric synthesis method applies to SIRs. The principle of the method consists of two previously established patterns:

1. There is a minimum order of the reactance function of the input impedance (5.35), which makes it possible to perform the required placement of some critical frequencies associated with the coupling conditions (5.32).

2. The characteristic impedances Z_{0i} of the optimal SIR take only two values (5.36), and the explicit dependence of the coefficients $Z^N(S)$ on Z_{0i} is used to determine them.

To obtain the elements of the SIR transfer matrix $\left[A^{(N)} \quad B^{(N)} \quad C^{(N)} \quad D^{(N)}\right]$ of order N, we need the SIR transfer matrix $\left[A^{(N-1)} \quad B^{(N-1)} \quad C^{(N-1)} \quad D^{(N-1)}\right]$ of order $(N-1)$ multiply on the left by the transfer matrix of the TL segment with characteristic impedance Z_{0N}

$$\begin{bmatrix} A^{(N)} & B^{(N)} \\ C^{(N)} & D^{(N)} \end{bmatrix} = \left(1 - S^2\right)^{-1/2} \begin{bmatrix} 1 & SZ_{0N} \\ S/Z_{0N} & 1 \end{bmatrix} \begin{bmatrix} A^{(N-1)} & B^{(N-1)} \\ C^{(N-1)} & D^{(N-1)} \end{bmatrix} \tag{5.37}$$

The expression (5.37) is recurrent. It is used to determine the polynomial coefficients a_i, b_i (5.19) of the transfer matrix for SIRs with different values of N. The coefficients a_i, b_i for $N \leq 4$ are presented in Table 5.5a.

Note that the terms of the coefficients a_i and b_i are mutually opposite quantities, so that we can move simply from the expressions for a_i to the expressions for b_i, and vice versa.

The coefficients a_i for $N = 5, 6$ are presented in Table 5.5b.

At the second stage, the values of Z_{0i} are determined, which act as parameters. Therefore, this sequence of optimal SIRs construction is called parametric synthesis.

Let us consider the application of the method using various examples. The results will be summarized in Table 5.6 (quarter-wave SIRs) and Table 5.7 (half-wave SIRs). These tables contain topologies, critical frequencies, functions $Z^N(S)$, and the ratio of critical frequencies. The arrow shows the input of each resonator. A narrower strip corresponds to a higher characteristic impedance.

For $N = 2$, we have:

No. 1 (Table 5.6): Let us consider the problem of removing the resonance frequencies ω_{p1} and ω_{p2} of quarter-wave SIRs from each other. In a uniform resonator, $\omega_{p2}/\omega_{p1} = 3$. The minimum order of the function $Z^N(S)$, which allows solving this problem, equals $N_{\min} = 2$. This function is written as

$$Z^2(S) = \frac{a_1{}^2 S}{b_0^{(2)} + b_2^{(2)} S^2} = \frac{(Z_{01} + Z_{02})\, S}{1 + (Z_{01}/Z_{02})\, S^2}, \tag{5.38}$$

where $S = j\Omega = j\tan(\omega t_d/2)$, and coefficients $a_1^{(2)}$, $b_0^{(2)}$, $b_2^{(2)}$ are contained in Table 5.5a.

The pole of function (5.38) is located at the frequency $\Omega_{p1}^2 = b_0^{(2)}/b_2^{(2)} = Z_{02}/Z_{01}$. Its decrease leads to an increase in the ratio ω_{p2}/ω_{p1}. The maximum effect will be provided when $Z_{01} = Z_{0\max} = mZ_0$ and $Z_{02} = Z_{0\min} = Z_0$. This resonator is the most demanded, and it has been used for a long time [14]. Along with the increased

Table 5.5a Transfer matrix polynomial coefficients of SIRs with $N \leq 4$.

N	$a_0^{(N)}, b_0^{(N)}$	$a_1^{(N)}, b_1^{(N)}$	$a_2^{(N)}, b_2^{(N)}$	$a_3^{(N)}, b_3^{(N)}$	$a_4^{(N)}, b_4^{(N)}$
1	$a_0^{(1)} = 1, b_0^{(1)} = 1$	$a_1^{(1)} = Z_{01}$ $b_1^{(1)} = 1/Z_{01}$			
2	$a_0^{(2)} = 1, b_0^{(2)} = 1$	$a_1^{(2)} = Z_{01} + Z_{02}$ $b_1^{(2)} = 1/Z_{01} + 1/Z_{02}$	$a_2^{(2)} = Z_{02}/Z_{01}$ $b_2^{(2)} = Z_{01}/Z_{02}$		
3	$a_0^{(3)} = 1, b_0^{(3)} = 1$	$a_1^{(3)} = Z_{01} + Z_{02} + Z_{03}$ $b_1^{(3)} = \frac{1}{Z_{01}} + \frac{1}{Z_{02}} + \frac{1}{Z_{03}}$	$a_2^{(3)} = \frac{Z_{02}}{Z_{01}} + \frac{Z_{03}}{Z_{01}} + \frac{Z_{03}}{Z_{02}}$ $b_2^{(3)} = \frac{Z_{01}}{Z_{02}} + \frac{Z_{01}}{Z_{03}} + \frac{Z_{02}}{Z_{03}}$	$a_3^{(3)} = \frac{Z_{01}Z_{03}}{Z_{02}}$ $b_3^{(3)} = \frac{Z_{02}}{Z_{01}Z_{03}}$	
4	$a_0^{(4)} = 1, b_0^{(4)} = 1$	$a_1^{(4)} = Z_{01} + Z_{02} + Z_{03} + Z_{04}$ $b_1^{(4)} = \frac{1}{Z_{01}} + \frac{1}{Z_{02}} + \frac{1}{Z_{03}} + \frac{1}{Z_{04}}$	$a_2^{(4)} = \frac{Z_{02}}{Z_{01}} + \frac{Z_{03}}{Z_{01}} + \frac{Z_{04}}{Z_{01}}$ $+\frac{Z_{03}}{Z_{02}} + \frac{Z_{04}}{Z_{02}} + \frac{Z_{04}}{Z_{03}}$ $b_2^{(4)} = \frac{Z_{01}}{Z_{02}} + \frac{Z_{01}}{Z_{03}} + \frac{Z_{01}}{Z_{04}}$ $+\frac{Z_{02}}{Z_{03}} + \frac{Z_{02}}{Z_{04}} + \frac{Z_{03}}{Z_{04}}$	$a_3^{(4)} = \frac{Z_{01}Z_{03}}{Z_{02}} + \frac{Z_{01}Z_{04}}{Z_{02}}$ $+\frac{Z_{01}Z_{04}}{Z_{03}} + \frac{Z_{02}Z_{04}}{Z_{03}}$ $b_3^{(4)} = \frac{Z_{02}}{Z_{01}Z_{03}} + \frac{Z_{02}}{Z_{01}Z_{04}}$ $+\frac{Z_{03}}{Z_{01}Z_{04}} + \frac{Z_{03}}{Z_{02}Z_{04}}$	$a_4^{(4)} = \frac{Z_{02}Z_{04}}{Z_{01}Z_{03}}$ $b_4^{(4)} = \frac{Z_{01}Z_{03}}{Z_{02}Z_{04}}$

Table 5.5b Transfer matrix polynomial coefficients of SIRs with $N = 5, 6$.

$N = 5$

$$a_0^{(5)} = 1,\ a_1^{(5)} = Z_{01} + Z_{02} + Z_{03} + Z_{04} + Z_{05}$$

$$a_2^{(5)} = \frac{Z_{02}}{Z_{01}} + \frac{Z_{03}}{Z_{01}} + \frac{Z_{04}}{Z_{01}} + \frac{Z_{05}}{Z_{01}} + \frac{Z_{03}}{Z_{02}} + \frac{Z_{04}}{Z_{02}} + \frac{Z_{05}}{Z_{02}} + \frac{Z_{04}}{Z_{03}} + \frac{Z_{05}}{Z_{03}} + \frac{Z_{05}}{Z_{04}}$$

$$a_3^{(5)} = \frac{Z_{01}Z_{03}}{Z_{02}} + \frac{Z_{01}Z_{04}}{Z_{02}} + \frac{Z_{01}Z_{05}}{Z_{02}} + \frac{Z_{01}Z_{04}}{Z_{03}} + \frac{Z_{01}Z_{05}}{Z_{03}} + \frac{Z_{01}Z_{05}}{Z_{04}} + \frac{Z_{02}Z_{04}}{Z_{03}} + \frac{Z_{02}Z_{05}}{Z_{03}} + \frac{Z_{02}Z_{05}}{Z_{04}} + \frac{Z_{03}Z_{05}}{Z_{04}}$$

$$a_4^{(5)} = \frac{Z_{02}Z_{04}}{Z_{01}Z_{03}} + \frac{Z_{02}Z_{05}}{Z_{01}Z_{03}} + \frac{Z_{02}Z_{05}}{Z_{01}Z_{04}} + \frac{Z_{03}Z_{05}}{Z_{01}Z_{04}} + \frac{Z_{03}Z_{05}}{Z_{02}Z_{04}}$$

$$a_5^{(5)} = \frac{Z_{01}Z_{03}Z_{05}}{Z_{02}Z_{04}}$$

$N = 6$

$$a_0^{(6)} = 1;\ a_1^{(6)} = Z_{01} + Z_{02} + Z_{03} + Z_{04} + Z_{05} + Z_{06}$$

$$a_2^{(6)} = \frac{Z_{02}}{Z_{01}} + \frac{Z_{03}}{Z_{01}} + \frac{Z_{04}}{Z_{01}} + \frac{Z_{05}}{Z_{01}} + \frac{Z_{06}}{Z_{01}} + \frac{Z_{03}}{Z_{02}} + \frac{Z_{04}}{Z_{02}} + \frac{Z_{05}}{Z_{02}} + \frac{Z_{06}}{Z_{02}} + \frac{Z_{04}}{Z_{03}} + \frac{Z_{05}}{Z_{03}} + \frac{Z_{06}}{Z_{03}} + \frac{Z_{05}}{Z_{04}} + \frac{Z_{06}}{Z_{04}} + \frac{Z_{06}}{Z_{05}}$$

$$a_3^{(6)} = \frac{Z_{01}Z_{03}}{Z_{02}} + \frac{Z_{01}Z_{04}}{Z_{02}} + \frac{Z_{01}Z_{05}}{Z_{02}} + \frac{Z_{01}Z_{06}}{Z_{02}} + \frac{Z_{01}Z_{04}}{Z_{03}} + \frac{Z_{01}Z_{05}}{Z_{03}} + \frac{Z_{01}Z_{06}}{Z_{03}} + \frac{Z_{02}Z_{04}}{Z_{03}} + \frac{Z_{02}Z_{05}}{Z_{03}} + \frac{Z_{02}Z_{06}}{Z_{03}} + \frac{Z_{01}Z_{05}}{Z_{04}} + \frac{Z_{01}Z_{06}}{Z_{04}} + \frac{Z_{02}Z_{05}}{Z_{04}} + \frac{Z_{02}Z_{06}}{Z_{04}} + \frac{Z_{03}Z_{05}}{Z_{04}} + \frac{Z_{03}Z_{06}}{Z_{04}} + \frac{Z_{01}Z_{06}}{Z_{05}} + \frac{Z_{02}Z_{06}}{Z_{05}} + \frac{Z_{03}Z_{06}}{Z_{05}} + \frac{Z_{04}Z_{06}}{Z_{05}}$$

$$a_4^{(6)} = \frac{Z_{02}Z_{04}}{Z_{01}Z_{03}} + \frac{Z_{02}Z_{05}}{Z_{01}Z_{03}} + \frac{Z_{02}Z_{06}}{Z_{01}Z_{03}} + \frac{Z_{02}Z_{05}}{Z_{01}Z_{04}} + \frac{Z_{02}Z_{06}}{Z_{01}Z_{04}} + \frac{Z_{03}Z_{05}}{Z_{01}Z_{04}} + \frac{Z_{03}Z_{06}}{Z_{01}Z_{04}} + \frac{Z_{02}Z_{06}}{Z_{01}Z_{05}} + \frac{Z_{03}Z_{06}}{Z_{01}Z_{05}} + \frac{Z_{04}Z_{06}}{Z_{01}Z_{05}} + \frac{Z_{03}Z_{05}}{Z_{02}Z_{04}} + \frac{Z_{03}Z_{06}}{Z_{02}Z_{04}} + \frac{Z_{03}Z_{06}}{Z_{02}Z_{05}} + \frac{Z_{04}Z_{06}}{Z_{02}Z_{05}} + \frac{Z_{04}Z_{06}}{Z_{03}Z_{05}}$$

$$a_5^{(6)} = \frac{Z_{01}Z_{03}Z_{05}}{Z_{02}Z_{04}} + \frac{Z_{01}Z_{03}Z_{06}}{Z_{02}Z_{04}} + \frac{Z_{01}Z_{03}Z_{06}}{Z_{02}Z_{05}} + \frac{Z_{01}Z_{04}Z_{06}}{Z_{02}Z_{05}} + \frac{Z_{01}Z_{04}Z_{06}}{Z_{03}Z_{05}} + \frac{Z_{02}Z_{04}Z_{06}}{Z_{03}Z_{05}}$$

$$a_6^{(6)} = \frac{Z_{02}Z_{04}Z_{06}}{Z_{01}Z_{03}Z_{05}}$$

ω_{p2}/ω_{p1} ratio, it greatly lowers the main resonant frequency ω_{p1}. The expressions given in Table 5.6 allow us to estimate the achieved effects. So at $m = 5$ we obtain the ratio $\omega_{p2}/\omega_{p1} = 6.47$, and the resonant frequency ω_{p1} appears lower than the resonant frequency of the quarter-wave resonator ω°_{p1} by 1.87 times.

No. 2 (Table 5.6): Improving the tunability of the resonator by tuning the main resonant frequency with a variable capacitance. For this, it is necessary to expand the resonance region by increasing the value of ω_{p1} [28]. The maximum upward displacement of the pole Ω_{p1} of function (5.38) is attained when $Z_{01} = Z_{0\min} = Z_0$ and $Z_{02} = Z_{0\max} = mZ_0$. The resonant region in a tunable quarter-wave resonator covers the electrical length range from 0 to 90°. The expression from Table 5.6 shows that the upper boundary of the SIR resonance region expands to 131.8° at $m = 5$.

For $N = 3$, we have:

No. 1 (Table 5.7): Removing the resonance frequencies ω_{p1} and ω_{p2} of a half-wave SIR from each other. In a uniform resonator, $\omega_{p2}/\omega_{p1} = 2$. The minimum order of the input impedance function (5.35) is $N_{\min} = 3$. This function is written as

$$Z^3(S) = \frac{a_0^{(3)} + a_2^{(3)} S^2}{b_1^{(3)} S + b_3^{(3)} S^3}. \tag{5.39}$$

where $S = j\Omega = j\tan(\omega t_d/3)$ and the coefficients of the numerator and denominator polynomials are given in Table 5.5a. The pole of function (5.39) is located at the frequency $\Omega^2_{p1} = \frac{b_1^{(3)}}{b_3^{(3)}}$. Its decreasing leads to an increase in the ratio ω_{p2}/ω_{p1}. Using the formulas in Table 5.5 for $N = 3$, we determine

$$\Omega^2_{p1} = \frac{b_1^{(3)}}{b_3^{(3)}} = \frac{Z_{01}}{Z_{02}} + \frac{Z_{01}Z_{03}}{Z^2_{02}} + \frac{Z_{03}}{Z_{02}}, \tag{5.40}$$

whence it follows that the value of Ω^2_{p1} is minimal if $Z_{01} = Z_{03} = Z_{0\min} = Z_0$ and $Z_{02} = Z_{0\max} = mZ_0$. We obtained the same result laboriously presented in Figure 5.10 based on multiple solutions of the inverse spectral problem for SIR.

No. 2 (Table 5.7): Convergence of resonant frequencies ω_{p1} and ω_{p2} (decreasing ω_{p2}/ω_{p1} ratio). If the oscillations with the indicated frequencies are separated from each other by "small" frequency intervals, all of them take part in the wide passband formation. This resonator can be used to create an ultra-wideband BPF. An increase in the pole ω_{p1} (5.40) leads to the convergence of ω_{p1} and ω_{p2}. The maximum magnification effect is achieved when $Z_{01} = Z_{03} = Z_{0\max} = mZ_0$ and $Z_{02} = Z_{0\min} = Z_0$. The expression from Table 5.7 shows that the degree of convergence ω_{p1} and ω_{p2} is characterized by the ratio $\omega_{p2}/\omega_{p1} = 1.24$ for $m = 5$.

No. 3 (Table 5.7): Increasing the tunability of a half-wave resonator, tunable by a capacitor. An increase in the frequency interval between ω_{01} and ω_{p1}

Table 5.6 Optimal SIRs of quarter-wave type

No.	Topology and critical frequencies	Input impedance	Ratio of critical frequencies
No. 1	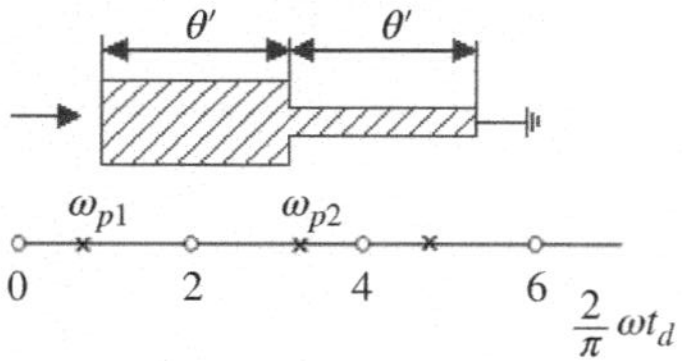	$Z^2(S) = Z_0 \dfrac{(1+m)S}{1+mS^2}$	$\dfrac{\omega_{p2}}{\omega_{p1}} = \dfrac{\pi}{\tan^{-1}\sqrt{1/m}} - 1,$ $\dfrac{\omega_{p1}}{\omega^{\circ}_{p1}} = \dfrac{4\tan^{-1}\sqrt{1/m}}{\pi}$
No. 2	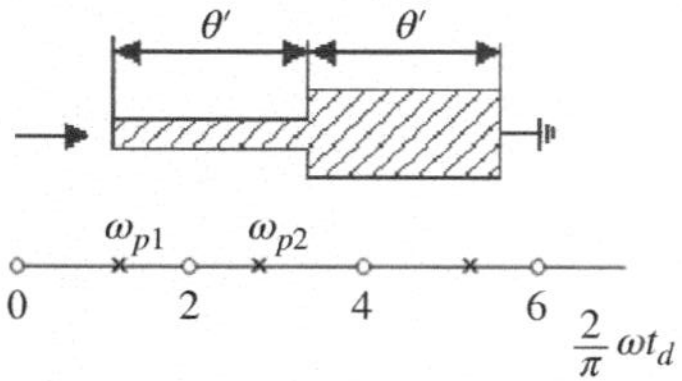	$Z^2(S) = Z_0 \dfrac{(1+m)S}{1+\frac{1}{m}S^2}$	$\dfrac{\omega_{p1}}{\omega^{\circ}_{p1}} = \dfrac{4\tan^{-1}\sqrt{m}}{\pi}$
No. 3	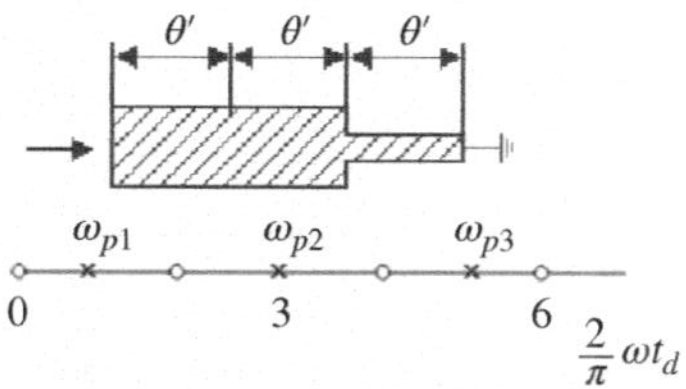	$Z^3(S) = Z_0 \dfrac{(2+m)S + mS^3}{1+(1+2m)S^2}$	$\dfrac{\omega_{p3}}{\omega_{p1}} = \dfrac{\pi}{\tan^{-1}\sqrt{\frac{1}{1+2m}}} - 1$

(Continued)

Table 5.6 (Continued)

No.	Topology and critical frequencies	Input impedance	Ratio of critical frequencies
No. 4	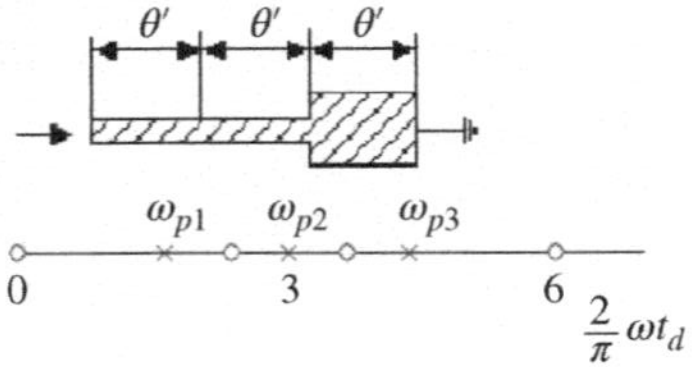	$Z^3(S) = Z_0 \dfrac{(1+2m)S + S^3}{1+\left(1+\frac{2}{m}\right)S^2}$	$\dfrac{\omega_{p3}}{\omega_{p1}} = \dfrac{\pi}{\tan^{-1}\sqrt{\frac{m}{2+m}}} - 1$
No. 5	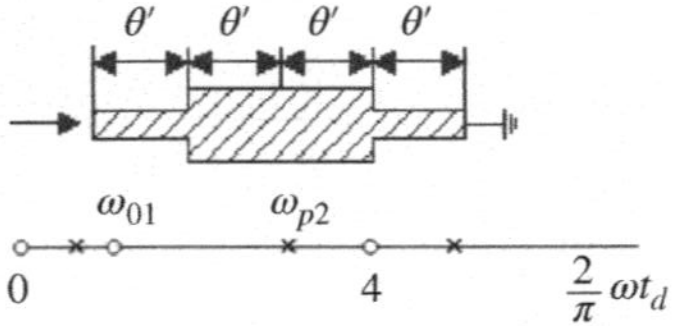	$Z^4(S) = Z_0 \dfrac{2(1+m)S + 2m(1+m)S^3}{1+2\left(1+m+\frac{1}{m}\right)S^2 + S^4}$	$\dfrac{\omega_{p2}}{\omega_{01}} = \dfrac{\tan^{-1}\sqrt{1+m+\frac{1}{m}+\sqrt{\left(1+m+\frac{1}{m}\right)^2-1}}}{\tan^{-1}\sqrt{\frac{1}{m}}}$
No. 6	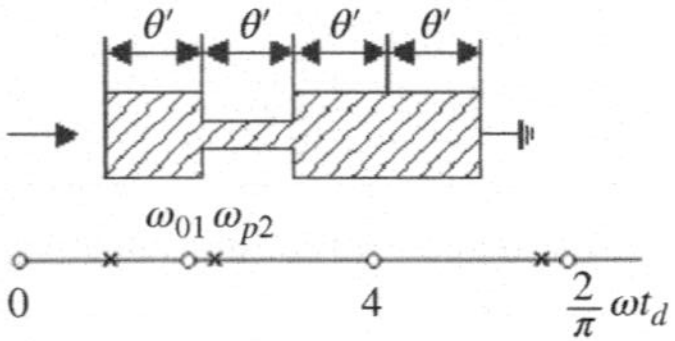	$Z^4(S) = Z_0 \dfrac{(3+m)S + \left(1+m+\frac{2}{m}\right)S^3}{1+\left(3+m+\frac{2}{m}\right)S^2 + mS^4}$	$\dfrac{\omega_{p2}}{\omega_{01}} = \dfrac{\tan^{-1}\sqrt{\frac{\left(1+\frac{3}{m}+\frac{2}{m^2}\right)}{2}+\sqrt{\frac{\left(1+\frac{3}{m}+\frac{2}{m^2}\right)^2}{4}-\frac{1}{m}}}}{\tan^{-1}\sqrt{\frac{3+m}{(1+m+2/m)}}}$

No.	Topology and critical frequencies	Input impedance	Ratio of critical frequencies
No. 1	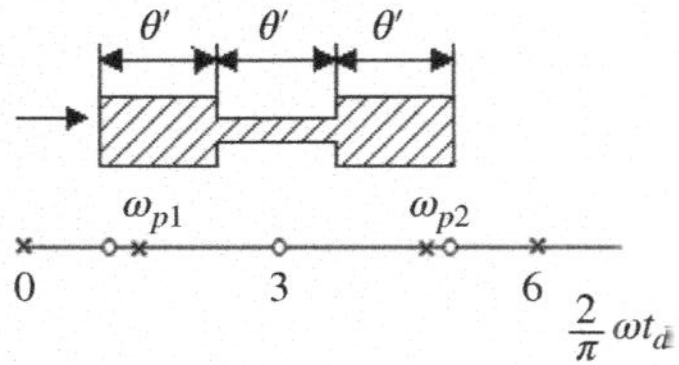	$Z^3(S) = Z_0 \frac{1 + \left(1 + m + \frac{1}{m}\right) S^2}{\left(2 + \frac{1}{m}\right) S + mS^2}$	$\frac{\omega_{p2}}{\omega_{p1}} = \frac{\pi}{\tan^{-1}\sqrt{\frac{2}{m} + \frac{1}{m^2}}} - 1$
No. 2	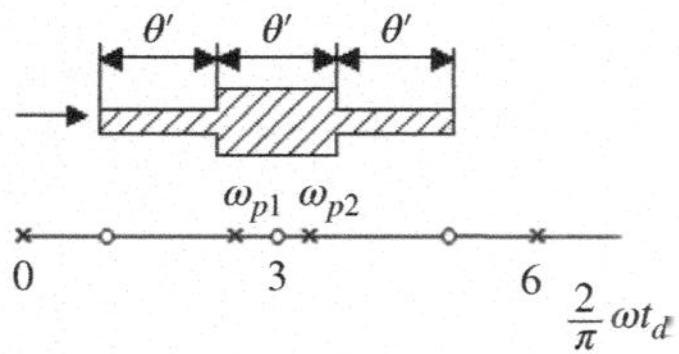	$Z^3(S) = Z_0 \frac{1 + \left(1 + m + \frac{1}{m}\right) S^2}{\left(1 + \frac{2}{m}\right) S + \frac{1}{m^2} S^3}$	$\frac{\omega_{p2}}{\omega_{p1}} = \frac{\pi}{\tan^{-1}\sqrt{2m + m^2}} - 1$
No. 3	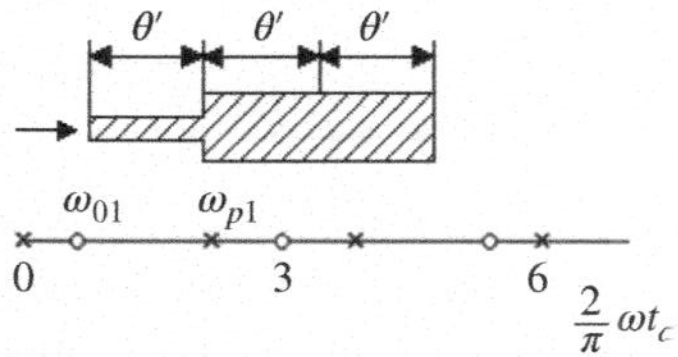	$Z^3(S) = Z_0 \frac{1 + (1 + 2m) S^2}{\left(2 + \frac{1}{m}\right) S + \frac{1}{m} S^3}$	$\frac{\omega_{p1}}{\omega_{01}} = \frac{\tan^{-1}\sqrt{1 + 2m}}{\tan^{-1}\sqrt{\frac{1}{1+2m}}}$
No. 4	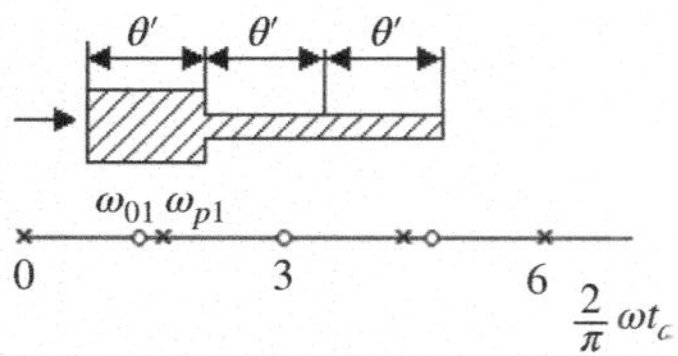	$Z^3(S) = Z_0 \frac{1 + \left(1 + \frac{2}{m}\right) S^2}{\left(1 + \frac{2}{m}\right) S + S^3}$	$\frac{\omega_{p1}}{\omega_{01}} = \frac{\tan^{-1}\sqrt{1 + \frac{1}{m}}}{\tan^{-1}\sqrt{\frac{1}{1+\frac{2}{m}}}}$

(*Continued*)

Table 5.7 (Continued)

No.	Topology and critical frequencies	Input impedance	Ratio of critical frequencies
No. 5	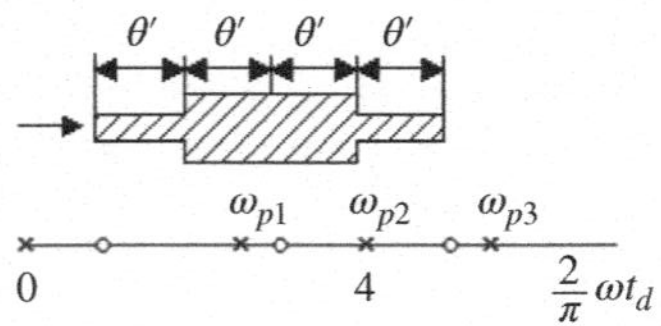	$Z^4(S) = Z_0 \frac{1 + 2\left(1 + m + \frac{1}{m}\right) S^2 + S^4}{2\left(1 + \frac{1}{m}\right) S + 2\left(\frac{1}{m} + \frac{1}{m^2}\right) S^3}$	$\frac{\omega_{p3}}{\omega_{p1}} = \frac{\pi}{\tan^{-1}\sqrt{m}} - 1$
No. 6	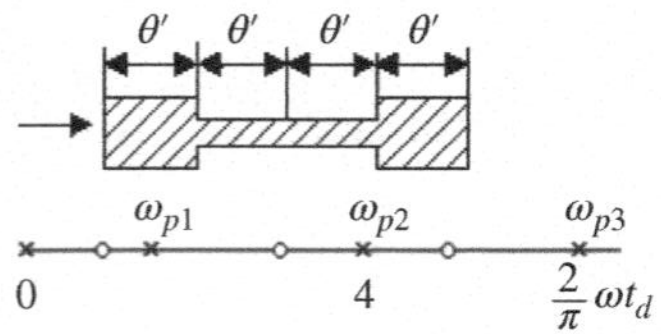	$Z^4(S) = Z_0 \frac{1 + 2\left(1 + m + \frac{1}{m}\right) S^2 + S^4}{2\left(1 + \frac{1}{m}\right) S + 2(1 + m) S^3}$	$\frac{\omega_{p3}}{\omega_{p1}} = \frac{\pi}{\tan^{-1}\sqrt{\frac{1}{m}}} - 1$

(increase ω_{p1}/ω_{01} ratio) leads to an expansion of the tuning range of the resonator with a lumped capacitor at the main resonance frequency ω_1: $\omega_{01} \leq \omega_1 \leq \omega_{p1}$ [28]. The half-wave resonator has a ratio $\omega_{p1}/\omega_{01} = 2$. This problem can be solved using the third-order reactance function (5.39) ($N = 3$). The pole of this function Ω_{p1} is expressed by equality (5.40), and the zero Ω_{01} of this function is determined by the expression

$$\Omega_{01}^2 = \frac{a_0^{(3)}}{a_2^{(3)}} = \frac{1}{\frac{Z_{02}}{Z_{01}} + \frac{Z_{03}}{Z_{01}} + \frac{Z_{03}}{Z_{02}}} = \frac{Z_{01}Z_{02}}{Z_{02}^2 + Z_{01}Z_{03} + Z_{02}Z_{03}}. \tag{5.41}$$

where $a_0^{(3)}$ and $a_2^{(3)}$ are taken from Table 5.5a.

It follows from (5.40) and (5.41) that the maximum value of the ratio Ω_{p1}/Ω_{01} is achieved at $Z_{01} = Z_{02} = Z_{0\min} = Z_0$ and $Z_{03} = Z_{0\max} = mZ_0$. The expression from Table 5.7 gives the ratio $\omega_{p1}/\omega_{01} = 4.36$ at $m = 5$.

No. 4 (Table 5.7): Increased stability of the main frequency of the half-wave resonator in the case of variation of the capacitive load. It can be achieved by decreasing the frequency interval between zero ω_{01} and pole ω_{p1}. The smaller the ω_{p1}/ω_{01} ratio, the higher the stability. Using (5.40) and (5.41), we determine that the minimum ratio Ω_{p1}/Ω_{01} is ensured at $Z_{01} = Z_{02} = Z_{0\max} = mZ_0$, and $Z_{03} = Z_{0\min} = Z_0$. The expression from Table 5.7 shows that the degree of convergence ω_{p1} and ω_{01} is characterized by the ratio $\omega_{p1}/\omega_{01} = 1.25$ at $m = 5$.

No. 3 (Table 5.6): Removing the resonance frequencies ω_{p1} and ω_{p3} of a quarter-wave SIR from each other. A wide rejection band can be achieved using this resonator in a BPF. This is achieved if the spurious passband at the frequency ω_{p2} is suppressed due to the zero coupling coefficient between resonators. In a uniform resonator, we have $\omega_{p3}/\omega_{p1} = 5$. The minimum order of the input impedance function (5.35) is $N_{\min} = 3$. This function is written as

$$Z^3(S) = \frac{a_1^{(3)}S + a_3^{(3)}S^3}{b_0^{(3)} + b_2^{(3)}S^2}, \tag{5.42}$$

where $S = j\Omega = j\tan(\omega t_d/3)$, and the coefficients of polynomials are given in Table 5.5a.

Lowering the pole Ω_{p1} of the function (5.42)

$$\Omega_{p1}^2 = \frac{b_0^{(3)}}{b_2^{(3)}} = \frac{1}{\frac{Z_{01}}{Z_{02}} + \frac{Z_{01}}{Z_{03}} + \frac{Z_{02}}{Z_{03}}} = \frac{Z_{02}Z_{03}}{Z_{02}^2 + Z_{01}(Z_{02} + Z_{03})} \tag{5.43}$$

automatically leads to an increase in Ω_{p3} and ω_{p3}/ω_{p1} ratio. From (5.43), it follows that the minimum value of Ω_{p1} is provided at $Z_{01} = Z_{0\max} = mZ_0$ and $Z_{02} = Z_{03} = Z_{0\min} = Z_0$. The formula from Table 5.6 gives the value of the ratio $\omega_{p3}/\omega_{p1} = 9.73$ at $m = 5$.

No. 4 (Table 5.6): The mutual convergence of resonant frequencies ω_{p1}, ω_{p2}, ω_{p3} of quarter-wave SIR allows realizing a multimode operation and using such a resonator in wideband BPFs. The ω_{p3}/ω_{p1} ratio can characterize this convergence. The smaller it is, the closer to each other the indicated frequencies. For this, the pole Ω_{p1} (5.43) of the resonator considered earlier must be moved up in frequency. This automatically leads to an increase in Ω_{p3} and a decrease in the ω_{p3}/ω_{p1} ratio. From (5.43), the maximum value of Ω_{p1} is achieved at $Z_{01} = Z_{0\min} = Z_0$ and $Z_{02} = Z_{03} = Z_{0\max} = mZ_0$. The expression from Table 5.6 gives the ratio $\omega_{p3}/\omega_{p1} = 3.48$ at $m = 5$.

For $N = 4$, we have:

No. 5 (Table 5.7): The mutual convergence of resonant frequencies ω_{p1}, ω_{p2}, ω_{p3} of half-wave SIR allows the use of such a resonator in broadband BPFs [29]. The ratio ω_{p3}/ω_{p1} characterizes the degree of convergence of these frequencies. In a uniform resonator, we have $\omega_{p3}/\omega_{p1} = 3$. The minimum order of the input impedance function (5.35), which makes it possible to reduce the ω_{p3}/ω_{p1} ratio, is equal to $N_{\min} = 4$. This function is written as

$$Z^4(S) = \frac{a_0^{(4)} + a_2^{(4)}S^2 + a_4^{(4)}S^4}{b_1^{(4)}S + b_3^{(4)}S^3}, \tag{5.44}$$

where $S = j\Omega = j\tan(\omega t_d/4)$, and the numerator and denominator polynomials' coefficients are given in Table 5.5a.

Note that the zeros of the input impedance (5.44) do not play any role in the problem under consideration, and they are not considered. Increase the frequency of the pole Ω_{p1} of the function (5.44)

$$\Omega_{p1}^2 = \frac{b_1^{(4)}}{b_3^{(4)}}, \tag{5.45}$$

where

$$b_1^{(4)} = \frac{1}{Z_{01}} + \frac{1}{Z_{02}} + \frac{1}{Z_{03}} + \frac{1}{Z_{04}}, \tag{5.46}$$

$$\begin{aligned} b_3^{(4)} &= \frac{Z_{02}}{Z_{01}Z_{03}} + \frac{Z_{02}}{Z_{01}Z_{04}} + \frac{Z_{03}}{Z_{01}Z_{04}} + \frac{Z_{03}}{Z_{02}Z_{04}} \\ &= \frac{1}{Z_{01}Z_{04}}\left[\frac{Z_{02}(Z_{03} + Z_{04})}{Z_{03}} + \frac{Z_{03}(Z_{01} + Z_{02})}{Z_{02}}\right] \end{aligned} \tag{5.47}$$

automatically leads to a decrease in Ω_{p3} and ω_{p3}/ω_{p1} ratio. To increase Ω_{p1} (5.45), it is necessary to increase the coefficient $b_1^{(4)}$ (5.46) and decrease $b_3^{(4)}$ (5.47). The maximum effect of increasing Ω_{p1} (5.45) is achieved when

$Z_{01} = Z_{04} = Z_{0\max} = mZ_0$ and $Z_{02} = Z_{03} = Z_{0\min} = Z_0$. The expression from Table 5.7 gives the ratio $\omega_{p3}/\omega_{p1} = 1.73$ at $m = 5$.

No. 6 (Table 5.7): Removing the resonance frequencies ω_{p1} and ω_{p3} of a half-wave SIR from each other. A wide rejection band can be achieved in BPF if the spurious passband at the frequency ω_{p2} is suppressed due to the zero-coupling coefficient between resonators. This problem is the opposite of the one considered earlier. A fourth-order reactance function is also used (5.44) to solve it. The expressions (5.45)–(5.47) show that the maximum ratio ω_{p3}/ω_{p1} is achieved at $Z_{01} = Z_{04} = Z_{0\min} = Z_0$ and $Z_{02} = Z_{03} = Z_{0\max} = mZ_0$. If the parameter $m = 5$, we obtain $\omega_{p3}/\omega_{p1} = 6.47$ from the expression in Table 5.7. To illustrate the effects achieved in the last two resonators, we use the linear simulator of Microwave Office software. The schematic and simulation results are shown in Figure 5.12 for $m = 5$.

No. 5 (Table 5.6): Increasing the tunability of a quarter-wave resonator with a capacitor at the second resonant frequency ω_2. An increase in the frequency interval between ω_{01} and ω_{p2} (increase ω_{p2}/ω_{01} ratio) leads to an expansion of the cavity tuning range with a lumped capacitor at the second resonance frequency

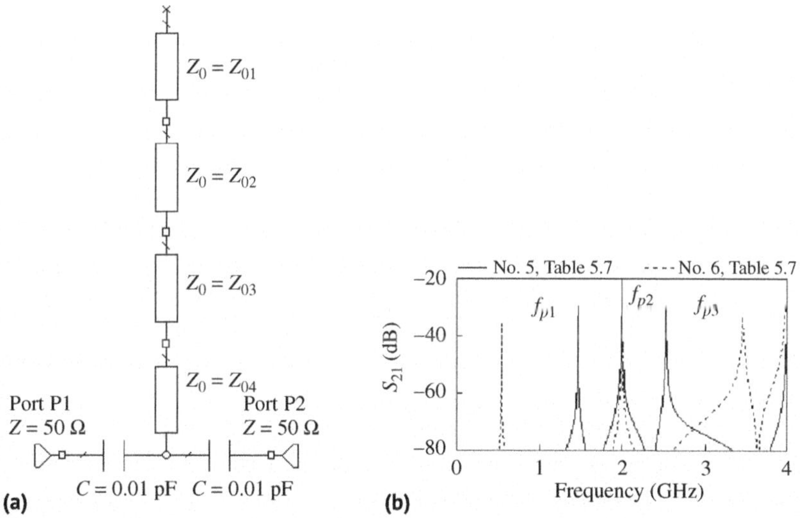

Figure 5.12 Half-wave SIR with four TL segments. (a) Microwave Office schematic. (b) Resonant frequencies.

ω_2: $\omega_{01} \leq \omega_2 \leq \omega_{p2}$ [28]. The quarter-wave resonator has the ratio $\omega_{p2}/\omega_{01} = 3/2$. This problem can be solved using the fourth-order reactance function (5.35) ($N = 4$):

$$Z^4(S) = \frac{a_1^{(4)} S + a_3^{(4)} S^3}{b_0^{(4)} + b_2^{(4)} S^2 + b_4^{(4)} S^4}, \tag{5.48}$$

where $S = j\Omega = j\tan(\omega t_d/4)$, and the coefficients of polynomials are given in Table 5.5a.

The expressions determine zero Ω_{01} and pole Ω_{p2} of the function (5.48):

$$\begin{aligned}\Omega_{01}^2 &= \frac{a_1^{(4)}}{a_3^{(4)}} = \frac{Z_{01} + Z_{02} + Z_{03} + Z_{04}}{\frac{Z_{01}Z_{03}}{Z_{02}} + \frac{Z_{01}Z_{04}}{Z_{02}} + \frac{Z_{01}Z_{04}}{Z_{03}} + \frac{Z_{02}Z_{04}}{Z_{03}}}\\ &= Z_{02}Z_{03}\frac{Z_{01} + Z_{02} + Z_{03} + Z_{04}}{Z_{01}(Z_{03} + Z_{04}) + Z_{04}(Z_{01} + Z_{02})},\end{aligned} \tag{5.49}$$

$$\Omega_{p2}^2 = \frac{b_2^{(4)}}{2b_4^{(4)}} + \sqrt{\left(\frac{b_2^{(4)}}{2b_4^{(4)}}\right)^2 - \frac{b_0^{(4)}}{b_4^{(4)}}}. \tag{5.50}$$

The analysis of expressions (5.49) and (5.50) shows that the maximum ratio Ω_{p2}/Ω_{01} is achieved at $Z_{02} = Z_{03} = Z_{0\min} = Z_0$ and $Z_{01} = Z_{04} = Z_{0\max} = mZ_0$. The expression from Table 5.6 gives the ratio $\omega_{p2}/\omega_{01} = 3.075$ at $m = 5$.

No. 6 (Table 5.6): Increased stability of the second resonant frequency of a quarter-wave resonator in the case of variation of the capacitive load can be achieved by decreasing the frequency interval between zero ω_{01} and pole ω_{p2}. To solve this problem, the input impedance function (5.48) can be used. Using expressions (5.49) and (5.50), we establish that the minimum ratio Ω_{p2}/Ω_{01} is ensured at $Z_{01} = Z_{02} = Z_{04} = Z_{0\min} = Z_0$ and $Z_{03} = Z_{0\max} = mZ_0$. The expression from Table 5.7 shows that the degree of convergence ω_{p2} and ω_{01} is characterized by the ratio $\omega_{p2}/\omega_{01} = 1.063$ at $m = 5$.

The proposed parametric synthesis method is quite simple. It avoids the time consumption. Using it, we can design SIR with different distributions of critical frequencies, which are helpful for practical use.

5.7 Structural-parametric Synthesis

Let us consider distributed Richards circuits [27], which contain serial and parallel stubs, and connecting line segments of the same length. Their input impedance at the Richards variable S (5.18) is described by the reactance function $Z^N(S)$. The implementation of the function $Z^N(S)$ by the Cauer method [30] is based on the separation of its poles at $S = 0$ and $S = \infty$, which transforms the function

$Z^N(S)$ into $Z^{N-1}(S)$. This method results in only two open- or short-circuit stubs connected in series or parallel. Decreasing the order of the function $Z^N(S)$ by one unit is also caused by the separation in the circuit of a cascaded line segment based on Richards' theorem [20].

Determining the number of distributed circuits realizing the given $Z^N(S)$ is not difficult. So $Z^1(S)$ has a single realization, $Z^2(S)$ has three realizations established by direct calculation. With a further increase in N (from $N-1$ to N), the number of realizations triples each time. Therefore, the total number of distributed circuits n realizing the given $Z^N(S)$ [31]:

$$n = 3^{(N-1)}. \tag{5.51}$$

For each distributed structure (resonator), it is necessary to solve the problem of parametric synthesis considered earlier. Since there are many such structures (5.51), the method for their optimization is called structural-parametric synthesis. The Kuroda identities [27, 32] are helpful when considering such structures. They allow us to change the stub position while changing its characteristic impedance.

We use this method to synthesize stub structure resonators with different distributions of critical frequencies, which are helpful for practical use. Let us compare them with SIRs, synthesized in the previous section.

1. The convergence of three resonant frequencies ω_{p3}, ω_{p2}, and ω_{p1}, realized in a quarter-wave SIR (No. 4, Table 5.6).

The input impedance of this SIR is described by a third-order reactance function (5.42). Now and in the future, we consider resonators with parallel stubs, which are most in demand in planar TL design. Using only parallel stubs significantly reduces the number of structures (5.51) implementing the $Z^N(S)$ function. Figure 5.13 shows the distributed structures, the input impedance of which is

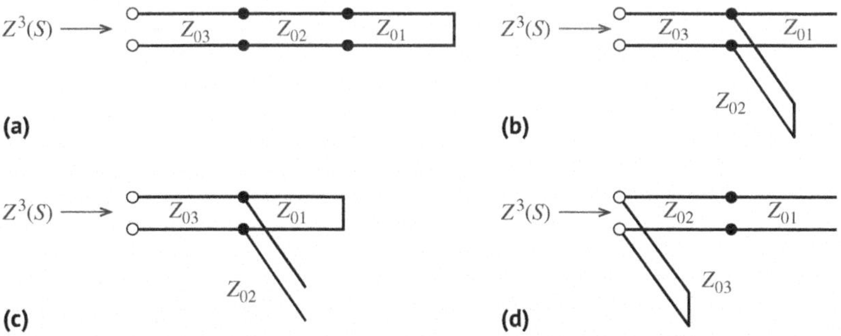

Figure 5.13 Distributed circuits implementing the third-order reactance function $Z^3(S)$ (5.42). (a) SIR. (b) Short-circuited stub in the middle. (c) Open-circuited stub in the middle. (d) Short-circuited segment connected to the input.

described by the function $Z^3(S)$ (5.42). Solving the problem of parametric synthesis for each of these resonators, we establish that the resonator shown in Figure 5.13b is the best. Its coefficients of $Z^3(S)$ (5.42) are determined by the expressions:

$$a_1^{(3)} = Z_{01}Z_{02}Z_{03} + Z_{01}Z_{03}^2, \quad a_3^{(3)} = Z_{02}Z_{03}^2,$$
$$b_0^{(3)} = Z_{01}Z_{03}, \quad b_2^{(3)} = Z_{01}Z_{02} + Z_{02}Z_{03}.$$

The pole Ω_{p1}^2 of function (5.42) is equal to the ratio

$$\Omega_{p1}^2 = \frac{b_0^{(3)}}{b_2^{(3)}} = \frac{Z_{01}Z_{03}}{Z_{02}\left(Z_{01} + Z_{03}\right)},$$

which shows that the value Ω_{p1} is maximum, and the Ω_{p3}/Ω_{p1} ratio is minimum if $Z_{01} = Z_{03} = Z_{0\max} = mZ_0$ and $Z_{02} = Z_{0\min} = Z_0$. In this case

$$Z^3(S) = Z_0 \frac{(1+m)S + S^3}{1 + (2/m)S^2}. \tag{5.52}$$

From (5.52) it follows

$$\frac{\omega_{p3}}{\omega_{p1}} = \frac{\pi}{\tan^{-1}\sqrt{m/2}} - 1. \tag{5.53}$$

The topology of the resonator obtained as a result of structural-parametric synthesis is shown in Figure 5.14a. The solid line in Figure 5.14b shows the ω_{p3}/ω_{p1} ratio (5.53) dependence on the parameter m. The dotted line in the figure shows a similar dependence for SIR (No. 4, Table 5.6). The stub resonator has the advantage.

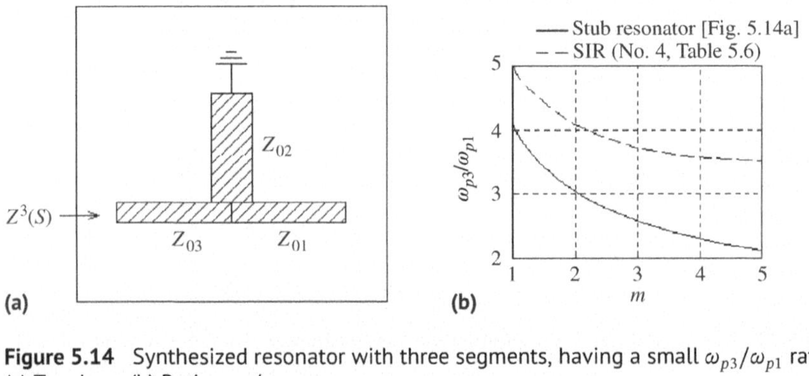

Figure 5.14 Synthesized resonator with three segments, having a small ω_{p3}/ω_{p1} ratio. (a) Topology. (b) Ratio ω_{p3}/ω_{p1}.

2. Convergence of three resonant frequencies ω_{p3}, ω_{p2}, and ω_{p1}, realized in a half-wave SIR (No. 5, Table 5.7).

The input impedance of this SIR is described by a fourth-order reactance function (5.44). We will not present all distributed structures with parallel stubs, the input impedance of which is described by the function $Z^4(S)$ (5.44). Note that the best one is the resonator shown in Figure 5.15a. Its coefficients of $Z^4(S)$ (5.44) are determined by the expressions

$$a_0^{(4)} = Z_{01}, \quad a_2^{(4)} = Z_{04}\left(1 + \frac{Z_{01}}{Z_{02}} + \frac{Z_{01}}{Z_{03}}\right) + \frac{Z_{01}Z_{03}}{Z_{02}},$$

$$a_4^{(4)} = \frac{Z_{03}Z_{04}}{Z_{02}}, \quad b_1^{(4)} = 1 + \frac{Z_{01}}{Z_{02}} + \frac{Z_{01}}{Z_{03}} + \frac{Z_{01}}{Z_{04}} + \frac{Z_{01}Z_{03}}{Z_{02}Z_{04}},$$

$$b_3^{(4)} = \frac{Z_{03}}{Z_{02}}.$$

The pole Ω_{p1} of the function (5.44) is equal to the ratio $\Omega_{p1}^2 = \frac{b_1^{(4)}}{b_3^{(4)}}$, which shows that the ratio Ω_{p3}/Ω_{p1} is minimum, if $Z_{01} = Z_{02} = Z_{04} = Z_{0\max} = mZ_0$ and $Z_{03} = Z_{0\min} = Z_0$. In this case

$$Z^4(S) = mZ_0 \frac{1 + \left(2 + m + \frac{1}{m}\right) S^2 + \left(\frac{1}{m}\right) S^4}{\left(1 + 3m + m^2\right) S + \left(\frac{1}{m}\right) S^3}. \tag{5.54}$$

From (5.54), it follows

$$\frac{\omega_{p3}}{\omega_{p1}} = \frac{\pi}{\tan^{-1}\sqrt{m\left(1 + 3m + m^2\right)}} - 1. \tag{5.55}$$

Figure 5.15b shows the topology of the synthesized resonator, and Figure 5.15c shows the ratio ω_{p3}/ω_{p1} (5.55) by a solid line. For comparison, the dotted line

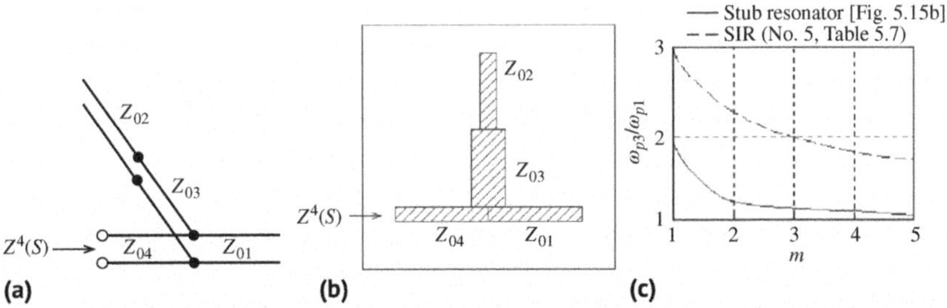

Figure 5.15 Synthesized resonator with four segments, having small ω_{p3}/ω_{p1} ratio. (a) Symbol. (b) Topology. (c) Ratio ω_{p3}/ω_{p1}.

shows the similar dependence for SIR (No. 5, Table 5.7), which is used in a wide-band multimode microstrip BPF [29]. The synthesized resonator with a stub has a significant advantage over SIR. For $m = 5$ it provides a ratio $\omega_{p3}/\omega_{p1} = 1.09$, while SIR has $\omega_{p3}/\omega_{p1} = 1.73$.

3. Increase in ω_{p2}/ω_{p1} ratio implemented in half-wave SIR (No. 1, Table 5.7), which leads to expansion of BPF stopband.

The input impedance of this SIR is described by a third-order reactance function (5.39). As before, we are not presenting all distributed structures with parallel stubs, the input impedance of which is described by the function $Z^3\,(S)$ (5.39). The best resonator is shown in Figure 5.16a. The coefficients of $Z^3\,(S)$ (5.39) of this resonator depend on the characteristic impedances Z_{0i} as follows:

$$a_0^{(3)} = Z_{04},\ a_2^{(3)} = \frac{Z_{02}Z_{04}}{Z_{01}} + \frac{Z_{03}Z_{04}}{Z_{01}} + \frac{Z_{03}Z_{04}}{Z_{02}},$$

$$b_1^{(3)} = 1 + \frac{Z_{04}}{Z_{01}} + \frac{Z_{04}}{Z_{02}} + \frac{Z_{04}}{Z_{03}},\ b_3^{(3)} = \frac{Z_{02}}{Z_{01}} + \frac{Z_{03}}{Z_{01}} + \frac{Z_{03}}{Z_{02}} + \frac{Z_{02}Z_{04}}{Z_{01}Z_{03}}.$$

It is essential to note the following characteristic feature. If an open line segment is connected in parallel at the input to a half-wave SIR with an odd number of composite segments, then the order of the input reactance function $Z^N\,(S)$ does not change. Only the coefficients of the function $Z^N\,(S)$ are changed. This applies to the resonator in Figure 5.16a.

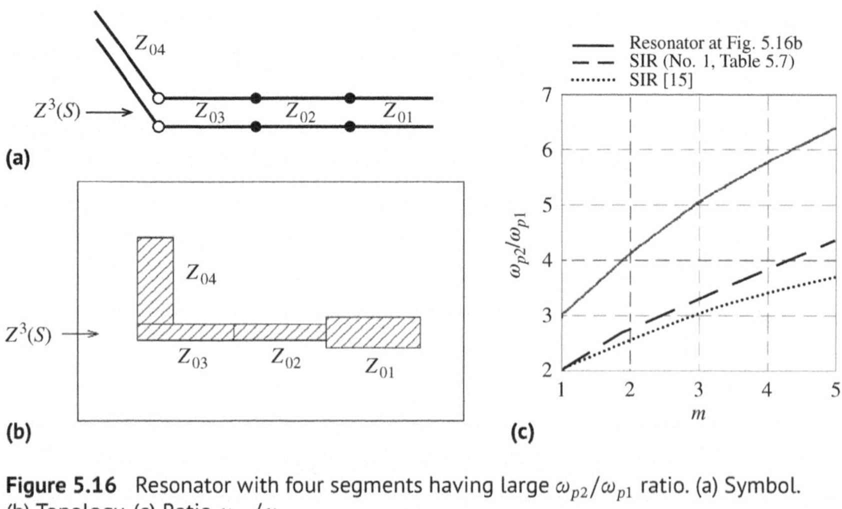

Figure 5.16 Resonator with four segments having large ω_{p2}/ω_{p1} ratio. (a) Symbol. (b) Topology. (c) Ratio ω_{p2}/ω_{p1}.

The pole Ω_{p1} of the function (5.39) is equal to the ratio

$$\Omega_{p1}^2 = \frac{b_1^{(3)}}{b_3^{(3)}} = \frac{Z_{01}Z_{04}\left(\frac{1}{Z_{01}} + \frac{1}{Z_{02}} + \frac{1}{Z_{03}} + \frac{1}{Z_{04}}\right)}{Z_{02} + Z_{03} + \frac{Z_{01}Z_{03}}{Z_{02}} + \frac{Z_{02}Z_{04}}{Z_{03}}},$$

which shows that the value of Ω_{p1} is minimal, and the ratio Ω_{p2}/Ω_{p1} is maximal if $Z_{01} = Z_{04} = Z_{0\min} = Z_0$ and $Z_{02} = Z_{03} = Z_{0\max} = mZ_0$. In this case

$$Z^3(S) = Z_0 \frac{1 + (1 + 2m)S^2}{2\left(1 + \frac{1}{m}\right)S + 2(1 + m)S^3}. \tag{5.56}$$

From (5.56) it follows

$$\frac{\omega_{p2}}{\omega_{p1}} = \frac{\pi}{\tan^{-1}\sqrt{\frac{1}{m}}} - 1. \tag{5.57}$$

Figure 5.16b shows the topology of the synthesized resonator. The dependence of ω_{p3}/ω_{p1} on the parameter m (5.57) is presented in Figure 5.16c. The dashed line in this figure shows a similar dependence for SIR (No. 1, Table 5.7). For comparison, in the same figure, the dotted line plots the ratio ω_{p3}/ω_{p1} for SIR [15], which is used in BPFs with a wide stopband [33]. The synthesized stub resonator has a significant advantage over SIRs.

4. Increase in ω_{p2}/ω_{p1} ratio implemented in quarter-wave SIR (No. 1, Table 5.6), which leads to stopband expansion of BPF.

The input impedance of this SIR is described by a second-order reactance function (5.38). The best one among the various resonators is shown in Figure 5.17a. Note that the parallel connection of an open section to the quarter-wave SIR input with an even number of composite sections does not increase the order of the input reactance function $Z^N(S)$. This resonator (Figure 5.17a) differs in that the number of its segments is greater than the order of the reactance input function. The coefficients of $Z^2(S)$ (5.38) of this resonator depend on the characteristic impedances Z_{0i}: $a_1^{(2)} = Z_{03}(Z_{01} + Z_{02})$, $b_0^{(2)} = Z_{03}$, $b_2^{(2)} = Z_{01} + Z_{02} + Z_{01}Z_{03}/Z_{02}$.

The pole Ω_{p1} of the function (5.38) is equal to the ratio

$$\Omega_{p1}^2 = \frac{b_0^{(2)}}{b_2^{(2)}} = \frac{Z_{03}}{Z_{01}(1 + Z_{03}/Z_{02}) + Z_{02}},$$

which shows that the value of Ω_{p1} is minimal, and the ratio Ω_{p2}/Ω_{p1} is maximal if $Z_{03} = Z_{0\min} = Z_0$ and $Z_{01} = Z_{0\max} = mZ_0$. The Z_{02} value can be either minimum or maximum. The intermediate value Z_{02} ($Z_0 < Z_{02} < mZ_0$) worsens the result.

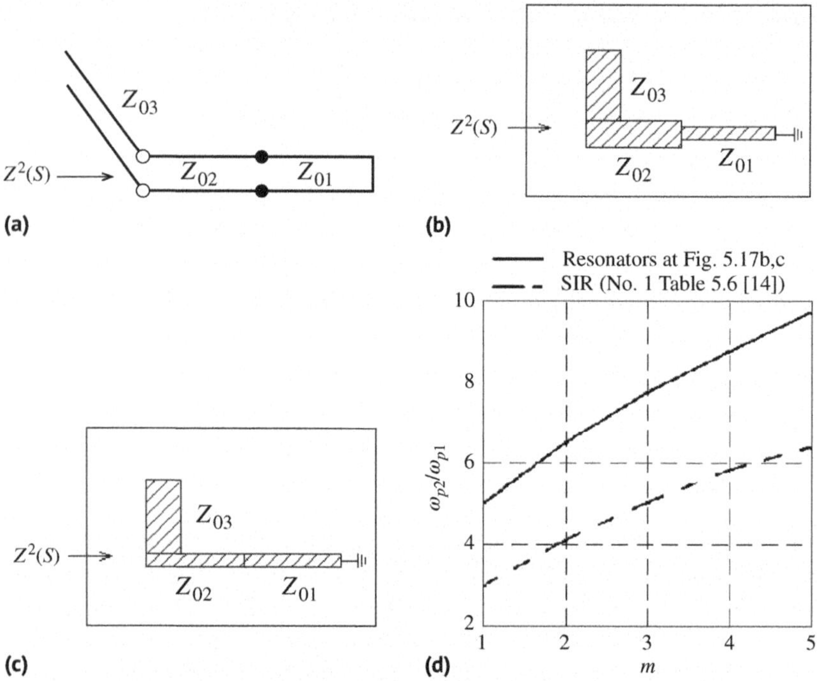

Figure 5.17 Resonator with three segments having large ω_{p2}/ω_{p1} ratio. (a) Symbol. (b) Topology with $Z_{02} = Z_{0\min}$. (c) Topology with $Z_{02} = Z_{0\max}$. (d) Ratio ω_{p2}/ω_{p1}.

If $Z_{02} = Z_{0\min} = Z_0$, then

$$Z^2(S) = Z_0 \frac{(1+m)S}{1+(1+2m)S^2}. \tag{5.58}$$

If $Z_{02} = Z_{0\max} = mZ_0$, then

$$Z^2(S) = Z_0 \frac{2mS}{1+(1+2m)S^2}. \tag{5.59}$$

In both cases (5.58) and (5.59), the topologies of which are shown in Figure 5.17b, c, respectively, the ratio ω_{p2}/ω_{p1} is the same

$$\frac{\omega_{p2}}{\omega_{p1}} = \frac{\pi}{\tan^{-1}\sqrt{\frac{1}{(1+2m)}}} - 1. \tag{5.60}$$

In Figure 5.17d, the solid line shows the dependence of ω_{p3}/ω_{p1} on the parameter m (5.60). The dashed line in this figure shows the similar dependence for SIR (No. 1, Table 5.6), which has remained a leader for a long time, starting from the work of Makimoto and Yamashita [14].

In all the cases considered earlier, the results of structural-parametric synthesis exceed those of parametric synthesis. This is because structural-parametric synthesis has wider possibilities for choosing the best solution.

The results of structural-parametric synthesis, particularly the solutions presented in Figure 5.17, indicate that the condition (5.36) holds in these cases. The characteristic impedances Z_{0i} take only two values $Z_{0\min}$ and $Z_{0\max}$. Complying with condition (5.36) significantly reduces labor costs in structural-parametric synthesis. A linear circuit simulator (Figure 5.12) lets us quickly enumerate all possible options and choose the best one. More laborious is the analytical justification of the solutions' optimality, the establishment of expressions for the input impedance Z^N (S) of the synthesized resonators, and the relationship between critical frequencies of these resonators, as was done earlier.

5.8 Microstrip BPF with Resonator Obtained by Structural-parametric Synthesis

Consider a microstrip second-order BPF with resonator (Figure 5.17c) obtained by structural-parametric synthesis. The parameters of this BPF were determined as follows: center frequency $f_0 = 910$ MHz; BW = 26 MHz (FBW = 0.0286), passband ripple $L_{\text{Ar}} = 0.2$ dB $(L_R = 13.467$ dB). Table 2.6 shows g parameters: $g_0 = 1$, $g_1 = 1.0378$, $g_2 = 0.6745$, $g_3 = 1.5386$. Using expressions (3.8), we determine the coupling coefficient between the resonators $K_{12} = 0.034$ and the external Q-factor $Q_e = 36.3$. Topology of the BPF is shown in Figure 5.18a. The filter is implemented on a substrate of TMM-10i (Rogers), $\varepsilon_r = 9.8$, $\tan\delta = 0.002$, $h = 1.905$ mm, and it has a size of 34 mm × 15.2 mm. The resonators are tuned to 910 MHz. The coupling coefficient $K_{12} = 0.034$ between resonators is provided by the gap $\Delta_{1,2} = 3.2$ mm. The input and output are coupled with loads by the tapped lines with a width of 1.2 mm and a series capacitor with capacitance $C = 0.5$ pF. The tapped lines are connected to the points with coordinate $L' = L_1/2$ (Figure 5.18a). The capacitance value $C = 0.5$ pF provides the value $Q_e = 36.3$. The $\Delta_{1,2}$ and C values are determined using the formula (3.10) and the electromagnetic simulator of Microwave Office software.

The photograph of this filter is shown in Figure 5.18b, and its frequency responses are depicted in Figure 5.18c, d. The measured parameters are as follows: $f_0 = 910$ MHz; BW = 26 MHz; the insertion loss at the center frequency $L_{A0} = 1.6$ dB; the return loss in bandwidth $L_R > 13$ dB. The measured value of ratio $f_{p2}/f_{p1} = 6.23/0.91$ GHz $= 6.84$. The width of the sections of microstrip lines in the resonators is $w_1 = 3$ mm $(Z_{01} \approx 36\ \Omega)$ and $w_2 = 0.6$ mm $(Z_{02} \approx 79\ \Omega)$, and the parameter m of the resonator is $m = 2.2$. Using (5.60), we determine the calculated value $f_{p2}/f_{p1} = 6.73$. The error between measured and calculated f_{p2}/f_{p1} values

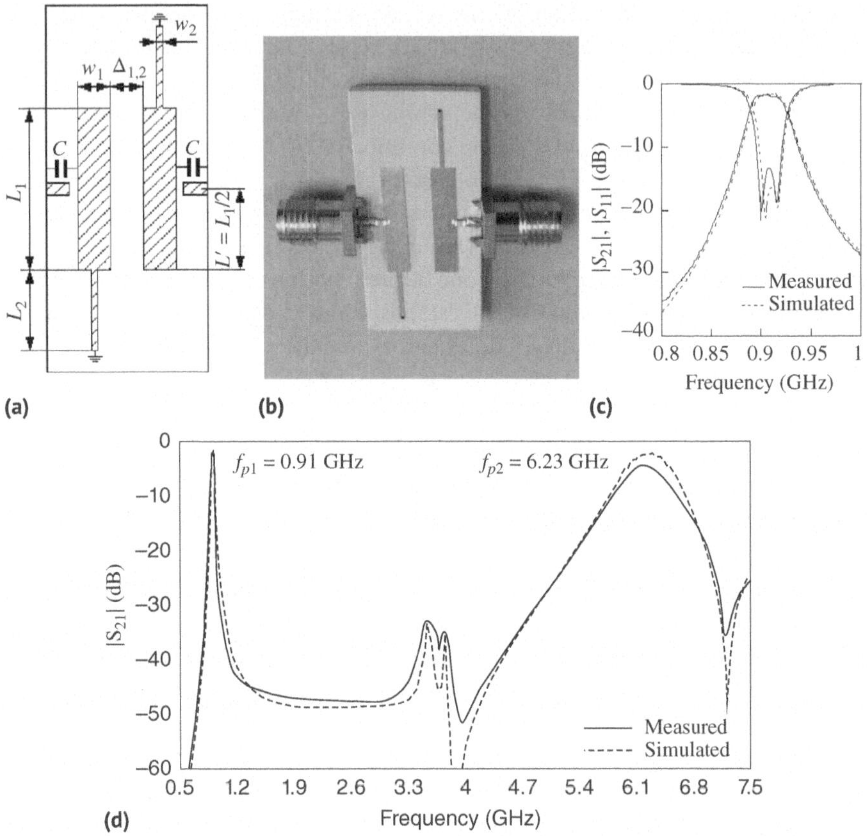

Figure 5.18 Microstrip interdigital BPF. (a) Topology: $L_1 = 14.8$ mm, $w_1 = 3$ mm, $L_2 = 7.6$ mm, $w_2 = 0.6$ mm, $\Delta_{1,2} = 3.2$ mm, $L' = 3.8$ mm. (b) Photograph of fabricated filter. *Source*: [22]/with permission of IEEE. (c) Frequency responses in a narrow frequency band. (d) Insertion loss in a wide frequency band.

equals 1.6%. For $m = 2.2$, the optimal SIR No. 1 in Table 5.6 is characterized by the value $f_{p2}/f_{p1} = 4.3$.

Because the resonant frequencies f_{p2} and f_{p1} of the synthesized resonators are significantly removed from each other, the microstrip filter in Figure 5.18 has an extended stopband. The number of resonators of the BPF under consideration can be increased, increasing its selectivity.

5.9 Transmission Line Resonators with Mutually Opposite Critical Frequencies

In some problems of TL resonator synthesis, the question arises: What should be done with resonator to change its critical frequencies to the opposite ones? Let us take a resonator of a nonuniform TL segment with a characteristic impedance

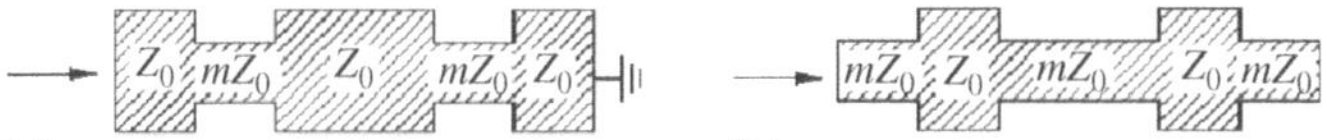

(a) (b)

Figure 5.19 SIRs with mutually opposite critical frequencies. (a) SIR with short-circuited far end. (b) SIR with open-circuited far end.

$Z_0\,(x)$ open at the far end. It is characterized by resonant $\{\omega_{0i}\}$ and antiresonant $\{\omega_{pi}\}$ frequencies. If we change the characteristic impedance $Z_0\,(x)$ of the segment according to the formula

$$Z_0^*\,(x) = \frac{R^2}{Z_0(x)} \tag{5.61}$$

where R [Ω] is a constant, and simultaneously change the boundary conditions at the far end to short-circuited ones, then the sets of resonant $\{\omega_{0i}\}$ and antiresonant $\{\omega_{pi}\}$ will swap places.

This equality follows from expressions (5.10)–(5.13) for CIR and from the SIR input immittances' coefficients formation (Table 5.5a). Optimal SIRs are two-level and contain TL segments with characteristic impedances Z_0 and mZ_0. In these resonators, condition (5.61) manifests itself so that the indicated characteristic impedances change places.

The indicated rule is illustrated in Figure 5.19 using the example of two-level SIRs. The considered regularity is helpful and will be used further.

References

1 G. Megla, *Dezimeterwellentechnik: Theorie Und Technik Der Dezimeterschaltungen*, 4th ed. Leipzig: Fachbuchverlag, 1955.

2 E. A. Guillemin, *Synthesis of Passive Networks: Theory and Methods Appropriate to the Realization and Approximation Problems.* New York: Wiley, 1957.

3 A. V. Zakharov, M. Y. Ilchenko, and L. S. Pinchuk, Microstrip U-shaped resonators, *Radioelectron. Commun. Syst.*, vol. 55, no. 8, pp. 368–374, 2012, https://doi.org/10.3103/S0735272712080043.

4 R. N. Ghose, Exponential transmission lines as resonators and transformers, *IEEE Trans. Microw. Theory Tech.*, vol. 5, no. 3, pp. 213–217, 1957, https://doi.org/10.1109/TMTT.1957.1125143.

5 C. P. Womack, The use of exponential transmission lines in microwave components, *IEEE Trans. Microw. Theory Tech.*, vol. 10, no. 2, pp. 124–132, 1962, https://doi.org/10.1109/TMTT.1962.1125468.

6 I. Sugai, The solutions for nonuniform transmission line problems (Correspondence), *Proc. IRE*, vol. 48, no. 8, pp. 1489–1490, 1960, https://doi.org/10.1109/JRPROC.1960.287576.

7 C. Nwoke, An exact solution for the nonuniform transmission line problem, *IEEE Trans. Microw. Theory Tech.*, vol. 38, no. 7, pp. 944–946, 1990, https://doi.org/10.1109/22.55789.

8 D. B. Large, Synthesis of a transmission-line resonator with a specified frequency spectrum, *Proc. IEEE*, vol. 52, no. 5, pp. 633–634, 1964, https://doi.org/10.1109/PROC.1964.3034.

9 O. N. Litvinenko and V. I. Soshnikov, *Theory of Nonuniform Lines and Their Application in Radio Engineering* [in Russian]. Moscow: Sov. Radio, 1964.

10 A. Kachalov, Y. Kurylev, and M. Lassas, *Inverse Boundary Spectral Problems (Monographs and Surveys in Pure and Applied Mathematics)*, 1st ed. Chapman and Hall/CRC, 2001.

11 M. Y. Il'chenko and A. V. Zakharov, Construction of resonators with a mutually reversed distribution of natural frequencies, *Sov. J. Commun. Technol. Electron.*, vol. 35, no. 9, pp. 128–131, 1990.

12 M. Y. Il'chenko, A. V. Zakharov, and E. E. Karyakin, Series resonators for microwave filters with single passband, *Radiotekhnika i Elektron.*, vol. 36, no. 7, pp. 1386–1389, 1991.

13 A. V. Zakharov, M. E. Il'chenko, and V. A. Syzranov, Filters on oscillations of the form $n\lambda/2$ with single transmission band, *Izv. VUZ Radioelektronika*, vol. 39, no. 6, pp. 9–15, 1996.

14 M. Makimoto and S. Yamashita, Compact bandpass filters using stepped impedance resonators, *Proc. IEEE*, vol. 67, no. 1, pp. 16–19, 1979, https://doi.org/10.1109/PROC.1979.11196.

15 M. Makimoto and S. Yamashita, Bandpass filters using parallel coupled stripline stepped impedance resonators, *IEEE Trans. Microw. Theory Tech.*, vol. 28, no. 12, pp. 1413–1417, 1980, https://doi.org/10.1109/TMTT.1980.1130258.

16 B. Kapilevich, A varactor-tunable filter with constant bandwidth and loss compensation, *Microw. J.*, no. 4, pp. 106–114, 2007, https://www.microwavejournal.com/articles/4575-a-varactor-tunable-filter-with-constant-bandwidth-and-loss-compensation.

17 W. Qin, J. Cai, Y.-L. Li, and J.-X. Chen, Wideband tunable bandpass filter using optimized varactor-loaded SIRs, *IEEE Microw. Wirel. Components Lett.*, vol. 27, no. 9, pp. 812–814, 2017, https://doi.org/10.1109/LMWC.2017.2734848.

18 J. X. Chen, Y. Ma, J. Cai, L. H. Zhou, Z. H. Bao, and W. Che, Novel frequency-agile bandpass filter with wide tuning range and spurious suppression, *IEEE Trans. Ind. Electron.*, vol. 62, no. 10, pp. 6428–6435, 2015, https://doi.org/10.1109/TIE.2015.2427122.

19 A. Zakharov, S. Litvintsev, and M. Ilchenko, Transmission line tunable resonators with intersecting resonance regions, *IEEE Trans. Circuits Syst. II Express Briefs*, vol. 67, no. 4, pp. 660–664, 2020, https://doi.org/10.1109/TCSII.2019.2922429.

20 P. I. Richards, Resistor-transmission-line circuits, *Proc. IRE*, vol. 36, no. 2, pp. 217–220, 1948, https://doi.org/10.1109/JRPROC.1948.233274.

21 H. J. Riblet, General synthesis of quarter-wave impedance transformers, *IRE Trans. Microw. Theory Tech.*, vol. 5, no. 1, pp. 36–43, 1957, https://doi.org/10.1109/TMTT.1957.1125088.

22 A. Zakharov, Parametric and structural-parametric synthesis of nonuniform transmission line resonators, *IEEE Trans. Circuits Syst. I Regul. Pap.*, vol. 68, no. 3, pp. 1055–1067, 2021, https://doi.org/10.1109/TCSI.2020.3044925.

23 G. A. Korn and T. M. Korn, *Mathematical Handbook for Scientists and Engineers: Definitions, Theorems, and Formulas for Reference and Review*. New York, NY: McGraw-Hill, 1961.

24 H. J. Carlin, Distributed circuit design with transmission line elements, *Proc. IEEE*, vol. 59, no. 7, pp. 1059–1081, 1971, https://doi.org/10.1109/PROC.1971.8334.

25 A. V. Zakharov, Synthesis of oscillator systems from segments of stepped lines, *Radio Eng. Electron. Phys. (English Transl. Radiotekhnika i Elektron.)*, vol. 29, no. 9, pp. 139–144, 1984.

26 S. Litvintsev, S. Rozenko, and L. Pinchuk, Synthesis of transmission line resonators for harmonic suppression, in *2022 IEEE 2nd Ukrainian Microwave Week (UkrMW)*, 2022, pp. 379–382, https://doi.org/10.1109/UkrMW58013.2022.10037139.

27 A. Matsumoto (Ed.), *Microwave Filters and Circuits*, 1st ed. Academic Press, 1970, https://www.elsevier.com/books/microwave-filters-and-circuits/matsumoto/978-0-12-027961-6.

28 A. Zakharov and M. Ilchenko, Circuit function characterizing tunability of resonators, *IEEE Trans. Circuits Syst. I Regul. Pap.*, vol. 67, no. 1, pp. 98–107, 2020, https://doi.org/10.1109/TCSI.2019.2940066.

29 L. Zhu, S. Sun, and W. Menzel, Ultra-wideband (UWB) bandpass filters using multiple-mode resonator, *IEEE Microw. Wirel. Components Lett.*, vol. 15, no. 11, pp. 796–798, 2005, https://doi.org/10.1109/LMWC.2005.859011.

30 W. Cauer, W. Klein, F. M. Pelz, K. Cauer, and G. E. Knausenberger, *Synthesis of Linear Communications Networks*. New York, NY: McGraw-Hill, 1958.

31 M. Y. Il'chenko and A. V. Zakharov, Construction of oscillatory systems based on distributed stub structures having T-waves, *Radioelectron. Commun. Syst.*, vol. 33, no. 1, pp. 9–15, 1990.

32 H. Ozaki and J. Ishii, Synthesis of a class of strip-line filters, *IRE Trans. Circuit Theory*, vol. 5, no. 2, pp. 104–109, 1958, https://doi.org/10.1109/TCT.1958.1086441.

33 J. T. Kuo and E. Shih, Microstrip stepped impedance resonator bandpass filter with an extended optimal rejection bandwidth, *IEEE Trans. Microw. Theory Tech.*, vol. 51, no. 5, pp. 1554–1559, 2003, https://doi.org/10.1109/TMTT.2003.810138.

6

Bandpass Filters with Cross-couplings

The use of cross-coupling in bandpass filters (BPFs) results in more varied frequency responses. This occurs because transmission zeros (TZs) are introduced into the transfer function of the BPF. The presence of these TZs leads to an increase in BPF selectivity and stopband level and also allows for obtaining linear phase and flat group delay. The TZs could be situated on the complex plane $s = \sigma + j\omega$, including real frequencies ($j\omega$-axis), which are also called attenuation poles. In N-order BPF, the TZ number can reach the value $(N - 2)$ [1–5]. Using a source/load-multiresonator coupling in BPF increases the TZ number from $(N - 2)$ to N [6]. However, such couplings reduce the filter rejection level [7], which is not always desirable. The era of BPFs with cross-couplings began with the works of R. M. Kurzrok [1, 2], related to triplet and quadruplet BPFs with one cross-coupling, which require more detailed consideration.

6.1 Triplet Bandpass Filters

When describing BPF with center frequency f_0 and fractional bandwidth (FBW), the normalized frequency is often used [8]

$$\Omega = \frac{1}{\text{FBW}}\left(\frac{f}{f_0} - \frac{f_0}{f}\right). \tag{6.1}$$

Equality (6.1) can be replaced by an approximate one

$$\Omega \approx \frac{f - f_0}{\text{BW}/2}, \tag{6.2}$$

where BW is the bandwidth. From (6.2), it follows that Ω is the number of bandwidth halves that fit into the frequency interval from f to f_0. Half of the bandwidth is used as the measurement unit of Ω. If $f = f_0$, then $\Omega = 0$.

The normalized coupling coefficients are defined by the equality

$$m_{ij} = \frac{K_{ij}}{\text{FBW}} \tag{6.3}$$

The complex frequency $s = \sigma + j\Omega$ is used for BPF description.

The coupling structure of triplet BPF with cross-coupling is shown in Figure 6.1a. The dotted line indicates the cross-coupling K_{13}. Each of the specified couplings can be either positive or negative. The mandatory condition is the equality $|K_{12}| = |K_{23}|$. Cross-coupling leads to the TZ appearance at the real frequency f_z, which corresponds to the normalized value Ω_z (6.1).

The first R. M. Kurzrok formula [1] expresses the position of Ω_z of the triplet BPF using normalized coupling coefficients

$$\Omega_z = \frac{m_{12}m_{23}}{m_{13}}. \tag{6.4}$$

From (6.4), it follows that $\Omega_z > 0$ ($f_z > f_0$) if one or three coupling coefficients are positive. If one or three coupling coefficients have a "–" sign, then $\Omega_z < 0$ ($f_z < f_0$).

Formula (6.4) together with classical expressions (3.6) and (3.8) form the basis of the design of the considered triplet BPF.

To illustrate the design sequence, we will use the schematic (Figure 6.1b) with quarter-wave resonators and J inverters (Figure 3.3), the transverse branches of which are not shown. A J inverter with inductors is used to implement a positive coupling coefficient between resonators, but if specified coupling is negative, a J inverter with capacitors is used. J inverters with quarter-wave segments are used to couple end resonators to the loads. The resonators are characterized by $f_0 = 1$ GHz, $Z_0 = 10\ \Omega$, and susceptance slope parameter $b = \pi/4Z_0$.

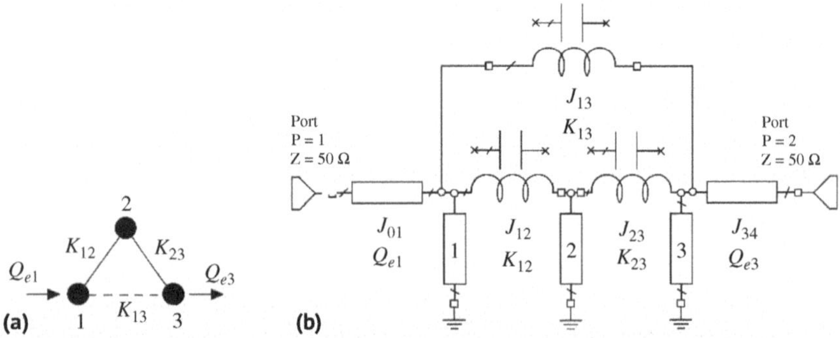

Figure 6.1 Triplet BPF. (a) Coupling structure. (b) Schematic with quarter-wave resonators and J inverters.

Suppose a Chebyshev prototype with $f_0 = 1$ GHz, BW = 50 MHz (FBW = 0.05), and $L_{Ar} = 0.2$ dB ($L_R = 13.467$ dB) is chosen. The prototype parameters (Table 2.6) are $g_0 = 1$, $g_1 = g_3 = 1.2275$, $g_2 = 1.525$. Using formulas (3.8), we determine the main coupling coefficients $K_{12} = K_{23} = 0.0420$ and the external quality factor of end resonators $Q_{e1} = Q_{e3} = 24.55$.

According to (3.6), the specified coupling values are provided by J inverters: $J_{12} = J_{23} = 0.0033\ \Omega^{-1}$; $J_{01} = J_{34} = 0.007999\ \Omega^{-1}$. The J_{12}, J_{23} inverters contain an inductor $L = 48.25$ nH (3.7a), and the J_{01}, J_{34} inverters contain a quarter-wave segment with characteristic impedance $Z_c = 125.02\ \Omega$ (3.7c). The parameters of the filter under consideration without cross-coupling are determined.

In the triplet BPF considered, two possible options exist for the location of one TZ at real frequencies (the $j\Omega$-axis of the complex plane). They are shown in Figure 6.2a. We require that TZ be located in the upper s-half-plane and take the value $\Omega_{z1} = 2$. To implement this value, it is necessary to determine the normalized cross-coupling coefficient m_{13} using formula (6.4). Since $m_{12} = m_{23} = 0.84$, we obtain $m_{13} = 0.3528$ from (6.4), corresponding to the cross-coupling coefficient $K_{13} = 0.01764$. According to (3.7a), this coupling coefficient value is implemented by the J_{13} inverter with the inductance $L_{13} = 113.59$ nH. All parameters of the triplet BPF are determined based on the initial data. In Figure 6.2b, the solid line shows the frequency responses simulated by the linear circuit simulator.

It is well known that introducing TZs into the Chebyshev inline filter leads to some change in the pulsation level, and their equally pulsating character also changes. Return loss and VSWR are also very sensitive when introducing TZs. It is

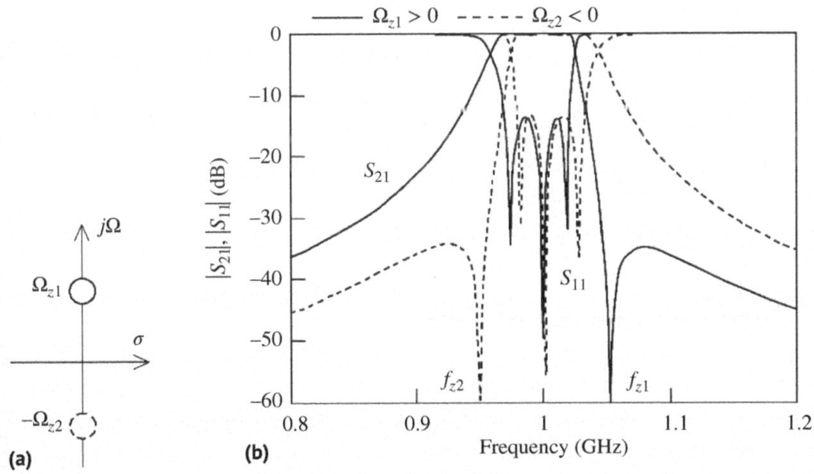

Figure 6.2 Performance of triplet BPF. (a) Situated TZ. (b) Simulated frequency responses.

necessary to perform optimization to return the original BPF parameters to their original state, which is used in various design methods [9–11].

The case considered earlier is not an exception: the characteristic impedances of the quarter-wave segments connected with the loads were reduced by 6%, and the resonant frequencies of the resonator took the values $f_{01} = f_{03} = 0.094$ GHz, $f_{02} = 1.005$ GHz. The optimization is carried out to ensure a given (initial) level of ripple (return loss) and bandwidth.

Now we require that for the same filter TZ is located in the lower s-half-plane and takes the value $\Omega_{z2} = -2$. Substituting this value into (6.4) yields $m_{13} = -0.3528$, corresponding to the cross-coupling coefficient $K_{13} = -0.01764$. According to (3.7b), this coupling coefficient value is realized by the J_{13} inverter with capacitance $C_{13} = 0.2205$ pF. All parameters of the triplet BPF are determined from the initial data. The dotted line shows its simulated frequency responses in Figure 6.2b. In this case, after optimization, the absolute value of the negative coupling coefficient K_{13} was increased by 12%, and the resonant frequencies of the resonator took the values $f_{01} = f_{03} = 1.009$ GHz, $f_{02} = 0.9945$ GHz.

If a positive value of the normalized frequency Ω_{z1} is given, then the corresponding attenuation pole f_{z1} is determined by the equality [11]

$$f_{z1} = f_0 \frac{\sqrt{(\mathrm{FBW} \cdot \Omega_{z1})^2 + 4} + \mathrm{FBW} \cdot \Omega_{z1}}{2}. \tag{6.5}$$

If the value of Ω_{z2} is negative, then the corresponding attenuation pole f_{z2} is determined by the expression

$$f_{z2} = f_0 \frac{\sqrt{(\mathrm{FBW} \cdot \Omega_{z2})^2 + 4} - \mathrm{FBW} \cdot \Omega_{z2}}{2}. \tag{6.6}$$

The frequency responses in Figure 6.2b show that after the attenuation pole, there is a rise, resulting in the insertion loss reaching the maximum value L_{Amax}. In Kurzrok [1], an expression allows one to calculate the L_{Amax} value. The closer f_z to f_0, the greater the L_{Amax} value and lower the rejection level of a triplet BPF. Thus, an increase in the one-sided selectivity of the triplet BPF is accompanied by a decrease in the rejection level in a stopband.

6.2 Quadruplet Bandpass Filter

The coupling structure of a quadruplet BPF with cross-coupling K_{14} is shown in Figure 6.3a. The dotted line indicates the cross-coupling K_{14}. Each of the specified couplings can be positive or negative. The mandatory condition fulfills the equality $|K_{12}| = |K_{34}|$. The cross-coupling K_{14} leads to two equally spaced TZs, which can be located at real frequencies at $S_{1,2} = \pm j\Omega_{1,2}$ $(\Omega_1 = -\Omega_2)$ or on the σ-axis at $S_{1,2} = \pm\sigma_{1,2}$ $(\sigma_1 = -\sigma_2)$.

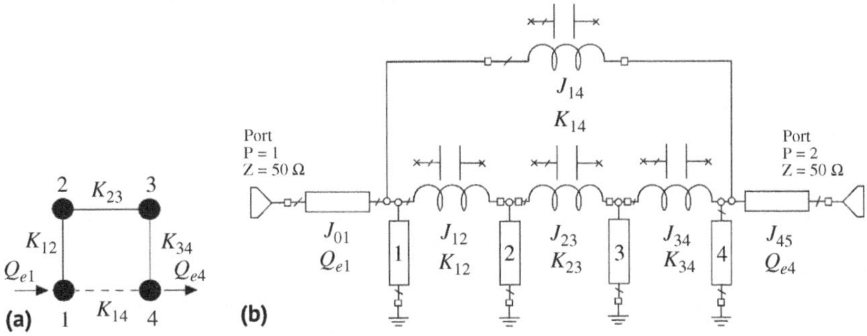

Figure 6.3 Quadruplet BPF with cross-coupling K_{14}. (a) Coupling structure. (b) Schematic with quarter-wave resonators and J inverters.

The second R. M. Kurzrok formula [2] expresses the position of the two zeros of the transmission $S_{1,2}$ of a quadruplet BPF through the normalized coupling coefficients

$$S_{1,2}^2 = \frac{m_{12}m_{23}m_{34}}{m_{14}} - m_{23}^2. \tag{6.7}$$

From (6.7), it follows that if the product of the signs of all the coupling coefficients gives "+," then TZs are located on the $j\Omega$-axis. If the product of the signs of all the coupling coefficients gives "–," then TZs are located on the σ-axis. Expression (6.7) together with formulas (3.6) and (3.8) form the basis of the design of the considered quadruplet BPF.

To illustrate the design sequence, we will use the schematic (Figure 6.3b) with quarter-wave resonators and J inverters, the transverse branches of which are not shown. A J inverter with inductors is used to implement a positive coupling coefficient between resonators, but if the specified coupling is negative, a J inverter with capacitors is used. J inverters with quarter-wave segments are used to couple the end resonators to the loads. The resonators are characterized by $f_0 = 1$ GHz, $Z_0 = 10\ \Omega$, and susceptance slope parameter $b = \pi/4Z_0$.

Suppose a Chebyshev prototype with $f_0 = 1$ GHz, BW = 50 MHz (FBW = 0.05), and $L_{\text{Ar}} = 0.1$ dB ($L_R = 16.426$ dB) is chosen. The prototype parameters (Table 2.6) are $g_0 = 1$, $g_1 = 1.1088$, $g_2 = 1.3061$, $g_3 = 1.7703$, $g_4 = 0.8180$, and $g_5 = 1.3554$. We use formulas (3.8) to determine the main coupling coefficients $K_{12} = K_{34} = 0.0415$, $K_{23} = 0.0329$, and the external quality factor of end resonators $Q_{e1} = Q_{e4} = 22.18$.

According to (3.6), the specified coupling values are provided by J inverters: $J_{12} = J_{34} = 0.00326\ \Omega^{-1}$; $J_{23} = 0.00258\ \Omega^{-1}$; $J_{01} = J_{45} = 0.00842\ \Omega^{-1}$. To simplify the consideration, all couplings between resonators will be considered positive and represented by inductors (3.7a): $L_{12} = L_{34} = 48.83$ nH; $L_{23} = 61.59$ nH. The J_{01}, J_{45} inverters contain a quarter-wave segment with characteristic impedance

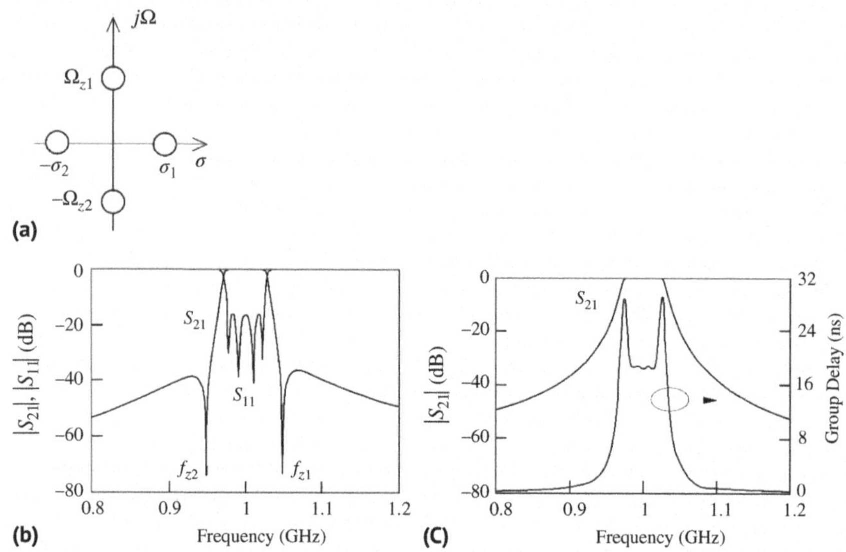

Figure 6.4 Performance of a quadruplet BPF. (a) Situated TZ. (b) Simulated frequency responses of a quadruplet BPF with TZs on the $j\Omega$-axis. (c) Simulated frequency responses of a quadruplet BPF with TZs on the σ-axis.

$Z_c = 118.83\ \Omega$ (3.7c). The parameters of the filter under consideration without cross-coupling are defined.

As already noted, two arrangement variants of two equidistant TZs in quadruplet BPF are shown in Figure 6.4a. Let us consider one of them, when TZs are located at real frequencies, and their normalized value $|\Omega_{1,2}| = 2$. To implement such an arrangement of TZs, it is necessary to determine the normalized cross-coupling coefficient m_{14} by formula (6.7). Using the normalization of the coupling coefficients (6.3), we obtain $m_{12} = m_{34} = 0.83$, $m_{23} = 0.658$. Substituting these values and value of $S_{1,2}^2 = -4$ into (6.7), we obtain $m_{14} = -0.1271$. This value corresponds to the cross-coupling coefficient $K_{14} = -0.00635$, which J inverter realizes with capacitance $C_{14} = 0.0794$ pF (3.7b). All parameters of the quadruplet BPF are determined based on the initial data. As a result of optimization, the characteristic impedances of the quarter-wave sections of the coupling to the loads were reduced by 6%, and the main coupling coefficient K_{23} was increased by 10%. Figure 6.4b shows the simulated frequency responses of the quadruplet BPF with TZs on the $j\Omega$-axis.

Let us consider the second variant, when TZs are located on the σ-axis and take the values $|\sigma_{1,2}| = 2$. The values of the main coupling coefficients remained positive and unchanged, as well as the characteristic impedances $Z_{01} = Z_{45}$. The required normalized cross-coupling coefficient m_{14}, providing the values

$|\sigma_{1,2}| = 2$, is still determined from expression (6.7). In this case, $S_{1,2}^2 = 4$ and formula (6.7) leads to $m_{14} = 0.1022$. This value corresponds to the coupling coefficient $K_{14} = 0.00511$, which J inverter realizes with inductance $L_{14} = 396.6$ nH (3.7a). All parameters of the quadruplet BPF are determined based on the initial data. As a result of optimization, the characteristic impedances of the quarter-wave sections of the coupling to the loads were reduced by 4%, and the main coupling coefficients $K_{12} = K_{34}$ were increased by 2%. Figure 6.4c shows the simulated frequency responses of the quadruplet BPF with TZs on the σ-axis. This quadruplet BPF is a delay line filter with a value of $\tau = 8.7 \pm 0.15$ ns in a frequency band of 28 MHz.

Formulas (6.4) for triplet BPF and (6.7) for quadruplet BPF, which are of considerable interest, are given without proof in Kurzrok [1, 2]. The derivation of these formulas will be presented in Section 7.2.

6.3 Quadruplet BPF with Two Cross-couplings

Figure 6.5a shows the coupling structure of the quadruplet BPF with two cross-couplings. Like the previous quadruplet BPF with one cross-coupling, it has two TZs. To determine the possible arrangements of these TZs, we use the approach [12] that represents the filter under consideration as a union of two triplet BPFs No. 1 and No. 2 (Figure 6.5a). Each of these triplet BPFs has one TZ located to the right or left of f_0. It is determined by the sign rule of their coupling coefficients (6.4). The arrangement of the two TZs of the quadruplet BPF will be the same as that of the two triplet BPFs No. 1 and No. 2.

Let all the main coupling coefficients of this quadruplet BPF have the sign "+" (Figure 6.5a). If both cross-couplings are characterized by the sign "+," then according to the sign rule (6.4), each triplet BPF will have a right-handed TZ. In this case, two TZs of quadruplet BPF will also be right-handed (Figure 6.5b). If the sign of the cross-coupling K_{13} is changed to negative "–," then both TZs will become left-handed (Figure 6.5c). These TZs will be opposite-handed and asymmetrically located (Figure 6.5d) in two cases: $K_{13} > 0$, $K_{14} < 0$ and $K_{13} < 0$, $K_{14} > 0$.

If the signs of the main coupling coefficients change, the number of combinations determining the nature of the TZs' placement becomes quite large. Thus, if the main coupling coefficient K_{23} is negative, the two TZs' right-hand arrangement will require that the cross-couplings K_{13} and K_{14} are also negative.

Each cross-coupling coefficient affects the position of both TZ1 and TZ2. To illustrate this, we remove the cross-coupling K_{14}. In this case, the coupling structure in Figure 6.5a will contain only one triplet BPF, including resonators 1–2–3. The cross-coupling m_{13} defines one TZ, according to (6.4). We use the main coupling coefficients of the preceding quadruplet BPF $m_{12} = m_{34} = 0.83$, $m_{23} = 0.658$.

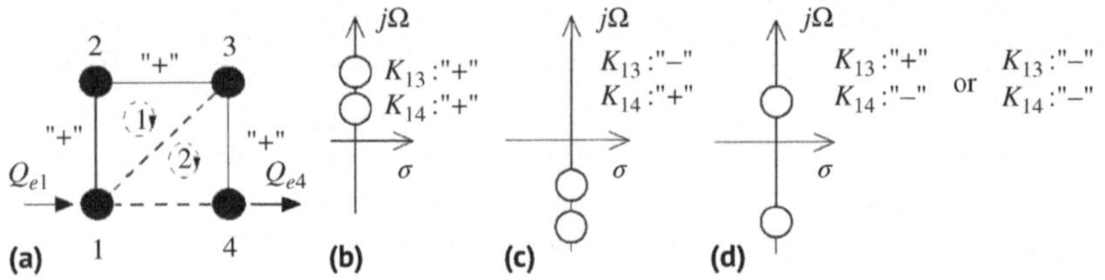

Figure 6.5 Quadruplet BPF with two cross-couplings and its TZs. (a) Coupling structure with all positive main couplings. (b) Two right-sided TZs. (c) Two left-sided TZs. (d) Two both-sided asymmetric TZs.

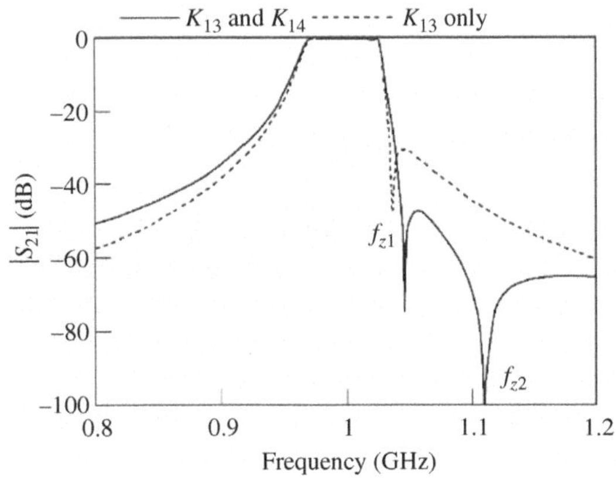

Figure 6.6 Influence of cross-couplings of quadruplet BPF (Figure 6.5a) on insertion loss function.

We set the value of the TZ $\Omega_{z1} = 1.5$, generated by triplet BPF No. 1. Substituting the given values into (6.4), we find the value of the normalized cross-coupling $m_{13} = 0.364$, which corresponds to the value of $K_{13} = 0.0182$ (6.3) and the value of the inductance $L_{13} = 111.3$ nH (3.11a), used in the J inverter. The dotted line shows the simplified insertion loss corresponding to this case in Figure 6.6.

Let us introduce into the filter under consideration the cross-coupling $m_{14} = 0.06$ ($K_{14} = 0.003$, $L_{14} = 675.5$ nH). This will lead to the appearance of the second right-hand TZ, which is reflected by the solid line on the simulated insertion loss curve (Figure 6.6). It is noteworthy that the introduction of cross-couplings m_{14} affected the position of the first TZ, and the frequency f_{z1} shifted to the right. The value of m_{14} affects both f_{z1} and f_{z2}. The same applies to the cross-coupling m_{13}. In this filter, it is impossible to regulate the position of TZ1 or TZ2 separately; they are interconnected.

6.4 Cascaded Quadruplet Filters

One option to increase the TZ number is to cascade several triplet or quadruplet BPFs. The advantage of such a connection is that it allows independent adjustment of the TZs. Cascade connection of two triplet BPFs leads to a cascaded trisection (CT) filter with five or six resonators. The CT filters are considered in Refs. [11, 13]. However, such filters are unattractive because with $N = 5$ resonators, they provide only two real TZs. The quadruplet BPF also provides two TZs, but with fewer resonators, $N = 4$. Cascade connection of two quadruplet BPFs with one cross-coupling K_{14} leads to a cascaded quadruplet (CQ) filter [14–16] with seven or eight resonators (Figure 6.7). One of the latest methods for designing CQ filters [16] involves optimization.

The design method of CQ filters that are considered as follows also includes optimization, but it is based on the second R. M. Kurzrok formula (6.7) and classical expressions (3.6) and (3.8). For example, we will design a seventh-order CQ filter, the circuit diagram of which is shown in Figure 6.8a. As before, this filter contains quarter-wave resonators and J inverters; the transverse branches are not shown. The main connections between resonators will be considered positive. They are realized by J inverters with inductances. The J inverters with quarter-wave segments connect the end resonators to the loads. The resonators are characterized by $f_0 = 1$ GHz, $Z_0 = 10\ \Omega$, and susceptance slope parameter $b = \pi/4Z_0$.

The initial design is a Chebyshev prototype with $f_0 = 1$ GHz, BW = 50 MHz (FBW = 0.05), and $L_{\text{Ar}} = 0.1$ dB ($L_R = 16.426$ dB). The prototype parameters (Table 2.6) are $g_0 = 1$, $g_1 = g_7 = 1.1811$, $g_2 = g_6 = 1.4228$, $g_3 = g_5 = 2.0966$, $g_4 = 1.5733$. Using formulas (3.8), we determine the main coupling coefficients $K_{12} = K_{67} = 0.03857$, $K_{23} = K_{56} = 0.02895$, $K_{34} = K_{45} = 0.02753$, and the external quality factor of end resonators $Q_{e1} = Q_{e7} = 23.622$.

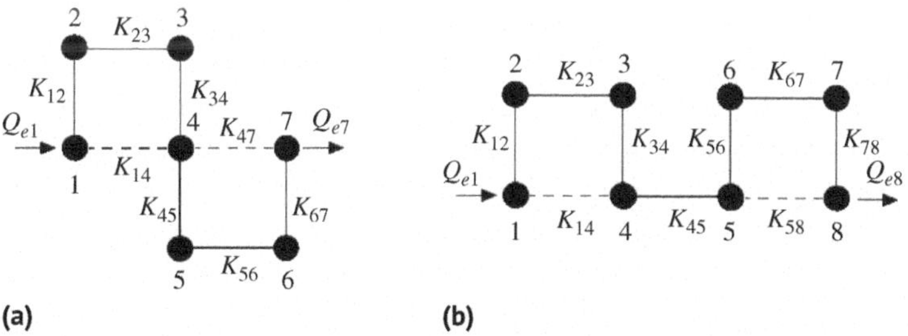

Figure 6.7 Coupling structures of CQ filters. (a) With seven resonators. (b) With eight resonators.

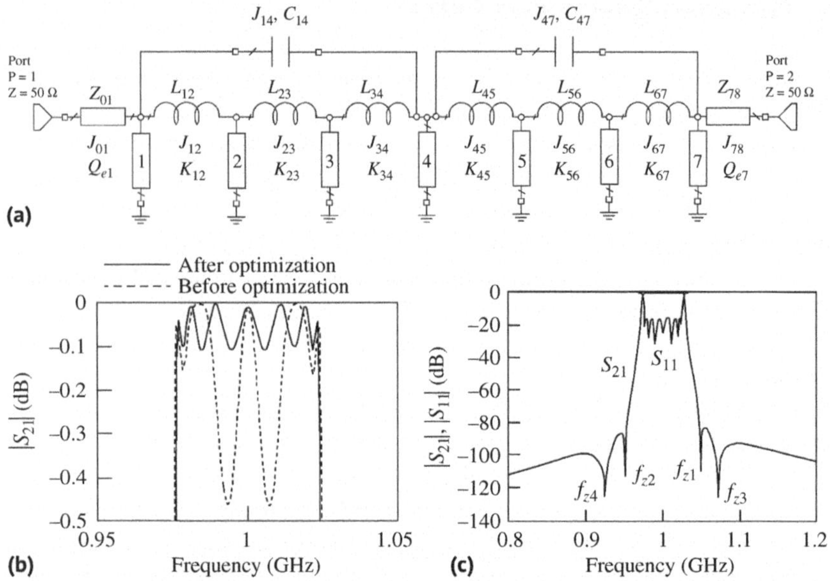

Figure 6.8 Seventh-order CQ filter. (a) Schematic with quarter-wave resonators and J inverters. (b) Pulsations in BW. (c) Frequency responses after optimization.

According to (3.6), the specified coupling values are provided by J inverters: $J_{12} = J_{67} = 0.003029\ \Omega^{-1}$, $J_{23} = J_{56} = 0.002274\ \Omega^{-1}$, $J_{34} = J_{45} = 0.002186\ \Omega^{-1}$, $J_{01} = J_{78} = 0.008154\ \Omega^{-1}$. The inverters between the resonators are presented by inductors (3.7a): $L_{12} = L_{67} = 52.54$ nH, $L_{23} = L_{56} = 67.0$ nH, $L_{34} = L_{45} = 73.61$ nH. The J_{01}, J_{78} inverters contain a quarter-wave segment with characteristic impedance $Z_{01} = Z_{78} = 122.63\ \Omega$ (3.7c).

Let us require that the first quadruplet BPF, including resonators 1–2–3–4, has TZs located at real frequencies, and their normalized value $S_{1,2} = j\Omega_{1,2} = \pm j2$. To implement such an arrangement of TZs, it is necessary to determine the normalized cross-coupling coefficient m_{14} using formula (6.7). Using the normalization of the coupling coefficients (6.3), we obtain $m_{12} = 0.7714$, $m_{23} = 0.5790$, and $m_{34} = 0.5506$. Substituting these values and the value of $S_{1,2}^2 = -4$ into (6.7), we obtain $m_{14} = -0.0671$. This value corresponds to the cross-coupling coefficient $K_{14} = -0.00336$, which J_{14} inverter realizes with capacitance $C_{14} = 0.042$ pF (3.7b).

We require that the second quadruplet BPF, including resonators 4–5–6–7, generates a pair of TZs: $S_{3,4} = j\Omega_{3,4} = \pm j3$. Substituting the normalized coupling coefficients of this quadruplet BPF and the value of $S_{1,2}^2 = -9$ into (6.7), we obtain $m_{47} = -0.02838$. This value corresponds to the cross-coupling coefficient $K_{47} = -0.00142$, which the J_{47} inverter realizes with capacitance $C_{14} = 0.0177$ pF (3.7b).

All coupling elements of the filter in question are defined. In Figure 6.8b, the dotted line shows the filter ripple in the passband. They are not equal, and their level is 0.46 dB. As a result of optimization, the main coupling coefficients changed: $K_{12} = K_{67} = 0.03930$ (1.9%), $K_{23} = K_{56} = 0.03039$ (5%), $K_{34} = K_{45} = 0.02643$ (−4.2%), and the external quality factor of the end resonators $Q_u = 23.115$ (−2.2%). The difference from the original values is indicated in parentheses. The pulsations became equal-ripple, and their level amounted to $L_{Ar} = 0.1$ dB, which is reflected in Figure 6.8b by a solid line. Figure 6.8c depicts the simulated frequency responses of the considered CQ filter with two pairs of predicted TZs on the real frequencies after optimization. This filter provides independent control of each TZ pair.

6.5 General Coupling Matrix

To describe BPF with cross-couplings, the coupling matrix $[m]$ is widely used [3–5, 11]. The BPF with N resonators is characterized by coupling coefficients K_{ij} (i, $j = 1, 2, \ldots, N$), external quality factor of the end resonators Q_{e1}, Q_{eN}, central frequency of the bandwidth f_0, bandwidth BW, and fractional bandwidth FBW=BW/f_0.

6.5.1 Coupling Matrix Definition

To describe BPF, we use the normalized coupling coefficients m_{ij} (6.3), external Q-factors $q_{e1} = Q_{e1}\mathrm{FBW}$, $q_{eN} = Q_{eN}\mathrm{FBW}$, and complex frequency $s = \sigma + j\Omega$, where Ω is the normalized frequency (6.1).

Expressions for S parameters of the N-order BPF can be represented in matrix form [11]

$$S_{21} = \frac{2}{\sqrt{q_{e1}q_{eN}}}[A]_{N1}^{-1}, \tag{6.8}$$

$$S_{11} = 1 - \frac{2}{q_{e1}}[A]_{11}^{-1}, \tag{6.9}$$

where $[A]_{N1}^{-1}$ and $[A]_{11}^{-1}$ are minors of the inverse matrix $[A]^{-1}$. A square $N \times N$ matrix $[A]$ is the normalized impedance matrix $[\tilde{Z}]$ or admittance matrix $[\tilde{Y}]$. The matrix $[A]$ has the form

$$[A] = [q] + s\,[U] - j\,[m], \tag{6.10}$$

where $[q]$ is the $(N \times N)$ matrix with all zeros, except for $q_{11} = 1/q_{e1}$ and $q_{NN} = 1/q_{eN}$, $[U]$ is the $N \times N$ unit or identity matrix. The matrix $[m]$ in (6.10) is called the general coupling matrix

$$[m] = \begin{bmatrix} m_{11} & m_{12} & \cdots & m_{1N} \\ m_{21} & m_{22} & \cdots & m_{2N} \\ \vdots & \vdots & \vdots & \vdots \\ m_{N1} & m_{N2} & \cdots & m_{NN} \end{bmatrix}. \qquad (6.11)$$

The properties of the elements of the matrix $[m]$ are discussed in detail in Cameron et al. and Hong [5, 11]. In particular, the matrix (6.11) is symmetric $m_{ij} = m_{ji}$. The elements m_{ii} of the main diagonal represent the conductivity of the resonators at the central frequency of the passband f_0. The elements are expressed as $m_{ii} = -b_i\Omega_{0i}$, where Ω_{0i} is the normalized resonant frequency of the resonators and b_i is the susceptance slope parameter (3.2). The values m_{ii} and Ω_{0i} have opposite signs. This is explained in Figures 3.6b and 3.7b. If $f_{01} = f_{02} = ... = f_{0N} = f_0$, then all diagonal elements (6.11) are equal to zero: $m_{ii} = 0$. This is a condition of the synchronous filter tuning. This can be realized by using inverters as filter coupling elements since inverters do not shift the resonant frequencies of resonators.

6.5.2 Canonical Filters

The so-called canonical filters [17] have become widespread. The coupling structure of such a filter for $N = 6$ is shown in Figure 6.9a. The TZ number of such filters reaches the value $(N - 2)$. They can be located at real frequencies and on the complex plane. The most practical case is when the canonical filter is symmetrical, which for $N = 6$ is expressed by the equalities: $K_{12} = K_{56}$, $K_{23} = K_{45}$, and $Q_{e1} = Q_{e6}$. Synthesis methods [3, 4, 17] can be used to design such filters.

The canonical filter in Figure 6.9a with all positive main couplings can implement two TZs on the $j\Omega$-axis and two TZs located on the σ-axis (Figure 6.4a), if the cross-couplings $K_{25} > 0$ and $K_{16} < 0$. The TZs on the $j\Omega$-axis provide a quasi-elliptic transfer function, and the TZs on the σ-axis provide a flat group delay. The circuit of such a filter, containing quarter-wave resonators ($f_0 = 1$ GHz, $Z_0 = 10\ \Omega$) and J inverters, is shown in Figure 6.9b. The transverse branches of the inverters are not shown. Quarter-wave sections with characteristic impedances Z_{01} and Z_{67} carry out the coupling of the end resonators to the loads.

With the pulsation value $L_{Ar} = 0.1$ dB ($L_R = 16.426$ dB), FBW $= 0.05$, and a pair of TZs at real frequencies $j\Omega_{1,2} = \pm j1.6$, the filter has the coupling coefficients $K_{12} = K_{56} = 0.04221$, $K_{23} = K_{45} = 0.03093$, $K_{34} = 0.02994$, $K_{25} = 0.0007$, $K_{16} = -0.001962$, and external quality factor $Q_{e1} = Q_{e6} = 22.23$. Using expressions

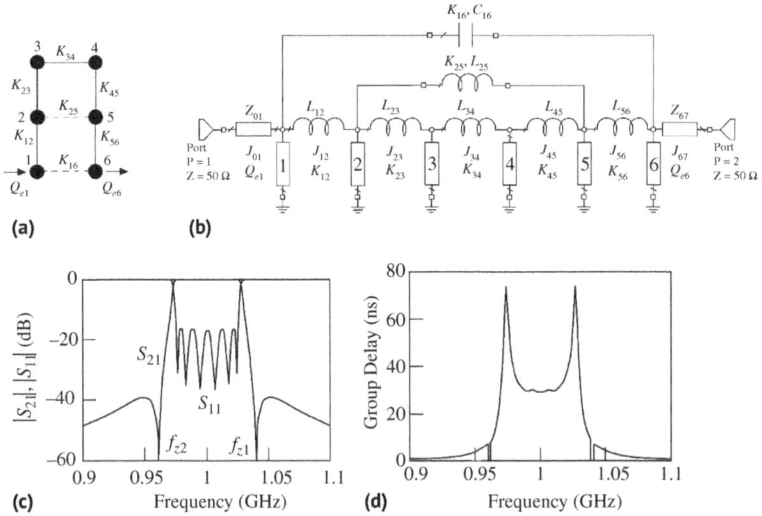

Figure 6.9 Sixth-order canonical filter. (a) Coupling structure. (b) Schematic. (c) Simulated frequency responses. (d) Simulated group delay.

(3.11a) and (3.11b), we make the transition from the coupling coefficients to the values of the lumped elements in Figure 6.9b: $L_{12} = L_{67} = 48.01$ nH, $L_{23} = L_{56} = 65.51$ nH, $L_{34} = 67.68$ nH, $L_{25} = 2894.86$ nH, and $C_{16} = 0.0245$ pF. The extreme quarter-wave sections provide the required values $Q_{e1} = Q_{e6} = 22.23$ with the characteristic impedance $Z_{01} = Z_{67} = 116.03\ \Omega$. All circuit parameters in Figure 6.9b are defined. Simulated frequency responses of the BPF are shown in Figure 6.9c. Figure 6.9d shows the group delay, which is sufficiently flat in the frequency band slightly less than 50% of the passband.

By normalizing the coupling coefficients and $Q_{e1} = Q_{e6}$ value, we obtain $m_{12} = m_{56} = 0.8442$, $m_{23} = m_{45} = 0.6186$, $m_{34} = 0.5988$, $m_{25} = 0.014$, $m_{16} = -0.03924$, $q_{e1} = q_{e6} = 1.112$. The general coupling matrix of the canonical BPF for synchronous tuning is written as

$$[m] = \begin{bmatrix} 0 & 0.8442 & 0 & 0 & 0 & -0.03924 \\ 0.8442 & 0 & 0.6186 & 0 & 0.0140 & 0 \\ 0 & 0.6186 & 0 & 0.5988 & 0 & 0 \\ 0 & 0 & 0.5988 & 0 & 0.6186 & 0 \\ 0 & 0.0140 & 0 & 0.6186 & 0 & 0.8442 \\ -0.03924 & 0 & 0 & 0 & 0.8442 & 0 \end{bmatrix}. \quad (6.12)$$

$$q_e = 1.112$$

Despite the wide possibilities of the $[m]$ matrix in describing BPF, it has certain limitations. The matrix $[m]$ is obtained from the impedance matrix $[Z]$ or admittance matrix $[Y]$ of an electric circuit with lumped elements [18]. The matrices $[Z]$ and $[Y]$ express the Kirchhoff laws for electrical circuits. Therefore, they apply to BPFs with reflective-type resonators, which are included in BPF as a one-port network with input admittance Y_{in} or input impedance Z_{in}. The through-type resonator is a two-port network [19]. It is described by three parameters: input admittance Y_{in} (or impedance Z_{in}), output admittance Y_{out} (or impedance Z_{out}), and transfer admittance Y_{12} (or impedance Z_{12}). The equivalent circuit of a through-type resonator with lumped elements is presented in Cohn [19]. This resonator has two open ends, and in addition to the resonance phenomenon, it implements the immittance transformation. Introducing through-type resonators into electrical circuits like a two-port network does not allow for their description by Kirchhoff laws and $[Z]$ or $[Y]$ matrix.

Special techniques are used to work with these circuits [20, 21]. Adapting the $[m]$ matrix to BPFs containing through-type resonators is advisable. In the following, we consider a modified coupling matrix [22], which can describe BPFs with resonators of reflection type and through type. The consideration is carried out for a matrix $[m]$ of size $N \times N$ with simple couplings.

6.6 Coupling Matrix Modification Using Through-type Resonators

6.6.1 Cases of Inapplicability of Coupling Matrix

Figure 6.10a shows a photograph of the microstrip triplet BPF. Its end resonators 1 and 3 are quarter-wave, and half-wave through-type central resonator *2* is connected in the filter as a two-port network. The BPF was fabricated on a TMM-10i substrate (Rogers) 37.6 mm × 15.6 mm size, $\varepsilon_r = 9.8$, $\tan\delta = 0.002$, and $h = 1.905$ mm. Metalized holes ensure a short circuit at the ends of the quarter-wave resonators. All couplings of the BPF are positive. A conductive connection is used to connect the filter to the loads. There is a positive cross-coupling between *1* and *3* resonators. In Figure 6.10b, the solid lines show their measured frequency responses.

The experimental filter is characterized by the following parameters: center frequency $f_0 = 2008$ MHz, bandwidth BW = 92 MHz (FBW = 0.0458), left-handed TZ located at $f_z = 1876$ MHz, insertion loss at the center frequency $L_{A0} = 1.5$ dB, return loss $L_R > 16.5$ dB. In this case, the details of the filter topology are not essential. The measured main coupling coefficients $K_{12} = K_{23} = 0.04$, the cross-coupling $K_{13} = 0.0123$, and the external quality factor of end resonators $Q_e = 22.92$ are important.

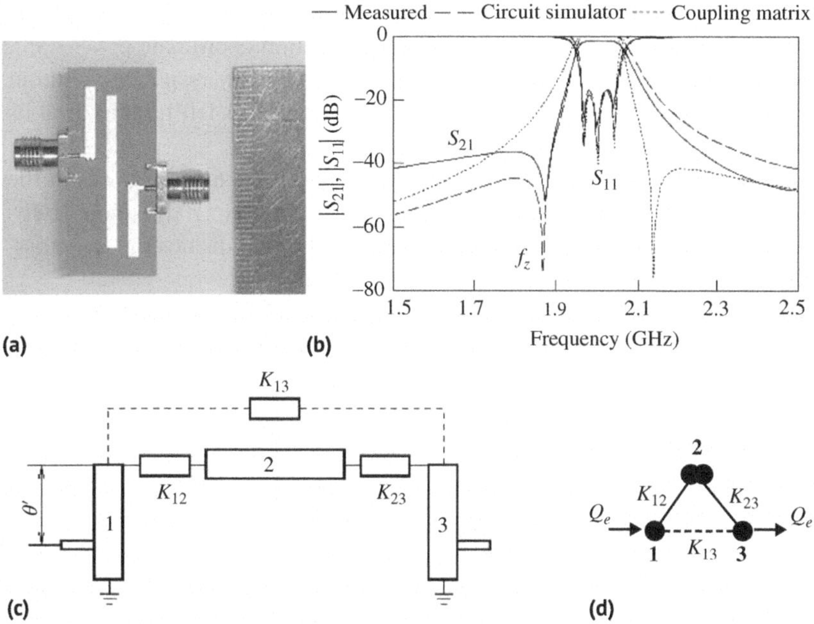

Figure 6.10 Microstrip third-order BPF with one cross-coupling. (a) Photograph. *Source:* [22]/Fig. 1/with permission of IEEE. (b) Frequency responses. (c) Schematic for linear circuit simulator. (d) Coupling structure.

We use the linear circuit simulator of Microwave Office software to obtain the frequency responses of the considered filter (Figure 6.10c). The resonators 1 and 3 are quarter-wave at $f_0 = 2008$ MHz and their characteristic impedance $Z_0 = 10\ \Omega$. The through-type resonator 2 is a half-wave at the same frequency, and its characteristic impedance $Z_0 = 20\ \Omega$. Quarter-wave open-ended TL segments with characteristic impedance Z_{ij} are used as admittance inverters. In Figure 6.10c, they are indicated by smaller rectangles. Since J_{12} inverter implements the coupling between the resonators $K_{12} = J_{12}/\ (b_1 b_2)^{1/2}$ (3.4), and the characteristic impedance Z_{12} of a quarter-wave segment is characterized by the value $Z_{12} = 1/J_{12}$, then the value $Z_{12} = 1/K_{12}\ (b_1 b_2)^{1/2}$ provides a coupling coefficient K_{12} between two resonators. Here, b_1 and b_2 are the susceptance slope parameters of the resonators, $b_1 = \pi/4Z_0$ for $\lambda/4$ resonator and $b_2 = \pi/2Z_0$ for $\lambda/2$ resonator [23]. The main couplings $K_{12} = K_{23} = 0.04$ are provided by segments with $Z_{12} = Z_{23} = 318.31\ \Omega$, and the cross-coupling $K_{13} = 0.0123$ is realized by a segment with $Z_{13} = 1035.15\ \Omega$.

The external Q-factor of the end resonators is calculated as follows: $Q_e = R_L b = R_L \pi/4Z_0 \cos^2 \theta'$ [23], where θ' is the coordinate of a conductive connection of the input and output loads; $R_L = 50\ \Omega$. The formula shows that the value $Q_e = 22.92$ of the end resonators is provided at the load connection coordinate $\theta' = 65.55°$. All filter elements in Figure 6.10c are defined. Its frequency responses, without losses, are shown in Figure 6.10b by dashed lines. The TZ is left-handed, and its value $f_z = 1876$ MHz is the same as measured.

We use the coupling matrix (6.11) to construct the frequency responses of the considered filter. Its coupling structure is shown in Figure 6.10d. A double circle indicates the through-type resonator. The dashed line indicates the K_{13} cross-coupling.

Normalization leads to the values $m_{12} = m_{23} = 0.873$, $m_{13} = 0.268$, and $q_e =$ 1.049. The matrix $[m]$ of the experimental filter at synchronous tuning is defined

$$[m] = \begin{bmatrix} 0 & m_{12} & m_{13} \\ m_{12} & 0 & m_{23} \\ m_{13} & m_{23} & 0 \end{bmatrix} = \begin{bmatrix} 0 & 0.873 & 0.268 \\ 0.873 & 0 & 0.873 \\ 0.268 & 0.873 & 0 \end{bmatrix}, \quad q_e = 1.049 \tag{6.13}$$

The frequency responses corresponding to the coupling matrix (6.13) are shown in Figure 6.10b by dotted lines. The "Filters designer" of CTS Microwave Studio software was used to calculate and plot these curves. Ensuring "equal-ripple" return loss in the passband is achieved by adjusting the main diagonal elements: $m_{11} = m_{33} = 0.1$, $m_{22} = -0.25$. The values of m_{ii} play a secondary role in this case, and they will not be given further. Then the TZ f_z is right-handed ($f_z > f_0$), which does not match the measured value f_z.

A similar discrepancy occurs in another triplet BPFs (Figure 6.11) with a through-type resonator. For them, K_{13} is not cross-coupling, but is provided by nearby resonators 1 and 3. The triplet with open-loop resonators (Figure 6.11a) has been known for a long time [24]. With all positive couplings, it has a left-hand TZ, which has not yet been explained. The triplet in Figure 6.11b [25] has $\lambda/4$ and $\lambda/2$ resonators. The task is to eliminate the existing discrepancy.

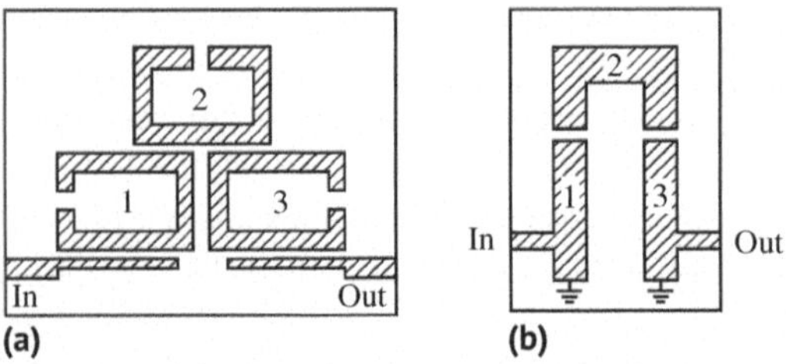

Figure 6.11 Microstrip triplet BPFs with through-type resonator. (a) With open-loop resonators. (b) With $\lambda/4$ and $\lambda/2$ resonators.

6.6.2 Problem Solution

Let us consider the through-type resonator in more detail. The voltage resonance of the half-wave resonator is observed for electric lengths $\theta_n = (n+1)\pi$, $n = 0,\ 1,\ 2, \ldots$. In this case, the transfer matrix of this resonator takes the form

$$\begin{bmatrix} A & B \\ C & D \end{bmatrix} = \begin{bmatrix} \cos\theta_n & jZ_0 \sin\theta_n \\ jZ_0^{-1} \sin\theta_n & \cos\theta_n \end{bmatrix} = \begin{bmatrix} \pm 1 & 0 \\ 0 & \pm 1 \end{bmatrix}, \tag{6.14}$$

where $A = D = 1$ for odd n and $A = D = -1$ for even n. At different values of n, the through-type resonator transforms impedance differently. To determine these differences, we use the transfer matrix of admittance inverter [11]

$$\begin{bmatrix} A & B \\ C & D \end{bmatrix} = \begin{bmatrix} 0 & \pm j/J \\ \pm jJ & 0 \end{bmatrix},$$

where positive values of J correspond to positive coupling, and negative values of J correspond to negative coupling.

A cascade connection of an inverter with a positive parameter J and a through-type resonator in the resonance regime leads to the following result

$$\begin{bmatrix} 0 & j/J \\ jJ & 0 \end{bmatrix} \begin{bmatrix} 1 & 0 \\ 0 & 1 \end{bmatrix} = \begin{bmatrix} 0 & j/J \\ jJ & 0 \end{bmatrix}, \quad n \text{ is odd}, \tag{6.15}$$

$$\begin{bmatrix} 0 & j/J \\ jJ & 0 \end{bmatrix} \begin{bmatrix} -1 & 0 \\ 0 & -1 \end{bmatrix} = \begin{bmatrix} 0 & -j/J \\ -jJ & 0 \end{bmatrix}, \quad n \text{ is even}. \tag{6.16}$$

If n is odd, the nature of the coupling does not change (6.15); if n is even, the magnetic coupling is perceived as an electric one (6.16) and vice versa. In this case, the voltages at the ends of a through-type resonator have opposite phases.

The through-type resonator is used in a BPF at a fundamental resonant frequency ($n = 0$); therefore, it changes the sign of the coupling coefficient.

6.6.3 Rules for Changing Sign of Coupling Coefficient

Figure 6.12 shows the coupling structure of BPF with one through-type resonator, which has two ends. The left end of the resonator is cross-coupled with left-side resonators; the right end is cross-coupled with right-side resonators. The main coupling coefficients K_r or K_l are located to the right and left of the through-type resonator. Figure 6.12 shows three different cross-couplings to a through-type resonator. Each case corresponds to specific rules for changing signs K_r or K_l, which are set using the linear circuit simulator.

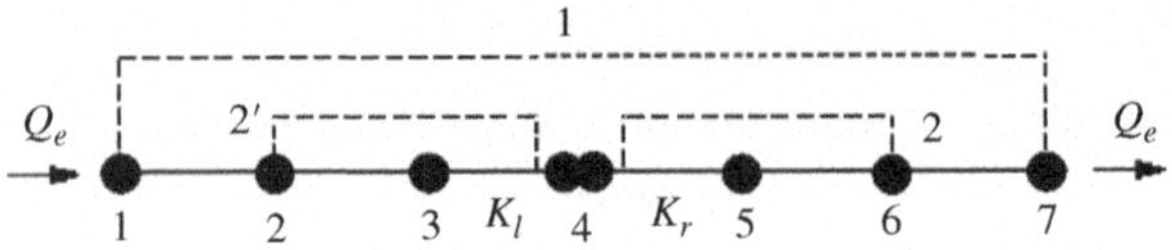

Figure 6.12 Coupling structure of BPF with one through-type resonator.

Case 1: The cross-couplings are not connected to the through-type resonator. There is only one cross-coupling 1 in Figure 6.12. *Rule 1*: Changing the sign of one of the two main couplings K_r or K_l is necessary.

Case 2: One cross-coupling (2 or 2') is connected to one end of the through-type resonator. *Rule 2*: Changing the sign of the coupling coefficient adjacent to the "free" end of the through-type resonator is necessary. If cross-coupling 2 is used, then the sign of K_l must be changed; if cross-coupling 2' is used, then the sign of K_r must be changed.

Case 3: Two cross-couplings 2 and 2' are connected to the right and left ends of the through-type resonator. *Rule 3*: The signs of the coupling coefficients adjacent to the through-type resonator remain the same.

These rules make it possible to modify the coupling matrix of various BPFs with a single through-type resonator. Note that this modification does not require changing the signs of the cross-couplings adjacent to the through-type resonator. The examples of modifying the matrix $[m]$ of some BPFs are presented as follows.

6.6.4 Modification of Matrix [*m*]

The triplet BPFs in Figure 6.13 are equivalent. The two filters on the right side of these figures are modified, since they contain all resonators of the reflection type. This equivalence is a consequence of Rule 1. Changing the sign of m_{12} or m_{23} to the opposite in (6.13), we obtain the modified matrix $[m]$. In this case, the character of the dotted curve for $|S_{21}|$ in Figure 6.10b becomes the same as the other two curves, and the TZ is left-handed. In this case, the Kurzrok formula (6.4), which determines the position of the TZ Ω_z, is changed and takes the form

$$\Omega_z = -\frac{m_{12}m_{23}}{m_{13}}. \tag{6.17}$$

On the right side of (6.17), the sign "−" appeared, which converts the right-hand TZ into the left-hand one, and vice versa.

Figure 6.14a shows a photograph of the microstrip filter in question, which is made on the same substrate and occupies an area of 21.6 mm × 20.8 mm. Resonators 1, 3, and 4 in the filter are quarter-wave. The through-type half-wave resonator 2 has a U-shape. The filter has three cross-couplings: K_{13},

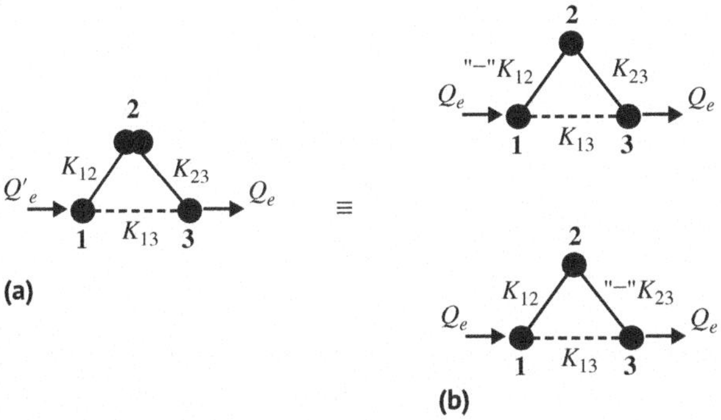

Figure 6.13 Equivalent triplet BPFs. (a) Triplet BPF with through-type resonator. (b) Triplet BPFs without through-type resonator.

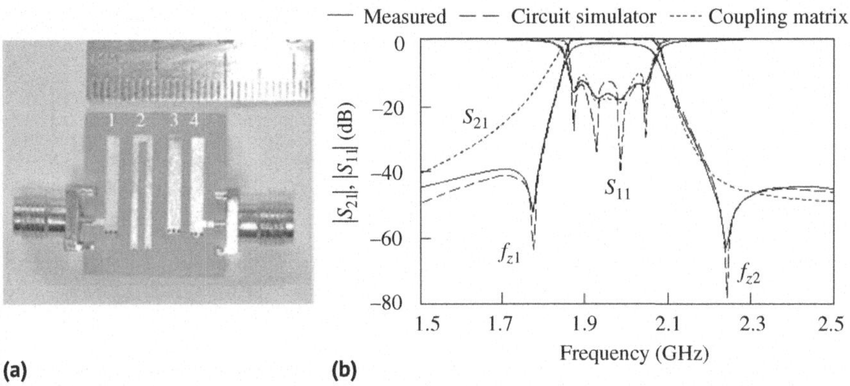

Figure 6.14 Microstrip fourth-order BPF. (a) Photograph. *Source*: [22]/Fig. 5/with permission of IEEE. (b) Frequency responses.

K_{14}, and K_{24}. The measured frequency responses of the filter are shown in Figure 6.14b by solid lines. The considered microstrip BPF has the following parameters: $f_0 = 1960$ MHz, BW = 160 MHz (FBW = 0.0816), $L_{A0} = 1.2$ dB, $L_R >$ 12.5 dB. The filter has two TZs at $f_{z1} = 1774$ MHz and $f_{z2} = 2240$ MHz.

As before, the filter topology details are unimportant to us. Of interest are the measured coupling coefficients between the resonators, the main $K_{12} = K_{34} =$ 0.08, $K_{23} = 0.065$, and cross-couplings $K_{13} = 0.0245$, $K_{14} = 0.0081$, $K_{24} = 0.0149$, and the external quality factor of the end resonators $Q_e = 12.5$. As a result of normalization, we get $m_{12} = m_{34} = 0.980$, $m_{23} = 0.796$, $m_{13} = 0.30$, $m_{14} = 0.0993$, $m_{24} = 0.182$, and $q_e = 1.02$.

The coupling structure of the filter under consideration is shown in Figure 6.15a. We transfer it to an equivalent structure with all reflective-type resonators. To do

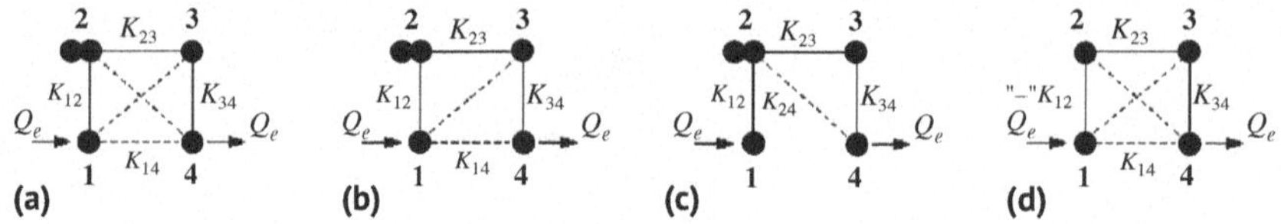

Figure 6.15 Defining equivalent quadruplet. (a) Initial quadruplet. (b) Coupling structure for Rule 1. (c) Coupling structure for Rule 2. (d) Equivalent quadruplet.

this, consider two auxiliary quadruplets (Figure 6.15b, c), to which the established rules of sign change apply. Rule 1 applies to the filter in Figure 6.15b, and it needs to change the sign of K_{12} or K_{23}. Rule 2 applies to the filter in Figure 6.15c, and the sign of K_{12} must be changed for it. Combining these solutions, we conclude that the equivalent filter should change the sign of K_{12}. Its coupling structure is shown in Figure 6.15d, which corresponds to the modified coupling matrix in synchronous tuning

$$[m] = \begin{bmatrix} 0 & -m_{12} & m_{13} & m_{14} \\ -m_{12} & 0 & m_{23} & m_{24} \\ m_{13} & m_{23} & 0 & m_{34} \\ m_{14} & m_{24} & m_{34} & 0 \end{bmatrix} = \begin{bmatrix} 0 & -0.980 & 0.300 & 0.0993 \\ -0.980 & 0 & 0.796 & 0.182 \\ 0.300 & 0.796 & 0 & 0.980 \\ 0.0993 & 0.182 & 0.980 & 0 \end{bmatrix}$$
$$q_e = 1.02 \tag{6.18}$$

The modified matrix (6.18) corresponds to the frequency responses shown in Figure 6.14b by dashed lines. They coincide with measured frequency responses. If the matrix $[m]$ is not modified, it will correspond to the frequency responses shown in Figure 6.14b by dotted lines, which is false.

If the through-type resonator occupies position 3, then using Rule 1 and Rule 2, we establish that in the modified matrix $[m]$, the sign of m_{34} will be changed.

The most common is the BPF quadruplet with one m_{14} cross-coupling. The position of two TZs of the filter with all reflective-type resonators is determined by the Kurzrok formula (6.7). If the resonator *2* or *3* of the quadruplet is replaced with a through-type resonator, then this formula is changed

$$S_{1,2}^2 = -\frac{m_{12}m_{23}m_{34}}{m_{14}} - m_{23}^2. \tag{6.19}$$

Expression (6.19) indicates that two TZs of this filter with all positive couplings are located at the real frequencies $S_{12} = j\Omega_{12}$ ($j\Omega_1 = -j\Omega_2$), and this quadruplet BPF is a quasi-elliptic filter.

6.6.5 Fifth-order Microstrip BPF with Six Cross-couplings

Figure 6.16a shows a photograph of the microstrip filter in question, which is made on the same substrate and occupies an area of 23.8 mm × 21.6 mm. The half-wave U-shape resonator 3 occupies a central position. The remaining resonators are quarter-wave. The filter has six positive cross-couplings: $K_{13} = K_{35}$, $K_{14} = K_{25}$, K_{15}, K_{24}. The solid lines in Figure 6.16b show the measured frequency responses of this filter. This microstrip BPF is characterized by the following parameters: $f_0 = 1965$ MHz, BW = 230 MHz (FBW = 0.117), $L_{A0} = 1.3$ dB, $L_R > 14$ dB. The filter has a left-handed TZ located at $f_z = 1784$ MHz.

The main couplings are $K_{12} = K_{45} = 0.086$, $K_{23} = K_{34} = 0.065$; and the cross-couplings are $K_{13} = K_{35} = 0.014$, $K_{14} = K_{25} = 0.0047$, $K_{24} = 0.02$, $K_{15} = 0.0027$. The measured external quality factor of the end resonators is $Q_e = 11.26$. As a result of normalization, we get $m_{12} = m_{45} = 0.735$, $m_{23} = m_{34} = 0.556$, $m_{13} = m_{35} = 0.120$, $m_{14} = m_{25} = 0.040$, $m_{24} = 0.171$, $m_{15} = 0.023$, and $q_e = 1.317$. Using Rule 1 and Rule 3, we arrive at an equivalent structure with the opposite sign of m_{23}. The modified matrix $[m]$ is written as

$$[m] = \begin{bmatrix} 0 & 0.735 & 0.120 & 0.040 & 0.023 \\ 0.735 & 0 & -0.556 & 0.171 & 0.040 \\ 0.120 & -0.556 & 0 & 0.556 & 0.120 \\ 0.040 & 0.171 & 0.556 & 0 & 0.735 \\ 0.023 & 0.040 & 0.120 & 0.735 & 0 \end{bmatrix}, \tag{6.20}$$

$$q_e = 1.31$$

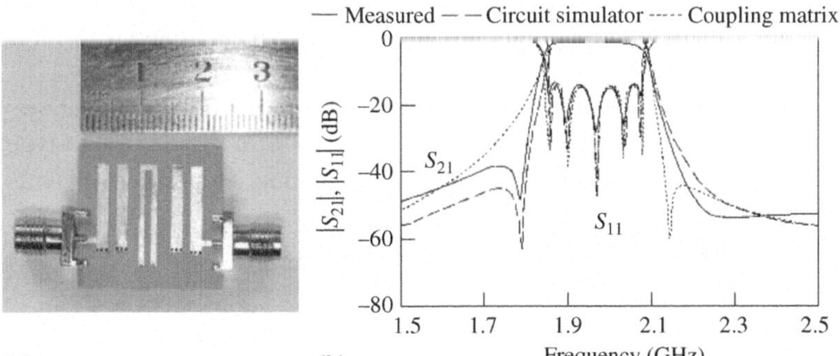

(a) (b)

Figure 6.16 Microstrip fifth-order BPF. (a) Photograph. *Source*: [22]/Fig. 7/with permission of IEEE. (b) Frequency responses.

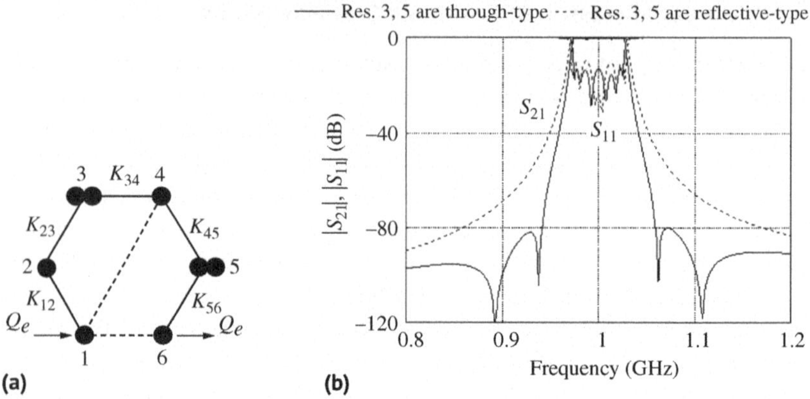

Figure 6.17 Sixth-order BPF with two through-type resonators and two cross-couplings. (a) Coupling structure. (b) Simulated frequency responses.

It corresponds to the frequency responses shown in Figure 6.16b by dashed lines. They coincide with measured frequency responses. If matrix (6.20) is not modified, it will correspond to the frequency responses shown in Figure 6.16b by dotted lines, which is false. The established rules for modifying the matrix $[m]$ are also valid for the extended matrix $[m]$ of size $(N+2)\times(N+2)$ and for several through-type resonators in a BPF.

In the latter case, through-type resonators should not be located next to each other, as shown in Figure 6.17 for a sixth-order BPF with two through-type resonators and two cross-couplings. During the study, it was also found that replacing reflective-type resonators with through-type resonators leads to a change in frequency responses: the right-hand TZ becomes left-hand (Figures 6.10 and 6.11); the insertion loss curve becomes more symmetrical (Figures 6.14 and 6.16); TZs appear in BPF (Figure 6.17b).

Three rules solve the problem of the coupling matrix $[m]$ modification. It allows us to correctly present the frequency responses of BPFs with some through-type resonators. The modification consists of sign-changing some coupling coefficients m_{ij} according to established rules.

References

1 R. M. Kurzrok, General three-resonator filters in waveguide, *IEEE Trans. Microw. Theory Tech.*, vol. 14, no. 1, pp. 46–47, 1966, https://doi.org/10.1109/TMTT.1966.1126154.

2 R. M. Kurzrok, General four-resonator filters at microwave frequencies, *IEEE Trans. Microw. Theory Tech.*, vol. 14, no. 6, pp. 295–296, 1966, https://doi.org/10.1109/TMTT.1966.1126254.

3 A. E. Atia and A. E. Williams, Narrow-bandpass waveguide filters, *IEEE Trans. Microw. Theory Tech.*, vol. 20, no. 4, pp. 258–265, 1972, https://doi.org/10.1109/TMTT.1972.1127732.

4 A. Atia, A. Williams, and R. Newcomb, Narrow-band multiple-coupled cavity synthesis, *IEEE Trans. Circuits Syst.*, vol. 21, no. 5, pp. 649–655, 1974, https://doi.org/10.1109/TCS.1974.1083913.

5 R. J. Cameron, C. M. Kudsia, and R. R. Mansour, *Microwave Filters for Communication Systems*, 2nd ed. Hoboken, NJ: John Wiley & Sons, Inc., 2018, https://doi.org/10.1002/9781119292371.

6 S. Amari, U. Rosenberg, and J. Bornemann, Adaptive synthesis and design of resonator filters with source/load-multiresonator coupling, *IEEE Trans. Microw. Theory Tech.*, vol. 50, no. 8, pp. 1969–1978, 2002, https://doi.org/10.1109/TMTT.2002.801348.

7 R. A. Johnson, *Mechanical Filters in Electronics*. John Wiley & Sons, 1983, https://www.amazon.com/Mechanical-Filters-Electronics-Wiley-filters/dp/0471089192.

8 G. C. Temes and S. K. Mitra (Eds.), *Modern Filter Theory and Design*. New York: Wiley, 1973.

9 W. A. Atia, K. A. Zaki, and A. E. Atia, Synthesis of general topology multiple coupled resonator filters by optimization, in *1998 IEEE MTT-S International Microwave Symposium Digest (Cat. No.98CH36192)*, 1998, vol. 2 pp. 821–824, https://doi.org/10.1109/MWSYM.1998.705116.

10 W. Meng, H.-M. Lee, K. A. Zaki, and A. E. Atia, Synthesis of wideband multicoupled resonators filters, *IEEE Trans. Microw. Theory Tech.*, vol. 59, no. 3, pp. 593–603, 2011, https://doi.org/10.1109/TMTT.2010.2095032.

11 J.-S. Hong, *Microstrip Filters for RF/Microwave Applications*, 2nd ed. New Jersey: Wiley, 2011, https://doi.org/10.1002/9780470937297.

12 J. B. Thomas, Cross-coupling in coaxial cavity filters-a tutorial overview, *IEEE Trans. Microw. Theory Tech.*, vol. 51, no. 4, pp. 1368–1376, 2003, https://doi.org/10.1109/TMTT.2003.809180.

13 R. Hershtig, R. Levy, and K. Zaki, Synthesis and design of cascaded trisection (CT) dielectric resonator filters, in *27th European Microwave Conference, 1997*, 1997, vol. 2 pp. 784–791, https://doi.org/10.1109/EUMA.1997.337890.

14 R. J. Cameron and J. D. Rhodes, Asymmetric realizations for dual-mode bandpass filters, *IEEE Trans. Microw. Theory Tech.*, vol. 29, no. 1, pp. 51–58, 1981, https://doi.org/10.1109/TMTT.1981.1130286.

15 R. Levy, Direct synthesis of cascaded quadruplet (CQ) filters, *IEEE Trans. Microw. Theory Tech.*, vol. 43, no. 12, pp. 2940–2945, 1995, https://doi.org/10.1109/22.476634.

16 R. Levy and P. Petre, Design of CT and CQ filters using approximation and optimization, *IEEE Trans. Microw. Theory Tech.*, vol. 49, no. 12, pp. 2350–2356, 2001, https://doi.org/10.1109/22.971620.

17 G. Pfitzenmaier, Synthesis and realization of narrow-band canonical microwave bandpass filters exhibiting linear phase and transmission zeros, *IEEE Trans. Microw. Theory Tech.*, vol. 30, no. 9, pp. 1300–1311, 1982, https://doi.org/10.1109/TMTT.1982.1131252.

18 S. Seshu and N. Balabanian, *Linear Network Analysis*. Wiley, 1959.

19 S. B. Cohn, Direct-coupled-resonator filters, *Proc. IRE*, vol. 45, no. 2, pp. 187–196, 1957, https://doi.org/10.1109/JRPROC.1957.278389.

20 P. I. Richards, Resistor-transmission-line circuits, *Proc. IRE*, vol. 36, no. 2, pp. 217–220, 1948, https://doi.org/10.1109/JRPROC.1948.233274.

21 H. J. Riblet, General synthesis of quarter-wave impedance transformers, *IRE Trans. Microw. Theory Tech.*, vol. 5, no. 1, pp. 36–43, 1957, https://doi.org/10.1109/TMTT.1957.1125088.

22 A. V. Zakharov and S. Litvintsev, Coupling matrix modification for bandpass filters with through-type resonators and simple couplings, *IEEE Trans. Circuits Syst. II Express Briefs*, vol. 71, no. 4, pp. 1864–1868, 2024, https://doi.org/10.1109/TCSII.2023.3337439.

23 G. L. Matthaei, L. Young, and E. M. T. Jones, *Microwave Filters, Impedance-Matching Networks, and Coupling Structures*. New York: Artech House Books, 1980.

24 J.-S. Hong and M. J. Lancaster, Microstrip cross-coupled trisection bandpass filters with asymmetric frequency characteristics, *IEE Proc. - Microwaves, Antennas Propag.*, vol. 146, no. 1, p. 84, 1999, https://doi.org/10.1049/ip-map:19990146.

25 A. Zakharov, S. Rozenko, and M. Ilchenko, Two types of trisection bandpass filters with mixed cross-coupling, *IEEE Microw. Wirel. Components Lett.*, vol. 28, no. 7, pp. 585–587, 2018, https://doi.org/10.1109/LMWC.2018.2837905.

7

Bandpass Filters with Mixed Couplings

Using mixed coupling coefficients between resonators containing magnetic and electric components results in the appearance of transmission zeros (TZs) in a bandpass filter (BPF) without cross-couplings.

This effect was discovered long before 2000 [1–3], when they were not yet called mixed. These couplings increase the TZ number by one compared to BPF, which uses only simple couplings. After 2000, a systematic study and development of the theory of BPFs with these couplings began [4–8]. In Levy [4], a BPF with "resonant" coupling was considered, which is a special, but important, case of mixed coupling. The term "mixed" coupling was first used in 2006 [5].

Mixed coupling coefficients are also called frequency-dependent [9–16]. Note that the coupling coefficient K characterizes the electromagnetic interaction between two resonators with the same resonant frequency f_0 and refers specifically to this frequency f_0. If the value of f_0 changes, then the K value will also change. The concept of frequency-dependent coupling was first proposed in 1992 [9] and was quite fruitful. The same idea was repeated seven years later in Amari and Bornemann [10]. Then it took another nine years to express the coefficient of mixed coupling as a linear dependence on the normalized frequency [11].

Considerable attention in the scientific and technical literature began to be paid to BPFs with frequency-dependent couplings after publication in 2012–2013 of a series of papers [12–16]. In these articles, parametric synthesis, also called optimization, is used. The filter's coupling structure is specified during parametric synthesis, while the coupling matrix elements serve as variables. Special algorithms are used to modify the variable elements, ensuring the specified TZ position. This synthesis method has considerable generality, allowing many essential TZ placements for BPFs with different coupling structures to be established. Using mixed couplings between adjacent resonators leads to TZs located only at real frequencies ($j\omega$-axis) and being attenuation poles. Cross-couplings are necessary for such filters to obtain TZs located on the complex plane [17–19].

Note that many simple couplings, including those defined by J inverters with inductances (Figure 3.3a) and capacitances (Figure 3.3b), depend on frequency, either positive or negative. Mixed coupling takes on both positive and negative values as frequency changes, and in the process of its change, it crosses the $K = 0$ axis. In the second-order BPF, the value of $K = 0$ corresponds to a particular value of frequency f_z, which is the TZ or attenuation pole. The term "frequency-dependent coupling" is applied exclusively to mixed coupling, for which the condition $K = 0$ is ensured.

It should be noted that industry experts did not expect the development of the theory of mixed coupled BPFs. Thus, the large European company Epcos AG in the early 2000s serially produced microwave ceramic filters with mixed couplings (second-, third-, and fourth-order), which ensured its commercial success. The development of these filters used EM simulation programs that appeared at that time.

7.1 Representation Forms of Mixed Coupling Between Resonators

Let us look at some features of mixed coupling. An admittance inverter is used to implement this (Figure 7.1a). It is characterized by the J parameter, containing the magnetic J_m and electric J_e components [20]

$$J = J_m + J_e = 1/\omega L - \omega C. \tag{7.1}$$

The expression for the mixed coupling coefficient is obtained by substituting (7.1) into (3.4)

$$K = K_m + K_e = \left.\frac{1/\omega L - \omega C}{\sqrt{b_1 b_2}}\right|_{\omega=\omega_0}. \tag{7.2}$$

Dependence (7.2) is represented in Figure 7.1b by a solid line, where the resonant frequency of resonators is equal to $f_0 = 1$ GHz. The components of the mixed coupling coefficient K (7.2) are defined as follows:

$$K_m = \left.\frac{1/\omega L}{\sqrt{b_1 b_2}}\right|_{\omega=\omega_0}, \tag{7.3a}$$

$$K_e = \left.\frac{-\omega C}{\sqrt{b_1 b_2}}\right|_{\omega=\omega_0}. \tag{7.3b}$$

They are simple couplings depending on frequency. Moreover, the slope of their frequency dependence is always negative. The difference between a mixed and simple coupling is that the graph of its frequency dependence $K = K(\omega)$ intersects

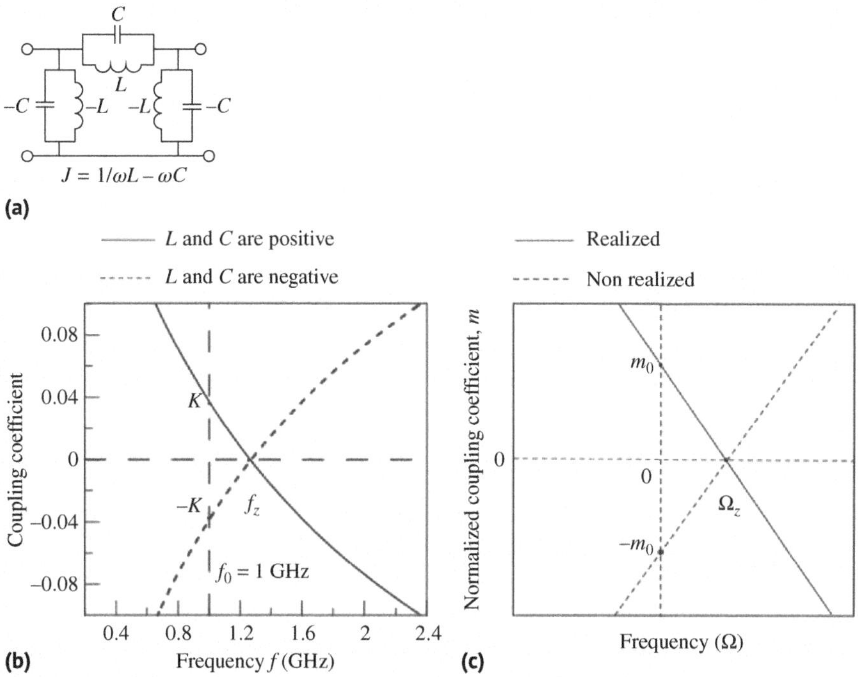

Figure 7.1 Admittance inverter for mixed coupling. (a) Schematic. (b) Coupling coefficient versus frequency f. (c) Linear representation of mixed coupling versus Ω.

the $K = 0$ axis. In other words, the mixed coupling takes the value $K = 0$ at a certain frequency ω_z, which is TZ.

The frequency ω_z is the resonant frequency of the LC chain of the J inverter $\omega_z = 1/\sqrt{LC}$. If a J inverter connects two resonators with a resonant frequency ω_0, then the last equality can be represented as [20]:

$$\omega_z = \omega_0\sqrt{K_m/\ |K_e|}. \tag{7.4}$$

To measure the coefficient K of mixed coupling between resonators, their weak coupling with loads is used. In this case, two peaks corresponding to two coupling frequencies f_e and f_o are visible on the transfer function. These frequencies and formula (3.10) allow us to determine the modulus of the coupling coefficient $|K|$. The transfer function also clearly shows the TZ frequency f_z, which determines the sign of K.

From (7.4) it follows that with right-handed TZ ($f_z > f_0$), the mixed coupling coefficient is positive $K > 0$, but if TZ is left-handed ($f_z < f_0$), then $K < 0$. Expressions (7.2) and (7.4) allow us to separate the mixed coupling coefficient K into magnetic and electric components [20]

$$K_m = \frac{K}{1 - (f_0/f_z)^2}, \tag{7.5a}$$

$$K_e = \frac{K}{1 - (f_z/f_0)^2}, \tag{7.5b}$$

where $f_0 = (f_o + f_e)/2$, f_e and f_o are the even- and odd-mode frequencies.

To introduce a mixed coupling into the coupling matrix, we use a linear representation form [11]

$$m(\Omega) = m_0 - a\Omega, \tag{7.6}$$

where $m_0 = (K_m + K_e)/\mathrm{FBW}$, $a = \sqrt{K_m |K_e|}$ is the slope parameter, and $\Omega = (f/f_0 - f_0/f)/\mathrm{FBW}$ is the normalized frequency.

The frequency dependence (7.6) is depicted by a solid line in Figure 7.1c. The slope of the mixed coupling remains negative for the more general case when it is carried out by an arbitrary LC circuit with a reactive input admittance $B(\omega)$. The fact is that the slope of the frequency dependence $K(\omega)$ is opposite to the slope of the function $B(\omega)$. The value of $dB(\omega)/d\omega$ is always positive, which was established based on the use of energy functions [21].

Some works use the BPF design method with mixed couplings between adjacent resonators, sometimes leading to a positive frequency slope of the mixed coupling (Figure 7.1b, c, dotted lines). This is physically impossible since it requires negative values of L and C. For negative L and C, the value of ω_z will remain the same (Figure 7.1b, c, solid lines), and expression (7.6) will contain the term $a\Omega$ with the "+" sign.

The question arises: Can a physically impossible coupling with a positive frequency slope be converted into a physically feasible coupling with a negative slope? To do this, we must transform the dotted lines into solid lines (Figure 7.1b, c). This transformation for the linear representation form (7.6) of the mixed coupling is carried out as follows.

Rule: To make a physically impossible mixed coupling, in the representation form (7.6) with positive slope, physically feasible, it is necessary to replace the "+" sign in front of the term $a\Omega$ with the "–" sign, and also change the sign in front of m_0 to the opposite.

Note that this rule does not apply to mixed cross-coupling, nor to a mixed coupling between adjacent resonators, which is additionally covered by simple or mixed cross-coupling.

The forms of a mixed coupling representation (7.2) and (7.6) are interconnected and complement each other. The transition from (7.2) to (7.6) is carried out according to the formulas:

$$m_0 = (K_m + K_e)/\text{FBW}, \tag{7.7a}$$

$$a = \sqrt{K_m\,|K_e|}. \tag{7.7b}$$

The expressions that carry out the transition from (7.6) to (7.2) are as follows:

$$K = m_0\text{FBW}, \tag{7.8}$$

$$\begin{matrix} K_m \\ K_e \end{matrix} = K/2 \pm \sqrt{(K/2)^2 + a^2}, \tag{7.9}$$

where "+" sign corresponds to K_m, and "−" sign to K_e.

It is important to note that if $K = 0$, then $m_0 = 0$ (7.6); however, $m\,(\Omega) \neq 0$. Such a mixed coupling is called resonant coupling [4] and is written in the form

$$m\,(\Omega) = -a\Omega. \tag{7.10}$$

The resonant coupling (7.10) plays a special role when creating mixed coupled BPFs with improved performance.

7.2 Direct and Inverse Problems of Mixed Coupled BPFs

The matrix representation form can be used to calculate the parameter S_{21} of the N-order BPF [22].

Formula (6.8), expressing the S_{21} parameter of BPF in matrix form, will be presented in a slightly different form

$$S_{21} = 2[\widetilde{Z}]^{-1}_{N1}/\sqrt{q_{e1}q_{en}}, \tag{7.11}$$

where $[\widetilde{Z}]^{-1}_{N1}$ is the minor of the element located at the intersection of the first column and the Nth row of the inverse normalized impedance matrix $[\widetilde{Z}]^{-1}$. A rather labor-intensive procedure is required to calculate the inverse matrix [23]. Since the normalized admittance matrix $[\widetilde{Y}]$ and the normalized impedance matrix $[\widetilde{Z}]$ are mutually reversible $[\widetilde{Y}] = [\widetilde{Z}]^{-1}$ [24], this makes it possible to simplify (7.11):

$$S_{21} = 2M_{N1}/\sqrt{q_{e1}q_{en}}, \tag{7.12}$$

where M_{N1} is the minor of the admittance matrix $[\widetilde{Y}]$, which can be represented in the form [22]

$$[\widetilde{Y}] = \begin{bmatrix} 1/q_1 + S & -jm_{12} & \dots & -jm_{1N} \\ -jm_{21} & S & \dots & -jm_{2N} \\ \vdots & \vdots & \ddots & \vdots \\ -jm_{N1} & -jm_{N2} & \cdots & 1/q_1 + S \end{bmatrix}. \tag{7.13}$$

To determine the TZs of BPF by using the admittance matrix (7.13), we have to equate to zero its minor M_{N1}, which is the determinant of a square matrix of size $(N-1) \times (N-1)$

$$M_{N1} = \begin{vmatrix} m_{12} & m_{13} & \cdots & m_{1N} \\ m_{22} & S' & \cdots & m_{2N} \\ \vdots & \vdots & \ddots & \vdots \\ m_{(N-1)2} & m_{(N-1)3} & \cdots & m_{(N-1)N} \end{vmatrix} = 0, \tag{7.14}$$

where $S' = S/j$ and $S'^2 = -S^2$. Note that the impedance matrix $[\widetilde{Z}]$ (7.11) was chosen as the initial matrix, and the subsequent transition was to the admittance matrix $[\widetilde{Y}]$. However, if we had initially chosen matrix $[\widetilde{Y}]$ (7.13) and then moved to the matrix $[\widetilde{Z}]$, the result would have been the same. The TZs in both cases are determined by expression (7.14).

We use expression (7.14) to analyze some BPFs with simple cross-coupling. The trisection BPF with simple cross-coupling has a minor (7.14) written as

$$M_{31} = \begin{vmatrix} m_{12} & m_{13} \\ S' & m_{23} \end{vmatrix} = m_{12}m_{23} - m_{13}S'. \tag{7.15}$$

Equating minor M_{31} (7.15) to zero, we obtain the expression for the TZ $S_1' = m_{12}m_{23}/m_{13}$ or $\Omega_1 = m_{12}m_{23}/m_{13}$, representing the first formula of R. Kurzrok (6.4).

The minor (7.14) for the quadruplet BPF with one cross-coupling m_{14} has the form

$$M_{41} = \begin{vmatrix} m_{12} & m_{13} & m_{14} \\ S' & m_{23} & m_{24} \\ m_{32} & S' & m_{34} \end{vmatrix} = m_{14}S'^2 + m_{12}m_{23}m_{34} - m_{14}m_{23}m_{32}. \tag{7.16}$$

Equating (7.16) to zero, and taking into account that $S'^2 = -S^2$, we successively obtain the second formula $S_{12} = \pm\left[(m_{12}m_{23}m_{34}/m_{14}) - m_{23}^2\right]^{1/2}$ of R. Kurzrok (6.7).

If the N-order BPF contains all mixed couplings, then in the determinant (7.14), it is necessary to replace everything m_{ik} with the linear form of the representation of the mixed couplings (7.6): $m_{0ik} - a_{ik}S'$. Equation $M_{N1} = 0$ (7.14) will in this case be an algebraic equation of degree $(N - 1)$

$$S'^{N-1} + A_2 S'^{N-2} + \ldots + A_{N-1}S' + A_N = 0, \tag{7.17}$$

where A_i are the coefficients of the polynomial, which depend on the BPF coupling coefficients, and are real numbers.

Equation (7.17) defines $(N - 1)$ TZs if the N-order BPF contains simple and mixed couplings, or all mixed couplings. If all couplings of the N-order BPF are simple, then the equation $M_{N1} = 0$ is an algebraic equation of degree $(N - 2)$. In this case, the number of TZs will be smaller; it is equal to $(N - 2)$. In general, determining the TZs of the N-order BPF involves solving an algebraic equation of degree $(N - 1)$ or $(N - 2)$, the number of roots of which will be equal to $(N - 1)$ or $(N - 2)$, respectively.

The inverse problem for the BPFs under consideration must be solved in the case where it is necessary to obtain the predicted TZs $S_1', S_2', \ldots, S_{N-1}'$. We use the equivalent form of the polynomial representation (7.17) to solve this problem

$$S^{N-1} + A_2 S'^{N-2} + \ldots + A_{N-1}S' + A_N = \left(S' - S_1'\right)\left(S' - S_2'\right)\ldots\left(S' - S_{N-1}'\right). \tag{7.18}$$

The relationship between the coefficients A_i of the polynomial (7.18) and the zeros S_i' is expressed by the Vieta's formulas [23].

$$\begin{aligned} A_2 &= -\left(S_1' + S_2' + \cdots + S_{N-1}\right) \\ A_3 &= S_1'S_2' + S_1'S_3' + \cdots + S_{N-2}'S_{N-1}' \\ &\ldots\;\ldots\;\ldots \\ A_N &= (-1)^{N-1}\; S_1'S_2'S_{N-1}' \end{aligned} \tag{7.19}$$

Solving the inverse problem for a BPF is possible because, using (7.19), we can determine the coupling coefficients from the given values of $S_1', S_2', \ldots, S_{N-1}'$. This is possible for cases of simple and mixed coupling in BPFs of different orders. In the following, we consider third- and fourth-order BPFs with mixed cross-coupling, for which equations (7.19) will be used.

7.3 Triplet BPF with Mixed Cross-coupling

The coupling structure of triple BPF with mixed cross-coupling is shown in Figure 7.2a. The filter scheme is shown in Figure 7.2b. It contains quarter-wave resonators and J inverters. The J inverters in the form of quarter-wave sections, which are indicated by small rectangles, are used for coupling between resonators.

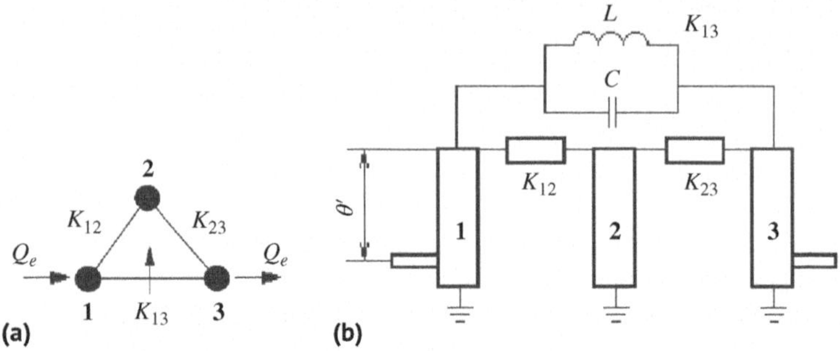

Figure 7.2 Trisection BPF with mixed cross-coupling. (a) Coupling structure. (b) Schematic with J inverters.

This J inverter performs cross-coupling for mixed coupling (Figure 7.1a) and transverse branches are not shown.

The use of a mixed cross-coupling in the triplet BPF (Figure 7.2a) leads to the need to replace in (7.15) m_{13} with $m_{13}(\Omega) = m_{013} - a_{13}S'$, which will lead to a second-degree equation

$$S'^2 - (m_{013}/a_{13})\,S' + m_{12}m_{23}/a_{13} = 0. \tag{7.20}$$

The analysis problem presented by equation (7.20) involves determining two TZs S'_1 and S'_2, based on given coupling coefficients.

It is easy to determine that the roots S'_1 and S'_2 of equation (7.20) are related to its coefficients by the formulas

$$S'_1 + S'_2 = m_{013}/a_{13}, \tag{7.21a}$$

$$S'_1 S'_2 = m_{12}m_{23}/a_{13}. \tag{7.21b}$$

The inverse problem is represented by equalities (7.21). In this case, the coefficients m_{12} and m_{23} are known. It is necessary to determine the mixed cross-coupling $m_{13}(\Omega) = m_{013} - a_{13}S'$ based on (7.21). The solution to this problem is represented as

$$a_{13} = m_{12}m_{23}/S'_1 S'_2, \tag{7.22a}$$

$$m_{013} = m_{12}m_{23}\,(S'_1 + S'_2)/S'_1 S'_2. \tag{7.22b}$$

Some examples are considered as follows. If the TZs are located on the σ-axis and are equidistant from $S = 0$: $S_1 = \sigma_1$ $(S'_1 = \sigma_1/j)$; $S_2 = -\sigma_1$ $(S'_2 = -\sigma_1/j)$, then from (7.22b) it will follow that $m_{013} = 0$ $(K_{m13} = |K_{e13}|)$. In this case, equality (7.22a) is satisfied if $m_{23} = m_{12}$, i.e., the signs of the main couplings coincide, and the equality is written in the form

$$a_{13} = K_{m13} = |K_{e13}| = m_{12}^2/\sigma_1^2. \tag{7.23}$$

Note that the TZs on the σ-axis are equidistant from $S = 0$ if the trisection BPF is symmetric ($K_{12} = K_{23}$) with resonant cross-coupling (7.10): $K_{13} = 0$; $m_{13}(\Omega) = -a_{13}\Omega$.

If the TZs are located on the $j\Omega$-axis and are equidistant from $S = 0$: $S_1 = j\Omega_1$ $(S_1' = \Omega_1)$, $S_2 = -j\Omega_1$ $(S_2' = -\Omega_1)$, then from (7.22b) it will follow that $m_{013} = 0$ $(K_{m13} = |K_{e13}|)$. In this case, equality (7.22a) is satisfied if $m_{23} = -m_{12}$, i.e., the signs of the main couplings have different signs, and the equality is written in the form

$$a_{13} = K_{m13} = |K_{e13}| = m_{12}^2/\Omega_1^2. \tag{7.24}$$

The TZs on the $j\Omega$-axis are equidistant from $S = 0$ if the trisection BPF is asymmetric ($K_{12} = -K_{23}$) with resonant cross-coupling (7.10): $K_{13} = 0$.

The TZs on the $j\Omega$-axis can coincide $S_1 = S_2 = j\Omega_1$ $(S_1' = S_2' = \Omega_1)$, which leads to a second-degree TZ. This can only happen under condition $m_{23} = m_{12}$ (7.22a). In this case, equality (7.22) can be represented as

$$a_{13} = m_{12}^2/\Omega_1^2, \tag{7.25a}$$

$$m_{013} = 2m_{12}^2/\Omega_1. \tag{7.25b}$$

Only in a symmetric ($K_{12} = K_{23}$) trisection BPF can the TZ of the second degree be realized. Moreover, its approach to f_0 is accompanied by an increase in m_{013} and a_{13} (7.25).

In the following, we consider, based on (7.22), a set of different options for the placement of the two TZs of the trisection BPF, which has a mixed cross-coupling. In performing this analysis, we will use a filter circuit (Figure 7.2b) whose resonators have a characteristic impedance $Z_0 = 10\ \Omega$ and a resonant frequency $f_0 = 1$ GHz. In this case, the expression for the coupling coefficients between adjacent resonators (3.5c) will take the form

$$K_{i,i+1} = 4Z_0/\pi Z_{i,i+1}. \tag{7.26}$$

The magnetic K_m (7.3a) and electric K_e (7.3b) components of the mixed coupling are expressed by the formulas

$$K_m = 2Z_0/\pi^2 f_0 L, \tag{7.27a}$$

$$K_e = -8Z_0 f_0 C. \tag{7.27b}$$

Consider the Chebyshev filter prototype with $f_0 = 1$ GHz, FBW = 0.05, $n = 3$, and ripple $L_{Ar} = 0.2$ dB. The values $n = 3$ and $L_{Ar} = 0.2$ dB correspond to the g parameters (Table 2.6): $g_0 = 1$, $g_1 = q_3 = 1.2275$, $g_2 = 1.525$. Substituting these parameters and FBW = 0.05 into (3.8), we determine the coupling coefficients $K_{12} = K_{23} = 0.0420$ ($m_{12} = 0.84$) and the value Q_e of end resonators of $Q_e = 24.55$. The $\lambda/4$ resonators of the BPF have characteristic impedances $Z_0 = 10\ \Omega$ and slope parameter $b = \pi/4Z_0$. The coupling coefficients are provided by $\lambda/4$ segments with $Z_{12} = Z_{23} = 303.15\ \Omega$ (3.11c). The end $\lambda/4$ resonators are conductively connected to the loads of $R_L = 50\ \Omega$. The value of Q_e is provided by the connection coordinate θ', which is measured from the resonator's open end. The value of $Q_e = 24.55$ is provided at coordinate $\theta' = 66.42°$.

7.3.1 Various Arrangements of Transmission Zeros

Table 7.1 presents the frequency responses corresponding to different arrangements of the two TZs. The first seven positions correspond to the symmetrical ($K_{12} = K_{23}$) trisection BPF.

Case 1: TZs of the trisection BPF are placed equidistantly on the σ-axis (No. 1, Table 7.1). For this, the cross-coupling must be resonant $K_{13} = 0$. Let $|\sigma_{1,2}| = 8$, then $a_{13} = K_{m13} = |K_{e13}| = 0.01102$ (7.23). The $K_{m13} = 0.01102$ component, according to (7.27a), corresponds to the value $L = 183.8$ nH, and the $K_{e13} = -0.01102$ component, according to (7.27b), corresponds to the value $C = 0.1378$ pF. The frequency response for this case is shown by the dotted line (position 1), and the BPF has a group delay $\tau = 11.8 \pm 0.1$ ns. The simulation was performed by the linear simulator of the Microwave Office (AWR). We bring the TZs closer together by a factor of four, setting them equal to $|\sigma_{1,2}| = 2$. The calculations give: $a_{13} = K_{m13} = |K_{e13}| = 0.1764$, $L = 11.488$ nH, $C = 2.205$ pF. The corresponding frequency response is shown by the solid line (position 1), and the BPF has a group delay $\tau = 13.3 \pm 0.15$ ns. Note that the mixed coupling components increased by 16 times when the TZs were brought four times closer to each other.

Case 2: TZs are located in the upper half-plane of s and equidistant relative to the $j\Omega$-axis (No. 2, Table 7.1). Let $S_1 = 2 + j0.4$ ($S_1' = -2/j + 0.4$) and $S_2 = -2 + j0.4$ ($S_2' = -2/j + 0.4$). For the TZs, we obtain the following values from (7.22): $a_{13} = 0.1696$, $m_{013} = 0.1357$. These values allow us to determine the mixed coupling coefficient (7.8): $K_{13} = m_{013}\text{FBW} = 0.006785$, and its components (7.9) $K_{m13} = 0.173$ ($L = 11.713$ nH) and $K_{e13} = -0.1662$ ($C = 2.0775$ pF). The simulated frequency responses for this case are shown in position No. 2 (Table 7.1).

Case 3: TZs are located in the lower half-plane of s and equidistant relative to the $j\Omega$-axis (No. 3, Table 7.1). Let $S_1 = 2 - j0.4$ and $S_2 = -2 - j0.4$. Performing

Table 7.1 TZs and frequency responses of triplet BPF with mixed cross-coupling.

No.	TZs	Frequency responses	K_{13}	K_{12}, K_{23}
1	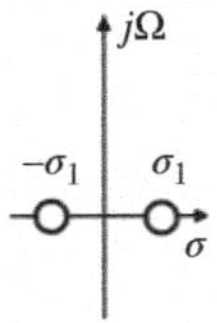	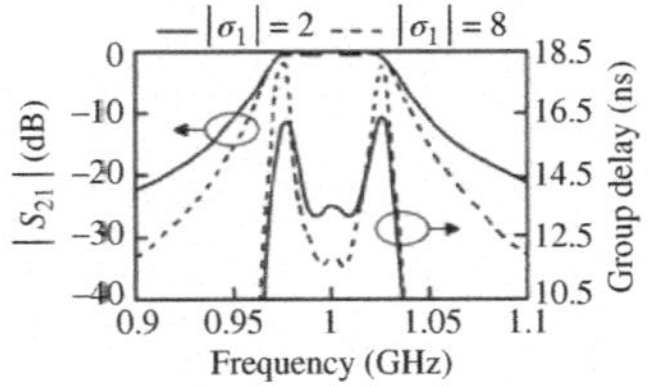	$K_{13} = 0$	$K_{12} = K_{23}$
2	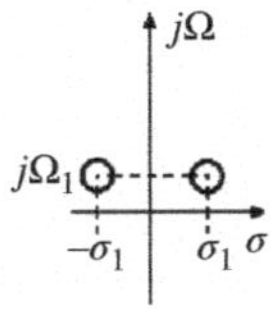	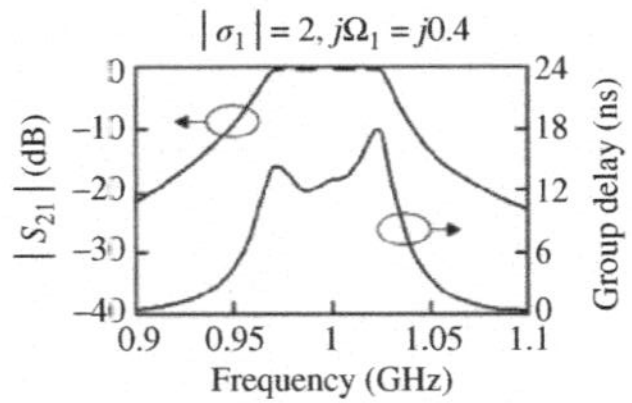	$K_{13} > 0$	$K_{12} = K_{23}$
3	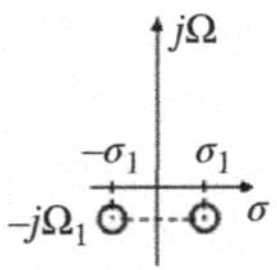	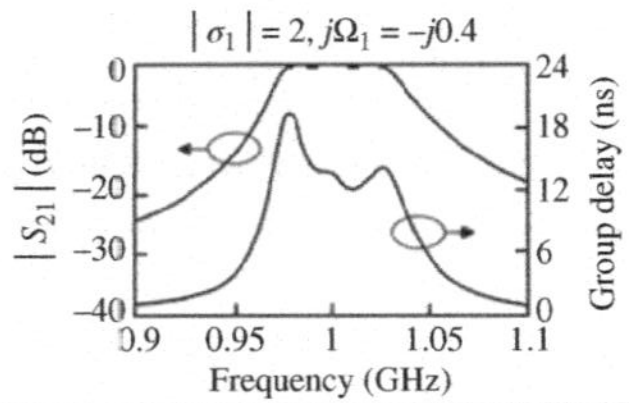	$K_{13} < 0$	$K_{12} = K_{23}$

(Continued)

Table 7.1 (Continued)

No.	TZs	Frequency responses	K_{13}	K_{12}, K_{23}
4	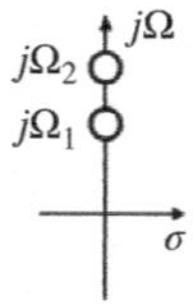	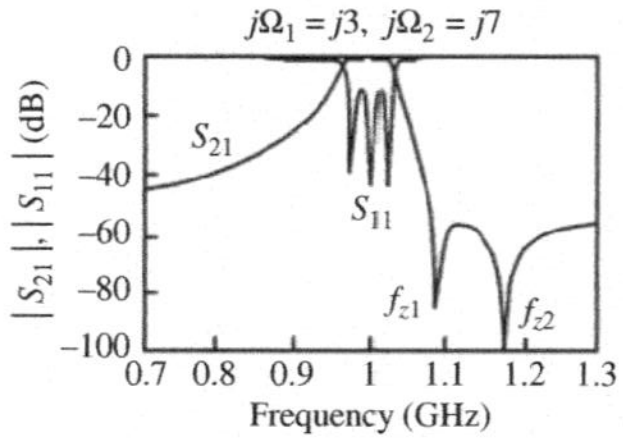	$K_{13} > 0$	$K_{12} = K_{23}$
5	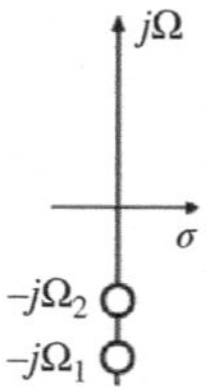	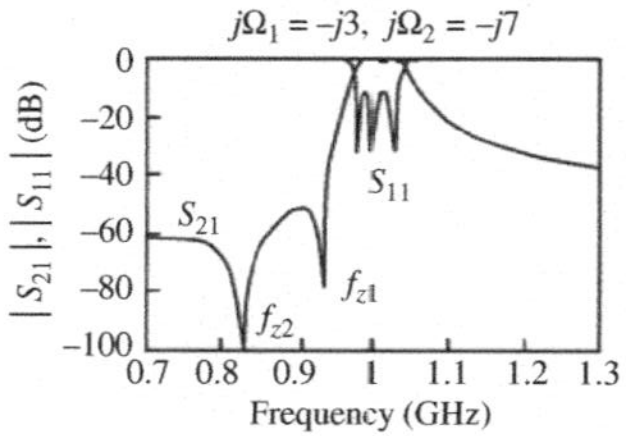	$K_{13} < 0$	$K_{12} = K_{23}$

6	$j\Omega_1 = j\Omega_2$ 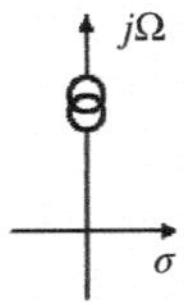	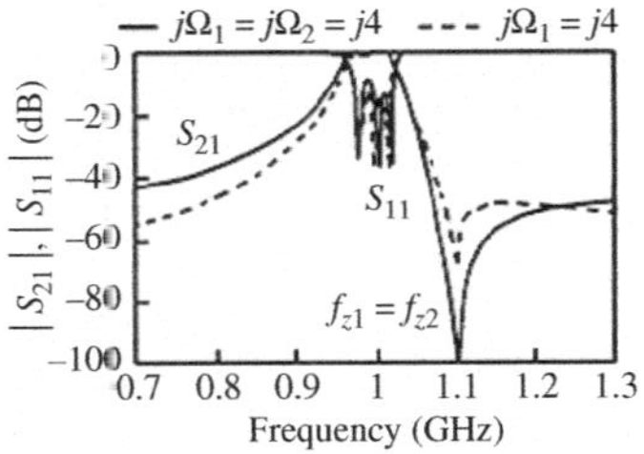	$K_{13} > 0$	$K_{12} = K_{23}$
7	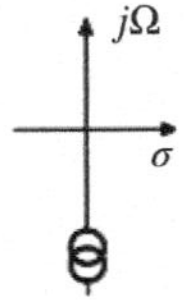 $-j\Omega_1 = -j\Omega_2$	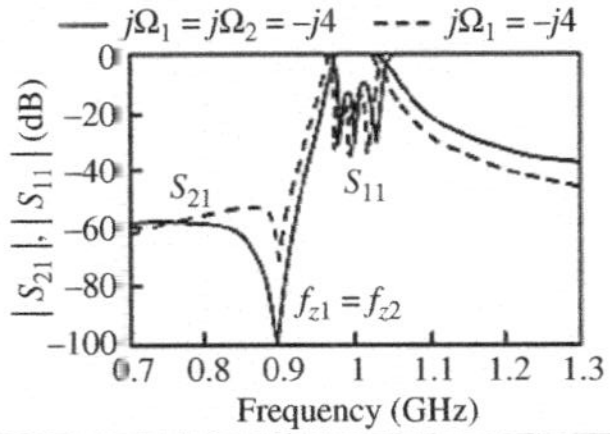	$K_{13} < 0$	$K_{12} = K_{23}$

(*Continued*)

Table 7.1 (Continued)

No.	TZs	Frequency responses	K_{13}	K_{12}, K_{23}
8	$j\Omega_2 = -j\Omega_1$ 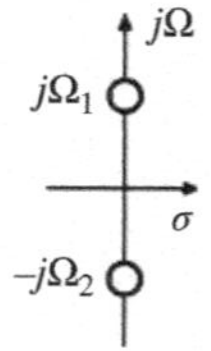	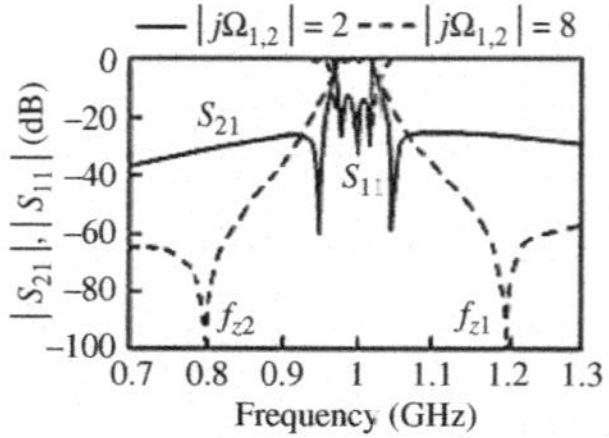	$K_{13} = 0$	$K_{12} = -K_{23}$
9	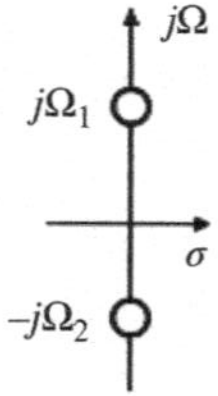	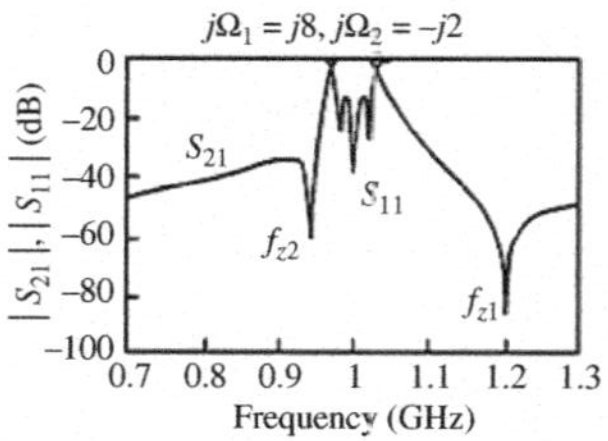	$K_{13} > 0$	$K_{12} = -K_{23}$
10	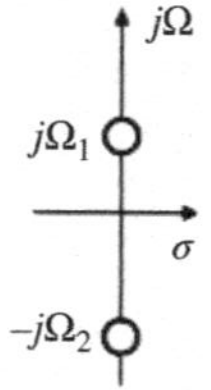	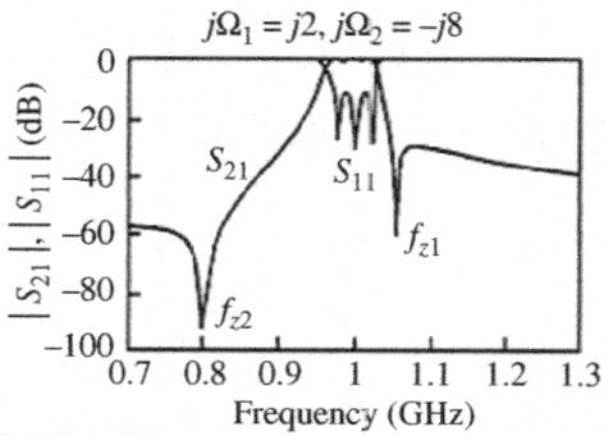	$K_{13} < 0$	$K_{12} = -K_{23}$

the previous calculations, we get $a_{13} = 0.1696$, $m_{013} = -0.1357$, $K_{13} = -0.006785$, $K_{m13} = 0.1662$ ($L = 12.192$ nH), and $K_{e13} = -0.1730$ ($C = 2.162$ pF). The simulated frequency responses for this case are shown in Table 7.1 (position No. 3). The two equidistant relative to the $j\Omega$-axis TZs of the BPF No. 2 and No. 3 moved from the σ-axis to the upper or lower half-plane of s. Therefore they do not have a flat group delay.

Case 4: Two TZs are located on the $j\Omega$-axis in the upper s-half-plane (No. 4, Table 7.1). Let $S_1 = j3$ ($f_{z1} = 1.075$ GHz), $S_2 = j7$ ($f_{z2} = 1.175$ GHz) or $S_1' = 3$ and $S_2' = 7$. Substituting these values into (7.22), we obtain the parameters of the linear form of the mixed cross-coupling representation $a_{13} = 0.0336$, $m_{013} = 0.336$. Using (7.8) and (7.9), we sequentially determine $K_{13} = 0.0168$, $K_{m13} = 0.0430$ ($L = 47.126$ nH), and $K_{e13} = -0.0262$ ($C = 0.3275$ pF). The frequency responses of this trisection BPF are presented in position No. 4 (Table 7.1).

Case 5: Two TZ are located on the $j\Omega$-axis in the lower s-half-plane (No. 5, Table 7.1). Let $S_1 = -j3$ ($f_{z1} = 0.925$ GHz), $S_2 = -j7$ ($f_{z2} = 0.825$ GHz). Substituting these values into (7.22), we obtain the parameters of the linear form of the mixed cross-coupling representation $a_{13} = 0.0336$, $m_{013} = -0.336$. Using (7.8) and (7.9), we sequentially determine $K_{13} = -0.0168$, $K_{m13} = 0.0262$ ($L = 77.344$ nH), and $K_{e13} = -0.043$ ($C = 0.5375$ pF). The frequency responses of this trisection BPF are presented in position No. 5 (Table 7.1).

Case 6: Second-order TZ is located on the $j\Omega$-axis in the upper s-half-plane (No. 6, Table 7.1). Let $S_1 = S_2 = j4$ ($f_z = 1.100$ GHz) or $S_1' = S_2' = 4$. From (7.25), we get: $a_{13} = 0.0441$, $m_{013} = 0.3528$. Using (7.8) and (7.9), we sequentially determine $K_{13} = 0.01764$, $K_{m13} = 0.0538$ ($L = 37.665$ nH), and $K_{e13} = -0.0362$ ($C = 0.4525$ pF). The frequency responses of this BPF are presented in position No. 6 (Table 7.1) by a solid line. If the trisection BPF has a simple TZ, its frequency responses will be displayed by a dashed line in the same figure.

Case 7: Second-order TZ is located on the $j\Omega$-axis in the lower s-half-plane (No. 7, Table 7.1). Let $S_1 = S_2 = -j4$ ($f_z = 0.9$ GHz) and using (7.25), we get: $a_{13} = 0.0441$, $m_{013} = -0.3528$. Using (7.8) and (7.9) leads to values: $K_{13} = -0.01764$, $K_{m13} = 0.0362$ ($L = 55.978$ nH), $K_{e13} = -0.0538$ ($C = 0.6725$ pF). The frequency responses of this BPF are presented in position No. 7 (Table 7.1) by a solid line. The second-order TZ produces a deeper attenuation pole on the transfer characteristic than the first-order TZ (dashed line).

For trisection BPF with the second-order TZ, equalities (7.22) and (7.25) are satisfied, from which the condition follows:

$$K_{13}^2/4a_{13} = K_{12}^2. \tag{7.28}$$

The values: $K_{12} = 0.042$, $K_{13} = -0.01764$, $a_{13} = 0.0441$ correspond to Case 7. These values also satisfy condition (7.28).

If the trisection BPF is asymmetric ($K_{12} = -K_{23}$), then its two TZs can have three possible positions Nos. 8–10.

Case 8: The TZs of the trisection BPF are placed equidistantly on the $j\Omega$-axis (No. 8, Table 7.1). Let $|\Omega_{1,2}| = 8$, and performing calculations using formulas (7.24) and (7.27), we successively obtain $K_{13} = 0$, $K_{m13} = 0.01102$, $K_{e13} = -0.01102$, and $L =$ 183.8 nH, $C = 0.1378$ pF. The dotted curves in position No. 8 depict the frequency responses of the BPF. If $|\Omega_{1,2}| = 2$, then the corresponding values are $K_{13} = 0$, $K_{m13} = 0.1764$ ($L = 11.488$ nH), and $K_{e13} = -0.1764$ ($C = 2.205$ pF). The frequency responses for this case are depicted by solid lines in the same figure. Note that the components $K_m = |K_e|$ increased 16 times when the two TZs approached four times.

Case 9: Two TZ are located on the $j\Omega$-axis, shifted upward (No. 9, Table 7.1). Let $S_1 = j8$ ($f_{z1} = 1.2$ GHz), $S_2 = -j2$ ($f_{z2} = 0.95$ GHz) or $S_1' = 8$ and $S_2' = -2$. Performing calculations using formulas (7.22) and (7.9), we obtain: $a_{13} = 0.0441$, $m_{013} = 0.2646$, $K_{13} = 0.01323$, $K_{m13} = 0.0512$ ($L = 39.58$ nH), and $K_{e13} = -0.038$ ($C = 0.475$ pF). The plots in position No. 9 (Table 7.1) display the frequency responses for the case of BPF under consideration.

Case 10: Two TZ are located on the $j\Omega$-axis, shifted downward (No. 10, Table 7.1). Let $S_1 = j2$ ($f_{z1} = 1.05$ GHz), $S_2 = -j8$ ($f_{z2} = 0.8$ GHz). Performing calculations using formulas (7.22) and (7.9), we obtain: $a_{13} = 0.0441$, $m_{013} = -0.2646$, $K_{13} = -0.01323$, $K_{m13} = 0.038$ ($L = 53.33$ nH), and $K_{e13} = -0.0512$ ($C = 0.64$ pF). The plots in position No. 10 (Table 7.1) display the frequency responses for this filter's case.

Ten different arrangements of two TZs (Table 7.1) are provided in the trisection BPF under consideration. All other arrangements of TZs are physically unrealizable. In these cases, the coefficients (7.21) of polynomial (7.20) will not be real numbers. A deeper attenuation pole on the transfer function is provided by the second-order TZ (positions Nos. 6 and 7).

Following is an experimental confirmation of the existence of 10 variants of the arrangement of two TZs in a trisection BPF with mixed cross-coupling.

7.3.2 Microstrip Quasi-inline Trisection BPFs with $\lambda/4$ Resonators

The topology and photograph of the quasi-inline microstrip trisection BPF with $\lambda/4$ resonators is shown in Figure 7.3. Its size is 21.6 mm × 16.8 mm. The substrate is TMM-10i (Rogers), $\varepsilon_r = 9.8$, $h = 1.905$ mm, $\tan\delta = 0.002$. Metallized

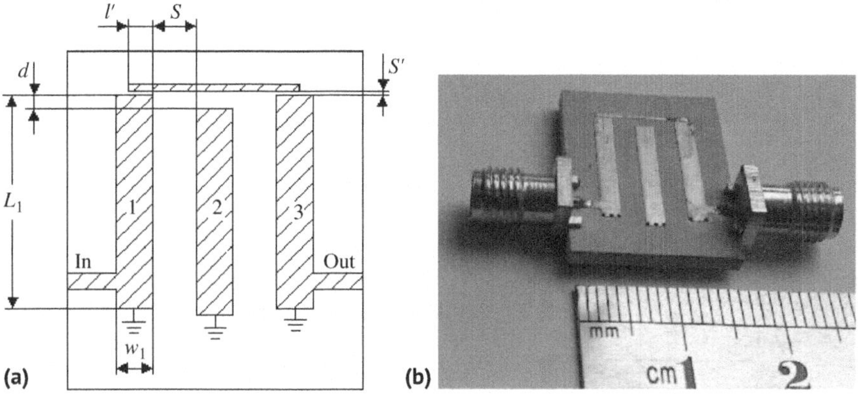

Figure 7.3 Microstrip quasi-inline trisection BPF with $\lambda/4$ resonators. (a) Topology: $L_1 = 13.6$ mm, $w_1 = 2$ mm, $S = 2.4$ mm, $d = 0.9$ mm, $S' = 0.2$ mm, $l' = 1.8$ mm. (b) Photograph of fabricated filter. *Source*: [19] / with permission of IEEE.

holes with a diameter of 0.4 mm provide a short circuit at the ends of the resonators. Resonators 1 and 3 are shifted upward relative to the middle resonator 2 by $d = 0.9$ mm, which makes this BPF a quasi-inline filter. The resonators are tuned to a resonant frequency of 2 GHz. Mixed cross-coupling exists between resonators 1 and 3. Parasitic magnetic interaction between these resonators forms the magnetic component K_{m13}. The geometric parameters of the filter provide the following values of the coupling coefficients $K_{12} = K_{23} = 0.056$ and $K_{m13} = 0.0139$. The procedure for determining the coupling coefficients is well known [25]. We used the EM simulator of the computer program Microwave Office (AWR) to determine these coefficients.

The external Q-factor of the end resonators, $Q_e = 17.06$, is provided by the connecting coordinate of the I/O segments. The electrical component K_{e13} of the mixed cross-coupling is provided by the strip connecting resonators 1 and 3 through capacitive gaps. Decreasing the gap S' and increasing the length l' lead to an increase in $|K_{e13}|$. For values of $S' = 0.2$ mm and $l' = 1.8$ mm, the condition $K_{13} = 0$ is satisfied, which means that $|K_{e13}| = K_{m13} = 0.0139$. The BPF has a flat group delay, and its two TZs are located equidistantly on the σ-axis.

Position No. 1 (Table 7.2) shows the simulated and measured frequency responses of the BPF with two different geometric parameters. They are shown in the insets. With a gap $S = 2.4$ mm between the resonators and $d = 0.9$ mm, we have the following measured date: $f_0 = 2$ GHz, BW = 151 MHz, $L_{A0} = 1.0$ dB, $L_R > 14.2$ dB. In this case, the group delay is $\tau = 4.3 \pm 0.1$ ns within the frequency range of 74 MHz. If $S = 1$ mm and $d = 2.7$ mm, then K_{12} is the same, but K_{m13} increases to the value of $K_{m13} = 0.0271$. To compensate for this value, an electrical

Table 7.2 Frequency responses of microstrip quasi-inline trisection BPF with mixed cross-coupling.

No.	Frequency response	K_{13}	Parameters
1	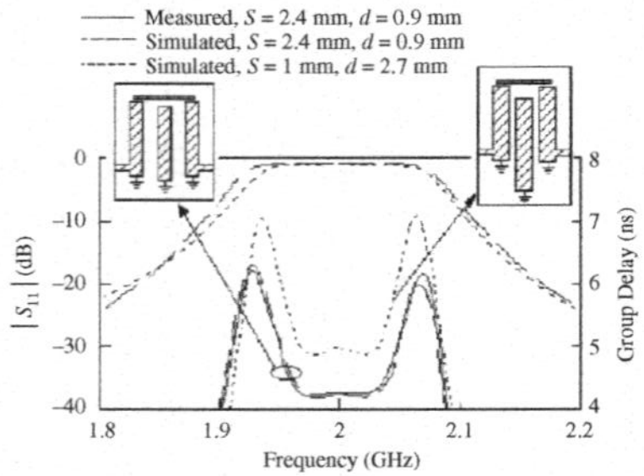		$S = 2.4$ mm $d = 0.9$ mm
2	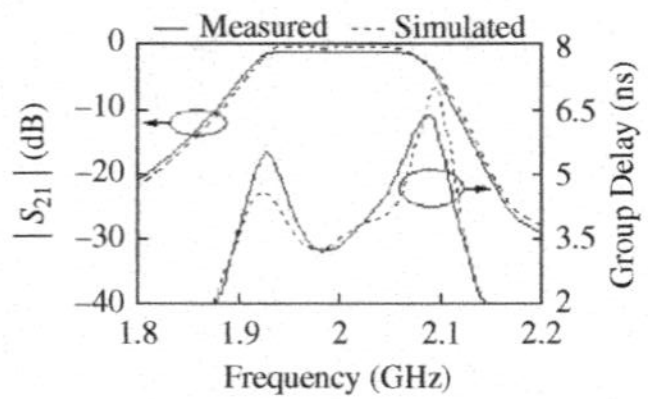	$K_{13} > 0$	$S' = 0.2$ mm $l' = 1.6$ mm

3	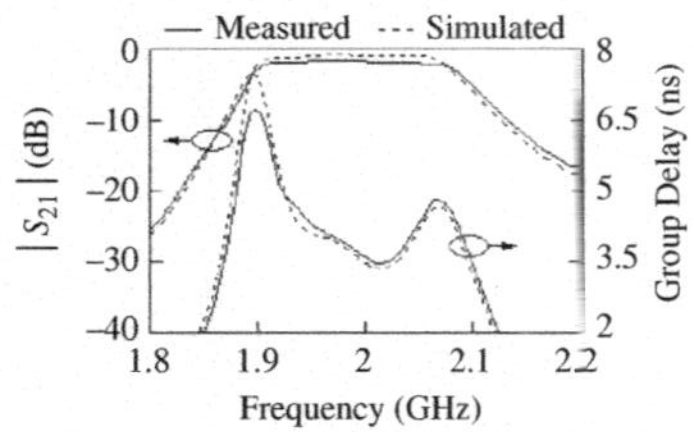	$K_{13} < 0$	$S' = 0.025$ mm $l' = 1.0$ mm
4	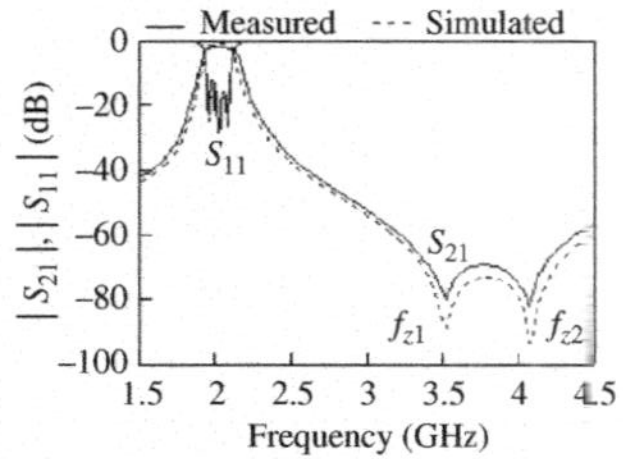	$K_{13} > 0$	$S' = 0.2$ mm $l' = 1.4$ mm

(*Continued*)

Table 7.2 (Continued)

No.	Frequency response	K_{13}	Parameters
5	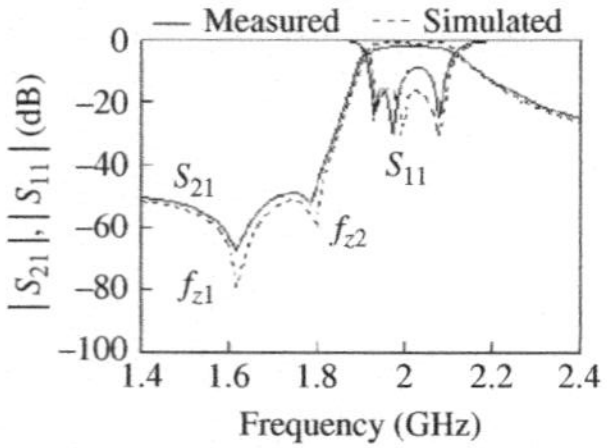	$K_{13} < 0$	$S = 0.025$ mm $l' = 2$ mm
6	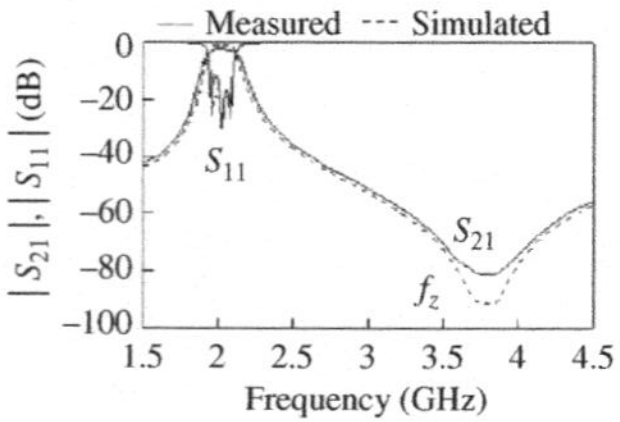	$K_{13} > 0$	$S' = 0.2$ mm $l' = 1.5$ mm

7	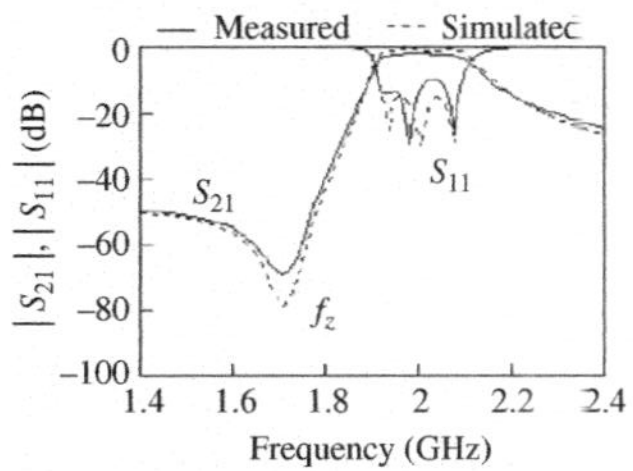	$K_{13} < 0$	$S' = 0.025$ mm $l' = 1.8$ mm
8	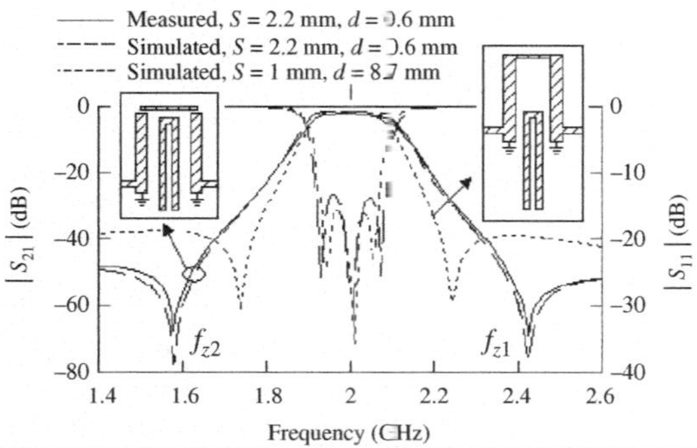		$S = 2.2$ mm $d = 0.6$ mm

(*Continued*)

Table 7.2 (Continued)

No.	Frequency response	K_{13}	Parameters
9	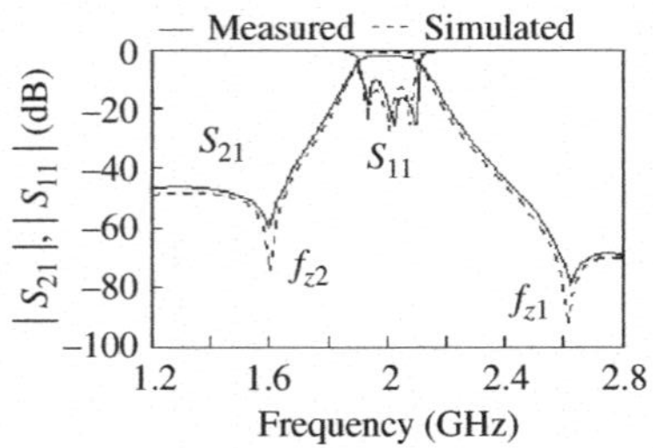	$K_{13} > 0$	$S' = 0.2$ mm $l' = 1.7$ mm
10	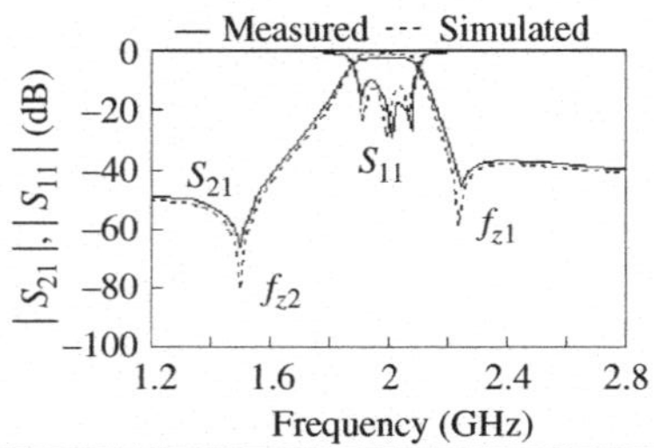	$K_{13} < 0$	$S' = 0.05$ mm $l' = 2$ mm

component $K_{e13} = -0.0271$ is required, which is provided with the parameters of the connecting strip $S' = 0.05$ mm and $l' = 1.8$ mm. In this case, the group delay increased from 4.3 to 4.9 ns.

The symmetrical ($K_{12} = K_{23}$) microstrip trisection BPF shown in Figure 7.3 has seven different arrangements of the two TZs and the corresponding frequency responses. All of them are listed in Table 7.2 (items Nos. 1–7). BPF No. 1 has the resonant cross-coupling $K_{13} = 0$. Trisection BPFs Nos. 2, 4, 6 have $K_{13} > 0$, and filters Nos. 3, 5, 7 are characterized by the $K_{13} < 0$ value. The gap $S' = 0.2$ is used to realize $K_{13} > 0$, and the gap $S' = 0.025$ mm is used to obtain the $K_{13} < 0$ values. Filter parameters such as $S = 2.4$ mm, $d = 0.9$ mm, and $K_{m13} = 0.0139$ remain constant in all cases. The only variable is the electrical component K_{e13}, which is provided by changing the value of l', at a given gap S' value.

The cases of arrangement of two TZs, equidistant with respect to the $j\Omega$-axis, in the upper S half-plane or the lower half-plane of s, correspond to the frequency responses shown in positions No. 2 and No. 3 (Table 7.2), respectively. These two variants of arrangement of complex TZs do not provide the flat group delay. Two one-sided TZs on the $j\Omega$-axis are characterized by the frequency responses shown in positions No. 4 and No. 5. If $K_{13} > 0$ (No. 4), then these TZs ($f_{z1} = 3.51$ GHz and $f_{z2} = 4.08$ GHz) are located at higher frequencies than f_0. If $K_{13} < 0$ (No. 5), then the indicated TZs ($f_{z1} = 1.798$ GHz and $f_{z2} = 1.618$ GHz) are left-sided.

Positions No. 6 and No. 7 show the frequency responses corresponding to one second-order TZ. At $K_{13} > 0$ (No. 6), the TZ ($f_z = 3.8$ GHz) is located to the right of f_0. At $K_{13} < 0$ (No. 7), the TZ ($f_z = 1.708$ GHz) is left-handed.

7.3.3 Microstrip Quasi-inline Trisection BPFs with $\lambda/4$ and $\lambda/2$ Resonators

An asymmetric ($K_{12} = -K_{23}$) trisection BPF can implement three possible arrangements of the two TZs and their corresponding frequency responses, Nos. 8–10 (Table 7.1). However, these BPFs can be made symmetrical. To achieve this, the middle $\lambda/4$ resonator, which is a reflective type, should be replaced with a through-type resonator, which can be, in particular, a half-wave resonator. A through-type resonator is indicated by a double circle (Figure 6.13).

A microstrip quasi-inline trisection BPF with $\lambda/4$ and $\lambda/2$ resonators is shown in Figure 7.4. It occupies an area of 21.6 mm × 17.4 mm. The substrate and $\lambda/4$ resonator parameters are the same as in the previous case. The through-type resonator *2* is a hairpin resonator with a microstrip line width of 1 mm. The selected geometric parameters of the filter provide coupling coefficients $K_{12} = K_{23} = 0.065$ and $K_{m13} = 0.0235$. The external Q-factor of the end resonators, $Q_e = 14.67$, is provided by the coordinate connecting the I/O short line. As before, the values of S' and l' of the band connecting resonators 1 and 3 through capacitive gaps determine the required component K_{e13} of the mixed coupling K_{13}. If $S' = 0.1$ mm and

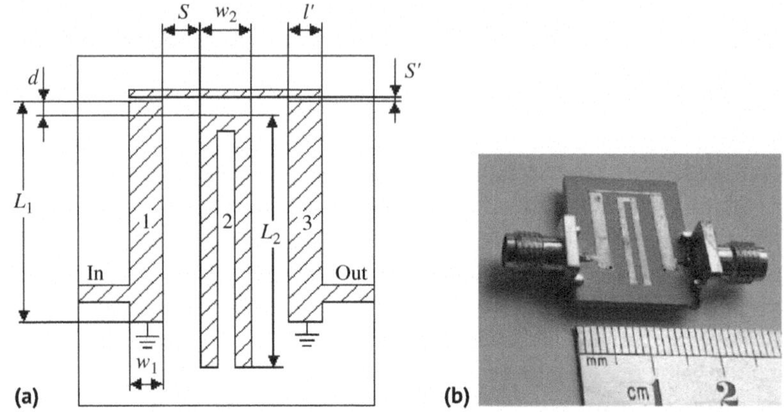

Figure 7.4 Microstrip quasi-inline trisection BPF with $\lambda/4$ and $\lambda/2$ resonators. (a) Topology: $L_1 = 13.6$ mm, $w_1 = 2$ mm, $L_2 = 15.7$ mm, $w_2 = 3$ mm, $S = 2.2$ mm, $d = 0.6$ mm, $S' = 0.1$ mm, $l' = 2$ mm. (b) Photograph of fabricated filter. *Source*: [19] / with permission of IEEE.

$l' = 2.0$ mm, then $K_{13} = 0$ and $|K_{e13}| = K_{m13} = 0.0235$. In this case, the two TZs are located equidistant regarding f_0. This corresponds to a quasi-elliptic frequency response.

Position No. 8 (Table 7.2) shows the simulated and measured frequency responses of the symmetric trisection BPF with two different geometric parameters corresponding to the insert topologies. If $S = 2.2$ mm and $d = 0.6$ mm, the measured data are: $f_0 = 2$ GHz, BW = 171 MHz, $L_{A0} = 0.9$ dB, $L_R > 13$ dB. The two TZs are: $f_{z1} - f_0 = 425$ MHz, $f_0 - f_{z2} = 420$ MHz. They are almost equidistant with respect to f_0. Using formula (7.24), we obtain the calculated value of $\left|f_{z1,2} - f_0\right| = 424$ MHz. The index 35 dB ($f_0 \pm 280$ MHz) characterizes the selectivity of the BPF. When using the normalized frequency $j\Omega$, the selectivity index is written as 35 dB ($\pm j3.3$).

If $S = 1$ mm and $d = 8.7$ mm, then the magnetic parasitic cross-coupling increased by $K_m = 0.0385$, while the value of K_{12} remains the same. The value of the magnetic component of the mixed cross-coupling is compensated by selecting the coupling strip parameters $S' = 0.05$ mm and $l' = 2$ mm. Position No. 8 (Table 7.2) illustrates the change in the frequency responses of the BPF under consideration, associated with a change in its geometric parameters.

If $K_{13} > 0$, then the two both-sided TZs are shifted up in frequency: $f_{z1} = 2.616$ GHz and $f_{z2} = 1.60$ GHz. If $K_{13} < 0$, then the two opposite TZs are shifted down in frequency: $f_{z1} = 2.237$ GHz and $f_{z2} = 1.50$ GHz. Positions No. 9 and No. 10 (Table 7.2) show the frequency responses of the triplet BPF under consideration when its two TZs are not equidistant with respect to f_0.

7.4 Quadruplet BPF with Mixed Cross-coupling

Figure 7.5a shows the coupling structure of a symmetrical $K_{12} = K_{34}$ $(m_{12} = m_{34})$ quadruplet BPF. Its cross-coupling K_{14} is mixed. In expression (7.16), which defines the minor of quadruplet BPF, we must replace m_{14} with $m_{14}(S') = m_{014} - a_{14}S'$, which will bring this minor to the form

$$M_{41} = (m_{014} - a_{14}S')\,S'^2 + m_{12}^2 m_{23} - (m_{014} - a_{14}S')\,m_{23}^2. \tag{7.29}$$

The third-degree algebraic equation defining the TZs of the quadruplet BPF is obtained by equating the minor (7.29) to zero:

$$S'^3 - \frac{m_{014}}{a_{14}}S'^2 - m_{23}^2 S' - \frac{m_{12}^2 m_{23}}{a_{14}} = 0. \tag{7.30}$$

The relationship between the signs of the coupling coefficients m_{014} and m_{23} determines the position of the roots of the algebraic equation (7.30).

For the quadruplet BPF under consideration, described by equation (7.30), the inverse problem (7.19) is represented by a system of three equations for determining the coefficients of the mixed coupling $m_{14}(\Omega) = m_{014} - a_{14}\Omega$:

$$-S'_1 S'_2 S'_3 = -m_{12}^2 m_{23}/a_{14}, \tag{7.31a}$$

$$-(S'_1 + S'_2 + S'_3) = -m_{014}/a_{14}, \tag{7.31b}$$

$$S'_1 S'_2 + S'_1 S'_3 + S'_2 S'_3 = -m_{23}^2. \tag{7.31c}$$

The following equalities represent the solution of system (7.31):

$$a_{14} = m_{12}^2 m_{23}/S'_1 S'_2 S'_3, \tag{7.32a}$$

$$m_{014} = m_{12}^2 m_{23}\,(S'_1 + S'_2 + S'_3)\,/S'_1 S'_2 S'_3, \tag{7.32b}$$

$$S'_3 = -\left(m_{23}^2 + S'_1 S'_2\right)/(S'_1 + S'_2). \tag{7.32c}$$

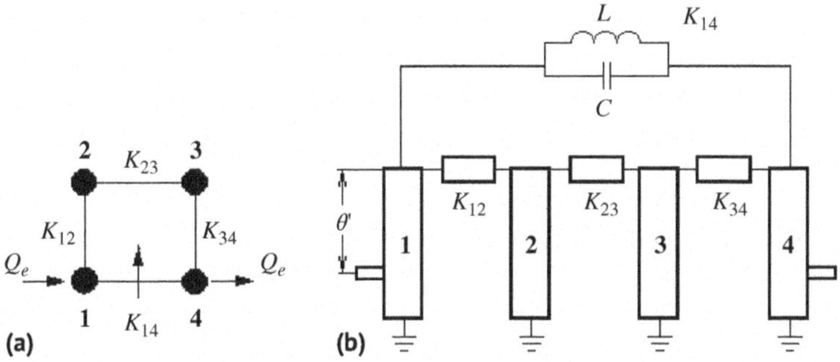

Figure 7.5 Quadruplet BPF with mixed cross-coupling. (a) Coupling structure. (b) Schematic.

The parameters m_{014} and a_{14} of the mixed coupling are not included in the third equality (7.32c) because the three TZs of the quadruplet BPF are interdependent. Two of the three TZs are independent, and the third TZ is dependent on the first two and is determined from equality (7.32c).

The two TZs (S_1' and S_2') in (7.32c) are independent. Constraints on some placement options for the three TZs follow from the solution of the inverse problem (7.32).

- The location of one TZ on the $S = \sigma$ ($S' = j\sigma$) axis is not allowed, since the left-hand side of (7.32b) will not be a real number.
- Equidistant location of two TZs on the $S = \sigma$ ($S' = j\sigma$) axis is also not allowed, since in this case the denominator of the right-hand side of (7.32c) will vanish.
- Three TZs on the $S = j\Omega$ ($S' = \Omega$) axis cannot be unilateral. If two TZs have the same sign, then, according to (7.32c), the third TZ must have the opposite sign.
- Two of the three TZs located on the $S = j\Omega$ ($S' = \Omega$) axis cannot be equidistant. The denominator of the right-hand side of (7.32c) vanishes if $j\Omega_2 = -j\Omega_1$ ($\Omega_2 = -\Omega_1$).

7.4.1 Various Arrangements of Transmission Zeros

Following are the admissible placements of three TZs of a quadruplet BPF with mixed cross-coupling K_{14}, which are established based on the solution of the inverse problem (7.32).

The quadruplet BPF schematic is shown in Figure 7.5b. It contains the same elements as the trisection BPF shown in Figure 7.2b. We set the following parameters for this filter: $f_0 = 1$ GHz, FBW = 0.05, $n = 4$, and ripple $L_{Ar} = 0.2$ dB. From Table 2.6, we determine the g parameters corresponding to the values $n = 4$ and $L_{Ar} = 0.2$ dB: $g_0 = 1$, $g_1 = 1.3028$, $g_2 = 1.2814$, $g_3 = 1.9761$, $g_4 = 0.8468$. Using the g parameters, FBW value, and formulas (3.8), we find $K_{12} = K_{34} = 0.0387$ ($m_{12} = 0.774$), $|K_{23}| = 0.0314$ ($|m_{23}| = 0.628$), $Q_e = 26.056$. Coupling between the resonators is accomplished by the quarter-wave segments with the characteristic impedances (7.26): $Z_{12} = Z_{34} = 329\ \Omega$; $|Z_{23}| = 409.49\ \Omega$. The signs of the main coupling coefficients $K_{12} = K_{34}$ do not affect the TZs. As follows from (7.30) and (7.31), we can assign them the "+" sign. The sign of K_{23} significantly affects the TZs, so we determined the modulus of this coupling coefficient. The value of $Q_e = 26.056$ corresponds to the load connection coordinate of $\theta' = 67.15°$.

Table 7.3 presents various options for the arrangement of the three TZs of the quadruplet BPF and their corresponding frequency responses.

Case 1: All three TZs are located on the $S = j\Omega$ ($S' = \Omega$) axis. Two TZs (No. 1, Table 7.3) are located in the lower S half-plane. The two both-sided TZs will be defined as follows: $S_1' = 2$ ($f_{z1} = 1.05$ GHz), $S_2' = -8$ ($f_{z2} = 0.8$ GHz). We determine the third TZs from (7.32c): $S_3' = -2.6$ ($f_{z3} = 0.935$ GHz). The values characterizing the BPF under consideration: $a_{14} = 0.009044$, $m_{014} = -0.0778$, $K_{14} = -0.0039$, $K_{m14} = 0.0073$ ($L = 277.59$ nH), and $K_{e14} = -0.0112$ ($C = 0.14$ pF) were obtained by successively using formulas (7.32a), (7.32b), and (7.9). The simulated frequency responses corresponding to these TZs are shown in position No. 1.

Case 2: All three TZs are located on the $S = j\Omega$ ($S' = \Omega$) axis. Two TZs (No. 2, Table 7.3) are located in the upper S half-plane. The two both-sided TZs will be defined as follows: $S_1' = -2$ ($f_{z1} = 0.95$ GHz), $S_2' = 8$ ($f_{z2} = 1.2$ GHz). The third TZs we determine from (7.32c): $S_3' = 2.6$ ($f_{z3} = 1.065$ GHz). The values characterizing the quadruplet BPF: $a_{14} = 0.009044$, $m_{014} = 0.0778$, $K_{14} = 0.0039$, $K_{m14} = 0.0112$ ($L = 180.9$ nH), and $K_{e14} = -0.0073$ ($C = 0.091$ pF) were obtained by formulas (7.32a), (7.32b), and (7.9). The position No. 2 represents the simulated frequency responses for the case.

Case 3: The TZs are located on the $S = j\Omega$ ($S' = \Omega$) axis; the second-order TZ is located in the lower S half-plane, and the first-order TZ is located in the upper S half-plane (No. 3, Table 7.3). The second-degree TZ is set to $S_1' = S_2' = -6$ ($f_z = 0.85$ GHz). The values characterizing the quadruplet BPF: $S_3' = 3.033$ ($f_{z3} = 1.0758$ GHz), $a_{14} = 0.003445$, $m_{014} = -0.0309$, $K_{14} = -0.001545$, $K_{m14} = 0.00276$ ($L = 734.2$ nH) and $K_{e14} = -0.0043$ ($C = 0.05375$ pF) were obtained by formulas (7.32) and (7.9). The simulated frequency responses corresponding to these TZs are shown in position No. 3.

Case 4: The TZs are located on the $S = j\Omega$ ($S' = \Omega$) axis: the second-order TZ is located in the upper S half-plane, and the first-order TZ is located in the lower S half-plane (No. 4, Table 7.3). The second-degree TZ is set to $S_1' = S_2' = 6$ ($f_z = 1.15$ GHz). The values characterizing the quadruplet BPF: $S_3' = -3.033$ ($f_{z3} = 0.9242$ GHz), $a_{14} = 0.003445$, $m_{014} = 0.0309$, $K_{14} = 0.001545$, $K_{m14} = 0.0043$ ($L = 471.26$ nH), and $K_{e14} = -0.00276$ ($C = 0.0345$ pF) were obtained by formulas (7.32) and (7.9). The position No. 4 represents the simulated frequency responses for the case.

Table 7.3 Transmission zeros and frequency responses of quadruplet BPF with mixed cross-coupling.

No.	TZ	Frequency responses	K_{14}	K_{23}
1	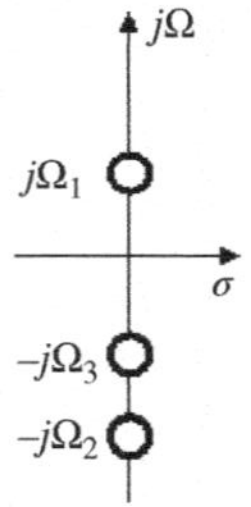	$j\Omega_1 = j2, j\Omega_2 = -j8, j\Omega_3 = -j2.6$ 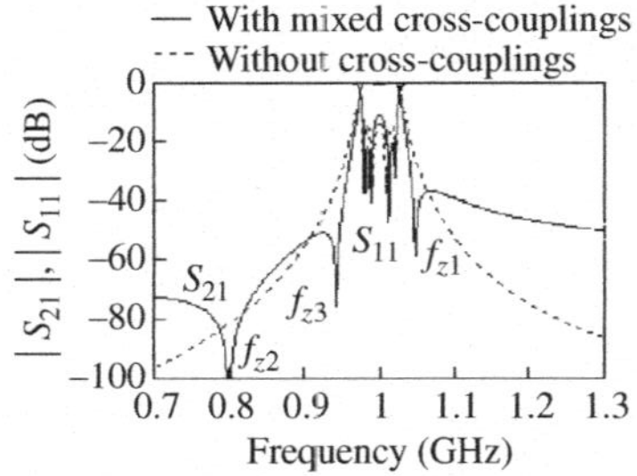	$K_{14} < 0$	$K_{23} > 0$
2	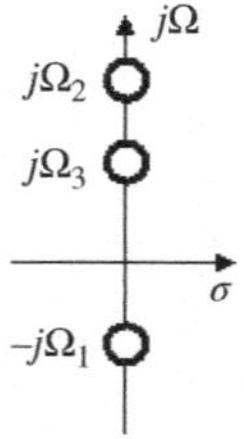	$j\Omega_1 = -j2, j\Omega_2 = j8, j\Omega_3 = j2.6$ 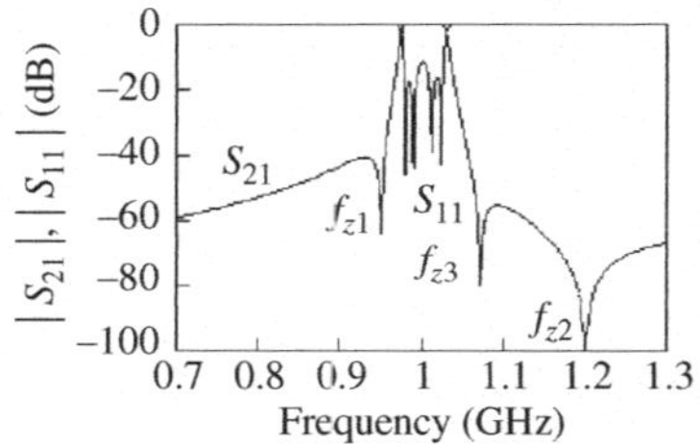	$K_{14} > 0$	$K_{23} < 0$

3	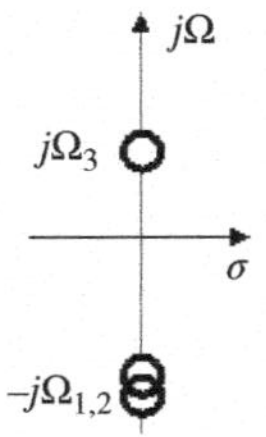	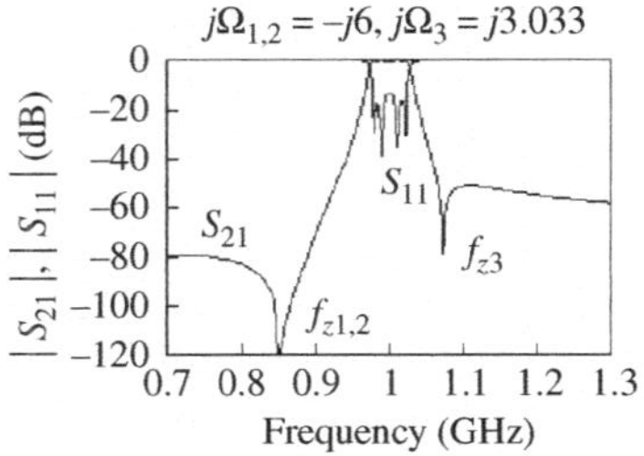	$K_{14} < 0$	$K_{23} > 0$
4	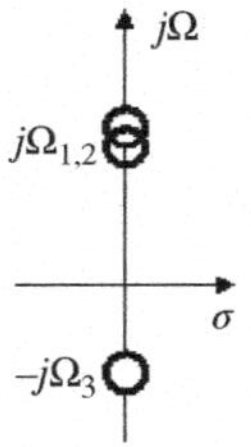	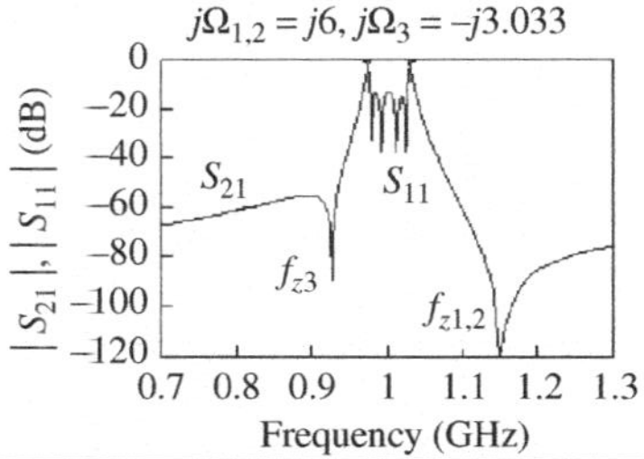	$K_{14} > 0$	$K_{23} < 0$

(*Continued*)

Table 7.3 (Continued)

No.	TZ	Frequency responses	K_{14}	K_{23}
5	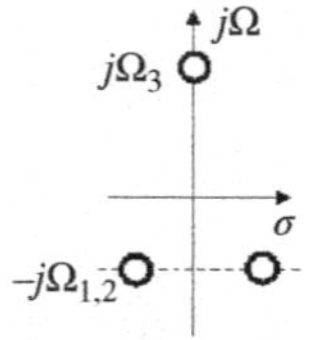 $\lvert j\Omega_{1,2}\rvert = j\Omega_3/2$ $\lvert\sigma_{1,2}\rvert = \lvert\sigma_{1,2}'\rvert$	$j\Omega_3 = -j2.5, j\Omega_{1,2} = -j1.25, \lvert\sigma_{1,2}\rvert = \lvert\sigma_{1,2}'\rvert$ 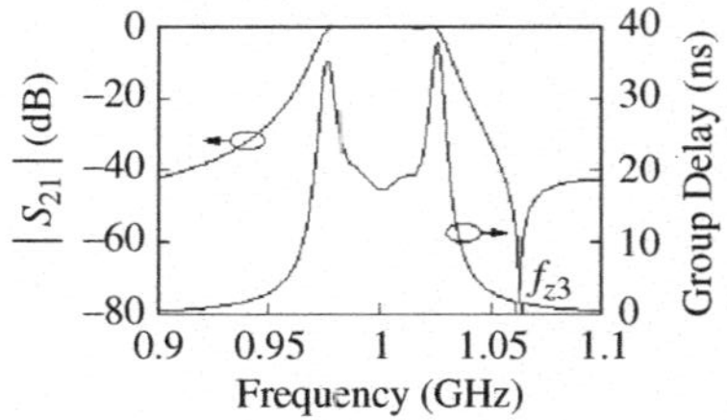	$K_{14} = 0$	$K_{23} > 0$
6	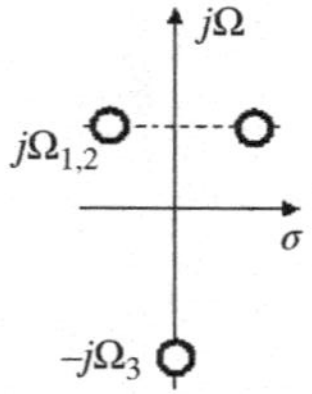 $\lvert j\Omega_{1,2}\rvert = j\Omega_3/2$ $\lvert\sigma_{1,2}\rvert = \lvert\sigma_{1,2}'\rvert$	$j\Omega_3 = -j2.5, j\Omega_{1,2} = j1.25, \lvert\sigma_{1,2}\rvert = \lvert\sigma_{1,2}'\rvert$ 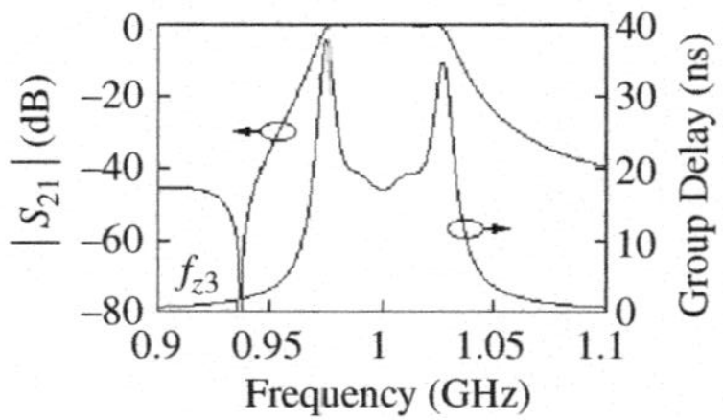	$K_{14} = 0$	$K_{23} < 0$

7	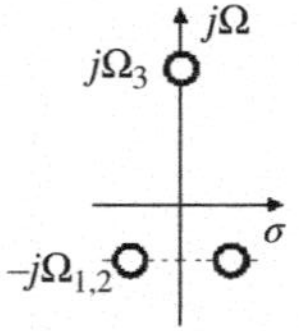 $\lvert j\Omega_{1,2} \rvert < j\Omega_3/2$ $\lvert \sigma_{1,2} \rvert < \lvert \sigma_{1,2}' \rvert$	$j\Omega_3 = j2.5, j\Omega_{1,2} = -j0.6, \lvert \sigma_{1,2} \rvert = 1.5$ 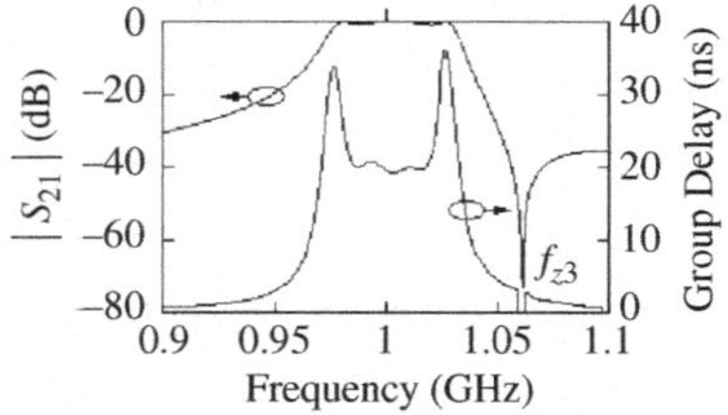	$K_{14} > 0$	$K_{23} > 0$
8	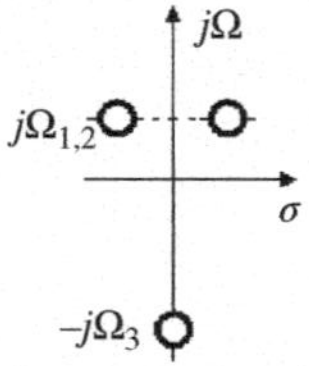 $\lvert j\Omega_{1,2} \rvert < j\Omega_3/2$ $\lvert \sigma_{1,2} \rvert < \lvert \sigma_{1,2}' \rvert$	$j\Omega_3 = -j2.5, j\Omega_{1,2} = j0.6, \lvert \sigma_{1,2} \rvert = 1.5$ 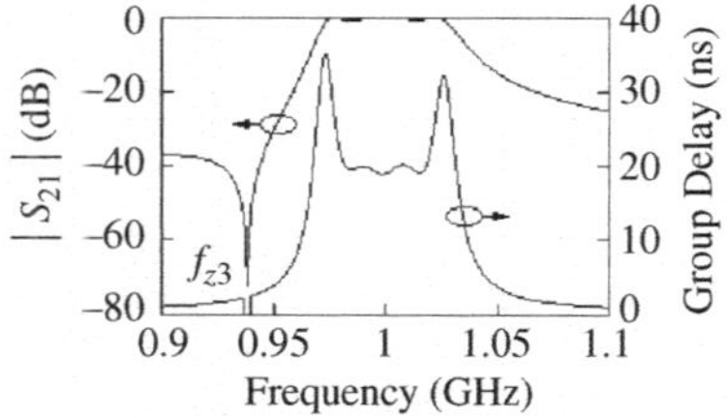	$K_{14} < 0$	$K_{23} < 0$

Case 5: The three TZs (No. 5, Table 7.3) correspond to conditions $K_{14} = 0$ and $S_3 = j\Omega_3$. In equation (7.30), the term $m_{014}S_2'/a_{14}$ disappears when condition $K_{14} = 0$ is satisfied, and it becomes the so-called reduced equation [23]. This equation has two complex conjugate roots $S_{1,2}' = \Omega_{1,2} \pm j\left|\sigma_{1,2}\right|$ and one real root $S_3' = \Omega_3$. From the solution of the inverse problem (7.32), it follows that one real root defines two complex conjugate roots with negative imaginary part

$$\Omega_{1,2} = \frac{\Omega_3}{2}, \tag{7.33}$$

and real

$$\sigma_{1,2}^2 = \frac{3\Omega_3^2}{4} - m_{23}^2 \tag{7.34}$$

in parts.

Let us define one TZ at a real frequency of $f_{z3} = 1.0625$ GHz ($\Omega_3 = 2.5$). Using expressions (7.33) and (7.34), we determine the components of two complex TZs: $\Omega_{1,2} = -1.25$ and $\left|\sigma_{1,2}\right| = 2.329$. The sign m_{23} in (7.32a) must be positive since $S_1'S_2'S_3' = 17.467 > 0$. Substituting the values of three TZs into (7.32a) yields $a_{14} = K_{m14} = |K_{e14}| = 0.0215$, $L = 98.90$ nH, $C = 0.2688$ pF. Position No. 5 represents the simulated frequency responses for this case.

Case 6: The three TZs (No. 6, Table 7.3) correspond to conditions $K_{14} = 0$ and $S_3 = -j\Omega_3$. Let us definc one TZ at a real frequency of $f_{z3} = 0.9375$ GHz ($\Omega_3 = -2.5$). Using expressions (7.33), (7.34), and (7.32a), we get: $\Omega_{1,2} = 1.25$, $m_{23} < 0$, $a_{14} = K_{m14} = |K_{e14}| = 0.0215$, $L = 98.90$ nH, $C = 0.2688$ pF. The position No. 6 represents the simulated frequency responses for the case. In cases No. 5 and No. 6, condition $K_{14} = 0$ did not lead to flat group delay, like trisection BPF with $K_{13} = 0$.

Case 7: The three TZs (No. 7, Table 7.3) correspond to flat group delay and $S_3 = j\Omega_3$. The value is $\Omega_3 = 2.5$. To obtain a flat group delay, it is necessary to reduce the values of the imaginary (7.33) and real (7.34) parts of the complex TZs for case No. 5. We will take the value $\left|\sigma_{1,2}\right| = 1.5$ (in contrast to $\left|\sigma_{1,2}\right| = 2.329$ in No. 5). Condition (7.32c) determines the value $\Omega_{1,2} = -0.6$. Note that as $\left|\sigma_{1,2}\right|$ decreases, $\left|\Omega_{1,2}\right|$ decreases too. All three TZs are given; solving the inverse problem (7.32) and using (7.9) yield: $m_{23} > 0$, $a_{14} = 0.0809$, $m_{014} = 0.1052$, $K_{14} = 0.00526$, $K_{m14} = 0.08357$ ($L = 24.25$ nH), $K_{e14} = -0.0783$ ($C = 0.979$ pF). The position No. 7 represents the simulated frequency responses with flat group delay for this case.

Case 8: The three TZs (No. 8, Table 7.3) correspond to flat group delay and $S_3 = -j\Omega_3$. The value is $\Omega_3 = -2.5$. In this case, the components and the two complex TZs are smaller than in case No. 5: $\left|\sigma_{1,2}\right| = 1.5$, $\Omega_{1,2} = 0.6$. All three TZs are given; solving the inverse problem (7.32) and using (7.9) yield: $m_{23} < 0$, $a_{14} = 0.0809$, $m_{014} = -0.1052$, $K_{14} = -0.00526$, $K_{m14} = 0.0783$ $(L = 25.88\ \text{nH})$, $K_{e14} = -0.08357$ ($C = 1.045$ pF). Position No. 8 represents the simulated frequency responses with flat group delay for this case.

To obtain a flat group delay (cases Nos. 7, 8), the TZs should be located at the corners of an isosceles triangle; the vertex lies on the $j\Omega$-axis; and the sides intersect the σ-axis. In this case, a small value of the mixed cross-coupling $|K_{14}|$ is realized at sufficiently large values of the components K_{m14} and $|K_{e14}|$.

7.4.2 Stripline Quadruplet BPF

We implement some positions of Table 7.3 by using stripline quadruplet BPF, the topology of which is presented in Figure 7.6a. The coupling structure of the BPF is shown in Figure 7.5a. The resonators (Nos. 1, 4) of the BPF are SIRs, and the resonators (Nos. 2, 3) are quarter-wave resonators, which are connected to SIRs through capacitive gaps. The main coupling K_{23} is simple and positive, and the main couplings $K_{12} = K_{34}$ are simple and negative. The mixed cross-coupling K_{14} can be either positive or negative [26], which correspond to positions No. 7 and No. 1 (Table 7.3).

The BPF was constructed using substrates with parameters $\varepsilon_r = 36$, $\tan\delta = 0.0003$, and $h = 1$ mm. The BPF dimensions are 17.8 mm × 10.2 mm × 2 mm. The resonators are tuned to a frequency $f_0 = 1.425$ GHz. A metal strip with a perimeter width of 1.2 mm is used to connect a metallized end surface to the short-circuited ends of the resonators. The resonator lengths, counted from the metal strip, are $L_1 = 7.4$ mm and $L_2 = 7.6$ mm. All resonators have a width of $w = 2.4$ mm. Gaps between resonators are $S_{14} = 1$ mm, $S_{12} = 0.6$ mm, and $S_{23} = 1.2$ mm.

Figure 7.6b shows a photograph of the interior of the stripline BPF. Position 2 in Figure 7.6a shows the conductive pads at the ends of the filter that connect its inner and outer parts. They are shown in the photograph of the filter (Figure 7.6c), which depicts the reverse side of the stripline BPF. Short zig-zag segments 0.4 mm wide connect the end resonators to the input and output.

The quadruplet BPF with the given parameters has a mixed cross-coupling $K_{14} = 0.00294$ with components $K_m = 0.03529$, $K_e = -0.03235$, and main simple couplings $K_{12} = K_{34} = -0.0399$, $K_{23} = 0.0287$. Its external Q-factor $Q_e = 23.1$ is determined by the position of the short zig-zag segment. The signs of the coupling coefficients K_{23} and K_{14} coincidences are positive. Due to this stripline quadruplet BPF is a delay line filter with right-handed attenuation poles (No. 7, Table 7.3).

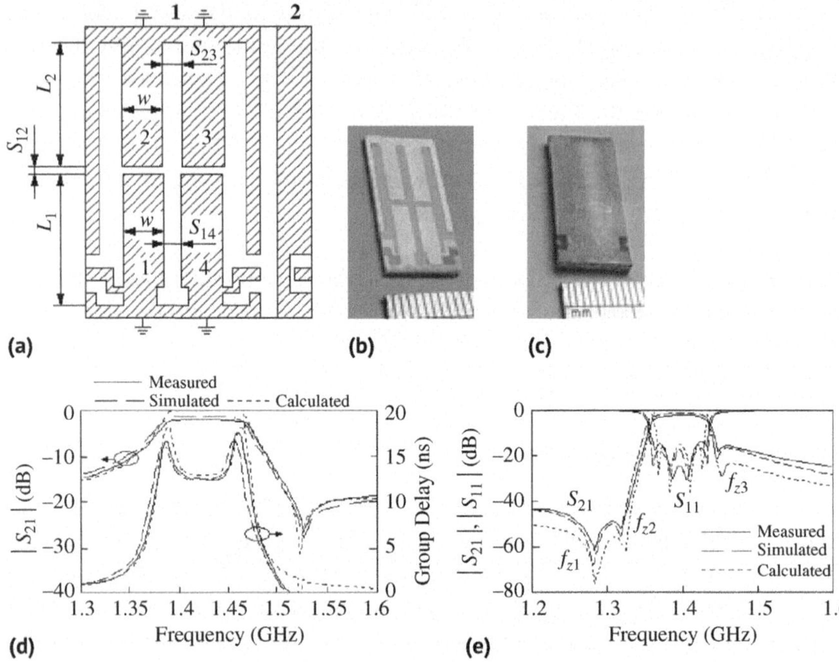

Figure 7.6 Stripline quadruplet BPF with mixed cross-coupling. (a) Topology. (b) Photograph of inside view. (c) Photograph of outside view. (d) Frequency responses for case $K_{23} > 0$, $K_{14} > 0$. (e) Frequency responses for case $K_{23} > 0$, $K_{14} < 0$. *Source*: [19] / with permission of IEEE.

Figure 7.6d shows the simulated and measured frequency responses of the BPF. The delay line BPF has the following measured date: $f_0 = 1425$ MHz; BW = 72 MHz (FBW = 0.0505); the group delay $\tau = 12.7 \pm 0.2$ ns within the frequency range of 44 MHz; $L_{A0} = 1.7$ dB; and $L_R \geq 17$ dB. The attenuation pole of this filter is due to the right-hand TZ (1525 MHz). The normalized admittance matrix Y (7.13) of the quadruplet BPF contains the following normalized coupling coefficients: $m_{12} = m_{34} = -0.79$, $m_{23} = 0.568$, $m_{14}(\Omega) = 0.0582 - 0.0338\ \Omega$, and a normalized external Q-factor $q_e = 1.17$.

If we keep the basic coupling coefficients the same and change the K_{14} sign to the opposite, this filter will have three TZs at real frequencies (No. 1, Table 7.3). To change the sign of K_{14}, the gap S_{14} between SIRs was reduced from 1 to 0.6 mm, and the rectangles removed from SIRs were increased from 1.4 mm × 0.4 mm to

2.0 mm × 0.5 mm. As a result, the negative value of $K_{14} = -0.015$ with the components $K_{m14} = 0.0517$ and $K_{e14} = -0.0667$ was obtained. In this case, $m_{14}(\Omega) = -0.0297 - 0.0537\ \Omega$.

Figure 7.6e shows the simulated and measured frequency responses of this filter. The measured values are $f_0 = 1395$ MHz, BW = 72 MHz, $L_{A0} = 1.7$ dB, and $L_R \geq 14$ dB. Three TZs are located at the frequencies of $f_0 = 1395$ MHz, BW = 72 MHz, $L_{A0} = 1.7$ dB, and $L_R \geq 14$ dB. The frequency is reduced by 30 MHz compared to the previous case (Figure 7.6d), which is due to a decrease in the resonant frequency of SIRs.

The analysis showed that the main advantage of triplet and quadruplet BPFs with mixed cross-coupling lies not in the number of TZs increased by one, but in the many different options for their location. This provides significant diversity in frequency responses.

7.5 Triplet BPF with All Mixed Couplings

Figure 7.7a, b shows the coupling structure of the triplet BPF with all mixed couplings. The circles in this figure indicate reflective-type resonators, such as quarter-wave ones. Suppose mixed coupling coefficients between adjacent resonators K_{12} and K_{23} have the same sign, the arrows in Figure 7.7a are oriented in the same direction. In this case, the triplet BPF is symmetrical. If K_{12} and K_{23} have different signs, the corresponding arrows in Figure 7.7b will be oriented in opposite directions, corresponding to an asymmetric triplet BPF.

Figure 7.7c shows the schematic of a mixed coupled triplet BPF with J inverters, which does not show the transverse branches of the inverters. The filter uses $\lambda/4$ resonators with characteristic impedance $Z_0 = 10\ \Omega$, susceptance slope parameter $b = \pi/4Z_0$, and $f_0 = 1$ GHz. The external Q-factor of the end resonators is determined by the coordinate of the load connection ($R_L = 50\ \Omega$). Let us analyze this filter using the inverse problem method, as was done in Litvintsev and Zakharov [27].

Using the linear representation form (7.6) of mixed coupling coefficients $m_{12}(\Omega) = m_{012} - a_{12}S'$, $m_{23}(\Omega) = m_{023} - a_{23}S'$, $m_{13}(\Omega) = m_{013} - a_{13}S'$, we write down the minor (7.15) for the trisection BPF with all mixed couplings as follows:

$$M_{31} = (m_{012} - a_{12}S')(m_{023} - a_{23}S') - (m_{013} - a_{13}S')S' = 0. \tag{7.35}$$

Equation (7.35) is an algebraic equation of the second degree, defining two roots S'_1 and S'_2, representing two TZs. We use the equation for two critical cases, when the triplet BPF with all mixed couplings is symmetric and asymmetric.

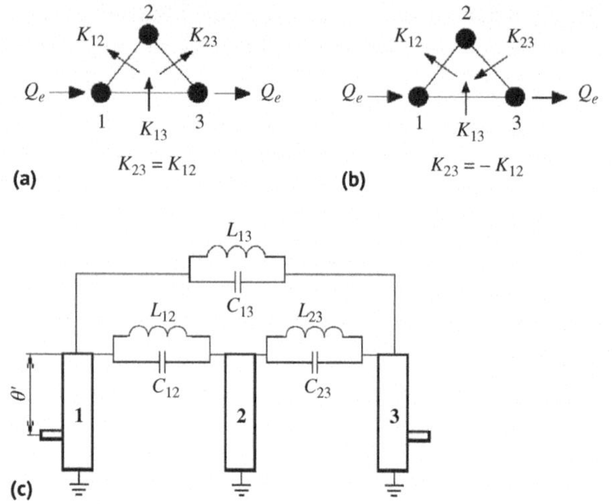

Figure 7.7 Triplet BPF with all mixed couplings. (a) Coupling structure for symmetric BPF. (b) Coupling structure for asymmetric BPF. (c) Schematic.

7.5.1 Symmetric triplet BPF

The symmetric triplet BPF has $m_{23} = m_{12}$ $(m_{023} - a_{23}S' = m_{012} - a_{12}S')$, and equation (7.35) becomes

$$S'^2 - \frac{2m_{012}a_{12} + m_{013}}{a_{13} + a_{12}^2}S' + \frac{m_{012}^2}{a_{13} + a_{12}^2} = 0. \tag{7.36}$$

In this case, the system of equations (7.21) will take the form

$$\frac{2m_{012}a_{12} + m_{013}}{a_{13} + a_{12}^2} = S'_1 + S'_2, \tag{7.37a}$$

$$\frac{m_{012}^2}{a_{13} + a_{12}^2} = S'_1 S'_2. \tag{7.37b}$$

Note that the main coupling coefficients of the BPF under consideration are known. The mixed cross-coupling is determined by two values m_{013} and a_{13}. The solution for the equation system (7.37) toward the unknowns m_{013} and a_{13} has the form

$$m_{013} = -2m_{012}a_{12} + m_{012}^2\frac{S'_1 + S'_2}{S'_1 S'_2}, \tag{7.38a}$$

$$a_{13} = \frac{m_{012}^2}{S'_1 S'_2} - a_{12}^2. \tag{7.38b}$$

Expressions (7.38) represent a solution to the inverse problem for a symmetric triplet BPF with all mixed couplings. Analysis based on (7.38) shows that there are seven different options for placing two TZs: two options for one-sided TZs on the $j\Omega$-axis; two second-order TZ options on the $j\Omega$-axis; two options for TZs on the complex plane; and one option for equidistant TZs on the σ-axis. These placements of two TZs are the same as those of triplet BPF with simple main couplings (Figure 7.2). They are presented in Table 7.1, and we will not repeat them.

The most significant interest is the equidistant TZs on the σ-axis. If the condition

$$m_{013} = -2m_{012}a_{12} \tag{7.39}$$

is met, then the second term in equation (7.36) equals zero and its solution is written as $S'^2 = -m_{012}^2/\left(a_{13} + a_{12}^2\right)$. Passing to the usual variable $S = \sigma + j\Omega$, we obtain

$$\sigma_{1,2} = \pm\frac{m_{012}}{\sqrt{a_{13} + a_{12}^2}}. \tag{7.40}$$

The TZs σ_1 and σ_2 in (7.40) are equidistant on the σ-axis ($\sigma_2 = -\sigma_1$), providing the filter with a flat group delay.

If the triplet BPF only has a mixed cross-coupling, then the second term $\left(a_{12}^2\right)$ under the radical (7.40) is absent, and this formula becomes

$$\sigma_{1,2} = \pm\frac{m_{012}}{\sqrt{a_{13}}}. \tag{7.41}$$

Comparison of (7.40) and (7.41) shows that adding mixed couplings between adjacent resonators to the triplet BPF leads to a convergence of TZs on the σ-axis.

The effect of using all mixed couplings in triplet BPF can be considered within the scheme (Figure 7.7c) with admittance inverters for mixed couplings. Let us set the initial parameters of the third-order BPF without cross-coupling: $f_0 = 1$ GHz, FBW = 0.05, ripple $L_{Ar} = 0.2$ dB. The prototype parameters for these data are $g_0 = 1$, $g_1 = g_3 = 1.2275$, $g_2 = 1.1525$. Using classic formulas (3.8), we obtain coupling coefficients $K_{12} = K_{23} = 0.042$ and external quality factor of the end resonators $Q_{e1} = Q_{e3} = Q_e = 24.55$. Normalizing these values, we get $m_{12} = K_{12}/\text{FBW} = 0.84$, $q_e = \text{FBW}\ Q_e = 1.228$.

Let us introduce into this filter a mixed cross-coupling $m_{13}\ (\Omega) = m_{013} - a_{13}\ \Omega$, defining it in such a way as to provide a pair of equidistant TZs: $\left|\sigma_{1,2}\right| = 8$. Substituting the indicated value into (7.41), we determine the slope parameter

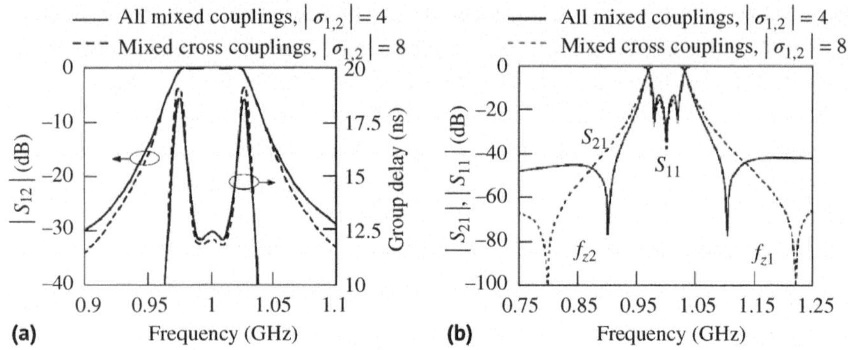

Figure 7.8 Frequency responses of mixed coupled triplet BPFs with equidistant TZs. (a) Symmetrical triplet BPF with TZs at σ-axis. (b) Asymmetric triplet BPF with TZs at $j\Omega$-axis.

$a_{13} = 0.0105$. The mixed cross-coupling is defined as $m_{13}(\Omega) = -0.0105\,\Omega$. The frequency responses for this case are presented in Figure 7.8a by dotted lines. The group delay of the filter is $\tau = 11.9 \pm 0.1$ ns.

Let us introduce into this filter the mixed couplings between adjacent resonators $m_{12}(\Omega) = m_{012} - a_{12}\Omega = 0.84 - a_{12}\Omega$, while ensuring a double approach of equidistant TZs: $|\sigma_{1,2}| = 4$. Substituting the indicated value into (7.40), we determine the slope parameter: $a_{12} = 0.1833$. Mixed couplings between adjacent resonators are defined as $m_{12}(\Omega) = m_{23}(\Omega) = 0.84 - 0.1833\,\Omega$. For the mixed cross-coupling m_{13}, the slope parameter a_{13} will remain the same, and the value of m_{013} will change to $m_{013} = -0.3079$, which follows from (7.39). The mixed cross-coupling is defined: $m_{13}(\Omega) = -0.3079 - 0.0105\,\Omega$. The frequency responses of the symmetric triplet BPF with all mixed couplings are presented in Figure 7.8a with solid lines. The group delay of the filter is $\tau = 12.2 \pm 0.15$ ns. The introduction of mixed couplings between adjacent resonators into the symmetric triplet BPF led to the convergence of TZs by two times (from $|\sigma_{1,2}| = 8$ to $|\sigma_{1,2}| = 4$) and to an increase in the group delay from 11.9 to 12.2 ns.

7.5.2 Asymmetric Triplet BPF

The asymmetric triplet BPF (Figure 7.7b) has $m_{023} - a_{23}S' = -m_{012} - a_{12}S'$, and equation (7.35) takes the form

$$\left(a_{12}^2 + a_{13}\right) S'^2 - m_{013}S' - m_{012}^2 = 0. \tag{7.42}$$

The inverse problem solution of the filter has the form

$$m_{013} = -m_{012}^2 \frac{S_1' + S_2'}{S_1' S_2'}, \tag{7.43a}$$

$$a_{13} = -\frac{m_{012}^2}{S_1' S_2'} - a_{12}^2. \tag{7.43b}$$

Analysis based on (7.43) shows three different options for placing two TZs on the $j\Omega$-axis, on both sides of the center frequency. These placements of two TZs are the same as for triplet BPF with simple main couplings (Nos. 8, 9, 10 from Table 7.1). We will not repeat them.

Of most significant interest to us is the equidistant location of TZs, which is ensured for the condition

$$m_{013} = 0. \tag{7.44}$$

The mixed cross-coupling m_{13} is written as $m_{13} = -a_{13}\Omega$ in this case. If condition (7.44) is met, the solution of equation (7.42) is written in the form

$$\Omega_{1,2} = \pm \frac{m_{012}}{\sqrt{a_{13} + a_{12}^2}}. \tag{7.45}$$

Solution (7.45) shows that an increase in slope parameters a_{13} and a_{12} of mixed couplings leads to a decrease in $|\Omega_{1,2}|$ and an increase in BPF selectivity. If the triplet BPF has a mixed cross-coupling only, then the second term (a_{12}^2) under the radical (7.45) is absent, which brings this formula to the form

$$\Omega_{1,2} = \pm \frac{m_{012}}{\sqrt{a_{13}}}. \tag{7.46}$$

Comparison of (7.45) and (7.46) shows that the addition of mixed couplings between adjacent resonators and the asymmetric triplet BPF leads to a convergence of TZs on the $j\Omega$-axis and an increase in the selectivity of quasi-elliptic BPF.

Let us illustrate the effect of using all mixed couplings in an asymmetric triplet BPF (Figure 7.7b). Set the initial parameters of the third-order BPF without cross-coupling as in the previous example: $f_0 = 1$ GHz, FBW $= 0.05$, ripple $L_{Ar} = 0.2$ dB. This will lead to coupling coefficients $m_{12} = 0.84$, $m_{23} = -0.84$.

Let us introduce into this filter a mixed cross-coupling $m_{13}(\Omega) = -a_{13}\Omega$ ($m_{013} = 0$), defining it in such a way as to provide a pair of equidistant TZs: $|\Omega_{1,2}| = 8$. Substituting the indicated value into (7.46), we determine $a_{13} = 0.0105$ and $m_{13}(\Omega) = -0.0105\ \Omega$. The frequency responses for this case are presented in Figure 7.8b by dotted lines.

Let us introduce into this filter mixed couplings between adjacent resonators $m_{12}(\Omega) = 0.84 - a_{12}\Omega$, while ensuring a double approach of equidistant TZs:

$|\Omega_{1,2}| = 4$. Substituting the indicated value into (7.45), we determine the slope parameter: $a_{12} = 0.1833$. Mixed couplings between adjacent resonators are defined as $m_{12}\ (\Omega) = 0.84 - 0.1833\ \Omega$ and $m_{23}\ (\Omega) = -0.84 - 0.1833\ \Omega$. The mixed cross-coupling m_{13} remains the same $m_{13}\ (\Omega) = -0.0105\ \Omega$. The asymmetric triplet BPF frequency responses with all mixed couplings are presented by solid lines in Figure 7.8b. Introducing mixed couplings between adjacent resonators into the asymmetric triplet BPF led to the convergence of TZs twice (from $|\Omega_{1,2}| = 8$ to $|\Omega_{1,2}| = 4$) and increased selectivity.

Formulas (7.40) and (7.45) express the TZ position through the coupling coefficients of symmetric and asymmetric triplet BPFs with all mixed couplings. They indicate that using all mixed couplings in triplet BPFs leads to an increase in flat group delay of symmetric BPF and an increase in selectivity of asymmetric BPF.

7.5.3 Microstrip Realization

Let us look at examples of symmetrical and asymmetrical microstrip triplet BPFs.

1. *Microstrip symmetric triplet BPF*: The condition expressed by formula (7.39) is the basis of the design of a symmetric triplet BPF with all mixed couplings and flat group delay.

Let us set the following initial parameters for BPF: $f_0 = 2015$ MHz, BW = 100 MHz (FBW = 0.0496), $L_{\text{Ar}} = 0.1$ dB (return loss $L_R = 16.426$ dB). The Chebyshev low-pass prototype parameters are $g_0 = 1$, $g_1 = g_3 = 1.0315$, $g_2 = 1.1474$. Using classic formulas (3.8), we determine the main coupling coefficients $K_{12} = K_{23} = 0.0450$ ($m_{012} = 0.887$) and the external quality factor $Q_{e1} = Q_{e2} = Q_e = 20.8$.

Figure 7.9a shows the topology of the designed microstrip triplet BPF, which covers an area of 9 mm × 6 mm. The filter contains two varieties of SIRs with resonant frequency $f_0 = 2.015$ GHz. The filter uses the substrate with the following parameters: $\varepsilon_r = 92$, $\tan\delta = 0.0004$, thickness $h = 1$ mm. Short circuit at the ends of SIRs is provided by metallized end faces of the substrate.

The procedure for determining the mixed coupling coefficient and its components is well known [28]. For this purpose, the AWR EM simulator is used. With the geometric dimensions shown in Figure 7.9a, we determine the even ($f_e = 2.044$ GHz) and odd ($f_o = 2.107$ GHz) modes of coupled resonators 1 and 2, and the TZ ($f_z = 2.621$ GHz). According to (3.10), the frequencies f_e and f_o are tuned to provide the required mixed coupling coefficient K_{12}. Using the TZ value determined by EM simulation ($f_z = 2.621$ GHz) and expressions (7.5), we extract the magnetic $K_{m12} = 0.174$ and electric $K_{12} = -0.129$ components of mixed coupling K_{12}. We also determine the slope parameter (7.7b): $a_{12} = \sqrt{K_{m12}\ |K_{e12}|} = 0.1498$. Substituting m_{012} and a_{12} into (7.39) yields the required normalized mixed cross-coupling value for the flat group delay $m_{013} = -0.266$, or $K_{13} = -0.0132$.

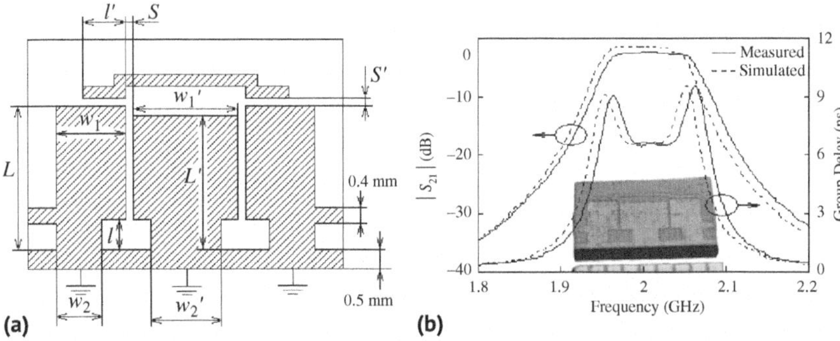

Figure 7.9 Microstrip symmetric triplet BPF with all mixed couplings and flat group delay. (a) Topology: $L = 3.7$ mm, $w_1 = 2$ mm, $w_2 = 1.3$ mm, $L' = 3.4$ mm, $w_1' = 3$ mm, $w_2' = 2$ mm, $l = 0.8$ mm, $S = 0.2$ mm, $S' = 0.2$ mm, $l' = 1.1$ mm. (b) Measured and simulated frequency responses. *Source*: [28]/Fig. 4/with permission of IEEE.

The cross-coupling K_{13} between resonators 1 and 3 is also mixed. Its magnetic component K_{m13} is a parasitic coupling between these resonators. It is determined by measurement, $K_{m13} = 0.0263$. The strip connecting resonators 1 and 3 through the capacitive gaps insert the electric component K_{e13} into the mixed cross-coupling. An increase in $|K_{e13}|$ occurs with an increase in the length l' and a decrease in the gap S'. Since $K_{13} = -0.0132$ and $K_{m13} = 0.0263$, it is necessary to implement the value $K_{e13} = -0.0395$ in the filter, which is provided by the following strip parameters: $S' = 0.2$ mm, $l' = 1.1$ mm. The external Q-factor $Q_e = 20.8$ is ensured by choosing the load connection coordinate and using expression (3.20).

Figure 7.9b shows the frequency responses of this filter and a filter photo in the inset. The measured parameters are as follows: $f_0 = 2015$ MHz, BW = 100 MHz, $L_{A0} = 1.5$ dB, $L_R > 16$ dB; the group delay $\tau = 6.6 \pm 0.1$ ns within the frequency range of 47 MHz, which is slightly less than 50% of the bandwidth.

2. *Microstrip asymmetric triplet BPF*: The basis of the microstrip symmetric triplet BPF design with all mixed couplings and improving selectivity is the condition for the cross-coupling (7.44), which must be resonant.

Let us set the following initial parameters for BPF: $f_0 = 2160$ MHz, BW = 140 MHz (FBW = 0.0648), $L_{\text{Ar}} = 0.2$ dB; TZs location $\left|f_{z1,2} - f_0\right| < 315$ MHz. Determining g parameters from these data and using formulas (3.8), we define $K_{12} = |K_{23}| = 0.0545$, $Q_{e1} = Q_{e2} = Q_e = 18.94$.

To create an asymmetric triplet BPF, we propose to use one asymmetric SIR of $\lambda/4$ type, which results in two pairs of resonators (Figure 7.10a) with mixed couplings of different signs. In these pairs of resonators, we used a TMM-10i dielectric substrate (Rogers) with parameters $\varepsilon_r = 9.8$, $h = 1.016$ mm, and $\tan\delta = 0.002$.

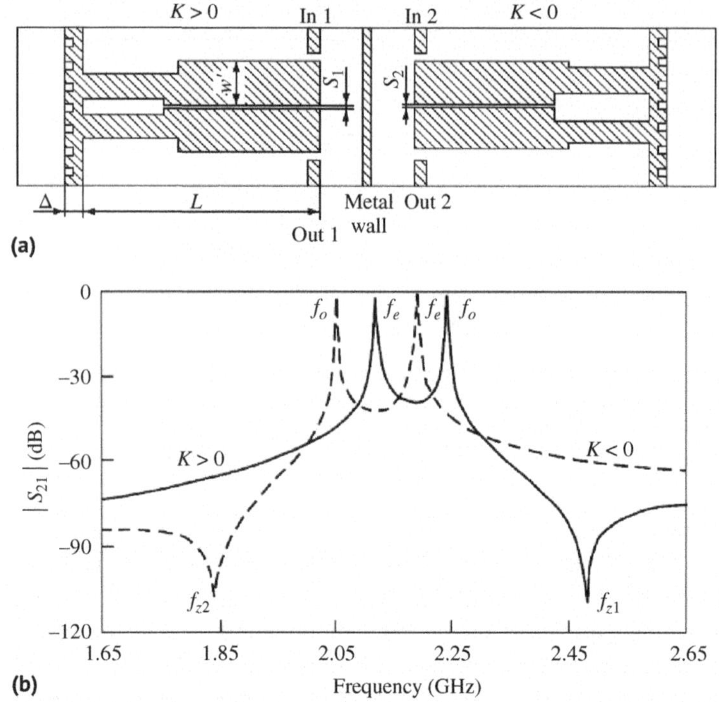

Figure 7.10 Two pairs with asymmetric microstrip SIR. (a) Topology: $L = 11$ mm, $w = 2$ mm, $\Delta = 0.8$ mm, $S_1 = 0.2$ mm, $S_2 = 0.1$ mm. (b) Simulated characteristic frequencies.

Pairs of resonators in Figure 7.10a are grounded through the strip of width Δ with metallized holes of 0.4 mm diameter. The asymmetric SIR is a rectangular strip of size L mm $\times$ w mm, in the short circuit region of which two different rectangles are extracted. By changing the size of these rectangles and the gap S between the SIRs, we can achieve different arrangements of the characteristic frequencies f_e, f_o, and f_z (Figure 7.10b).

The orientation of the resonators of the left pair in Figure 7.10a leads to a positive mixed coupling ($K > 0$). If these resonators are rotated 180° relative to the horizontal axis, then the right pair in Figure 7.10b is formed ($K < 0$). The smaller gap between SIRs and larger extractable rectangles provides the values of $K < 0$. Each pair should have a coupling coefficient $|K| = 0.0545$. Using the expression (3.10), we determine the frequencies of two modes $\left|f_{o,e} - f_0\right| = 60$ MHz, providing the specified value of $|K|$. We require that f_{z1} and f_{z2} of the considered pairs of SIRs be equidistant concerning f_0: $\left|f_{z1,2} - f_0\right| = 315$ MHz. The interval between these TZs is: $f_{z1} - f_{z2} = 630$ MHz.

Tuning to the specified characteristic frequencies by optimizing the SIR shape was carried out simultaneously for two pairs of resonators. In this case, an insulating metal wall (Figure 7.10a) and the EM simulator of MWO software were used. Figure 7.10a shows the optimization results of the SIR shape and its geometric dimensions. In the short circuit region of each SIR pair, two rectangles with dimensions of 4.4 mm × 0.6 mm and 3.8 mm × 0.3 mm were extracted. The coupling between the resonators is a real electromagnetic coupling and is not carried out by the mixed admittance inverter (Figure 7.1a). As a result, the coupling frequencies f_o and f_e are shifted (Figure 7.10b) relative to the calculated values $|f_{o,e} - f_0| = 60$ MHz. At $K > 0$, these frequencies are increased and reduced at $K < 0$.

Combining SIR pairs (Figure 7.10a) results in an asymmetric BPF triplet [27], the topology and photograph of which are shown in Figure 7.11a, b. The filter covers an area of 14.4 mm × 9.5 mm. Since a quasi-TEM wave propagates in a microstrip filter, there is a magnetic cross-coupling $K_{m13} = 0.015$ between resonators 1 and 3. To realize the resonant cross-coupling ($K_{13} = 0$), we must use

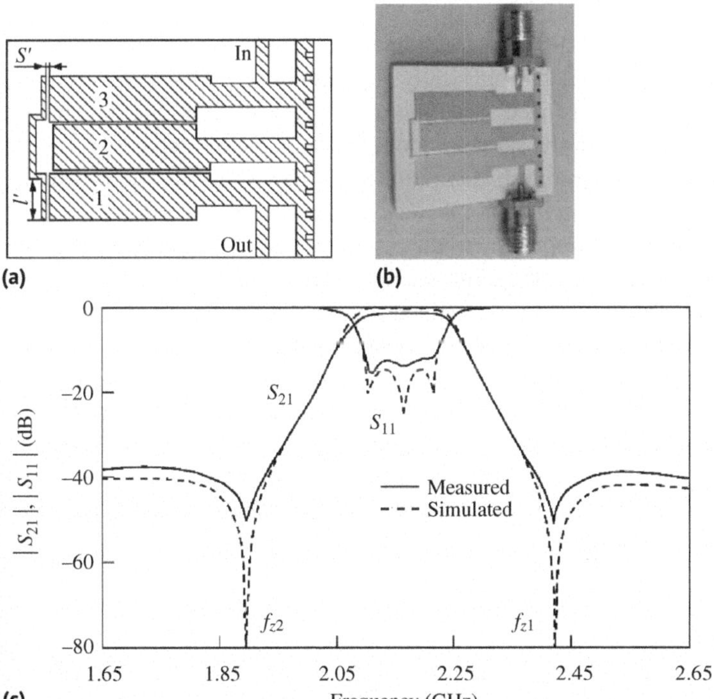

(a) (b) (c)

Figure 7.11 Microstrip asymmetric triplet BPF with all mixed couplings. (a) Topology: $S' = 0.05$ mm; $l' = 2$ mm. (b) Photograph. (c) Frequency responses. *Source:* [27] / with permission of Springer Nature.

the electric component $|K_{e13}| = K_{m13} = 0.015$. The electrical component of K_{e13} is provided by a thin strip connecting resonators 1 and 3 through the capacitive gaps S'. The required condition $|K_{e13}| = K_{m13} = 0$ is provided at $S' = 0.05$ mm and $l' = 2$ mm. The special coordinate provides the required value of $Q_e = 18.94$ I/O taped line connection.

The BPF frequency responses are shown in Figure 7.11c. Measured data: BW = 140 MHz, insertion loss $L_{A0} = 1.1$ dB, return loss $L_R > 13$ dB. The TZs are located at frequencies $f_{z1} = 2419$ MHz and $f_{z2} = 1893$ MHz. The interval between them, $f_{z1} - f_{z2} = 526$ MHz, has decreased by 104 MHz compared to the initial interval of 630 MHz (Figure 7.11b). The selectivity of this filter 35 dB ($f_0 \pm 210$ MHz) can be expressed at the normalized frequency $j\Omega$: 35 dB ($\pm j3.0$). This is a fairly high selectivity indicator.

7.6 Comments on BPFs with Mixed Couplings

Following are comments that focus on what the authors consider essential aspects of the theory of mixed coupled BPFs.

1. In some works, the linear representation of the mixed coupling $m = m_0 - a\Omega$ has a positive slope, which is physically impossible. Changing the signs of m_0 and a simultaneously is necessary to make it physically possible. The mixed coupling equals zero at the same frequency $\Omega_z = m_0/a$ in both cases.
2. In some works, a different representation form $M = K/x - J/b$ of the mixed coupling was used, which combines J and K inverters. Note that J inverter refers to the resonant frequency of the resonator, and K inverter refers to the antiresonant frequency of the same resonator. Since the indicated frequencies differ, this representation form of mixed coupling is invalid.
3. A widespread idea is that a pair of two equidistant TZs on the σ-axis, being shifted into the complex plane $s = \sigma + j\Omega$, provides a linear phase and flat group delay in a triplet BPF with mixed cross-coupling. This idea is erroneous, as confirmed by positions Nos. 2 and 3 (Table 7.1). Adding another TZ located at a real frequency is necessary to provide a flat group delay with the indicated arrangement of two TZs. Four resonators are required to implement a filter with three TZs, not three.

References

1 J. L. Allen, Inhomogeneous coupled-line filters with large mode-velocity ratios, *IEEE Trans. Microw. Theory Tech.*, vol. 22, no. 12, pp. 1182–1186, 1974, https://doi.org/10.1109/TMTT.1974.1128461.

2 T. Sugiura, Analysis of distributed-lumped strip transmission lines, *IEEE Trans. Microw. Theory Tech.*, vol. 25, no. 8, pp. 656–661, 1977, https://doi.org/10.1109/TMTT.1977.1129181.

3 T. Ishizaki, M. Fujita, H. Kagata, T. Uwano, H. Miyake, A very small dielectric planar filter for portable telephones, in *1993 IEEE MTT-S International Microwave Symposium Digest*, 1993, pp. 177–180, https://doi.org/10.1109/MWSYM.1993.276916.

4 R. Levy, New cascaded trisections with resonant cross-couplings (CTR Sections) applied to the design of optimal filters, in *IEEE MTT-S International Microwave Symposium Digest*, 2004, vol. 2 pp. 447–450, https://doi.org/10.1109/mwsym.2004.1336007.

5 K. Ma, J.-G. Ma, K. S. Yeo, M. A. Do, A compact size coupling controllable filter with separate electric and magnetic coupling paths, *IEEE Trans. Microw. Theory Tech.*, vol. 54, no. 3, pp. 1113–1119, 2006, https://doi.org/10.1109/TMTT.2005.864118.

6 Q.-X. Chu, H. Wang, A compact open-loop filter with mixed electric and magnetic coupling, *IEEE Trans. Microw. Theory Tech.*, vol. 56, no. 2, pp. 431–439, 2008, https://doi.org/10.1109/TMTT.2007.914642.

7 H. Wang, Q.-X. Chu, An inline coaxial quasi-elliptic filter with controllable mixed electric and magnetic coupling, *IEEE Trans. Microw. Theory Tech.*, vol. 57, no. 3, pp. 667–673, 2009, https://doi.org/10.1109/TMTT.2009.2013290.

8 W. Shen, L.-S. Wu, X.-W. Sun, W.-Y. Yin, J.-F. Mao, Novel substrate integrated waveguide filters with mixed cross coupling (MCC), *IEEE Microw. Wirel. Components Lett.*, vol. 19, no. 11, pp. 701–703, 2009, https://doi.org/10.1109/LMWC.2009.2032007.

9 B. A. Belyaev, V. V. Tyurnev, Frequency-dependent coupling coefficients of microstrip resonators, *Electron. Eng. Ser. Microw. Eng.*, no. 4, pp. 23–28, 1992.

10 S. Amari, J. Bornemann, Using frequency-dependent coupling to generate finite attenuation poles in direct-coupled resonator bandpass filters, *IEEE Microw. Guid. Wave Lett.*, vol. 9, no. 10, pp. 404–406, 1999, https://doi.org/10.1109/75.798030.

11 S. Amari, M. Bekheit, F. Seyfert, Notes on bandpass filters whose inter-resonator coupling coefficients are linear functions of frequency, in *2008 IEEE MTT-S International Microwave Symposium Digest*, 2008, pp. 1207–1210, https://doi.org/10.1109/MWSYM.2008.4633275.

12 L. Szydlowski, A. Lamecki, M. Mrozowski, Coupled-resonator filters with frequency-dependent couplings: coupling matrix synthesis, *IEEE Microw. Wirel. Components Lett.*, vol. 22, no. 6, pp. 312–314, 2012, https://doi.org/10.1109/LMWC.2012.2197386.

13 L. Szydlowski, A. Lamecki, M. Mrozowski, Coupled-resonator waveguide filter in quadruplet topology with frequency-dependent coupling – a design based on coupling matrix, *IEEE Microw. Wirel. Components Lett.*, vol. 22, no. 11, pp. 553–555, 2012, https://doi.org/10.1109/LMWC.2012.2225604.

14 L. Szydlowski, N. Leszczynska, A. Lamecki, M. Mrozowski, A substrate integrated waveguide (SIW) bandpass filter in a box configuration with frequency-dependent coupling, *IEEE Microw. Wirel. Components Lett.*, vol. 22, no. 11, pp. 556–558, 2012, https://doi.org/10.1109/LMWC.2012.2221690.

15 L. Szydlowski, A. Jedrzejewski, M. Mrozowski, A trisection filter design with negative slope of frequency-dependent crosscoupling implemented in substrate integrated waveguide (SIW), *IEEE Microw. Wirel. Components Lett.*, vol. 23, no. 9, pp. 456–458, 2013, https://doi.org/10.1109/LMWC.2013.2272611.

16 L. Szydlowski, N. Leszczynska, M. Mrozowski, Generalized Chebyshev bandpass filters with frequency-dependent couplings based on stubs, *IEEE Trans. Microw. Theory Tech.*, vol. 61, no. 10, pp. 3601–3612, 2013, https://doi.org/10.1109/TMTT.2013.2279777.

17 S. Tamiazzo, G. Macchiarella, Synthesis of cross-coupled filters with frequency-dependent couplings, *IEEE Trans. Microw. Theory Tech.*, vol. 65, no. 3, pp. 775–782, 2017, https://doi.org/10.1109/TMTT.2016.2633258.

18 M. Y. Sandhu, A. Lamecki, R. Gómez-García, M. Mrozowski, Compact quasi-elliptic-type inline waveguide bandpass filters with nonlinear frequency-variant couplings, *IEEE Trans. Microw. Theory Tech.*, vol. 71, no. 11, pp. 4933–4946, 2023, https://doi.org/10.1109/TMTT.2023.3269518.

19 A. Zakharov, Transmission zeros of trisection and quadruplet bandpass filters with mixed cross coupling, *IEEE Trans. Microw. Theory Tech.*, vol. 69, no. 1, pp. 89–100, 2021, https://doi.org/10.1109/TMTT.2020.3034663.

20 A. V. Zakharov, S. Litvintsev, Lumped-distributed resonators providing multiple transmission zeros in bandpass filters with simple and mixed couplings, *IEEE Trans. Circuits Syst. I Regul. Pap.*, vol. 71, no. 8, pp. 3502–3513, 2024, https://doi.org/10.1109/TCSI.2024.3375961.

21 E. A. Guillemin, *Synthesis of Passive Networks: Theory and Methods Appropriate to the Realization and Approximation Problems*. New York: Wiley, 1957.

22 J.-S. Hong, *Microstrip Filters for RF/Microwave Applications*, 2nd ed. New Jersey: Wiley, 2011, https://doi.org/10.1002/9780470937297.

23 G. A. Korn, T. M. Korn, *Mathematical Handbook for Scientists and Engineers: Definitions, Theorems, and Formulas for Reference and Review*. New York, NY: McGraw-Hill, 1961.

24 S. Seshu, N. Balabanian, *Linear Network Analysis*. Wiley, 1959.

25 A. Zakharov, S. Rozenko, M. Ilchenko, Two types of trisection bandpass filters with mixed cross-coupling, *IEEE Microw. Wirel. Components Lett.*, vol. 28, no. 7, pp. 585–587, 2018, https://doi.org/10.1109/LMWC.2018.2837905.

26 A. Zakharov, M. Ilchenko, Coupling coefficients between resonators in stripline combline and pseudocombline bandpass filters, *IEEE Trans. Microw. Theory Tech.*, vol. 68, no. 7, pp. 2679–2690, 2020, https://doi.org/10.1109/TMTT.2020.2988866.

27 S. Litvintsev, A. Zakharov, Inverse problem of triplet BPF with all mixed couplings, *Radioelectron. Commun. Syst.*, vol. 67, no. 4, pp. 196–211, 2024, https://doi.org/10.3103/S0735272724050017.

28 S. Litvintsev, A. Zakharov, M. Ilchenko, Trisection bandpass filters with all mixed couplings, *IEEE Microw. Wirel. Components Lett.*, vol. 29, no. 9, pp. 592–594, Sep. 2019, https://doi.org/10.1109/LMWC.2019.2929650.

8

Mixed Coupled High-order Inline and Cascaded Filters

The transmission zero (TZ) number can be increased by increasing the order of inline bandpass filter (BPF) with mixed couplings between adjacent resonators. The TZ number of such a filter with the number of resonators N can be $(N + 1)$ [1]. Each mixed coupling generates one TZ. In such a filter, the number of mixed couplings is $(N + 1)$, which leads to the number of TZs equal to $(N + 1)$. They consist of $(N - 1)$ couplings between resonators and two couplings between end resonators and loads. Mixed coupled inline BPFs with the specified number of TZs are considered in Refs. [2–4]. The disadvantage of these filters is that their TZs are located only at real frequencies.

The TZ placement on the complex plane is possible in more complex mixed coupled BPFs that use cross-couplings. Increasing the TZ number in such BPFs is possible by cascading lower-order filters [5–9]. As in filters with all simple couplings, such filters allow the TZs to be adjusted independently. Two design methods related to these two varieties of mixed coupled BPFs are discussed as follows.

8.1 Design Method of BPF with Mixed Couplings Between Adjacent Resonators

Let us look at some features of mixed coupling. For ease of presentation, we will write down formula (7.1) once again, which defines J parameter of admittance inverter (Figure 7.1a) for mixed coupling

$$J = J_m + J_e = 1/\omega_0 L - \omega_0 C, \tag{8.1}$$

where $J_m = 1/\omega_0 L$ is the magnetic component of J inverter, $J_e = -\omega_0 C$ is the electric component, and ω_0 is the central frequency of the BPF passband. The J values are determined by classical expressions (3.6). The task is to extract the magnetic J_m and electric J_e components from the obtained value of J.

The resonant frequency of the LC chain of the inverter (Figure 7.1a)

$$\omega_z = 1/\sqrt{LC} \tag{8.2}$$

corresponds to the TZ of BPF since the signal does not pass through the inverter at this frequency.

Since $L = 1/\omega_0 J_m$ and $C = |J_e|/\omega_0$, then $LC = |J_e|/J_m\omega_0^2$. Substituting this quantity into (8.2) leads to the equality

$$\omega_z = \omega_0\sqrt{J_m/|J_e|}. \tag{8.3}$$

Let us create a system of equations (8.1) and (8.3), and solve it concerning the components J_m and J_e of the J inverter. As a result, we obtain the required division of J into magnetic and electric components:

$$J_m = \frac{J}{1-(\omega_0/\omega_z)^2}, \tag{8.4a}$$

$$J_e = \frac{J}{1-(\omega_z/\omega_0)^2}. \tag{8.4b}$$

We have $J > 0$ in (8.4) if $\omega_z > \omega_0$. In this case, the magnetic component dominates in the mixed coupling. We have $J < 0$ in (8.4) if $\omega_z < \omega_0$. In this case, the electric component dominates in the mixed coupling.

The design method still uses classical expressions (3.6). However, in this case, they determine the absolute values of $\left|J_{i,i+1}\right|$, $|J_{01}|$, and $\left|J_{n,n+1}\right|$. The usual values of J parameters, considering the signs "+" or "–," are determined by the TZ position relative to ω_0. The proposed design method of BPF with mixed couplings between adjacent resonators always provides physically realizable parameters.

For example, consider a fifth-order inline BPF with mixed couplings between adjacent resonators. The filter circuit with admittance inverters is shown in Figure 8.1a. The transverse branches of the inverters are not shown. The quarter-wave resonators of the filter are tuned to the frequency of $f_0 = 1$ GHz and have a characteristic impedance of $Z_0 = 10\ \Omega$.

Let us set the parameters of this filter as follows: $f_0 = 1$ GHz, BW = 50 MHz (FBW = 0.05), $L_{Ar} = 0.1$ dB. We require that the filter has six TZs generated by mixed couplings: $f_{z1} = 0.7$ GHz, $f_{z2} = 0.8$ GHz, $f_{z3} = 0.9$ GHz, $f_{z4} = 1.1$ GHz, $f_{z5} = 1.2$ GHz, $f_{z6} = 1.3$ GHz.

The corresponding g parameters for this case are written out from Table 2.6: $g_0 = 1$, $g_1 = g_5 = 1.1146$, $g_2 = g_4 = 1.3712$, $g_3 = 1.9750$. Using formulas (3.6), we determine the moduli of J parameters of the inverters used in BPF:

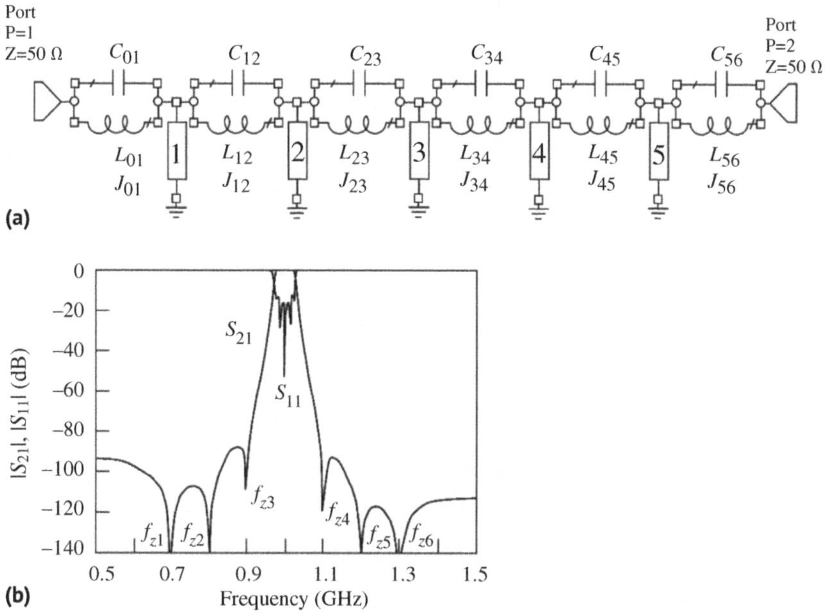

Figure 8.1 Inline fifth-order BPF with mixed couplings between adjacent resonators. (a) Schematic with J inverters. (b) Simulated frequency responses.

$$|J_{01}| = 0.008394\ \Omega^{-1}\ (f_{z3} = 0.9\ \text{GHz}), \quad |J_{12}| = 0.003176\ \Omega^{-1}\ (f_{z2} = 0.8\ \text{GHz}),$$
$$|J_{23}| = 0.002386\ \Omega^{-1}\ (f_{z1} = 0.7\ \text{GHz}), \quad |J_{34}| = 0.002386\ \Omega^{-1}\ (f_{z6} = 1.3\ \text{GHz}),$$
$$|J_{45}| = 0.003176\ \Omega^{-1}\ (f_{z5} = 1.2\ \text{GHz}), \quad |J_{56}| = 0.008394\ \Omega^{-1}\ (f_{z4} = 1.1\ \text{GHz}). \tag{8.5}$$

The TZs generated by each inverter are shown in parentheses in expressions (8.5). Let the inverter with J_{01} generates left-handed TZ at the frequency of $f_{z3} = 0.9$ GHz. Then the value of J_{01} should be negative $J_{01} = -0.008394\ \Omega^{-1}$. Substituting this value and $f_{z5} = 0.9$ GHz into (8.4), we obtain components $J_{01m} = 0.03578\ \Omega^{-1}$, and $J_{01e} = -0.04418\ \Omega^{-1}$. Using the obtained values and expression (8.1), we get the inductance and capacitance values of the inverter for mixed coupling: $L_{01} = 4.45$ nH, $C_{01} = 7.03$ pF.

The calculated parameters of the J_{01} inverter are contained in the second row of Table 8.1. This table contains the parameters of the remaining J inverters used in the filter under consideration. All elements of this filter are defined, and its simulated frequency responses are shown in Figure 8.1b. All TZs are located at the prescribed frequencies.

Table 8.1 *J* inverter parameters of BPF in Figure 8.1a.

J_{ij}, Ω^{-1}	f_z, GHz	J_{ijm}, Ω^{-1}	J_{ije}, Ω^{-1}	L_{ij}, nH	C_{ij}, pF
$J_{01} = -0.008394$	0.9	$J_{01m} = 0.03578$	$J_{01e} = -0.04418$	$L_{01} = 4.45$	$C_{01} = 7.03$
$J_{12} = -0.003176$	0.8	$J_{12m} = 0.005646$	$J_{12e} = -0.008822$	$L_{12} = 28.19$	$C_{12} = 1.40$
$J_{23} = -0.002386$	0.7	$J_{23m} = 0.002292$	$J_{23e} = -0.004678$	$L_{23} = 69.44$	$C_{23} = 0.744$
$J_{34} = 0.002386$	1.3	$J_{34m} = 0.005844$	$J_{34e} = -0.003458$	$L_{34} = 27.23$	$C_{34} = 0.55$
$J_{45} = 0.003176$	1.2	$J_{45m} = 0.01039$	$J_{45e} = -0.007218$	$L_{45} = 15.32$	$C_{45} = 1.15$
$J_{56} = 0.008394$	1.1	$J_{56m} = 0.04836$	$J_{56e} = -0.03997$	$L_{56} = 3.29$	$C_{56} = 6.36$

8.2 Cascaded Two Triplet BPFs with Mixed Cross-couplings

The basis of the proposed design method is the feasible TZ locations (Tables 7.1 and 7.3) and exact solutions of the inverse problem for triplet BPFs (7.22) and quadruplet BPFs (7.32). The proposed method consists of three stages.

Stage 1: Using initial data and classical formulas (3.8), we determine the coupling coefficients between adjacent resonators $K_{i,i+1}$ and the external quality factor of the end resonators Q_{e1}, Q_{eN} of the Chebyshev filter, which corresponds to a given level of ripple L_{Ar} (return loss L_R) in the passband.

Stage 2: We set the required TZ location, and using solutions to the inverse problem (7.22) and (7.32), we determine mixed cross-couplings of cascaded triplet and quadruplet BPFs.

Stage 3: Optimization of the coupling coefficients between adjacent resonators $K_{i,i+1}$ and the external quality factor of the end resonators Q_{e1}, Q_{eN} to ensure a given (initial) level of ripple (return loss) and bandwidth.

It is well known [10] that introducing TZs into a Chebyshev filter affects the level of ripple L_{Ar} in the passband. To restore the previous ripple level, it is necessary to carry out optimization [11, 12], which is typical for most design methods. In this method, optimization is carried out using a linear simulator. The variable values are the coupling coefficients and the external quality factor of the end resonators.

Consider using the proposed design method for various triplet and quadruplet BPFs cascading options.

A cascade connection of two triplet BPFs is called a cascade triplet (CT) filter for short [13]. Figure 8.2 shows the coupling structure of a CT filter with mixed cross-couplings, which contains six resonators. This filter schema combines the triplet BPFs shown in Figure 7.2b. It contains quarter-wave resonators ($f_0 = 1$ GHz, $Z_0 = 10\ \Omega$) and J inverters in quarter-wave sections, which are indicated by small rectangles. A J inverter with an LC chain carries out mixed cross-couplings K_{13} and K_{46}; the transverse branches of the inverter are not shown. The external quality factor Q_e of the end resonators is ensured by choosing the coordinate θ' for connecting loads to them. As in the previous section, the coordinate θ' is measured from the open end of the quarter-wave resonator.

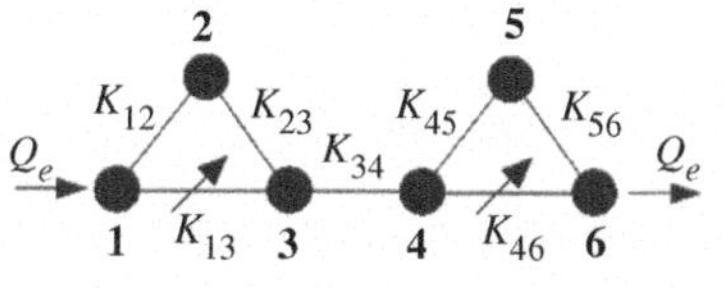

Figure 8.2 Coupling structure CT filter with mixed cross-couplings.

Let it be required to implement a CT BPF design with ripple $L_{Ar} = 0.1$ dB ($L_R = 16.426$ dB), FBW = 0.05, and $f_0 = 1$ GHz, which has two pairs of equidistant TZs: $S_{1,2} = \pm j5$ ($S'_{1,2} = \pm 5$) and $S_{3,4} = \pm j2.5$ ($S'_{3,4} = \pm 2.5$) shown in Figure 8.3a. We use position No. 8 from Table 7.1 in each triplet BPF with resonant cross-coupling $m_{13} = m_{46} = 0$.

At Stage 1, we write the g parameters of the prototype with $N = 6$ and equal-ripple $L_{Ar} =$ 0.1 dB (Table 2.6): $g_0 = 1$, $g_1 = 1.1681$, $g_2 = 1.4039$, $g_3 = 2.0562$, $g_4 = 1.5170$, $g_5 = 1.9029$, $g_6 = 0.8618$, $g_7 = 1.3554$. Using (3.8), we determine the main coupling coefficients $K_{12} = K_{56} = 0.03904$ ($m_{12} = m_{56} = 0.7808$), $K_{23} = K_{45} = 0.02943$ ($m_{23} = m_{45} = 0.5886$), $K_{34} = 0.02831$ ($m_{34} = 0.5662$), and the external quality factor $Q_{e1} = Q_{e6} = Q_e = 23.362$.

Based on (3.11c), we determine the characteristic impedances of quarter-wave sections used as J inverters: $Z_{12} = Z_{56} = 326.05\ \Omega$; $Z_{23} = Z_{45} = 432.71\ \Omega$; $Z_{34} = 449.64\ \Omega$. The coordinate for connecting the loads to the end resonators $\theta' = 65.796°$ provides an external quality factor $Q_e = 23.362$.

At Stage 2, we will use position No. 8 from Table 7.1 with two equidistant TZs. This position requires the use of asymmetric triplet BPFs with opposite signs of

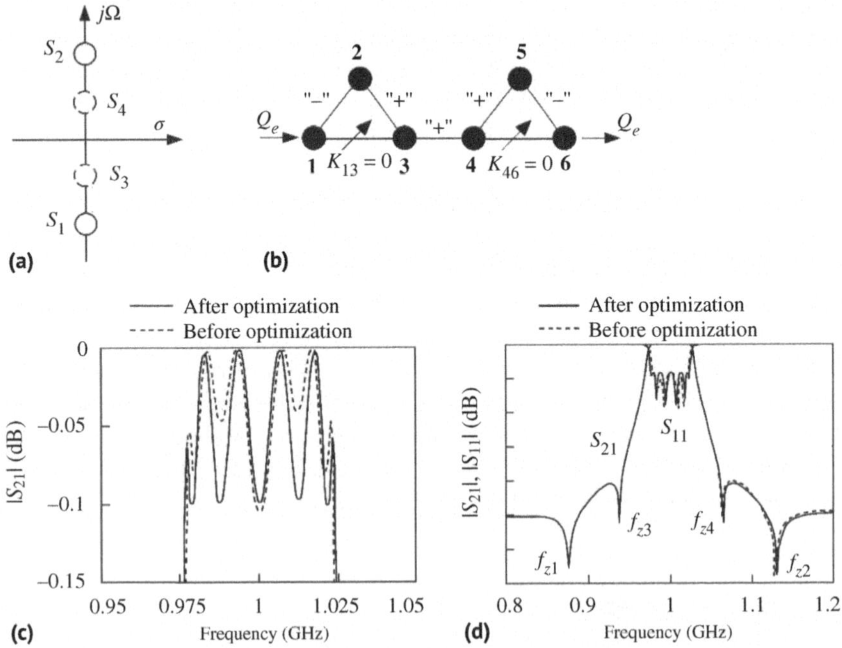

Figure 8.3 CT filter with mixed cross-couplings and two equidistant pairs of TZs on the $j\Omega$-axis. (a) TZ placement. (b) Coupling structure. (c) Ripple level in bandpass. (d) Simulated frequency responses.

the main couplings, which is reflected in the coupling structure of the CT filter in Figure 8.3b.

Let the first triplet BPF, formed by resonators 1–2–3, implement TZs: $S_{1,2} = \pm j5$, which are shown in Figure 8.3a with solid circles. Substituting these TZs values into formulas (7.22), we determine $a_{13} = 0.01838$, $m_{013} = 0$. The mixed cross-coupling m_{13} in the linear form (7.6) is defined as $m_{13} = -0.01838\ \Omega$. Using expressions (7.8) and (7.9), we transit to a mixed coupling in the form (7.2): $K_{m13} = 0.01838$; $K_{e13} = -0.01838$. These values and (7.3) lead to the inductance value $L_{13} = 110.25$ nH and capacitance $C_{13} = 0.23$ pF of the used J_{13} inverter.

The second triplet BPF, formed by resonators 4–5–6, implements TZs: $S_{3,4} = \pm j2.5$, which are depicted in Figure 8.3a by dotted circles. Substituting the indicated values in (7.22) gives $a_{46} = 0.07353$, $m_{046} = 0$. The mixed cross-coupling m_{46} in form (7.6) is defined as $m_{46} = -0.07353\ \Omega$. Using expressions (7.8) and (7.9), we transit to a mixed coupling in the form (7.2): $K_{m46} = 0.07353$, $K_{e46} = -0.07353$. The resulting mixed coupling components and formulas (7.3) determine the inductance $L_{46} = 27.56$ nH and capacitance $C_{46} = 0.919$ pF of the used J_{46} inverter.

All parameters of the considered CT filter (Figure 8.3b) are defined. The second column of Table 8.2 contains the values $K_{i,i+1}$ and Q_e. The dotted lines in Figure 8.3c, d depict the frequency responses of this CT filter. The ripple level (Figure 8.3c) was 0.105 dB, slightly exceeding the specified value of 0.1 dB. The pulsations are not equal ripples.

At Stage 3, optimizing the $K_{i,i+1}$ and Q_e values will ensure a pulsation level of 0.1 dB and their equal ripple. The optimized values of the coupling coefficients $K_{i,i+1}$ are extracted from the optimized characteristic impedances of the quarter-wave coupling segments according to (3.11c), as was in Ref. [13], [Chapter 8.3.4]. The optimization results are presented in the third column of Table 8.2. The solid lines in Figure 8.3c, d show the frequency responses of the CT filter after optimization. There are practically no differences in the insertion loss curves (Figure 8.3d) and there are slight differences in return loss.

Table 8.2 Values of $K_{i,i+1}$ and Q_e of CT filter in Figure 8.3b with TZs: $S_{1,2} = \pm j5$ and $S_{3,4} = \pm j2.5$ before and after optimization.

$K_{i,i+1}, Q_e$	Before optimization	After optimization	Difference, %
$K_{12} = K_{56}$	-0.03904	-0.03881	0.59
$K_{23} = K_{45}$	0.02943	0.03003	2.04
K_{34}	0.01940	0.01947	0.36
Q_e	23.362	24.50	4.87

If we do not require equal-ripple pulsations, since this does not have much practical significance, but set only their maximum level $L_{Ar} = 0.1$ dB, then the optimization process will be significantly simplified. It will only come down to a slight change in the external quality factor Q_e of the end resonators.

Generally, a CT filter with two mixed cross-couplings can have a significant variety of frequency responses, representing different combinations of the 10 TZs placement positions from Table 7.1.

8.3 Cascaded Quadruplet and Triplet BPFs with Mixed Cross-couplings

Like the previous one, the filter in question is a sixth-order BPF. Its coupling structure is shown in Figure 8.4a. Let us design this BPF with the same primary parameters: $L_{Ar} = 0.1$ dB, FBW = 0.05, and $f_0 = 1$ GHz. Its g parameters, main coupling coefficients $K_{i,i+1}$ (characteristic impedances $Z_{i,i+1}$), and external quality factor of end resonators Q_e (load connection coordinate θ') will be the same as the previous filter. This filter must combine No. 1 (Table 7.3) of quadruplet BPF and No. 8 (Table 7.1) of triplet BPF and has five TZs at real frequencies. The quadruplet BPF provides TZs $S_1 = -j6$ $(S_1' = -6)$, $S_2 = j2$ $(S_2' = 2)$. The value of S_3 has not yet been determined. Its TZs are depicted in Figure 8.4b with solid circles. The triplet BPF has TZs $S_{4,5} = \pm j4$ $(S_{45}' = \pm 4)$, which are depicted in Figure 8.4b with dotted circles.

The quadruplet BPF is formed by resonators 1–2–3–4. Using equality (7.32c), we further determine its TZ $S_3' = -2.913$ $(S_3 = -j2.913)$. Then from (7.32a), we find the mixed cross-coupling slope parameter a_{14} =0.007443, and from (7.32b), we determine the value $m_{014} = -0.05147$. The linear representation of the mixed cross-coupling K_{14} is defined as $m_{14} = -0.05147 - 0.007445\ \Omega$. Using (7.8) and (7.9), we determine the components of this mixed coupling: $K_{m14} = 0.00627$, $K_{e14} = -0.008844$. These values and formulas (7.3) lead to the inductance value $L_{14} = 323.19$ nH and capacitance value $C_{14} = 0.1106$ pF of J_{14} inverter.

The triplet BPF is formed by 4–5–6 resonators. The mixed cross-coupling K_{46} must be defined in such a way as to provide equidistant TZs $S_{4,5} = \pm j4$ $(S_{4,5}' = \pm 4)$. In this case, the coupling coefficients K_{45} and K_{56} should have opposite signs; we will assign a "–" sign to the coefficient K_{45}. To determine the cross-coupling m_{46}, we use equalities (7.22). As a result, we obtain $a_{46} = 0.02872$, $m_{046} = 0$. The linear representation form of the mixed cross-coupling m_{46} is defined by $m_{46} = -0.02872\ \Omega$. The components of this mixed coupling, $K_{m46} = 0.02872$, $K_{e46} = 0.02872$, are established using (7.8) and (7.9). Using these values and formulas (7.3), we determine the inductance $L_{14} = 70.56$ nH and the capacitance $C_{14} = 0.359$ pF of the J_{46} inverter.

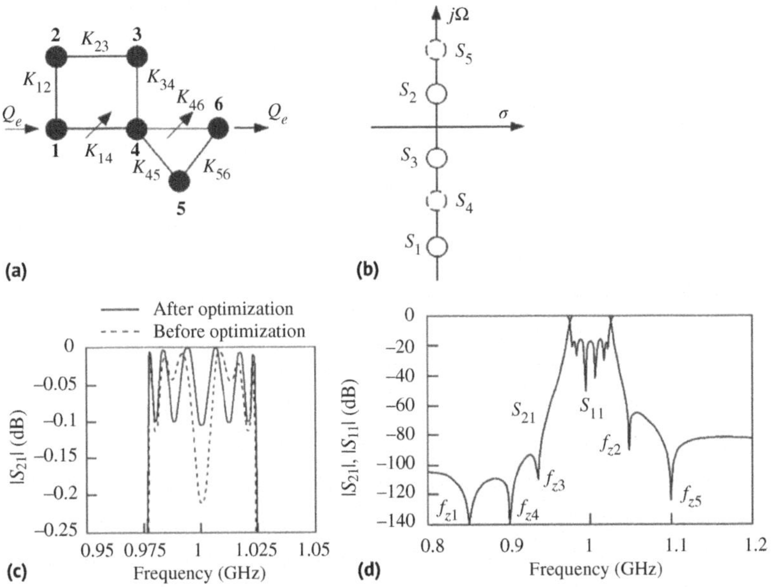

Figure 8.4 Cascaded quadruplet and triplet BPFs with mixed cross-couplings. (a) Coupling structure. (b) TZ placement. (c) Pulsation in BW, (d) Frequency responses after optimization.

All coupling elements of the filter in question are defined. In Figure 8.4c, the dotted line shows the filter ripple in the passband. They are not equal, and their level slightly exceeds 0.2 dB. As a result of optimization, the main coupling coefficients changed: $K_{12} = 0.03816$ (2.3%), $K_{23} = 0.0303$ (2.96%), $K_{34} = 0.02959$, $K_{45} = -0.0346$ (17.7%), $K_{56} = 0.02563$ (52.3%), and the external quality factor of the extreme resonators $Q_u = 22.49$ (3.9%). The difference from the original values is indicated in parentheses. The pulsations became equal-ripple, and their level amounted to $L_{Ar} = 0.1$ dB, which is reflected in Figure 8.4c by a solid line. Figure 8.4d depicts the simulated frequency responses of the considered BPF after optimization.

8.4 Cascaded Two Quadruplet Bandpass Filters with Mixed Cross-couplings

A cascade connection of two quadruplet BPFs is called a cascaded quadruplet (CQ) filter for short [13]. Figure 8.5 shows the structure of connections of a CQ filter with mixed cross-couplings, which contains seven resonators. Two quadruplet BPFs can have eight possible arrangements of three TZs, as shown in Table 7.3. Let us consider two typical cascading options of the quadruplet BPF.

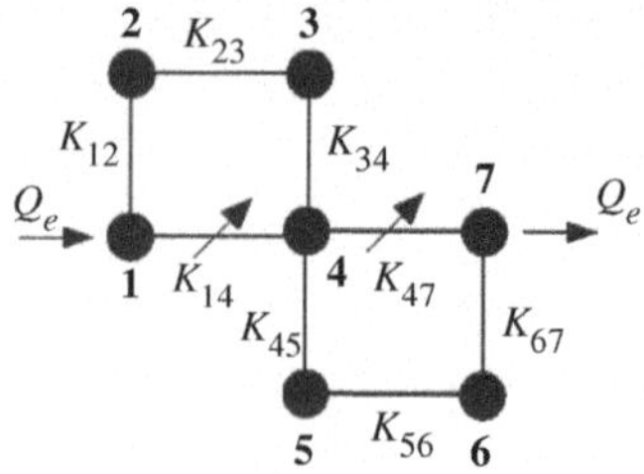

Figure 8.5 Coupling structure of CQ filter with mixed cross-couplings.

Option 1: The first option involves implementing a mixed coupled CQ filter design with ripple $L_{Ar} = 0.1$ dB, FBW = 0.05, and $f_0 = 1$ GHz, with three pairs of equidistant TZs at real frequencies (Figure 8.6a): $S_{1,2} = \pm j2$ $(S'_{1,2} = \pm 2)$, $S_{3,4} = \pm j6$ $(S'_{3,4} = \pm 6)$. The values of $S_{5,6}$ have not yet been determined. This filter combines No. 1 and No. 2 from Table 7.3.

Let us write g parameters (Table 2.6) of the prototype with $N = 7$ and $L_{Ar} = 0.1$ dB: $g_0 = 1$, $g_1 = g_7 = 1.1811$, $g_2 = g_6 = 1.4228$, $g_3 = g_5 = 2.0966$, $g_4 = 1.5733$. Using (3.8), we determine the main coupling coefficients $K_{12} = K_{67} = 0.03857$ $(m_{12} = m_{67} = 0.7714)$, $K_{23} = K_{56} = 0.02895$ $(m_{23} = m_{56} = 0.5790)$, $K_{34} = K_{45} = 0.02753$ $(m_{34} = m_{45} = 0.5506)$ and the external quality factor $Q_e = 23.622$. The main coupling coefficients are realized by transmission line (TL) sections with characteristic impedances: $Z_{12} = Z_{67} = 330.11\ \Omega$, $Z_{23} = Z_{56} = 439.81\ \Omega$, and $Z_{34} = Z_{45} = 449.75\ \Omega$, which are established using (3.11c). The load connection coordinate $\theta' = 67.03°$ provides the external quality factor $Q_e = 23.622$. Using expression (7.32c), we define two TZs $S_{5,6} = \pm j2.9162$ $(S'_{5,6} = \pm 2.9162)$.

A quadruplet BPF containing resonators 1–2–3–4 provides TZs with odd numbers $S_1 = j2$ $(S'_1 = 2)$, $S_3 = -j6$ $(S'_3 = -6)$, $S_5 = -j2.9162$ $(S'_5 = -2.9162)$, which are indicated in Figure 8.6a by solid circles. From (7.32a), we find the slope parameter of the mixed cross-coupling $a_{14} = 0.007028$, and from (7.32b), we determine the value $m_{014} = -0.0486$. The mixed cross-coupling m_{14} in the linear form (7.6) is defined as $m_{14} = -0.0486 - 0.007028\ \Omega$. Using expressions (7.8) and (7.9), we transit to a mixed coupling in the form (7.2): $K_{m14} = 0.005917$, $K_{e14} = -0.00835$. These values and formulas (7.3) lead to the inductance value $L_{14} = 342.47$ nH and the capacitance value $C_{14} = 0.1044$ pF of the used J_{14} inverter.

Quadruplet BPF containing 4–5–6–7 resonators provides TZs with even numbers $S_2 = -j2$ $(S'_2 = 2)$, $S_4 = j6$ $(S'_4 = 6)$, $S_6 = j2.9162$ $(S'_6 = 2.9162)$, which are indicated in Figure 8.6a by dotted circles. For a given TZ location, the main coupling coefficient m_{56} must be negative $m_{56} < 0$ (Table 7.3). Using the aforementioned calculation sequence, we obtain the mixed cross-coupling m_{47} in the linear form (7.6): $m_{47} = 0.0486 - 0.007028\ \Omega$. Using expressions (7.8) and (7.9), we transit to a mixed coupling in the form (7.2): $K_{m47} = 0.00835$, $K_{e47} = -0.005917$. Using (7.3), we determine the inductance and capacitance of J_{47} inverter: $L_{47} = 242.68$ nH, $C_{47} = 0.0740$ pF.

The pulsations of this filter in the passband are not equal-ripple. After optimization, they become equal-ripple with a level of $L_{Ar} = 0.1$ dB. The coupling coefficients after optimization take the following values: $K_{12} = K_{67} = 0.03789$ (1.79%), $K_{23} = K_{56} = 0.03105$ (7.25%), $K_{34} = K_{45} = 0.02726$ (0.99%). The difference from the original values is indicated in parentheses. Figure 8.6b shows the simulated frequency responses of this filter after optimization. The position of the TZs corresponds to the prescribed values.

Option 2: The second option involves implementing a mixed coupled CQ filter design with four quadrantal symmetry complex TZs and two equidistant TZs at real frequencies (Figure 8.7a). As before, we will assume $L_{Ar} = 0.1$ dB, FBW = 0.05, and $f_0 = 1$ GHz. Therefore, the coupling coefficients between adjacent resonators and the value of the external quality factor of the end resonators Q_e are the same as for the previous filter. Let us set the following TZs: $S_{1,3} = \pm 1.5 - j0.3$ $(S'_{1,3} = -0.3 \pm j1.5)$, $S_{2,4} = \pm 1.5 + j0.3$ $(S'_{2,4} = 0.3 \pm j1.5)$. The values of $S_{5,6}$ have not yet been determined. This CQ filter combines No. 7 and No. 8 from Table 7.3, which provide the flat group delay of each quadruplet BPF. Using the expression (7.32c), we define two TZs $S_{5,6} = \pm j4.459$ $(S'_{5,6} = \pm 4.459)$.

A quadruplet BPF containing resonators 1–2–3–4 provides TZs with odd numbers $S_{1,3} = \pm 1.5 - j0.3$ $(S'_{1,3} = -0.3 \pm j1.5)$, $S_5 = j4.459$ $(S'_5 = 4.459)$, which are indicated in Figure 8.7a by solid circles. From (7.32a), we find the slope parameter of the mixed cross-coupling $a_{14} = 0.02357$, and from (7.32b), we determine the value of $m_{014} = 0.09094$. The linear representation of the mixed cross-coupling m_{14} is determined as $m_{14} = 0.09094 - 0.02357$ Ω. Using expressions (7.8) and (7.9), we determine the components of this mixed coupling, $K_{m14} = 0.02595$ and

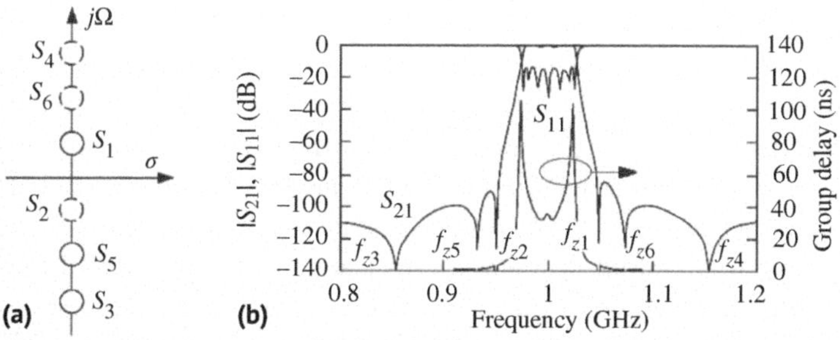

Figure 8.6 Mixed coupled CQ filter with three pairs of equidistant TZs at real frequencies. (a) Transmission zeros. (b) Simulated frequency responses.

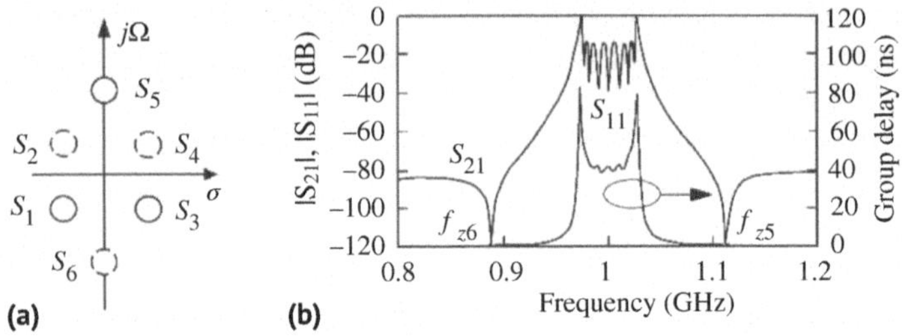

Figure 8.7 Mixed coupled CQ filter with four quadrantal symmetry complex zeros and two equidistant zeros at real frequencies. (a) Transmission zeros. (b) Simulated frequency responses.

$K_{e14} = -0.0214$. These values and formulas (7.3) lead to the inductance value $L_{14} = 78.09$ nH and capacitance $C_{14} = 0.2675$ pF of the J_{14} inverter.

Quadruplet BPF containing 4–5–6–7 resonators provides TZs with even numbers $S_{2,4} = \pm 1.5 + j0.3$ $(S'_{2,4} = 0.3 \pm j1.5)$, $S_6 = -j4.459$ $(S'_6 = -4.459)$, which are indicated in Figure 8.7a by dotted circles. For a given TZ location, the main coupling coefficient m_{56} must be negative $m_{56} < 0$ (Table 7.3). Using the aforementioned calculation sequence, we obtain a linear representation form of the mixed cross-coupling $m_{47} = -0.09094 - 0.02357\ \Omega$. Using (7.8) and (7.9), we determine the components of this mixed coupling: $K_{m47} = 0.0214$, $K_{e47} = -0.02595$. Using (7.3), we determine the inductance and capacitance of J_{47} inverter: $L_{47} = 94.69$ nH and $C_{47} = 0.3244$ pF.

The pulsations of this filter in the passband are not equal-ripple. After optimization, they become equal-ripple with a level of $L_{Ar} = 0.1$ dB. The coupling coefficients after optimization take the following values: $K_{12} = K_{67} = 0.03894$ (0.95%), $K_{23} = K_{56} = 0.02652$ (9.2%), $K_{34} = K_{45} = 0.02894$ (4.9%). The difference from the original values is indicated in parentheses. Figure 8.7b shows the simulated frequency responses of this filter after optimization.

In Ref [13, Fig. 9.26], two options for the TZ's location are considered, which lead to linearization of the phase and provide a flat group delay of the BPF. In the first most common case, two TZs are located equidistantly on the σ-axis (No. 1 in Table 7.1). In this case, BPF can provide a flat group delay in a frequency band of approximately 40–50% of the bandwidth. In the second case, four TZs are placed on the complex plane in quadrant symmetry. This leads to an expansion of the frequency band with flat group delay. At the same time, this is accompanied by a decrease in BPF selectivity. Due to the quadrant symmetry of the four TZs, the mixed coupled CQ filter in Figure 8.7 has a frequency band extended to 65% with a flat group delay at the level of 41 ns. This filter does not exhibit any deterioration in selectivity due to the two TZs located on the $j\Omega$-axis. The filter in Figure 8.7 is a mixed coupled quasi-elliptical BPF with flat group delay in an extended frequency band.

8.5 Cascaded Two Quadruplets and One Triplet BPFs with Mixed Cross-couplings

Figure 8.8a shows the coupling structure of the filter under consideration, which contains nine resonators and includes two quadruplet BPFs and one triplet BPF with mixed cross-couplings. Each quadruplet BPF can have 8 options for the location of three TZs (Table 7.3), and the triplet BPF can have 10 options for two TZs (Table 7.1).

Let us implement the design of the considered filter with ripple $L_{Ar} = 0.1$ dB, FBW = 0.05, and $f_0 = 1$ GHz, which has eight TZs shown in Figure 8.8b. The first six TZs are located as in the previous CQ filter (Figure 8.7a): $S_{1,3} = \pm 1.5 - j0.3$ $(S'_{1,3} = -0.3 \pm j1.5)$, $S_{2,4} = \pm 1.5 + j0.3$ $(S'_{2,4} = 0.3 \pm j1.5)$,

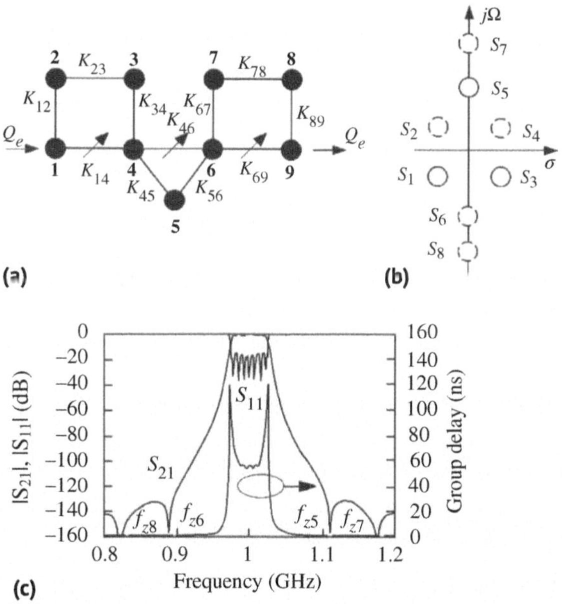

Figure 8.8 Cascaded two quadruplets and one triplet BPFs with mixed cross-couplings. (a) Coupling structure. (b) TZ placement. (c) Frequency responses after optimization.

$S_{5,6} = \pm j4.459$ $(S'_{5,6} = \pm 4.441)$. The remaining two TZs, located equidistantly at real frequencies $S_{7,8} = \pm j7$ $(S'_{7,8} = \pm 7)$, are implemented by a triplet BPF with resonators 4–5–6.

Let us write g parameters (Table 2.6) of the prototype with $N = 9$ and $L_{Ar} = 0.1$ dB: $g_1 = g_9 = 1.1956$, $g_2 = g_8 = 1.4425$, $g_3 = g_7 = 2.1345$, $g_4 = g_6 = 1.6167$, $g_5 = 2.2053$. Using (3.8), we determine the main coupling coefficients $K_{12} = K_{89} = 0.03808$ $(m_{12} = m_{89} = 0.7616)$, $K_{23} = K_{78} = 0.02849$ $(m_{23} = m_{78} = 0.5698)$, $K_{34} = K_{67} = 0.02692$ $(m_{34} = m_{67} = 0.5384)$, $K_{45} = K_{56} = 0.02648$ $(m_{45} = m_{56} = 0.5296)$, and the external quality factor $Q_e = 23.912$. The main coupling coefficients are realized by TL sections with characteristic impedances $Z_{12} = Z_{89} = 334.36\ \Omega$, $Z_{23} = Z_{78} = 446.91\ \Omega$, $Z_{34} = Z_{67} = 472.97\ \Omega$, $Z_{45} = Z_{56} = 480.83\ \Omega$, which are established using (3.11c). The load connection coordinate $\theta' = 66.09°$ provides the external quality factor $Q_e = 23.912$.

The first quadruplet BPF containing resonators 1–2–3–4 provides odd-numbered TZs $S_{1,3} = \pm 1.5 - j0.3$ $(S'_{1,3} = -0.3 \pm j1.5)$, $S_5 = j4.441$ $(S'_5 = 4.441)$, which are indicated in Figure 8.8b by solid circles. From (7.32a), we find the slope parameter of the mixed cross-coupling $a_{14} = 0.02307$, and from (7.32b), we determine the value $m_{014} = 0.08861$. The linear representation form of the mixed cross-coupling m_{14} is determined as $m_{14} = 0.08861 - 0.02307\ \Omega$. Using expressions (7.8) and (7.9), we determine the components of this mixed coupling: $K_{m14} = 0.02539$, $K_{e14} = -0.02096$. These values and formulas (7.3) lead to the inductance value $L_{14} = 79.81$ nH and capacitance $C_{14} = 0.2620$ of the J_{14} inverter.

The second quadruplet BPF containing 6–7–8–9 resonators provides TZs with even numbers $S_{2,4} = \pm 1.5 + j0.3$ $(S'_{2,4} = 0.3 \pm j1.5)$, $S_6 = -j4.441$ $(S'_6 = -4.441)$, which are indicated in Figure 8.8b by dotted circles. For a given TZ location, the main coupling coefficient m_{78} must be negative $m_{78} < 0$ (Table 7.3). Using the aforementioned calculation sequence, we obtain a linear representation of the mixed cross-coupling $m_{69} = -0.08861 - 0.02357\ \Omega$. Using (7.8) and (7.9), we determine the components of this mixed coupling: $K_{m69} = 0.02096$, $K_{e69} = -0.02539$. Using (7.3), we determine the inductance and capacitance of J_{69} inverter: $L_{69} = 96.68$ nH and $C_{69} = 0.3174$ pF.

The triplet BPF, formed by resonators 4–5–6, provides equidistant TZs at real frequencies $S_{7,8} = \pm j7$ $(S'_{7,8} = \pm 7)$, which are depicted in Figure 8.8b by dotted circles. The indicated arrangement of TZs corresponds to position No. 8 from Table 7.1 and requires using the main couplings K_{45} and K_{56} with opposite signs. Substituting these TZs values into formulas (7.22), we determine $a_{46} = 0.005724$, $m_{046} = 0$. The mixed cross-coupling m_{46} in the linear form is defined as $m_{46} = -0.005724\ \Omega$. Using expressions (7.8) and (7.9), we transit to a mixed coupling in the form (7.2): $K_{m46} = 0.005724$, $K_{e46} = -0.005724$. These values and formulas (7.3) lead to the inductance value $L_{46} = 354.02$ nH and the capacitance value $C_{46} = 0.07155$ pF of the used J_{46} inverter.

The pulsations of this filter in the passband are not equal-ripple. After optimization, they become equal-ripple with a level of $L_{Ar} = 0.1$ dB. The coupling coefficients after optimization take the following values: $K_{12} = 0.03784$ (0.6%), $K_{23} = 0.02530$ (12.6%), $K_{34} = 0.02668$ (0.9%), $K_{45} = -0.02598$ (1.9%), $K_{56} = 0.02598$ (1.9%), $K_{67} = 0.02666$ (1.0%), $K_{78} = -0.02472$ (15.2%), $K_{89} = 0.03730$ (2.1%). The difference from the original values is indicated in parentheses. Figure 8.8c shows the simulated frequency responses of this filter after optimization. The filter is characterized by a flat group delay, and its attenuation poles are located at prescribed frequencies.

8.6 Stripline Realization of CT Filter with Mixed Cross-couplings

Let us design a fully shielded stripline CT filter with two pairs of equidistant TZs at real frequencies (Figure 8.3d). However, the coupling structure presented in Figure 8.3a is unsuitable for implementation in stripline construction, since it contains asymmetric triplet BPFs. The main coupling coefficients K_{12} and K_{23} have opposite signs; the same applies to K_{45} and K_{56}.

To eliminate this drawback, we will use the results of Section 6.6 presented in Figure 8.9. The triplet BPFs shown in Figure 8.9 are equivalent, since they have the same frequency responses. Two asymmetric triplet BPFs on the right side of Figure 8.9 are equivalent to a symmetric triplet BPF on the left.

This equivalence is ensured by replacing the middle resonator of the reflection-type (left side) with a through-type resonator (right side). A double circle indicates a through-type resonator in Figure 8.9b. A half-wave resonator with voltages of different signs at opposite ends can be used as a through-type resonator. Using this pattern, we can modify a CT filter with asymmetric triplet BPFs (Figure 8.9a) into a CT filter with symmetric triplet BPFs (Figure 8.9b), in which resonators *2* and *5* are half-wave through-type resonators. The choice of the signs of the coupling coefficients indicated in Figure 8.9 will become evident upon further consideration of the stripline implementation of the CT filter.

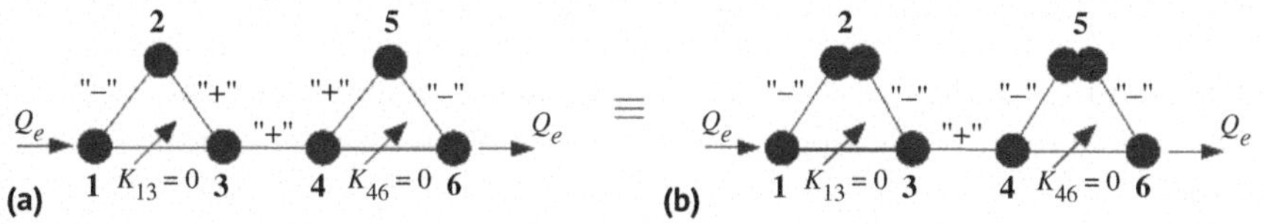

Figure 8.9 Modification of CT filter. (a) CT filter with asymmetric triplet BPFs. (b) Modified CT filter with symmetric triplet BPFs.

8.6.1 Modified CT Filter with Two Through-type Resonators

In the circuit of each triplet BPF (Figure 8.9b), replacing the $\lambda/4$ resonators 2 and 5 with half-wave through-type resonators is necessary. Since the susceptance slope parameter of a half-wave resonator is $b = \pi/2Z_0$ [13], which is twice as large as that of a quarter-wave resonator, its characteristic impedance is chosen to be equal to $Z_0 = 20\ \Omega$, which is also twice as large as that of a quarter-wave resonator.

Let it be required to design a modified CT filter with $f_0 = 1.765$ GHz, BW = 80 (FBW = 0.0453), $L_{Ar} = 0.1$ dB, which has two pairs of equidistant TZs: $f_{z1,2} = f_0 \pm 75$ MHz and $f_{z3,4} = f_0 \pm 205$ MHz.

As initial ones, we choose the parameters of the prototype with $N = 6$ and equal-ripple $L_{Ar} = 0.1$ dB (Table 2.6): $g_0 = 1$, $g_1 = 1.1681$, $g_2 = 1.4039$, $g_3 = 2.0562$, $g_4 = 1.5170$, $g_5 = 1.9029$, $g_6 = 0.8618$, $g_7 = 1.3554$. Using (3.8), we determine the coupling coefficients $|K_{12}| = |K_{56}| = 0.03537$ ($|m_{12}| = |m_{56}| = 0.7809$), $|K_{23}| = |K_{45}| = 0.02666$ ($|m_{23}| = |m_{45}| = 0.5886$), $K_{34} = 0.02565$ ($m_{34} = 0.5662$) and the external quality factor $Q_e = 25.78$. The main coupling coefficients are realized by sections of TLs with characteristic impedances: $|Z_{12}| = |Z_{56}| = 326.14\ \Omega$, $|Z_{23}| = |Z_{45}| = 432.63\ \Omega$, and $Z_{34} = 449.75\ \Omega$, which is established using (3.11c). The value of $Q_e = 25.78$ is provided by the load connection coordinate $\theta' = 67.03°$.

Let us set the TZ position on the $j\Omega$-axis: $S_{1,2} = \pm j1.838$, $S_{3,4} = \pm j4.86$. The first triplet BPF, formed by resonators 1–2–3, will implement TZs $S_{1,2}$, and the second triplet BPF, formed by resonators 4–5–6, will implement TZs $S_{3,4}$. Since the TZs are located equidistantly, it follows from (7.22b) $m_{013} = m_{46} = 0$. Substituting the values of $S_{1,2}$ into (7.22a), we determine the dispersion part of the mixed coupling of the first triplet BPF: $a_{13} = K_{m13} = |K_{e13}| = 0.1360$. Substituting the values of $S_{3,4}$ into (7.22a) gives $a_{46} = K_{m46} = |K_{e46}| = 0.01946$. The values of lumped inductances and capacitances of mixed cross-couplings are determined based on (7.3): $L_{13} = 8.442$ nH, $C_{13} = 0.963$ pF, and $L_{46} = 58.998$ nH, $C_{46} = 0.1378$ pF. All elements of the CT filter in question are defined.

In Figure 8.10a, the dotted line shows the simulated insertion loss of the filter before optimization. The pulsations are not equal-ripple, and the bandwidth is narrower. After optimization, the coupling coefficients took the following values: $K_{12} = -0.04375$, $K_{23} = -0.03374$, $K_{34} = 0.02895$, $K_{45} = -0.02889$, $K_{56} = -0.03979$, $K_{m13} = |K_{e13}| = 0.2130$, $K_{m46} = |K_{e46}| = 0.02371$. In Figure 8.10a, the solid line shows that after optimization, the ripples became equal-ripple with level $L_{Ar} = 0.1$ dB and the bandwidth expanded. The frequency responses of this filter are shown in Figure 8.10b. The ripple level is $L_{Ar} = 0.1$ dB, and TZs have the prescribed values.

8.6.2 Modified Stripline CT Filter

We use a coupling structure with through-type resonators for stripline implementation of mixed coupled CT filter (Figure 8.9b). The stripline construction is realized by connecting two substrates with topologically conductive patterns

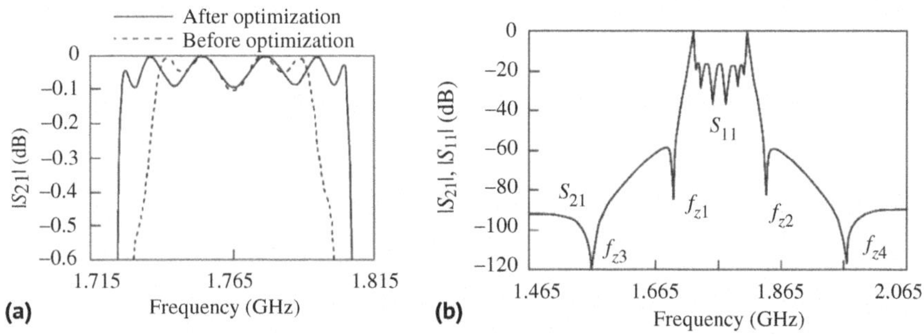

Figure 8.10 Simulated frequency responses of modified CT filter with mixed cross-couplings. (a) Pulsation in BW. (b) Frequency responses after optimization.

according to the "face-to-face" principle. Substrate parameters: $h = 1$ mm, $\varepsilon_r = 36$, $\tan\delta = 0.0003$. The filter's size is 18 mm × 15.5 mm × 2 mm.

The topology of the central conductors of the filter is shown in Figure 8.11a (position 1). The filter uses four stepped-impedance resonators (SIRs) and two through-type half-wave resonators with a resonance frequency of 1.765 GHz. The SIRs are short-circuited on a metallized end face. The metal strip located on the perimeter of the filter does not allow copper to enter the interior of the filter.

The greatest width of all SIRs is $w_1 = 2.4$ mm. The width of the strip of the hairpin resonator is $w_3 = 1.6$ mm, and the width of the entire hairpin resonator is $w_3' = 4.6$ mm. Conductive strips on the two ends of the filter, which are shown in Figure 8.11a (position 2), connect the inner and outer parts of the filter. The latter are contact pads. They are shown in the filter photograph in Figure 8.11b, where the filter is shown on the reverse side. The filter design is intended for surface mounting. The dielectric substrates were coated with copper by vacuum deposition. The thickness of the copper deposition is 6 μm.

The main couplings of the filter K_{12}, K_{23}, K_{45}, and K_{56} act between the open ends of the resonators. They are electrical and are assigned a minus sign (Figure 8.9b). The values of these couplings are realized using the corresponding gaps Δ_{12}, Δ_{23}, Δ_{45}, and Δ_{56}. The coupling coefficient K_{34} between resonators 3 and 4 is positive, and its value is provided by the gap Δ_{34}.

The first triplet BPF formed by resonators 1–2–3 provides two TZs: $S_{1,2} = \pm j1.838$. To do this, the mixed cross-coupling coefficient K_{13} between resonators 1 and 3 must be zero, $K_{13} = 0$, and have components $K_{m13} = |K_{e13}| = 0.2130$. The question of ensuring the specified values $K_{m13} = |K_{e13}|$ of SIRs is discussed in Lu et al. [7] and Zakharov and Ilchenko [14], and we will not repeat it. The aforementioned value of the magnetic and electric components will be achieved if, in the short-circuit area, a rectangle of 0.6 mm × 0.4 mm is extracted from the center strip of each SIR, and the gap between these resonators is $\Delta_{13} = 0.2$ mm.

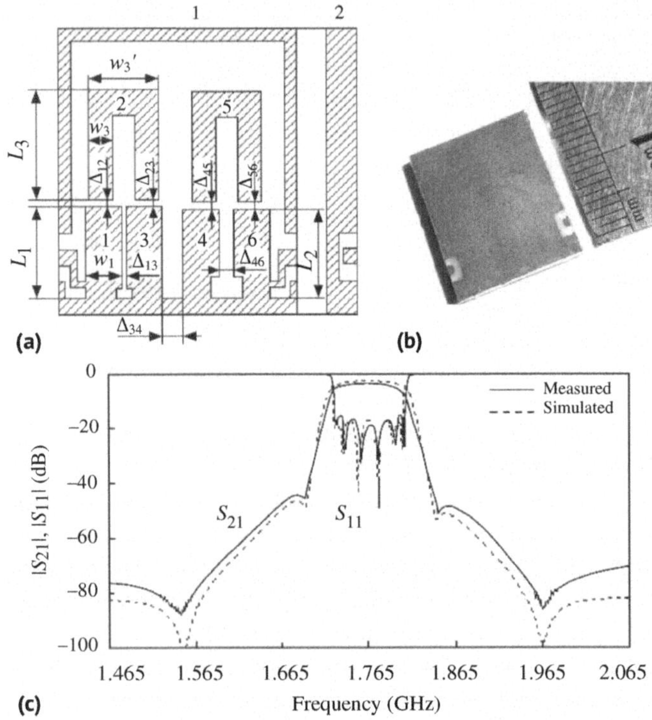

Figure 8.11 Stripline modified CT filter with mixed cross-couplings. (a) Topology: $w_1 = 2.4$ mm, $L_1 = 5.8$ mm, $L_2 = 5.6$ mm, $L_3 = 7.05$ mm, $w_3 = 1.6$ mm, $w_3' = 4.6$ mm, $\Delta_{12} = 0.35$ mm, $\Delta_{23} = 0.5$ mm, $\Delta_{45} = 0.55$ mm, $\Delta_{56} = 0.45$ mm, $\Delta_{34} = 1.3$ mm, $\Delta_{13} = 0.2$ mm, $\Delta_{46} = 1$ mm. (b) Photograph of fabricated filter. (c) Frequency responses.

The second triplet BPF formed by resonators 4–5–6 provides two TZs: $S_{3,4} = \pm j4.86$. To do this, the mixed cross-coupling coefficient must satisfy the conditions $K_{46} = 0$ and $K_{m46} = |K_{e46}| = 0.02371$. These conditions are ensured if resonators 4 and 6 are separated by the gap $\Delta_{46} = 1$ mm, and in the short circuit region, a rectangle of 1.4 mm × 0.6 mm was extracted from the center strip of these SIRs. The position of the zigzag tapped line determines the external Q-factor $Q_e = 25.8$.

This filter's measured and simulated responses are depicted in Figure 8.11c. The TZs $f_{z1} = 1691$ MHz, $f_{z2} = 1842$ MHz, $f_{z3} = 1555$ MHz, and $f_{z4} = 1966$ MHz are almost equidistant with respect to f_0.

Other measured parameters are center frequency $f_0 = 1.762$ GHz, BW = 80 MHz, insertion loss at center frequency $L_{A0} = 2.4$ dB, and return loss $R_L > 16$ dB. The selectivity of this filter, 45 dB ($f_0 \pm 71$ MHz), can be expressed at the normalized frequency $j\Omega$: 45 dB ($\pm j1.78$). This is a reasonably high selectivity indicator.

The analytical method for designing a cascade connection of triplet and quadruplet BPFs with mixed cross-coupling is considered. It is based on the exact solution of the inverse problem using formulas (7.22) and (7.32), and it is supplemented with optimization. These formulas allow us to accurately determine the mixed cross-coupling coefficients for a given location of TZs. This method made it possible to implement, using mixed couplings, a quadrant arrangement of four TZs on the complex plane, which provides the filter with a flat group delay.

The method guarantees the physical feasibility of the specified TZs and allows various frequency responses to be obtained in cascaded filters. If there is no requirement for equal-ripple pulsations, which is the case in many technical applications, then the optimization process is simplified as much as possible. It comes down to changing only the external quality factor of the end resonators Q_e. The considered approach does not limit the number of cascaded triplets and quadruplets with mixed cross-couplings.

References

1 R. Levy, New cascaded trisections with resonant cross-couplings (CTR Sections) applied to the design of optimal filters, in *IEEE MTT-S International Microwave Symposium Digest*, 2004, vol. 2 pp. 447–450, https://doi.org/10.1109/mwsym.2004.1336007.

2 D. Deslandes, Bandpass filter with N+1 transmission zeroes using frequency-variant couplings, in *Proceedings of the 8th IEEE International NEWCAS Conference 2010*, 2010, pp. 217–220, https://doi.org/10.1109/NEWCAS.2010.5603777.

3 J. J. Vague, D. Rubio, M. A. Fuentes, S. Cogollos, M. Baquero, V. E. Boria, and M. Guglielmi, Inline combline filters of order N with up to N + 1 transmission zeros, *IEEE Trans. Microw. Theory Tech.*, vol. 69, no. 7, pp. 3287–3297, 2021, https://doi.org/10.1109/TMTT.2021.3072370.

4 M. Mul, M. Jasinski, A. Lamecki, R. Gomez-Garcia, and M. Mrozowski, Inline microwave filters with N+1 transmission zeros generated by frequency-variant couplings: coupling-matrix-based synthesis and design, *IEEE Trans. Circuits Syst. II Express Briefs*, vol. 69, no. 3, pp. 824–828, 2022, https://doi.org/10.1109/TCSII.2021.3125460.

5 P. Zhao, and K. Wu, Cascading fundamental building blocks with frequency-dependent couplings in microwave filters, *IEEE Trans. Microw. Theory Tech.*, vol. 67, no. 4, pp. 1432–1440, 2019, https://doi.org/10.1109/TMTT.2019.2895532.

6 Y. He, G. Macchiarella, Z. Ma, and N. Yoshikawa, Synthesis and design of quasi-canonical planar filters comprising cascaded frequency-variant blocks, *IEEE Trans. Microw. Theory Tech.*, vol. 69, no. 1, pp. 671–681, 2021, https://doi.org/10.1109/TMTT.2020.3039814.
7 D. Lu, Y. Yang, Y. Wu, and M. Yu, Planar/waveguide quasi-elliptic linear-phase filters using mixed cascaded frequency-dependent triplets with extracted pole sections, *IEEE Trans. Microw. Theory Tech.*, vol. 69, no. 6, pp. 3070–3083, 2021, https://doi.org/10.1109/TMTT.2021.3074602.
8 D. Lu, B. Zhu, and M. Yu, Synthesis approach to planar filter design using new cascade hybrid triplet block with arbitrary-placed TZ pair, *IEEE Trans. Microw. Theory Tech.*, vol. 70, no. 4, pp. 2196–2206, 2022, https://doi.org/10.1109/TMTT.2022.3148361.
9 Y. Zhang, F. Seyfert, S. Amari, M. Olivi, and K.-L. Wu, General synthesis method for dispersively coupled resonator filters with cascaded topologies, *IEEE Trans. Microw. Theory Tech.*, vol. 69, no. 2, pp. 1378–1393, 2021, https://doi.org/10.1109/TMTT.2020.3041223.
10 G. C. Temes and S. K. Mitra (Eds.), *Modern Filter Theory and Design*. New York: Wiley, 1973.
11 W. A. Atia, K. A. Zaki, and A. E. Atia, Synthesis of general topology multiple coupled resonator filters by optimization, in *1998 IEEE MTT-S International Microwave Symposium Digest (Cat. No.98CH36192)*, 1998, vol. 2 pp. 821–824, https://doi.org/10.1109/MWSYM.1998.705116.
12 W. Meng, H.-M. Lee, K. A. Zaki, and A. E. Atia, Synthesis of wideband multicoupled resonators filters, *IEEE Trans. Microw. Theory Tech.*, vol. 59, no. 3, pp. 593–603, 2011, https://doi.org/10.1109/TMTT.2010.2095032.
13 J.-S. Hong, *Microstrip Filters for RF/Microwave Applications*, 2nd ed. New Jersey: Wiley, 2011, https://doi.org/10.1002/9780470937297.
14 A. Zakharov and M. Ilchenko, Coupling coefficients between resonators in stripline combline and pseudocombline bandpass filters, *IEEE Trans. Microw. Theory Tech.*, vol. 68, no. 7, pp. 2679–2690, 2020, https://doi.org/10.1109/TMTT.2020.2988866.

9

Resonators with One or Two Antiresonant Frequencies Close to Operating Resonant Frequency

Throughout the entire period of filter development, there has been a general tendency to increase the transmission zero (TZ) number for a given number of resonators N. The approach to forming TZs, based on cross-coupling between resonators [1–5], began to develop earlier. This approach can reach the TZ number $(N-2)$. If source/load-multiresonator couplings [6] are used in a bandpass filter (BPF), the specified TZ number can be increased to N.

The next approach to generating TZs is using mixed couplings between resonators in BPF [7–12]. These couplings are also called frequency-dependent [10–12] and can operate between adjacent and nonadjacent resonators. The TZ number in BPF can reach $(N+1)$ when using mixed couplings.

The most recent approach is that each resonator contributes one or two TZs to the BPF transfer function [13–29]. The operation of such resonators is based on a physical process associated with the propagation of electromagnetic waves along the resonator itself. This process significantly differs from electromagnetic interaction between two resonators, leading to mixed coupling.

TZs generated by the resonators are located at real frequencies, also called attenuation poles. The formation of TZs by the resonators themselves makes it possible to significantly increase the total TZ number in the BPF. If simple couplings are used in BPF, then the TZ number can equal $2N$ [16]. If mixed couplings are used in BPF, the TZ number can reach the value $(3N+1)$ [25].

Several recently developed microwave resonators and BPFs with multiple TZs are described as follows.

9.1 Features of Resonators with Multiple Critical Frequencies

The generalized circuit of the N-order BPF we are considering is shown in Figure 3.1. Such filters contain parallel-type resonators and admittance inverters as coupling elements. Microwave resonators contain at least one distributed element (transmission line [TL] segment), so their input admittance is a reactance function of infinite order on the frequency variable $s = \sigma + j\omega$ [30]:

$$Y_{\text{in}}(s) = H \frac{s\left(s^2 + \omega_{01}^2\right)\left(s^2 + \omega_{02}^2\right)\ldots}{\left(s^2 + \omega_{p1}^2\right)\left(s^2 + \omega_{p2}^2\right)\ldots}, \tag{9.1}$$

where H is the admittance coefficient. Zeros ω_{0n} ($n = 1, 2, \ldots$) and poles ω_{pn} of the reactance function (9.1) are located on the $j\omega$-axis. They are simple and alternate. These frequencies are called resonant and antiresonant, respectively. It is known that at the antiresonant frequency ω_p, the input admittance Y_{in} of the resonator goes to infinity, and the resonator blocks the signal passing through the filter. This leads to the appearance of TZ at the antiresonant frequency ω_p.

If the antiresonant frequency ω_{p1} or ω_{p2} is shifted closer to ω_{01}, TZ will appear in a BPF, located to the left or right of the passband. If ω_{p1} and ω_{p2} are simultaneously approached to ω_{01}, then two TZs will appear from different sides of the passband, increasing the BPF selectivity. This way of growing selectivity can be considered an alternative to cross- and mixed couplings. In addition, the predicted location of ω_{pn} (near ω_{01}) results in additional TZs at prescribed frequencies. This can increase the BPF attenuation in specific frequency regions, effectively suppressing unwanted signals in those bands. Therefore, a special (prescribed) arrangement of antiresonant frequencies ω_{pn} near ω_{01} is considered as a means of improving the filter's performance.

In traditional TL resonators, the antiresonant frequencies ω_{p1} and ω_{p2} are far from the resonant frequency ω_{01}. To confirm the aforementioned, consider a half-wave resonator (Figure 9.1a) with characteristic impedance Z_0 and length l. Its input admittance is written as

$$Y_{\text{in}} = jZ_0^{-1} \tan(\omega l/v) = jZ_0^{-1} \tan(\theta), \tag{9.2}$$

where $\theta = \omega l/v$ is the electrical length of the resonator, and v is the velocity of the electromagnetic wave in the resonator.

From (9.2) it follows that the resonant frequencies of the considered resonator ω_{0n} correspond to the electrical lengths $\theta_{0n} = n\pi$. The antiresonant frequencies ω_{pn} correspond to electrical lengths $\theta_{pn} = (2n - 1)\pi/2$. These critical electrical lengths are shown in Figure 9.1b. Two antiresonant frequencies ω_{p1} and ω_{p2}

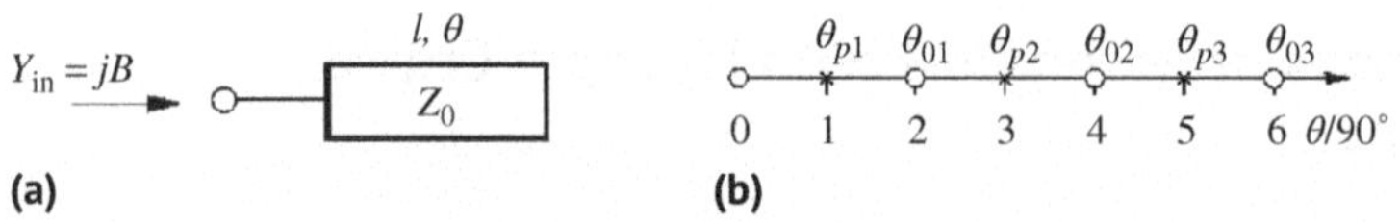

Figure 9.1 $\lambda/2$ resonator. (a) Schematic. (b) Critical electrical length.

are located to the left and right of ω_{01} at a considerable distance, characterized by the ratios $\omega_{01}/\omega_{p1} = 2$ and $\omega_{p2}/\omega_{01} = 1.5$. Since these antiresonant frequencies are located significantly from ω_{01}, they practically do not affect the filter's performance.

The problem is to be able to control individual antiresonant frequencies of the resonators, adjusting their position relative to the operating resonant frequency.

9.2 Transmission Line Resonators with One Antiresonant Frequency Close to Main Resonance Frequency

Such resonators are presented in Table 9.1. They occupy the first three positions in this table.

9.2.1 Resonator No. 1 from Table 9.1: $\lambda/4$ Resonator with Right-sided ω_{p1} Relative to ω_{01}

Figure 9.2a shows the $\lambda/4$ resonator with electric length θ and characteristic impedance Z_0. The electric length $\theta' = q\theta$ $(0 < q \leq 1)$ is the point at which the input admittance $B(\omega)$ is monitored. The function $B(\omega)$ is expressed by formula (3.14a).

This resonator has a set of resonant frequencies ω_{0i} and their corresponding resonant electric lengths $\theta_{0i} = (2i-1)(\pi/2), i = 1, 2 \ldots$, at which its input admittance (9.3) is equal to zero. The electric lengths θ_{0i} do not depend on the coordinate q and are related to each other as odd numbers 1:3:5.... The antiresonant frequencies ω_{pn} correspond to the antiresonant electric lengths θ_{pi}, which rely on the connection coordinate q.

If q changes within $0 < q \leq 2/3$, the position of the first antiresonant electric length θ_{p1} is described by the expression

$$\theta_{p1} = \frac{\pi/2}{1-q}. \tag{9.3}$$

The smaller q, the closer θ_{p1} to θ_{01}. From (9.3) it follows that the ratio θ_{p1}/θ_{01} is determined by the equality

$$\theta_{p1}/\theta_{01} = 1/(1-q). \tag{9.4}$$

Figure 9.2b shows the critical electrical lengths of the resonator for a small q value. They are arranged in pairs. Input admittance for this case is shown in Figure 9.2c.

Table 9.1 Resonators generating one TZ.

Res.	No.	Ref.	Type	Schematic	$Y_{in} = jB$ graph	$Y_{in} = jB$ function
TL	1	[13]	$\lambda/4$	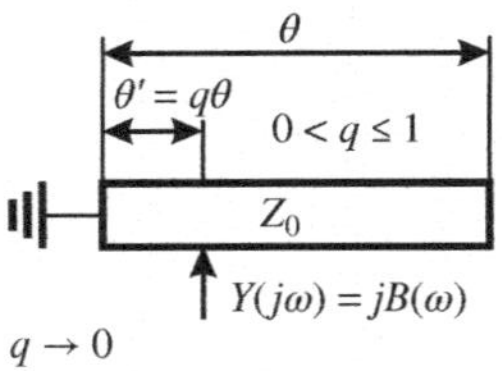	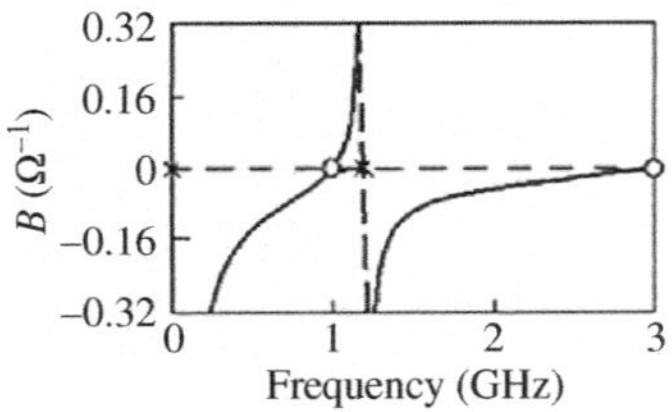	$Y_L = jB_L = \dfrac{-jZ_0^{-1}\cos\theta}{\sin q\theta\cos(1-q)\theta}$
	2	[14]	$\lambda/4$	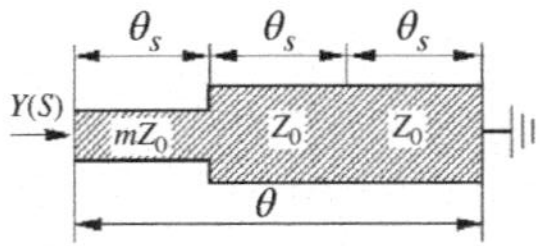	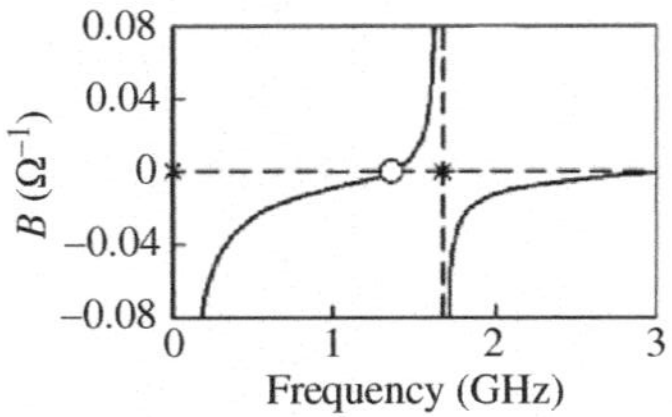	$Y^{(3)}(S) = Z_0^{-1}\dfrac{1+(1+2/m)S^2}{(2+m)S+mS^3}$ $S = j\tan(\theta/3)$
	3	[14]	$\lambda/2$	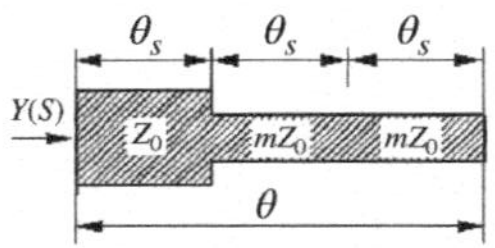	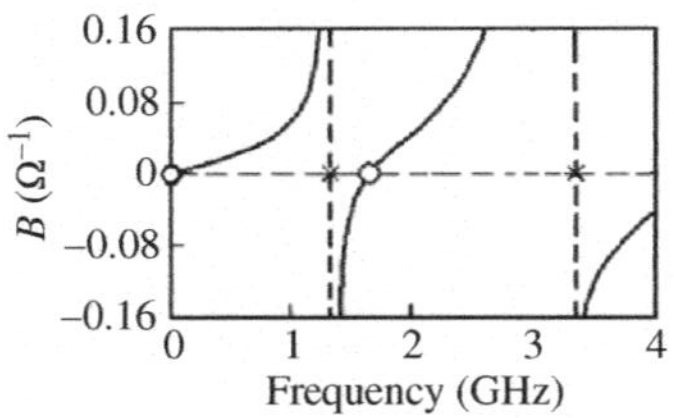	$Y^{(3)}(S) = Z_0^{-1}\dfrac{(1+2/m)S+S^3}{1+(1+2/m)S^2}$ $S = j\tan(\theta/3)$

LD	4	[16]	$\lambda/2$	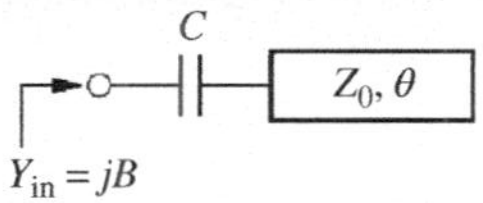	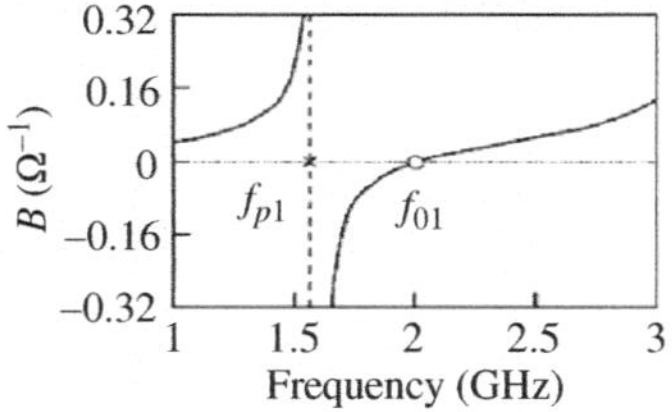	$Y_{in}(j\omega) = \dfrac{[\ldots]}{\cos\theta + \sin\theta/(Z_0\omega C)}$
	5	[16]	$\lambda/2$	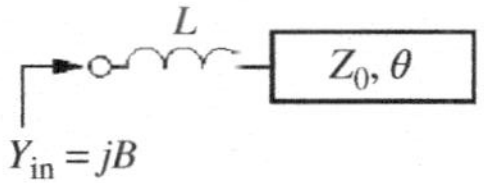	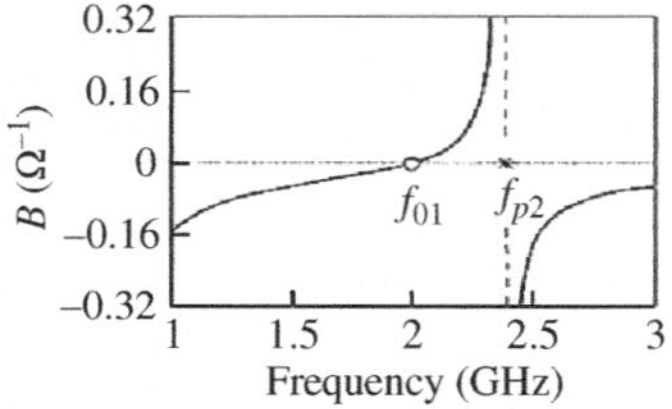	$Y_{in}(j\omega) = \dfrac{jZ_0^{-1}\sin\theta}{\cos\theta - (\omega L/Z_0)\sin\theta}$
	6	[17, 18]	$\lambda/4$	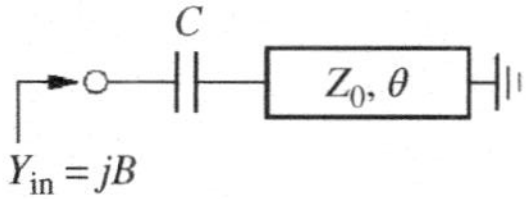	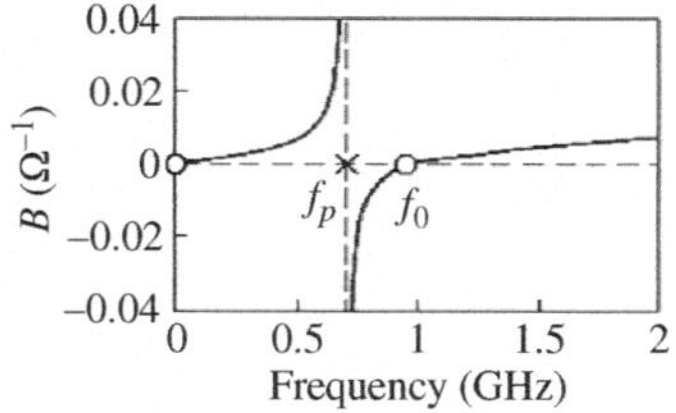	$Y_{in}(j\omega) = \dfrac{-jZ_0^{-1}\cos\theta}{\sin\theta - \cos\theta/(Z_0\omega C)}$
	7	[17, 18]	$\lambda/4$	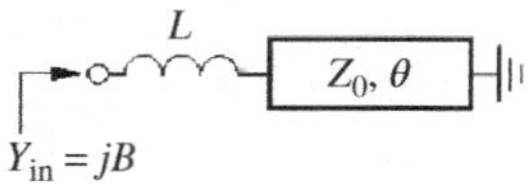	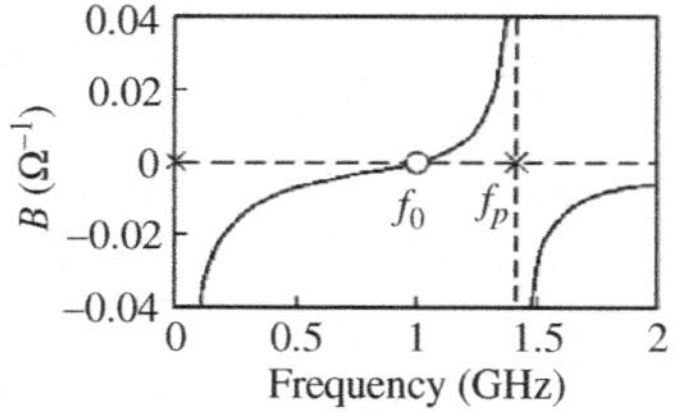	$Y_{in}(j\omega) = \dfrac{-jZ_0^{-1}\cos\theta}{\sin\theta + (\omega L/Z_0)\cos\theta}$

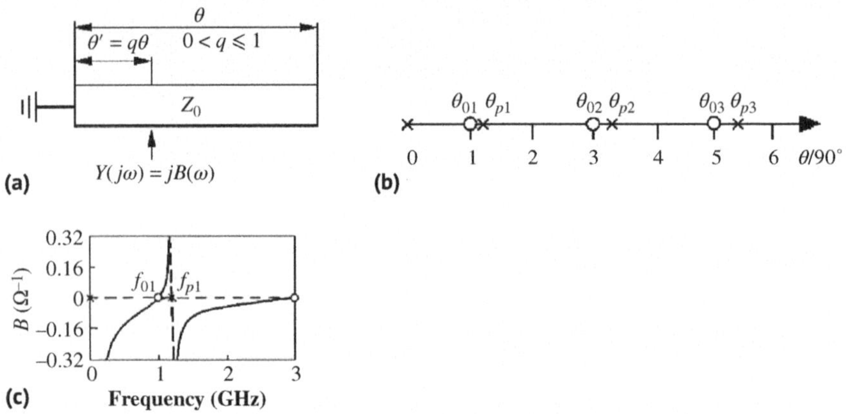

Figure 9.2 $\lambda/4$ resonator with different connection options. (a) Topology. (b) Critical electrical length for small q value. (c) Input admittance $B(\omega)$ for small q value.

Expression (9.4) can be presented as

$$q = 1 - (\theta_{p1}/\theta_{01})^{-1}. \tag{9.5}$$

Equality (9.5) determines the connection coordinate q, at which the required ratio θ_{p1}/θ_{01} is realized. Let $f_{01} = 1$ GHz ($\theta_{01} = 90°$) and $f_{p1} =$ 1.2 GHz ($\theta_{01} = 108°$). Using (9.5), we obtain $q = 1/6$. Figure 9.2c shows the input impedance B for this case.

Susceptance slope parameter b_i of the considered resonator at the main resonant frequency ω_{01} is expressed by the formula (3.15a). It is presented in a more visual representation as follows

$$b = \pi/4Z_0 \sin^2(q\pi/2). \tag{9.6}$$

Expression (9.6) indicates that b increases due to a decrease in q.

In the resonator under consideration, the antiresonant frequency ω_{p1} can be located arbitrarily close to the resonant frequency ω_{01}. However, in this case, the slope parameter (9.6) becomes quite large, and the coupling coefficient $K = J/b_1$ between two such resonators becomes relatively small. Therefore, it is advisable to use this resonator in narrowband filters.

9.2.2 Resonator No. 2 from Table 9.1: $\lambda/4$-type SIR with Right-sided ω_{p1} Relative to ω_{01}

We will use the parametric synthesis method to construct SIR with closely spaced resonant and antiresonant frequencies (Section 5.6). This method uses the polynomial coefficients (Table 5.5) of the numerator and denominator of the SIR input immittance. It is suitable for implementing input impedance Z and input admittance Y. In subsequent consideration, we will use the function Y, for which the zeros ω_{0i} are resonant frequencies and the poles ω_{pi} are antiresonant frequencies.

We implement a case [14] when the pole ω_{p1} of the Y function is located to the right of ω_{01}.

According to the parametric synthesis method, to implement critical frequencies ω_{01} and ω_{p1} close to each other, it is necessary to use SIR with the number of component segments $N = 3$. The input admittance function of such an SIR is written as

$$Y^{(3)}(S) = \frac{b_0^{(3)} + b_2^{(3)} S^2}{a_1^{(3)} S + a_3^{(3)} S^3}, \tag{9.7}$$

where $S = j\Omega = j\tan\theta_s = j\tan(\theta/3)$.

The numerator and denominator coefficients (9.7) are defined in Table 5.5a: $b_0^{(3)} = 1$, $b_2^{(3)} = Z_{01}/Z_{02} + Z_{01}/Z_{03} + Z_{02}/Z_{03}$, $a_1^{(3)} = Z_{01} + Z_{02} + Z_{03}$, $a_3^{(3)} = Z_{01}Z_{03}/Z_{02}$. Remember that the segments are numbered from the far (short-circuited) end. From (9.7) it follows that the critical frequencies Ω_{01} and Ω_{p1} on the Ω-axis are expressed by the equalities:

$$\Omega_{01}^2 = b_0^{(3)}/b_2^{(3)} = 1/\left(Z_{01}/Z_{02} + Z_{01}/Z_{03} + Z_{02}/Z_{03}\right), \tag{9.8a}$$

$$\Omega_{p1}^2 = a_1^{(3)}/a_3^{(3)} = Z_{02}\left(Z_{01} + Z_{02} + Z_{03}\right)/Z_{01}Z_{03}. \tag{9.8b}$$

The equalities (9.8a) and (9.8b) show that the requirement to minimize the ratio Ω_{p1}/Ω_{01} is provided for the characteristic impedances $Z_{01} = Z_{02} = Z_0$, $Z_{03} = mZ_0$.

The topology of the synthesized SIR is shown in Figure 9.3a. Its input admittance (9.7) takes the form

$$Y^{(3)}(S) = Z_0^{-1}\frac{1 + (1 + 2/m) S^2}{(2 + m) S + mS^3}. \tag{9.9}$$

The critical frequencies on the Ω-axis and the critical electrical length of the resonator are determined by the following equations:

$$\Omega_{01}^2 = (1 + 2/m)^{-1}, \quad \theta_{01} = 3\tan^{-1}\sqrt{(1 + 2/m)^{-1}}, \tag{9.10a}$$

$$\Omega_{p1}^2 = 1 + 2/m, \quad \theta_{p1} = 3\tan^{-1}\sqrt{1 + 2/m}. \tag{9.10b}$$

The location of the critical electrical lengths of this SIR is shown in Figure 9.3b, and the input admittance $Y^{(3)} = jB(\omega)$ for the case $m = 5$ is demonstrated in Figure 9.3c.

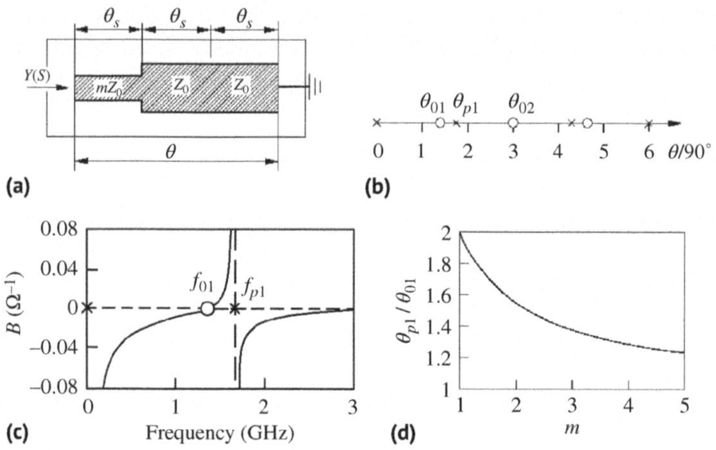

Figure 9.3 $\lambda/4$-type SIR with right-sided ω_{p1} relative to ω_{01}. (a) Topology. (b) Critical electrical length. (c) Input admittance for $m = 5$. (d) Ratio θ_{p1}/θ_{01}.

The aforementioned equalities allow us to determine the ratio ω_{p1}/ω_{01}, which is the same as θ_{p1}/θ_{01}:

$$\frac{\omega_{p1}}{\omega_{01}} = \frac{\theta_{p1}}{\theta_{01}} = \frac{\tan^{-1}\sqrt{1+2/m}}{\tan^{-1}\sqrt{(1+2/m)^{-1}}}. \tag{9.11}$$

From (9.11) it follows that $\omega_{p1}/\omega_{01} = 1.24$ at $m = 5$; $\omega_{p1}/\omega_{01} = 1.123$ at $m = 10$; and $\omega_{p1}/\omega_{01} = 1.042$ at $m = 30$. In the latter case, the convergence degree of ω_{01} and ω_{p1} is 4.2%. The practically feasible parameter value for SIR fabricated on a single substrate is $m \approx 10$ (Table 5.3). If the low-impedance part of the SIR is placed on a substrate with a high dielectric constant, and the high-impedance part is placed on a nearby substrate with a low dielectric constant, then the parameter value $m = 30$ can be realized. So, if $\varepsilon_r = 92$ and $h = 0.508$ mm, then the width of the metal strip $w = 3.2$ mm has a characteristic impedance $Z_0 = 5\ \Omega$. If $\varepsilon_r = 2.2$ and $h = 0.508$ mm, the characteristic impedance $Z_0 = 150\ \Omega$ is provided with a strip width $w = 0.15$ mm. Such an SIR on two adjacent different substrates has a value of $m = 30$. In coaxial SIRs, implementing the parameter $m = 30$ does not cause difficulties.

Substituting (9.9) into (3.2) gives the susceptance slope parameter b of the considered resonator

$$b = (2Z_0)^{-1}(1+2/m)\tan^{-1}\sqrt{(1+2/m)^{-1}}. \tag{9.12}$$

As m increases, this SIR's slope parameter b (9.12) decreases. When m changes from 1 to 5, the parameter b decreases by a factor of 1.6.

9.2.3 Resonator No. 3 from Table 9.1: $\lambda/2$-type SIR with Left-sided ω_{p1} Relative to ω_{01}

SIR with such an arrangement of critical frequencies differs from the previous SIR (Figure 9.3) in that its zero ω_{01} and pole ω_{p1} have swapped places $(\omega_{p1} < \omega_{01})$. We do not need to use the parametric synthesis method to build it. In this case, it is enough to make the far end of the SIR open-circuit and replace the characteristic impedances of the constituent segments with the opposite ones, according to equality (5.61), illustrated in Figure 5.19. As a result of these changes, we arrive at SIR, the topology of which is shown in Figure 9.4a, and the critical electrical lengths are shown in Figure 9.4b. The input admittance (9.9) is converted as follows

$$Y^{(3)}(S) = Z_0^{-1} \frac{(1 + 2/m)\, S + S^3}{1 + (1 + 2/m)\, S^2}, \tag{9.13}$$

where $S = j\Omega = j \tan \theta_s = j \tan(\theta/3)$. The input admittance $Y^{(3)} = jB(\omega)$ for case $m = 5$ is shown in Figure 9.4c.

The critical frequencies on the Ω-axis and the critical electrical length of the resonator are determined by obvious equalities:

$$\Omega_{01}^2 = \left(1 + \frac{2}{m}\right), \quad \theta_{01} = 3\tan^{-1}\sqrt{\left(1 + \frac{2}{m}\right)}$$

$$\Omega_{p1}^2 = \left(1 + \frac{2}{m}\right)^{-1}, \quad \theta_{p1} = 3\tan^{-1}\sqrt{\left(1 + \frac{2}{m}\right)^{-1}}.$$

The ratio ω_{01}/ω_{p1} coincides with the ratio of electrical lengths θ_{01}/θ_{p1}:

$$\frac{\omega_{01}}{\omega_{p1}} = \frac{\theta_{01}}{\theta_{p1}} = \frac{\tan^{-1}\sqrt{1 + 2/m}}{\tan^{-1}\sqrt{(1 + 2/m)^{-1}}}. \tag{9.14}$$

The ratio (9.14) on the parameter m of the resonator is depicted in Figure 9.4d.

As one would expect, the right-hand sides of equalities (9.11) and (9.14) coincide, as the graphs in Figures 9.3d and 9.4d. The same values as a previous SIR characterize the convergence degree of ω_{01} and ω_{p1}: $\omega_{01}/\omega_{p1} = 1.24$ at $m = 5$; $\omega_{01}/\omega_{p1} = 1.123$ at $m = 10$; and $\omega_{01}/\omega_{p1} = 1.042$ at $m = 30$. In the latter case, the convergence degree of ω_{01} and ω_{p1} is 4.2%.

A natural question arises: Can we improve the results using SIRs with different electrical lengths of the component segments? Consider SIR (Figure 9.4) with $m = 5$ in detail to answer this. It has one normalized characteristic impedance

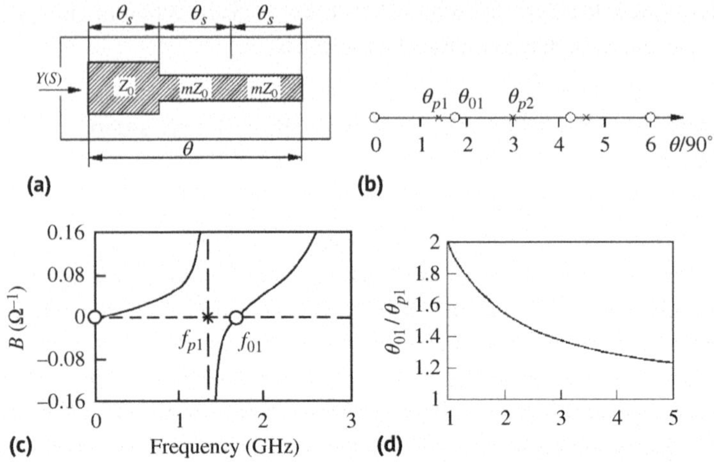

Figure 9.4 $\lambda/2$-type SIR with left-sided ω_{p1} relative to ω_{01}. (a) Topology. (b) Critical electrical length. (c) Input admittance for $m = 5$. (d) Ratio θ_{01}/θ_{p1}.

jump coordinate $\eta = 1/3$, which requires no explanation. If the electrical length of this resonator is 90° at the frequency of 1 GHz, then it has the values f_{01} and f_{p1} presented by the following values:

$$\eta = \frac{1}{3}: \; f_{01} = 1.659 \text{ GHz}, \quad f_{p1} = 1.341 \text{ GHz}, \quad \frac{f_{01}}{f_{p1}} = 1.237. \tag{9.15}$$

The similar values correspond to a small change in the coordinate η ($\eta = 0.32$ and $\eta = 0.34$), which are defined using a linear circuit simulator:

$$\begin{aligned} \eta = 0.32: \; f_{01} &= 1.642 \text{ GHz}, \quad f_{p1} = 1.324 \text{ GHz}, \quad \frac{f_{01}}{f_{p1}} = 1.240 \\ \eta = 0.34: \; f_{01} &= 1.669 \text{ GHz}, \quad f_{p1} = 1.347 \text{ GHz}, \quad \frac{f_{01}}{f_{p1}} = 1.239 \end{aligned} \tag{9.16}$$

Comparison of (9.15) and (9.16) shows that deviations η from the value $\eta = 1/3$ did not improve results. The minimum ratio f_{01}/f_{p1} is achieved at $\eta = 1/3$. The same is typical for other optimal SIRs.

For the considered SIR with input admittance (9.13), expression (3.2) gives

$$b = Z_0^{-1}\left(1 + \frac{m}{2}\right)\tan^{-1}\sqrt{1 + \frac{2}{m}}. \tag{9.17}$$

As m increases, the slope parameter (9.17) increases, in contrast to the previous resonator (9.12).

9.3 Lumped-distributed Resonator with One Antiresonant Frequency Close to Main Resonance Frequency

Such resonators are presented in Nos. 4–7 of Table 9.1. In Ref. [14], lumped-distributed (LD) resonators (No. 4, No. 5) were considered, representing a half-wave TL segment to which a capacitance or inductance is cascaded. The input of these resonators is located on the side of the lumped element. To obtain the Y_{in} function of the LD resonators, it is necessary to transform the input admittance of the half-wave resonator (9.2) using the transfer matrix $[ABCD]$ of the lumped circuit. In this case, expression (2.19) will take the form

$$Y_{\text{in}} = \frac{C + jDZ_0^{-1} \tan\theta}{A + jBZ_0^{-1} \tan\theta}. \tag{9.18}$$

9.3.1 Resonator No. 4 from Table 9.1: $\lambda/2$ Resonator with C

Figure 9.5a shows the resonator circuit with C. The transfer matrix of capacitance C has the form

$$\begin{bmatrix} A & B \\ C & D \end{bmatrix} = \begin{bmatrix} 1 & 1/j\omega C \\ 0 & 1 \end{bmatrix}. \tag{9.19}$$

Substituting (9.19) into (9.18), we obtain

$$Y_{\text{in}}(j\omega) = jZ_0^{-1} \frac{\sin\theta}{\cos\theta + \sin\theta/(Z_0\omega C)}. \tag{9.20}$$

The zeros of function Y_{in} (9.20) are the same as the zeros of function (9.2). This means that the resonant frequencies ω_{0i} of the compound resonator (Figure 9.5a) are the same as the $\lambda/2$ resonator. The corresponding electrical lengths are as follows: $\theta_{0i} = i\pi$, $i = 1, 2, \ldots$. Equating the denominator (9.20) to zero, we obtain an equation for determining the poles ω_p:

$$\cos\theta_p + \frac{\sin\theta_p}{Z_0\omega_p C} = 0.$$

The electrical length θ_p of the resonator in the last equality corresponds to the antiresonant frequency ω_p. Figure 9.5b shows the electrical lengths θ_{0i} and θ_{pi}, corresponding to the critical frequencies of the resonator ω_{0i} and ω_{pi}. With an increase in the serial number n, the difference between the critical frequencies ω_{0n} and ω_{pn} decreases; if $n \to \infty$, then $(\omega_{0n} - \omega_{pn}) \to 0$. Connecting the capacitance C to a half-wave resonator leads to a left-sided pole ω_p close to ω_0. This is reflected in the function of the input admittance $Y(j\omega)$ of this resonator (Figure 9.5c), which has a resonant frequency $f_0 = 2$ GHz. On the transfer function $|S_{21}|$, TZ appears at the frequency ω_p (Figure 9.5d). The smaller the C value, the closer ω_p to ω_0. The antiresonant frequency ω_p and the corresponding TZ are controlled by capacitance C. If the value ω_p (θ_p) is given, then from the last equality it follows

$$C = \frac{-\tan\theta_p}{\omega_p Z_0,} \quad (\pi/2 < \theta_p < \pi). \tag{9.21}$$

If the value ω_p is set in (9.21), then the value θ_p corresponding to it becomes known automatically.

Table 9.2 shows the calculated values of the capacitance C (9.21) when the antiresonant frequency of the resonator under consideration approaches the fundamental resonant frequency $f_{01} = 2$ GHz. The characteristic impedance Z_0 of the half-wave segment is equal to 50, 10, and 5 Ω. The data in Table 9.2 indicate that at practical values of capacitance C (0.5 pF and 0.25 pF), a high degree of convergence between the antiresonant and resonant frequencies $f_{01}/f_{p1} = 1.01$ is achieved. The difference between these frequencies is 1%.

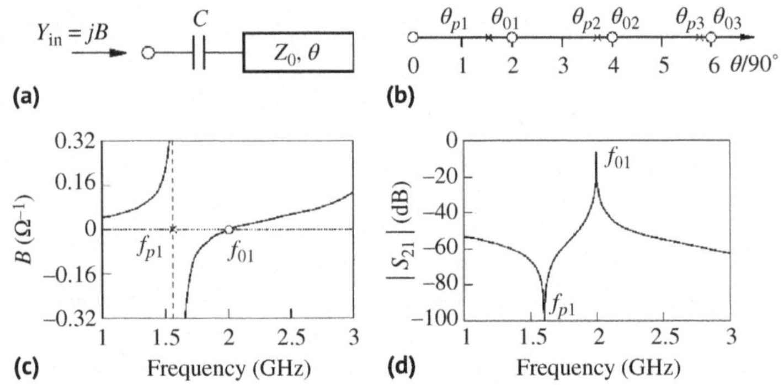

Figure 9.5 $\lambda/2$ resonator with capacitance C. (a) Schematic. (b) Critical electrical lengths. (c) Input admittance. (d) Frequency response.

Table 9.2 Values of capacitance C when left-hand antiresonant frequency f_{p1} approaches main resonant frequency f_{01} for resonator in Figure 9.5 at $f_{01} = 2$ GHz.

f_{p1}, GHz		1.5	1.7	1.9	1.95	1.98
θ_{p1}, °		135	153	171	175.5	178.2
$Z_0 = 5$ Ω	C, pF	21.23	9.54	2.66	1.28	0.5
$Z_0 = 10$ Ω	C, pF	10.61	4.77	1.33	0.64	0.25
$Z_0 = 50$ Ω	C, pF	2.12	0.95	0.26	0.13	0.05

Substituting (9.20) into (3.2), we obtain

$$b = \pi/2Z_0. \tag{9.22}$$

This parameter b is the same as for the $\lambda/2$ resonator. This is a significant result used in the filter development.

9.3.2 Resonator No. 5 from Table 9.1: $\lambda/2$ Resonator with L

Figure 9.6a shows the resonator circuit with L. The transfer matrix of inductance L has the form

$$\begin{bmatrix} A & B \\ C & D \end{bmatrix} = \begin{bmatrix} 1 & j\omega L \\ 0 & 1 \end{bmatrix}. \tag{9.23}$$

Substituting (9.23) into (9.18), we obtain

$$Y_{\text{in}}(j\omega) = jZ_0^{-1} \frac{\sin\theta}{\cos\theta - (\omega L/Z_0)\sin\theta}. \tag{9.24}$$

The resonant frequencies of this resonator ω_{0i}, which are zeros of the input function Y_{in} (9.24), coincide with the resonant frequencies of the $\lambda/2$ resonator. Equating the denominator (9.24) to zero, we obtain the equation for determining the poles ω_p:

$$\cos\theta_p - \left(\frac{\omega_p L}{Z_0}\right)\sin\theta_p = 0.$$

Figure 9.6b shows the electrical lengths θ_{0n} and θ_{pn} corresponding to the critical frequencies ω_{0n} and ω_{pn} of this resonator. With an increase in the serial number n, the difference between the critical frequencies ω_{pn+1} and ω_{0n} decreases; if

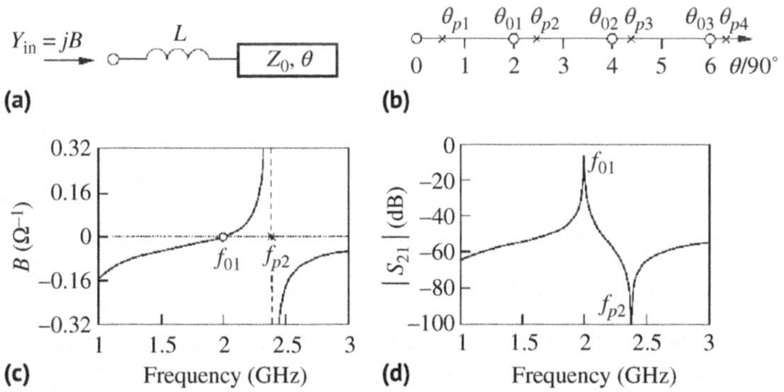

Figure 9.6 $\lambda/2$ resonator with inductance L. (a) Schematic. (b) Critical electrical lengths. (c) Input admittance. (d) Frequency response.

$n \to \infty$, then $(\omega_{pn+1} - \omega_{0n}) \to 0$. Connecting to a $\lambda/2$ resonator an inductance L leads to a right-hand pole ω_p close to ω_0. This is reflected in the input admittance $Y(j\omega)$ of this resonator (Figure 9.6c), which has a resonant frequency $f_0 = 2$ GHz. On the transfer function $|S_{21}|$, TZ appears at the frequency ω_p (Figure 9.6d). The larger the value of L, the closer ω_p to ω_0. For the resonator under consideration, the antiresonant frequency ω_p is controlled by inductance L. If the value ω_p (θ_p) is given, then from the last equality it follows

$$L = \left(\frac{Z_0}{\omega_p}\right) \cot\theta_p, \quad (3\pi/2 > \theta_p > \pi). \tag{9.25}$$

Table 9.3 presents the calculated values of the inductance L (9.25) when the antiresonant frequency f_{p2} of the considered resonator approaches the fundamental resonant frequency $f_{01} = 2$ GHz. As in Table 9.2, the characteristic impedance Z_0 of the half-wave segment is equal to 50, 10, and 5 Ω. The data in Table 9.3 indicate that at practical values of the inductance L (12.8 nH and 25.6 nH) [31, 32], a high degree of convergence between the antiresonant and resonant frequencies $f_{p2}/f_{01} = 1.01$ is achieved. The difference between these frequencies is 1%.

Substituting (9.24) into (3.2), we obtain $b = \pi/2Z_0$. The slope parameter b is the same as for $\lambda/2$ resonator.

The LD resonators (No. 6 and No. 7 in Table 9.1) are a quarter-wave resonator with input admittance (5.2a) to which a capacitance or inductance is cascaded [17, 18].

To obtain the Y_{in} function of the resonators on the co-side of lumped elements, it is necessary to transform the input admittance (5.2a) of the quarter-wave resonator using the transfer matrix $[ABCD]$ of the lumped circuit.

In this case, expression (2.19) takes the form

$$Y_{in} = \frac{C - jDZ_0^{-1}\cot\theta}{A - jBZ_0^{-1}\cot\theta}. \tag{9.26}$$

Table 9.3 Values of inductance L when the right-hand antiresonant frequency f_{p2} approaches the main resonant frequency f_{01} for resonator in Figure 9.6 at $f_{01} = 2$ GHz.

f_{p2}, GHz		2.5	2.3	2.1	2.05	2.02
θ_{p2}, °		225	207	189	184.5	181.8
$Z_0 = 5$ Ω	L, nH	0.32	0.68	2.39	4.93	12.8
$Z_0 = 10$ Ω	L, nH	0.64	1.36	4.78	9.86	25.6
$Z_0 = 50$ Ω	L, nH	3.18	6.79	23.8	49.3	–

9.3.3 Resonator No. 6 from Table 9.1: $\lambda/4$ Resonator with C

Figure 9.7a shows the resonator circuit with capacitance C. Substituting the transfer matrix of capacitance C (9.19) into (9.26), we obtain the input admittance of the LD resonator

$$Y_{\text{in}}(j\omega) = -Z_0^{-1}\frac{\cos\theta}{\sin\theta - \cos\theta/(Z_0\omega C)}. \tag{9.27}$$

The resonant frequencies of this resonator, ω_{0i}, which are the zeros of the input function Y_{in} (9.27), coincide with the resonant frequencies of the quarter-wave resonator (5.2a). They correspond to electrical lengths $\theta_{0i} = (2i-1)\pi/2$ and are related to each other as odd numbers 1:3:5. Equating the denominator (9.27) to zero, we obtain the equation for determining the poles ω_p:

$$\sin\theta_p - \cos\theta_p/(Z_0\omega_p C) = 0.$$

In the last expression, θ_p is the electrical length of the resonator corresponding to the antiresonant frequency ω_p. Figure 9.7b shows the electrical lengths θ_{0i} and θ_{pi} corresponding to the critical frequencies of the resonator ω_{0i} and ω_{pi}. Connecting a capacitance C to a quarter-wave resonator leads to a left-sided pole ω_p close to ω_0. This is reflected in the input admittance function $Y(j\omega)$ of this resonator (Figure 9.7c), which has a resonant frequency $f_0 = 1$ GHz. The TZ appears on the transfer function $|S_{21}|$ at the frequency ω_p (Figure 9.7d). The smaller the C value, the closer ω_p to ω_0.

The antiresonant frequency ω_p and the corresponding TZ are controlled by a capacitance C. If the value ω_p (θ_p) is given, then from the last equality it follows

$$C = \frac{1}{\omega_p Z_0 \tan\theta_p} \qquad (\omega_p < \omega_0, \theta_p < \pi/2). \tag{9.28}$$

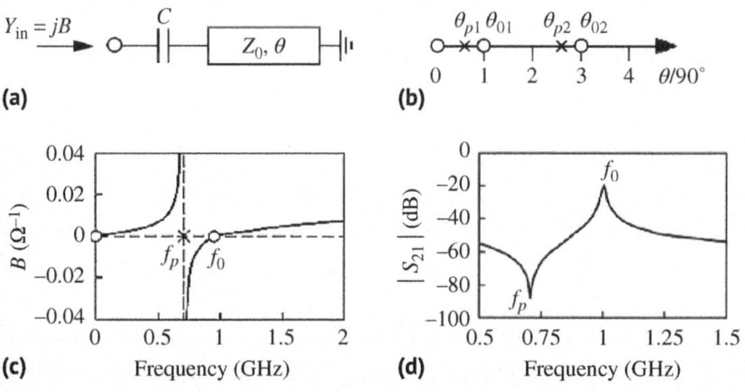

Figure 9.7 $\lambda/4$ resonator with capacitance C. (a) Schematic. (b) Critical electrical length. (c) Input admittance. (d) Frequency response.

Substituting the input admittance (9.27) into (3.2), we obtain the susceptance slope parameter of the considered resonator $b = \pi/4Z_0$. It is the same as for the $\lambda/4$ resonator.

This resonator and the resonator with a half-wave TL segment (Figure 9.5) provide a left-hand position of the antiresonant frequency ω_{p1} to the main resonance frequency ω_{01}. The convergence degree of these frequencies can be 1% with a practically realizable value of the capacitance C.

9.3.4 Resonator No. 7 from Table 9.1: $\lambda/4$ Resonator with L

Figure 9.8a shows the resonator circuit with inductance L. Substituting the transfer matrix of inductance L (9.23) into (9.26), we obtain the input admittance of the LD resonator

$$Y_{\text{in}}(j\omega) = \frac{-jZ_0^{-1}\cos\theta}{\sin\theta + (\omega L/Z_0)\cos\theta}. \tag{9.29}$$

The resonant frequencies of this resonator ω_{0i} are zeros of the input function Y_{in} (9.25) and coincide with the resonant frequencies of the quarter-wave resonator (5.2a). Equating the denominator (9.29) to zero, we obtain the equation for determining the poles ω_p:

$$\sin\theta_p + \left(\frac{\omega L}{Z_0}\right)\cos\theta_p = 0. \tag{9.30}$$

Figure 9.8b shows the electrical lengths θ_{0i} and θ_{pi}, corresponding to the critical frequencies of the resonator ω_{0i} and ω_{pi}. Connecting an inductance L to a quarter-wave resonator leads to a right-hand pole ω_p close to ω_0. This is reflected in the input admittance function $Y(j\omega)$ of this resonator (Figure 9.8c). The attenuation pole appears at the frequency ω_p on the transfer function $|S_{21}|$ (Figure 9.8d). The larger the value of L, the closer ω_p to ω_0.

For the resonator under consideration, the antiresonant frequency ω_p is controlled by an inductance L. If the value ω_p (θ_p) is given, then from (9.30) we have

$$L = -\left(\frac{Z_0}{\omega_p}\right)\tan\theta_p \quad (\omega_p > \omega_0, \theta_p > \pi/2). \tag{9.31}$$

Substituting input admittance (9.29) into (3.2), we obtain the susceptance slope parameter of this resonator $b = \pi/(4Z_0)$, which is the same as for the $\lambda/4$ resonator.

This resonator and a resonator with a half-wave TL segment (Figure 9.6) provide a right-hand position of the antiresonant frequency ω_{p1} to the main resonance frequency ω_{01}. The convergence degree of these frequencies can be 1% at a practically achievable value of the inductance L.

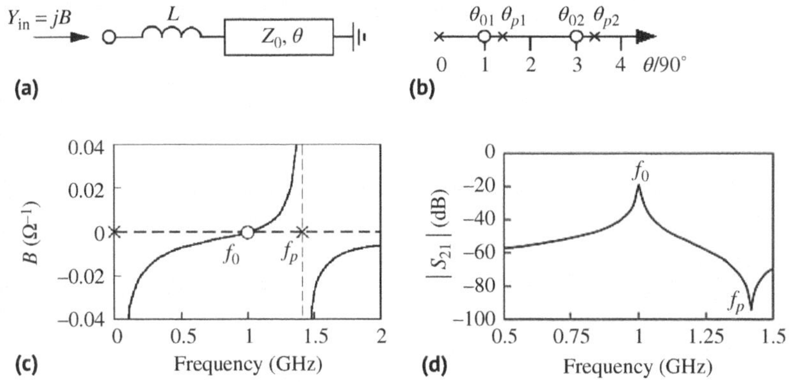

Figure 9.8 $\lambda/4$ resonator with inductance L. (a) Schematic. (b) Critical electrical length. (c) Input admittance. (d) Frequency response.

9.4 TL Resonators with Two Antiresonant Frequencies Close to Operating Resonant Frequency

Such resonators are presented in Table 9.4a (Part 1: TL resonators). They occupy six positions in this table.

9.4.1 Resonator No. 1 from Table 9.4: Uniform Resonator on $3\lambda/4$ Oscillations

This is a well-known resonator [19, 20] with a normalized connection coordinate q (Figure 9.9a). In this case, it is used at the highest resonant frequency ω_{02}, corresponding to $3\lambda/4$ oscillations, which allows it to be used at high frequencies without excessive reduction in its length.

To analyze the operation of this resonator, it is advisable to use trajectories (Figure 9.9b) of their antiresonant frequencies [19], which clearly explains the principle of their operation. The electric length $\theta' = q\theta$ $(0 < q \leq 1)$ is the point at which the input admittance Y_{in} is monitored. The function Y_{in} (3.14a) of the resonator is conveniently represented in the form:

$$B = \frac{-Z_0^{-1} \cos\theta}{\sin(q\theta)\cos(1-q)\theta}. \tag{9.32}$$

The condition $\cos\theta = 0$ shows that the resonant electrical lengths of this resonator $\theta_{0i} = (2i-1)(\pi/2), i = 1, 2 ...,$ at which its input admittance (9.32) is equal to zero, do not depend on the coordinate q and are related to each other as odd numbers 1:3:5.... They are marked with circles in Figure 9.9b. The antiresonant

Table 9.4a Resonators generating two TZs (TL resonators).

Res	No.	Ref.	Type	Schematic	$Y_{in} = jB$ graph	$Y_{in} = jB$ function
TL	1	[19, 20]	$3\lambda/4$	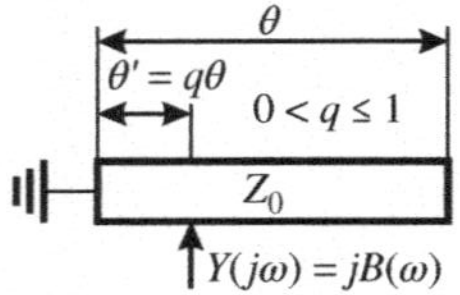 $q \to 2/3$	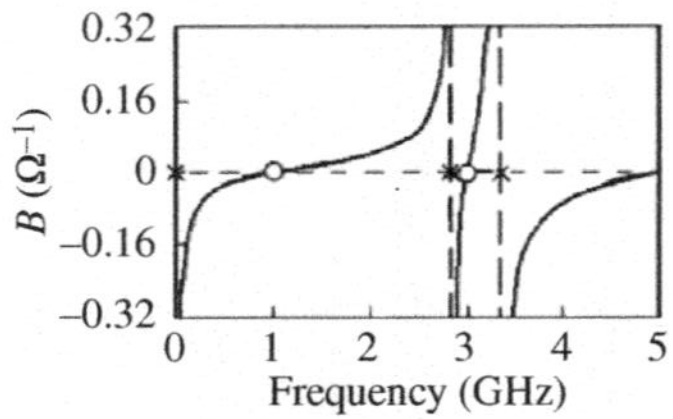	$B = Z_0^{-1}\,[\tan(1 - q\theta) - \cot(q\theta)]$
	2	[14, 20, 21]	$\lambda/2$	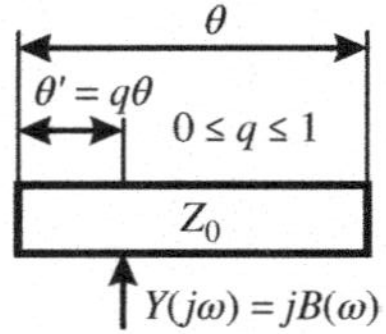 $q \to 1/2$	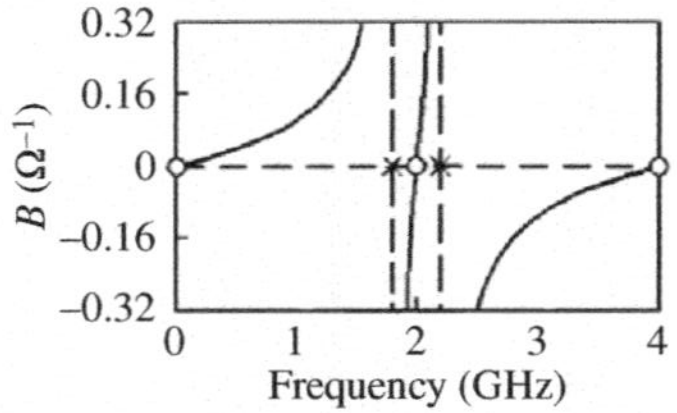	$B = Z_0^{-1}\,[\tan(1 - q\theta) + \tan(q\theta)]$
	3	[20]	λ	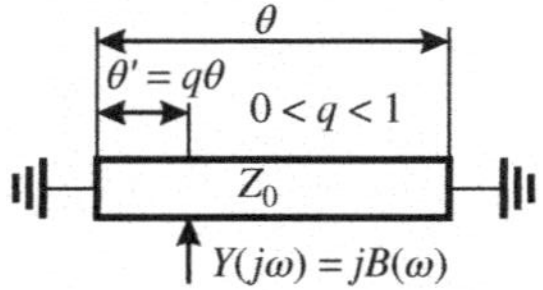 $q \to 1/2$	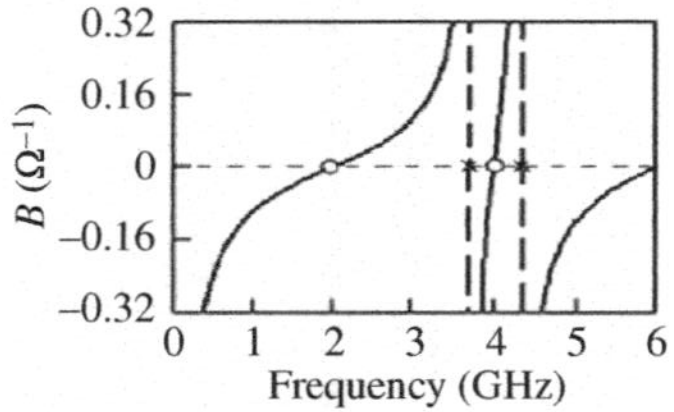	$B = -Z_0^{-1}\,[\cot(1 - q\theta) + \cot(q\theta)]$

4	[14]	$\lambda/2$	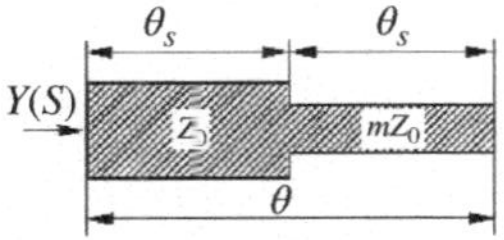	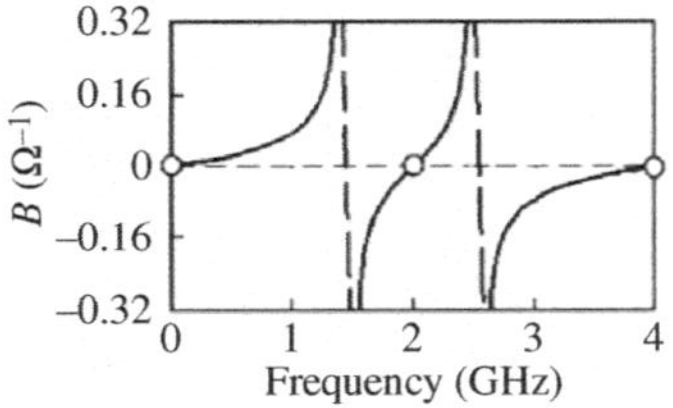	$Y^{(2)}(S) = Z_0^{-1}\dfrac{(1+1/m)S}{1+(1/m)S^2} =$ $= jZ_0^{-1}\dfrac{(1+1/m)\tan(\theta/2)}{1-(1/m)\tan^2(\theta/2)},$ $S = j\tan\left(\dfrac{\theta}{2}\right)$
5	[22, 23]	$3\lambda/4$	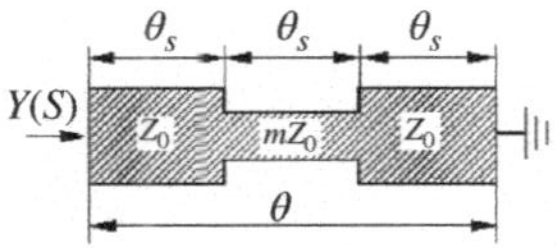	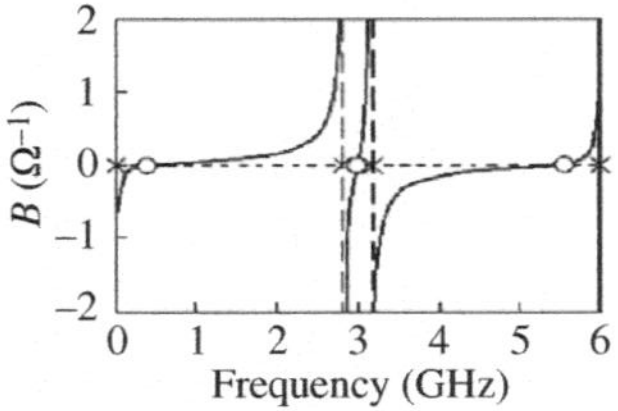	$Y^{(3)}(S) =$ $= Z_0^{-1}\dfrac{1+(1+m+1/m)S^2}{(2+m)S+(1/m)S^3},$ $S = j\tan\left(\dfrac{\theta}{3}\right)$
6	[22, 24]	$5\lambda/4$	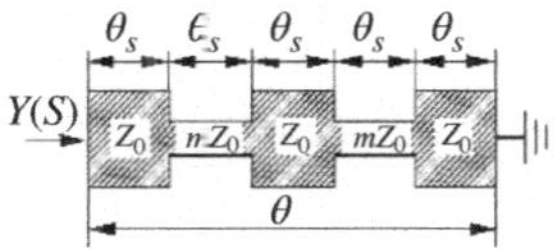	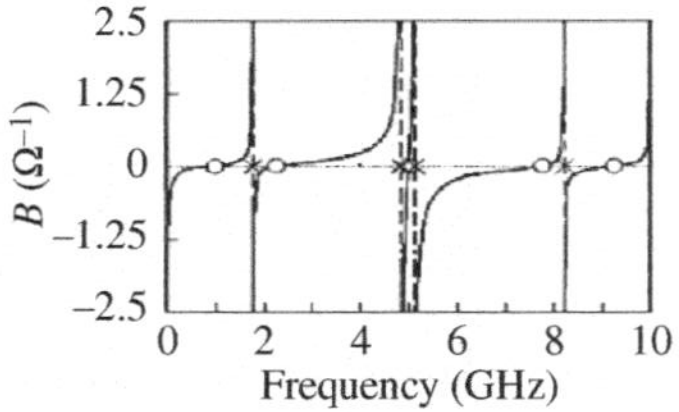	$Y^{(5)}(S) = \dfrac{b_0+b_2S^2+b_4S^4}{a_1S+a_3S^3+a_5S^5},$ $b_0 = 1,$ $b_2 = 4+3m+\dfrac{3}{m},$ $b_4 = 1+m+m^2+\dfrac{1}{m}+\dfrac{1}{m^2},$ $a_1 = 3+2m,$ $a_3 = 3+2m+m^2+\dfrac{4}{m},$ $a_5 = \dfrac{1}{m^2}, S = j\tan\left(\dfrac{\theta}{5}\right)$

Table 9.4b Resonators generating two TZs (LD resonators).

Res	No.	Ref.	Type	Schematic	$Y_{in} = jB$ graph	$Y_{in} = jB$ function
LD	7	[16]	$\lambda/2$	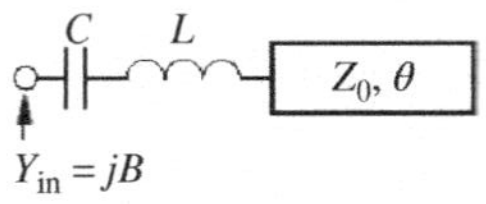	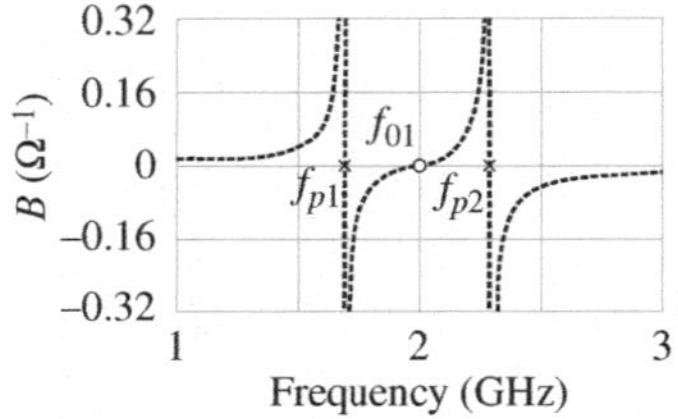	$Y_{in}(j\omega) = \dfrac{jZ_0^{-1}\sin\theta}{\cos\theta - Z_0^{-1}\left(\omega L - \frac{1}{\omega C}\right)\sin\theta}$
	8	[17, 18]	$\lambda/4$	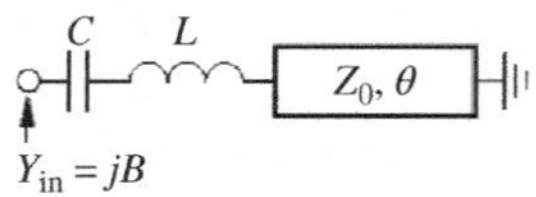	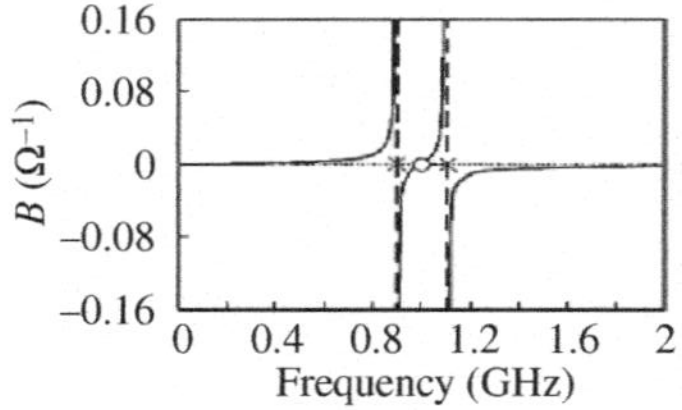	$Y_{in}(j\omega) = \dfrac{-jZ_0^{-1}\cos\theta}{\sin\theta + Z_0^{-1}\left(\omega L - \frac{1}{\omega C}\right)\cos\theta}$
	9	[25]	$\lambda/4$	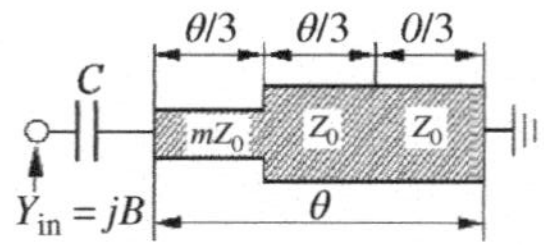	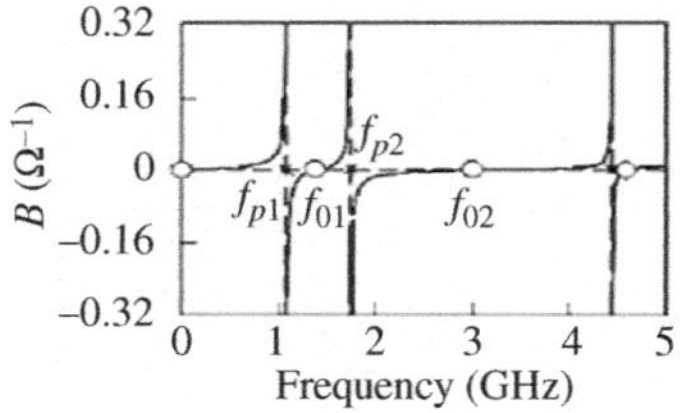	$Y_{in} = -jZ_0^{-1}\dfrac{1-(1+2/m)\,\Omega^2}{A+B}$, $A = (2+m)\,\Omega - m\Omega^3$, $B = -Z_0^{-1}\dfrac{1-(1+2/m)\,\Omega^2}{\omega C}$, $\Omega = \tan\left(\dfrac{\theta}{3}\right)$

10	[25]	$\lambda/2$	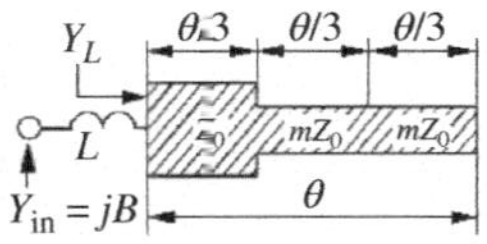	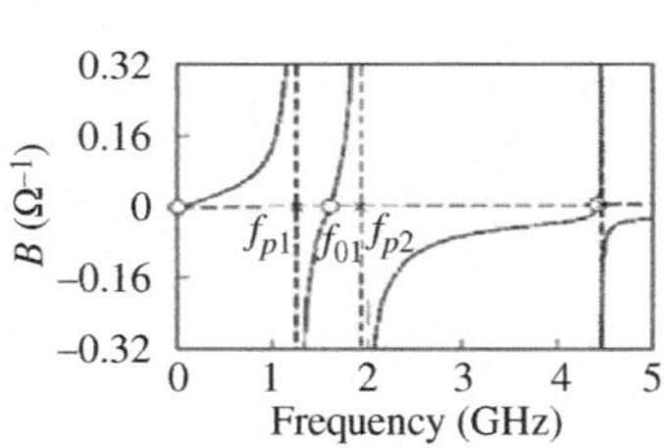	$Y_{in} = jZ_0^{-1}\dfrac{(1+2/m)\,\Omega - \Omega^3}{A+B}$, $A = 1 - (1+2/m)\,\Omega^2$, $B = -Z_0^{-1}\left[(1+2/m)\,\Omega - \Omega^3\right]\omega L$, $\Omega = \tan\left(\dfrac{\theta}{3}\right)$
11	[25]	$\lambda/4$	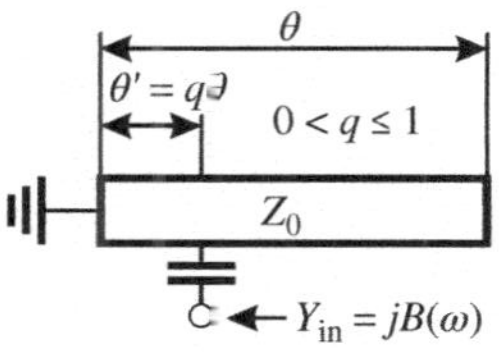	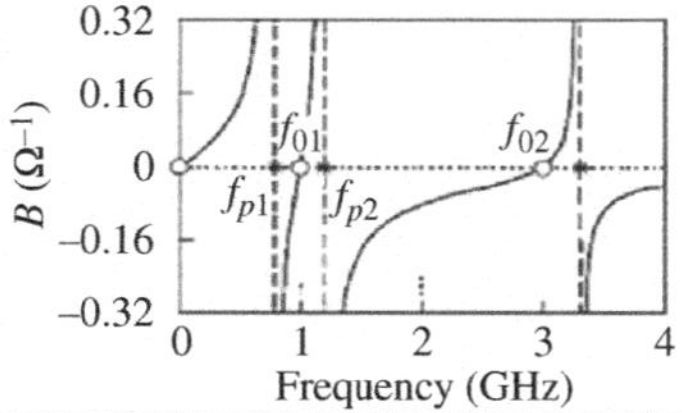	$Y_{in} = \dfrac{-jZ_0^{-1}\cos\theta}{\sin q\theta\cos(1-q)\,\theta - \frac{\cos\theta}{Z_0\omega C}}$

frequencies ω_{pn} correspond to the antiresonant electric lengths θ_{pi}, which can be determined by equating the denominator (9.32) to zero:

$$\theta_{pi} = \frac{i\pi}{q}, \tag{9.33a}$$

$$\theta_{pi} = \frac{(2i-1)(\pi/2)}{1-q}. \tag{9.33b}$$

The trajectories of the antiresonant electric lengths θ_{pi} (9.33) versus q are shown in Figure 9.9b. The intersection points of the curves with an arbitrary horizontal line ($0 < q \leq 1$) correspond to the antiresonant electric lengths θ_{pi} of the resonator at the connection coordinate q.

If we use the second resonant frequency of the resonator ω_{02}, which corresponds to $3\lambda/4$-type oscillations, then for values of q close to 2/3, the antiresonant frequencies ω_{p1} and ω_{p2} will be quite close to ω_{02}. The closer to the coordinate $q = 2/3$ the resonator input is located, the closer to ω_{02} the ω_{p1} and ω_{p2} are located. The trajectories in Figure 9.9b also indicate that the antiresonant frequencies ω_{p1} and ω_{p2} are located at different distances from ω_{02}, i.e., they are not equidistant. If $q < 2/3$, then $\theta_{02} - \theta_{p1} > \theta_{p2} - \theta_{02}$; if $q > 2/3$, then $\theta_{02} - \theta_{p1} < \theta_{p2} - \theta_{02}$. Let $q = 0.7 > 2/3$. Substituting this value into (9.33), we get: $\theta_{p1} = 257.14°$ and $\theta_{p2} = 300°$. If the electrical length $\theta_{02} = 3\pi/2$ corresponds to the resonant frequency $f_{02} = 3$ GHz, then $f_{p1} = 2.857$ GHz $(f_{02} - f_{p1} = 143\ \text{MHz})$ and $f_{p2} = 3.333$ GHz $(f_{p2} - f_{02} = 333\ \text{MHz})$. The f_{p2} pole is more distant from zero f_{02} than the f_{p1} pole.

From (9.32) and (3.2), it follows that the susceptance slope parameter of the considered resonator at the second resonant frequency ω_{02} is expressed by the formula

$$b_2 = \frac{3\pi}{4Z_0 \sin^2(q3\pi/2)}. \tag{9.34}$$

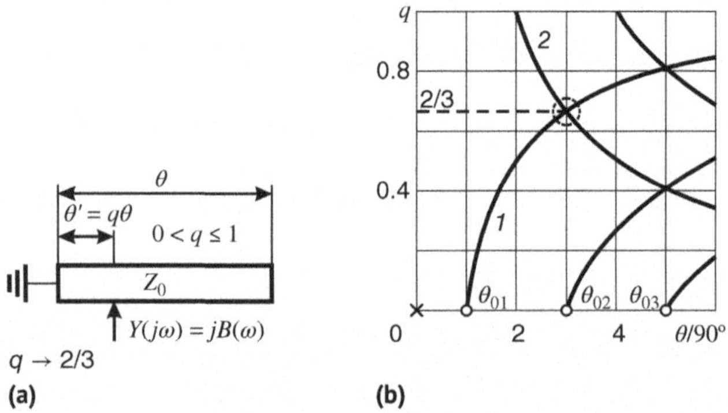

Figure 9.9 Uniform resonator operating on $3\lambda/4$-type oscillations. (a) Schematic. (b) Trajectories of antiresonant electric lengths θ_{pi} (9.33) versus q.

If q approaches from the left or right to 2/3, then the parameter b_2 (9.34) increases. For $q \to 2/3$, the value of $b_2 \to \infty$.

9.4.2 Resonator No. 2 from Table 9.4: Uniform Resonator on $\lambda/2$ Oscillations

This is a well-known resonator [20, 21] with two open ends and a normalized connection coordinate q (Figure 9.10a). It is used at the fundamental resonant frequency ω_{01}, corresponding to $\lambda/2$ oscillations. When the coordinate q is close to 1/2, its two antiresonant frequencies ω_{p1} and ω_{p2} are located next to ω_{01}. The electric length $\theta' = q\theta$ $(0 \le q \le 1)$ is the point at which the input admittance $B(\omega)$ is monitored. The function Y_{in} (3.14b) of the resonator is conveniently represented in the form

$$B = \frac{Z_0^{-1} \sin\theta}{\cos(q\theta)\cos(1-q)\theta}. \tag{9.35}$$

The resonant frequencies ω_{0i} $(i = 1, 2, \ldots)$ and their corresponding electric lengths $\theta_{0i} = i\pi$ are the zeros of the function B (9.35). Equating the denominator (9.35) to zero, we determine the antiresonant electric lengths of the resonator

$$\theta_{pi} = \frac{(2i-1)(\pi/2)}{1-q}, \tag{9.36a}$$

$$\theta_{pi} = \frac{(2i-1)(\pi/2)}{q}, \tag{9.36b}$$

The trajectories of the antiresonant electrical lengths θ_{pi} (9.36) versus q are shown in Figure 9.10b. The intersection points of the curves with an arbitrary

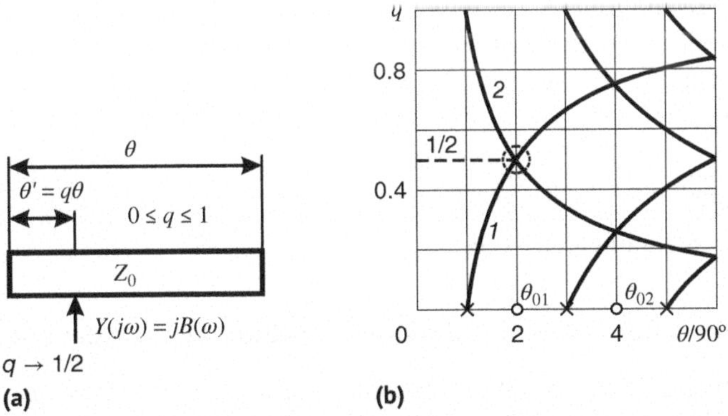

Figure 9.10 Uniform resonator operating on $\lambda/2$-type oscillations. (a) Schematic. (b) Trajectories of antiresonant electric lengths θ_{pi} (9.36) versus q.

horizontal line ($0 \le q \le 1$) correspond to the antiresonant electrical lengths θ_{pi} of the resonator at the connection coordinate q. These trajectories (Figure 9.10b) show that it is possible to bring the antiresonant frequencies ω_{p1} and ω_{p2} closer to the fixed resonant frequencies ω_{01} by appropriate choice of the connection coordinate q.

The q values adjacent to the left or right of the value 1/2 are of practical interest. In this case, the antiresonant electrical lengths θ_{p1} and θ_{p2} are close to the resonant electrical length $\theta_{01} = \pi$, corresponding to the main resonant frequency ω_{01}. If $q < \frac{1}{2}$, then the trajectory θ_{p1} corresponds to curve *1*, and the trajectory θ_{p2} corresponds to curve *2* (Figure 9.10b). The corresponding equalities describe the curves for θ_{p1} and θ_{p2}

$$\theta_{p1} = \frac{\pi/2}{1-q}, \quad \text{for} \quad q < \frac{1}{2}, \tag{9.37a}$$

$$\theta_{p2} = \frac{\pi/2}{q}, \quad \text{for} \quad q < \frac{1}{2}. \tag{9.37b}$$

If $q > 1/2$, then the right parts of (9.37) change places

$$\theta_{p1} = \frac{\pi/2}{q} \quad \text{for} \quad q > \frac{1}{2}, \tag{9.38a}$$

$$\theta_{p2} = \frac{\pi/2}{1-q}, \quad \text{for} \quad q > \frac{1}{2}. \tag{9.38b}$$

The values θ_{p1} and θ_{p2} in both cases (9.37) and (9.38) are not equidistant from θ_{01}. Thc valuc of θ_{p2} is always away from θ_{01} than θ_{p1}. Lct $q = 0.45 < 1/2$. Substituting this value into (9.37), we get: $\theta_{p1} = 163.54°$ and $\theta_{p2} = 200°$. If the electrical length $\theta_{01} = \pi$ corresponds to the resonance frequency $f_{01} = 2$ GHz, then $f_{p1} =$ 1.817 GHz $(f_{01} - f_{p1} = 183\ \text{MHz})$ and $f_{p2} = 2.222$ GHz $(f_{p2} - f_{02} = 222\ \text{MHz})$. The f_{p2} pole is more distant from f_{01} zero than the f_{p1} pole.

The susceptance slope parameter b_i of the considered resonator at the main resonant frequency ω_{01} is expressed by the formula (3.15b). It is presented in a more visual form as follows

$$b = \frac{\pi}{2Z_0 \cos^2(q\pi)}. \tag{9.39}$$

If q tends to 1/2, then parameter $b \to \infty$.

9.4.3 Resonator No. 3 from Table 9.4: Uniform Resonator on λ Oscillations

This resonator (Figure 9.11a) has two short-circuited ends and a normalized connection coordinate q [20]. It was used at the second resonant frequency ω_{02}, corresponding to λ oscillations. Using λ-type oscillations makes it possible to use this resonator at sufficiently high frequencies with an acceptable geometric length.

When the coordinate q is close to 1/2, its two antiresonant frequencies ω_{p1} and ω_{p2} are located next to ω_{02}. The electric length $\theta' = q\theta$ $(0 < q < 1)$ is the point at which the input admittance $B(\omega)$ is monitored. The function Y_{in} (3.14c) of the resonator is conveniently represented in the form

$$B = \frac{-Z_0^{-1}\sin\theta}{\sin(q\theta)\sin[(1-q)\theta]}. \tag{9.40}$$

The resonant frequencies ω_{0i} $(i = 1, 2, \ldots)$ and their corresponding electric lengths $\theta_{0i} = i\pi$ are the zeros of the function B (9.40). Equating the denominator (9.40) to zero, we determine the antiresonant electric lengths of the resonator

$$\theta_{pi} = \frac{i\pi}{q}, \tag{9.41a}$$

$$\theta_{pi} = \frac{i\pi}{1-q}. \tag{9.41b}$$

The trajectories of the antiresonant electrical lengths θ_{pi} (9.41) versus q are shown in Figure 9.11b. These trajectories show that it is possible to bring the antiresonant frequencies ω_{p1} and ω_{p2} closer to the fixed resonant frequencies ω_{02} $(\theta_{02} = 2\pi)$ by an appropriate choice of the connection coordinate q. If $q = 0.45$, then from (9.41), we get $\theta_{p1} = 1.818°$ and $\theta_{p2} = 2.222°$. Similar to the previous resonator, the antiresonant frequency ω_{p2} is further removed from the resonant frequency ω_{02} than ω_{p1}.

Substitution of (9.40) into (3.2) leads to the susceptance slope parameter of the considered resonator, operating at the second resonant frequency ω_{02}

$$b_2 = \frac{\pi}{Z_0\sin^2(q\pi)}. \tag{9.42}$$

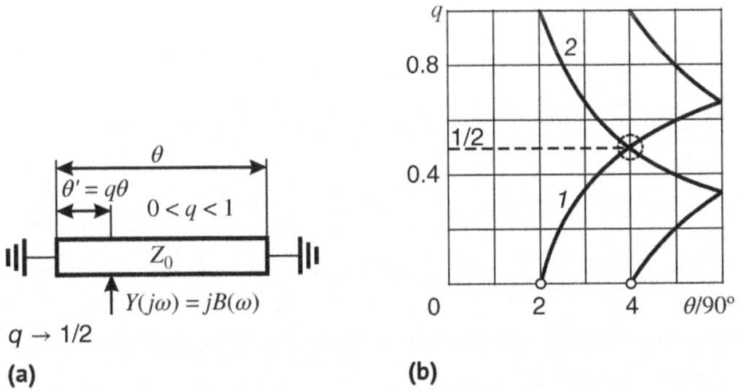

Figure 9.11 Uniform resonator operating on λ-type oscillations. (a) Schematic. (b) Trajectories of antiresonant electric lengths θ_{pi} (9.41) versus q.

If q approaches from the left or right to 1/2, the parameter b_2 (9.42) increases. For $q \to 1/2$, the value of $b_2 \to \infty$.

9.4.4 Resonator No. 4 from Table 9.4: SIR on $\lambda/2$ Oscillations

We implement a case when both poles ω_{p1} and ω_{p2} of the Y function of SIR are located at a small distance on both sides from the main resonant frequency ω_{01}, corresponding to $\lambda/2$ oscillations [14]. The parametric synthesis method (Section 5.6) indicates that using SIR with two open ends and the number of composite segments $N = 2$ is necessary. Using the coefficients of the numerator and denominator polynomials for the input admittance of SIR (Table 5.5a), we write the function $Y^{(2)}(S)$ in the general form

$$Y^{(2)}(S) = \frac{b_1^{(2)} S}{a_0^{(2)} + a_2^{(2)} S^2} = \frac{(1/Z_{01} + 1/Z_{02})\, S}{1 + (Z_{02}/Z_{01})\, S^2}, \tag{9.43}$$

where $S = j\Omega = j \tan \theta_s = j \tan(\theta/2)$.

Zero ω_{01} of the function (9.43) occurs at the value of the frequency variable $S = j\infty$, which corresponds to the electrical length of the composite segment $\theta_s = \pi/2$ and the electrical length of the entire resonator $\theta_{01} = \pi$, which corresponds to the oscillations of $\lambda/2$ type.

The pole Ω_{p1} of function (9.43) on the Ω-axis is expressed by the equality $\Omega_{p1}^2 = a_0^{(2)}/a_2^{(2)} = Z_{01}/Z_{02}$, from which it follows that to increase Ω_{p1}, it is necessary to use the characteristic impedances: $Z_{01} = mZ_0$, $Z_{02} = Z_0$. In this case, function (9.42) takes the form

$$Y^{(2)}(S) = Z_0^{-1} \frac{(1 + 1/m)\, S}{1 + (1/m)\, S^2} = jZ_0^{-1} \frac{(1 + 1/m) \tan(\theta/2)}{1 - (1/m) \tan^2(\theta/2)}, \tag{9.44}$$

and the electrical length θ_{p1} is determined by the equality

$$\theta_{p1} = 2 \tan^{-1} \sqrt{m}. \tag{9.45}$$

Since the function (9.44) is periodic with period $\theta = 2\pi$, then

$$\theta_{p2} = 2\pi - \theta_{p1} = 2\pi - 2 \tan^{-1} \sqrt{m}. \tag{9.46}$$

The electrical lengths θ_{p1} (9.45) and θ_{p2} (9.46) are located symmetrically with respect to the resonant electrical length $\theta_{01} = \pi$, which does not depend on the parameter m of the resonator. The synthesized SIR topology and critical electrical lengths are shown in Figure 9.12a, b. Using (9.45), we obtain the ratio of critical frequencies ω_{01}/ω_{p1} of the considered SIR

$$\frac{\omega_{01}}{\omega_{p1}} = \frac{\theta_{01}}{\theta_{p1}} = \frac{\pi}{\theta_{p1}} = \frac{\pi}{2 \tan^{-1} \sqrt{m}}. \tag{9.47}$$

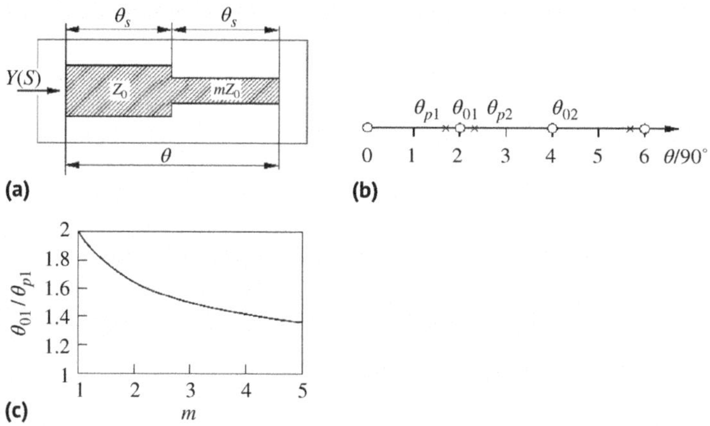

Figure 9.12 Synthesized SIR with $\lambda/2$ oscillation and antiresonant frequencies ω_{p1} and ω_{p2} near resonant frequency ω_{01}. (a) Topology. (b) Critical electrical length. (c) Ratio θ_{01}/θ_{p1}.

The dependence (9.47) is shown in Figure 9.12c. From (9.47), it follows that $\omega_{01}/\omega_{p1} = 1.36$ at $m = 5$, $\omega_{01}/\omega_{p1} = 1.24$ at $m = 10$, and $\omega_{01}/\omega_{p1} = 1.13$ at $m = 30$. In the latter case, the convergence degree of ω_{01} and ω_{p1} is 13%.

For the considered SIR with input admittance (9.44), expression (3.2) gives the susceptance slope parameter

$$b = \frac{(1+m)\,\pi}{4Z_0}. \tag{9.48}$$

As m increases, the slope parameter (9.48) increases too. When m changes from 1 to 5, parameter b (9.48) increases by 3.

9.4.5 Resonator No. 5 from Table 9.4: SIR on $3\lambda/4$ Oscillations

We implement a case when both poles ω_{p1} and ω_{p2} of the Y function of SIR are located at a small distance on both sides from the second resonant frequency ω_{02}, corresponding to $3\lambda/4$ oscillations [22, 23]. The parametric synthesis method (Section 5.6) indicates that using SIR with one short-circuited end and the number of composite segments $N = 3$ is necessary. The input admittance of the SIR has the form

$$Y_{\text{in}}^{3}(S) = \frac{b_0^{(3)} + b_2^{(3)}S^2}{a_1^{(3)}S + a_3^{(3)}S^3}, \tag{9.49}$$

where $S = j\Omega = j\tan\theta_s = j\tan\left(\dfrac{\theta}{3}\right)$.

The coefficients of the numerator and denominator polynomials (9.49) are determined from Table 5.5a

$$b_0^{(3)} = 1, \ b_2^{(3)} = \frac{Z_{01}}{Z_{02}} + \frac{Z_{01}}{Z_{03}} + \frac{Z_{02}}{Z_{03}},$$
$$a_1^{(3)} = Z_{01} + Z_{02} + Z_{03}, \ a_3^{(3)} = Z_{01}Z_{03}/Z_{02}. \tag{9.50}$$

At $S = \pm j\infty$, the input admittance (9.49) is zero, which is ensured when the electrical length of the resonator is $\theta_{02} = 3\pi/2$. This corresponds to $3\lambda/4$-type oscillations, occurring at the second resonant frequency ω_{02}. The pole ω_{p1} of function (9.49) is located at the frequency $\Omega_{p1}^2 = a_1^{(3)}/a_3^{(3)}$. Its increase leads to a decrease in the ratio ω_{p2}/ω_{p1}. Using the coefficients (9.50), we determine

$$\Omega_{p1}^2 = \frac{a_1^{(3)}}{a_3^{(3)}} = Z_{02}\left(\frac{1}{Z_{01}} + \frac{1}{Z_{03}} + \frac{Z_{02}}{Z_{01}Z_{03}}\right). \tag{9.51}$$

From (9.51), the maximum value of Ω_{p1}^2 is achieved at $Z_{01} = Z_{03} = Z_0$ and $Z_{02} = Z_{0\max} = mZ_0$. The parametric synthesis is completed. The synthesized SIR and its characteristics are presented in Figure 9.13. Taking into account (9.50), its input admittance (9.49) takes the form:

$$Y_{\text{in}}(S) = Z_0^{-1}\frac{1 + (1 + m + 1/m)\,S^2}{(2 + m)\,S + (1/m)\,S^3}. \tag{9.52}$$

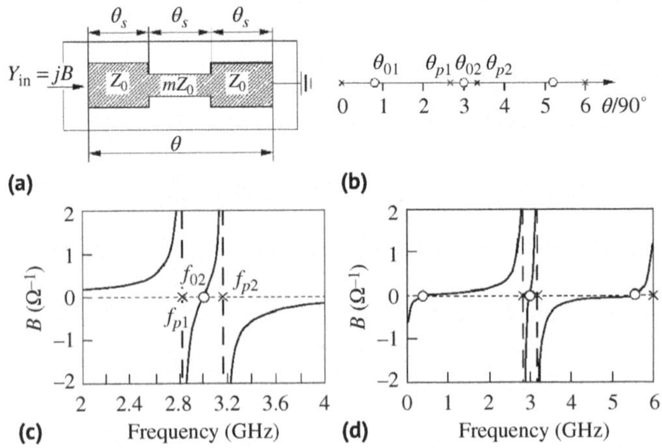

Figure 9.13 Synthesized SIR with oscillation $3\lambda/4$ and antiresonant frequencies ω_{p1} and ω_{p2} closed to resonant frequency ω_{02}. (a) Topology. (b) Critical electrical length. (c) Input admittance in narrow frequency band for case $Z_0 = 10\ \Omega$, $m = 10$, $f_{02} = 3$ GHz. (d) Input admittance over a wide frequency band for the same case.

Equating the denominator (9.52) to zero, we determine the electrical length θ_{p1} corresponding to the antiresonant frequency ω_{p1}

$$\theta_{p1} = 3\tan^{-1}\sqrt{(m+2)\,m}. \tag{9.53}$$

Since function (9.52) is periodic with period $\theta = 3\pi$, then the electrical length θ_{p2}, corresponding to the antiresonant frequency ω_{p2}, is determined by the equality

$$\theta_{p2} = 3\pi - \theta_{p1} = 3\pi - 3\tan^{-1}\sqrt{(m+2)\,m}. \tag{9.54}$$

The electrical lengths θ_{p1} (9.53) and θ_{p2} (9.54) are located equidistantly with respect to the electrical length at resonance $\theta_{02} = 3\pi/2$, which does not depend on the parameter m of the resonator. Using (9.53) and (9.54), we obtain the ratio of antiresonant frequencies ω_{p2}/ω_{p1} of the considered SIR

$$\frac{\omega_{p2}}{\omega_{p1}} = \frac{\theta_{p2}}{\theta_{p1}} = \frac{\pi}{\tan^{-1}\sqrt{(m+2)\,m}} - 1. \tag{9.55}$$

The dependence (9.55) is displayed in Table 9.5. The antiresonant frequencies ω_{p2} and ω_{p1} significantly approach each other as parameter m increases. At $m = 30$, the convergence degree is 4%.

For the considered SIR, with input admittance (9.52), expression (3.2) gives the susceptance slope parameter at resonant frequency ω_{02}

$$b_2 = \frac{(1+m+m^2)\,\pi}{4Z_0}. \tag{9.56}$$

When m increases, then b_2 (9.56) increases too.

9.4.6 Resonator No. 6 from Table 9.4: SIR on $5\lambda/4$ Oscillations

We implement a case when both poles ω_{p2} and ω_{p3} of the Y function of SIR are located at a small distance on both sides from the third resonant frequency ω_{03}, corresponding to $5\lambda/4$ oscillations [22, 24]. The parametric synthesis method

Table 9.5 Ratio ω_{p2}/ω_{p1} versus m (9.55).

m	1	5	10	20	30
$\frac{\omega_{p2}}{\omega_{p1}}$	3	1.24	1.12	1.06	1.04

(Section 5.6) indicates that using SIR with one short-circuited end, the number of composite segments $N = 5$ is necessary. The input admittance of the SIR has the form

$$Y^5(S) = \frac{b_0^{(5)} + b_2^{(5)}S^2 + b_4^{(5)}S^4}{a_1^{(5)}S + a_3^{(5)}S^3 + a_5^{(5)}S^5}, \tag{9.57}$$

where

$$S = j\Omega = j\tan\left(\frac{\theta}{5}\right), \tag{9.58}$$

$$a_5^{(5)} = Z_{01}Z_{03}Z_{05}/Z_{02}Z_{04}. \tag{9.59}$$

The remaining coefficients of function (9.57) are contained in Table 5.5b.

At $S = \pm j\infty$, the input admittance (9.57) is zero, which is ensured when the electrical length of the resonator is $\theta_{03} = 5\pi/2$. It corresponds to $5\lambda/4$-type oscillations, occurring at the third resonant frequency ω_{03}. From (9.57) it follows that the leading root Ω_{p2}^2 of the algebraic equation of the second degree determines the pole ω_{p2} of function (9.57)

$$\Omega_{p2}^2 = \frac{a_3^{(5)}}{2a_5^{(5)}} + \sqrt{\left[\frac{a_3^{(5)}}{2a_5^{(5)}}\right]^2 - \frac{a_1^{(5)}}{a_5^{(5)}}}. \tag{9.60}$$

The maximum value of Ω_{p2}^2 is ensured at the minimum value of the coefficient (9.59) $a_5^{(5)} = \dfrac{Z_{01}Z_{03}Z_{05}}{Z_{02}Z_{04}}$, which leads to the final solution: $Z_{01} = Z_{03} = Z_{05} = Z_{0\min} = Z_0$ and $Z_{02} = Z_{04} = Z_{0\max} = mZ_0$. The synthesized SIR and its characteristics are presented in Figure 9.14. Its input admittance (9.57) takes the form:

$$Y_{\text{in}}(S) = jB = Z_0^{-1}\frac{1 + (4 + 3m + 3/m)S^2 + \left(1 + m + m^2 + 1/m + 1/m^2\right)S^4}{(3 + 2m)S + (3 + 2m + m^2 + 4/m)S^3 + (1/m^2)S^5}. \tag{9.61}$$

Equating the denominator (9.61) to zero, we determine the electrical length θ_{p2} corresponding to the antiresonant frequency ω_{p2}

$$\theta_{p2}(m) = 5\tan^{-1}\left[A/2 + \sqrt{(A/2)^2 - m^2(3 + 2m)}\right]^{\frac{1}{2}}, \tag{9.62}$$

where $A = 4m + 3m^2 + 2m^3 + m^4$.

Since function (9.61) is periodic with period $\theta = 5\pi$, then the electrical length θ_{p3}, corresponding to the antiresonant frequency ω_{p3}, is determined by the equality

$$\theta_{p3}(m) = 5\pi - \theta_{p2}(m). \tag{9.63}$$

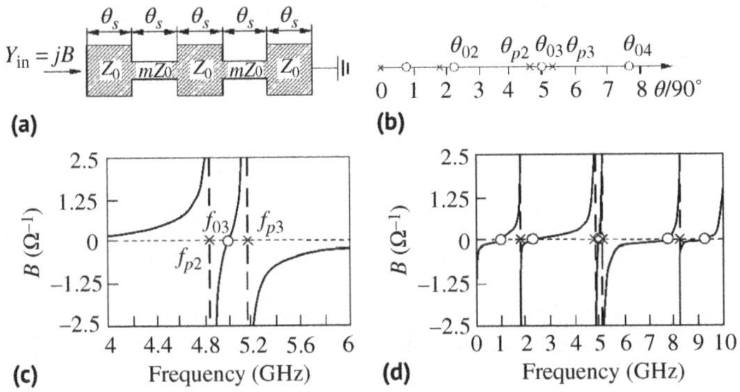

Figure 9.14 Synthesized SIR with $5\lambda/4$ oscillation and antiresonant frequencies ω_{p2} and ω_{p3} close to resonant frequency ω_{03}. (a) Topology. (b) Critical electrical length. (c) Input admittance in narrow frequency band at $Z_0 = 10\ \Omega$, $m = 4$, $f_{03} = 5$ GHz. (d) Input admittance over a wide frequency band for the same case.

Table 9.6 Ratio ω_{p3}/ω_{p2} versus m (9.64).

m	1	2	3	4	5	10	30
$\dfrac{\omega_{p3}}{\omega_{p2}}$	1.5	1.19	1.10	1.06	1.04	1.01	1.001

The electrical lengths θ_{p2} (9.62) and θ_{p3} (9.63) are located equidistantly with respect to the electrical length at resonance $\theta_{03} = 5\pi/2$, which does not depend on the parameter m of the resonator. Using (9.62) and (9.63), we obtain the ratio

$$\frac{\omega_{p3}}{\omega_{p2}} = \frac{\theta_{p3}}{\theta_{p2}} = \frac{\pi}{\tan^{-1}\left[A/2 + \sqrt{(A/2)^2 - m^2\,(3+2m)}\right]^{\frac{1}{2}}} - 1, \tag{9.64}$$

where A is the same as in (9.62).

The dependence (9.64) is displayed in Table 9.6. The antiresonant frequencies ω_{p3} and ω_{p2} significantly approach each other as parameter m increases. At $m = 5$, the degree of convergence is 4%.

For the considered SIR, the susceptance slope parameter (3.2) at ω_{03} is expressed by the formula

$$b_3 = \frac{(1 + m + m^2 + m^3 + m^4)\,\pi}{4Z_0}. \tag{9.65}$$

As m increases, the susceptance slope parameter b_3 (9.65) increases as well.

9.5 LD Resonators with Two Antiresonant Frequencies Close to Operating Resonant Frequency

Such resonators are presented in Table 9.4b (LD resonators). They occupy five positions in this table.

9.5.1 Resonator No. 7 from Table 9.4: Half-wave Resonator with C and L

This LD resonator (Figure 9.15a) is a half-wave resonator to which an LC chain is cascaded [16]. The resonator input is located on the side of the lumped elements.

The transfer matrix of a series LC circuit has the form

$$\begin{bmatrix} A & B \\ C & D \end{bmatrix} = \begin{bmatrix} 1 & j\left(\omega L - 1/\omega C\right) \\ 0 & 1 \end{bmatrix}. \tag{9.66}$$

Substituting (9.66) into (9.18), we obtain the input admittance of the resonator

$$Y_{\text{in}}\left(j\omega\right) = jZ_0^{-1} \frac{\sin\theta}{\cos\theta - Z_0^{-1}\left(\omega L - 1/\omega C\right)\sin\theta}. \tag{9.67}$$

Expression (9.67) indicates that the resonant frequencies ω_{0i} of the considered and $\lambda/2$ resonators coincide. They correspond to electrical lengths $\theta_{0n} = n\pi$. Equating the denominator of the input admittance Y_{in} (9.67) to zero, we obtain the equation for determining the poles ω_p:

$$\omega_p L - \frac{1}{\omega_p C} = Z_0 \cot\theta_p. \tag{9.68}$$

Figure 9.15b shows the electrical lengths θ_{0n} and θ_{pn}, corresponding to the critical frequencies ω_{0n} and ω_{pn} of the resonator.

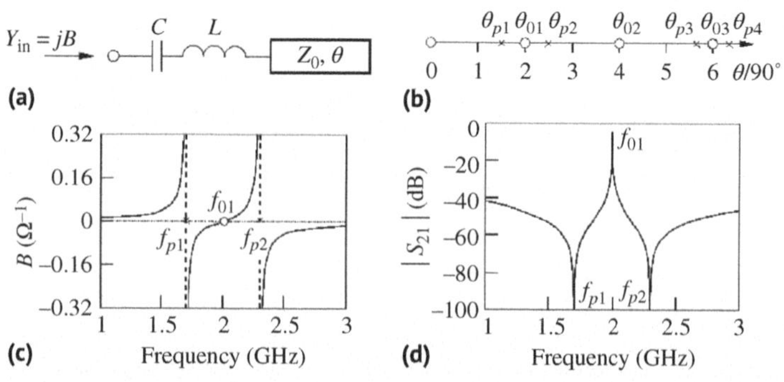

Figure 9.15 $\lambda/2$ resonator with capacitance C and inductance L. (a) Schematic. (b) Critical electrical lengths. (c) Input admittance. (d) Frequency response.

Connecting capacitance C and inductance L to the $\lambda/2$ resonator leads to the approach of the antiresonant frequencies ω_{p1} and ω_{p2} near the resonant frequency ω_{01} from both sides. Along with the main resonant frequency ω_{01}, other resonant frequencies with odd numbers n are surrounded by closely spaced antiresonant frequencies (Figure 9.15b). As n increases, the convergence degree of the indicated frequencies increases. This is reflected in the function of the input admittance $Y(j\omega)$ of this resonator (Figure 9.15c), which has the resonant frequency $f_0 = 2$ GHz. In this case, the transfer function $|S_{21}|$ has two TZs close to ω_{01} (Figure 9.15d). Their positions can be adjusted by changing L and C.

Table 9.7 presents the calculated values of C and L (9.68) with the equidistant approaches f_{p1} and f_{p2} of the resonator under consideration to the fundamental resonant frequency $f_{01} = 2$ GHz. As before, the Z_0 takes the values 50, 10, and 5 Ω. At $Z_0 = 5$ Ω, a relatively high convergence degree of f_{p1}, f_{p2}, and f_{01} is provided, characterized by the ratio $f_{p2}/f_{p1} = 1.07$ (7%) at practically realizable values of $C = 0.12$ pF and $L = 52.8$ nH. Substituting (9.67) into (3.2), we obtain the susceptance slope parameter of the LD resonator $b = \pi/2Z_0$. It is the same as for the half-wave resonator.

9.5.2 Resonator No. 8 from Table 9.4: Quarter-wave Resonator with *C* and *L*

This LD resonator (Figure 9.16a) is a quarter-wave resonator to which an LC chain is cascaded [17, 18]. The resonator input is located on the side of the lumped elements. Figure 9.16b shows its critical electrical lengths and antiresonant frequencies ω_{p1} and ω_{p2} near the fundamental resonant frequency ω_{01}. This resonator is a cascade connection of a short-circuited quarter-wave segment and a series LC circuit.

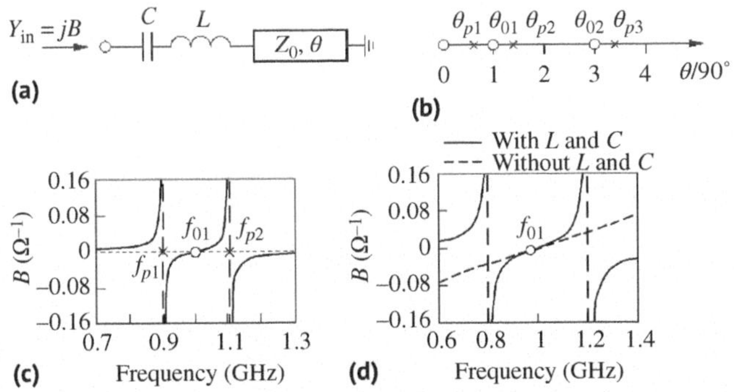

Figure 9.16 $\lambda/4$ resonator with capacitance C and inductance L. (a) Schematic. (b) Critical electrical length. (c) Input admittance for case $Z_0 = 10$ Ω, $C = 0.519$ pF, $L = 49.4$ nH, and $f_{01} = 1$ GHz. (d) Illustration of property $b = \text{const}$.

Table 9.7 Values of C and L as antiresonant frequencies f_{p1} and f_{p2} approach fundamental resonant frequency f_{01} for resonator in Figure 9.15a at $f_{01} = 2$ GHz.

No.		**1**	**2**	**3**	**4**	**5**	**6**	**7**
$\frac{f_{p1}}{f_{p2}}$, GHz		1.5/2.5	1.7/2.3	1.75/2.25	1.8/2.2	1.85/2.15	1.9/2.1	1.93/2.07
$\frac{\theta_{p1}}{\theta_{p2}}$, °		135/225	153/207	157.5/202.5	162/198	166.5/193.5	171/189	173.7/186.3
$Z_0 = 5\ \Omega$	C, pF	8.54	2.5	1.66	1.02	0.58	0.26	0.12
	L, nH	0.79	2.58	3.86	6.25	10.95	24.4	52.8
$Z_0 = 10\ \Omega$	C, pF	4.27	1.25	0.83	0.51	0.29	0.13	0.06
	L, nH	1.58	5.17	7.72	12.5	21.9	48.8	105.6
$Z_0 = 50\ \Omega$	C, pF	0.854	0.25	0.166	0.106	0.058	0.026	–
	L, nH	7.9	25.8	38.6	60.2	109.5	244	–

Substituting (9.66) into (9.27), we obtain the input admittance of the resonator

$$Y_{\text{in}}(j\omega) = -jZ_0^{-1}\frac{\cos\theta}{\sin\theta + Z_0^{-1}(\omega L - 1/\omega C)\cos\theta}. \tag{9.69}$$

where Z_0 is the characteristic impedance of $\lambda/4$ TL segment.

Expression (9.69) indicates that the resonant frequencies ω_{0i} of the considered resonator are the same as those of the $\lambda/4$ resonator. These resonant frequencies are related to each other as odd numbers. Equating the denominator of input admittance Y_{in} (9.69) to zero, we obtain an equation for determining poles ω_p:

$$\omega_p L - 1/\omega_p C = -Z_0 \tan\theta_p. \tag{9.70}$$

Expression (9.70) indicates that connection to a quarter-wave resonator series LC circuit leads to antiresonant frequencies ω_{p1} and ω_{p2} approaching the resonant frequency ω_{01} from both sides. From (9.70) it follows that changing the values of L and C makes it possible to change the position of the antiresonant frequencies ω_{p1} and ω_{p2} over a wide range.

The most useful case is when ω_{p1} and ω_{p2} are equidistant relative to ω_{01}. Table 9.8 presents the calculated values of C and L (9.70) with the equidistant approaches f_{p1} and f_{p2} of the resonator under consideration to the main resonant frequency $f_{01} = 1$ GHz. The characteristic impedance Z_0 of the quarter-wave segment equals 20, 10, and 5 Ω. The data in Table 9.8 indicate that at practical values of capacitance C and inductance L, the high convergence degree between the antiresonant frequencies f_{p1} and f_{p2} is achieved.

Using a TL segment with low characteristic impedance is promising in such resonators. The value $Z_0 = 5$ Ω presented in Table 9.8 is realized by the width of the microstrip line $w = 3$ mm on a substrate with a dielectric constant $\varepsilon_r = 92$, the

Table 9.8 Parameters of LD resonators in Figure 9.16a with $f_{01} = 1$ GHz and equidistant antiresonant frequencies f_{p1} and f_{p2}.

No.		1	2	3	4	5
f_{p1}/f_{p2}, GHz		0.6/1.4	0.7/1.3	0.8/1.2	0.9/1.1	0.975/1.025
Ratio f_{p2}/f_{p1}		2.333	1.857	1.5	1.222	1.051
θ_{p1}/θ_{p2}, °		54/126	63/117	72/108	81/99	87.75/92.25
$Z_0 = 5$ Ω	C, pF	21.76	10.72	4.26	1.038	0.0635
	L, nH	1.38	2.6	6.19	24.7	400
$Z_0 = 10$ Ω	C, pF	10.89	5.36	2.13	0.519	0.0318
	L, nH	2.76	5.2	12.38	49.4	800
$Z_0 = 20$ Ω	C, pF	5.44	2.68	1.065	0.260	0.0159
	L, nH	5.52	10.4	24.76	98.48	1600

thickness of which is $h = 0.5$ mm. In this case, the value $C = 0.0635$ pF can be provided by a lumped capacitor [31] and the value $L = 400$ nH by a lumped inductor [32], which leads to the convergence degree of ω_{p1} and ω_{p2} equal to 5.1%. This is somewhat better than the resonator in Figure 9.15a.

The input admittance (9.69) of the considered resonator with $Z_0 = 10\ \Omega$, $C = 0.519$ pF, $L = 49.4$ nH, and $f_{01} = 1$ GHz is shown in Figure 9.16c.

Substituting (9.69) into (3.2), we obtain the susceptance slope parameter of the LD resonator $b = \pi/4Z_0$. It is the same as a quarter-wave resonator. Parameter b is not dependent on the values of L and C. The two poles of the function Y_{in} (9.69) of this resonator can be located as close as we like to the frequency of the main resonance, which does not increase parameter b. This result is unexpected because the parameter b was expected to increase. Figure 9.16d compares the input admittance of the considered resonator with and without lumped elements. The presence of lumped elements L and C leads to the poles ω_{p1} and ω_{p2} being close to ω_0, however, the slope of the input admittance at the frequency ω_0 and the parameter b do not change. The property $b =$ const significantly simplifies the BPF design with such resonators.

9.5.3 Resonator No. 9 from Table 9.4: SIR with *C*

All resonators discussed as follows use a cascade connection of one lumped element L or C to the TL segment [25]. The resonator input is located on the lumped element side. To obtain the Y_{in} function of LD resonators, it is necessary to transform the input admittance Y_L of the TL segment using the transfer matrix $[ABCD]$ of the lumped element:

$$Y_{\text{in}} = \frac{C + Y_L D}{A + Y_L B}. \tag{9.71}$$

The considered LD resonator (Figure 9.17a) is a stepped-impedance segment with a short-circuited end and capacitance C. The input admittance Y_L of the TL segment is defined by expression (9.9), which we write in the form

$$Y_L = jB_L = -jZ_0^{-1}\frac{1-(1+2/m)\,\Omega^2}{(2+m)\,\Omega - m\Omega^3}. \tag{9.72}$$

where $\Omega = \tan(\theta/3)$.

Transforming the admittance Y_L (9.72) via the transfer matrix of capacitance (9.19), we obtain the input admittance Y_{in} of the LD resonator under consideration

$$Y_{\text{in}} = jB_{\text{in}} = -jZ_0^{-1}\frac{1-(1+2/m)\,\Omega^2}{(2+m)\,\Omega - m\Omega^3 - Z_0^{-1}\left[1-(1+2/m)\,\Omega^2\right]/\omega C}. \tag{9.73}$$

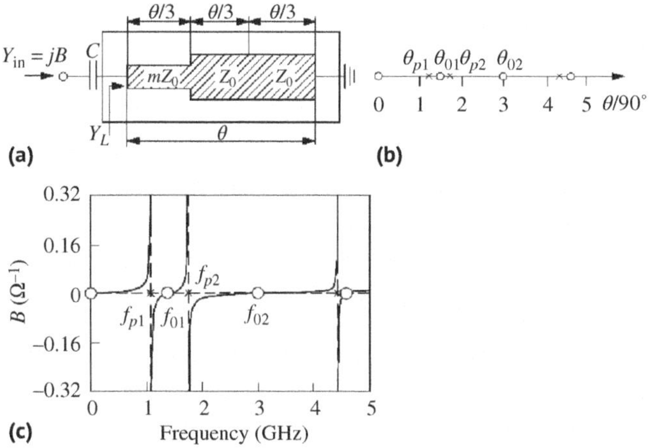

Figure 9.17 Stepped-impedance segment with capacitance. (a) Schematic (b) Critical electrical length. (c) Input admittance for $m = 10$, $Z_0 = 10\ \Omega$, $C = 1.51$ pF.

A comparison of (9.73) and (9.72) shows that the resonant frequencies ω_{0i} and the corresponding electrical lengths θ_{0i} of the LD resonator and stepped-impedance segment coincide, and their antiresonant frequencies ω_{pi} and the corresponding electrical lengths θ_{pi} differ from each other. Figure 9.17b shows the electrical lengths θ_{0i} and θ_{pi}, corresponding to the critical frequencies ω_{0i} and ω_{pi} of the considered resonator. Equating the numerator (9.73) to zero, we determine the electrical length

$$\theta_{01} = 3\tan^{-1}(1 + 2/m)^{-1/2}, \tag{9.74}$$

corresponding to the main resonant frequency ω_{01}.

Equality (9.74) indicates that θ_{01} depends on the parameter m. As m increases, the value of θ_{01} also increases. Equating the denominator (9.73) to zero, we obtain the equation that determines the antiresonant frequencies ω_{pi} (θ_{pi}):

$$\omega_p C = -B_L(\omega_p) = Z_0^{-1}\frac{1 - (1 + 2/m)\,\Omega_p^2}{(2 + m)\,\Omega_p - m\Omega_p^3}. \tag{9.75}$$

For the resonator under consideration, ω_{p1} and ω_{p2} can be located close to ω_{01}. The convergence degree of ω_{p2} and ω_{01} is characterized by the ratio $\omega_{p2}/\omega_{01} > 1$. The smaller this ratio, the closer ω_{p2} to ω_{01}. From (9.74) and (9.75), it follows that the lower bound of this ratio is determined by the equality

$$\mathrm{Inf}\left\{\frac{\omega_{p2}}{\omega_{01}}\right\} = \frac{\tan^{-1}(1+2/m)^{1/2}}{\tan^{-1}(1+2/m)^{-1/2}}, \tag{9.76}$$

which is provided at $C \to \infty$. The maximum approach (9.76) depends only on the parameter m of the stepped-impedance segment. The larger m, the closer ω_{p2} and ω_{01} are to each other. The stepped-impedance segment in the resonator was synthesized before (Figure 9.2b).

From (9.74) and (9.75), the antiresonant frequency ω_{p1} can be located arbitrarily close to ω_{01}. This is ensured by a decrease in the value of C for any values of the parameter m, $\omega_{p1} \to \omega_{01}$ if $C \to 0$.

The resonator makes it possible to ensure an equidistant arrangement of ω_{p1} and ω_{p2} relative to ω_{01}. Table 9.9 presents the calculated values of ratio θ_{p2}/θ_{01} and C (9.75) for equidistant arrangement of ω_{p1} and ω_{p2} for different values of the parameter m and $Z_0 = 10\ \Omega$. As m increases, the convergence degree of ω_{p2} and ω_{01} increases. With $m = 10$, it is characterized by the ratio $\omega_{p2}/\omega_{01} = 1.23$. This case corresponds to the input admittance function of the resonator, which is shown in Figure 9.17c.

At $m = 30$, the convergence degree of ω_{p1} and ω_{p2} is characterized by the ratio $\omega_{p2}/\omega_{01} = 1.13$. Table 9.9 also contains the electrical length θ_{01} (9.74) at the main resonance frequency.

The peculiarity of this resonator is that its distributed part ensures that the right-handed antiresonant frequency ω_{p2} is close to ω_{01}, and the capacitance C is the left-handed one ω_{p1}.

Substituting (9.73) into (3.2), we determine the susceptance slope parameter of the LD resonator

$$b = \frac{(1+2/m)\tan^{-1}(1+2/m)^{-1/2}}{2Z_0}. \tag{9.77}$$

The parameter b (9.77) decreases as m increases (Table 9.9). It does not depend on the value of capacitance C and coincides with (9.12).

Table 9.9 Ratio θ_{p2}/θ_{02} at equidistant arrangement of antiresonant frequencies in resonator in Figure 9.17 with $Z_0 = 10\ \Omega$.

m	3	5	7	10
θ_{p2}/θ_{01}	1.47	1.34	1.28	1.23
C, pF	11.0	4.20	2.50	1.51
θ_{01}, °	113.28	120.61	124.23	127.18
b, Ω^{-1}	0.05492	0.04912	0.04646	0.04439

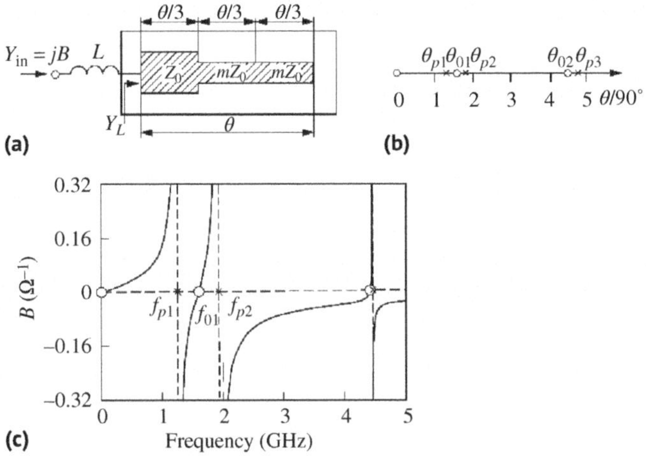

Figure 9.18 Stepped-impedance segment with inductance. (a) Schematic. (b) Critical electrical length. (c) Input admittance for case $m = 10$, $Z_0 = 10\ \Omega$, $L = 0.8$ nH.

9.5.4 Resonator No. 10 from Table 9.4: SIR with *L*

This LD resonator is a stepped-impedance segment with open-circuited ends and inductance L (Figure 9.18a). The input admittance Y_L of the TL is defined by (9.13), which we rewrite as follows:

$$Y_L = jB_L = jZ_0^{-1}\frac{(1+2/m)\,\Omega - \Omega^3}{1-(1+2/m)\,\Omega^2}, \tag{9.78}$$

where $\Omega = \tan(\theta/3)$.

Modifying admittance Y_L (9.78) through the transfer matrix of inductance (9.23), we obtain the input admittance Y_{in} of the LD resonator under consideration

$$Y_{\text{in}} = jB_{\text{in}} = jZ_0^{-1}\frac{(1+2/m)\,\Omega - \Omega^3}{1-(1+2/m)\,\Omega^2 - Z_0^{-1}\left[(1+2/m)\,\Omega - \Omega^3\right]\omega L}. \tag{9.79}$$

The resonant frequencies ω_{0i} and the corresponding electrical lengths θ_{0i} of the LD resonator and stepped-impedance segment coincide, and their antiresonant frequencies ω_{pi} and the corresponding electrical lengths θ_{pi} differ from each other, as follows from the comparison of (9.79) and (9.78). Figure 9.18b shows the electrical lengths θ_{0i} and θ_{pi}, corresponding to the critical frequencies ω_{0i} and ω_{pi} of the resonator.

Equating the numerator (9.79) to zero, we determine the electrical length θ_{01} at the fundamental resonant frequency of the resonator

$$\theta_{01} = 3\tan^{-1}(1+2/m)^{1/2}. \tag{9.80}$$

Equality (9.80) indicates that θ_{01} depends on the parameter m. If m increases, the value of θ_{01} decreases. Equating the denominator (9.79) to zero, we obtain the equation that determines the antiresonant frequencies $\omega_{pi}\left(\theta_{pi}\right)$:

$$\omega_p L = \frac{1}{B_L\left(\omega_p\right)} = Z_0 \frac{1-(1+2/m)\,\Omega^2}{(1+2/m)\,\Omega-\Omega^3}. \tag{9.81}$$

The antiresonant frequencies ω_{p1} and ω_{p2} of this resonator can be located close to ω_{01}. The ratio $\omega_{01}/\omega_{p1} > 1$ characterizes the convergence degree of ω_{p1} and ω_{01}. The smaller the ratio ω_{01}/ω_{p1}, the closer ω_{p1} to ω_{01}. From (9.80) and (9.81), it follows that the lower limit of this ratio is determined by equality

$$\operatorname{Inf}\left\{\frac{\omega_{01}}{\omega_{p1}}\right\} = \frac{\tan^{-1}(1+2/m)^{1/2}}{\tan^{-1}(1+2/m)^{-1/2}}, \tag{9.82}$$

which is held at $L \to \infty$ and coincides with (9.76). The maximum approach (9.82) depends on the parameter m of the stepped-impedance segment. The larger m, the closer ω_{p1} and ω_{01} are to each other. The stepped-impedance segment with both open ends in the resonator is synthesized before (Figure 9.4a).

From (9.80) and (9.81), the antiresonant frequency ω_{p2} can be located arbitrarily close to ω_{01}. This is ensured by increasing the values of L for any values of the parameter m ($\omega_{p2} \to \omega_{01}$, if $L \to \infty$).

This resonator can also provide the equidistant arrangement of ω_{p1} and ω_{p2} relative to ω_{01}. Table 9.10 presents the calculated values of the ratio θ_{p2}/θ_{01} and L (9.81) for equidistant location ω_{p1} and ω_{p2} for different values of m and $Z_0 = 10\ \Omega$. We can see from Table 9.10 that as m increases, the convergence degree of ω_{p2} and ω_{01} increases. At $m = 10$, it is characterized by the ratio $\omega_{p2}/\omega_{01} = 1.21$. This case corresponds to the input admittance shown in Figure 9.18c. For $m = 30$, the convergence degree of ω_{01} and ω_{p2} is characterized by the ratio $\omega_{p2}/\omega_{01} = 1.12$.

Table 9.10 Ratio θ_{p2}/θ_{01} at equidistant arrangement of antiresonant frequencies in resonator in Figure 9.18a with $Z_0 = 10\ \Omega$.

m	3	5	7	10
θ_{p2}/θ_{01}	1.35	1.29	1.25	1.21
L, nH	0.46	0.62	0.72	0.80
θ_{01}, °	156.72	149.39	145.77	142.82
b, Ω^{-1}	0.2279	0.3042	0.3816	0.4985

Substituting (9.79) into (3.2), we determine the susceptance slope parameter of this LD resonator

$$b = \frac{(1 + m/2)\tan^{-1}(1 + 2/m)^{1/2}}{Z_0}. \tag{9.83}$$

The parameter b (9.83) increases with increasing m (Table 9.10). It does not depend on the inductance L and coincides with (9.17).

9.5.5 Resonator No. 11 from Table 9.4: Quarter-wave Segment with Nontraditional Connection of Capacitance

Figure 9.19a shows a schematic of resonator No. 11. It consists of a uniform quarter-wave segment with characteristic impedance Z_0 and electrical length θ, to which capacitance C is connected at a distance $q\theta$ ($0 < q \leq 1$) from the short-circuited end. The input admittance Y_L of the quarter-wave segment at the point $q\theta$ is written as

$$Y_L = jB_L = -jZ_0^{-1}\frac{\cos\theta}{\sin(q\theta)\cos[(1-q)\theta]}. \tag{9.84}$$

Substituting (9.84) into (9.71), we determine the input admittance of the LD resonator:

$$Y_{\text{in}} = jB_{\text{in}} = -jZ_0^{-1}\frac{\cos\theta}{\sin(q\theta)\cos[(1-q)\theta] - \cos\theta/(Z_0\omega C)}. \tag{9.85}$$

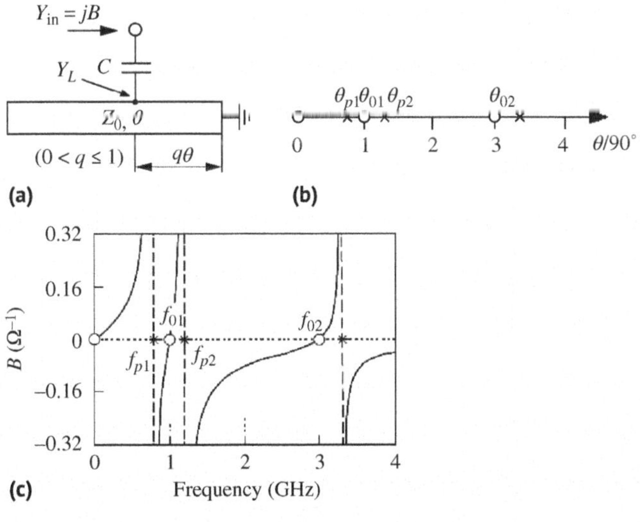

Figure 9.19 Uniform quarter-wave segment with capacitance C. (a) Schematic. (b) Critical electrical length. (c) Input admittance for $Z_0 = 30\ \Omega$, $q = 0.074$, $C = 56$ pF, $f_{01} = 1$ GHz.

Equating the numerator (9.85) to zero, we determine the resonant frequencies of the resonator ω_{0i}, $i = 1, 2, \ldots$, which correspond to the resonant electrical lengths $\theta_{0i} = (2i - 1)\,\pi/2$. These frequencies are the same as those of a quarter-wave resonator. They are related to each other as odd numbers 1:3:5.... The main resonant frequency ω_{01} corresponds to the electrical length

$$\theta_{01} = \frac{\pi}{2}. \tag{9.86}$$

Equating the denominator (9.85) to zero, we obtain the equation that determines the antiresonant frequencies ω_{pi} (θ_{pi}):

$$\omega_p C = -B_L\left(\omega_p\right) = Z_0^{-1} \frac{\cos\theta_p}{\sin\left(q\theta_p\right)\cos\left[(1-q)\,\theta_p\right]}. \tag{9.87}$$

The electrical length θ_p in (9.87) corresponds to the antiresonant frequency ω_p.

Analysis of expression (9.87) shows that at small q values, two antiresonant frequencies ω_{p1} and ω_{p2} are close to the resonant frequency ω_{01} (Figure 9.19b). For a given value of Z_0, the values of q and C can be determined from (9.87) so that the antiresonant frequencies ω_{p1} and ω_{p2} are located equidistant relative to ω_{01}.

The second row of Table 9.11 shows various options for the equidistant arrangement of the antiresonant frequencies f_{p1} and f_{p2} of the resonator at $f_{01} = 1$ GHz and $Z_0 = 30\ \Omega$. Rows 3 and 4 of Table 9.11 show the values of q and C, which provide the indicated values of antiresonant frequencies. We can see from Table 9.11 that the approach of f_{p1} and f_{p2} to f_{01} is accompanied by a decrease in q and an increase in C. Figure 9.19c shows the input admittance B_{in} (9.85) of the resonator from Table 9.11, which has antiresonant frequencies $f_{p1} = 0.8$ GHz and $f_{p2} = 1.2$ GHz. In this case, the convergence degree of f_{p2} and f_{01} is characterized by the ratio $f_{p2}/f_{01} = 1.20$. With a further decrease in q and an increase in C, we can achieve the convergence degree of f_{p2} and f_{01} equal to $f_{p2}/f_{01} = 1.15$.

Table 9.11 Parameters of LD resonator in Figure 9.19a with equidistant antiresonant frequencies f_{pk} $(k = 1, 2)$ at $f_{01} = 1$ GHz and $Z_0 = 30\ \Omega$.

Res. No.	1	2	3
f_{p1}, GHz	0.6	0.7	0.8
f_{p2}, GHz	1.4	1.3	1.2
q_i	0.2222	0.1444	0.07444
C_i, pF	34.5	36.8	56
b_i	0.2238	0.07296	1.923

Substituting (9.85) into (3.2), we determine the susceptance slope parameter of the resonator:

$$b = \frac{\pi}{4Z_0 \sin^2 (q\pi/2)}. \tag{9.88}$$

The parameter b (9.88) increases with decreasing connection coordinate q (Table 9.11). It does not depend on the value of capacitance C and coincides with (9.6).

9.6 Systematization and Basic Properties of Resonators with One or Two Antiresonant Frequencies Close to Operating Resonant Frequency

9.6.1 Resonators with One Antiresonant Frequency Close to the Main Resonant Frequency

1. These seven resonators are contained in Table 9.1. Three of them (Nos. 1–3) are TL resonators, the remaining four resonators (Nos. 4–7) are LD ones. All of them operate at the main resonant frequency ω_{01}.
2. Quarter-wave resonator No. 1 provides a right-hand approach of the antiresonant frequency ω_{p1} to ω_{01}. The convergence degree of the indicated critical frequencies is not limited. As they approach, the susceptance slope parameter b increases significantly.
3. Stepped-impedance resonators No. 2 and No. 3 provide a right-hand and left-hand approach of ω_{p1} to ω_{01}, respectively. The convergence degree of these critical frequencies can be approximately 4%. As ω_{p1} and ω_{01} get closer, the susceptance slope parameter b of these resonators changes oppositely. For resonator No. 2, parameter b decreases; for resonator No. 3, this parameter increases.
4. Half-wave resonators with capacitance C (No. 4) and inductance L (No. 5) provide a left-hand and right-hand approach of ω_{p1} to ω_{01}, respectively. The convergence degree of these critical frequencies can be approximately 1%. Resonant frequencies and the susceptance slope parameter b of these resonators are independent of lumped elements. They are the same as those of a half-wave resonator.
5. Quarter-wave resonators with capacitance C (No. 6) and inductance L (No. 7) provide a left-hand and right-hand approach of ω_{p1} to ω_{01}, respectively. The convergence degree of these frequencies can be approximately 1%. The susceptance slope parameter b and resonant frequencies of these resonators are independent of lumped elements. They are the same as those of a quarter-wave resonator.

9.6.2 Resonators with Two Antiresonant Frequencies Close to the Operating Resonant Frequency

These 11 resonators are contained in Table 9.4. Six of them (Nos. 1–6) are TL resonators, and the remaining five resonators (Nos. 7–11) are LD ones. These resonators use main resonant frequencies, corresponding to oscillations $\lambda/4$ and $\lambda/2$, and higher resonant frequencies, corresponding to oscillations $3\lambda/4$, λ, $5\lambda/4$ (Table 9.12).

1. The use of higher oscillation types makes it possible to increase the geometric length of the resonators when they operate at very high frequencies, including the millimeter range.
2. Uniform resonators No. 1, No. 2, and No. 3 operate on oscillations $3\lambda/4$, $\lambda/2$, and λ, respectively. The effect of approaching the operating resonant frequency of two double-side antiresonant frequencies is achieved by choosing a connection coordinate q close to the normalized values: $q = 2/3$ for resonator No. 1, $q = 1/2$ for resonator No. 2, and $q = 1/2$ for resonator No. 3. The convergence degree of the indicated critical frequencies is not limited. As they approach, the susceptance slope parameter b increases significantly. The main disadvantage of these resonators is the nonequidistant location of two antiresonant frequencies relative to the operating resonant frequency. The right-hand antiresonant frequency is always further away than the left-hand one.
3. Synthesized stepped-impedance resonators No. 4, No. 5, No. 6 are arranged in increasing order of the used oscillation type: $\lambda/2$, $3\lambda/4$, $5\lambda/4$. Their two antiresonant frequencies are equidistant relative to the operating resonant frequency, which is a definite advantage. The greater the irregularity parameter m of these resonators, the closer to each other their critical frequencies are located, and the greater the susceptance slope parameter b of the resonators. The convergence degree of these critical frequencies increases when higher types of oscillations are used. At $m = 30$, the convergence degree is 13% for resonator No. 4 with $\lambda/2$ oscillations, 4% for resonator No. 5 with $3\lambda/4$ oscillations, and 0.1% for resonator No. 6 with $5\lambda/4$ oscillations. In the latter case, the critical frequencies practically merge.

Table 9.12 Oscillation types used in resonators from Table 9.4.

Oscillation	$\lambda/4$	$\lambda/2$	$3\lambda/4$	λ	$5\lambda/4$
Resonator No.	8, 9, 11	2, 4, 7, 10	1, 5	3	6

4. LD resonators No. 7 and No. 8 contain half-wave (No. 7) or quarter-wave (No. 8) resonators and a cascaded LC chain. These resonators operate at the main resonant frequency ω_{01}, with antiresonant frequencies ω_{p1} and ω_{p2} close to it. The appropriate choice of elements L and C allows us to place the antiresonant frequencies equidistant relative to ω_{01}. The convergence degree of antiresonant frequencies ω_{p1} and ω_{p2} can be 7% for resonator No. 7 and 5% for resonator No. 8. The susceptance slope parameter b of these resonators is constant. It does not depend on the convergence degree of the critical frequencies. The value of b is the same as that of the corresponding half-wave or quarter-wave resonators. This result was somewhat unexpected (Figure 9.16d).
5. LD resonators No. 9 and No. 10 contain SIR and one lumped element. They also operate at the main resonant frequency ω_{01}, with antiresonant frequencies ω_{p1} and ω_{p2} close to it. For a given parameter m of the SIRs used, the equidistant location of ω_{p1} and ω_{p2} relative to ω_{01} is ensured by the value of the lumped element. Resonator No. 9 uses SIR with a short-circuited end, and tuning is carried out by capacitance C. Resonator No. 10 uses SIR with an open end, and inductance L is used for tuning. For both resonators, with equidistant tuning of antiresonant frequencies, a convergence degree of critical frequencies can be realized, expressed by the approximate equality $\omega_{p2}/\omega_{01} \approx 1.2$. Resonant frequencies and the susceptance slope parameter b of these resonators are independent of the lumped elements. These values are exactly the same as those of the synthesized SIRs No. 3 and No. 4 from Table 9.1, which form the basis of the LD resonators No. 9 and No. 10 from Table 9.4.
6. LD resonator No. 11 contains a uniform quarter-wave segment and capacitance C. This resonator operates at the main resonant frequency ω_{01} with antiresonant frequencies ω_{p1} and ω_{p2} close to it. The equidistant location of ω_{p1} and ω_{p2} relative to ω_{01} is ensured by the q coordinate of the connection of capacitance C and its value. With equidistant tuning of antiresonant frequencies, the convergence degree of critical frequencies can be realized, expressed by the approximate equality $\omega_{p2}/\omega_{01} \approx 1.2$. The resonant frequencies of this resonator and the susceptance slope parameter b do not depend on the value of capacitance C.
7. A common property for all LD resonators in Tables 9.1 and 9.4 is that their resonant frequencies and susceptance slope parameter b do not depend on the lumped elements. These values are exactly the same as those of the distributed part of these resonators.

References

1 R. M. Kurzrok, General three-resonator filters in waveguide, *IEEE Trans. Microw. Theory Tech.*, vol. 14, no. 1, pp. 46–47, 1966, https://doi.org/10.1109/TMTT.1966.1126154.

2 R. M. Kurzrok, General four-resonator filters at microwave frequencies, *IEEE Trans. Microw. Theory Tech.*, vol. 14, no. 6, pp. 295–296, 1966, https://doi.org/10.1109/TMTT.1966.1126254.

3 A. E. Atia and A. E. Williams, Narrow-bandpass waveguide filters, *IEEE Trans. Microw. Theory Tech.*, vol. 20, no. 4, pp. 258–265, 1972, https://doi.org/10.1109/TMTT.1972.1127732.

4 A. Atia, A. Williams, and R. Newcomb, Narrow-band multiple-coupled cavity synthesis, *IEEE Trans. Circuits Syst.*, vol. 21, no. 5, pp. 649–655, 1974, https://doi.org/10.1109/TCS.1974.1083913.

5 R. J. Cameron, C. M. Kudsia, and R. R. Mansour, *Microwave Filters for Communication Systems*, 2nd ed. Hoboken, NJ: John Wiley & Sons, Inc., 2018, https://doi.org/10.1002/9781119292371.

6 S. Amari, U. Rosenberg, and J. Bornemann, Adaptive synthesis and design of resonator filters with source/load-multiresonator coupling, *IEEE Trans. Microw. Theory Tech.*, vol. 50, no. 8, pp. 1969–1978, 2002, https://doi.org/10.1109/TMTT.2002.801348.

7 K. Ma, J.-G. Ma, K. S. Yeo, and M. A. Do, A compact size coupling controllable filter with separate electric and magnetic coupling paths, *IEEE Trans. Microw. Theory Tech.*, vol. 54, no. 3, pp. 1113–1119, 2006, https://doi.org/10.1109/TMTT.2005.864118.

8 Q.-X. Chu and H. Wang, A compact open-loop filter with mixed electric and magnetic coupling, *IEEE Trans. Microw. Theory Tech.*, vol. 56, no. 2, pp. 431–439, 2008, https://doi.org/10.1109/TMTT.2007.914642.

9 R. Levy, New cascaded trisections with resonant cross-couplings (CTR Sections) applied to the design of optimal filters, in *IEEE MTT-S International Microwave Symposium Digest*, 2004, vol. 2 pp. 447–450, https://doi.org/10.1109/mwsym.2004.1336007.

10 S. Amari, M. Bekheit, and F. Seyfert, Notes on bandpass filters whose inter-resonator coupling coefficients are linear functions of frequency, in *2008 IEEE MTT-S International Microwave Symposium Digest*, 2008, pp. 1207–1210, https://doi.org/10.1109/MWSYM.2008.4633275.

11 L. Szydlowski, A. Lamecki, and M. Mrozowski, Coupled-resonator filters with frequency-dependent couplings: coupling matrix synthesis, *IEEE Microw. Wirel. Components Lett.*, vol. 22, no. 6, pp. 312–314, 2012, https://doi.org/10.1109/LMWC.2012.2197386.

12 D. Lu, B. Zhu, and M. Yu, Synthesis approach to planar filter design using new cascade hybrid triplet block with arbitrary-placed TZ pair, *IEEE Trans. Microw. Theory Tech.*, vol. 70, no. 4, pp. 2196–2206, 2022, https://doi.org/10.1109/TMTT.2022.3148361.

13 S. Litvintsev and A. Zakharov, Bandpass filters with 2N+1 transmission zeros at real frequencies, *Radioelectron. Commun. Syst.*, vol. 66, no. 6, pp. 315–321, 2023, https://doi.org/10.3103/S0735272723100035.

14 A. V. Zakharov and S. Litvintsev, Transmission line bandpass filters with multiple attenuation poles and small number of resonators, *AEU - Int. J. Electron. Commun.*, p. 155131, 2024, https://doi.org/10.1016/j.aeue.2024.155131.

15 A. Zakharov, Parametric and structural-parametric synthesis of nonuniform transmission line resonators, *IEEE Trans. Circuits Syst. I Regul. Pap.*, vol. 68, no. 3, pp. 1055–1067, 2021, https://doi.org/10.1109/TCSI.2020.3044925.

16 A. V. Zakharov and S. Litvintsev, Lumped-distributed resonators providing N or 2N transmission zeros at real frequencies in bandpass filters without cross and mixed couplings, *IEEE Trans. Microw. Theory Tech.*, vol. 72, no. 6, pp. 3682–3691, 2024, https://doi.org/10.1109/TMTT.2023.3332260.

17 S. Litvintsev and A. Zakharov, Bandpass filters with increased to 3N+1 number of attenuation poles, *Radioelectron. Commun. Syst.*, vol. 66, no. 8, pp. 385–402, 2023, https://doi.org/10.3103/S073527272311002X.

18 J.-S. Lim and D. C. Park, A modified Chebyshev bandpass filter with attenuation poles in the stopband, *IEEE Trans. Microw. Theory Tech.*, vol. 45, no. 6, pp. 898–904, 1997, https://doi.org/10.1109/22.588597.

19 A. Zakharov, S. Rozenko, and S. Litvintsev, Combline filters with increased stopband and one-sided selectivity, *IEEE Microw. Wirel. Technol. Lett.*, vol. 33, no. 4, pp. 407–410, 2023, https://doi.org/10.1109/LMWT.2022.3221269.

20 C. Quendo, E. Rius, and C. Person, Narrow bandpass filters using dual-behavior resonators based on stepped-impedance stubs and different-length stubs, *IEEE Trans. Microw. Theory Tech.*, vol. 52, no. 3, pp. 1034–1044, 2004, https://doi.org/10.1109/TMTT.2004.823582.

21 K. Wada and O. Hashimoto, Fundamentals of open-ended resonators and their application to microwave filters, *IEICE Trans. Electron.*, vol. E83-C, no. 11, pp. 1763–1775, 2000, https://search.ieice.org/bin/summary.php?id=e83-c_11_1763&category=C&year=2000&lang=E&abst=.

22 A. Zakharov and S. Litvintsev, Planar bandpass filters based on resonators generating transmission zeros, *IEEE Circuits Syst. Mag.*, vol. 25, no. 2, pp. 57–73, 2025, https://doi.org/10.1109/MCAS.2025.3543781.

23 S. Litvintsev and A. Zakharov, Synthesis of TL resonators with 3λ/4-mode oscillations for bandpass filters with multiple transmission zeros, in *2024 IEEE 29th International Seminar/Workshop on Direct and Inverse Problems of Electromagnetic and Acoustic Wave Theory (DIPED)*, 2024, pp. 145–149, https://doi.org/10.1109/DIPED63529.2024.10706048.

24 S. Litvintsev and A. Zakharov, Design of bandpass filters with high mode 5λ/4 oscillations and equidistant transmission zeros pairs, in *2024 IEEE 29th International Seminar/Workshop on Direct and Inverse Problems of Electromagnetic and Acoustic Wave Theory (DIPED)*, 2024, pp. 150–154, https://doi.org/10.1109/DIPED63529.2024.10706170.

25 A. V. Zakharov and S. Litvintsev, Lumped-distributed resonators providing multiple transmission zeros in bandpass filters with simple and mixed couplings, *IEEE Trans. Circuits Syst. I Regul. Pap.*, vol. 71, no. 8, pp. 3502–3513, 2024, https://doi.org/10.1109/TCSI.2024.3375961.

26 W. J. Feng, W. Q. Che, Y. M. Chang, S. Y. Shi, and Q. Xue, High selectivity fifth-order wideband bandpass filters with multiple transmission zeros based on transversal signal-interaction concepts, *IEEE Trans. Microw. Theory Tech.*, vol. 61, no. 1, pp. 89–97, 2013, https://doi.org/10.1109/TMTT.2012.2227785.

27 W. Feng, X. Gao, and W. Che, Bandpass filters with multiple transmission zeros using open/shorted stubs, *IET Microwaves, Antennas Propag.*, vol. 9, no. 8, pp. 769–774, 2015, https://doi.org/10.1049/iet-map.2014.0402.

28 C. Quendo, E. Rius, C. Person, and M. Ney, Integration of optimized low-pass filters in a bandpass filter for out-of-band improvement, *IEEE Trans. Microw. Theory Tech.*, vol. 49, no. 12, pp. 2376–2383, 2001, https://doi.org/10.1109/22.971624.

29 A. Manchec, C. Quendo, E. Rius, C. Person, and J.-F. Favennec, Synthesis of dual behavior resonator (DBR) filters with integrated low-pass structures for spurious responses suppression, *IEEE Microw. Wirel. Components Lett.*, vol. 16, no. 1, pp. 4–6, 2006, https://doi.org/10.1109/LMWC.2005.861361.

30 E. A. Guillemin, *Synthesis of Passive Networks: Theory and Methods Appropriate to the Realization and Approximation Problems*. New York: Wiley, 1957.

31 AVX Corporation, RF/Microwave Products, *Products*, 2023. https://www.kyocera-avx.com/products/.

32 Coilcraft. Inc., Chip Inductors – 0603HP Series (1608), *Web-site*, 2022. https://www.coilcraft.com.

10

Filters with Simple Couplings and Multiple Transmission Zeros Generated by Resonators

Transmission zeros (TZ) generated by the resonators are located at real frequencies, also called attenuation poles. Formation of TZs by the resonators themselves significantly increases their total number in the bandpass filter (BPF). The advantage of this approach is clearly illustrated in Figure 10.1, which shows the frequency responses of a second-order BPF with TZs at real frequencies. The BPF has no TZ (dotted curve) if the coupling between its resonators is simple. The BPF has one TZ (dashed curve) if the coupling between its resonators is mixed. The filter has four TZs (solid curve) if each resonator generates two TZs.

This section presents numerous examples of inline BPFs containing transmission line (TL) and lumped-distributed (LD) resonators, as presented in Tables 9.1 and 9.4. The BPF properties of such resonators were discussed in Refs. [1–16].

10.1 BPF with Simple Couplings and Resonators with One Antiresonance Frequency Close to Main Resonance Frequency

Using N-order inline BPF resonators with one antiresonance frequency ω_{p1} located near the fundamental resonant frequency ω_{01} (Table 9.1) will lead to the TZ number equal to N. In this case, analytical expressions describing the properties of resonators, and the classical method of G. Matthaei, expressed by formulas (3.6) and (3.8), are used for BPF design. Table 9.1 contains seven resonators; in the following examples, we use only some of them.

10.1.1 Fourth-order BPF with LD Resonators and Quasi-elliptic Transfer Function

The circuit diagram of the filter under consideration is shown in Figure 10.2a. This filter contains LD resonators No. 6 and No. 7 (Table 9.1) and quarter-wave TL segments as a J inverter.

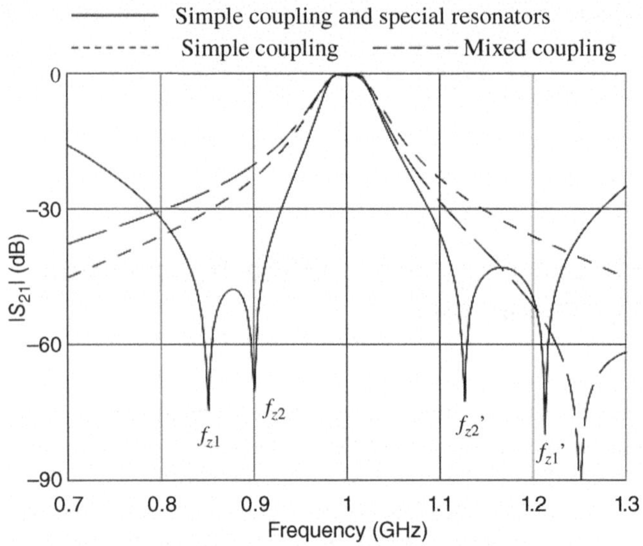

Figure 10.1 Insertion loss of a second-order BPF.

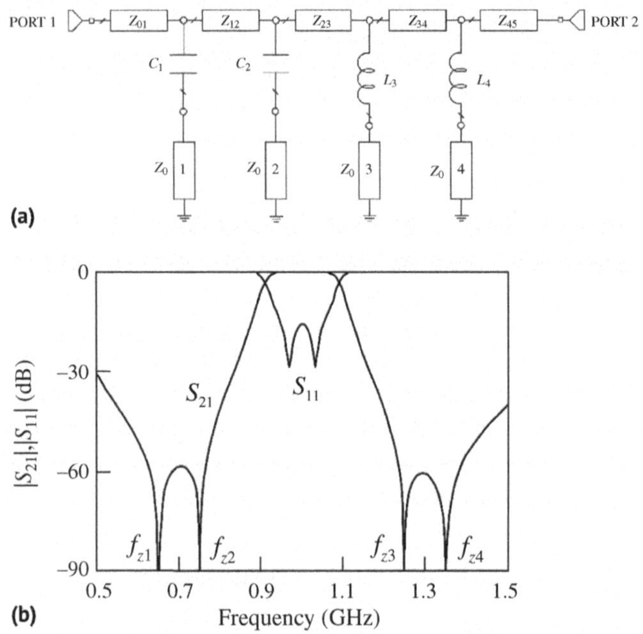

Figure 10.2 Fourth-order BPF with LD resonators and four TZs. (a) Schematic. (b) Simulated frequency responses.

Table 10.1 Resonator parameters for BPF in Figure 10.2a.

Res. No.	1	2	3	4
f_p, GHz	0.65	0.75	1.25	1.35
θ_p, °	58.5	67.5	112.5	121.5
C_i, pF	15	8.79	—	—
L_i, nH	—	—	3.07	1.92

As an example, let us set the following parameters of the fourth-order BPF shown in Figure 10.2a: center frequency $f_0 = 1$ GHz, bandwidth BW = 200 MHz (FBW = 0.20), passband ripple $L_{Ar} = 0.1$ dB. From Table 2.6, we find g parameters: $g_1 = 1.1088$, $g_2 = 1.3061$, $g_3 = 1.7703$, $g_4 = 0.8180$, $g_5 = 1.3554$. Substituting these g parameters and the initial data into (3.8), we determine $K_{12} = K_{34} = 0.1660$, $K_{23} = 0.1316$, $Q_{e1} = Q_{e4} = Q_e = 5.544$. In the composite resonators of the filter, we will use quarter-wave segments at the frequency of 1 GHz with a characteristic impedance $Z_0 = 10\ \Omega$.

Let us set different values, f_{p1}, f_{p2}, f_{p3}, and f_{p4}, of the antiresonant frequencies of each of the four resonators, as indicated in the second row of Table 10.1. They correspond to the antiresonant electrical lengths of the resonators θ_{p1}, θ_{p2}, θ_{p3}, θ_{p4} (row 3). Antiresonance frequencies f_{p1}, $f_{p2} < f_0$ are realized by resonators with capacitance C (No. 6, Table 9.1). The values of C of these two resonators are determined using expression (9.28). The calculated values of C_i are listed in the fourth row of Table 10.1. The antiresonance frequencies f_{p3}, $f_{p4} > f_0$ are realized by resonators with inductance L (No. 7, Table 9.1). We determine the L values of these two resonators using expression (9.31) and put them in the last row of Table 10.1.

As J inverters, we use a quarter-wave segment with a characteristic impedance $Z_{i,i+1}$. In this case, we use expression (3.11c) $Z_{i,i+1} = 4Z_0/\pi K_{i,i+1}$.

Substituting the values $K_{12} = K_{34} = 0.1660$, $K_{23} = 0.1316$, and $Z_0 = 10\ \Omega$ into this expression, we obtain $Z_{12} = Z_{34} = 76.68\ \Omega$; $Z_{23} = 96.74\ \Omega$. For the leftmost quarter-wave segment to provide resonator 1 with an external quality factor Q_{e1}, its characteristic impedance Z_{01} must satisfy the equality

$$Z_{01} = \sqrt{\frac{Q_{e1}}{bG_A}}, \tag{10.1}$$

which follows from (3.6b) and (3.8b). The characteristic impedance Z_{45} of the rightmost segment is determined similarly. Substituting the values $Q_e = 5.544$ and $G_A = 0.02\ \Omega^{-1}$ into (10.1), we obtain $Z_{01} = Z_{45} = 59.41\ \Omega$. All the filter parameters in Figure 10.2a are defined. Figure 10.2b shows the frequency responses of the considered filter, simulated by the linear circuit simulator of Microwave Office software.

Figure 10.3 illustrates the influence of the quality factor of lumped elements C_i (Q_{Ci}) and L_i (Q_{Li}), $i = 1, 2$, on the frequency responses of the considered BPF. Let us note some regularities:

- Quality factor of lumped elements Q_{Ci} and Q_{Li} does not affect the passband insertion loss.
- Quality factor of lumped elements affects the "depth" of the attenuation poles that accompany it.
- A decrease in the quality factor leads to a decrease in the "depth" of the attenuation poles, while its frequency does not change.

All four TZs are located at the prescribed frequencies in the filter under consideration. Note that to implement four TZs in canonical BPF [17], six resonators and two cross-couplings are required. The advantage of the considered filter with LD resonators is that the lumped elements do not affect the resonant frequency of the resonators and, consequently, the position of the central frequency, and the

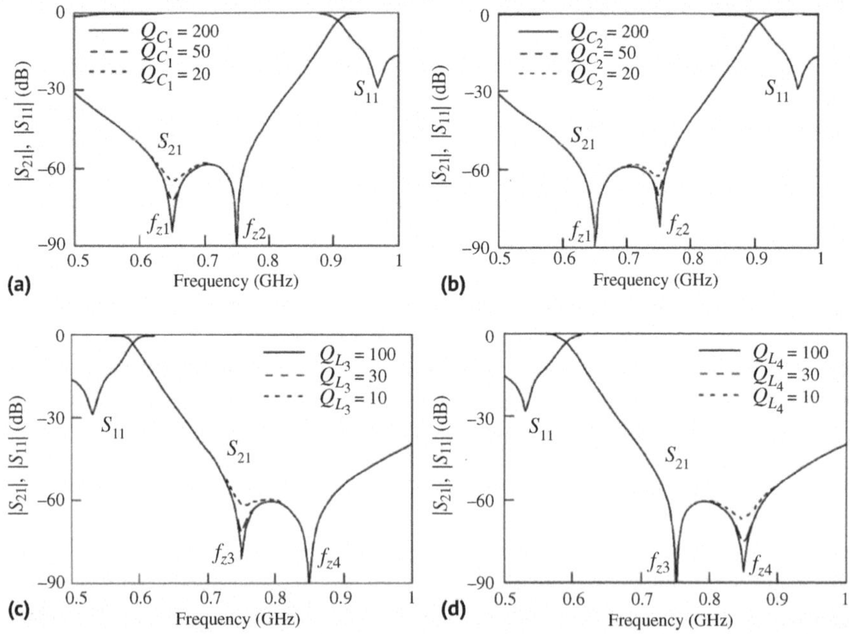

Figure 10.3 Influence of quality factor of lumped elements on frequency responses of BPF in Figure 10.2. (a) Influence Q_{C1}. (b) Influence Q_{C2}. (c) Influence Q_{L3}. (d) Influence Q_{L4}.

finite quality factor of these elements does not affect the filter's insertion loss at the central frequency. The design of this filter is relatively simple because the lumped elements do not affect the susceptance slope parameter b of resonators.

10.1.2 Second-order BPF with Two SIRs and Two Right-sided TZs

Figure 10.4a shows the circuit of a second-order BPF with resonator No. 2 (Table 9.1) and admittance inverters containing capacitances. The transverse branches of the J inverters are not shown. The resonators used are SIRs. They operate at the fundamental resonant frequency ω_{01} with a closely spaced right-sided antiresonant frequency ω_{p1}. The upper part of each SIR (Figure 10.4a) contains a segment with increased characteristic impedance mZ_0 and electrical length $\theta_s = \pi/3$. The lower part of the SIR includes a segment with lower characteristic impedance Z_0 and doubled electrical length $2\theta_s$. The value of Z_0 is set equal to 5 Ω. The properties of these resonators are described by formulas (9.9)–(9.12).

Let us set the following parameters of this BPF: center frequency $f_0 = 1$ GHz, FBW $= 0.03$, passband ripple $L_{Ar} = 0.2$ dB, which corresponds to the return loss $L_R = 13.467$ dB.

We will define two right-side antiresonant frequencies for a sharp right slope: f_{p1} =1.05 GHz $(f_{p1}/f_{01} = 1.05)$ for resonator 1 and $f_{p1} = 1.08$ GHz $(f_{p1}/f_{01} = 1.08)$ for resonator 2. These data are contained in Table 10.2 (rows 1 and 2).

From (9.11), we determine the parameter m, which provides the required ratio f_{p1}/f_{01} (row 3). Knowing m, we determine the electrical length θ_{01} of the resonator from (9.10a) and find the susceptance slope parameter b from (9.12) at the $Z_0 = 5$ Ω. The indicated values are contained in Table 10.2 (rows 4 and 5).

From Table 2.6, we find g parameters: $g_0 = 1$, $g_1 = 1.0378$, $g_2 = 0.6745$, $g_3 = 1.5386$. Substituting these values, susceptance slope parameters, and

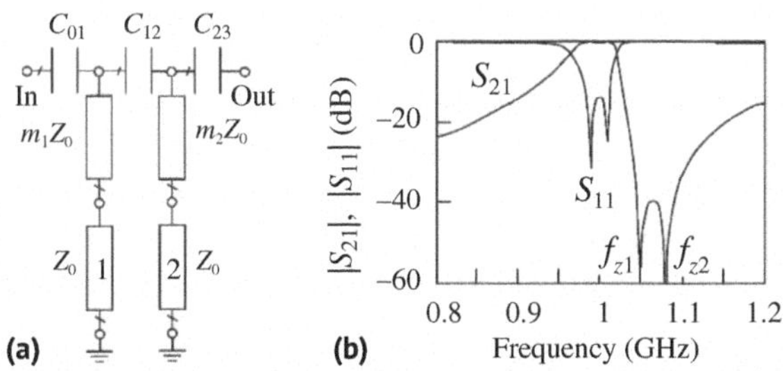

Figure 10.4 Second-order BPF with SIRs and two right-sided TZs. (a) Schematic: $Z_0 = 5$ Ω, $m_1 = 25.1$, $m_2 = 15.5$, $C_{12} = 0.48$ pF, $C_{01} = 1.10$ pF, and $C_{23} = 1.12$ pF. (b) Simulated frequency responses.

Table 10.2 SIR parameters of BPF in Figure 10.4a.

		Resonator	
No.	**Parameter**	**SIR-1**	**SIR-2**
1	f_{p1}, GHz	1.05	1.08
2	f_{p1}/f_{01}	1.05	1.08
3	m	25.1	15.5
4	θ_{01}, °	131.71	129.79
5	b, Ω^{-1}	0.08273	0.08525

FBW = 0.03 into (3.6), we determine the J parameters of the inverters used in BPF: $J_{12} = 0.00301\ \Omega^{-1}$; $J_{01} = 0.006916\ \Omega^{-1}$; $J_{23} = 0.00702\ \Omega^{-1}$. If using inverters with capacitances, then according to (3.7b): $C_{12} = 0.48$ pF, $C_{01} = 1.10$ pF, and $C_{23} = 1.12$ pF. All parameters of the considered filter are defined. Figure 10.4b shows its simulated frequency responses. The second-order filter has increased right-hand selectivity due to two TZs at prescribed frequencies.

10.1.3 Second-order BPF with LD Resonators and Two Left-sided TZs

The circuit diagram of the filter under consideration is shown in Figure 10.5a. The filter contains LD resonator No. 4 (Table 9.1) and J inverters, containing inductances. The transverse branches of the J inverters are not shown. The resonators are half-wave TL segments with characteristic impedance $Z_0 = 10\ \Omega$ and capacitance C. The properties of these resonators are described by formulas (9.20)–(9.22).

Let us set the following parameters of this BPF: center frequency $f_0 = 2$ GHz, FBW = 0.02, passband ripple $L_{Ar} = 0.2$ dB. To have a sharp left slope, we define two left-sided TZs: $f_{z1} = 1.93$ GHz and $f_{z2} = 1.95$ GHz. To provide these TZs, we use $\lambda/2$ resonators with capacitance C. From Table 2.6, we find g parameters: $g_0 = 1$, $g_1 = 1.0378$, $g_2 = 0.6745$, $g_3 = 1.5386$. Substituting these values and FBW = 0.02 into (3.6), we determine the J parameters of the inverters used in BPF: $J_{12} = 0.00376\ \Omega^{-1}$; $J_{01} = J_{23} = 0.00778\ \Omega^{-1}$.

If inverters with inductances are used, then according to (3.7a), $L_{12} = 21.19$ nH and $L_{01} = L_{23} = 10.23$ nH. If inverters with capacitances are used, then according to (3.7b), $C_{12} = 0.299$ pF and $C_{01} = C_{23} = 0.619$ pF. The required capacitances of LD resonators (Figure 10.5a) are determined using (9.21). The specified value $f_{z1} = 1.93$ GHz is provided at the capacitance value $C_1 = 0.91$ pF, and the value $f_{z2} = 1.95$ GHz is supplied at the capacitance value $C_2 = 0.64$ pF. All parameters of the considered filter are defined. Figure 10.5b shows its simulated frequency responses with the prescribed location of two TZs.

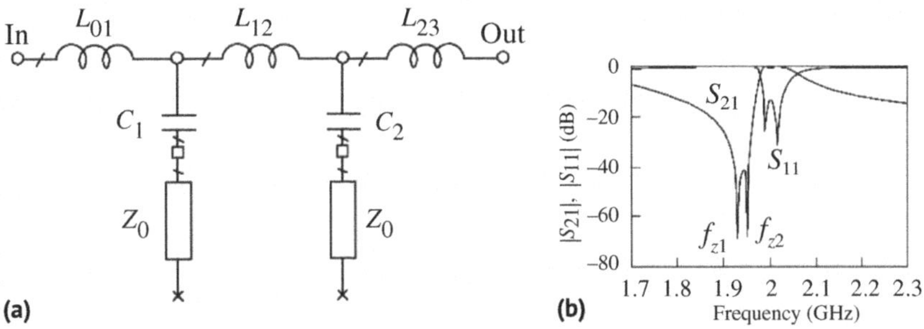

Figure 10.5 Second-order BPF with LD resonators and two left-sided TZs. (a) Schematic. (b) Simulated frequency responses.

The filters in Figures 10.4 and 10.5 have increased one-sided selectivity, which is in demand when using BPFs in duplexers. This effect is achieved by using two one-sided TZs.

Note that to implement two one-side TZs through cross-couplings, it is necessary to use the fourth-order BPF with two cross-couplings [17].

10.2 BPF with Simple Couplings and Resonators with Two Antiresonance Frequencies Close to Operating Resonant Frequency

Using N-order inline BPF resonators with two antiresonance frequencies near the operating resonant frequency (Table 9.4) will result in TZ number equal to $2N$. In this case, analytical expressions describing the properties of these resonators are used for BPF design and general method expressed by formulas (3.6) and (3.8). Table 9.4 contains 11 resonators. In the following examples, we use most of these resonators, which can provide equidistant and nonequidistant arrangements of two antiresonant frequencies relative to the operating resonant frequency. The consideration is not carried out in the order of resonators in Table 9.4.

10.2.1 BPF with Resonators Nos. 9, 10, and 11 (Table 9.4)

These resonators [2] are LD resonators containing a TL segment and one lumped element.

10.2.1.1 Second-order BPF with Resonator No. 9 (Table 9.4) and Sharp Left Slope

Figure 10.6a shows the circuit of a second-order BPF with resonator No. 9 (Table 9.4) and admittance inverters with capacitances. This resonator contains a stepped-impedance segment with a short-circuited end and capacitance C. Let us set the following parameters of this BPF: $f_0 = 2$ GHz, BW = 100 MHz

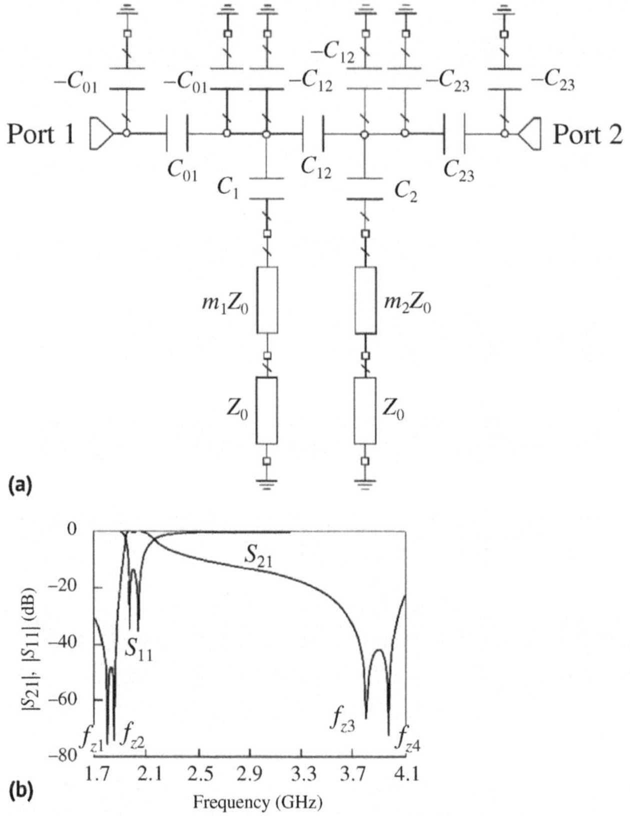

Figure 10.6 Second-order BPF with LD resonator No. 9 (Table 9.4) and increased left-sided selectivity. (a) Schematic ($Z_0 = 10\ \Omega$, $m_1 = m_2 = m = 3$, $C_1 = 0.87$ pF, $C_2 = 0.65$ pF, $C_{01} = C_{23} = 0.579$ pF, $C_{12} = 0.261$ pF). (b) Simulated frequency responses.

(FBW = 0.05), passband ripple $L_{Ar} = 0.2$ dB, which corresponds to the return loss $L_R = 13.467$ dB. To have a sharp left slope, we define two left-sided TZs: $f_{z1} = 1.8$ GHz and $f_{z2} = 1.85$ GHz. To provide these TZs, we use two such resonators with parameters: $Z_0 = 10\ \Omega$ and $m_1 = m_2 = m = 3$.

Using (9.74) and (9.77), we determine the electrical length of these resonators $\theta_{01} = 113.28°$ at the main resonant frequency $f_{01} = 2$ GHz and their susceptance slope parameter $b = 0.05492\ \Omega^{-1}$. These values are presented in the second column of Table 9.9.

From Table 2.6, we find g parameters: $g_0 = 1$, $g_1 = 1.0378$, $g_2 = 0.6745$, $g_3 = 1.5386$. Substituting these values, FBW = 0.05 and $b = 0.05492\ \Omega^{-1}$, into (3.6), we determine the J parameters of the inverters used in BPF: $J_{12} = 0.00328\ \Omega^{-1}$;

$J_{01} = J_{23} = 0.00728\ \Omega^{-1}$. If inverters with capacitances are used, then according to (3.7b), $C_{12} = 0.261$ pF and $C_{01} = C_{23} = 0.579$ pF.

The specified TZ value $f_{z1} = 1.8$ GHz is provided by resonator 1 with capacitance C_1. Using equation (9.75), we determine the value of $C_1 = 0.87$ pF. The specified TZ value $f_{z2} = 1.85$ GHz is provided by resonator 2 with capacitance $C_2 = 0.65$ pF. From (9.75), it follows that the frequency $f_{z1} = 1.8$ GHz will be accompanied by a frequency $f_{z3} = 3.8$ GHz, and the frequency $f_{z2} = 1.85$ GHz will be accompanied by a frequency $f_{z4} = 3.97$ GHz.

All parameters of the filter (Figure 10.6a) are defined. Figure 10.6b shows its frequency responses simulated with the linear circuit simulator of Microwave Office software. The second-order BPF has increased left-handed selectivity due to two left-sided TZs located near the passband and generated by resonator No. 9 (Table 9.4).

10.2.1.2 Second-order BPF with Resonator No. 10 (Table 9.4) and Sharp Right Slope

Figure 10.7a shows the circuit of a second-order BPF with resonator No. 10 (Table 9.4) and admittance inverters with capacitances. This resonator contains a stepped-impedance segment with an open-circuited end and an inductance L. Filter parameters f_0, BW (FBW), $L_{Ar} = 0.2$ dB, and g parameters are the same, as in the previous filter.

To have a sharp right slope, we will define two right-sided TZs: $f_{z1} = 2.2$ GHz and $f_{z2} = 2.15$ GHz. To provide these TZs, we use two such resonators with parameters: $Z_0 = 10\ \Omega$ and $m_1 = m_2 = m = 3$.

Using (9.80) and (9.83), we determine the electrical length of these resonators $\theta_{01} = 156.72°$ at the main resonant frequency $f_{01} = 2$ GHz, and their susceptance slope parameter $b = 0.2279\ \Omega^{-1}$. These values are presented in the second column of Table 9.10.

Substituting g parameters, FBW = 0.05 and $b = 0.2279\ \Omega^{-1}$ into (3.6), we determine the J parameters of the inverters used in BPF: $J_{12} = 0.01362\ \Omega^{-1}$; $J_{01} = J_{23} = 0.01482\ \Omega^{-1}$. If inverters with capacitances are used, then according to (3.7b), $C_{12} = 1.084$ pF and $C_{01} = C_{23} = 1.084$ pF. The first resonator provides the specified TZ value $f_{z1} = 2.2$ GHz with inductance L_1. Using equation (9.81), we determine the value $L_1 = 1.8$ nH.

The specified TZ value $f_{z2} = 2.15$ GHz is provided by the second resonator with inductance $L_2 = 2.4$ nH. This filter has two more TZs generated by these resonators. The first resonator generates $f_{z3} = 0.914$ GHz and second resonator generates $f_{z4} = 0.824$ GHz. The indicated frequencies solve equation (9.81) at $L_1 = 1.8$ nH and $L_2 = 2.4$ nH, respectively. All parameters of the considered BPF are defined. The simulated frequency responses of the BPF are shown in Figure 10.7b, in which f_{zi} are numbered from right to left. The second-order BPF has

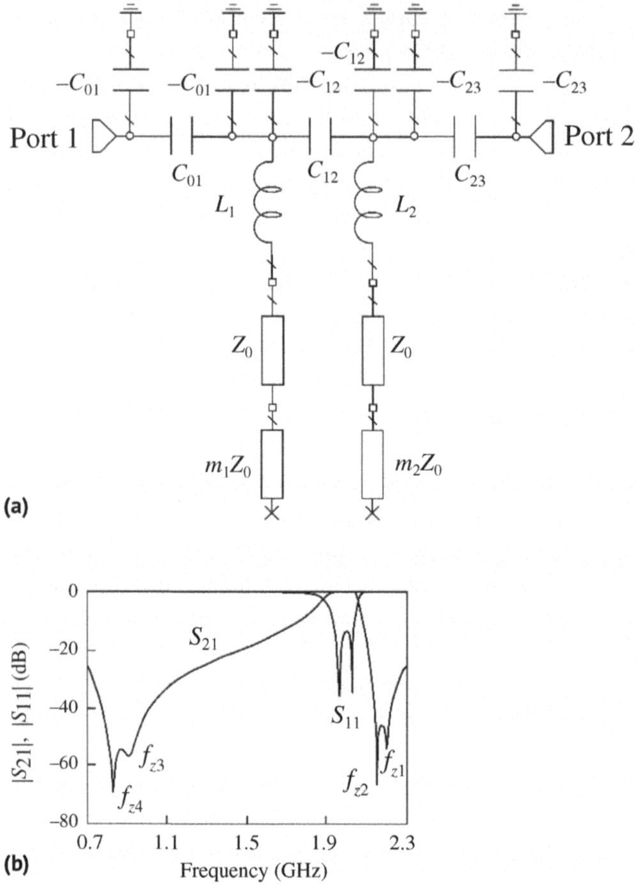

Figure 10.7 Second-order BPF with LD resonator No. 10 (Table 9.4) and increased right-sided selectivity. (a) Schematic: $Z_0 = 10\ \Omega$, $m_1 = m_2 = m = 3$, $L_1 = 1.8$ nH ($Q_{L1} = 30$), $L_2 = 2.4$ nH ($Q_{L2} = 200$), $C_{01} = C_{23} = 1.179$ pF, $C_{12} = 1.084$ pF. (b) Simulated frequency responses.

increased right-handed selectivity due to two right-sided TZs generated by resonator No. 10 (Table 9.4) and located near the passband.

The frequency responses in Figure 10.7b are simulated with different quality factors of inductors: $Q_{L1} = 30$ and $Q_{L2} = 200$. Resonator 1 with inductor L_1 forms TZs at frequencies f_{z1} and f_{z3}. Resonator 2 with inductor L_2 forms TZs at frequencies f_{z2} and f_{z4}. A decrease in Q_{L1} reduces the "depth" of the attenuation poles at frequencies f_{z1} and f_{z3}. At the same time, reducing Q_{L1} does not affect the insertion loss in the passband. The same principles are true when using the previous resonator No. 9 (Table 9.4) with capacitance. These two established features of BPFs with LD resonators discussed are practically useful.

The second-order BPFs shown in Figures 10.6 and 10.7 have increased one-sided selectivity and can be helpful for use in duplexers. Each BPF has four TZs, requiring six resonators using cross-couplings [17]. The asymmetry of the considered BPFs is due to the asymmetric location of the TZs.

The used LD resonators No. 9 and No. 10 (Table 9.4) can also generate a pair of TZs equidistant relative to f_0, as reflected in Tables 9.9 and 9.10. The BPFs will have increased two-sided selectivity and symmetrical frequency responses in this case.

10.2.1.3 Third-order BPF with Resonator No. 11 (Table 9.4) and Three Equidistant Pairs of TZs

Figure 10.8a shows the circuit of a third-order BPF with resonator No. 11 (Table 9.4) and admittance inverters containing capacitors. The transverse branches of the inverters are not shown. This resonator contains a quarter-wave TL segment and capacitance C. Let us set the following parameters of this BPF: $f_0 = 1$ GHz, BW = 60 MHz (FBW = 0.06), passband ripple $L_{Ar} = 0.2$ dB (return loss $L_R = -13.467$ dB). We require that the filter has three equidistant pairs of

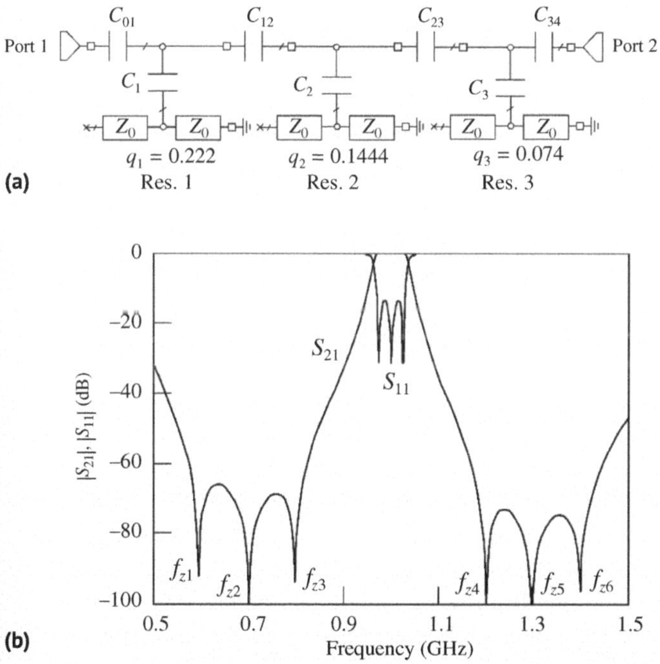

Figure 10.8 Third-order BPF with resonator No. 11 (Table 9.4) and three equidistant pairs of TZs. (a) Schematic ($Z_0 = 30\ \Omega$, $q_1 = 0.2222$, $C_1 = 34.5$ pF; $q_2 = 0.1444$, $C_2 = 36.8$ pF; $q_3 = 0.07444$, $C_3 = 56$ pF). (b) Simulated frequency responses.

TZs: $f_{z1,6} = 0.6$ and 1.4 GHz; $f_{z2,5} = 0.7$ and 1.3 GHz; $f_{z3,4} = 0.8$ and 1.2 GHz. To implement the prescribed TZs, we use resonator No. 11 (Table 9.4) with $Z_0 = 30\ \Omega$ and other parameters q_i and C_i (Table 9.11). As in Table 9.11, we arrange the resonators in the sequence 1–2–3 with the corresponding normalized coordinate connection of the capacitance to the resonator: $q_1 = 0.2222$, $q_2 = 0.1444$, $q_3 = 0.07444$.

From Table 2.6, we find g parameters: $g_1 = 1.2275$, $g_2 = 1.1525$, $g_3 = 1.2275$, $g_4 = 1.0$. Substituting these parameters, FBW = 0.06 and susceptance slope parameters b_i from Table 9.11 into (3.6), we determine the J parameters of the inverters used in BPF: $J_{01} = 0.01480\ \Omega^{-1}$, $J_{12} = 0.01716\ \Omega^{-1}$, $J_{23} = 0.05032\ \Omega^{-1}$, $J_{34} = 0.04336\ \Omega^{-1}$. If inverters with capacitances are used, then according to (3.7b): $C_{01} = 2.355$ pF, $C_{12} = 2.731$ pF, $C_{23} = 8.008$ pF, $C_{34} = 6.90$ pF.

All parameters of the BPF are defined. The simulated frequency responses are shown in Figure 10.8b. The third-order BPF without cross- and mixed couplings has three pairs of equidistant TZs at prescribed frequencies. Eight resonators would be required to implement six TZs in a cross-coupled BPF [17].

10.2.2 BPF with TL Resonators Nos. 4, 5, and 6 (Table 9.4)

These resonators are SIRs and do not contain lumped elements. They are considered as follows in ascending order of the type of oscillation used in them: $\lambda/2$, $3\lambda/4$, $5\lambda/4$.

10.2.2.1 Third-order BPF with SIRs No. 4 (Table 9.4) Operating on $\lambda/2$ Oscillations

Figure 10.9a shows the circuit of a third-order BPF with SIRs No. 4 (Table 9.4) and quarter-wave segments as admittance inverters. The SIR has two sections and an open-circuited end. Its properties are described by formulas (9.44)–(9.48). It has two antiresonant frequencies, ω_{p1}, ω_{p2}, located equidistant relative to ω_{01} [4].

For example, let us set the following parameters of a third-order BPF: center frequency $f_0 = 2$ GHz, bandwidth BW = 60 MHz, FBW = 0.03, passband ripple $L_{Ar} = 0.1$ dB.

Let us set different values, f_{p1} and f_{p2}, of the antiresonant frequencies for each of the three SIRs, as presented in the second row of Table 10.3. They correspond to the antiresonant electrical lengths of the resonators θ_{p1} and θ_{p2} (row 3). From expressions (9.45) and (9.46), we determine the parameter m of SIR (row 4), which ensures the given locations θ_{p1} and θ_{p2}. Let us take the value Z_0 of each resonator $Z_0 = 20\ \Omega$. The values of the high-impedance segment mZ_0 of each SIR will be different; they are given in the fifth row of Table 10.3. The last row of Table 10.3 shows the values of the susceptance slope parameter b of the resonators, which are calculated by formula (9.48).

From Table 2.6, we find g parameters for $n = 3$ and $L_{Ar} = 0.1$ dB: $g_0 = g_4 = 1$, $g_1 = g_3 = 1.0315$, $g_2 = 1.1474$. Substituting these g parameters, the susceptance

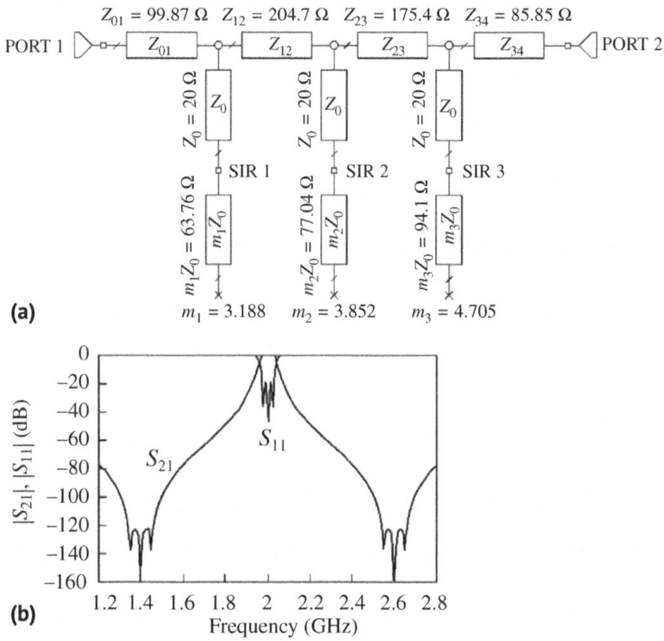

Figure 10.9 Third-order BPF with SIRs No. 4 (Table 9.4) operating on $\lambda/2$ oscillations with three pairs of equidistant TZs. (a) Schematic. Electrical lengths of all segments are 90° at 2 GHz. (b) Frequency responses.

Table 10.3 SIR parameters of BPF in Figure 10.9a.

	Resonator		
Parameter	**SIR-1**	**SIR-2**	**SIR-3**
f_{p1}, f_{p2}, GHz	1.35, 2.65	1.4, 2.6	1.45, 2.55
θ_{p1}, θ_{p2}, °	121.5, 238.5	126, 234	130.5, 229.5
m	3.188	3.852	4.705
mZ_0, Ω	63.76	77.04	94.1
b_i, Ω^{-1}	0.1645	0.1905	0.2240

slope parameters b_i from Table 10.3 and FBW value into (3.6), we determine J parameters of admittance inverters: $J_{01} = 0.01001\ \Omega^{-1}$, $J_{12} = 0.00489\ \Omega^{-1}$, $J_{23} = 0.00570\ \Omega^{-1}$, $J_{34} = 0.01165\ \Omega^{-1}$. Using (3.7c), we determine the characteristic impedances of the quarter-wave sections: $Z_{01} = 99.87\ \Omega$, $Z_{12} = 204.7\ \Omega$, $Z_{23} = 175.40\ \Omega$, $Z_{34} = 85.85\ \Omega$.

All filter parameters in Figure 10.9a are defined. Figure 10.9b shows the simulated frequency responses of this filter. This third-order BPF with simple couplings has six regulated TZs. Due to these TZs, this filter has a high attenuation level in the frequency bands 1.3–1.5 GHz and 2.5–2.7 GHz, allowing us to suppress unwanted signals in these bands effectively. Moreover, frequency bands with high attenuation can be adjusted by the appropriate TZ location.

10.2.2.2 Fourth-order BPF with SIRs No. 5 (Table 9.4) Operating on $3\lambda/4$ Oscillations

Figure 10.10a shows the circuit of a fourth-order BPF with SIRs No. 5 (Table 9.4) and admittance inverters containing capacitors. The transverse branches of the inverters with negative capacitances are not shown. The SIR has three sections and a short-circuited end. Its properties are described by formulas (9.52)–(9.56). It operates on $3\lambda/4$ oscillations and has two antiresonant frequencies, ω_{p1} and ω_{p2}, which are equidistant relative to the second resonant frequency ω_{02}.

Let us set the following parameters of this BPF: center frequency $f_0 = 6$ GHz, FBW = 0.05, passband ripple $L_{Ar} = 0.1$ dB (return loss $L_R = 16.426$ dB). We require that the filter has four equidistant TZ pairs: $f_{z1,8} = 5.4$ and 6.6 GHz; $f_{z2,7} = 5.5$ and 6.5 GHz; $f_{z3,6} = 5.6$ and 6.4 GHz; $f_{z4,5} = 5.7$ and 6.3 GHz. To implement the prescribed TZs, we use SIRs (Figure 9.13) of $3\lambda/4$ type with $f_{01} = 6$ GHz and $Z_0 = 10\ \Omega$. The other parameters f_{p1}, f_{p2}, and θ_{p1} of the resonators are indicated in Table 10.4 (rows 2–4).

To determine the parameter m of each resonator, we use expression (9.53), which we present in the form

$$m = \left[1 + \tan^2\left(\frac{\theta_{p1}}{3}\right)\right]^{1/2} - 1. \tag{10.2}$$

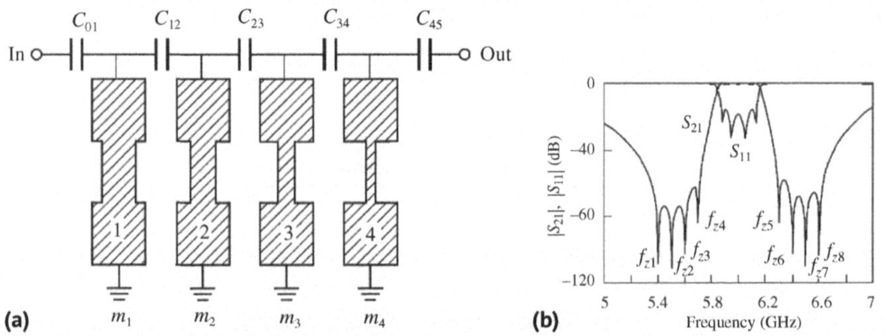

Figure 10.10 Fourth-order BPF with SIRs No. 5 (Table 9.4) operating on $3\lambda/4$ oscillations with four pairs of equidistant TZs. (a) Schematic. (b) Simulated frequency responses with resonators 1–2–3–4 from Table 10.4.

Table 10.4 Parameters of SIRs of $3\lambda/4$ type, with $f_{02} = 6$ GHz and $Z_0 = 10\,\Omega$ for BPF in Figure 10.10.

No.	1	2	3	4
f_{p1}, GHz	5.4	5.5	5.6	5.7
f_{p2}, GHz	6.6	6.5	6.4	6.3
θ_{p1}, °	243	247.5	252	256.5
m	5.392	6.661	8.57	11.745
b, Ω^{-1}	2.785	4.086	6.52	11.835

Substituting θ_{p1} from Table 10.4 into (10.2) leads to the corresponding values of m presented in the same table (row 5). The last row of this table contains the susceptance slope parameter of each resonator calculated by formula (9.56). All parameters of SIRs in BPF in Figure 10.10a are defined.

From Table 2.6, we find g parameters for $n = 4$ and ripple $L_{Ar} = 0.1$ dB: $g_1 = 1.1088$, $g_2 = 1.3061$, $g_3 = 1.7703$, $g_4 = 0.8180$, $g_5 = 1.3554$. Substituting these parameters, FBW = 0.05 and susceptance slope parameter b_i from Table 10.4 into (3.6), we determine the J parameters of the inverters used in BPF: $J_{01} = 0.0501\,\Omega^{-1}$, $J_{12} = 0.1402\,\Omega^{-1}$, $J_{23} = 0.1697\,\Omega^{-1}$, $J_{34} = 0.3650\,\Omega^{-1}$, $J_{45} = 0.1033\,\Omega^{-1}$. If inverters with capacitances are used, then according to (3.7b): $C_{01} = 1.329$ pF, $C_{12} = 3.719$ pF, $C_{23} = 4.501$ pF, $C_{34} = 9.68$ pF, $C_{45} = 2.74$ pF. All parameters of the BPF are defined. Its simulated frequency responses are shown in Figure 10.10b. The filter has four pairs of equidistant TZs located at prescribed frequencies. Note that the location of the resonators in the BPF can be arbitrary.

10.2.2.3 Third-order BPF with SIRs No. 6 (Table 9.4) Operating on $5\lambda/4$ Oscillations

Figure 10.11a shows the circuit of a third-order BPF with SIRs No. 6 (Table 9.4) and admittance inverters containing capacitors. The transverse branches of the inverters with negative capacitances are not shown. The SIR has five sections and a short-circuited end. Its properties are described by formulas (9.61)–(9.65). It operates on $5\lambda/4$ oscillations and has two antiresonant frequencies, ω_{p2} and ω_{p3}, at an equidistant relative to the third resonant frequency, ω_{03}.

Let us set the following parameters of this BPF: center frequency $f_0 = 10$ GHz, FBW = 0.04, passband ripple $L_{Ar} = 0.1$ dB. We require that the filter has three equidistant pairs of TZs: $f_{z1,6} = 9.2$ and 10.8 GHz; $f_{z2,5} = 9.4$ and 10.6 GHz; $f_{z3,4} = 9.6$ and 10.4 GHz. To implement the prescribed TZs, we use SIRs of $5\lambda/4$ type with $f_{03} = 10$ GHz and $Z_0 = 10\,\Omega$. The other parameters f_{p2}, f_{p3}, and θ_{p2} of the resonators are indicated in Table 10.5 (rows 2–4). The parameter m of each resonator is determined from equality (9.62). The calculation results are presented in the same table (row 5). The last row of this table contains the susceptance slope

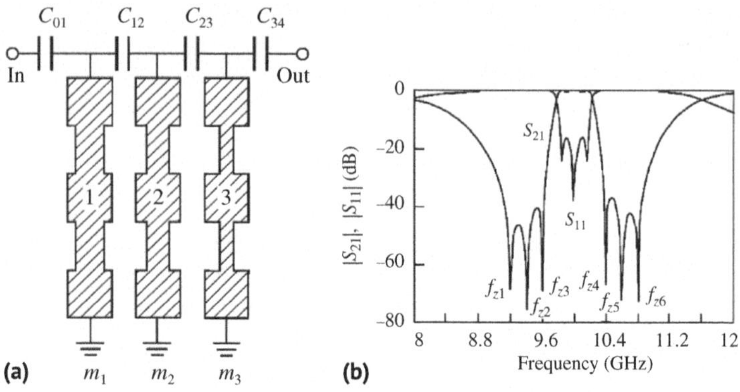

Figure 10.11 Third-order BPF with SIRs No. 6 (Table 9.4) operating on $5\lambda/4$ oscillations with three pairs of equidistant TZs. (a) Schematic. (b) Simulated frequency responses with resonators 1–2–3 from Table 10.5.

Table 10.5 Parameters of SIRs of $5\lambda/4$ type with $f_{03} = 10$ GHz and $Z_0 = 10\,\Omega$ for BPF in Figure 10.11(a).

No.	1	2	3
f_{p2}, GHz	9.2	9.4	9.6
f_{p3}, GHz	10.8	10.6	10.4
θ_{p2}, °	414	423	432
m	2.143	2.607	3.365
b, Ω^{-1}	3.037	5.836	14.295

parameter of each resonator calculated by formula (9.65). All parameters of SIRs in BPF in Figure 10.11a are defined.

From Table 2.6, we find g parameters for $n = 3$ and $L_{Ar} = 0.1$ dB: $g_1 = 1.0315$, $g_2 = 1.1474$, $g_3 = 1.0315$, $g_4 = 1$. Substituting these parameters, FBW = 0.04 and susceptance slope parameter b_i from Table 10.5 into (3.6), we determine the J parameters of the inverters used in BPF: $J_{01} = 0.0485\,\Omega^{-1}$, $J_{12} = 0.1548\,\Omega^{-1}$, $J_{23} = 0.3358\,\Omega^{-1}$, $J_{34} = 0.1053\,\Omega^{-1}$. If inverters with capacitances are used, then according to (3.7b): $C_{01} = 0.772$ pF, $C_{12} = 2.464$ pF, $C_{23} = 5.344$ pF, $C_{34} = 1.676$ pF. All parameters of the considered BPF are defined, and its simulated frequency responses are shown in Figure 10.11b.

The filter has three pairs of equidistant TZs located at prescribed frequencies.

The filters in Figures 10.10 and 10.11 use SIRs whose length is significantly longer than quarter-wave resonators, which allows them to be used at very high frequencies.

10.2.3 BPF with Microstrip Uniform TL Resonator No. 2 (Table 9.4) Operating on $\lambda/2$ Oscillations

Figure 10.12a shows the topology of a microstrip third-order BPF with uniform TL resonator No. 2 (Table 9.4) and quarter-wave segments as admittance inverters.

The resonator is a half-wave segment with a special normalized connection coordinate q. Its properties are described by formulas (9.35)–(9.39). It operates on $\lambda/2$ oscillations and has two antiresonant frequencies ω_{p1}, ω_{p2} located on both sides of the main resonant frequency ω_{01}.

Let us set the following parameters of the microstrip third-order BPF: $f_0 =$ 1 GHz, BW = 40 MHz (FBW = 0.04), $L_{\mathrm{Ar}} = 0.1$ dB, $N = 3$. The filter used a TMM-10i substrate (Rogers), with $\varepsilon_r = 9.8$, $h = 1.016$ mm, $\tan\delta = 0.002$, three $\lambda/2$ resonators of width $w = 1.4$ mm. ($Z_0 = 41.68\ \Omega$), and length $L = 57.7$ mm.

Antiresonant frequencies ω_{p1} and ω_{p2} of each resonator are not equidistant relative to ω_{01}. To simplify consideration, we will set only one antiresonant frequency f_{p1} and the electrical length θ_{p1} corresponding to it, as indicated in the second and third rows of Table 10.6. From (9.37a), we determine the connection coordinate q and the length l corresponding to it (rows 4, 5 in Table 10.6). The last row of this table shows the slope parameters b of the resonators, calculated by formula (9.39).

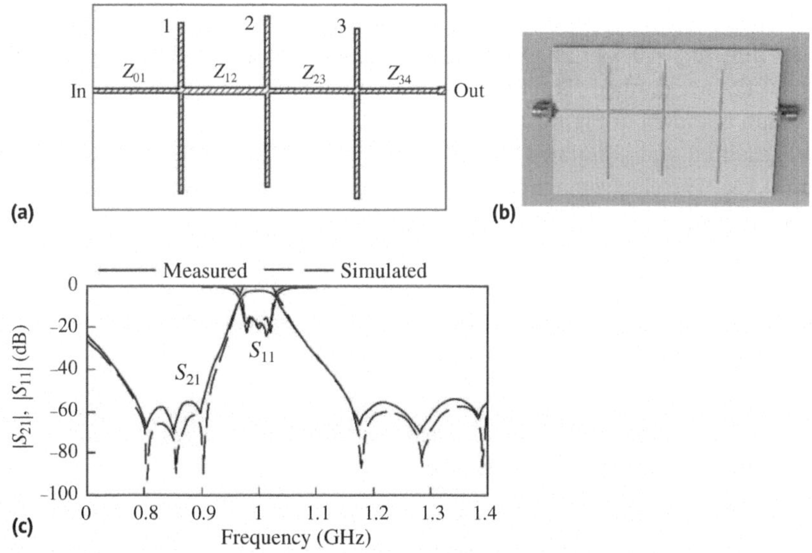

Figure 10.12 Microstrip third-order BPF with uniform resonator No. 2 (Table 9.4) operates on $\lambda/2$ oscillations with six TZs. (a) Topology. (b) Photograph. (c) Frequency responses.

Table 10.6 Parameters of resonators in Figure 10.12a.

Parameter	Resonator No. 1	2	3
f_{p1}, GHz	0.85	0.9	0.8
θ_{p1}, °	153	162	144
q	0.412	0.444	0.375
$l = qL$, mm	23.8	25.6	21.6
b, Ω^{-1}	0.506	1.230	0.257

From Table 2.6, we find g parameters for $n = 3$ and $L_{Ar} = 0.1$ dB: $g_1 = 1.0315$, $g_2 = 1.1474$, $g_3 = 1.0315$, $g_4 = 1$. Substituting these parameters, FBW = 0.04 and the susceptance slope parameter b_i from Table 10.6 into (3.6), we determine the J parameters of the inverters used in BPF: $J_{01} = 0.019810\ \Omega^{-1}$, $J_{12} = 0.02906\ \Omega^{-1}$, $J_{23} = 0.02067\ \Omega^{-1}$, $J_{34} = 0.01412\ \Omega^{-1}$. Using (3.7c), we determine the characteristic impedances of the quarter-wave sections used as admittance inverters: $Z_{01} = 50.48\ \Omega$, $Z_{12} = 34.47\ \Omega$, $Z_{23} = 48.38\ \Omega$, $Z_{34} = 70.83\ \Omega$.

It is easy to determine the length and width of the horizontal microstrip segments in Figure 10.12a with the aforementioned values $Z_{i,i+1}$: $L_{01} = 29.2$ mm, $w_{01} = 1$ mm; $L_{12} = 28.25$ mm, $w_{12} = 1.95$ mm; $L_{23} = 29.1$ mm, $w_{23} = 1.05$ mm; $L_{34} = 30$ mm, $w_{34} = 0.43$ mm. All geometric dimensions of the microstrip BPF in Figure 10.12a are defined. The filter occupies an area of 116.7 mm × 70 mm. Its photograph is shown in Figure 10.12b.

The measured and simulated frequency responses of the BPF are depicted in Figure 10.12c. The electromagnetic (EM) simulator of MWO (AWR) software was used for simulation. The measured data are as follows: $f_0 = 1$ GHz, BW = 40 MHz, the insertion loss at center frequency $L_{A0} = 1.9$ dB, the return loss in bandwidth $L_R > 14.5$ dB. The third-order microstrip BPF has six TZs. Three left-handed TZs are located at prescribed frequencies (Table 10.6). The considered third-order BPF has a high attenuation level (55 dB) in rejection bands.

10.2.4 Second-order BPF with LD Resonator No. 8 (Table 9.4), Containing LC Chain and Operating on $\lambda/4$ Oscillations

Figure 10.13a shows the circuit of a second-order BPF with LD resonator No. 8 (Table 9.4) and quarter-wave segments as admittance inverters. The resonator has $\lambda/4$ TL segment with a short-circuited end and an LC chain. Its properties are described by formulas (9.69) and (9.70). It has two antiresonant frequencies, ω_{p1} and ω_{p2} near ω_{01}, which can be located equidistantly.

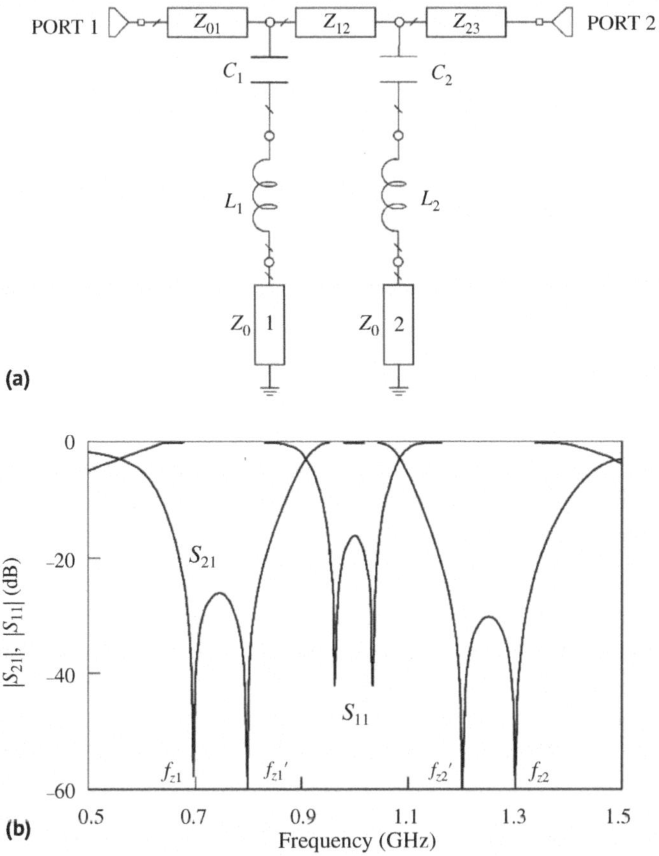

Figure 10.13 Second-order BPF with LD resonator No. 8 (Table 9.4) on $\lambda/4$ oscillations with two pairs of equidistant TZs. (a) Schematic. (b) Simulated frequency responses.

Let us design the second-order BPF (Figure 10.13a) with two pairs of equidistant TZs. The filter parameters are follows: $f_0 = 1$ GHz, BW = 120 MHz (FBW = 0.12), $L_{Ar} = 0.1$ dB. We require that this filter has two equidistant pairs of TZs: $f_{z1,2} =$ 0.7 and 1.3 GHz; $f_{z1',2'} = 0.8$ and 1.2 GHz. The filter uses quarter-wave line segments at the frequency of 1 GHz with characteristic impedance $Z_0 = 10\ \Omega$.

From Table 2.6, we find g parameters for $n = 2$ and $L_{Ar} = 0.1$ dB: $g_1 = 0.8430$, $g_2 = 0.6220$, $g_3 = 1.3554$. Substituting these parameters, FBW = 0.12 and susceptance slope parameter $b = \pi/4Z_0 = 0.07854\ \Omega^{-1}$ into (3.6), we determine the J parameters of the inverters used in BPF: $J_{01} = 0.01495\ \Omega^{-1}$, $J_{12} = 0.01302\ \Omega^{-1}$, $J_{23} = 0.01495\ \Omega^{-1}$. Using (3.7c), we determine the characteristic impedances of the quarter-wave sections used as admittance inverters: $Z_{01} = 66.87\ \Omega$, $Z_{12} = 76.83\ \Omega$, $Z_{23} = 66.87\ \Omega$.

Table 9.8 presents the calculated values of the lumped elements of the resonators used in Figure 10.13a, which provide the arrangement of two antiresonant frequencies mentioned earlier. For resonator 1, these values are $C_1 = 5.36$ pF, $L_1 = 5.2$ nH, and for resonator 2 we have $C_2 = 2.13$ pF and $L_2 = 12.38$ nH.

All BPF parameters are defined. Figure 10.13b shows the simulated frequency responses of this filter. As expected, the considered second-order BPF with simple couplings has two pairs of equidistant TZs at the predicted frequencies.

Figure 10.14 illustrates the influence of the quality factor of inductors with inductance L_1 (Q_{L1}) and L_2 (Q_{L2}) on the frequency responses of the considered BPF. Let us note some regularities:

- The finite values of Q_{L1} and Q_{L2} do not affect the insertion loss of the filter in the passband.
- The quality factor of each inductor affects the two TZs that accompany it.
- A decrease in the quality factor leads to decrease in the "depth" of the corresponding attenuation poles, but their frequency does not change. The quality factor of the capacitors has the same effect.

Let us implement the microstrip realization of the second-order BPF discussed earlier. To implement the microstrip BPF, a TMM-10i substrate (Rogers), with $\varepsilon_r = 9.8$, $h = 0.508$ mm, $\tan\delta = 0.002$, is used. The BPF topology is shown in Figure 10.15a. The following geometric parameters correspond to microstrip resonators and coupling $\lambda/4$ segments: $Z_0 = 10\ \Omega$ ($w = 5.1$ mm, $l = 25.8$ mm); $Z_{01} = Z_{23} = 66.87\ \Omega$ ($w = 0.25$ mm, $l = 14.9$ mm); $Z_{12} = 76.83\ \Omega$ ($w = 0.16$ mm, $l = 15.1$ mm). Microstrip segments of resonators are grounded by metallized holes of 0.6 mm in diameter. To reduce the dimensions of the BPF, the coupling

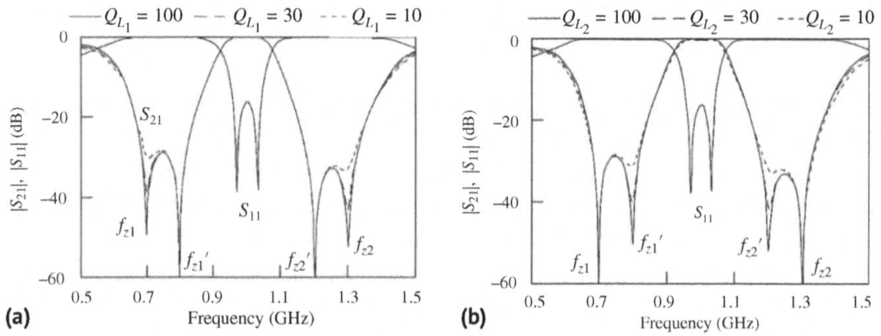

Figure 10.14 Influence of the quality factor of inductors with inductance L_1 and L_2 on BPF frequency responses in Figure 10.13(b). (a) Influence of Q_{L1}. (b) Influence of Q_{L2}.

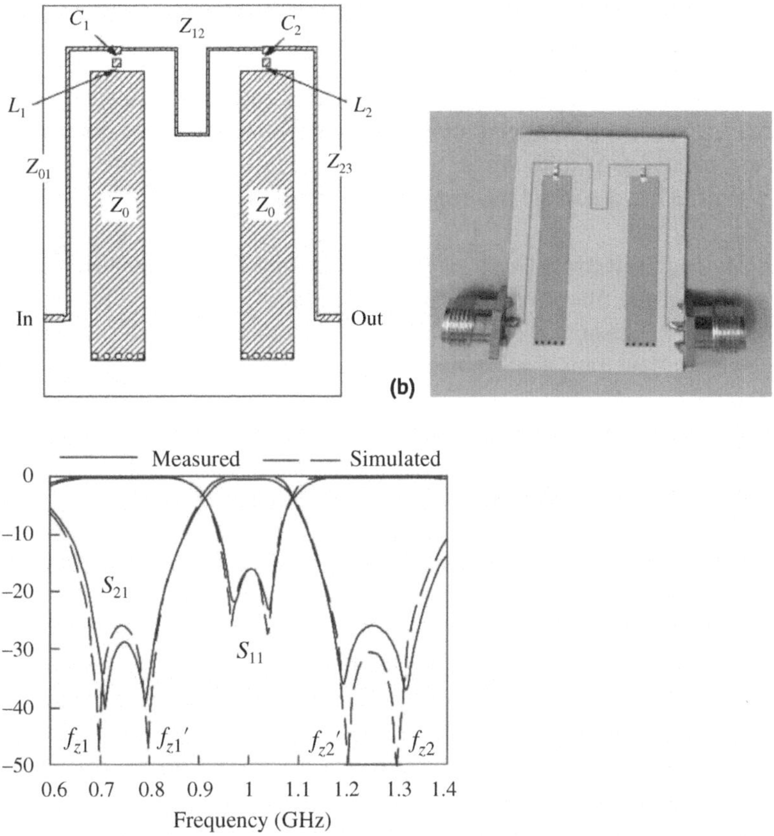

Figure 10.15 Microstrip second-order BPF with LD resonators No. 8 (Table 9.4) and four TZs. (a) Topology. (b) Photograph. (c) Frequency responses.

quarter-wave segments are bent. The filter occupies the area of 35 mm × 27 mm. Its photograph is shown in Figure 10.15b.

The EM simulator of Microwave Office software is used to tune the TZs to the desired frequencies by changing C_i and L_i. The manufactured filter used capacitors and inductors with size code 0402 with the following values: $C_1' = 3.9$ pF, $L_1' = 5.6$ nH; $C_2' = 2.0$ pF, $L_2' = 12$ nH.

The measured and simulated frequency responses of the BPF are depicted in Figure 10.15c. The measured data are $f_0 = 1$ GHz, BW = 120 MHz, $L_{A0} =$ 0.7 dB, and $L_R > 16$ dB in the bandwidth. The selectivity of this filter 30 dB ($f_0 \pm 185$ MHz) can be expressed at the normalized frequency $j\Omega$: 30 dB ($\pm j3.08$). This is a high selectivity for a two-order BPF.

10.2.5 BPF with LD Resonator No. 7 (Table 9.4) Containing LC Chain and Operating on $\lambda/2$ Oscillations

This resonator is a half-wave TL segment with a cascaded connection of an LC chain. It operates on $\lambda/2$ oscillations and has two antiresonant frequencies ω_{p1} and ω_{p2} located next to the main resonant frequency ω_{01}, including equidistant. Its properties are described by formulas (9.67) and (9.68).

10.2.5.1 High-selectivity Fourth-order BPF with Four Pairs of Equidistant TZs

The circuit of the filter under consideration is shown in Figure 10.16a. This filter contains LD resonator No. 7 (Table 9.4) and admittance inverters containing capacitances. The transverse branches of inverters with negative capacitances are not shown.

The filter parameters are follows: $f_0 = 2$ GHz, FBW = 0.05, $L_{Ar} = 0.1$ dB ($L_R = 16.426$ dB), and four pairs of TZs $f_{p1,8} = 1.75$ and 2.25 GHz; $f_{p2,7} = 1.8$ and 2.2 GHz; $f_{p3,6} = 1.85$ and 2.15 GHz; $f_{p4,5} = 1.9$ and 2.1 GHz.

To generate the fourth pair of equidistant TZs, four LD resonators are used (Figure 10.16a), which have characteristic impedance $Z_0 = 10\ \Omega$ and susceptance slope parameter $b = \pi/2Z_0 = 0.15708\ \Omega^{-1}$.

From Table 2.6, we find g parameters: $g_0 = 1$, $g_1 = 1.1088$, $g_2 = 1.3061$, $g_3 = 1.7703$, $g_4 = 0.8180$, $g_5 = 1.3554$. Substituting these values, parameter b and FBW = 0.05 into (3.6), we determine the J parameters of the inverters used in BPF: $J_{12} = J_{34} = 0.06525\ \Omega^{-1}$, $J_{23} = 0.05165\ \Omega^{-1}$, $J_{01} = J_{45} = 0.01190\ \Omega^{-1}$. Using inverters with capacitances, according to (3.7b): $C_{12} = C_{34} = 0.5192$ pF, $C_{23} = 0.411$ pF, and $C_{01} = C_{45} = 0.947$ pF.

Table 9.7 presents the calculated values of the lumped elements L_i and C_i used in Figure 10.16a of the resonators, which provide the aforementioned arrangement of four equidistant pairs of antiresonant frequencies. For resonator 1 with $f_{p1,8} = 1.75$ and 2.25 GHz, these values are equal to $C_1 = 0.83$ pF, $L_1 = 7.72$ nH; for

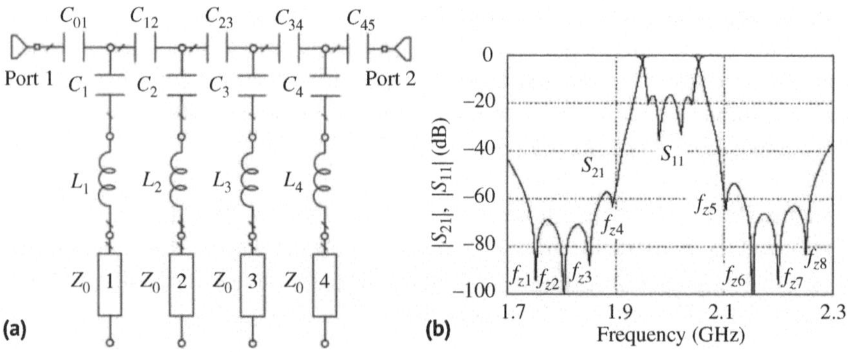

Figure 10.16 Fourth-order BPF with LD resonator No. 7 (Table 9.4) with four pairs of equidistant TZs. (a) Schematic. (b) Simulated frequency responses.

resonator 2 with $f_{p2,7} = 1.8$ and 2.2 GHz, we have $C_2 = 0.51$ pF and $L_2 = 12.5$ nH; the values $f_{p3,6} = 1.85$ and 2.15 GHz of the third resonator are provided at $C_3 = 0.29$ pF and $L_3 = 21.9$ nH; antiresonant frequencies $f_{p4,5} = 1.9$ and 2.1 GHz of the fourth resonator occur at $C_4 = 0.13$ pF and $L_4 = 48.8$ nH.

All parameters of the considered filter are defined. Figure 10.16b shows its simulated frequency responses. The four pairs of equidistant TZs of this BPF are located at the prescribed frequencies. The filter is characterized by a relatively high selectivity of 54 dB ($f_0 \pm 100$ MHz), which can be expressed at the normalized frequency $j\Omega$: 54 dB ($\pm j2$). To implement four TZ pairs in traditional cross-coupling filters, 10 resonators must be used [17].

10.2.5.2 BPFs with Equidistant Pairs of Higher Degree TZs

The TZ formation of higher degrees is particularly interesting, leading to increased filter attenuation at these frequencies. If in a third-order BPF with admittance inverters (Figure 10.17a) we use LD resonator No. 7 (Table 9.4), then we can implement a pair of equidistant TZs of the second or third degree. Admittance inverters in Figure 10.17a contain capacitances, and the transverse branches of inverters with negative capacitances are not shown.

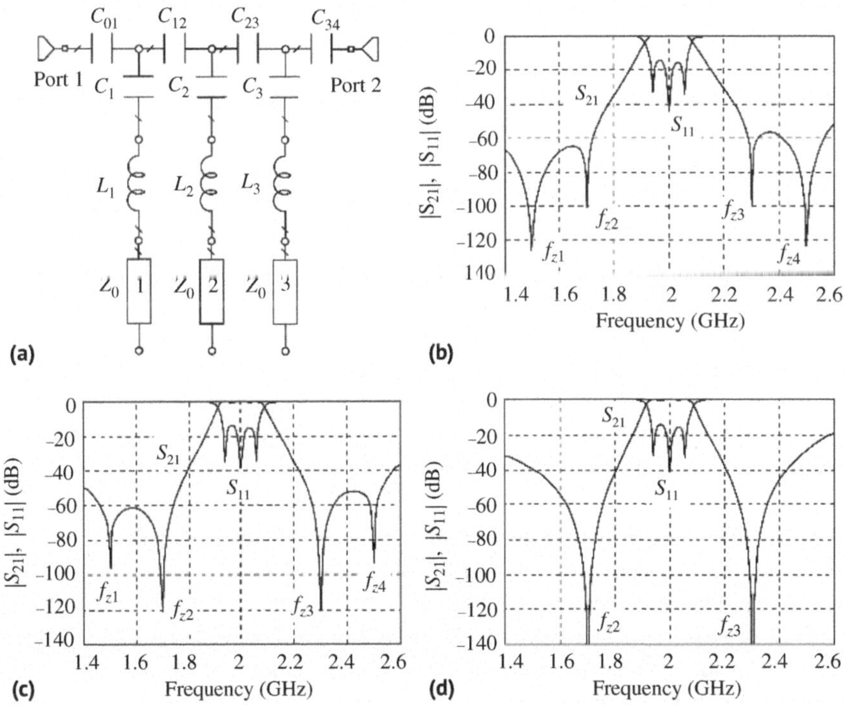

Figure 10.17 Third-order BPF with LD resonators No. 7 (Table 9.4) and higher degree TZs. (a) Schematic. (b) Second-degree TZs at the frequencies f_{z1} and f_{z4}. (c) Second-degree TZs at frequencies f_{z2} and f_{z3}. (d) Third-degree TZs at frequencies f_{z2} and f_{z3}.

The filter parameters will be set as follows: $f_0 = 2$ GHz, FBW = 0.075, $L_{Ar} = 0.2$ dB ($L_R = 13.467$ dB), and two pairs of TZs from Table 9.7: $f_{p1,4} = 1.5$ and 2.5 GHz; $f_{p2,3} = 1.7$ and 2.3 GHz. From Table 2.6, we find g parameters: $g_0 = g_4 = 1$, $g_1 = g_3 = 1.2275$, $g_2 = 1.525$. Resonators used in the filter have characteristic impedance $Z_0 = 10\ \Omega$ and susceptance slope parameter $b = \pi/2Z_0 = 0.15708\ \Omega^{-1}$.

Substituting g parameters, values of b and FBW = 0.075 into (3.6), we determine the J parameters of the inverters: $J_{01} = J_{34} = 0.01385\ \Omega^{-1}$, $J_{12} = J_{23} = 0.009540\ \Omega^{-1}$. If using inverters with capacitances, then according to (3.7b): $C_{01} = C_{34} = 1.102$ pF, $C_{12} = C_{23} = 0.7591$ pF.

To implement the prescribed equidistant TZ pairs, it is necessary to use resonators No. 1 and No. 2 (Table 9.7) with the corresponding L_i and C_i values. For resonator No. 1 with $f_{p1,4} = 1.5$ and 2.5 GHz, these values are equal to $C_1 = 4.27$ pF, $L_1 = 1.58$ nH; for resonator No. 2 with $f_{p2,3} = 1.7$ and 2.3 GHz, we have $C_2 = 1.25$ pF and $L_2 = 5.17$ nH.

Figure 10.17b–d shows the frequency responses of the considered third-order BPF for various options of used resonators from Table 9.7. If resonator No. 1 is used twice in a BPF, then the corresponding TZs ($f_{z1} = 1.5$ GHz and $f_{z4} = 2.5$ GHz) have a second degree (Figure 10.17b). If resonator No. 2 is used twice in a BPF, then the corresponding TZs ($f_{z2} = 1.7$ GHz and $f_{z3} = 2.3$ GHz) have a second degree (Figure 10.17c). The order of the resonators in the BPF can be arbitrary. If resonator No. 2 is used three times in a BPF, then its corresponding TZ pair has a third degree (Figure 10.17d). Generally, an N-order BPF will have a TZ pair of k degrees if it uses k identical LD resonators (Figure 9.15).

Let us evaluate the selectivity of the BPF for frequency responses in Figure 10.17: 43 dB ($f_0 \pm 250$ MHz) for Figure 10.17b; 50 dB ($f_0 \pm 250$ MHz) for Figure 10.17c; and 60 dB ($f_0 \pm 250$ MHz) for Figure 10.17d. The selectivity of BPF increases if the degree of the closest TZs to the passband is increased. This also applies to the fourth-order BPF in Figure 10.16. If the TZs with frequencies $f_{z4} = 1.9$ GHz and $f_{z5} = 2.1$ GHz are made to the second degree, then the BPF selectivity will increase from 54 dB ($\pm j2$) to 75 dB ($\pm j2$).

Thus, N-order BPFs with LD resonators and $2N$ TZs can have increased selectivity, which becomes higher if the degree of the closest TZs to the passband increases.

10.2.5.3 Microstrip Second-order BPF with Two Pairs of Equidistant TZs

Let us design a microstrip second-order BPF with two pairs of equidistant TZs. The filter parameters will be set as follows: $f_0 = 2$ GHz, FBW = 0.06, $L_{Ar} = 0.1$ dB ($L_R = 16.426$ dB), and two pairs of TZs from Table 9.7: $f_{p1,4} = 1.5$ and 2.5 GHz; $f_{p2,3} = 1.7$ and 2.3 GHz. We will use LD resonators (Figure 9.15) with $Z_0 = 10\ \Omega$ and $b = 0.15708\ \Omega^{-1}$ and quarter-wave TL segments as J inverters. The schematic of the BPF is shown in Figure 10.18a. From Table 2.6, we find the corresponding g parameters: $g_0 = 1$, $g_1 = 0.8430$, $g_2 = 0.6220$, $g_3 = 1.3554$. Substituting these values, slope parameter b and FBW = 0.06 into (3.6), we determine the J parameters of the inverters: $J_{01} = J_{23} = 0.01495\ \Omega^{-1}$, $J_{12} = 0.01301\ \Omega^{-1}$. Since $\lambda/4$ segments are used as J inverters, then from (3.7c), we determine their characteristic impedances: $Z_{01} = Z_{23} = 66.9\ \Omega$, $Z_{12} = 76.9\ \Omega$.

To implement the prescribed equidistant TZ pairs, it is necessary to use resonators No. 1 and No. 2 (Table 9.7) with the corresponding values of lumped elements: $L_1 = 1.58$ nH, $C_1 = 4.27$ pF; $L_2 = 5.17$ nH, $C_2 = 1.25$ pF.

The microstrip second-order BPF used a TMM-10i substrate (Rogers) with $\varepsilon_r = 9.8$, $h = 0.508$ mm, $\tan\delta = 0.002$. The following geometric parameters correspond to microstrip resonators and coupling $\lambda/4$ segments: $Z_0 = 10\ \Omega$ ($w = 5.1$ mm, $l = 25.8$ mm); $Z_{01} = Z_{23} = 66.9\ \Omega$ ($w = 0.25$ mm, $l = 14.9$ mm); $Z_{12} = 76.9\ \Omega$ ($w = 0.16$ mm, $l = 15.1$ mm). To reduce the dimensions of the BPF, the line segments with Z_{01} and Z_{23} are curved. The filter occupies the area of 45 mm × 35 mm. Its photograph is shown in Figure 10.18b.

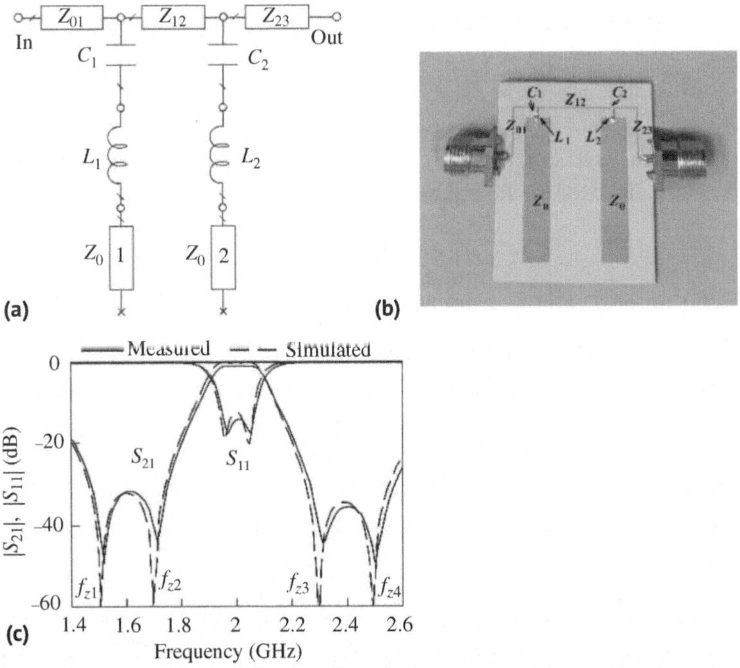

Figure 10.18 Second-order microstrip BPF with LD resonators No. 7 (Table 9.4) and four TZs. (a) Schematic. (b) Photograph. *Source*: [1] / with permission of IEEE. (c) Frequency responses.

The EM simulator of Microwave Office software is used to tune the TZs to the desired frequencies by changing C_i and L_i. The manufactured filter used capacitors and inductors with size code 0402 with the following values: $C_1' = 4.3$ pF, $L_1' = 0.8$ nH; $C_2' = 1$ pF, $L_2' = 4.7$ nH.

The measured data are as follows: $f_0 = 2$ GHz, BW = 120 MHz, the insertion loss $L_{A0} = 2.0$ dB, the return loss in bandwidth $L_R > 14$ dB. The selectivity of this filter 30 dB ($f_0 \pm 250$ MHz) can be expressed at the normalized frequency $j\Omega$: 30 dB ($\pm j4.17$). The considered BPF has significant potential for increasing selectivity. Its two closest TZs to the passband are at 1.7 and 2.3 GHz. These TZs can be located much closer to the passband and have a higher degree, further increasing the BPF selectivity.

The proposed BPFs can be significantly miniaturized using high dielectric constant substrates and inverters with lumped elements. When using a substrate with dielectric constant $\varepsilon_r = 92$, the length of the half-wave segment used in resonators at the frequency of 2 GHz will decrease to 9 mm, and its width will decrease to 1.2 mm.

10.3 Influence of Destabilizing Factors on BPF Frequency Responses

The destabilizing factors of the considered LD BPFs include deviation of C_i and L_i from nominal values, precision manufacturing, and temperature change. Appropriate sensitivities are used in Refs. [18, 19] to consider the influence of these factors.

10.3.1 Deviation of C_i and L_i from Nominal Values

In general, when C_i and L_i change, we obtain a shift of the central frequency ω_0 of the passband and TZs ω_{zi}, associated with the resonant ω_0 and antiresonant ω_{pi} frequencies of resonators. If the resonator in Figure 9.15 does not have inductance L, then equation (9.68), which determines the antiresonance frequencies of the resonator and the TZs ω_{zi} of the filter, takes the form

$$\omega C + B(\omega) = 0, \tag{10.3}$$

where $B(\omega) = Z_0^{-1} \tan\theta$ is the reactance function.

A relative change in capacitance $\Delta C/C$ leads to a relative shift in TZs $\Delta\omega_{zi}/\omega_{zi}$. The ratio of these quantities [19], expressed in differential form, is called sensitivity ω_{zi} to changes in C:

$$S_C = -\left.\frac{d\omega/\omega}{dC/C}\right|_{\omega=\omega_n} = -\left.\frac{d\ln\omega}{d\ln C}\right|_{\omega=\omega_n}. \tag{10.4}$$

The "–" sign on the right side of (10.4) makes the value $S_C^{\omega_{zi}}$ positive definite. Using equation (10.3), we can express the sensitivity $S_C^{\omega_{zi}}$ (10.4) in the form [20]

$$S_C^{\omega_{zi}} = \left(1 - \frac{dB/d\omega}{B/\omega}\right)^{-1}\Bigg|_{\omega=\omega_{zi}}. \tag{10.5}$$

The reactance functions satisfy the condition $dB/d\omega \geq |B|/\omega$; therefore, from (10.5), the next inequality follows

$$S_C^{\omega_{zi}} \leq \frac{1}{2}. \tag{10.6}$$

If the resonator in Figure 9.15 contains L only, then $S_L^{\omega_{zi}} \leq 1/2$. If the resonator in Figure 9.15 contains C and L, then the sensitivity will satisfy the condition

$$S_{C+L}^{\omega_{zi}} \leq 1. \tag{10.7}$$

Inequality (10.7) means that with a one-sided change in C and L by 2%, the values of ω_{zi} will change by no more than 2%.

Since the resonant frequency ω_0 of the resonators under consideration does not depend on C and L, the corresponding sensitivity is equal to zero

$$S_{C+L}^{\omega_0} = 0. \tag{10.8}$$

The property (10.8) is essential and practically applicable.

For the microstrip BPF (Figure 10.18), the fulfillment of conditions (10.7) and (10.8) is illustrated in Figure 10.19a, where the values of C_i and L_i are increased by 2%. Note that most capacitors and inductors with size code 0402 have a deviation from the nominal values not exceeding 2% [21]. The largest shift of f_{zi} (Figure 10.18a) occurred for $f_{z4} = 2.5$ GHz. It amounted to $\Delta f_{z4} = 27$ MHz (1.08%), which corresponds to the sensitivity of $S_{C+L}^{\omega_{z4}} = 0.54$ and agrees with (10.7).

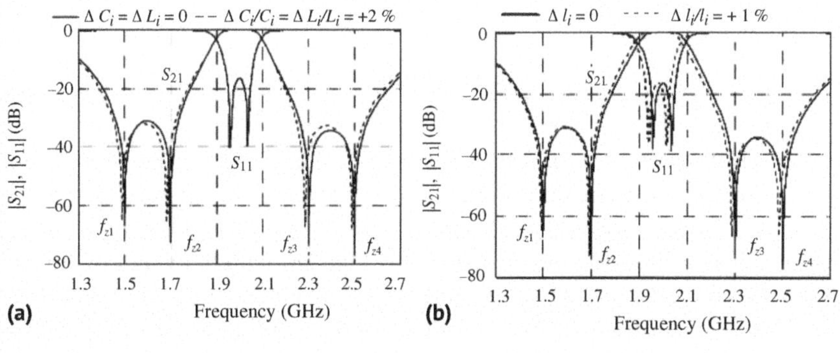

Figure 10.19 Influence of destabilizing factors on frequency responses. (a) Influence of C_i and L_i. (b) Influences of lengths l_i.

The value of ω_0 and the return loss curve in the BPF passband must remain unchanged.

10.3.2 Manufacturing Precision

Errors in the conductive patterning of microstrip BPFs also shift ω_0, ω_{zi}, and change the frequency responses. The change in the length l of the resonators is the main factor in the disturbance of frequency responses. Similar to (10.4), we denote the sensitivity of ω_0 and ω_{zi} to the length changes of the resonator as $S_l^{\omega_{0,zi}}$. It is easy to show that

$$S_l^{\omega_{0,zi}} = 1. \tag{10.9}$$

Equality (10.9) means that the change in l by 1% leads to the change in ω_0 and ω_{zi} also by 1%. Condition (10.9) is illustrated in Figure 10.19b for the microstrip BPF shown in Figure 10.18. The increase in l by 1% led to a decrease in ω_0 and ω_{zi} by 1%. Changing the width w of resonators has a much smaller effect on the frequency responses of BPF than changing l.

If conductive patterns on the substrate are formed by photolithography, then the accuracy of their reproduction is about 10 μm or less. With a reproduction accuracy of 10 μm, the length of the resonator used in the filter in Figure 10.18 will change by 0.08%, which leads to a shift of characteristic frequencies by 1.6 MHz (10.9).

Suppose BPF uses small gaps (e.g., 0.1 mm) between TL segments. In that case, the specified accuracy of 10 μm will change the gap width by 20%, inevitably changing the frequency responses.

10.3.3 Temperature Frequency Stability

The temperature coefficient of resonant frequency $\tau(f)$ for microstrip resonators with parallel connection of capacitance C and inductance L is determined by the approximate formula

$$\tau(f) \approx -\left[\tau(l) + \frac{\tau(\varepsilon)}{2} + \frac{\tau(C)}{2} + \frac{\tau(L)}{2}\right], \tag{10.10}$$

where $\tau(l)$ is the coefficient of dielectric substrate thermal expansion (max X, Y), $\tau(\varepsilon)$ is the temperature coefficient of dielectric constant, $\tau(C)$ is the temperature coefficient of capacitance, and $\tau(L)$ is the temperature coefficient of inductance.

In our case, the values of C and L do not affect the resonant frequency of the resonators; therefore, the terms $\tau(C)/2$ and $\tau(L)/2$ in (10.10) are absent. For thermoset dielectric substrate TMM-10i with $\varepsilon_r = 9.8$ according to data [22], $\tau(l) =$ 19 ppm/°C; $\tau(\varepsilon) = -43$ ppm/°C. Substituting these values into (10.10) gives $\tau(f) \approx 2.5$ ppm/°C. If the ambient temperature t changes by $\Delta t = 100$ °C, then the center frequency $f_0 = 2$ GHz of the BPF passband shifts by $\Delta f \approx \tau(f) f_0 \Delta t \approx$ 0.5 MHz. If we use alumina (Al_2O_3) substrate with the same dielectric constant and value $\tau(f) \approx -46.5$ ppm/°C, then with the exact temperature change, the frequency offset will decrease significantly ($\Delta f \approx -9.3$ MHz).

10.4 Basic Properties of BPF with Simple Couplings and Multiple TZs Generated by Resonators

It is advisable to systematize the main properties of the above-considered inline BPFs with simple couplings, which have multiple TZs located at real frequencies.

1. Using N-order inline BPF, which has resonators with one antiresonant frequency ω_{p1} located near the main resonant frequency ω_{01} (Table 9.1), results in TZ number equal to N.
2. Using N-order inline BPF, which has resonators with two antiresonant frequencies located near the operating resonant frequency (Table 9.4), results in several TZs equal to $2N$.
3. The filters used different types of oscillations $\lambda/4$, $\lambda/2$, $3\lambda/4$, λ, and $5\lambda/4$, allowing them to be used in an extensive frequency range.
4. The TZs may be equidistant or nonequidistant relative to the passband center frequency. The exception is the BPFs (Figures 10.9–10.11) that use SIRs Nos. 4–6 (Table 9.4), where the TZs are only equidistant.
5. The advantage of the considered filter with LD resonators (Tables 9.1 and 9.4) is that the lumped elements do not affect the resonant frequency of the resonators and, consequently, the position of the central frequency. These elements only affect the position of the TZs.
6. The lumped elements do not affect the susceptance slope parameter b of LD resonators from Tables 9.1 and 9.4, simplifying the design of BPFs with such resonators.
7. The quality factor of inductors Q_L and capacitors Q_C does not increase the BPF insertion loss in the passband; they only decrease the "depth" of the attenuation poles.
8. Filters with pairs of equidistant TZs of higher degree can be realized.
9. Using higher degree TZs results in improved selectivity of BPFs.
10. Using multiple TZs allows us to implement different frequency responses with few resonators. Thus, the second-order BPFs in Figures 10.4–10.7 have increased one-sided selectivity, which is highly demanded when using BPFs in duplexers. The third-order BPF in Figure 10.9 has a high attenuation level in adjustable frequency bands, allowing us to suppress unwanted signals in these bands effectively. Third-order BPF in Figure 10.17 with pairs of equidistant TZs of higher degree has increased selectivity.

11. The design sequence of the filters considered is quite simple. It is based on the classical design method (3.6), (3.8), and the formulas describing the position of the TZs generated by the resonators from Tables 9.1 and 9.4.
12. The coupling matrix [17] is not used to describe these filters since it does not reproduce the TZs generated by the resonators.

The disadvantages include a discrete series of nominal values of C_i, L_i of mass-produced lumped elements [21], which may not coincide with the calculated values. It leads to deviation of the positions of TZs from the prescribed values.

References

1 A. V. Zakharov and S. Litvintsev, Lumped-distributed resonators providing N or 2N transmission zeros at real frequencies in bandpass filters without cross and mixed couplings, *IEEE Trans. Microw. Theory Tech.*, vol. 72, no. 6, pp. 3682–3691, 2024, https://doi.org/10.1109/TMTT.2023.3332260.
2 A. V. Zakharov and S. Litvintsev, Lumped-distributed resonators providing multiple transmission zeros in bandpass filters with simple and mixed couplings, *IEEE Trans. Circuits Syst. I Regul. Pap.*, vol. 71, no. 8, pp. 3502–3513, 2024, https://doi.org/10.1109/TCSI.2024.3375961.
3 S. Litvintsev and A. Zakharov, Bandpass filters with increased to 3N+1 number of attenuation poles, *Radioelectron. Commun. Syst.*, vol. 66, no. 8, pp. 385–402, 2023, https://doi.org/10.3103/S073527272311002X.
4 A. V. Zakharov and S. Litvintsev, Transmission line bandpass filters with multiple attenuation poles and small number of resonators, *AEU - Int. J. Electron. Commun.*, p. 155131, 2024, https://doi.org/10.1016/j.aeue.2024.155131.
5 S. Litvintsev and A. Zakharov, Bandpass filters with 2N+1 transmission zeros at real frequencies, *Radioelectron. Commun. Syst.*, vol. 66, no. 6, pp. 315–321, 2023, https://doi.org/10.3103/S0735272723100035.
6 J.-S. Lim and D. C. Park, A modified Chebyshev bandpass filter with attenuation poles in the stopband, *IEEE Trans. Microw. Theory Tech.*, vol. 45, no. 6, pp. 898–904, 1997, https://doi.org/10.1109/22.588597.
7 M. E. Ilchenko and A. V. Zakharov, Wide band cutoff filters made of regular transmission line segments, *Radiotekhnika i Elektron.*, vol. 36, no. 8, pp. 1585–1589, 1991.
8 C. Quendo, E. Rius, and C. Person, Narrow bandpass filters using dual-behavior resonators based on stepped-impedance stubs and different-length stubs, *IEEE Trans. Microw. Theory Tech.*, vol. 52, no. 3, pp. 1034–1044, 2004, https://doi.org/10.1109/TMTT.2004.823582.

9 K. Wada and O. Hashimoto, Fundamentals of open-ended resonators and their application to microwave filters, *IEICE Trans. Electron.*, vol. E83-C, no. 11, pp. 1763–1775, 2000, https://search.ieice.org/bin/summary.php?id=e83-c_11_1763&category=C&year=2000&lang=E&abst=.

10 W. J. Feng, W. Q. Che, Y. M. Chang, S. Y. Shi, and Q. Xue, High selectivity fifth-order wideband bandpass filter with multiple transmission zeros based on transversal signal-interaction concepts, *IEEE Trans. Microw. Theory Tech.*, vol. 61, no. 1, pp. 89–97, 2013.

11 W. Feng, X. Gao, and W. Che, Bandpass filters with multiple transmission zeros using open/shorted stubs, *IET Microwaves, Antennas Propag.*, vol. 9, no. 8, pp. 769–774, 2015, https://doi.org/10.1049/iet-map.2014.0402.

12 C. Quendo, E. Rius, C. Person, and M. Ney, Integration of optimized low-pass filters in a bandpass filter for out-of-band improvement, *IEEE Trans. Microw. Theory Tech.*, vol. 49, no. 12, pp. 2376–2383, 2001, https://doi.org/10.1109/22.971624.

13 A. Manchec, C. Quendo, E. Rius, C. Person, and J.-F. Favennec, Synthesis of dual behavior resonator (DBR) filters with integrated low-pass structures for spurious responses suppression, *IEEE Microw. Wirel. Components Lett.*, vol. 16, no. 1, pp. 4–6, 2006, https://doi.org/10.1109/LMWC.2005.861361.

14 A. Zakharov and S. Litvintsev, Planar bandpass filters based on resonators generating transmission zeros, *IEEE Circuits Syst. Mag.*, vol. 25, no. 2, pp. 57–73, 2025, https://doi.org/10.1109/MCAS.2025.3543781.

15 S. Litvintsev and A. Zakharov, Synthesis of TL resonators with 3λ/4-mode oscillations for bandpass filters with multiple transmission zeros, in *2024 IEEE 29th International Seminar/Workshop on Direct and Inverse Problems of Electromagnetic and Acoustic Wave Theory (DIPED)*, 2024, pp. 145–149, https://doi.org/10.1109/DIPED63529.2024.10706048.

16 S. Litvintsev and A. Zakharov, Design of bandpass filters with high mode 5λ/4 oscillations and equidistant transmission zeros pairs, in *2024 IEEE 29th International Seminar/Workshop on Direct and Inverse Problems of Electromagnetic and Acoustic Wave Theory (DIPED)*, 2024, pp. 150–154, https://doi.org/10.1109/DIPED63529.2024.10706170.

17 R. J. Cameron, C. M. Kudsia, and R. R. Mansour, *Microwave Filters for Communication Systems*, 2nd ed. Hoboken, NJ: John Wiley & Sons, Inc., 2018, https://doi.org/10.1002/9781119292371.

18 H. W. Bode, *Network Analysis and Feedback Amplifier Design*. New Jersey: D. Van Nostrand Company, Inc., 1945.

19 K. Géher, *Theory of Network Tolerances*. Budapest: Akadémiai Kiadó, 1971, http://real-eod.mtak.hu/id/eprint/12705.

20 A. Zakharov and M. Ilchenko, Circuit function characterizing tunability of resonators, *IEEE Trans. Circuits Syst. I Regul. Pap.*, vol. 67, no. 1, pp. 98–107, 2020, https://doi.org/10.1109/TCSI.2019.2940066.

21 AVX Corporation, RF/Microwave Products, *Products*, 2023. https://www.kyocera-avx.com/products/.

22 Rogers Corporation, TMM® thermoset microwave materials. Data Sheet. http://www.rogerscorp.com.

11

Filters with Mixed Couplings and Multiple Transmission Zeros Generated by Resonators

It is well known [1–3] that the use of mixed couplings in inline N-order bandpass filters (BPFs) allows us to obtain $(N + 1)$ transmission zeros (TZs), which are located at real frequencies. Using resonators with one or two antiresonant frequencies close to the operating resonant frequency in such BPFs will significantly increase the TZ number. This section considers inline BPFs with mixed couplings [4–8], which contain resonators from Tables 9.1 and 9.4 and have many TZs. For the design of such BPFs, analytical expressions are used that describe the properties of the resonators, and the classical method of G. Matthaei, supplemented by expressions (8.4), which allow dividing the J parameter of the admittance inverter for mixed coupling into magnetic J_m and electric J_e components.

11.1 BPFs with Mixed Couplings and Resonators with One Antiresonance Frequency Close to Main Resonance Frequency

Suppose an inline N-order BPF with mixed couplings and $(N + 1)$ TZs uses resonators with one antiresonance frequency ω_{p1} near the fundamental resonant frequency ω_{01} (Table 9.1). In that case, the TZ number increases to $(2N + 1)$. To illustrate this, we use a quarter-wave resonator No. 1 (Table 9.1) in a second-order mixed coupled BPF [6] (Figure 11.1a). This resonator (Figure 9.2a) provides a right-hand TZ, and its properties are expressed by formulas (9.3)–(9.6). J inverters provide mixed couplings (Figure 7.1a), and transverse branches are not shown.

Let us design a second-order BPF with five TZs. Set the following parameters: center frequency $f_0 = 1$ GHz, bandwidth BW = 30 MHz (FBW = 0.03), passband ripple $L_{\text{Ar}} = 0.1$ dB ($L_R = 16.426$ dB), and TZs: $f_{z1} = 0.6$ GHz, $f_{z2} = 0.8$ GHz, $f_{z3} = 1.2$ GHz, $f_{z4} = 1.4$ GHz, $f_{z5} = 1.6$ GHz.

Two right-sided TZs, $f_{z3} = 1.2$ GHz and $f_{z4} = 1.4$ GHz, will be provided by appropriate coordinates q for connecting coupling elements to the resonators.

Using (9.5), we obtain the values of q_i for each resonator: $q_1 = 1/6$; $q_2 = 2/7$. We place these values in Table 11.1. Inverters connecting the resonators to the loads are also connected to the specified points of the resonators. Let us define the characteristic impedance of resonators: $Z_0 = 40\ \Omega$. The last column of Table 11.1 contains the susceptance slope parameter b of each resonator, calculated by (9.6).

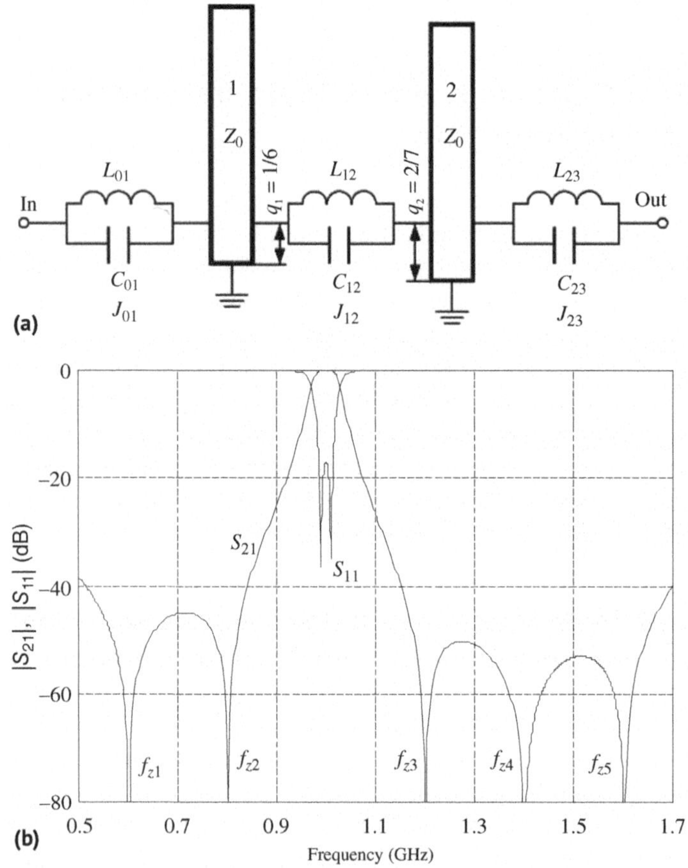

Figure 11.1 Second-order mixed coupled BPF with resonator No. 1 (Table 9.1) with five TZs. (a) Schematic. (b) Simulated frequency responses.

Table 11.1 Parameters of resonators in Figure 11.1a.

Resonator	f_{zi}, GHz	q_i	b, Ω^{-1}
No. 1	1.2	1/6	0.2931
No. 2	1.4	2/7	0.1043

From Table 2.6, we find g parameters for $n = 2$ and $L_{Ar} = 0.1$ dB: $g_0 = 1$, $g_1 = 0.8430$, $g_2 = 0.6220$, $g_3 = 1.3554$. Substituting these g parameters, FBW, and b from Table 11.1 into (3.6), we determine the module of the inverter's J parameters

$$
\begin{aligned}
|J_{01}| &= 0.01443\ \Omega^{-1}\ (f_{z1} = 0.6\ \text{GHz}),\\
|J_{12}| &= 0.00724\ \Omega^{-1}\ (f_{z2} = 0.8\ \text{GHz}),\\
|J_{23}| &= 0.00862\ \Omega^{-1}\ (f_{z5} = 1.6\ \text{GHz}).
\end{aligned}
\tag{11.1}
$$

The TZs generated by each inverter are shown in parentheses in expressions (11.1). Let the inverter with J_{01} generate a left-handed TZ at $f_{z1} = 0.6$ GHz. Then, the value of J_{01} should be negative $J_{01} = -0.01443\ \Omega^{-1}$. Substituting this value and $f_{z1} = 0.6$ GHz into (8.4), we obtain $J_{01m} = 0.00813\ \Omega^{-1}$, $J_{01e} = -0.02256\ \Omega^{-1}$. Using the obtained values and expressions (3.7a, b), we get the inductance and capacitance values of the inverter J_{01} for mixed coupling: $L_{01} = 19.6$ nH and $C_{01} = 3.59$ pF.

The J_{12} inverter generates left-handed TZ at $f_{z2} = 0.8$ GHz. Then, the value of J_{12} should be negative $J_{12} = -0.00724\ \Omega^{-1}$. Substituting this value and $f_{z2} = 0.8$ GHz into (8.4), we obtain $J_{12m} = 0.01288\ \Omega^{-1}$, $J_{12e} = -0.02012\ \Omega^{-1}$. Using the obtained values and expressions (3.7a, b), we get this inverter's inductance and capacitance values: $L_{12} = 12.36$ nH and $C_{12} = 3.2$ pF.

The inverter with $J_{23} = 0.00862\ \Omega^{-1}$ is positive because it generates right-handed TZ at frequency $f_{z5} = 1.6$ GHz. Carrying out calculations in the aforementioned sequence, we obtain $J_{23m} = 0.01414\ \Omega^{-1}$, $L_{23} = 11.26$ nH; $J_{23e} = -0.00552\ \Omega^{-1}$, $C_{23} = 0.88$ pF.

All BPF parameters in 11.1(a) are defined. Figure 11.1b shows the BPF frequency responses, simulated by the linear circuit simulator of Microwave Office. The filter is characterized by $f_0 = 1$ GHz, BW = 30 MHz, and five TZs at prescribed frequencies. Seven resonators would be required to implement five TZs in a BPF with simple couplings, including cross-coupling [9].

11.2 BPFs with Mixed Couplings and Resonators with Two Antiresonance Frequencies Close to Main Resonant Frequency

Suppose an inline N-order BPF with mixed couplings and $(N + 1)$ TZs uses resonators with two antiresonant frequencies near the operating resonant frequency (Table 9.4). In that case, the TZ number increases significantly to $(3N + 1)$. In the following section, we consider various inline mixed coupled BPFs with multiple TZs containing resonators from Table 9.4.

11.2.1 Third-order Mixed Coupled BPF with Resonator No. 4 (Table 9.4) with Eight TZs

Figure 11.2a shows the circuit of third-order BPF with resonators No. 4 (Table 9.4) and four equidistant TZ pairs [5], which have admittance inverters for simple and mixed couplings.

The resonators in the filter are stepped-impedance resonators with two sections and an open-circuit end. Its properties are described by (9.44)–(9.48). Each such resonator generates a pair of equidistant TZs in the BPF. These resonators have already been used in the third-order BPF (Figure 10.9) with simple couplings and $f_0 = 2$ GHz, FBW = 0.03. This filter had three equidistant TZ pairs, so it had a high attenuation level in the frequency bands 1.3–1.5 and 2.5–2.7 GHz, allowing us to suppress unwanted signals effectively. The frequency selectivity of this filter is 50 dB ($f_0 \pm 220$ MHz).

The filter in Figure 11.2a differs from the previous filter (Figure 10.9) in that mixed couplings replace the two simple couplings between resonators. They are represented by the admittance inverters for mixed coupling as a lumped *LC* circuit. The transverse branches of the inverters with negative values of elements are not shown. Such a *J* inverter replacement should lead to two additional TZs, increasing the filter selectivity. The remaining parameters of the filter in Figure 11.2a are the same as in Figure 10.9. We will leave the inverter modules J_{12} and J_{23} the same

$$\begin{aligned} |J_{12}| &= 0.00488\,\Omega^{-1}\ \ (f_z = 1.8\,\text{GHz}),\\ |J_{23}| &= 0.00570\,\Omega^{-1}\ \ (f_z = 2.2\,\text{GHz}), \end{aligned} \tag{11.2}$$

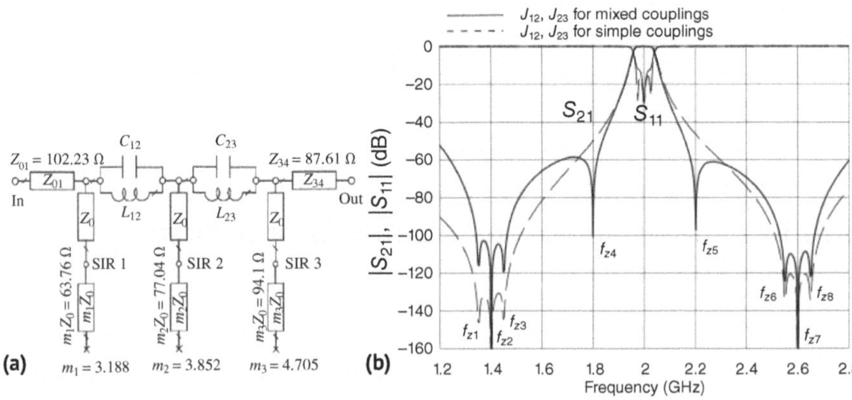

Figure 11.2 Third-order mixed coupled BPF with resonator No. 4 (Table 9.4) with eight TZs. (a) Schematic: $Z_0 = 20\ \Omega$, $C_{12} = 2.044$ pF, $L_{12} = 3.824$ nH, $C_{23} = 2.159$ pF, $L_{23} = 2.424$ nH. (b) Simulated frequency responses.

but we require that each inverter generates TZs shown in parentheses in expressions (11.2).

Let the inverter with J_{12} generate a left-handed TZ at $f_z = 1.8$ GHz. Then, the value of J_{12} should be negative $J_{12} = -0.00488\ \Omega^{-1}$. Substituting this value and $f_z = 1.8$ GHz into (8.4), we obtain $J_{01m} = 0.02081\ \Omega^{-1}$, $J_{01e} = -0.02569\ \Omega^{-1}$. Using the obtained values and expressions (3.7a) and (3.7b), we get the inductance and capacitance values of the inverter J_{12} for mixed coupling: $L_{12} = 3.824$ nH and $C_{12} = 2.044$ pF.

The inverter with $J_{23} = 0.00570\ \Omega^{-1}$ is positive because it generates right-handed TZ at frequency $f_z = 2.2$ GHz. Carrying out calculations in the aforementioned sequence, we obtain: $J_{23m} = 0.03283\ \Omega^{-1}$, $L_{23} = 2.424$ nH; $J_{23e} = -0.02713\ \Omega^{-1}$, $C_{23} = 2.159$ pF.

All filter parameters in Figure 11.2a are defined. Its simulated frequency responses are shown in Figure 11.2b by a solid line. The dashed lines in this figure show the frequency responses of the filter with simple couplings between resonators (Figure 10.9). Adding two TZs at frequencies of 1.8 and 2.2 GHz due to mixed couplings led to an increase in the filter selectivity from 50 dB ($f_0 \pm 220$ MHz) to 50 dB ($f_0 \pm 160$ MHz).

11.2.2 Third-order Mixed Coupled BPF with Lumped-Distributed Resonators No. 11 (Table 9.4) with 10 TZs

Let us confirm the possibility of obtaining $(3N + 1)$ TZs using the example of a third-order BPF, the schematic of which is shown in Figure 11.3a [4]. This filter contains lumped-distributed (LD) resonators No. 11 (Table 9.4) with $\lambda/4$ transmission line segment and admittance inverters for mixed couplings. The transverse branches of the inverters are not shown. The properties of the filter resonators are described by expressions (9.85)–(9.88).

Let us set the following parameters of this filter: $f_0 = 1$ GHz, FBW $= 0.06$, $L_{Ar} = 0.2$ dB. We require that the filter has 10 TZs, of which 6 TZs ($f_{z2,9} = 0.6$ and 1.4 GHz; $f_{z3,8} = 0.7$ and 1.3 GHz; $f_{z4,7} = 0.8$ and 1.2 GHz) are generated by resonators, and 4 TZs ($f_{z1} = 0.5$ GHz, $f_{z5} = 0.9$ GHz, $f_{z6} = 1.1$ GHz, $f_{z10} = 1.5$ GHz) are generated by mixed couplings.

To generate these six TZs, we use resonators in the BPF with $Z_0 = 30\ \Omega$ and parameters q_i and C_i indicated in Table 9.11. The resonators in Figure 11.3a are in sequence *1–3–2* from Table 9.11.

The g parameters for $n = 3$ and $L_{Ar} = 0.2$ dB can be found from Table 2.6: $g_1 = 1.2275$, $g_2 = 1.1525$, $g_3 = 1.2275$, $g_4 = 1.0$. Substituting these values, FBW $= 0.06$,

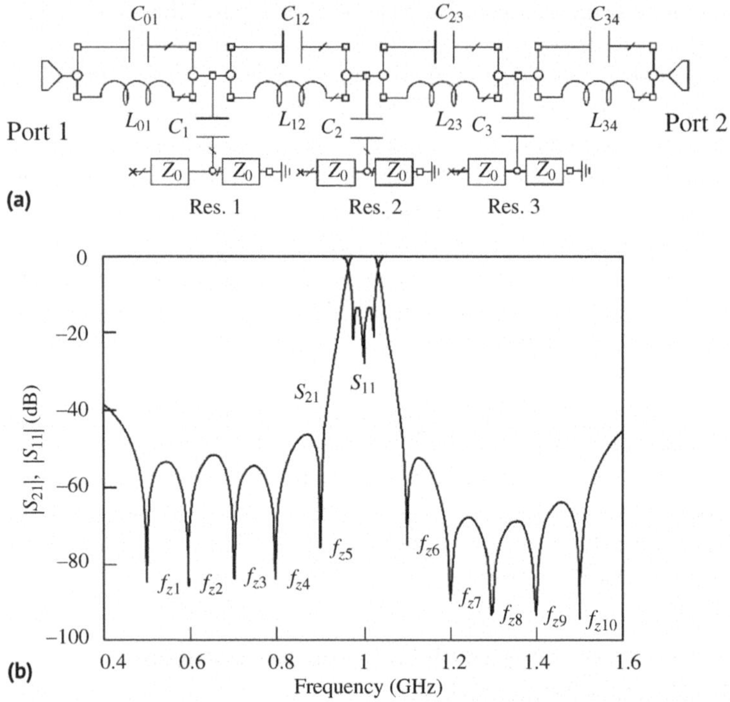

Figure 11.3 Third-order mixed coupled BPF with LD resonators No. 11 (Table 9.4) with 10 TZs. (a) Schematic ($Z_0 = 30\ \Omega$, $C_1 = 34.5$ pF, $C_2 = 56$ pF, $C_3 = 36.8$ pF). (b) Simulated frequency responses.

and susceptance slope parameters b_i from Table 9.11 into (3.6), we determine the moduli of J parameters of the inverters used in BPF:

$$
\begin{aligned}
|J_{01}| &= 0.01480\ \Omega^{-1}\ (f_{z5} = 0.9\ \text{GHz}), \\
|J_{12}| &= 0.03309\ \Omega^{-1}\ (f_{z1} = 0.5\ \text{GHz}), \\
|J_{23}| &= 0.05032\ \Omega^{-1}\ (f_{z10} = 1.5\ \text{GHz}), \\
|J_{34}| &= 0.02249\ \Omega^{-1}\ (f_{z6} = 1.1\ \text{GHz}).
\end{aligned}
\tag{11.3}
$$

The TZs generated by each inverter are shown in parentheses in expressions (11.3). Let the inverter with J_{01} generate a left-handed TZ at $f_{z5} = 0.9$ GHz. Then the value of J_{01} should be negative $J_{01} = -0.01480\ \Omega^{-1}$. Substituting this value and $f_{z5} = 0.9$ GHz into (8.4), we obtain $J_{01m} = 0.0631\ \Omega^{-1}$, $J_{01e} = -0.0779\ \Omega^{-1}$. Using these values and expressions (3.7a, b), we get the inductance and capacitance values of the inverter for the mixed coupling: $L_{01} = 2.522$ nH and $C_{01} = 12.398$ pF.

The J_{34} inverter generates right-handed TZ at $f_{z6} = 1.1$ GHz. Then the value of J_{34} should be positive $J_{34} = 0.0225\ \Omega^{-1}$. Substituting this value and $f_{z6} = 1.1$ GHz into (8.4), we obtain $J_{34m} = 0.1296\ \Omega^{-1}$, $J_{34e} = -0.1071\ \Omega^{-1}$. Using these values and expressions (3.7a, b), we get the inductance and capacitance values of this inverter: $L_{34} = 2.522$ nH and $C_{34} = 12.398$ pF.

The inverter with $J_{12} = -0.03309\ \Omega^{-1}$ generates left-handed TZ at $f_{z1} =$ 0.5 GHz. Carrying out calculations in the aforementioned sequence, we obtain $J_{12m} = 0.0110\ \Omega^{-1}$, $L_{12} = 14.47$ nH; $J_{12e} = -0.0441\ \Omega^{-1}$, $C_{12} = 7.02$ pF. The inverter with $J_{23} = 0.05032\ \Omega^{-1}$ generates right-handed TZ at $f_{z10} = 1.5$ GHz. In this case, expressions (8.4) and (3.7) give $J_{23m} = 0.0110\ \Omega^{-1}$, $L_{23} = 14.47$ nH; $J_{23e} = -0.0441\ \Omega^{-1}$, $C_{23} = 7.02$ pF.

All parameters of the BPF are defined. The simulated frequency responses are shown in Figure 11.3b. The third-order BPF with LD resonators and mixed couplings has 10 TZs at prescribed frequencies. To implement 10 TZs in a BPF with simple couplings, including cross-coupling, 12 resonators would be required [9].

11.2.3 Third-order Mixed Coupled BPF with LD Resonator No. 8 (Table 9.4) with 10 TZs

Consider the second example of a third-order mixed coupled BPF with 10 TZs. This filter is shown in Figure 11.4a. It contains LD resonators No. 8 (Table 9.4) with $\lambda/4$ transmission line segment and admittance inverters for mixed couplings [7]. The transverse branches of the inverters are not shown. The properties of the filter resonators are described by expressions (9.69) and (9.70).

Let us set the following parameters of this BPF: center frequency f_0 =1 GHz, FBW = 0.06, passband ripple $L_{Ar} = 0.2$ dB ($L_R = -13.467$ dB). We require the filter to have 10 TZs. Six of them ($f_{z2,9} = 0.6$ and 1.4 GHz; $f_{z3,8} = 0.7$ and 1.3 GHz; $f_{z4,7} = 0.8$ and 1.2 GHz) will be generated by LD resonators Nos. 1, 2, and 3 (Table 9.8). These resonators have $Z_0 = 10\ \Omega$ and the susceptance slope parameter $b = \pi/4Z_0 = 0.07854\ \Omega^{-1}$. The values of the lumped elements of these resonators are given in Table 9.8 (line $Z_0 = 10\ \Omega$). The four mixed couplings will generate the other four TZs ($f_{z1} = 0.5$ GHz, $f_{z5} = 0.9$ GHz, $f_{z6} = 1.1$ GHz, $f_{z10} = 1.5$ GHz).

The g parameters are the same as the previous filter: $g_1 = 1.2275$, $g_2 = 1.1525$, $g_3 = 1.2275$, and $g_4 = 1.0$. Substituting these values, together with susceptance slope parameters b and FBW, into (3.6), we determine the moduli of J parameters of the inverters used in BPF:

$$
\begin{aligned}
|J_{01}| &= 0.00876\ \Omega^{-1} \quad (f_{z5} = 0.9\ \text{GHz}), \\
|J_{12}| &= 0.00396\ \Omega^{-1} \quad (f_{z1} = 0.5\ \text{GHz}), \\
|J_{23}| &= 0.00396\ \Omega^{-1} \quad (f_{z10} = 1.5\ \text{GHz}), \\
|J_{34}| &= 0.00876\ \Omega^{-1} \quad (f_{z6} = 1.1\ \text{GHz}).
\end{aligned}
\tag{11.4}
$$

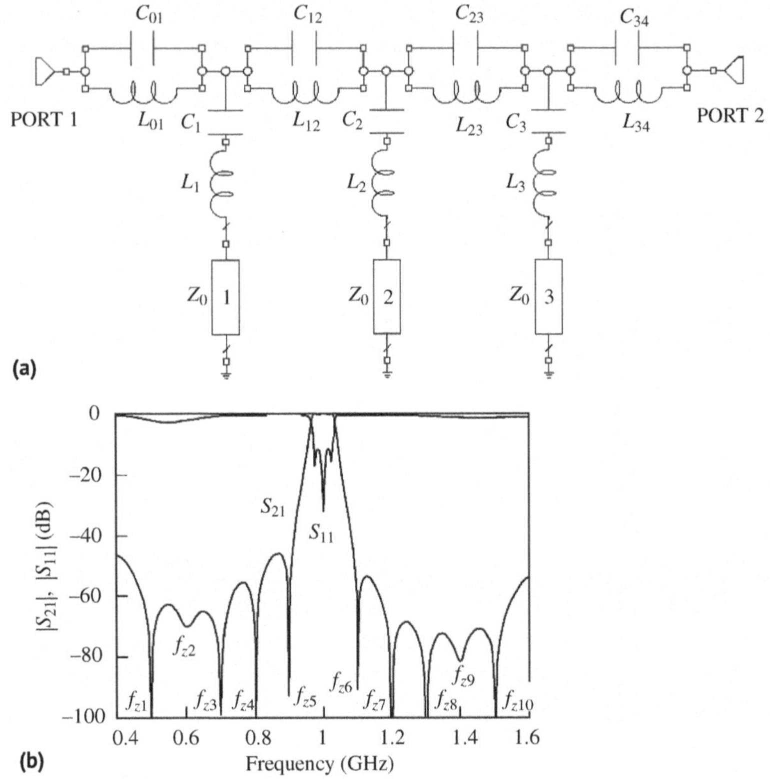

Figure 11.4 Third-order mixed coupled BPF with LD resonator No. 8 (Table 9.4) with 10 TZs. (a) Schematic. (b) Simulated frequency responses.

The TZs generated by each inverter are shown in parentheses in expressions (11.4). Let the inverter J_{01} generate a left-handed TZ at $f_{z5} = 0.9$ GHz. Then the value of J_{01} should be negative $J_{01} = -0.00876\ \Omega^{-1}$. Substituting this value and $f_{z5} = 0.9$ GHz into (8.4), we extract the components: $J_{01m} = 0.03735\ \Omega^{-1}$ and $J_{01e} = -0.04611\ \Omega^{-1}$. Using these values and expressions (3.7a, b), we obtain the inductance and capacitance values of the J_{01} inverter: $L_{01} = 4.26$ nH, $C_{01} = 7.34$ pF. The resulting calculated values of the components of the J_{01} inverter are contained in Table 11.2 (column J_{01}). Other columns of this table contain values similar to those of the other inverters (11.4).

All parameters of the considered BPF are defined. Its simulated frequency responses are shown in Figure 11.4b. The third-order BPF with LD resonators and mixed couplings has 10 TZs at prescribed frequencies.

When simulating the frequency responses in Figure 11.4b, the quality factors of capacitor C_1 and inductor L_1 were equal to $Q_{C1} = Q_{L1} = 20$. The first resonator

Table 11.2 Components of J inverters for BPF in Figure 11.4a.

Parameter	J_{ij}			
	J_{01}	J_{12}	J_{23}	J_{34}
f_z, GHz	0.9	0.5	1.5	1.1
J, Ω^{-1}	-0.00876	-0.00396	0.00396	0.00396
J_m, Ω^{-1}	0.03755	0.00132	0.00713	0.05049
J_e, Ω^{-1}	-0.04611	-0.00528	-0.00317	-0.04173
L, nH	4.26	120.57	22.315	3.152
C, pF	7.34	0.84	0.504	6.64

with C_1 and L_1 generates TZs (attenuation poles) at frequencies $f_{z2} = 0.6$ GHz and $f_{z9} = 1.4$ GHz. A decrease in Q_{C1} and Q_{L1} reduced the "depth" of the attenuation poles generated by the first resonator. The insertion loss in the filter passband did not increase. These features of BPF with the LD resonator (Figure 9.16) are positive and valuable for practice.

11.3 Microstrip Second-order Mixed Coupled BPF with Multiple TZs

The TZ number in second-order mixed coupled BPFs can be significant if they use the resonators from Tables 9.1 and 9.4. In the following section, we consider second-order microstrip BPFs with five, six, and seven TZs.

11.3.1 Microstrip BPF with Resonator No. 2 (Table 9.4) and One Mixed Coupling with Five TZs

Consider a microstrip second-order BPF with simple and mixed couplings (Figure 11.5a, b), which has five TZs [5]. This filter uses two resonators No. 2 (Table 9.4), each generating two TZs. The properties of the resonator are described by (9.35)–(9.39).

Let us set the following parameters: $f_0 = 1$ GHz, BW = 70 MHz (FBW = 0.07), $L_{Ar} = 0.2$ dB, corresponding to the return loss $L_R = 13.467$ dB. The filter should have five attenuation poles: two left-handed $f_{p1} = 0.7$ GHz, $f_{p2} = 0.8$ GHz and three right-handed $f_{p3} = 1.2$ GHz, f_{p4}, $f_{p5} > f_0$.

For design, we will use (3.8) to determine the coupling coefficients between resonators and the external Q-factors of end resonators. These formulas are preferable when some parameters in the design require measurement. The values

$L_{Ar} = 0.2$ dB, $N = 2$ correspond to g parameters: $g_0 = 1$, $g_1 = 1.0378$, $g_2 = 0.6745$, $g_3 = 1.5386$ (Table 2.6). Substituting these g parameters and the initial data into (3.8), we define $|K_{12}| = 0.0837$, $Q_{e1} = Q_{e3} = Q_e = 14.82$. The filter (Figure 11.5a, b) uses a TMM-10i substrate (Rogers), with $\varepsilon_r = 9.8$, $h = 1.016$ mm, $\tan\delta = 0.002$, and two resonators of width $w = 1.4$ mm ($Z_0 = 41.68\ \Omega$), which corresponds to $L = 57.7$ mm.

Capacitance C_{12} and inductance L_{12} between resonators provide the mixed coupling between resonators. In this case, the electromagnetic coupling between resonators is extremely weak. Matching the filter with the loads is carried out by simple couplings using capacitors C_{01} and C_{23}.

The ratio between the lengths of the short and long arms of the resonators controls the position of the left-handed TZs f_{z1} and f_{z2}. The shorter the short arm length, the more f_{z1} and f_{z2} were removed from f_0. The left resonator provides the value $f_{z1} = 0.7$ GHz with the length of the short arm $l = 18.2$ mm. The right resonator provides $f_{z2} = 0.8$ GHz at $l = 19.8$ mm.

The TZ $f_{z3} = 1.2$ GHz is generated by the mixed coupling $|K_{12}| = 0.0837$. Since the TZ is right-handed ($f_{z3} > f_0$), the mixed coupling coefficient K_{12} is positive $K_{12} = 0.0837$. Using (7.5) and $f_{z3} = 1.2$ GHz, we determined the magnetic $K_{m12} = 0.2739$ and electric $K_{e12} = -0.1902$ components of the mixed coupling K_{12}.

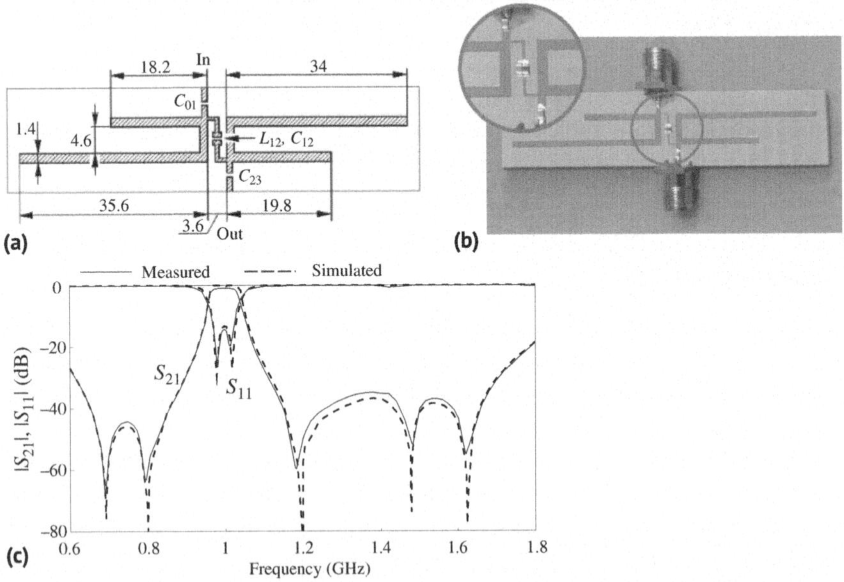

Figure 11.5 Microstrip second-order BPF with resonator No. 2 (Table 9.4) and one mixed coupling with five TZs. (a) Topology. (b) Photograph. *Source*: [5] / with permission of Elsevier. (c) Frequency responses.

From (3.10), $K = 0.0837$ is provided at even mode $f_e = 0.958$ GHz and odd mode $f_o = 1.042$ GHz coupled oscillations. Required values of $K = 0.0837$, $K_{m12} = 0.2739$, and $K_{e12} = -0.1902$ are provided when the specific frequencies (f_e, f_o, f_{z3}) take the values $f_e = 0.958$ GHz, $f_o = 1.042$ GHz, and $f_{z3} = 1.2$ GHz. The values of lumped capacitance C_{12} and inductance L_{12} are varied to realize the specific frequencies. To do this, we use the EM simulator of Microwave Office. This is ensured at $C_{12} = 8.2$ pF and $L_{12} = 2.14$ nH. The capacitances C_{01} and C_{23} provide $Q_e = 14.82$ and $L_R = 13.467$ dB at values $C_{01} = 2.1$ pF and $C_{23} = 2.65$ pF.

The microstrip BPF parameters in Figure 11.5a, b are defined. BPF occupies an area of 78 mm × 17.4 mm. Figure 11.5c shows the measured and simulated frequency responses. The capacitors and inductors with size code 0402 were used in the prototype filter. These lumped elements have nominal capacitances $C_{01} = 2.1$ pF, $C_{12} = 8.2$ pF, $C_{23} = 2.7$ pF, and inductance $L_{12} = 2.2$ nH. These values differ from the values of capacitances and inductances obtained by computer simulation, which leads to differences between the measured and simulated frequency responses.

The measured data are as follows: $f_0 = 1$ GHz, BW = 70 MHz, $L_{A0} = 0.9$ dB, $L_R > 14$ dB. The number of attenuation poles is 5. The attenuation poles f_{p1}, f_{p2}, and f_{p3} are at prescribed frequencies. The selectivity of this filter, 30 dB ($f_0 \pm 109$ MHz), can be expressed at the normalized frequency jΩ: 30 dB ($\pm$j3.1).

11.3.2 Microstrip BPF with Resonator No. 1 (Table 9.1) and Three Mixed Couplings with Five TZs

Let us consider a microstrip second-order BPF with three mixed couplings, which has five TZs [6]. The topology of this filter and its photograph are shown in Figure 11.6a, b. This filter uses two resonators No. 1 (Table 9.1), each generating one TZ. The properties of the resonator are described by (9.3)–(9.6).

Let us design a second-order BPF with five TZs. Set the following parameters: center frequency $f_0 = 1$ GHz, bandwidth BW = 40 MHz (FBW = 0.04), passband ripple $L_{Ar} = 0.1$ dB ($L_R = 16.426$ dB), and TZs: 0.6, 0.8, 1.2, 1.4, 1.6 GHz. The schematic of this filter and the initial data, except for the FBW, are the same as those of the BPF in Figure 11.1a. However, in this case, we will use formulas (3.8) rather than (3.6) for the filter design since they are preferable when some parameters in the design require measurement. Using (3.8) yields: $|K_{12}| = 0.0552$, $Q_{e1} = Q_{e3} = Q_e = 21.075$.

This filter uses a TMM-10i substrate (Rogers) with $\varepsilon_r = 9.8$, $h = 1.016$ mm, $\tan\delta = 0.002$, and two resonators of width $w = 1.4$ mm and length $L = 27.4$ mm. The width $w = 1.4$ mm corresponds to the characteristic impedance $Z_0 = 41.68\ \Omega$. The resonators are separated from each other by the gap of $\Delta_{12} = 2.6$ mm.

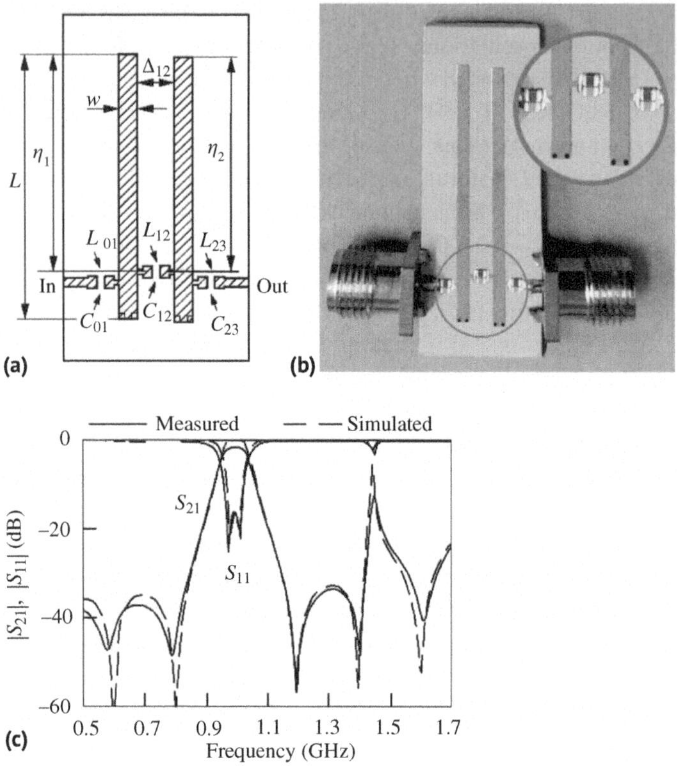

Figure 11.6 Microstrip second-order BPF with resonator No. 1 (Table 9.1) and three mixed couplings with five TZs. (a) Topology. (b) Photograph. *Source*: [6] / with permission of Springer Nature. (c) Frequency responses.

Two right-side TZs, 1.2 and 1.4 GHz, are determined by the coordinates of the points of the two $\lambda/4$ resonators defined in Table 11.1: $q_1 = 1/6, q_2 = 2/7$. The left-side TZ $f_z = 0.8$ GHz is generated by the mixed coupling $|K_{12}| = 0.0552$. Since the TZ is left-handed, the mixed coupling coefficient K_{12} is negative $K_{12} = -0.0552$. Using (8.4) and f_z =0.8 GHz, we determine the magnetic $K_{m12} = 0.0981$ and electric $K_{e12} = -0.1533$ components of the mixed coupling K_{12}. In this case, the mixed coupling between resonators is determined by C_{12}, L_{12}, and electromagnetic interaction between resonators, primarily of a magnetic nature.

From (3.10), the value of $K_{12} = -0.0552$ is provided at odd mode $f_o = 0.973$ GHz and even mode $f_e = 1.027$ GHz coupled oscillations. The required value of coupling K_{12} and its components K_{m12} and K_{e12} are provided when the specific frequencies (f_e, f_o, f_z) take the values $f_o = 0.973$ GHz, $f_e = 1.027$ GHz, and $f_z = 0.8$ GHz. The value of lumped capacitance C_{12} and inductance L_{12} are varied to realize the specific frequencies. To do this, we use the EM simulator of Microwave Office. This is ensured at $C_{12} = 6.0$ pF and $L_{12} = 8.2$ nH.

The coupling circuit of the first resonator with the load, formed by a parallel connection of C_{01} and L_{01}, is connected at 23.6 mm from the open end of the resonator. The matching coupling circuit formed by C_{23} and L_{23} is connected to the second resonator at 23.3 mm from the open end. These circuits generate TZs at frequencies of 0.6 and 1.6 GHz, respectively. They provide the required value $Q_e = 21.075$ with the following element values: $C_{01} = 29.8$ pF, $L_{01} = 2.36$ nH, $C_{23} = 11.1$ pF, $L_{23} = 0.89$ nH.

The measured and simulated frequency responses of the BPF are depicted in Figure 11.6c. The spurious peak at 1.44 GHz is due to the second resonant frequency of the resonators, which is lowered by the coupling elements.

The measured data are as follows: $f_0 = 1$ GHz, BW = 40 MHz, $L_{A0} = 1.5$ dB, $L_R > 15$ dB. The TZs are located at prescribed frequencies, and their number is $(2N + 1) = 5$. If we use only mixed couplings [1–3], the TZ number equals 3.

11.3.3 Microstrip BPF with Resonator No. 11 (Table 9.4) and Two Mixed Couplings with Six TZs

Consider a microstrip second-order BPF (Figure 11.7) with six TZs [4]. This filter uses two resonators No. 11 (Table 9.4), each generating two TZs. The resonator properties are described by (9.85)–(9.88). The filter uses two mixed couplings between resonators and loads. The resonators themselves are connected by simple coupling.

Let us set the filter parameters as follows: $f_0 = 1$ GHz, BW = 120 MHz (FBW = 0.12), $L_{Ar} = 0.2$ dB ($L_R = -13.467$ dB), six TZs at the frequencies $f_{z1,6} = 0.6$ and 1.4 GHz, $f_{z2,5} = 0.7$ and 1.3 GHz, $f_{z3,4} = 0.8$ and 1.2 GHz.

The filter used a TMM-10i substrate (Rogers), with $\varepsilon_r = 9.8$, $h = 1.016$ mm, $\tan\delta = 0.002$.

The filter topology is shown in Figure 11.7a. Both resonators contain a section of microstrip line short-circuited at one end with metallized holes with a diameter of 0.4 mm and lumped capacitances C_1 and C_2, respectively. The segment has a length of $l = 27.7$ mm and a width of $w = 2.4$ mm ($Z_0 = 30\ \Omega$). The coupling between these resonators is combined. A quarter-wave segment of width w_{12} carries it out and also depends on the size of the gaps Δ (Figure 11.7a). Matching of the filter with the loads is ensured by LC chains $L_{01}C_{01}$ and $L_{23}C_{23}$. The sections of short-circuited lines of the resonators are oriented toward each other, which ensures a fairly strong coupling between resonators. This feature makes significant adjustments to the BPF design process, which is carried out using the EM simulator of Microwave Office software and consists of three stages.

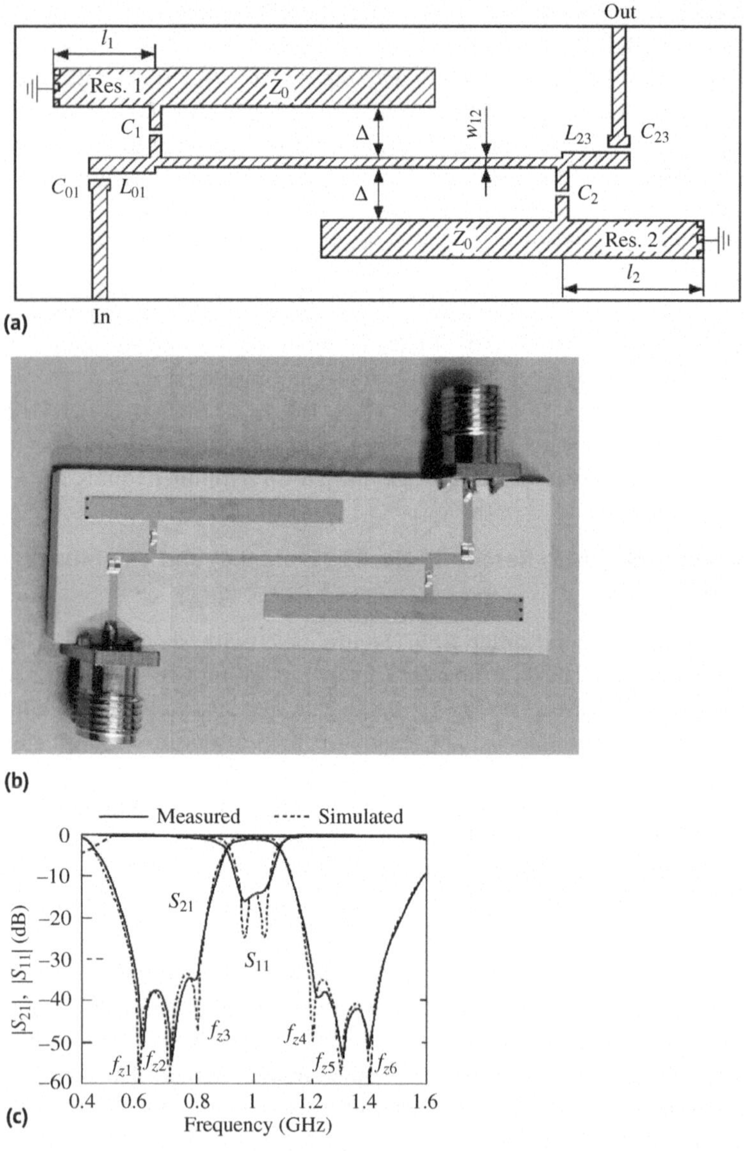

Figure 11.7 Microstrip second-order BPF with resonator No. 11 (Table 9.4) and two mixed couplings with six TZs. (a) Topology. (b) Photograph. *Source*: [4] / with permission of IEEE. (c) Frequency responses.

Stage 1: Implementing required coupling coefficient between resonators. The specified initial data corresponds to g parameters (Table 2.6): $g_0 = 1$, $g_1 = 1.0378$, $g_2 = 0.6745$, $g_3 = 1.5386$. The coupling coefficient K_{12} between resonators is determined by the equality (3.8a): $K_{12} = 0.1434$.

The coupling coefficient can also be determined from the values of even f_e and odd f_o modes of coupled oscillations by (3.10). The value $K_{12} = 0.1434$ is provided at $f_o = 1.072$ GHz and $f_e = 0.928$ GHz.

In the BPF under consideration, it is necessary to establish a weak coupling with the loads (for example, remove the inductances and take the values of capacitances C_{01} and C_{23} equal to 0.2 pF). At the same time, the frequencies f_o and f_e will be visible in the transfer function. The required values of f_o and f_e are provided with the width of the quarter-wave coupling section $w_{12} = 0.6$ mm ($Z_{12} = 62.3\ \Omega$), whose length $l_{12} = 29.7$ mm, and the size of the gaps $\Delta = 3.4$ mm.

Stage 2: Implementing TZs generated by LD resonators ($f_{z2,5} = 0.7$ and 1.3 GHz, $f_{z1,6} = 0.6$ and 1.4 GHz). For this purpose, the weak coupling of the resonators with the loads is also used. On the insertion loss curve, in addition to f_o and f_e, TZs are visible, which are regulated by the lengths l_1 and l_2 shown in Figure 11.7a and the corresponding values C_1 and C_2. The first resonator provides the TZs at frequencies $f_{z2,5} = 0.7$ and 1.3 GHz with $l_1 = 7.4$ mm, $C_1 = 11.2$ pF. Resonator *2* generates the TZs at frequencies $f_{z1,6} = 0.6$ and 1.4 GHz with $l_2 = 10.1$ mm, $C_2 = 13.0$ pF.

Stage 3: Implementing TZs generated by mixed couplings and matching filter with loads. The chain formed by L_{01} and C_{01} generates TZ at $f_{z3} = 0.8$ GHz, and the chain formed by L_{23} and C_{23} generates TZ at $f_{z4} = 1.2$ GHz. To do this, the product of the indicated elements must be constant and satisfy equality (8.2): $\omega_z = 1/\sqrt{LC}$. The elements of each pair are changed in such a way (while maintaining the LC product) to provide the required $L_R = -13.467$ dB. This is ensured with the following values: $L_{01} = 1.8$ nH, $C_{01} = 22$ pF; $L_{23} = 0.9$ nH, $C_{23} = 19.54$ pF.

The length of the segments connecting C_{01} with C_1 and C_{23} with C_2 is 4.8 mm, and their width is 1.2 mm.

All parameters of the considered microstrip BPF with lumped elements are determined. Figure 11.7b shows a photograph of the filter occupying the 52.7 mm × 17.8 mm area, and Figure 11.7c shows its frequency responses. The capacitors and inductors with size code 0402 are used in the prototype filter. The filter has six TZs, which are located at prescribed frequencies. Other measured data are as follows: $f_0 = 1$ GHz, BW = 120 MHz, the insertion loss $L_{A0} = 1$ dB, and the return loss in bandwidth $L_R > 13$ dB. The selectivity of this filter, 30 dB ($f_0 \pm 184$ MHz), can be expressed at the normalized frequency $j\Omega$: 30 dB ($\pm j3.07$).

In some cases, including operation at high frequencies, lumped capacitances can be replaced with microstrip capacitors in the form of interdigital fingers [9, Figure 4.6a] or capacitive gaps.

11.3.4 Microstrip BPF with Resonator No. 8 (Table 9.4) and Three Mixed Couplings with Seven TZs

Consider a microstrip second-order BPF (Figure 11.8) with seven TZs [7]. The filter uses two LD resonators No. 8 (Table 9.4), each generating two TZs. The resonator properties are described by (9.69)–(9.70). The filter uses three mixed couplings, including coupling between resonators and between resonators and loads.

Let us set the filter parameters: $f_0 = 1$ GHz, BW = 60 MHz (FBW = 0.06), $L_{Ar} = 0.2$ dB. We require that TZs be located at the frequencies:

$$\begin{gathered} f_{z1} = 0.65\,\text{GHz},\ f_{z2} = 0.75\,\text{GHz},\ f_{z3} = 0.85\,\text{GHz},\ f_{z4} = 1.15\,\text{GHz}, \\ f_{z5} = 1.25\,\text{GHz},\ f_{z6} = 1.35\,\text{GHz},\ f_{z7} = 1.45\,\text{GHz}. \end{gathered} \tag{11.5}$$

The topology and photograph of the microstrip BPF with the specified parameters are shown in Figure 11.8a, b, respectively. The filter used a TMM-10i substrate (Rogers), with $\varepsilon_r = 9.8$, $h = 1.016$ mm, $\tan\delta = 0.002$. The quarter-wave segments of resonator lines have length $L = 25.3$ mm and width $w = 2.2$ mm ($Z_0 = 20\ \Omega$). The short-circuits at the ends of the resonators are made of metallized holes with a diameter of 0.4 mm. The interval between resonators is $\Delta_{12} = 4.6$ mm. The filter occupies an area of 38 mm × 16 mm.

Two equidistant TZ pairs ($f_{z1,6} = 0.65$ and 1.35 GHz, $f_{z2,5} = 0.75$ and 1.25 GHz) are generated by LD resonators 1, 2 with $Z_0 = 20\ \Omega$ and corresponding lumped element values $L_1 = 7.45$ nH, $C_1 = 3.88$ pF, $L_2 = 15.4$ nH, $C_2 = 1.75$ pF. These values are determined using (9.70) and are listed in Table 11.3 (row 3). These resonators have the same susceptance slope parameter b, like a quarter-wave resonator, $b = \pi/4Z_0 = 0.03927\ \Omega^{-1}$. The three mixed couplings will generate the other three TZs ($f_{z3} = 0.85$ GHz, $f_{z4} = 1.15$ GHz, $f_{z7} = 1.45$ GHz) from the set (11.5). Table 2.6 shows g parameters: $g_1 = 1.2275$, $g_2 = 1.1525$, $g_3 = 1.2275$, $g_4 = 1.0$. Substituting these parameters, FBW = 0.06 and susceptance slope parameter $b = 0.03927\ \Omega^{-1}$ into (3.6), we determine the J parameters, which are contained in the first row of Table 11.3. Above these values are the TZs that are generated by these couplings.

The second row of Table 11.3 contains the components J_{ijm} and J_{ije} of J_{ij} inverters for mixed couplings obtained using (8.4). Using these values and expressions (3.7a) and (3.7b), we get the inductance and capacitance values of the J_{ij} inverters, which are contained in Table 11.3 (row 3).

In microstrip BPF (Figure 11.8), additional short microstrip sections with a characteristic impedance $Z_0 = 50\ \Omega$ are used to install lumped elements. To keep

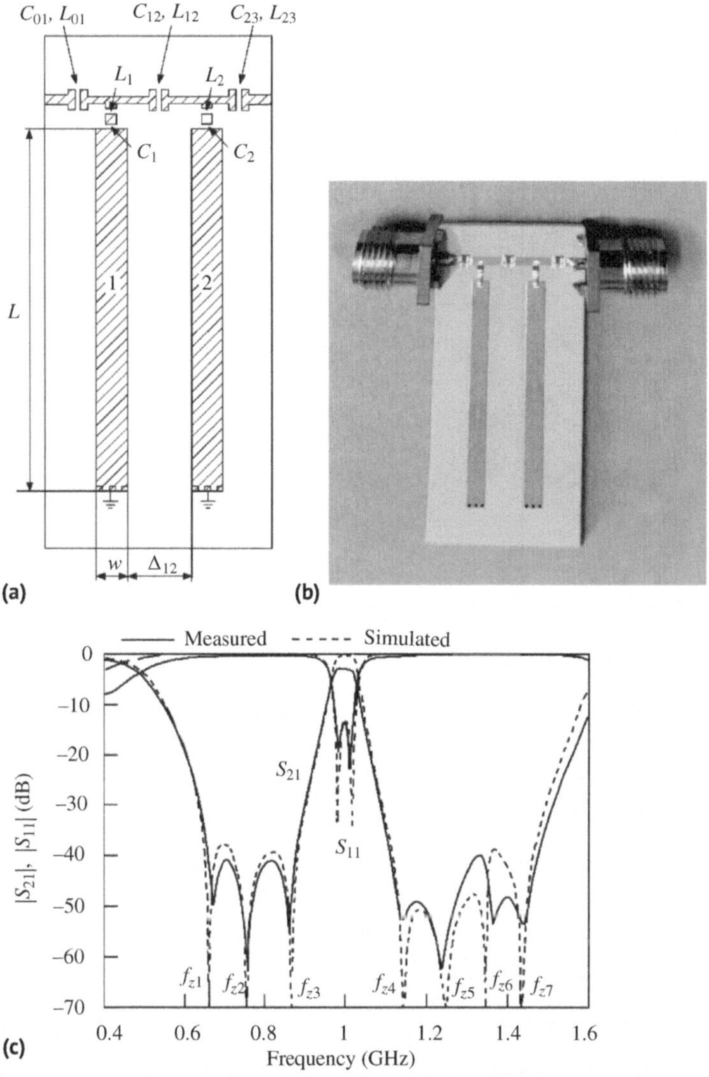

Figure 11.8 Microstrip second-order BPF with resonator No. 8 (Table 9.4) and three mixed couplings with seven TZs. (a) Topology: $L = 25.5$ mm, $w = 2.2$ mm, $\Delta_{12} = 4.6$ mm. (b) Photograph. (c) Frequency responses.

the previous positions of the TZs, it is necessary to optimize the C_{ij} and L_{ij} values, which are obtained by calculation and are contained in the third row of Table 11.3.

For optimization, the EM simulator of Microwave Office software was used. The optimized values of C'_{ij} and L'_{ij} differ slightly from the calculated ones. They are shown in the fourth row of Table 11.3 and correspond to the simulated frequency

Table 11.3 Parameters of microstrip second-order BPF (Figure 11.8) with seven TZs.

No.	f_{zk}, GHz	$f_{z7} = 1.45$	$f_{z1} = 0.65$ $f_{z6} = 1.35$	$f_{z3} = 0.85$	$f_{z2} = 0.75$ $f_{z5} = 1.25$	$f_{z4} = 1.15$
1	J_{ij}, Ω^{-1}	$J_{01} = 0.00674$		$J_{12} = -0.00282$		$J_{23} = 0.00674$
2	J_{ijm}, Ω^{-1} J_{ije}, Ω^{-1}	$J_{01m} = 0.01285$ $J_{01e} = -0.00611$		$J_{12m} = 0.00733$ $J_{12e} = -0.01015$		$J_{23m} = 0.02763$ $J_{23e} = -0.02089$
3	L_{ij}, nH C_{ij}, pF	$L_{01} = 12.38$ $C_{01} = 0.972$	$L_1 = 7.45$ $C_1 = 3.88$	$L_{12} = 21.71$ $C_{12} = 1.615$	$L_2 = 15.4$ $C_2 = 1.75$	$L_{23} = 5.76$ $C_{23} = 3.32$
4	L'_{ij}, nH C'_{ij}, pF	$L'_{01} = 12.3$ $C'_{01} = 0.97$	$L'_1 = 7.45$ $C'_1 = 3.88$	$L'_{12} = 23$ $C'_{12} = 1.55$	$L'_2 = 14.5$ $C'_2 = 1.67$	$L'_{23} = 5.76$ $C'_{23} = 3.32$
5	L''_{ij}, nH C''_{ij}, pF	$L''_{01} = 12.5$ $C''_{01} = 0.95$	$L''_1 = 6.9$ $C''_1 = 3.9$	$L''_{12} = 24$ $C''_{12} = 1.55$	$L''_2 = 15.9$ $C''_2 = 1.4$	$L''_{23} = 5.7$ $C''_{23} = 3.4$

responses presented in Figure 11.8c by dashed lines. The values C''_{ij} of capacitors and L''_{ij} of inductors with size code 0402 used in the filter differ from those obtained by simulation. They are shown in row 5 of Table 11.3 and correspond to the measured frequency responses presented in Figure 11.8c by solid lines.

The measured data are as follows: $f_0 = 1$ GHz, bandwidth at 3 dB level $\mathrm{BW}_{3\,\mathrm{dB}} =$ 60 MHz, the insertion loss $L_{A0} = 2.7$ dB, the return loss in bandwidth $L_R > 13$ dB, the left-hand rejection band by level 30 dB is 41.1%, the right-hand rejection band by level 30 dB is 38.8%. The selectivity of this filter, 30 dB ($f_0 \pm 96$ MHz), can be expressed at the normalized frequency $j\Omega$: 30 dB ($\pm j3.2$).

The rejection bands of the filters discussed earlier are not wide, but this does not reduce their practical significance. Methods for expanding these rejection bands are well known. For this purpose, BPF can be integrated into a single design with low- and high-pass sections [10–12]. The small-sized low- and high-pass filters with size codes 0402 and 0603 have been commercially produced [13]. Their use makes it possible to expand these bands without practically changing the size of original BPF.

In all the examples discussed earlier, a moderate return loss $L_R \approx 13.467$ dB was used, corresponding to ripple $L_{\mathrm{Ar}} = 0.2$ dB, and is quite acceptable in many applications. Increasing the L_R level is enough to reduce the external quality factor Q_e of the end resonators, which will require some change in the coupling coefficients between resonators.

11.4 Properties of BPF with Mixed Couplings and Multiple TZs Generated by Resonators

Below, we systematize the main properties of inline BPFs with mixed couplings, which have multiple TZs at real frequencies.

1. Using resonators with one antiresonance frequency ω_{p1} located near the fundamental resonant frequency ω_{01} (Table 9.1) in N-order inline mixed coupled BPF allows us to obtain $(2N + 1)$ TZs. Second-order microstrip BPF (Figure 11.6) has five TZs.
2. Using resonators with two antiresonant frequencies located near the operating resonant frequency (Table 9.4) in an N-order inline mixed coupled BPF allows us to obtain $(3N + 1)$ TZs. Second-order microstrip BPF (Figure 11.8) has 7 TZs, and third-order BPFs in Figures 11.3 and 11.4 have 10 TZs.
3. Many TZs allow us to have a high attenuation level in specific frequency bands, effectively suppressing unwanted signals in these bands while providing improved selectivity (Figure 11.2).
4. The TZs can be positioned equidistant from the center BPF frequency, allowing the implementation of quasi-elliptic filters.

5. In the considered BPFs, it is quite simple to implement equidistant TZ pairs of a higher degree, improving the selectivity of these filters.
6. The TZs can be positioned asymmetrically relative to the center frequency, resulting in BPFs with increased one-sided selectivity, which are in high demand in duplexer design.
7. Reducing Q-factors of inductors Q_{Li} and capacitors Q_{Ci} does not increase the BPF insertion loss in the passband. They only decrease the "depth" of the attenuation poles (Figure 11.4b). This practically significant effect is explained by the fact that the unloaded quality factor Q_u of the LD resonators under consideration does not depend on the quality factor of the lumped inductors and capacitors. It depends only on the unloaded quality factor Q_t of the distributed part of the resonators $Q_u = Q_t$.
8. The design sequence of the filters considered is simple. It is based on the classical design method (3.6), (3.8), formulas describing the position of the TZs generated by the resonators from Tables 9.1, 9.4, and expressions (8.4), which separate the electric and magnetic components from the admittance inverter for mixed coupling.
9. The simplest design is the BPF with LD resonators Nos. 4–7 (Table 9.1) and Nos. 7 and 8 (Table 9.4), which contain a uniform transmission line segment. This is because the L and C affect the TZ position but do not affect the susceptance slope parameter b. This parameter is constant and equal to $b = \pi/4Z_0$ for resonators with a quarter-wave segment and $b = \pi/2Z_0$ for resonators with a half-wave segment.
10. Using the coupling matrix [9, 14] to describe the considered filters with multiple TZs is impossible since it does not reproduce the TZs generated by the resonators.

References

1 R. Levy, New cascaded trisections with resonant cross-couplings (CTR Sections) applied to the design of optimal filters, in *IEEE MTT-S International Microwave Symposium Digest*, 2004, vol. 2 pp. 447–450, https://doi.org/10.1109/mwsym.2004.1336007.

2 J. J. Vague, D. Rubio, M. A. Fuentes, S. Cogollos, M. Baquero, V. E. Boria, and M. Guglielmi, Inline combline filters of order N with up to N + 1 transmission zeros, *IEEE Trans. Microw. Theory Tech.*, vol. 69, no. 7, pp. 3287–3297, 2021, https://doi.org/10.1109/TMTT.2021.3072370.

3 M. Mul, M. Jasinski, A. Lamecki, R. Gomez-Garcia, and M. Mrozowski, Inline microwave filters with N+1 transmission zeros generated by frequency-variant couplings: coupling-matrix-based synthesis and design, *IEEE Trans. Circuits Syst. II Express Briefs*, vol. 69, no. 3, pp. 824–828, 2022, https://doi.org/10.1109/TCSII.2021.3125460.

4 A. V. Zakharov and S. M. Litvintsev, Lumped-distributed resonators providing multiple transmission zeros in bandpass filters with simple and mixed couplings, *IEEE Trans. Circuits Syst. I Regul. Pap.*, vol. 71, no. 8, pp. 3502–3513, 2024, https://doi.org/10.1109/TCSI.2024.3375961.

5 A. V. Zakharov and S. M. Litvintsev, Transmission line bandpass filters with multiple attenuation poles and small number of resonators, *AEU - Int. J. Electron. Commun.*, p. 155131, 2024, https://doi.org/10.1016/j.aeue.2024.155131.

6 S. Litvintsev and A. Zakharov, Bandpass filters with 2N+1 transmission zeros at real frequencies, *Radioelectron. Commun. Syst.*, vol. 66, no. 6, pp. 315–321, 2023, https://doi.org/10.3103/S0735272723100035.

7 S. Litvintsev and A. Zakharov, Bandpass filters with increased to 3N+1 number of attenuation poles, *Radioelectron. Commun. Syst.*, vol. 66, no. 8, pp. 385–402, 2023, https://doi.org/10.3103/S073527272311002X.

8 A. Zakharov and S. Litvintsev, Planar bandpass filters based on resonators generating transmission zeros, *IEEE Circuits Syst. Mag.*, vol. 25, no. 2, pp. 57–73, 2025, https://doi.org/10.1109/MCAS.2025.3543781.

9 J.-S. Hong, *Microstrip Filters for RF/Microwave Applications*, 2nd ed. New Jersey: Wiley, 2011, https://doi.org/10.1002/9780470937297.

10 C. Quendo, E. Rius, C. Person, and M. Ney, Integration of optimized low-pass filters in a bandpass filter for out-of-band improvement, *IEEE Trans. Microw. Theory Tech.*, vol. 49, no. 12, pp. 2376–2383, 2001, https://doi.org/10.1109/22.971624.

11 J. Xu, Y.-X. Ji, W. Wu, and C. Miao, Design of miniaturized microstrip LPF and wideband BPF with ultra-wide stopband, *IEEE Microw. Wirel. Components Lett.*, vol. 23, no. 8, pp. 397–399, 2013, https://doi.org/10.1109/LMWC.2013.2269041.

12 L. Kumar and M. S. Parihar, Quasi-lumped analysis of wideband bandpass filter with high out-of-band rejection rate, *IEEE Trans. Components, Packag. Manuf. Technol.*, vol. 9, no. 8, pp. 1549–1558, 2019, https://doi.org/10.1109/TCPMT.2019.2918478.

13 AVX Corporation, RF/Microwave Products, *Products*, 2023. https://www.kyocera-avx.com/products/.

14 A. V. Zakharov and S. M. Litvintsev, Coupling matrix modification for bandpass filters with through-type resonators and simple couplings, *IEEE Trans. Circuits Syst. II Express Briefs*, vol. 71, no. 4, pp. 1864–1868, 2024, https://doi.org/10.1109/TCSII.2023.3337439.

12

Stripline Bandpass Filters

Small bandpass filters (BPFs) such as rectangular dielectric resonator filters [1], ceramic monoblock filters [2], surface acoustic wave (SAW) filters [3], and ceramic multilayer filters [4, 5] are in high demand. Stripline BPFs use dielectric substrates with a permittivity of $\varepsilon_r = 2\text{–}100$. This allows these filters to operate over a wide frequency range from 300 MHz to 100 GHz, making them very promising [1].

Low insertion loss in such filters can be achieved by using thick ($h = 2\text{–}3$ mm) substrates, which provide the resonators with a high unloaded quality factor. Stripline filters can be 1 mm thick or less if thin substrates ($h \leq 0.5$ mm) are used. Significant variety in the frequency responses of such filters is achieved by the variety of metallized topological patterns used.

In recent decades, there has been no significant development of stripline filters, despite their inherent advantages. This is apparently due to the statement made in Ref. [6] regarding the absence of electromagnetic coupling between stripline $\lambda/4$ and $\lambda/2$ resonators in combline and pseudocombline structures. This statement was duplicated in the authoritative publication [7], which slowed down the said development.

The structures identified earlier were considered unpromising. A 1983 review of stripline filters [8] did not mention them. The proposal for the absence of electromagnetic coupling between stripline resonators in combline or pseudocombline structures has been put forward for a long time and requires further revision.

Knowledge of the coupling coefficients between stripline resonators is used in the design of BPFs. Some stripline resonators exhibit mixed couplings, also called "frequency-dependent" [9], while some of them are called "resonant" [10]. Such couplings contain both magnetic K_m and electrical K_e components. They generate transmission zeros (TZs), leading to more varied frequency responses.

Mixed couplings are used in various transmission line designs: coaxial [11], multilayer [12], substrate integrated waveguide [13–15], and microstrip [16–20]. Each of these designs has its own patterns of dependence of the mixed coupling coefficients between resonators on the transmission line parameters. In some transmission line types, these patterns have been studied quite well, which cannot be said about stripline resonators.

This chapter studies the electromagnetic coupling between stripline resonators in combline and pseudocombline structures, which makes it possible to realize stripline BPFs with different frequency responses [21, 22].

12.1 Coupling Between Uniform Stripline Resonators

A pair of quarter-wave stripline resonators is shown in Figure 12.1. The stripline structure has a thickness of b and uses a dielectric with a permittivity of ε_r. The width of the central conductors of the resonators shown is w, and their length is L. A gap S separates these resonators from each other. All sides of the structure, including the top and bottom, are metallized, providing complete insulation. One end of the central conductors is shorted to the conducting wall, and the other is open.

A stripline resonator has multiple resonant frequencies, the lowest of which f_0 is considered as fundamental. At this frequency, the resonator is a quarter-wave resonator. The coupling coefficient K characterizes the electromagnetic interaction of two resonators at the resonant frequency. Even f_e and odd f_o modes of coupled oscillations are clearly visible in the transfer function when the resonators are weakly coupled to the loads. The coupling coefficient K is determined by formula (3.10) using the coupling frequencies. The phase of transfer function is used to determine the sign of K, as discussed in Section 3.2. Capacitive gaps were used to excite the resonators. The commercial program Microwave Office (AWR Company) was used to simulate the frequency responses.

The dependence of the coupling coefficient K on the dielectric constant ε_r and the gap value S, at a constant stripline thickness $b = 1$ mm, is shown in Figure 12.2a. The following parameters were used in the simulation: the gap value

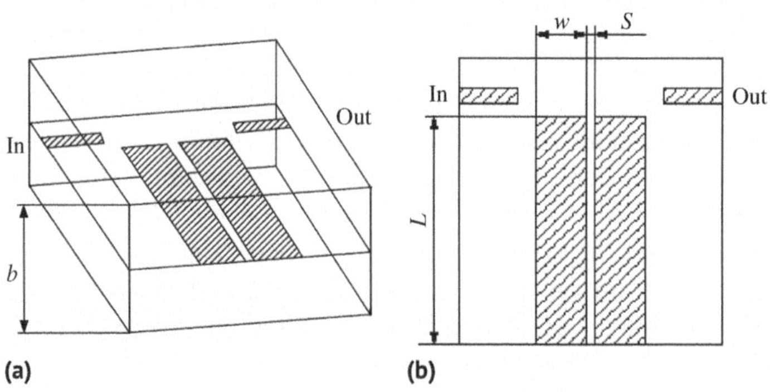

Figure 12.1 Stripline quarter-wave resonator. (a) 3D view. (b) Topology.

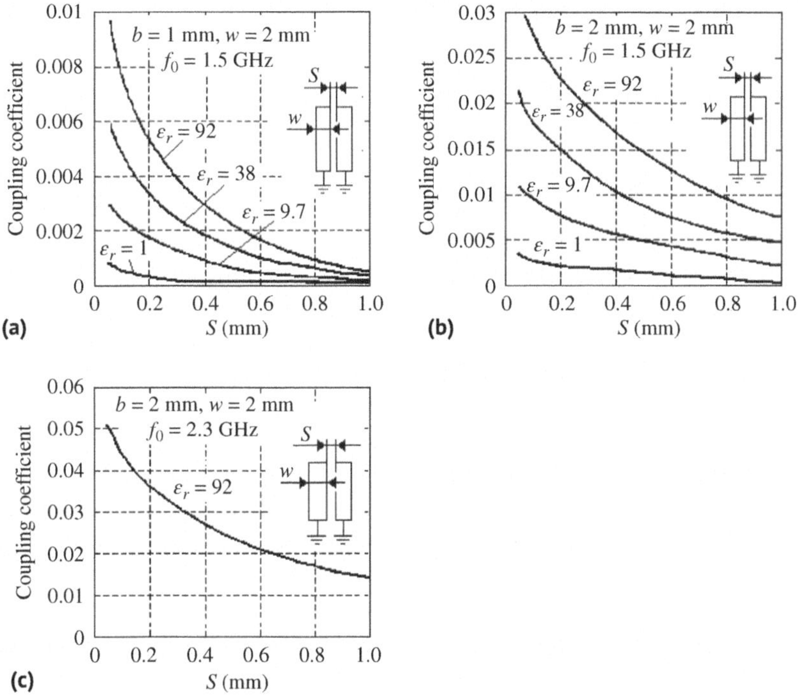

Figure 12.2 Coupling coefficient between stripline $\lambda/4$ resonators with wide $w = 2$ mm: (a) for $b = 1$ mm, $f_0 = 1.5$ GHz; (b) for $b = 2$ mm, $f_0 = 1.5$ GHz; (c) for $b = 2$ mm, $f_0 = 2.3$ GHz, $\varepsilon_r = 92$.

S varied from 0.05 to 1 mm; $w = 2$ mm; $\varepsilon_r = 92$, 38, 9.7, 1. The resonator length was taken equal to $L = 5$ mm at $\varepsilon_r = 92$, which provided a resonant frequency of $f_0 \approx 1508$ MHz. The length of the resonators for other dielectric constants was chosen to maintain the previous value of the resonant frequency f_0: $L = 7.9$ mm ($\varepsilon_r = 38$); $L = 15.8$ ($\varepsilon_r = 9.7$ mm); $L = 49.5$ mm ($\varepsilon_r = 1$). The simulation showed that for a given resonant frequency, the values of K increase with increasing ε_r.

The influence of the gap width S and the permittivity ε_r on K for a stripline thickness doubled $b = 2$ mm is shown graphically in Figure 12.2b. These dependencies show that the thickness b significantly influences the coupling coefficient K.

If at $\varepsilon_r = 92$, $b = 2$ mm, $w = 2$ we decrease the length of stripline resonators $L = 3$ mm, then their resonant frequency will increase to $f_0 \approx 2.3$ GHz. The K–S dependence for this case is shown in Figure 12.2c. Comparison of this dependence with the dependences in Figure 12.2b indicates that the values of K increase with increasing resonant frequency f_0 of the resonators. The dependences in Figure 12.2a–c display the patterns of change in K for stripline $\lambda/4$ resonators oriented in one direction. The values of K increase with increasing resonant frequency, stripline thickness b, and values of ε_r.

As an example, we will design a fourth-order Chebyshev stripline BPF with the following data: passband center frequency $f_0 = 2380$ MHz; passband ripple $L_{Ar} = 0.1$ dB; fractional bandwidth FBW = 0.035. From Table 2.6, we get the g parameters for the case $n = 4$ and $L_{Ar} = 0.1$ dB: $g_0 = 1$, $g_1 = 1.1088$, $g_2 = 1.3062$, $g_3 = 1.7704$, $g_4 = 0.8181$, $g_5 = 1.3554$. Substituting FBW and these parameters into (3.8), we obtain $K_{12} = K_{34} = 0.029$; $K_{23} = 0.023$; $Q_e = 31.7$.

The stripline BPF is formed by joining two aluminum oxide (Al_2O_3) substrates with a dielectric constant of $\varepsilon_r = 9.7$, $\tan\delta = 0.0002$, and a thickness of $h = 2$ mm [21]. All side surfaces of the filter are metallized, and it is fully shielded.

The relationships used in the BPF design are shown in Figure 12.3, where the filter elements are shown in the insets. The center conductors of the stripline resonators have a width of $w = 4$ mm. A 1 mm wide "protective" strip is placed around the filter perimeter, preventing tin from getting inside the filter during assembly. The "protective" strip slightly increases the value of K. The length of the resonators, measured from the "protective" strip, is $L = 8.6$ mm.

The K–S dependence for stripline resonators is shown by the curve in Figure 12.3a. The gaps of $S_{12} = S_{34} = 1.5$ mm and $S_{23} = 2.0$ mm correspond to coupling coefficients of $K_{12} = K_{34} = 0.029$ and $K_{23} = 0.023$.

The circuit that couples the end resonator to a 50 Ω load is shown in the inset of Figure 12.3b. It uses short stripline segments 1 mm and 0.5 mm wide. The distance of this segment from the "protective" strip is designated as l. The electromagnetic (EM) simulation of the group delay of parameter S_{11} for a resonator with a one-sided load and formula (3.20) are used to determine the external Q-factor Q_e of the end resonators. The curve in Figure 12.3b displays the dependence $Q_e = Q_e(l)$. A value of $Q_e = 31.7$ is achieved at $l = 0.7$ mm.

Vacuum deposition was used to coat alumina dielectric substrates with a copper layer. The layer thickness was 8 μm. The filter size is 26 mm × 12 mm × 4 mm. Position 1 in Figure 12.4a shows the topology of the filter's central conductors. To assemble the stripline BPF, its two substrates with conductive topologies are pressed together in a "face-to-face" principle. In this case, the metallized ends of

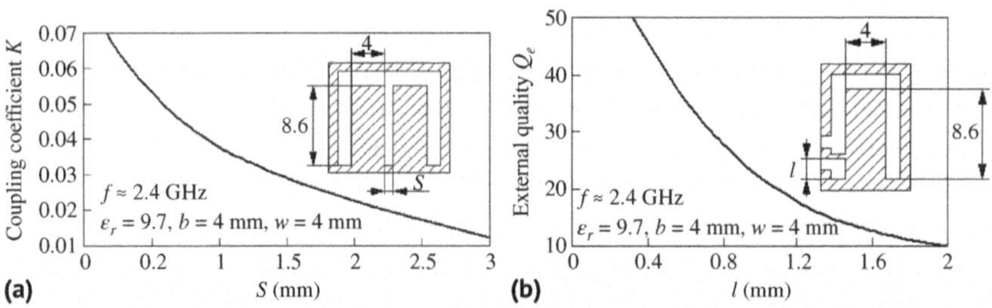

Figure 12.3 Dependencies used in the design of stripline BPF. (a) Coupling coefficient. (b) External quality factor.

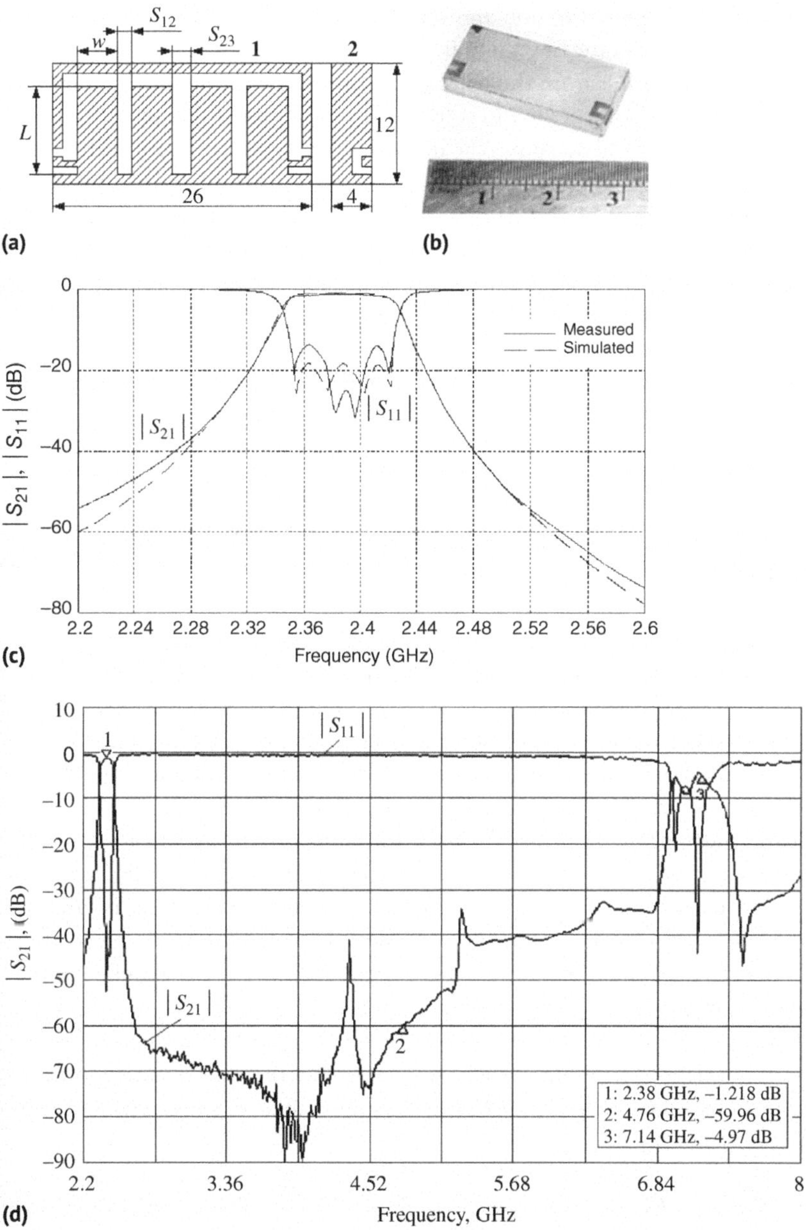

Figure 12.4 Stripline combline BPF. (a) Topology. (b) Photograph of a fabricated filter. (c) Frequency responses. (d) Measured frequency responses over wide frequency band.

the substrates are soldered to one another. Position 2 in Figure 12.4a shows the short strips that connect the outer and inner parts of the filter. Figure 12.4b shows a photograph of the filter from the back side. Surface mounting is used to install this BPF. Figure 12.4c shows the BPF's frequency responses in a relatively narrow frequency band. Figure 12.4d shows the measured frequency responses of this BPF over a wide frequency band.

The center frequency of the BPF is 2380 MHz, insertion loss at the center frequency is $L_{A0} = 1.22$ dB, bandwidth BW = 85 MHz, return loss $L_R > 13.0$ dB, and selectivity 40 dB ($f_0 \pm 115$ MHz). Filter attenuation at the frequency of $2f_0$ is 60 dB and at the frequency of $3f_0$ is less than 5 dB. Low attenuation at the frequency $3f_0$ is typical for filters with $\lambda/4$ resonators.

It has been confirmed experimentally and by EM simulation that there is electromagnetic coupling between the stripline $\lambda/4$ resonators and the combline structure is a BPF.

12.2 Coupling Between Stepped-impedance Stripline Resonators

It is useful to consider mixed coupling between stripline SIRs, which generate transmission zeros at given frequencies. The insets in Figure 2.15 show two different pairs of SIRs, one end of which is short-circuited. The figures themselves depict the K–S dependences obtained through computer simulation using formula (3.10). The following SIR parameters were chosen for the simulation: the dielectric constant $\varepsilon_r = 92$, $b = 2$ mm, the width of the wide part of the SIR $w_1 = 1.4$ mm, the narrow width of the SIR $w_2 = 0.9$ mm, and the resonator length $L = 3.6$ mm. The narrow section of the resonator has a variable length l. The solid curves in these figures correspond to a gap of $S = 0.2$ mm between the resonators, and the dashed curves correspond to a gap of $S = 0.4$ mm. The values of K can be positive, zero, or negative for the SIR pair in Figure 12.5a. If $K > 0$, the magnetic component dominates the interaction between the SIRs. If $K < 0$, the electrical component dominates. In this figure, the values of $K = K\,(l/L)$ first decrease, reach a minimum at $l/L \approx 1/2$, and then increase. The values of K are negative for most values of l. For a smaller gap ($S = 0.2$ mm), the values of K vary over a wider range: $-0.106 \leq K \leq 0.0385$.

The K values are always positive for the SIR pair in Figure 12.5b. The bell-shaped function $K = K\,(l/L)$ is also characteristic of this SIR pair. This function reaches a maximum at $l/L \approx 1/2$. The K values vary within the range $0.0385 \leq K \leq 0.0948$, with a gap between the resonators of $S = 0.2$ mm. Combining the two SIR variants considered earlier allows for a wide variation of K : $0.106 \leq K \leq 0.0948$.

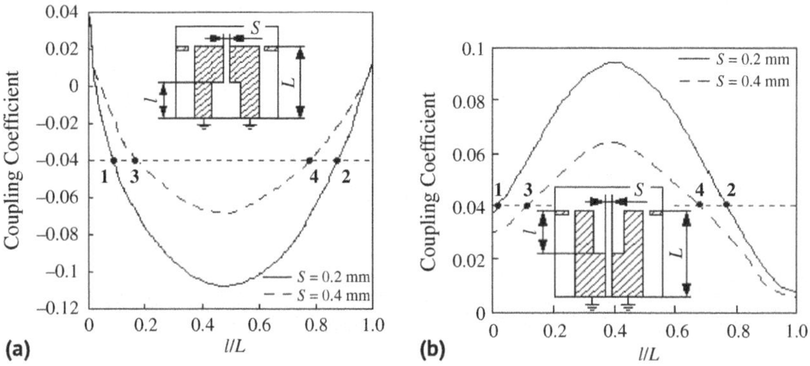

Figure 12.5 Coupling coefficient between stripline SIRs. (a) Short-circuited end of SIR is narrow. (b) Open-circuited end of SIR is narrow.

If the narrow part of the SIR is located near the open-circuited end, but in its "internal" part, then the dependences $K\ (l/L)$ will be similar to those shown in Figure 12.5a. If the narrow part of the SIR is located near the short-circuited end, but in its "internal" part, then the dependence $K\ (l/L)$ will be similar to that shown in Figure 12.5b.

Symmetrical fourth-order BPFs with two opposite-sided transmission zeros, which are caused by mixed coupling coefficients of different signs, are shown in Figure 12.6. The BPFs in Figure 12.6a, b have two negative mixed couplings and one positive mixed coupling. The filters in Figure 12.6c, d have two mixed couplings with $K > 0$ and one mixed coupling with $K < 0$.

The dependencies in Figure 12.5a, b display the patterns of change in the mixed K for SIRs. If the values of K and f_z are determined by measurements, then the components of the mixed coupling K_m and K_e can be determined using formulas (7.5). In Figure 12.5a, a horizontal line is drawn corresponding to the value $K = -0.04$. It intersects each of the constructed $K\ (l/L)$ curves at two points. If the gap between the resonators is $S = 0.2$ mm, then the corresponding intersection points are *1* and *2*. If $S = 0.4$ mm, then the corresponding points are *3* and *4*.

The indicated points correspond to the SIR topologies shown in Figure 12.7a. Table 12.1 show the values of K_m and K_e for the four points.

The values of K_m and K_e for the four points in Figure 12.5b with positive mixed coupling $K = 0.04$ are contained in Table 12.2. As in the previous case, the gap between the resonators $S = 0.2$ mm corresponds to points *1* and *2*, and the gap $S = 0.4$ mm corresponds to points *3* and *4*. The topologies of the SIR pairs corresponding to these points are shown in Figure 12.7b.

The topologies shown in Figure 12.7 and the data contained in Tables 12.1 and 12.2 indicate that a given value of K can be realized for mixed coupling with different gaps S between the SIRs, which also differ in topology. This value of K has

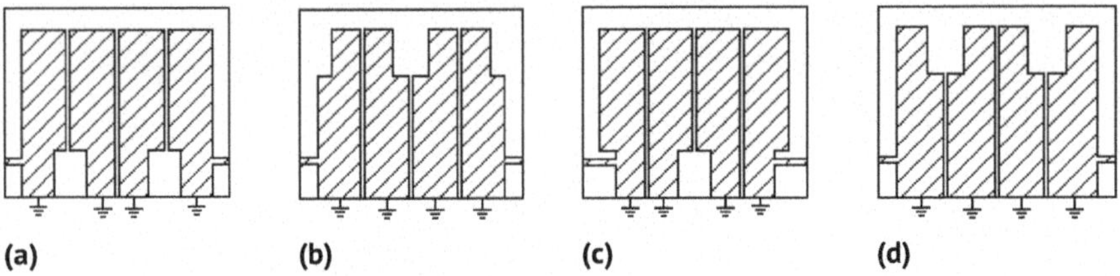

Figure 12.6 Topologies of stripline fourth-order BPFs with alternating mixed couplings. (a) Short-circuited end of SIRs is narrow, one positive and two negative couplings. (b) Open-circuited end of SIR is narrow, one positive and two negative couplings. (c) Short-circuited end of SIRs is narrow, two positive and one negative couplings. (d) Open-circuited end of SIR is narrow, two positive and one negative couplings.

Table 12.1 Components of mixed coupling $K = -0.04$ at points *1, 2, 3, 4* in Figure 12.5a.

No.	*1*	*2*	*3*	*4*
l/L	0.0903	0.8819	0.1806	0.7917
K_m	0.1295	0.0642	0.0658	0.0479
K_e	−0.1695	−0.1042	−0.1058	−0.0879
$K_m/\ \lvert K_e \rvert$	0.764	0.616	0.622	0.545

Table 12.2 Components of mixed coupling $K = 0.04$ at points *1, 2, 3, 4* in Figure 12.5b.

No.	*1*	*2*	*3*	*4*
l/L	0.0139	0.7917	0.0972	0.7917
K_m	0.2305	0.0829	0.1559	0.0699
K_e	−0.1905	−0.0429	−0.1159	−0.0299
$K_m/\ \lvert K_e \rvert$	1.210	1.932	1.345	2.338

different components K_m and K_e, which correspond to different transmission zero f_z positions. The position of f_z relative to f_0 for a given K can be controlled by the shape of the SIR and the gap S.

The analysis reveals the following patterns:

1. Decreasing the gap S between SIRs with a fixed value of mixed coupling K leads to an increase in the components K_m and K_e. At the same time, the interval between frequencies f_z and f_0 decreases.
2. If the gap S between SIRs is fixed, then to decrease the interval between frequencies f_z and f_0, it is necessary to decrease the ratio of lengths l/L.
3. The minimum permissible gap S between SIRs limits the degree of convergence of f_z and f_0.

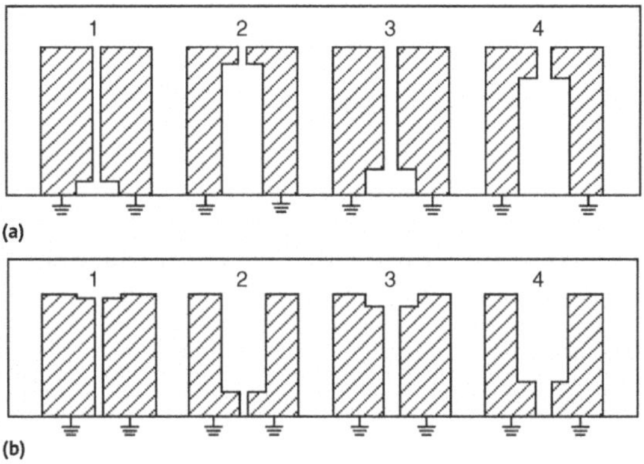

Figure 12.7 Pairs of SIRs with mixed coupling coefficient $|K| = 0.04$. (a) Pairs of SIRs corresponding to points *1, 2, 3, 4* in Figure 12.5a for $K = -0.04$. (b) Pairs of SIRs corresponding to points *1, 2, 3, 4* in Figure 12.5b for $K = 0.04$.

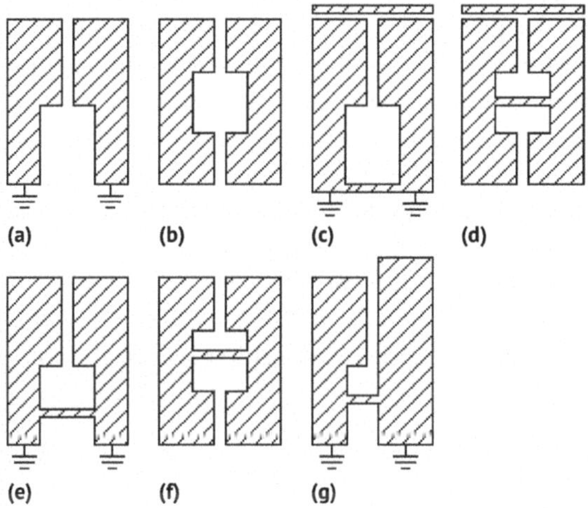

Figure 12.8 Mixed electromagnetic coupling between two stripline resonators: (a) quarter-wave type; (b) half-wave type; (c) quarter-wave type with two coupling strips; (d) half-wave type with two coupling strips; (e) quarter-wave type with moving coupling strip; (f) half-wave type with moving coupling strip; (g) stepped-impedance and quarter-wave resonators with coupling strip.

4. If $K_{1m} = K_{2m}$ and $K_{1e} = K_{2e}$, then the two mixed coupling coefficients K_1 and K_2 are equal to each other. The mixed coupling coefficients at points *1*, *2*, *3*, and *4* (Figure 12.5) do not satisfy these conditions.

In stripline BPF, various variants of implementing mixed coupling between two resonators can be used. Some of these variants are shown in Figure 12.8.

12.3 BPF with SIRs

Let us look at some examples of stripline BPFs with SIRs.

12.3.1 BPF Design Example

We realize a fourth-order stripline combline filter. This filter has a Chebyshev characteristic in the passband and the following specification:

- Center frequency: $f_0 = 1835$ MHz
- Bandwidth: BW = 90 MHz (FBW = 0.049)
- Passband ripple: $L_{\text{Ar}} = 0.2$ dB
- Transmission zeros: $f_{z1}, f_{z2} \in [f_0 \pm 200 \text{ MHz}]$

Note that the positions of the two transmission zeros are not precisely defined in this case. To ensure improved selectivity, they should be removed from the center frequency f_0 no more than 200 MHz. From Table 2.6 for $n = 4$ and $L_{\text{Ar}} = 0.2$ dB, we determine the g parameters: $g_0 = 1$, $g_1 = 1.3028$, $g_2 = 1.2844$, $g_3 = 1.9761$, $g_4 = 0.8468$, $g_5 = 1.5386$. Based on formula (3.8), for the obtained g parameters and the FBW = 0.049 value, we find the coupling coefficients $|K_{12}| = |K_{34}| = 0.0379$, $|K_{23}| = 0.0308$, and the value of $Q_e = 26.59$. The values of K_{12} and K_{34} will be considered positive, and the value of $K_{23} < 0$ will be taken as negative.

The components of the resulting mixed coupling coefficients must ensure the required location of the transmission zeros. The transmission zero f_{z1} corresponds to a positive coupling coefficient K_{12}. It must satisfy the condition $f_{z1} \leq (f_0 + 200 \text{ MHz})$. Inequality $f_{z1}/f_0 \leq 1.109$ follows from this condition. Using this inequality and expression (7.5), we determine the relationship between the components K_{m12} and K_{e12} of the mixed coupling K_{12}:

$$\frac{K_{m12}}{|K_{e12}|} \leq 1.23. \tag{12.1}$$

The transmission zero f_{z2} corresponds to a negative coupling K_{23}. From conditions $f_{z2} \geq (f_0 - 200 \text{ MHz})$, $f_{z2}/f_0 \geq 0.891$ and formulas (7.5) we get

$$\frac{K_{m23}}{|K_{e23}|} \geq 0.794. \tag{12.2}$$

The thermostable dielectric substrates with $\varepsilon_r = 92$, $h = 1$ mm, and $\tan\delta = 0.0003$ will be used in the developed stripline BPF. The thickness of the stripline BPF is $b = 2$ mm.

The topology of two pairs of stripline SIRs of the BPF under consideration is shown in Figure 12.9a, b. The resonators of these pairs are identical, but they are oriented differently in each pair. Figure 12.9a shows a pair of SIRs with $K < 0$, and Figure 12.9b shows a pair of SIRs with $K > 0$.

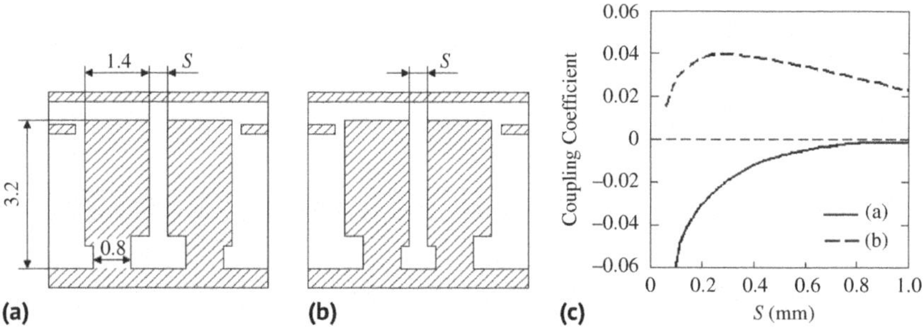

Figure 12.9 Resonator pairs of the developed BPF with mixed coupling coefficients. (a) Pairs of SIRs with $K < 0$. (b) Pairs of SIRs with $K > 0$. (c) Dependence of the value K on gap S.

The short-circuited ends of the SIR are adjacent to a 0.4 mm wide "protective" strip. The SIR length, measured from this strip, is $L = 3.2$ mm. The maximum SIR width is $w = 1.4$ mm. Two rectangles measuring 0.8 mm × 0.4 mm and 0.6 mm × 0.2 mm are removed from the stripline resonator in the short-circuit region, creating a stepped-impedance resonator. The magnetic component of the mixed coupling of each SIR pair is increased by the "protective" strip. This increase is greater as the strip becomes wider. To compensate for this additional positive coupling, the SIR width was reduced in the region of the short-circuited ends.

Computer simulation was used to construct the $K = K(S)$ dependence in Figure 12.9c for the SIR pairs under consideration. In the SIR pair in Figure 12.9b, with a gap of $S_{12} = S_{34} = 0.2$ mm, a positive coupling of $K_{12} = K_{34} = 0.0379$ is ensured. In the SIR pair in Figure 12.9a, with a gap of $S_{23} = 0.2$ mm, a negative coupling of $K_{23} = -0.0308$ is ensured.

The positive mixed coupling K_{12} generates a right-hand TZ at a frequency of $f_{z1} = 2034$ MHz. Using this value and K_{12}, based on (7.5), we determine the components of the mixed coupling: $K_{m12} = 0.2036$, $K_{e12} = -0.1657$. Condition (12.1) is satisfied for these values. The negative mixed coupling K_{23} generates a left-hand TZ at a frequency of $f_{z2} = 1668$ MHz. Formulas (7.5) allow us to determine the components of this coupling: $K_{m23} = 0.1547$, $K_{e23} = -0.1855$, which satisfy condition (12.2).

Figure 12.10 shows a stripline fourth-order BPF obtained by combining two pairs of SIRs shown in Figure 12.9. The dielectric substrates used in the filter with $\varepsilon_r = 92$, $h = 1$ mm, and $\tan\delta = 0.0003$ were coated with a layer of copper 8 μm thick. Vacuum deposition was used for the coating.

Position 1 in Figure 12.10a shows the topology of the filter's conductors, indicated in millimeter. The filter size is 7.4 mm × 4.2 mm × 2 mm. The topology of this filter differs from the topology of the filter in Figure 12.6c. This is due to the use of a 0.4 mm wide "protective" strip along the filter's perimeter. Position 2

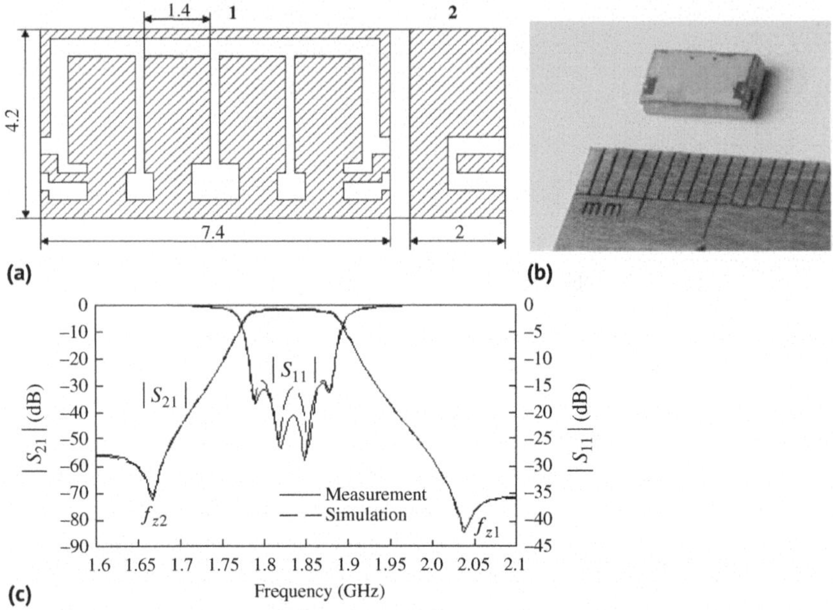

Figure 12.10 Miniaturized quasi-elliptic stripline BPF with mixed coupling. (a) Topology. (b) Photograph of fabricated filter. *Source*: [21] / with permission of IEEE. (c) Frequency responses.

in Figure 12.10a shows the short strips at both ends of the filter that connect the outer part of the filter to the inner part. Figure 12.10b shows a photograph of the reverse side of the filter, which is intended for surface mounting. The corresponding coordinate of the connection of the I/O strips to the outer resonators ensures the required external quality factor of the end resonators $Q_e = 26.59$.

Figure 12.10c simulated and measured frequency responses of the BPF. The filter has next characteristics: $f_0 = 1835$ MHz, insertion loss at f_0 is $L_{A0} = 1.9$ dB, insertion loss at the edges of the passband $L_{A\max} = 3.2$ dB, BW = 90 MHz, return loss $L_R > 14.0$ dB.

The TZs' position corresponds to the previously adopted specification: $f_{z1} = 2034$ MHz and $f_{z2} = 1668$ MHz. The BPF frequency selectivity is quite high: 45 dB ($f_0 \pm 135$ MHz). When using the normalized frequency $j\Omega = (f/f_0 - f_0/f)\,/\mathrm{FBW}$, it has the following notation 45 dB ($\pm j3$).

A very useful application is the effect of the mixed coupling coefficient transition through zero (Figure 12.5a), when the magnetic and electric components cancel each other $K_m = |K_e|$. If a half-wave hairpin resonator is placed over such a pair of SIRs, the resulting third-order BPF will be quasi-elliptical with two TZs equidistant with respect to f_0 [19]. The open ends of the hairpin resonator should be separated by a capacitive gap from the open ends of the SIRs. The larger the value of the mixed coupling component, the closer the two TZs are to each other. In this case, the mixed coupling is used as a cross-coupling.

If a $\lambda/4$ resonator is installed instead of a hairpin resonator, the resulting BPF (Figure 13.7) will be a delay line filter with a flat group delay [20]. These effects have been observed in microstrip BPFs [19, 20]. However, they can also be used in self-shielded stripline BPFs, whose resonators have a higher unloaded quality factor Q_u than microstrip resonators. The use of mixed cross-coupling $K_{13} = 0$ in the third-order filter provides it with two equidistant TZs, as in a quadruplet BPF with all simple couplings.

12.3.2 BPFs with Increased Attenuation at Frequencies Multiple of f_0

Bandpass microwave filters with quarter-wave resonators $\lambda/4$ have parasitic passbands (Figure 12.4d) at odd harmonics of the fundamental signal $3f_0$, $5f_0$, etc., which is their disadvantage. In transmitters, the useful and parasitic signals are emitted at the frequency f_0 and multiple harmonics $2f_0$, $3f_0$, respectively. To suppress them, it is sufficient to disrupt the multiplicity in the distribution of the filter passbands. In this case, achieving a wide rejection band is unnecessary. This problem can be solved by using filters with SIRs.

Experimental results of stripline BPF with SIRs made on alumina (Al_2O_3) substrates with a thickness of $h = 2$ mm and operating in the frequency range of 1500–2500 MHz are presented [23]. We omit the design details and limit ourselves to considering the design features and measurement data of the most significant practical interest.

12.3.2.1 Stripline Interdigital BPF with Center Frequency 1.5 GHz

Figure 12.11 shows a third-order stripline BPF of the interdigital type with SIRs, operating at the frequency of $f_0 = 1.5$ GHz. Vacuum deposition method was used to deposit an 8 μm thick copper layer on the substrates. Chemical etching was used to form the topological patterns (Figure 12.11a). The topology of the front side of the lower substrate is shown in position 1, and the front side of the upper substrate is shown in position 2. The conductors of the upper substrate are reduced by 0.2 mm on each side compared to the conductors of the lower substrate. This measure is due to the need to align the topological patterns during filter assembly.

Position 3 shows the contact pads etched on the outside of the bottom substrate, which are used for surface mounting of the filter. The ends of the bottom and top substrates are also metallized using vacuum deposition. Position 4 shows short conductive strips on the ends of the bottom substrate, which connect the contact pads to the filter's input and output center conductors. Both substrates are pressed together, and their ends are soldered together. Conductive patterns are located inside the BPF stripline, which must be protected from possible solder ingress

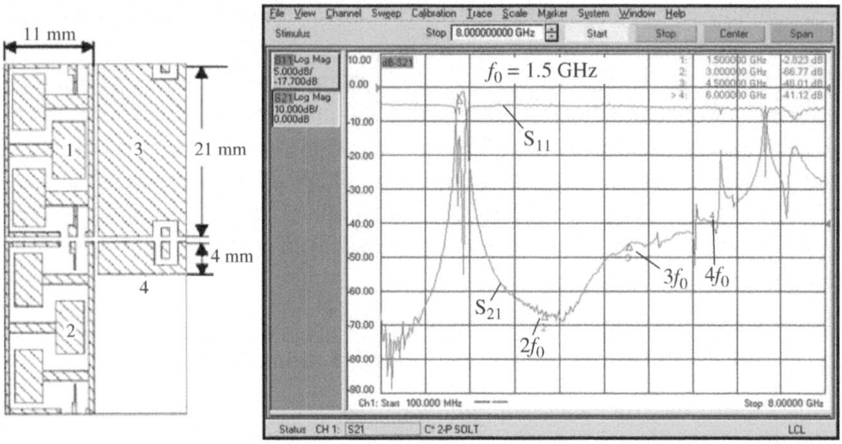

(a) (b)

Figure 12.11 Stripline interdigital BPF with SIRs and center frequency 1.5 GHz. (a) Topology. (b) Measured frequency responses.

during filter assembly. For this purpose, metallized strips are provided along the perimeter of the top and bottom substrates.

The thickness of the stripline BPF under consideration is $b = 4$ mm, its central conductor has a thickness of $t = 16$ μm, and the filter size is 21 mm × 11 mm × 4 mm. Each resonator is formed by two sections with low and high characteristic impedance. The high-impedance section contains a shorted end; the width of this section is $w = 1.6$ mm, and its characteristic impedance is $Z_{0\max} = 36.3\ \Omega$. Adjacent to the open end of the resonator is a section with low characteristic impedance ($Z_{0\min} = 14.1\ \Omega$); the conductor width of this section is $w = 6.8$ mm. The heterogeneity parameter of this resonator is $m = Z_{0\max}/Z_{0\min} = 2.57$.

The main passband (Figure 12.11b) has a center frequency of $f_0 = 1.5$ GHz, and the first parasitic passband has a center frequency of 6.84 GHz. The ratio of these frequencies is 4.5. The BPF is characterized by increased attenuation at frequency $3f_0$ (46 dB) due to the removal of the parasitic passband from the main one. At frequency $2f_0$, the attenuation is 66 dB, and at frequency $4f_0$, it is 41 dB. Over a wide frequency range, up to $4f_0$, the BPF has a high suppression level (at least 40 dB).

This filter is characterized by the following data: the bandwidth at 2 dB is $BW_{2\,dB}$ =105 MHz; $L_{A0} = 1$ dB; selectivity is characterized by the values of 30 dB ($f_0 - 160$ MHz) and 30 dB ($f_0 + 190$ MHz). The steepness of the left slope of this filter's frequency response is steeper than the right slope. This indicates that the electrical component dominates the electromagnetic interaction between the resonators.

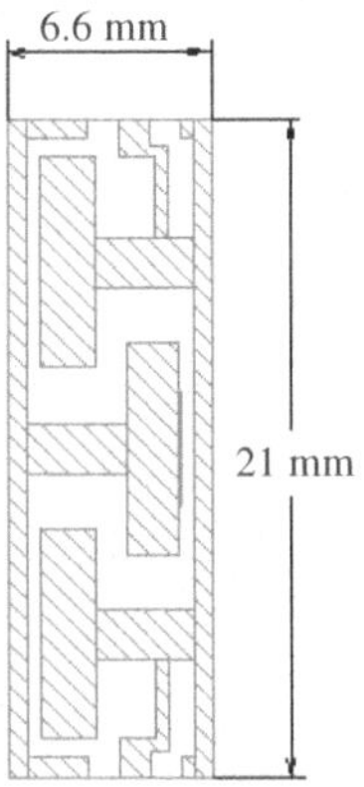

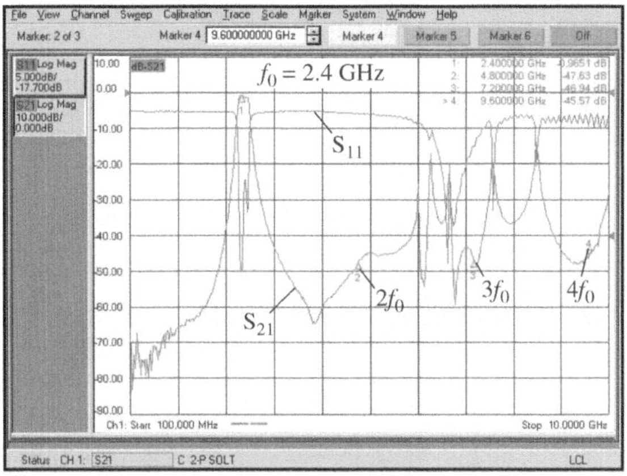

Figure 12.12 Stripline interdigital BPF with SIRs and center frequency 2.4 GHz. (a) Topology. (b) Measured frequency responses.

12.3.2.2 Stripline Interdigital BPF with Center Frequency 2.4 GHz

The third-order interdigital BPF with SIRs considered as follows (Figure 12.12a) has a center frequency of $f_0 = 2.4$ GHz and is characterized by dimensions of 21 mm × 6.6 mm × 4 mm. The width of its central conductors is the same as that of the previous filter: the low-impedance section ($Z_{0\min} = 14.1\ \Omega$) has a width of $w = 6.8$ mm and the high-impedance section ($Z_{0\max} = 36.3\ \Omega$) has a width of $w = 1.6$ mm. The parameter m of this resonator is $m = 2.57$.

The frequency responses of this filter are contained in Figure 12.12b. The filter attenuation at frequencies $2f_0$, $3f_0$, and $4f_0$ is approximately the same, equaling 46 dB. The filter loss at the central frequency: $L_{A0} = 0.8$ dB; the bandwidth at 0.5 dB: $\mathrm{BW}_{0.5\ \mathrm{dB}} = 175$ MHz. The filter selectivity is characterized by 30 dB ($f_0 - 260$ MHz) and 30 dB ($f_0 + 330$ MHz).

Unlike the previous filter (Figure 12.11), this filter does not have a continuous stopband in the frequency band, the right edge of which approximately corresponds to the value of $4f_0$. However, its resonators are characterized by the same parameter $m = 2.57$. Apparently, this is due to parasitic oscillations associated with the fact that the transverse dimensions of the resonator are larger than its longitudinal dimensions.

12.3.2.3 Stripline Combline BPF with Center Frequency 2.48 GHz

Figure 12.13a shows the fourth-order combline BPF with SIRs. Its size is 28 mm × 12 mm × 4 mm. The low-impedance section of the SIR, one end of which is short-circuited, is characterized by the values: $w = 5$ mm and $Z_0 = 17.9\ \Omega$.

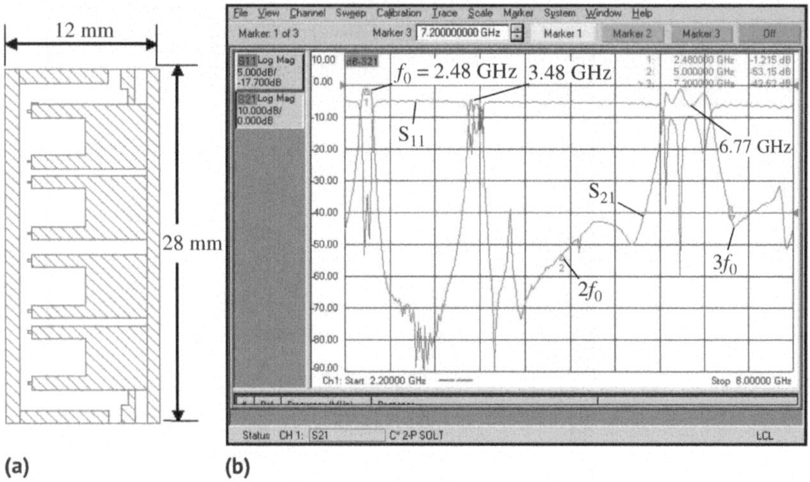

Figure 12.13 Stripline combline BPF with SIRs and center frequency 2.48 GHz. (a) Topology. (b) Measured frequency responses.

Two high-impedance sections with $w = 1$ mm and $Z_0 = 45\ \Omega$ adjoin it. The measured unloaded quality factor Q_u of these resonators is $Q_u \approx 600$.

Figure 12.13b shows the measured frequency responses of this filter. In is characterized by the following data: the central frequency is $f_0 = 2.48$ GHz; the bandwidth at the level of 0.5 dB is $\text{BW}_{0.5\text{ dB}} = 75$ MHz, which corresponds to FBW = 0.03; $L_{A0} = 1.2$ dB; the selectivity is characterized by the values of 30 dB ($f_0 + 95$ MHz) and 30 dB ($f_0 - 125$ MHz). The filter's frequency response is not symmetrical. Its right slope is steeper than the left one, indicating the dominance of the magnetic component in the electromagnetic interaction of the resonators. The filter attenuation at harmonics $2f_0$ and $3f_0$ frequencies is 53 and 42 dB, respectively.

Figure 12.13b shows that the nearest parasitic passband of the BPF is located at 3.48 GHz. This is due to parasitic oscillations. Computer simulation reveals the disappearance of this peak when replacing two high-impedance sections at the edges of the resonator with one section located in the middle of the resonator.

12.4 Coupling Coefficients at Higher Resonant Frequencies

Two different pairs of thin ($b = 1$ mm) SIRs with $\varepsilon_r = 9.7$ (alumina) and a length of $L = 10$ mm are shown in Figure 12.14a, b. The minimum width of the resonator is 0.8 mm ($Z_{0\max} = 24.26\ \Omega$), and the maximum width of the resonator is 1.6 mm ($Z_{0\min} = 14.75\ \Omega$). The gap S separating the resonators is $S = 0.1$ mm.

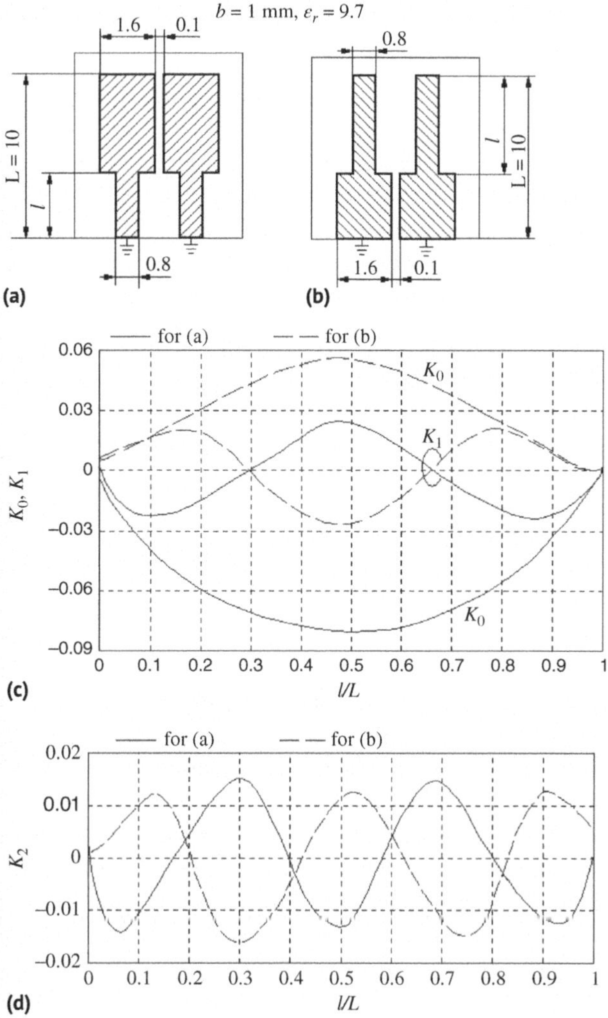

Figure 12.14 Coupling coefficients between low-profile stripline SIRs. (a) First-type SIRs. (b) Second-type SIRs. (c) Coupling coefficients K_0 and K_1. (d) Coupling coefficients K_2.

12.4.1 Coupling Coefficients Versus Shape of SIRs

A set of oscillations with resonant frequencies f_n, $n = 0, 1, \ldots$, which are electromagnetically coupled to each other, is inherent in these resonators. The coupling coefficients corresponding to resonance frequencies f_0, f_1, and f_2 are denoted as K_0, K_1, and K_2. The dependences of these coupling coefficients on the normalized value l/L of these SIRs are shown in Figure 12.14c, d.

The features of the change in the coupling coefficients between the resonators are noted as follows, which are shown in Figure 12.14c, d:

1. The alternating behavior is inherent to these functions K_n (l/L); they have $(2n + 1)$ extremes.
2. The function K_n (l/L) has $2n$ internal zeros. The coupling coefficients under consideration are mixed. Satisfaction of condition $K_1 = 0$ indicates the absence of coupling between the resonators at the frequency f_1. This leads to suppression of the first parasitic passband of the BPF and expansion of its rejection band. For this reason, condition $K_1 = 0$ has a practical interest.

The condition $K_1 = 0$ holds for $l/L \approx 1/3$ and $l/L \approx 2/3$ for both types of resonator pairs, as illustrated in Figure 12.14c. Further analysis is needed to determine which of the SIR pairs shown in Figure 12.14a, b is preferable.

For the SIRs shown in Figure 12.14a, b, the resonance equations hold

$$m \tan\left[\frac{\omega (L-l)}{v}\right] = -\cot\left(\frac{\omega l}{v}\right) \tag{12.3}$$

for the first type, and

$$m^{-1} \tan\left[\frac{\omega l}{v}\right] = -\cot\left[\frac{\omega (L-l)}{v}\right] \tag{12.4}$$

for the second type, where $m = Z_{0\max}/Z_{0\min}$ and v is the propagation speed of the electromagnetic wave.

The ratio of resonant frequencies f_2/f_0 characterizes the mutual distance of these frequencies from each other. The curves in Figure 12.15 show that the first type of SIR has a larger ratio f_2/f_0, so its use is preferable. This pair of resonators has two maxima in the dependence f_2/f_0 on the ratio l/L (Figure 12.15): at $l/L = 1/3$ and at $l/L = 2/3$. At these maxima, the ratio is determined by the formula

$$\frac{f_2}{f_0} = \frac{\pi}{\tan^{-1}\sqrt{1/(1+2m)}} - 1, \tag{12.5}$$

which follows from (12.3). From (12.5) we obtain $f_2/f_0 = 6.47$ for $m = 2$ and $f_2/f_0 = 7.69$ for $m = 3$.

At the maximum points $l/L \approx 1/3$ and $l/L \approx 2/3$, a double effect occurs: the maximum ratio f_2/f_0 and the zero value of $K_1 = 0$. These two effects make it possible to significantly expand the rejection band.

12.4.2 Low-profile Stripline BPF with Extended Stopband

The effect described earlier ($K_1 = 0$) is illustrated as follows, using the stripline third-order Chebyshev BPF as an example. The following parameters were adopted for the design: $f_0 = 2.45$ GHz, BW = 140 MHz (FBW = 0.057),

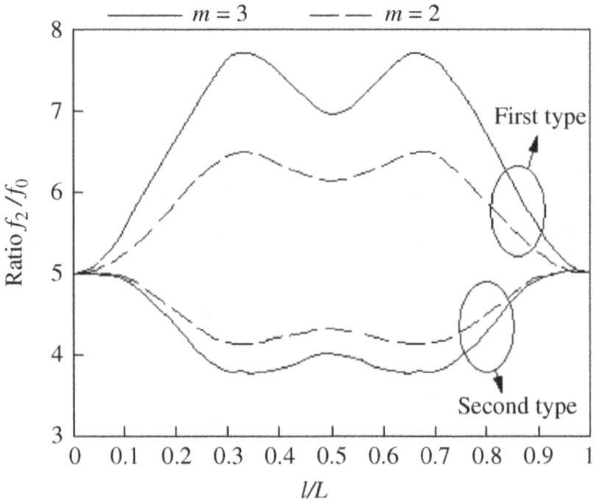

Figure 12.15 Ratio f_2/f_0 of stripline SIRs.

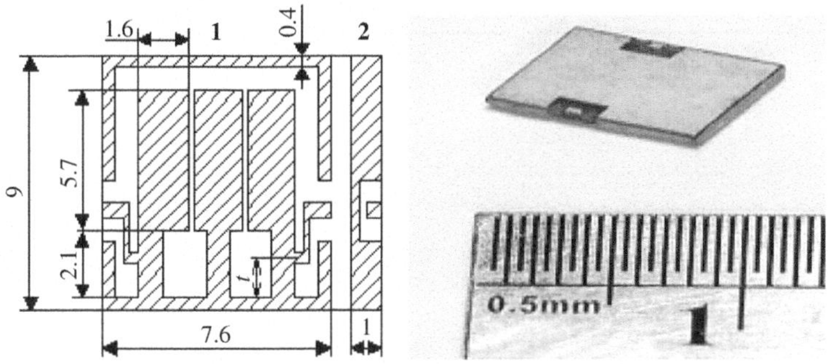

(a) (b)

Figure 12.16 Low-profile stripline BPF with $K_1 = 0$. (a) Topology. (b) Photograph of fabricated filter.

$L_{Ar} = 0.2$ dB. The following parameters correspond to the values $L_{Ar} = 0.2$ dB and $n = 3$ (Table 2.6): $g_0 = 1$, $g_1 = 1.2275$, $g_2 = 1.1525$, $g_3 = 1.2275$. Substituting the value FBW = 0.057 and the g parameters into (3.8), we determine the coupling coefficients $K_{12} = K_{23} = K_0 = 0.048$ and the Q_e value of the end resonators $Q_e = 21.5$.

Using thin substrates ($h = 0.5$ mm) from alumina ($\varepsilon_r = 9.7$), a thin ($b = 1$ mm) stripline BPF with SIRs was fabricated (Figure 12.16). Figure 12.16a, position 1, shows the topology of this filter, where the dimensions are indicated in millimeter: $L = 7.8$ mm, $l = 2.1$ mm ($l/L = 0.27$), $w_{min} = 0.8$ mm ($Z_{0max} = 24.26\ \Omega$), $w_{max} = 1.6$ mm ($Z_{0min} = 14.75\ \Omega$), $t = 1.2$ mm, $S = 0.2$ mm; the filter size is 9 mm × 7.6 mm × 1 mm. Short strips were formed at both ends of the filter (Figure 12.16a, position 2); they connect the outer and inner parts of the filter. The photograph of the filter (Figure 12.16b) shows its reverse side.

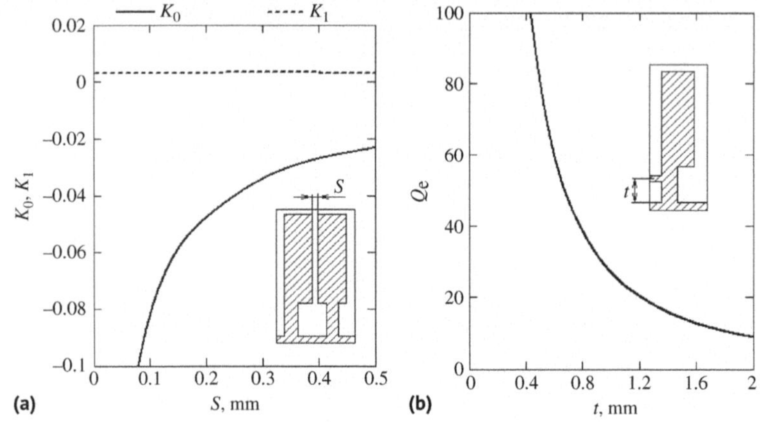

Figure 12.17 Design plots for low-profile stripline BPF. (a) Coupling coefficients K_0 and K_1. (b) External quality factor Q_e.

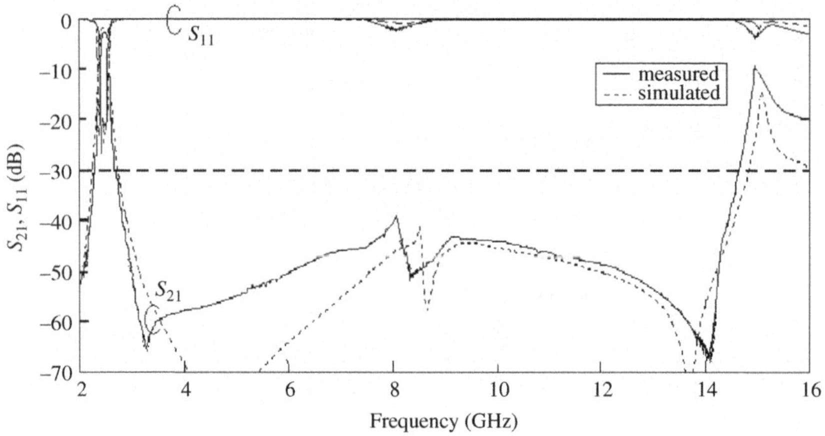

Figure 12.18 Measured and simulated frequency responses of low-profile stripline BPF with $K_1 = 0$.

The dependence of the coupling coefficients K_0 and K_1 on the gap size S between the SIRs is plotted graphically in Figure 12.17a. For all values of S, the coupling coefficient K_1 takes on a small value $K_1 = 0.0032$. The graph of the dependence of Q_e on t is shown in Figure 12.17b. It shows that the required value $Q_e = 21.5$ is achieved at $t = 1.2$ mm.

Figure 12.18 shows simulated and measured frequency responses of the BPF. The following values were measured: $f_0 = 2450$ MHz, $L_{A0} = 1.8$ dB, BW = 140 MHz, $L_R > 16$ dB. As predicted, the spurious passband associated with the resonant frequency f_1 was suppressed. The nearest spurious passband of this BPF is located at a frequency $6.06f_0$. The parameter m of the SIRs used in the filter is equal to $m = 24.26\ \Omega/14.75\ \Omega = 1.645$, substituting which into (12.5) yields the calculated value $f_2/f_0 = 5.98$. This value differs little from the measured value $f_2/f_0 = 6.06$.

This filter is characterized by a wide rejection band achieved by suppressing the spurious passband associated with the resonant frequency f_1. If f_{max} and f_{min} are designated as the cutoff frequencies of the rejection band with an attenuation level of 40 dB, we obtain $f_{max}/f_{min} = 5.44$. Note that the third-order BPF [24] with $\lambda/2$ and $\lambda/4$ resonators, using discriminating coupling, had a ratio of $f_{max}/f_{min} = 4.09$.

12.5 Invariance of Coupling Coefficients Relative to Dielectric Constant

Of scientific and practical interest is the influence of the dielectric constant ε_r on K between stripline resonators at the fundamental resonant frequency f_0. To eliminate the influence of short-circuited ends on K, we will consider half-wave resonators.

12.5.1 Different Location of Stripline Half-wave Resonators

The topology of a symmetric stripline construction of a pair of half-wave resonators is shown in Figure 12.19a. The stripline construction is completely isolated because its side surfaces are metallized. EM simulation was performed using Microwave Office (AWR) software, using the following parameters: $b = 2$ mm, $w = 1$ mm, $L = 6$ mm, $S = 0.2$ mm. The pair of stripline resonators occupies an area of 9.2 mm × 5.4 mm. The pair of resonators is separated from the metallized side surfaces by 1.6 mm. Using a dielectric constant $\varepsilon_r = 9.7$, the coupling frequencies are $f_e = 7220$ MHz and $f_o = 7604$ MHz, which corresponds to a coupling coefficient between the resonators of $K = 0.05177$ (3.10). If $\varepsilon_r = 2.2$, then the coupling frequencies will change significantly: $f_e = 15159$ MHz, $f_o = 15965$ MHz, but the coupling coefficient will retain its previous value: $K = 0.05176$ (3.10). Table 12.3 presents various values of ε_r and their corresponding K values. For different values of ε_r, only the f_e and f_o values change, while the K values remain virtually unchanged.

If the central conductors of the stripline resonators are shifted downward by 0.5 mm, as shown in Figure 12.19b, then with a change of ε_r from 2.2 to 92, the value of the coupling coefficient will be practically constant ($K = 0.0383$–0.0384). If the central conductors of the resonators are shifted downward and to the right, as shown in Figure 12.19c, then their resonant frequencies will be different. Due to the proximity of the metal, the resonant frequency of the right resonator

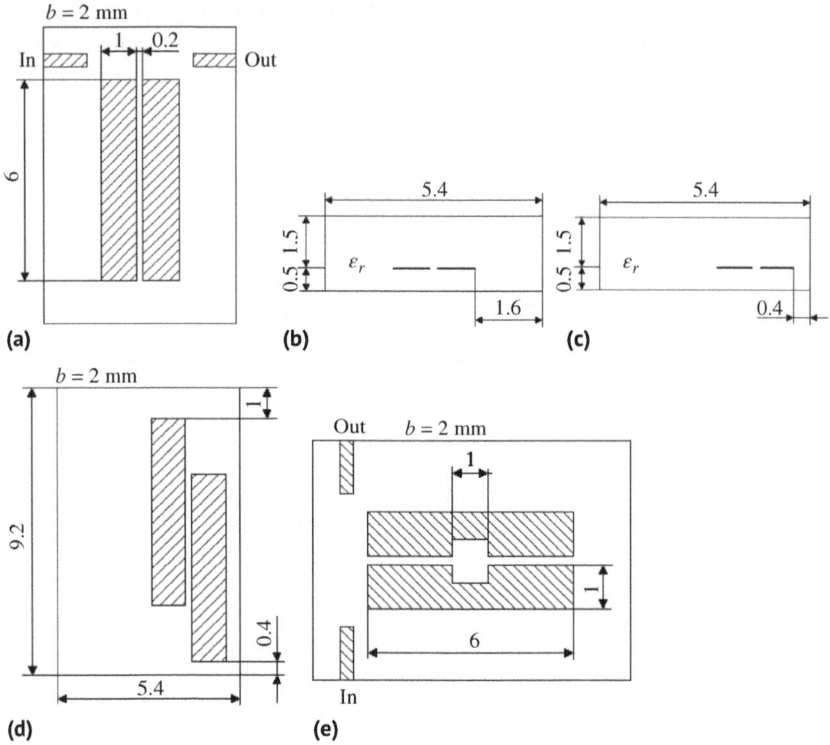

Figure 12.19 Pairs of stripline half-wave resonators. (a) Symmetrically positioned. (b) Offset down. (c) Offset down and to the right. (d) Displaced downward, to the right, and with respect to each other. (e) Half-wave type SIRs.

Table 12.3 Coupling coefficients of stripline resonators (Figure 12.19a).

ε_r	**2.2**	**9.7**	**21**	**38**	**92**
K	0.05176	0.05177	0.05178	0.05177	0.05170

will increase. For different resonant frequencies, the coefficient K is calculated using the formula [25]:

$$K = \pm\left(\frac{f_2}{f_1} + \frac{f_1}{f_2}\right)\sqrt{\left(\frac{f_o^2 - f_e^2}{f_o^2 + f_e^2}\right)^2 - \left(\frac{f_2^2 - f_1^2}{f_2^2 + f_1^2}\right)^2}. \tag{12.6}$$

In expression (12.6), f_1 and f_2 denote the resonance frequencies of the two resonators. When ε_r changes from 2.2 to 92 for the pair of resonators in Figure 12.19c, the coefficient K takes the values $K = 0.0352$–0.0355.

Figure 12.19d shows the most general case of the arrangement of the stripline resonators, when the center conductors are shifted additionally relative to each other. In this case, when ε_r changes from 2.2 to 92, the coefficient K takes the values $K = 0.1053$–0.1055.

Figure 12.19e shows a pair of SIRs with both open-circuited ends. The center conductor of each resonator is a 6 mm × 1 mm strip with a 1 mm × 0.4 mm rectangle removed from its center. The coupling coefficient of this pair is mixed and negative. It ranges from (−0.0613) to (−0.0614) as ε_r varies from 2.2 to 92.

Computer simulations reveal the following pattern. In a stripline structure with metallized side surfaces filled with a homogeneous dielectric, the electromagnetic coupling coefficient of the resonators is independent of the permittivity ε_r. The value of K depends only on the geometric parameters of the structure.

12.5.2 Dependence of Coupling Coefficient on Resonators' Length

In Section 12.1, the influence of several factors on the coupling coefficient K in stripline resonators was analyzed: geometric parameters, permittivity ε_r, operating frequency f_0. The pattern established in the previous paragraph dictates the need to reconsider the influence of these factors.

First, let us answer the question of how the length L of a stripline resonator affects K. Let us again consider a pair of half-wave resonators (Figure 12.19a). Any value of ε_r can be used in the simulation, since it does not affect the coupling coefficient K. We will leave the parameters of the resonator pair in Figure 12.19a unchanged: $w = 1$ mm, $b = 2$ mm, and $S = 0.2$ mm. Figure 12.20 shows the

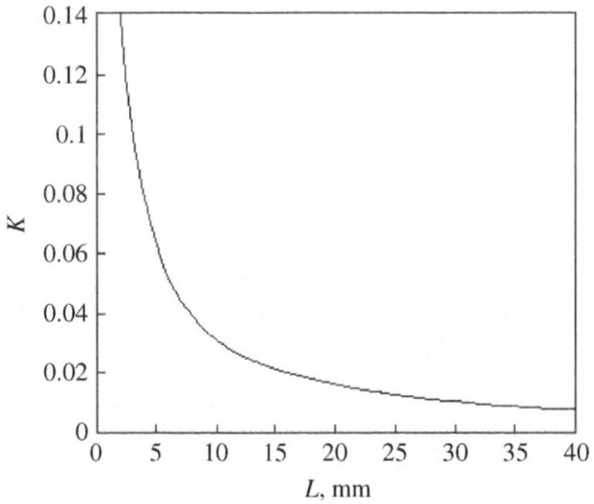

Figure 12.20 Dependence of coupling coefficient on the length of stripline half-wave resonators with parameters $b = 2$ mm, $w = 1$ mm, $S = 0.2$ mm.

dependence of the coupling coefficient K on the length L of the pair of stripline resonators under consideration. This dependence shows that the smaller L, the greater the K.

The dependence $K = K(L)$ (Figure 12.20) is universal, since it applies to stripline resonators using different values of ε_r. It is important to note that this dependence also applies to $\lambda/4$ resonators, since they have the same K value as $\lambda/2$ resonators, but their length is half as long.

Let us examine the dependencies in Figure 12.2 again. It seems that an increase in operating frequency f and ε_r leads to an increase in K. Now we can explain these dependencies. They are caused by a decrease in the resonator length L, which occurs with both an increase in f and an increase in ε_r.

To increase K in stripline resonators, along with decreasing the length L, other geometric parameters can be used. This is achieved by decreasing the gap S between the resonators, increasing the thickness b, and decreasing the width w of the resonator's center conductor.

12.5.3 Stripline BPF with Different ε_r

When changing the value of ε_r, only the coupling frequencies of the stripline structures change, while the ratios between the specified frequencies do not change. Figure 12.21a shows a stripline fourth-order BPF, which contains half-wave SIRs, and is weakly coupled to the loads. Its thickness $b = 1$ mm. The following parameters characterize the SIRs: the size of the central conductor $w = 1.1$ mm and $L = 4.4$ mm; rectangles of size 0.9 mm × 0.4 mm are removed from the central part of the conductors; the interresonator gap between resonators 1–2 and 3–4 is 0.3 mm, and between resonators 2–3 is 0.25 mm. The filter size is 7.05 mm × 5.9 mm × 1 mm. Coupling coefficients of the filter are mixed, with alternating signs.

The coupling frequencies of the BPF are clearly visible in its transfer characteristic in the case of weak coupling of the filter to the loads. We denote these frequencies as f_1, f_2, f_3, and f_4. If the filter uses a permittivity of $\varepsilon_r = 92$, then $f_1 = 2923$ MHz, $f_2 = 2982$ MHz, $f_3 = 3068$ MHz, and $f_4 = 3105$ MHz. These frequencies are shown in Figure 12.21b. If the value is reduced by a factor of four ($\varepsilon_r = 23$), then the coupling frequencies will increase exactly twofold: $f_1 = 5846$ MHz, $f_2 = 5964$ MHz, $f_3 = 6136$ MHz, and $f_4 = 6210$ MHz (Figure 12.21c). In a stripline high-order BPF with homogeneous dielectric filling and with all metallized surfaces, the relationship between the coupling frequencies does not depend on the value ε_r; it depends only on the geometric parameters.

This pattern allows the same stripline BPF design to be used to replicate frequency responses with different center frequencies f_0, changing only ε_r. The established pattern corresponds to the case of completely isolated stripline systems. Isolation is ensured by metallization of all side surfaces of the stripline structure.

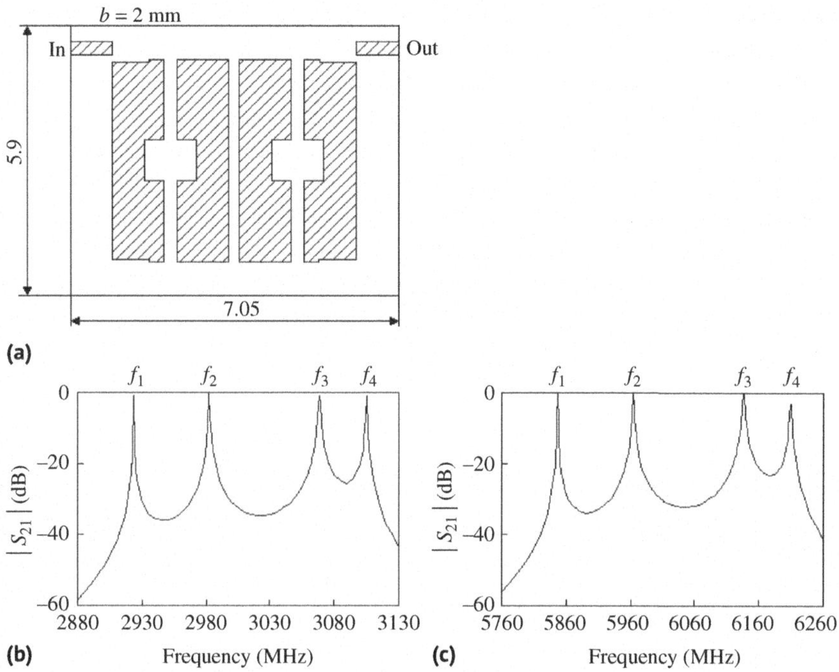

Figure 12.21 Stripline fourth-order BPF. (a) Topology. (b) Coupling frequencies for $\varepsilon_r = 92$. (c) Coupling frequencies for $\varepsilon_r = 23$.

The weak coupling of the end resonators to the loads is a certain approximation to an isolated stripline system. When applying real filter characteristics, it is necessary to increase the coupling of the end resonators to the loads. This violates the isolation condition, and the frequencies of the end resonators are detuned. The question naturally arises: what actions should be taken to keep the frequency response of the filter unchanged when ε_r changes?

The external Q-factor Q_e characterizes the coupling of the end resonators to the load; the lower the Q_e, the greater this coupling. Connecting a half-wave resonator to the load via a short conductive strip is the most common method. In this case, the value of Q_e can be calculated from expression (3.13). Let us write it in a form

$$Q_e = R_L b = \frac{R_L \pi}{2Z_0 \sin^2 \theta'}, \tag{12.7}$$

where $R_L = 50\ \Omega$ is the load resistance, and θ' is the coordinate of the load connection, which is counted from the resonator's middle. As the dielectric constant ε_r decreases, Z_0 increases, which in turn decreases Q_e (12.7). In this case, the connection coordinate θ' must be decreased to restore the previous value Q_e.

The stripline BPF in Figure 12.21a contains SIRs. For these resonators, expression (12.7) is not legitimate. However, the shift in the connection coordinate θ' associated with the change in the permittivity ε_r is preserved. Two load connection options and the BPF frequency responses for the corresponding values of $\varepsilon_r = 92$ and $\varepsilon_r = 23$ are shown in Figure 12.22. Also note that, simultaneously with shifting the load connection point θ' toward the middle of the SIR, it is necessary to elongate the resonator itself. This elongation is necessary to compensate for the frequency shift associated with the change in the coordinate θ'.

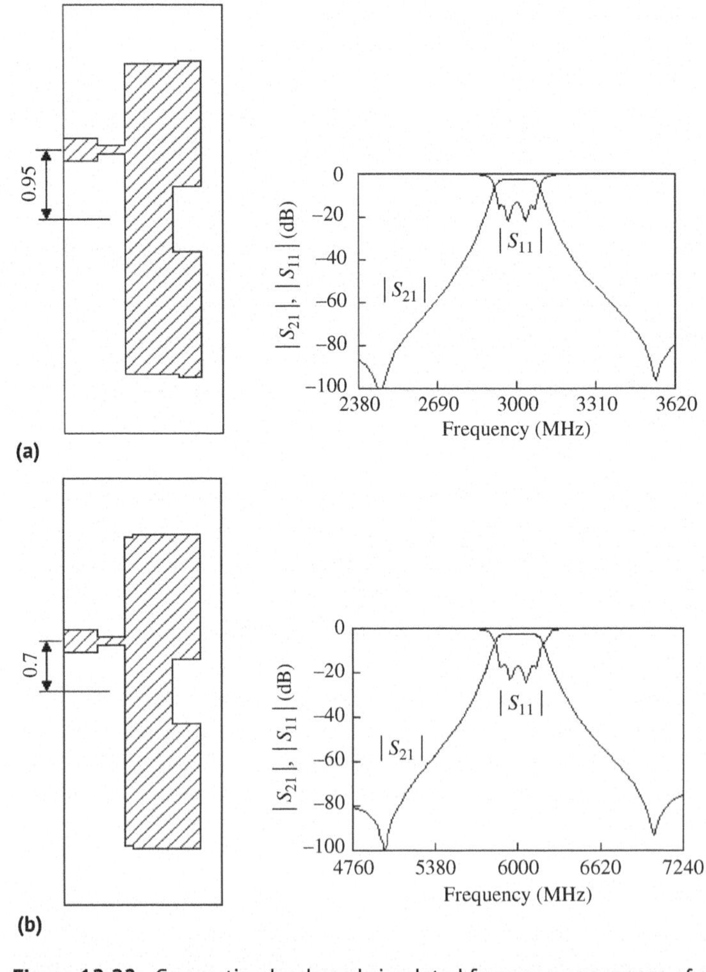

Figure 12.22 Connecting loads and simulated frequency responses of stripline BPF shown in Figure 12.21. (a) For case $\varepsilon_r = 92$, $f_0 = 3000$ MHz, BW = 140 MHz (b) For case $\varepsilon_r = 23$, $f_0 = 6000$ MHz, BW = 280 MHz.

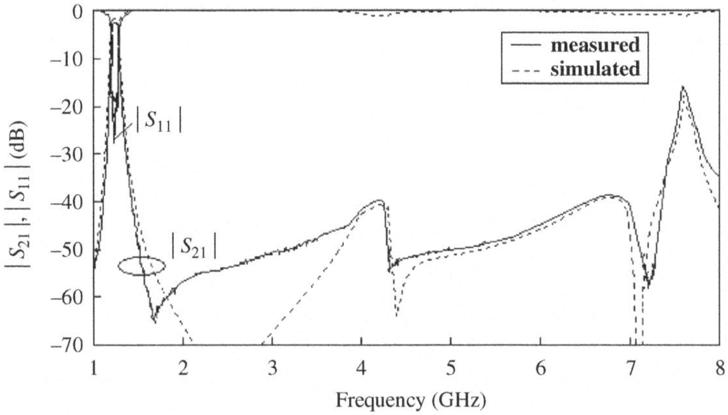

Figure 12.23 Frequency responses of a modified low-profile stripline BPF, shown in Figure 12.16.

The normalized frequency responses remained the same after changing the permittivity by a factor of four from $\varepsilon_r = 92$ to $\varepsilon_r = 23$. The f_0 value doubled from 3 to 6 GHz. The bandwidth also changed by a factor of two from 140 to 280 MHz. Key indicators change at $\sqrt{\varepsilon_{r2}/\varepsilon_{r1}}$ times.

If a stripline BPF with a dielectric constant ε_r has a good frequency response, it can be reproduced in another frequency band by appropriately changing ε_r. This requires minor adjustments to the end resonators and the θ' connection coordinates. This method is also applicable to stripline BPF with quarter-wave resonators.

The measured and simulated frequency responses of a low-profile BPF (Figure 12.16), whose dielectric constant ε_r was changed from 9.7 to 38, are shown in Figure 12.23. The connection coordinate of the loads t is increased from 1.2 to 1.8 mm, and the end resonators are somewhat shortened.

In this case, the square root is $\sqrt{\varepsilon_{r2}/\varepsilon_{r1}} \approx 1.98$. The filter's center frequency f_0 decreased by the same factor (from 2.45 to 1.24 GHz). Its BW also changed by the same factor (from 140 to 71 MHz). It is crucial that, with this change of $\sqrt{\varepsilon_{r2}/\varepsilon_{r1}} \approx 1.98$, the condition for suppressing the first parasitic passband of the BPF, expressed by the equality $K_1 = 0$, was maintained. The new filter also has an extended rejection band (Figure 12.23).

12.6 Millimeter-wave Stripline Bandpass Filters

Stripline BPFs can find wide application in the millimeter range, where it is advisable to use thin dielectric substrates ($h = 0.25 \ldots 0.5$ mm) with a low dielectric constant ($\varepsilon_r = 2 \ldots 3$). Figure 4.9 shows that stripline resonators in the specified frequency range have a relatively high unloaded quality factor Q_u. Therefore, it is necessary to answer some questions: What will the coupling coefficients K be at such high frequencies between stripline resonators with small values of ε_r and b? Will these values of K be sufficient for BPF design? These questions will be considered in this section [26].

12.6.1 Coupling Coefficients Between Half-wave Resonators

In the millimeter-wave range, we will analyze the values of K between half-wave stripline resonators. Of particular interest are dielectric materials with a small value of ε_r, which are widely available. These include metallized dielectric substrates manufactured by Rogers Corp. [27]. RT/Duroid 5880 substrates have a small permittivity of $\varepsilon_r = 2.2$. The substrates have a small thickness, including $h = 0.254$ mm and $h = 0.508$ mm. Stripline BPFs using such substrates will have double the thickness $b = 508$ mm and $b = 1.016$ mm. When analyzing stripline resonators with $\varepsilon_r = 2.2$ and $b = 1.016$ mm (508 mm), formula (3.10) and the Microwave Office (AWR) circuit design software will be used.

The topology of a pair of half-wave stripline resonators separated by a gap of $S = 0.05$ mm and the frequency dependence of their coupling coefficient K are shown in Figure 12.24. The resonant frequency is determined by the resonator length L, which is variable. Curves *1* and *2* are bounded on the right. This limitation is determined by the frequencies at which L values become extremely small, leading to longitudinal oscillations. The K value on curve *1* varies from 0.05 ($f = 30$ GHz) to 0.1 ($f = 62$ GHz). The K value on curve *2* varies from 0.03 ($f = 50$ GHz) to 0.08 ($f = 120$ GHz). These curves are plotted for a gap between the resonators of $S = 0.05$ mm. The K value decreases as the gap width increases.

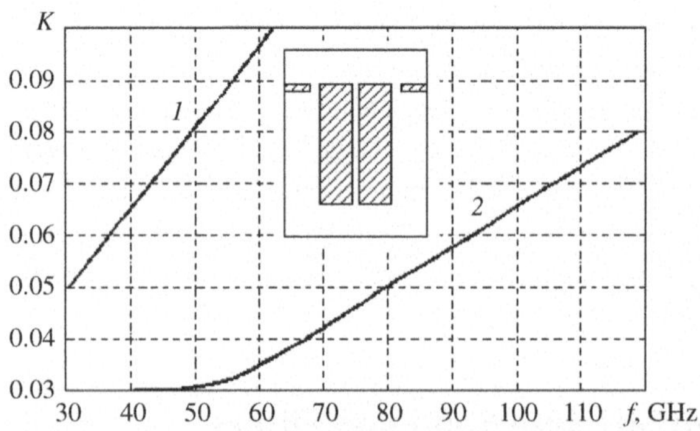

Figure 12.24 Frequency dependence of K between stripline half-wave resonators ($\varepsilon_r = 2.2$) with the parameters: $b = 1.016$ mm and $w = 0.8$ mm (curve *1*); $b = 0.508$ mm and $w = 0.4$ mm (curve *2*).

Since, for technological reasons, it is not practical to make the S value less than 0.05 mm, the K values shown in Figure 12.24 can be considered the maximum achievable for a given value of $\varepsilon_r = 2.2$.

The analysis shows that thin ($b = 1.016$ mm or less) half-wave stripline resonators with low permittivity ($\varepsilon_r = 2.2$) exhibit fairly strong electromagnetic coupling at frequencies above 30 GHz. This suggests that thin stripline pseudocombline structures with half-wave resonators represent BPFs in the millimeter range.

12.6.2 Coupling Coefficients Between SIRs of Half-wave type

Figure 12.25 shows two pairs of SIRs with open-circuited ends, which allows them to be classified as half-wave SIRs. Both pairs contain the same two resonators, but oriented differently. The thickness of the stripline resonators is b, and they are filled by dielectric with $\varepsilon_r = 2.2$. The length L and width w characterize the central conductor of the resonators used. A gap S separates the two resonators. Three rectangles are removed from the central conductor of each resonator. Two rectangles are located near the open-circuited ends of the resonator, and the third is in the middle of the opposite side of the resonator. The resonator is not symmetrical about the vertical axis that passes through its center. The electromagnetic coupling between the SIRs is mixed. The pair of resonators in Figure 12.25a is characterized by positive coupling ($K > 0$) and generates a right-handed TZ ($f_z > f_0$). The pair of resonators in Figure 12.25b is characterized by negative coupling ($K < 0$) and generates a left-handed TZ ($f_z < f_0$).

Both positive (Figure 12.25a) and negative (Figure 12.25b) coupling between resonators increases with the removal of larger rectangles from their central conductors.

The following parameters were used in the computer simulation: $b = 1.016$ mm, $w = 0.8$ mm ($Z_0 = 51.74\ \Omega$), and $L = 2.8$ mm. Rectangles of 0.2 mm × 0.1 mm in size were removed from the central conductors of the resonators near their open-circuited ends, and a rectangle of 0.6 mm × 0.2 mm in size was removed from the middle part. Each SIR has a resonant frequency of $f_0 = 31.2$ GHz. The dependence of K on the gap value S for each SIR pair is shown in Figure 12.25c. As S decreases, the value of $|K|$ for each pair increases monotonically. The K value of two SIR pairs varies in the range $-0.093 \leq K \leq 0.098$.

Knowledge of the maximum achievable K values is very useful when designing BPFs. It is already known that the K values of both signs increase as the rectangles made of the center conductors of the stripline resonators become larger. If we remove two 0.5 mm × 0.4 mm rectangles near the open-circuited ends of the resonator and one 1 mm × 0.4 mm rectangle from the middle of the resonator, we obtain new pairs of resonators as shown in Figure 12.26. These pairs of SIRs

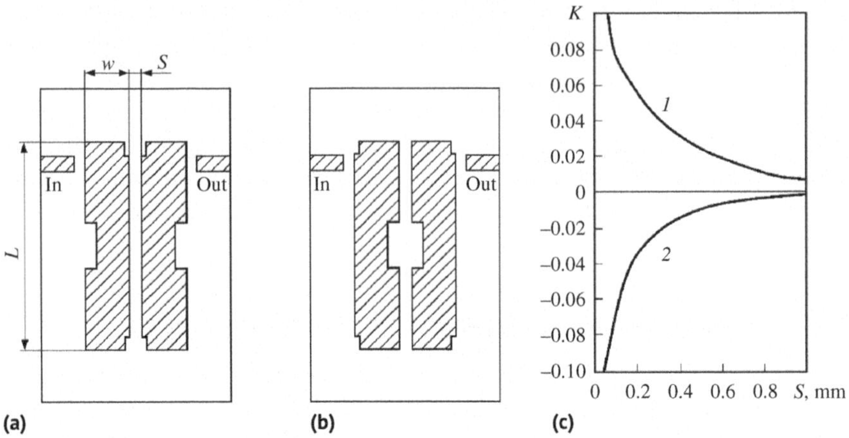

Figure 12.25 Stripline stepped-impedance resonators with (a) positive and (b) negative coupling coefficients; (c) K versus gap width S in Figure 12.25a (curve *1*) and Figure 12.25b (curve *2*).

have the maximum K values. At $S = 0.05$ mm, the pair in Figure 12.26a has a K value of 0.16, and the pair in Figure 12.26b is characterized by a K value of −0.236. The range of variation of the permissible K values is characterized by the interval $-0.236 \leq K \leq 0.16$. This limitation on the permissible values of K is valid for a pair of stripline SIRs with parameters: $\varepsilon_r = 2.2$, $b = 1.016$ mm, $w = 0.8$ mm, and $L = 2.8$ mm. The value of $|K|$ will increase as the stripline thickness of b increases.

The coupling coefficient of the considered resonators in the millimeter range can vary fairly widely, which allows the implementation of BPFs with different bandwidths. Mixed coupling coefficients vary in a broader range and can be positive or negative.

12.6.3 Characteristics of Millimeter-wave Stripline BPFs

In the specified frequency range, it is advisable to use BPFs with SIRs of the half-wave type (Figure 12.25), which have mixed coupling coefficients of different signs. Such resonators are longer than quarter-wave resonators. Alternation of signs of mixed coupling coefficients leads to different-sided TZs and allows the implementation of a quasi-elliptic transfer function.

Below is the design of a Chebyshev pseudocombline BPF with the following parameters: $f_0 = 31.2$ GHz, FBW = 0.054 (BW = 1.7 GHz), $L_{Ar} = 0.2$ dB, and $n = 4$. This filter will use two substrates with $\varepsilon_r = 2.2$ and $h = 0.5$ mm, as well as the SIRs shown in Figure 12.25. Using the values of $L_{Ar} = 0.2$ dB and $n = 4$, from Table 2.6, we determine the g parameters: $g_0 = 1$, $g_1 = 1.3028$, $g_2 = 1.2844$, $g_3 = 1.9761$, $g_4 = 0.8468$, and $g_5 = 1.5386$. Substituting the FBW and g parameters

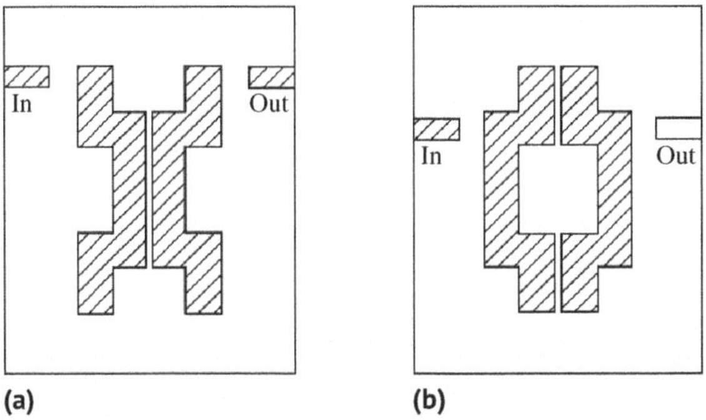

Figure 12.26 Pairs of stripline resonators with extremum (a) positive and (b) negative coupling coefficients.

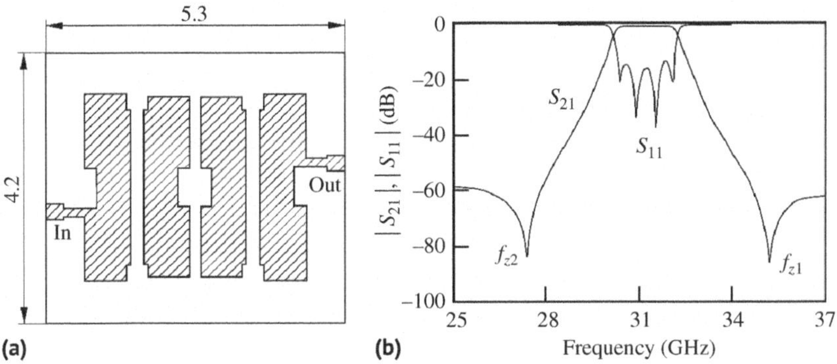

Figure 12.27 A fourth-order stripline BPF with mixed alternating couplings and $f_0 = 31.2$ GHz. (a) Topology. (b) Simulated frequency responses.

into formula (3.8) gives the coupling coefficients between adjacent resonators: $|K_{12}| = |K_{34}| = 0.0417$, $|K_{23}| = 0.0339$, and the external Q-factor: $Q_{e1} = Q_{e2} = Q_e = 24.12$. Two coupling coefficients will be considered positive: $K_{12} = K_{34} = 0.0417$ and one negative: $K_{23} = -0.0339$.

We use pairs of stripline resonators for this filter, as shown in Figure 12.25. The gap widths between resonators are determined by Figure 12.25c. To implement a positive coupling coefficient $K_{12} = K_{34} = 0.0417$, we use a pair of resonators, as shown in Figure 12.25a, and the gap width $S = 0.25$ mm. To implement a negative coupling coefficient $K_{23} = -0.0339$, we use a pair of resonators shown in Figure 12.25b with a gap width between resonators of $S = 0.175$ mm. The load connection coordinate provides the required external Q-factor $Q_e = 24.12$ of the end resonators, using expression (3.20). The topology of the central conductors of the stripline BPF (5.3 mm × 4.2 mm × 1.016 mm) with alternating mixed couplings is shown in Figure 12.27a.

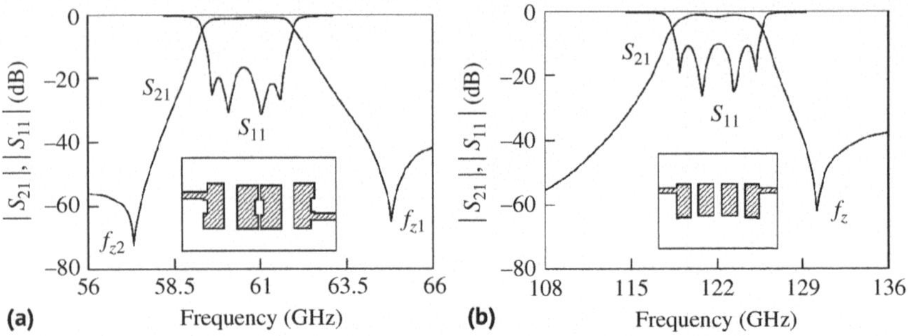

Figure 12.28 Millimeter-wave fourth-order stripline BPFs. (a) $f_0 = 60.66$ GHz, size 6.0 mm × 2.8 mm × 0.508 mm; (b) $f_0 = 122$ GHz, size 3.3 mm × 2.0 mm × 0.254 mm.

Simulated frequency responses of the filter are shown in Figure 12.27b. The filter has the following parameters: $f_0 = 31.2$ GHz, BW = 1.7 GHz (FBW = 0.054), $L_R > 14$ dB, $f_{z1} = 35.2$ GHz, $f_{z2} = 27.4$ GHz. Thanks to two-sided TZs, the filter has improved selectivity: 40 dB ($f_0 \pm 2.44$ GHz) or 40 dB ($\pm j2.87$).

Figure 12.28 shows the frequency responses of fourth-order stripline BPFs operating at higher frequencies. The insets show the topologies of the center conductors of these filters. The filter in Figure 12.28a operates at $f_0 = 60.66$ GHz; its bandwidth is BW = 2.28 GHz (FBW = 0.038), $L_R > 16.5$ dB, $f_{z1} = 64.7$ GHz, $f_{z2} = 57.3$ GHz. The filter selectivity is characterized by the values of 40 dB ($f_0 - 2.6$ GHz) and 40 dB ($f_0 + 3.45$ GHz). The filter size is 6.0 mm × 2.8 mm × 1.016 mm, and its resonators are 1.3 mm long and 0.8 mm wide.

The filter in Figure 12.28b operates at the frequency of $f_0 = 122$ GHz, and its bandwidth BW = 6.48 GHz (FBW = 0.053), $L_R > 10$ dB, $f_z = 130.2$ GHz. The filter selectivity is characterized by the values of 40 dB ($f_0 - 9.04$ GHz) and 40 dB ($f_0 + 7.05$ GHz). The filter size is 3.3 mm × 2.0 mm × 0.508 mm, and its resonators are 0.67 mm long and 0.4 mm wide.

Due to the small length of the stripline resonators of the half-wave type using $\lambda/2$ oscillations, the operating frequency $f_0 \approx 120$ GHz is the limit for stripline BPFs. To further increase the operating frequencies of stripline BPFs, it is necessary to use higher oscillation types $3\lambda/4$, λ, and $5\lambda/4$ in the resonators.

Thus, the coupling coefficient of the thin stripline resonators in the millimeter range can vary fairly, which allows for the implementation of BPFs with different bandwidths. Mixed coupling coefficients vary in a broader range and can be positive or negative. Using BPFs with alternating signs of mixed coupling coefficients in the specified frequency range leads to two-sided TZs and increased filter selectivity. Computer modeling has shown that operating frequencies $f_0 \approx 120$ GHz are the limiting ones for BPFs with half-wave resonators. To further increase the operating frequencies, it is necessary to use higher oscillation modes $3\lambda/4$, λ, and $5\lambda/4$.

12.7 Influence of Metal Layer Thickness on Coupling Coefficients and Frequency Responses

The cross-section of the stripline BPF design under consideration is shown in Figure 12.29. It is formed by connecting two substrates of thickness h, which have a dielectric constant ε_r. There are metal strips of thickness t between the substrates, which are the central conductors of stripline resonators. The gap t between the substrates is filled with air outside these strips. This air gap is absent in the widely used low-temperature cofired ceramic (LTCC) technology [4]. The thickness of the stripline structure is $b = 2h + t$.

The advantage of the stripline BPFs under consideration is that they are manufactured using a batch method. The manufacturing process is illustrated in Figure 12.30 with a particular example. On both sides of a metallized dielectric substrate ($\varepsilon_r = 92$) of size 30 mm × 24 mm × 1 mm, the topological drawings are applied by photolithography for fragments of 10 BPFs, 5 fourth-order BPFs, and 5 third-order BPFs. Then the substrate is cut into individual fragments using a diamond cutter, which is 160 μm thick. The fragments are collected in a package, and their ends are metallized. Then, the lower and upper parts of the BPF are superimposed face-to-face, and their ends are soldered to each other.

When using high dielectric constant substrates, the dimensions of the third-order and fourth-order quasi-stripline BPFs (Figure 12.30) at the frequency 2 GHz are 5.8 mm × 4.2 mm × 2 mm and 7.4 mm × 4.2 mm × 2 mm, respectively. The miniature monoblock ceramic filters have similar dimensions [2]. Note that in monoblock BPFs, the conducting topological pattern that forms the connections between the resonators is located on a thin edge (1.6–2 mm), which technologically limits their diversity. The topological pattern located "inside" the stripline BPF occupies almost its entire area. This allows the implementation of a significant variety of topological patterns and their corresponding frequency responses.

The basis for the BPF design is knowledge of the coupling coefficients K between resonators, which are determined by the parameters of the design used. When creating a stripline BPF by connecting two dielectric substrates, the thickness of the metal strips t can be different. The t value can vary from 1 to 20 μm if thin

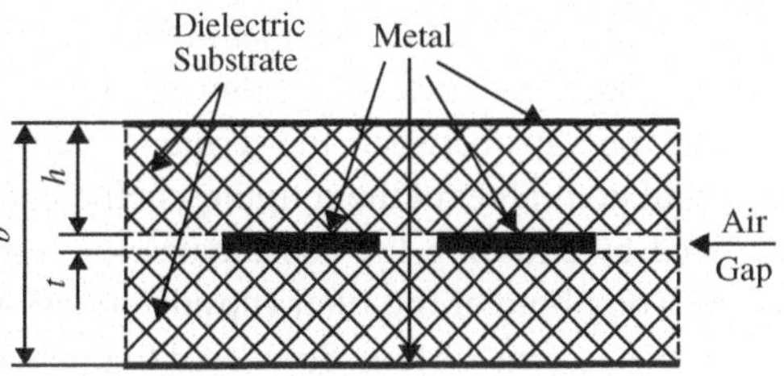

Figure 12.29 Stripline cross-section.

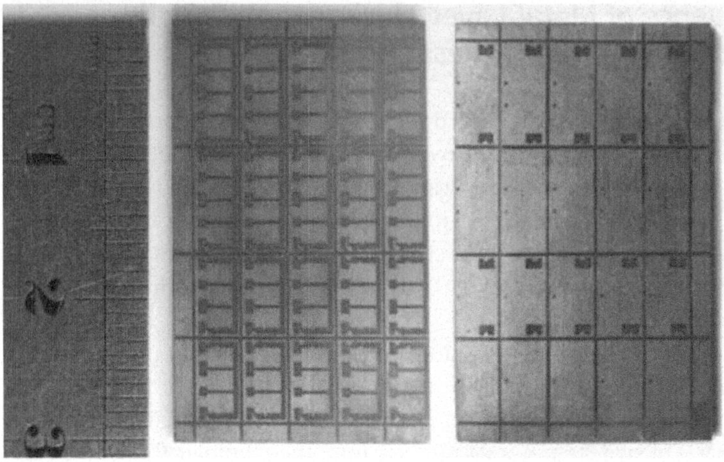

(a) (b)

Figure 12.30 Double-sided photolithography for stripline BPFs. (a) Front side of substrate. (b) Back side of substrate. *Source*: [22] / with permission of IEEE.

film technology is used. If laminated circuit board materials are used to create a stripline design, the t value can be up to 70 μm. Let us consider the influence of t on the frequency responses of two different BPFs. One has simple coupling coefficients, and the second uses mixed coupling coefficients.

12.7.1 BPF with Simple Couplings

Let us implement a ninth-order BPF with the following parameters: $f_0 = 2.4$ GHz, BW = 85 MHz (FBW = 0.0354), ripple $L_{Ar} = 0.1$ dB ($L_R = 16.426$ dB). It corresponds to g parameters (Table 2.6): $g_0 = q_{10} = 1$, $g_1 = g_9 = 1.1956$, $g_2 = g_8 = 1.4425$, $g_3 = g_7 = 2.1345$, $g_4 = g_6 = 1.6167$, $g_5 = 2.2053$. Substitution of these values and initial data into (3.8) leads to the coupling coefficients $K_{12} = K_{89} = 0.0270$, $K_{23} = K_{78} = 0.0202$, $K_{34} = K_{67} = 0.0191$, $K_{45} = K_{56} = 0.0187$, and the external quality factor of the end resonators $Q_{e1} = Q_{e9} = Q_e = 33.77$.

The filter will use combline topology (Figure 12.31a) with $\lambda/4$ resonators having the width $w = 4$ mm. To manufacture this filter, we use two alumina (Al_2O_3) substrates with $\varepsilon_r = 9.7$, $\tan\delta = 2 \times 10^{-4}$, and thickness $h = 2$ mm. Using EM simulator of Microwave Office software, the gaps between the resonators were determined, which correspond to the calculated values of the coupling coefficients $K_{i,i+1}$. The value of $Q_e = 33.77$ is realized by choosing the coordinates for connecting the loads to the end resonators. The photo of the filter (upside down) is shown in Figure 12.31b. Its size is 57.4 mm × 12 mm × 4 mm. Simulated frequency responses of this filter with $t = 1$ μm are shown in Figure 12.31c by dashed lines.

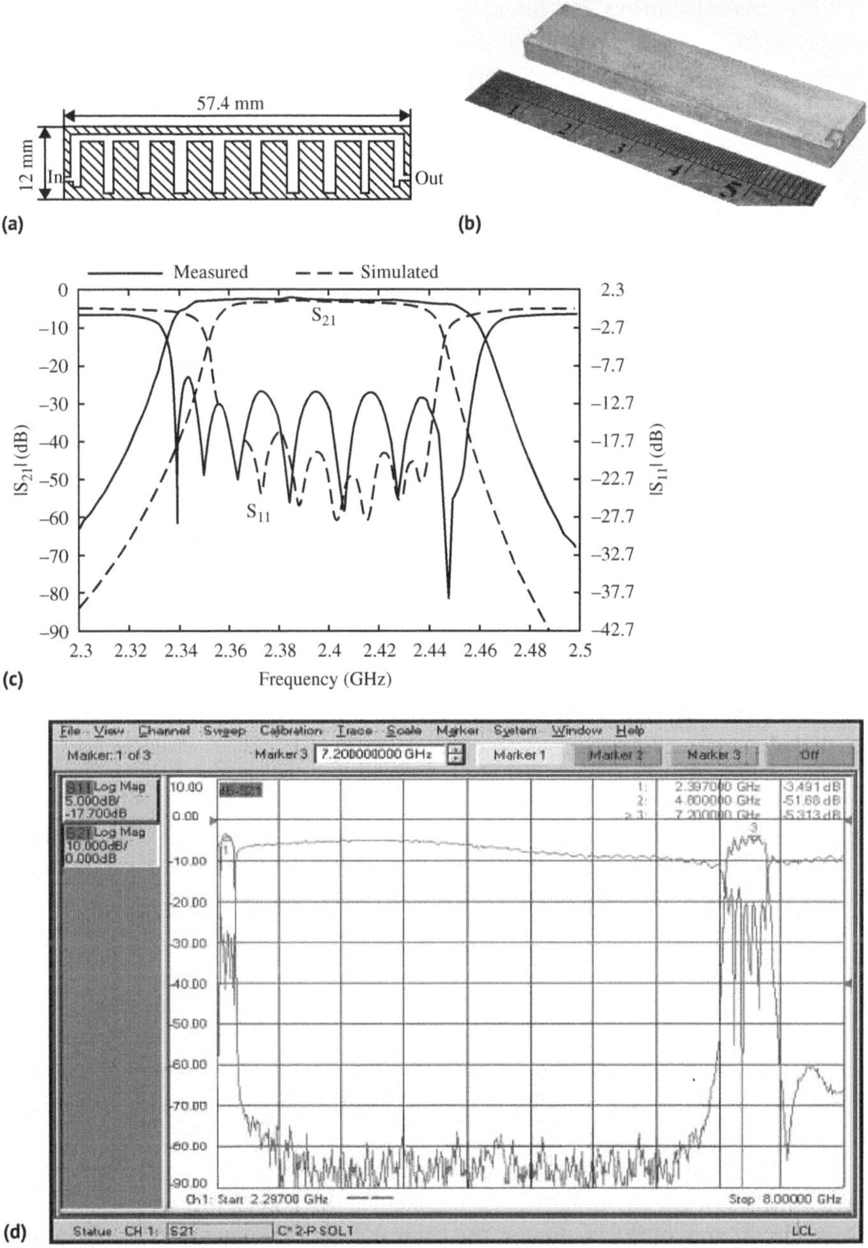

Figure 12.31 Stripline ninth-order BPF. (a) Topology. (b) Photograph. *Source*: [22] / with permission of IEEE. (c) Frequency responses in a narrow frequency band. (d) Measured frequency responses in a wide frequency band.

While manufacturing the filter, its substrates were coated with a layer of copper 8 μm thickness using thin film technology. After combining the patterns, both substrates were pressed against each other, and their ends were soldered together. The thickness of the central conductor is equal to $t = 16$ μm. The filter is designed for surface mounting. The measured filter characteristics include the following (Figure 12.31c): center frequency $f_0 = 2397$ MHz, the bandwidth by level 3 dB is $\mathrm{BW}_{3\ \mathrm{dB}} = 115$ MHz, insertion loss at the center frequency $L_{A0} = 2.3$ dB, return loss $L_R \geq 9.9$ dB, filter selectivity 63 dB ($f_0 \pm 100$ MHz). When using normalized frequency $\Omega = (f/f_0 - f_0/f)/\mathrm{FBW}$, this selectivity is 63 dB ($\pm j1.74$).

In addition to relatively low insertion loss and high selectivity, this filter has a high rejection level exceeding 80 dB (Figure 12.31d). Relatively small filter loss is due to stripline resonators' increased unloaded quality factor, which at $b \approx 4$ mm and $w = 4$ mm equals $Q_u \approx 590...610$.

However, the bandwidth of the manufactured filter has become wider, and the return loss level has decreased. These differences are because the simulation used the value $t = 1$ μm, but the manufactured filter had $t = 16$ μm. This fact leads to the need to study the influence of t on the coupling coefficient K between stripline resonators. We will conduct this study for a more general case, when the couplings between the resonators are mixed, and their signs alternate, providing quasi-elliptical frequency responses with two different-side TZs.

12.7.2 BPF with Mixed Couplings

We will return to the consideration of a stripline fourth-order BPF with mixed couplings of different signs, which uses two thermostable dielectric substrates with $\varepsilon_r = 92$, $h = 1$ mm, $\tan\delta = 0.0003$ (Figure 12.10). The filter size is 7.4 mm × 4.2 mm × 2 mm. The filter design was carried out for the initial data: $f_0 =$ 1835 MHz, BW = 90 MHz (FBW = 0.049), $L_{\mathrm{Ar}} = 0.2$ dB, $f_{z1,2} = (f_0 \pm 200$ MHz). Its coupling coefficients are $K_{12} = K_{34} = 0.0379$, $K_{23} = -0.0308$, and the external quality factor of the end resonator is $Q_e = 26.59$.

When performing EM simulation, the value $t = 16$ μm was used, which is the same as for the fabricated filter. Therefore, simulated and measured frequency responses agree with each other (Figure 12.10c). The measured filter characteristics: $f_0 = 1835$ MHz, BW = 90 MHz, $L_{A0} = 1.9$ dB, $L_R = 14.0$ dB, selectivity 45 dB ($f_0 \pm 135$ MHz), or 45 dB ($\pm j3.0$).

To implement two resonator pairs of the filter with $K > 0$ and $K < 0$ (Figure 12.32a), one asymmetric step-impedance resonator (SIR) was used. In the short circuit area of the resonator, the central strip, two rectangles of 0.8 mm × 0.4 mm and 0.6 mm × 0.2 mm were extracted. The gap between the resonators of each pair is 0.2 mm. Figure 12.32b, c shows the dependences of coupling coefficient K and its components K_m and K_e versus t for two pairs of resonators

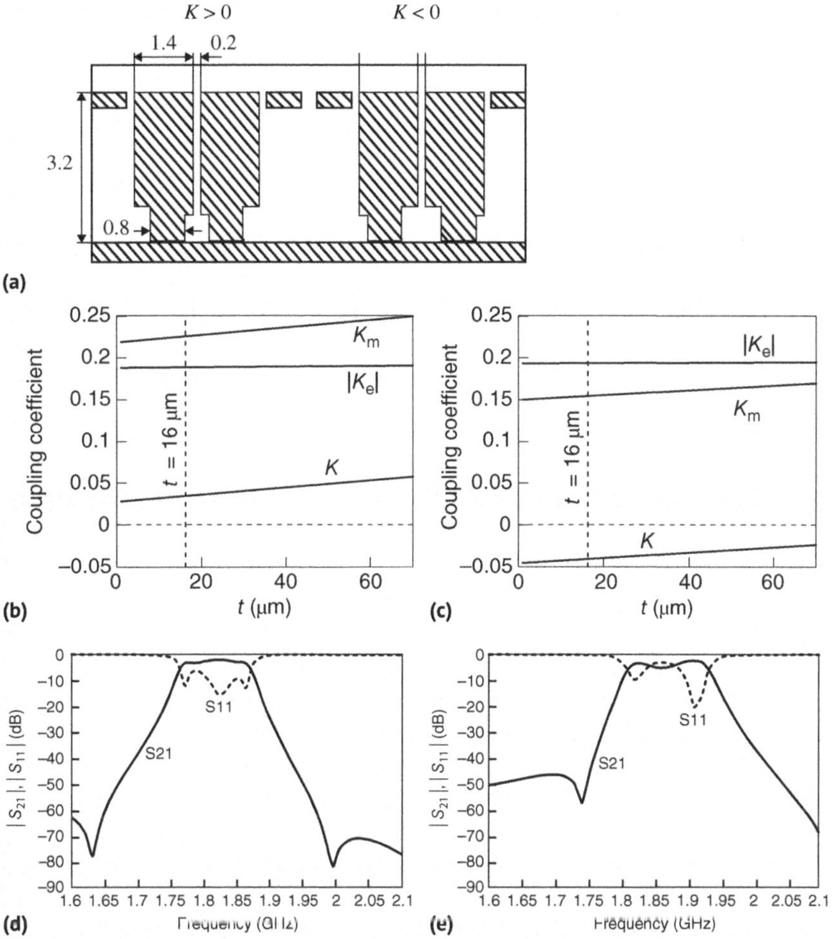

Figure 12.32 Influence of t on frequency responses of stripline mixed coupled BPF in Figure 12.10. (a) Pairs of mixed coupled resonators with $K > 0$ and $K < 0$. (b) Influence of t on positive coupling coefficient K and its components. (c) Influence of t on negative coupling coefficient K and its components. (d) Simulated frequency responses at $t = 1$ μm. (e) Simulated frequency responses at $t = 50$ μm.

shown in Figure 12.32a. The expressions (7.5) were used to determine K_m and K_e. As t increases, the magnetic component K_m increases linearly, while the electric component remains constant. This pattern leads to a linear change in the mixed coupling coefficient K. Moreover, the linearly changing functions $K(t)$ and $K_m(t)$ have the same slope. As t increases, the positive coupling K increases, whereas the modulus of negative coupling $|K|$ decreases.

This pattern affects the frequency responses of the stripline BPF with simple and mixed couplings. For a filter with simple couplings (Figure 12.31), a change in t is accompanied by a one-way change in $K_{i,i+1}$, which leads to a change in the bandwidth and a change in the level of return loss. At the same time, the "flat" top of the insertion losses and equal-wave ripples are preserved (Figure 12.31c). The required level of return losses can be restored by changing the coordinate of the load connection.

In a filter with mixed couplings of different signs, a change in t is accompanied by different changes in positive and negative couplings. This leads to the degradation of their frequency responses, which is illustrated in Figure 12.32d, e. This defect cannot be corrected. When simulating the frequency responses of a stripline BPF with mixed couplings, it is necessary to set the same value of t as that of the manufactured filter.

12.8 Stripline BPF with Two Different Dielectric Substrates

Two dielectric substrates with different thicknesses and dielectric constants ε_{r1} and ε_{r2} can be used in a stripline construction. In this case, the pure transverse electromagnetic wave (TEM) wave becomes a quasi-TEM wave, and cross-couplings occur between nonadjacent resonators, which is sometimes an advantage. Note that for stripline resonators with a uniform dielectric, the coupling coefficients do not depend on the dielectric constant ε_r (Section 12.5). They depend only on the geometrical parameters of the resonators. The features of the coupling coefficients between resonators in stripline structures with two different substrates, characterized by dielectric constants ε_{r1} and ε_{r2}, are of scientific and practical interest.

Figure 12.33 shows a pair of half-wave stripline resonators with different dielectric substrates, which are limited on all sides by metallized walls. The central conductors of the resonators are located between two dielectric substrates with ε_{r1} and ε_{r2} ($\varepsilon_{r1} < \varepsilon_{r2}$), which have the same thickness $h_1 = h_2 = 1$ mm. The resonators have the length $L = 10$ mm and width $w = 2$ mm and are separated by the gap $S = 4$ mm.

The coupling between these resonators is magnetic. Using EM simulation, it is possible to determine the even f_e and odd f_o modes of coupled oscillations and calculate the coupling coefficient K (3.10). Let $\varepsilon_{r1} = 5$, $\varepsilon_{r2} = 10$ ($\varepsilon_{r2}/\varepsilon_{r1} = 2$), then $f_e = 5.112$ GHz, $f_o = 6.332$ GHz, and $K = 0.211$. Let us change the dielectric constants $\varepsilon_{r1} = 10$, $\varepsilon_{r2} = 20$, leaving their ratio the same $\varepsilon_{r2}/\varepsilon_{r1} = 2$. In this case, the coupling frequencies decreased by $\sqrt{2}$ times: $f_e = 3.615$ GHz and $f_o = 4.478$ GHz. However, the coupling coefficient did not change, $K = 0.211$. The indicated values of the dielectric constants and the value of K are presented in Table 12.4

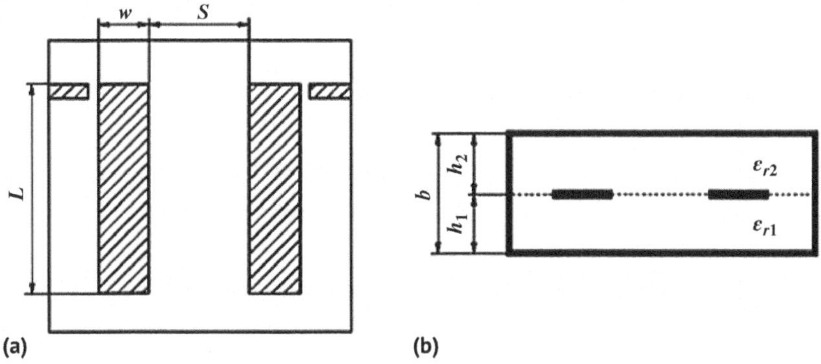

Figure 12.33 Pair of stripline $\lambda/2$ resonators. (a) Topology: $L = 10$ mm, $w = 2$ mm, $S = 4$ mm. (b) Cross-section: $h_1 = 1$ mm, $h_2 = 1$ mm, $b = 2$ mm.

Table 12.4 Coupling coefficients of stripline resonators at $\varepsilon_{r2}/\varepsilon_{r1} = 2$.

			$\varepsilon_{r2}/\varepsilon_{r1}$			
Note	**Coupling type**	**Figure**	**10/5**	**20/10**	**40/20**	**80/40**
K	Simple	12.33	0.211	0.211	0.211	0.215
K	Mixed	12.34	−0.0555	−0.0557	−0.0553	−0.0554
K_m	Mixed	12.34	0.0638	0.0640	0.0637	0.0638
K_e	Mixed	12.34	−0.1193	−0.1197	−0.1190	−0.1192

(line 2). This table also contains other values of K, corresponding to other dielectric constants characterized by the ratio $\varepsilon_{r2}/\varepsilon_{r1} = 2$. In all these cases, the coupling coefficient practically did not change.

Now consider the pair of SIRs shown in Figure 12.34. The resonators have the length $L = 10$ mm and width $w = 2$ mm and are separated by the $S = 0.2$ mm gap. The rectangles 2 mm × 1 mm are removed from the middle part of the central conductor of the resonators. The substrates have different thicknesses $h_1 = 1.5$ mm and $h_2 = 0.5$ mm.

The coupling between these resonators is mixed. Assuming $\varepsilon_{r1} = 5$, $\varepsilon_{r2} = 10$ ($\varepsilon_{r2}/\varepsilon_{r1} = 2$), by EM simulation, we determine the characteristic frequencies: $f_e = 4.348$ GHz, $f_o = 4.113$ GHz, $f_z = 3.094$ GHz. Using (3.10) and (7.5), we determine the coefficient of a mixed coupling and its components $K = -0.0555$, $K_m = 0.0638$, $K_e = -0.1193$. These values are given in Table 12.4, where they contain similar values for other dielectric constants with the ratio $\varepsilon_{r2}/\varepsilon_{r1} = 2$.

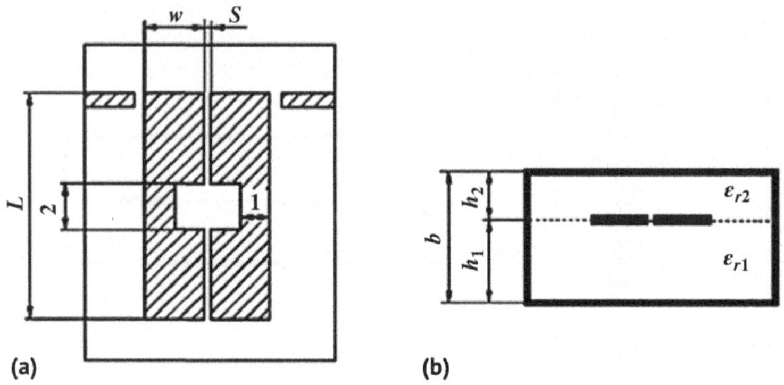

Figure 12.34 Pair of stripline SIRs. (a) Topology: $L = 10$ mm, $w = 2$ mm, $S = 0.2$ mm. (b) Cross-section: $h_1 = 1.5$ mm, $h_2 = 0.5$ mm, $b = 2$ mm.

Table 12.5 Influence of ratio $\varepsilon_{r2}/\varepsilon_{r1}$ on K for stripline $\lambda/2$ resonators.

		$\varepsilon_{r2}/\varepsilon_{r1}$	
Note	**Figure**	**10/5**	**20/5**
K	12.33	0.211	0.3597
K	12.34	-0.0555	-0.0339
K_m	12.34	0.0638	0.1496
K_e	12.34	-0.1193	-0.1835

Table 12.4 indicates that changes in ε_{r2} and ε_{r1}, while maintaining their ratio $\varepsilon_{r2}/\varepsilon_{r1}$ =const, do not lead to a change in K and components K_m and K_e. This pattern is valid in the more general case when a pair of stripline resonators (Figure 12.19c, d) is located asymmetrically relative to the vertical axis, and when the resonators are shifted relative to each other.

The influence of the ratio $\varepsilon_{r2}/\varepsilon_{r1}$ on the positive coupling coefficient between the stripline $\lambda/2$ resonators (Figure 12.33) is presented in Table 12.5 (line 1). The increase in the ratio $\varepsilon_{r2}/\varepsilon_{r1}$ leads to an increase in the value of K.

The values given in Table 12.5 (lines 2 and 4) characterize the influence of the ratio $\varepsilon_{r2}/\varepsilon_{r1}$ on the negative mixed coupling K and components between the pair of SIRs shown in Figure 12.34. An increase in the ratio $\varepsilon_{r2}/\varepsilon_{r1}$ increased the components K_m and $|K_e|$. However, the rise in K_m is more significant than $|K_e|$, which led to a decrease in the resulting value of $|K|$.

As a result, the following regularity was established. In stripline construction formed by two dielectric substrates with ε_{r2} and ε_{r1}, the electromagnetic coupling coefficient K between resonators depends only on the ratio $\varepsilon_{r2}/\varepsilon_{r1}$ and the geometric parameters of the structure. For a given ratio $\varepsilon_{r2}/\varepsilon_{r1}$, the value of K does not depend on the absolute values of ε_{r2} and ε_{r1}.

For stripline resonators with a uniform dielectric, the value of K does not depend on the dielectric constant ε_r at all (Section 12.5). Parameter K depends on the ratio $\varepsilon_{r2}/\varepsilon_{r1}$ in a two-layer stripline construction. The two-layer stripline BPFs, unlike microstrip BPFs, are self-shielded. At the same time, they have a magnetic interaction between nonadjacent resonators, which can be used in practice, particularly in delay line BPFs.

12.9 Stripline Delay Line Filter with Two Different Dielectric Substrates

Section 7.3 shows that the symmetric triplet BPF ($K_{12} = K_{23}$) with mixed cross-coupling is a delay line filter if its mixed coupling K_{13} is resonant, i.e., satisfies the condition $K_{m13} = |K_{e13}|$. In this case, its two TZs are equidistant on the σ-axis ($\sigma_2 = -\sigma_1$), which provides the filter with a flat group delay. The value of these TZs is determined by the equality:

$$\sigma_{1,2} = \frac{\pm m_{12}}{\sqrt{K_{m13}}} = \frac{\pm m_{12}}{\sqrt{|K_{e13}|}}, \tag{12.8}$$

where $m_{12} = K_{12}/\text{FBW}$.

The larger the magnetic component of the mixed cross-coupling K_{m13}, the closer the TZs σ_1 and σ_2 (12.8) are to each other, and the filter group delay is bigger (Table 7.1, position No. 1).

Let us set the following initial data for the third-order delay line filter: passband center frequency $f_0 = 2.15$ GHz, bandwidth BW = 100 MHz (FBW = 0.0465), ripple level $L_{\text{Ar}} = 0.1$ dB. The values $L_{\text{Ar}} = 0.1$ dB and $N = 3$ correspond to g parameters from Table 2.6: $g_1 = g_3 = 1.0315$, $g_2 = 1.474$, $g_4 = 1$. Using these parameters, FBW = 0.0465, and formulas (3.8), we define: $K_{12} = K_{23} = 0.038$ ($m_{12} = 0.817$), $Q_{e1} = Q_{e2} = Q_e = 22.2$.

Figure 12.35a, b shows the topology and photograph of a stripline BPF that implements the aforementioned parameters. This filter is based on two different dielectric substrates with $\varepsilon_{r1} = 9.7$ and $\varepsilon_{r2} = 38$ ($\varepsilon_{r2}/\varepsilon_{r1} = 3.918$), the thickness of which is $h_1 = h_2 = h = 1$ mm. The value of t is 16 μm. The filter size is 10 mm × 8.4 mm × 2 mm. Figure 12.35a presents the geometric dimensions of the filter elements. The required coupling coefficients between adjacent resonators are realized with the gap $S_1 = 0.6$ mm between resonators. The end resonators of the filter use the conductive connection with loads, which are presented by short segments. The required value of the external quality factor $Q_e = 22.2$ is provided by the corresponding connecting coordinate of a short segment to the resonator.

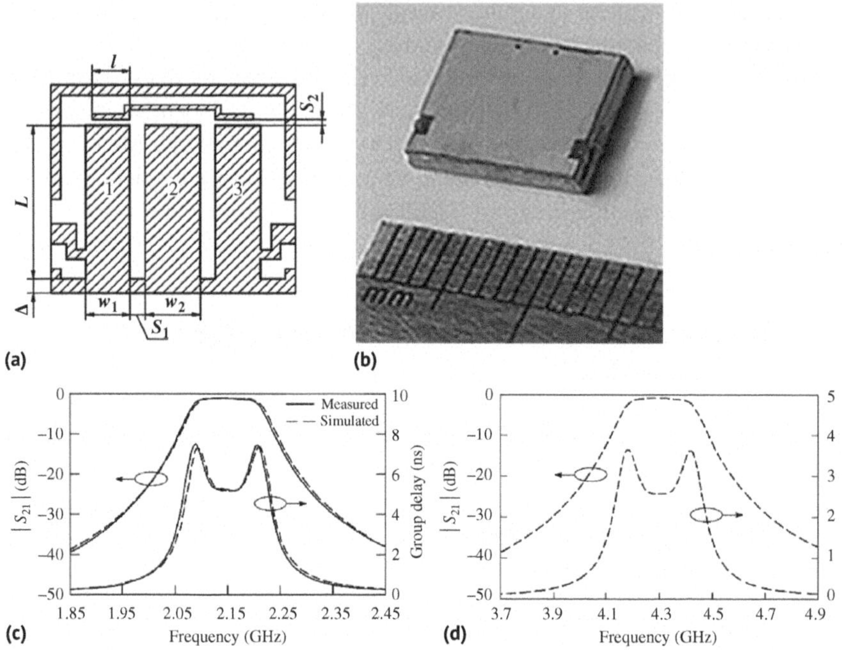

Figure 12.35 Stripline delay line filter with two different substrates. (a) Topology: $L = 6.2$ mm, $w_1 = 1.8$ mm, $w_2 = 2.2$ mm, $\Delta = 0.6$ mm, $S_1 = 0.6$ mm, $l = 1.5$ mm, $S_2 = 0.2$ mm. (b) Photograph. *Source*: [22] / with permission of IEEE. (c) Measured and simulated frequency responses for $\varepsilon_{r2}/\varepsilon_{r1} = 38/9.7 = 3.918$. (d) Simulated frequency responses for $\varepsilon_{r2}/\varepsilon_{r1} = 9.5/2.425 = 3.918$.

Between resonators 1 and 3, there is a mixed cross-coupling K_{13}. The measured magnetic component K_{m13} of this coupling coefficient is $K_{m13} = 0.0079$. The electric component K_{e13} is provided by a thin segment connecting the end resonators through the capacitive gaps S_2. The smaller S_2 and the greater the length l of the thin segment above each resonator, the greater the value of $|K_{e13}|$. Condition $K_{m13} = |K_{e13}|$, under which the electrical component compensates the magnetic component of mixed coupling, is provided for the values $S_2 = 0.2$ mm and $l = 1.5$ mm.

The frequency responses of the considered filter are shown in Figure 12.35c. The measured values are as follows: $f_0 = 2.15$ GHz, BW = 100 MHz, $L_{A0} = 1.2$ dB, $L_R > 16.5$ dB, group delay $\tau = 5.3 \pm 0.1$ ns in the frequency band of 50 MHz. Thus, the delay line filter can be implemented using a stripline construction with two different dielectric substrates and the simplest and most compact third-order combline structure with $\lambda/4$ resonators.

The pattern established earlier allows us to transfer the frequency response to the region of higher or lower frequencies for the given BPF topology. For the filter under consideration, let us decrease the dielectric constants of the used substrates by four times ($\varepsilon_{r1} = 2.425$, $\varepsilon_{r2} = 9.5$), while maintaining their ratio $\varepsilon_{r2}/\varepsilon_{r1} = 3.918$. The simulated frequency responses for this case are presented in Figure 12.35d. They are characterized by the values: $f_0 = 4.3$ GHz, BW = 200 MHz (FBW = 0.0465), group delay $\tau = 2.65 \pm 0.05$ ns in the frequency band of 100 MHz. The values of f_0 and BW increased by 2, and the group delay τ decreased by two times in the frequency band, which is two times larger than the original one.

12.10 Stripline Duplexers

Duplexers are an integral part of mobile communication systems, including cellular telephony. They separate signals in the receiving and transmitting channels [7]. The receiving channel is designated RX, and the transmitting channel is TX. BPFs are most often used in the RX and TX channels. Examples of two stripline duplexers are considered as follows.

12.10.1 Stripline Duplexer on Alumina Substrates

The considered duplexer (Figure 12.36) contains two fifth-order stripline combline filters in the transmission channel TX (2300–2370 MHz, $f_{0T} = 2335$ MHz) and receiving channel RX (2510–2580 MHz, $f_{0R} = 2545$ MHz). A frequency interval of 140 MHz separates the TX and RX channels, and the attenuation of each BPF at the frequencies of the adjacent channel is not less than 60 dB.

The BPFs used in the duplexer have fractional bandwidths $FBW_{TX} = 0.03$ and $FBW_{RX} = 0.0275$. The duplexer uses BPFs with quarter-wave resonators, which have a Chebyshev attenuation characteristic and are not characterized by

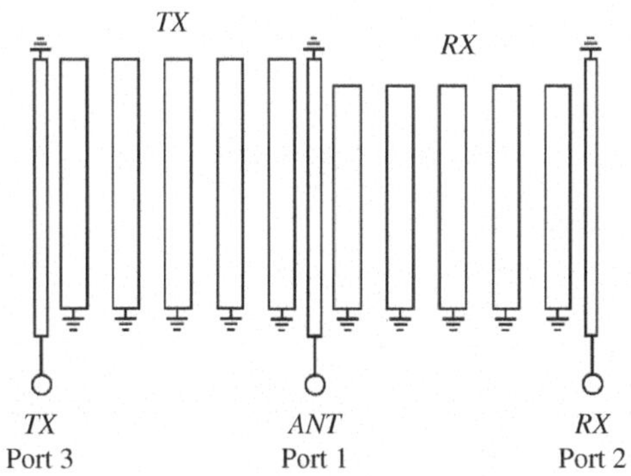

Figure 12.36 Schematic of duplexer with combline BPFs.

increased one-sided selectivity. To ensure the required attenuation (60 dB) at adjacent channel frequencies, the TX and RX filters must contain five resonators each, $n = 5$ [7]. We take the filter pulsation value to be $L_{Ar} = 0.1$ dB. For $L_{Ar} = 0.1$ dB and $n = 5$, we find the parameters of the low-frequency prototype from Table 2.6: $g_0 = g_6 = 1$; $g_1 = g_5 = 1.1468$; $g_2 = g_4 = 1.3712$; $g_3 = 1.9750$. Substituting these values and FBW into (3.8), we obtain for the TX filter: $K_{12} = K_{45} = 0.0239$, $K_{23} = K_{34} = 0.0182$, $Q_e = 38.22$, and for the RX filter: $K_{12} = K_{45} = 0.0219$, $K_{23} = K_{34} = 0.0167$, $Q_e = 41.71$.

The duplexer construction [28] is formed by connecting two dielectric substrates from alumina ($\varepsilon_r = 9.7$, $\tan\delta = 0.0002$) with $h = 2$ mm thickness. The vacuum deposition of copper was used to manufacture the duplexer, with a deposition thickness of 8 μm.

The topology of the central conductors of the duplexer is shown in Figure 12.37a, position 1, where the dimensions are given in millimeter. Position 2 shows the topology of the filter from the outside, which is mounted on the printed circuit board. The contact pads of position 2 are connected to three terminals of the duplexer, located in its "inner part," position 1. The connection is made by conductive strips located on the ends, which are not shown. The external appearance of the duplexer (from the back side) is shown in Figure 12.37b; its size is 57.5 mm × 11.8 mm × 4 mm. When assembling the filter, the substrates with conductive patterns are pressed against each other, and their metallized ends are

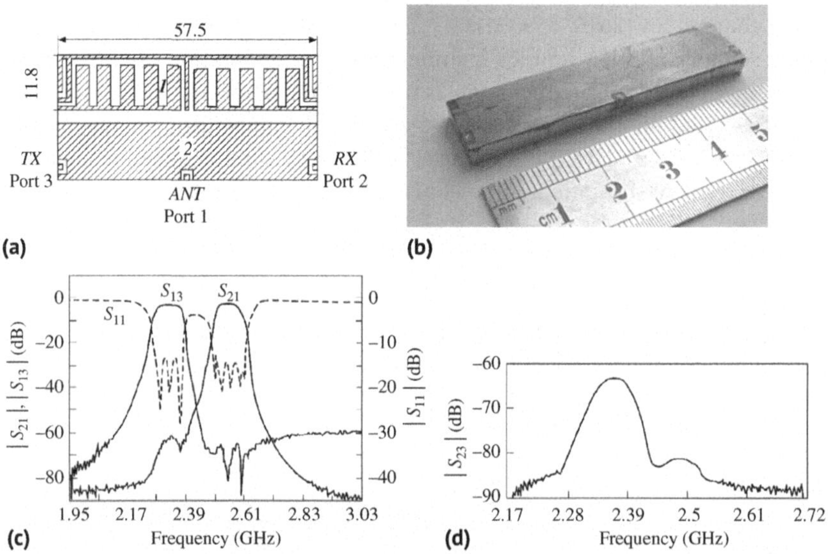

Figure 12.37 Stripline duplexer on alumina substrates. (a) Topology. (b) Photograph. *Source*: [28] / with permission of Springer Nature. (c) Measured insertion loss (S_{21}, S_{13}) and return loss (S_{11}). (d) Measured isolation (S_{23}).

soldered together. A metal strip 0.8 mm wide along the perimeter of the filter prevents tin solder from getting into the inner part of the filter during its assembly. The numbering of the duplexer ports in Figure 12.37a corresponds to the numbering in Figure 12.36.

The width of all duplexer resonators is $w = 3$ mm. The length of the TX filter resonators is 8.85 mm, and the RX filter resonators are 8.05 mm. The gaps between the TX filter resonators, which provide the aforementioned coupling coefficients, are $S_{12} = S_{45} = 1.8$ mm and $S_{23} = S_{34} = 2.3$ mm. The gaps between the RX filter resonators have the following values: $S_{12} = S_{45} = 1.9$ mm and $S_{23} = S_{34} = 2.35$ mm. To connect the end resonators of the TZ and RX filters to the loads, line segments with a width of 0.8 mm are used, one end of which is short-circuited. The gap value S between these segments and the filters' end resonators determines the end resonators' external quality factor Q_e. The value of the external quality factor $Q_e = 38.22$ of the end resonators in the TX filter is provided with the value $S = 1.1$ mm. The required value of external quality factor $Q_e = 41.71$ of the end resonators of the RX filter is provided with the same S value.

Figure 12.37c, d shows the measured frequency responses of the duplexer. The minimum losses in the frequency band $L_{A0} = 2.5$ dB of the TX and RX channels are the same. The maximum losses $L_{A\max} = 3.0$ dB are also the same. The ripple corresponds to the difference between the specified values and equals 0.5 dB. The return loss exceeds 13 dB.

An essential characteristic of the duplexer is the isolation between TX and RX channels. The transmitter signal enters port 3, passes the TX filter, and is fed to the antenna (port 1). Part of this signal enters the RX filter input, passes through it, and reaches port 2, which is connected to the receiver. The isolation between the TX and RX channels of the duplexer is described by the scattering parameter S_{23}. Figure 12.37d shows the measured isolation curve. The stripline duplexer under consideration has an isolation corresponding to 60 dB. The indicated values are systematized in Table 12.6.

Alumina has a relatively high thermal conductivity coefficient [25]: $K_T = 31.5$ W/m°C. The higher the K_T, the more intensely the heat is removed from the filter inside to the surface and dissipated. This allows the duplexer to operate at an increased power level.

Table 12.6 Specification of stripline duplexer on alumina substrates.

Frequency, MHz	BW, MHz	$L_{A\max}$, dB	L_{Ar}, dB	L_R, dB	Attenuation, dB (MHz)	Size, mm³
TX, 2300–2370	70	3.0	0.5	>13	60 (2510–2580)	57.5 × 11.8 × 4
RX, 2510–2580	70	3.0	0.5	>13	65 (2300–2370)	

12.10.2 Stripline Duplexer on High Dielectric Constant Substrates

Let us illustrate the use of thin ($b = 1$ mm) combline stripline structures using the example of a duplexer [29] for the W-CDMA (Wideband Code Division Multiple Access) communication system. Its TX (1920–1980 MHz) and RX (2110–2170 MHz) transmission channels have a width of 60 MHz and are separated by a frequency interval of 130 MHz.

The duplexer is made of two ceramic substrates with a dielectric constant $\varepsilon_r = 92$, the thickness of which is $h = 0.5$ mm. Figure 12.38a shows the topology of a stripline duplexer, which uses third-order BPFs, the end resonators connected to loads by short-circuited coupling at one end of the strips. The width of the resonators in the TX filter is 1.2 mm, and the RX channel filter has resonators 0.8 mm wide; the coupling segments are 0.4 mm wide. The size of the duplexer is 9.4 mm × 4.4 mm × 1 mm.

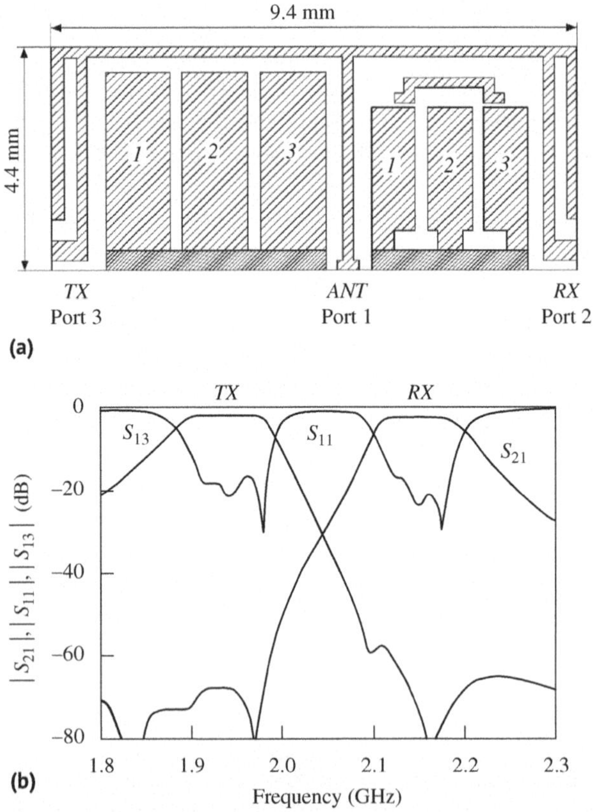

Figure 12.38 Stripline duplexer on high dielectric constant substrates. (a) Topology. (b) Simulated frequency responses.

Table 12.7 Specification of a thin (1 mm) stripline duplexer on dielectric substrates with $\varepsilon_r = 92$.

Frequency, MHz	BW, MHz	L_{Amax}, dB	L_{Ar}, dB	L_R, dB	Attenuation, dB (MHz)	Size, mm³
TX, 1920–1980	60	2.4	0.5	>14	55 (2110–2170)	$9.4 \times 4.4 \times 1$
RX, 2110–2170	60	3.2	0.6	>14	60 (1920–1980)	

The TX and RX filters have increased one-sided selectivity. The strip at the base of the TX filter resonators provides a positive cross-coupling K_{13} between resonators 1 and 3, which makes the right slope of the filter steeper. The RX channel uses SIRs that are all connected by mixed couplings. A thin strip connecting resonators 1 and 3 through capacitive gaps forms the cross-coupling K_{13}. As a result, the left slope of the filter is steeper.

Figure 12.38b shows the simulated frequency responses of the duplexer, and Table 12.7 presents its specification.

As a result of computer simulation, it was found that using thin substrates ($h = 0.5$ mm) with high dielectric constant ($\varepsilon_r = 92$) allows the design of miniature stripline duplexers.

12.11 Properties of Stripline BPFs with Two Dielectric Substrates and Metal Layer

It is advisable to systematize the properties of the stripline BPFs discussed earlier.

1. The stripline filters are designed to be fully shielded and can be manufactured using group technology (Figure 12.30).
2. By using substrates with different dielectric constants ($\varepsilon_r = 2.2$–100), the filters can operate in a very wide range, including the upper part of the meter range and the lower part of the millimeter range, up to frequencies of 120 GHz (Figure 12.28).
3. It is shown that there is electromagnetic coupling between the stripline resonators in combline ($\lambda/4$) and pseudocombline ($\lambda/2$) structures (Figures 12.2, 12.24), and they are BPFs.
4. The coupling coefficients between stripline uniform $\lambda/4$ and $\lambda/2$ resonators are positive and can vary within relatively wide limits. When using stripline SIRs, the coupling coefficients vary within wider limits (Figures 12.5, 12.25, 12.26). They are mixed and can be both positive and negative.
5. The use of mixed coupling coefficients generating TZs in stripline BPFs allows the implementation of a significant variety of frequency responses,

including quasi-elliptic frequency response (Figures 12.10, 12.27), flat group delay (Figure 12.35), and also to suppress multiple harmonics of the main signal (Figures 12.11–12.13).

6. The nature of the change in the SIRs coupling coefficients at higher resonant frequencies (Figure 12.14) and their passage through the zero value allows the implementation of stripline BPFs with a wide rejection band (Figure 12.18).
7. The coupling coefficient K between resonators in an isolated stripline structure with a homogeneous dielectric is invariant with respect to the permittivity constant ε_r. The value of K depends only on the geometric parameters of the stripline structure. The permittivity constant ε_r only transfers the coupling frequencies on the frequency axis (Figure 12.21), while maintaining the relationship between them. If there is a stripline BPF with a specific frequency response, it can be transferred in frequency by changing the value of ε_r.
8. In stripline construction formed by two dielectric substrates with ε_{r2} and ε_{r1}, the electromagnetic coupling coefficient K between resonators depends only on the ratio $\varepsilon_{r2}/\varepsilon_{r1}$ and geometric parameters of the structure. For a given ratio $\varepsilon_{r2}/\varepsilon_{r1}$, the value of K does not depend on the absolute values of ε_{r2} and ε_{r1}. When using two substrates with different dielectric constants, an electromagnetic interaction of a magnetic nature occurs between nonadjacent resonators, which expands the possibilities of forming different BPF frequency responses (Figure 12.35).
9. The coupling coefficients between stripline resonators depend on the thickness t of the metal layer. As t increases, the positive coupling K increases (Figure 12.32b), whereas the module of negative coupling $|K|$ decreases (Figure 12.32c). In a filter with mixed couplings of different signs, a change in t is accompanied by different changes in positive and negative couplings. This leads to the degradation of their frequency responses (Figure 12.32d, e).
10. High dielectric constant substrates ($\varepsilon_r = 92$) allow the design of miniature stripline filters and duplexers (Figures 12.10 and 12.38).
11. The use of thin substrates ($h = 0.5$ mm) results in thin stripline BPFs ($b = 1$ mm) that are completely shielded, and their resonators have an unloaded quality factor Q_u sufficient to realize relatively narrow passbands (Figure 12.16).
12. Using alumina substrates with a high thermal conductivity coefficient ($K_T = 31.5$ W/m°C) allows the design of stripline filters and duplexers with increased handling power.

References

1 M. Makimoto, S. Yamashita, *Microwave Resonators and Filters for Wireless Communication*, vol. 4. Berlin, Heidelberg: Springer Berlin Heidelberg, 2001, https://doi.org/10.1007/978-3-662-04325-7.

2 A. Fukasawa, Analysis and composition of a new microwave filter configuration with inhomogeneous dielectric medium, *IEEE Trans. Microw. Theory Tech.*, vol. 30, no. 9, pp. 1367–1375, 1982, https://doi.org/10.1109/TMTT.1982.1131262.

3 D. Morgan, *Surface Acoustic Wave Filters*. Academic Press, 2007, https://www.elsevier.com/books/surface-acoustic-wave-filters/morgan/978-0-12-372537-0.

4 L. K. Yeung, K.-L. Wu, Y. E. Wang, Low-temperature cofired ceramic LC filters for RF applications [Applications Notes], *IEEE Microw. Mag.*, vol. 9, no. 5, pp. 118–128, 2008, https://doi.org/10.1109/MMM.2008.927634.

5 T. Ishizaki, M. Fujita, H. Kagata, T. Uwano, H. Miyake, A very small dielectric planar filter for portable telephones, in *1993 IEEE MTT-S International Microwave Symposium Digest*, 1993, pp. 177–180, https://doi.org/10.1109/MWSYM.1993.276916.

6 J. Bolljahn, G. Matthaei, A study of the phase and filter properties of arrays of parallel conductors between ground planes, *Proc. IRE*, vol. 50, no. 3, pp. 299–311, 1962, https://doi.org/10.1109/JRPROC.1962.288322.

7 G. L. Matthaei, L. Young, E. M. T. Jones, *Microwave Filters, Impedance-Matching Networks, and Coupling Structures*. New York: Artech House Books, 1980.

8 L. Dworsky, Stripline filters - an overview, in *37th Annual Symposium on Frequency Control*, 1983.

9 L. Szydlowski, A. Lamecki, M. Mrozowski, Coupled-resonator filters with frequency-dependent couplings: Coupling matrix synthesis, *IEEE Microw. Wirel. Components Lett.*, vol. 22, no. 6, pp. 312–314, 2012, https://doi.org/10.1109/LMWC.2012.2197386.

10 R. Levy, New cascaded trisections with resonant cross-couplings (CTR Sections) applied to the design of optimal filters, in *IEEE MTT-S International Microwave Symposium Digest*, 2004, vol. 2 pp. 447–450, https://doi.org/10.1109/mwsym.2004.1336007.

11 H. Wang, Q.-X. Chu, An inline coaxial quasi-elliptic filter with controllable mixed electric and magnetic coupling, *IEEE Trans. Microw. Theory Tech.*, vol. 57, no. 3, pp. 667–673, 2009, https://doi.org/10.1109/TMTT.2009.2013290.

12 M. Hoft, T. Shimamura, Design of symmetric trisection filters for compact low-temperature Co-fired ceramic realization, *IEEE Trans. Microw. Theory Tech.*, vol. 58, no. 1, pp. 165–175, 2010, https://doi.org/10.1109/TMTT.2009.2035870.

13 W. Shen, L.-S. Wu, X.-W. Sun, W.-Y. Yin, J.-F. Mao, Novel substrate integrated waveguide filters with mixed cross coupling (MCC), *IEEE Microw. Wirel. Components Lett.*, vol. 19, no. 11, pp. 701–703, 2009, https://doi.org/10.1109/LMWC.2009.2032007.

14 K. Gong, W. Hong, Y. Zhang, P. Chen, C. J. You, Substrate integrated waveguide quasi-elliptic filters with controllable electric and magnetic mixed coupling, *IEEE Trans. Microw. Theory Tech.*, vol. 60, no. 10, pp. 3071–3078, 2012, https://doi.org/10.1109/TMTT.2012.2209437.

15 L. Szydlowski, A. Jedrzejewski, M. Mrozowski, A trisection filter design with negative slope of frequency-dependent crosscoupling implemented in substrate integrated waveguide (SIW), *IEEE Microw. Wirel. Components Lett.*, vol. 23, no. 9, pp. 456–458, 2013, https://doi.org/10.1109/LMWC.2013.2272611.

16 K. Ma, J.-G. Ma, K. S. Yeo, M. A. Do, A compact size coupling controllable filter with separate electric and magnetic coupling paths, *IEEE Trans. Microw. Theory Tech.*, vol. 54, no. 3, pp. 1113–1119, 2006, https://doi.org/10.1109/TMTT.2005.864118.

17 Q.-X. Chu, H. Wang, A compact open-loop filter with mixed electric and magnetic coupling, *IEEE Trans. Microw. Theory Tech.*, vol. 56, no. 2, pp. 431–439, 2008, https://doi.org/10.1109/TMTT.2007.914642.

18 F. Zhu, W. Hong, J.-X. Chen, K. Wu, Quarter-wavelength stepped-impedance resonator filter with mixed electric and magnetic coupling, *IEEE Microw. Wirel. Components Lett.*, vol. 24, no. 2, pp. 90–92, 2014, https://doi.org/10.1109/LMWC.2013.2290225.

19 A. Zakharov, S. Rozenko, M. Ilchenko, Two types of trisection bandpass filters with mixed cross-coupling, *IEEE Microw. Wirel. Components Lett.*, vol. 28, no. 7, pp. 585–587, 2018, https://doi.org/10.1109/LMWC.2018.2837905.

20 A. Zakharov, M. Ilchenko, Trisection microstrip delay line filter with mixed cross-coupling, *IEEE Microw. Wirel. Components Lett.*, vol. 27, no. 12, pp. 1083–1085, 2017, https://doi.org/10.1109/LMWC.2017.2759724.

21 A. Zakharov, M. Ilchenko, Coupling coefficients between resonators in stripline combline and pseudocombline bandpass filters, *IEEE Trans. Microw. Theory Tech.*, vol. 68, no. 7, pp. 2679–2690, 2020, https://doi.org/10.1109/TMTT.2020.2988866.

22 A. Zakharov, S. Litvintsev, Design features of fully shielded stripline bandpass filters affecting their frequency responses, *IEEE Trans. Components, Packag. Manuf. Technol.*, vol. 15, no. 6, pp. 1284–1291, 2025, https://doi.org/10.1109/TCPMT.2025.3566991.

23 A. V. Zakharov, M. Y. Ilchenko, V. Y. Karnauh, L. S. Pinchuk, Stripline bandpass filters with step-impedance resonators, *Radioelectron. Commun. Syst.*, vol. 54, no. 3, pp. 163–169, 2011, https://doi.org/10.3103/S0735272711030071.

24 Y. C. Li, X. Y. Zhang, Q. Xue, Bandpass filter using discriminating coupling for extended out-of-band suppression, *IEEE Microw. Wirel. Components Lett.*, vol. 20, no. 7, pp. 369–371, 2010, https://doi.org/10.1109/LMWC.2010.2049425.

25 J.-S. Hong, *Microstrip Filters for RF/Microwave Applications*, 2nd ed. New Jersey: Wiley, 2011, https://doi.org/10.1002/9780470937297.

26 A. V. Zakharov, M. E. Il'chenko, Millimiter-wave stripline bandpass filters, *J. Commun. Technol. Electron.*, vol. 62, no. 10, pp. 1184–1192, 2017, https://doi.org/10.1134/S1064226917100199.

27 Rogers Corporation RT/duroid® 5870 /5880, high frequency laminates datasheet. http://www.rogerscorp.com.

28 A. V. Zakharov, S. A. Rozenko, L. S. Pinchuk, Symmetric stripline duplexer, *Radioelectron. Commun. Syst.*, vol. 60, no. 11, pp. 495–502, 2017, https://doi.org/10.3103/S0735272717110048.

29 A. V. Zakharov, Stripline combline filters on substrates designed on high-permittivity ceramic materials, *J. Commun. Technol. Electron.*, vol. 58, no. 3, pp. 265–272, 2013, https://doi.org/10.1134/S1064226913030145.

13

Microstrip Resonators and Bandpass Filters

Microstrip bandpass filters (BPFs) are the most popular and are the subject of many publications in scientific and technical journals. The monograph considers these filters most fully and systematically [1]. Some new aspects concerning microstrip resonators and filters will be presented in the following sections, complementing the material presented in Ref. [1].

13.1 Coupling Between Stepped-impedance Microstrip Resonators

The resonators of the most compact BPFs are located next to each other. To change the coupling coefficient between two resonators with a fixed gap S between them, it is necessary to change the shape of the resonators. The coupling coefficients will be mixed in this case, containing magnetic and electric components $K = K_m + K_e$. The influence of the shape of closely spaced microstrip SIRs on the coupling coefficient between them is most fully described in Ref. [2]. Note that in microstrip designs, it is insufficient to control the interaction of only closely spaced resonators since there is a "strong" interaction between nonadjacent resonators.

To have a chance to create microstrip filters whose resonators are closely spaced and exhibit various characteristics, two issues deserve answers:

i. What is the range of variations in the coupling coefficient of adjacent resonators?

ii. Is it possible to control the interaction of nonadjacent resonators? If so, what is the range?

Let us consider these issues one after another.

13.1.1 Coupling Between Closely Spaced SIRs

The electromagnetic coupling coefficient K of two resonators can be calculated from the formula (3.10), in which f_e and f_o are the frequencies of even and odd modes in the system of two coupled resonators. The values of f_e and f_o can be determined via computer simulation of the frequency response of two coupled resonators if they are weakly loaded at the input and output. In this case, frequencies f_e and f_o correspond to pronounced peaks in the transfer function.

Figure 13.1 illustrates the dependence between the coupling coefficient and the shape of closely mounted stepped-impedance resonators (SIRs) in the microstrip embodiment with open-circuited ends. These dependences correspond to four typical modifications of their shapes. Frequency responses were simulated via the Microwave Office (AWR) computer software. The simulation was performed under the following assumptions: the substrate thickness $h = 1$ mm, the relative

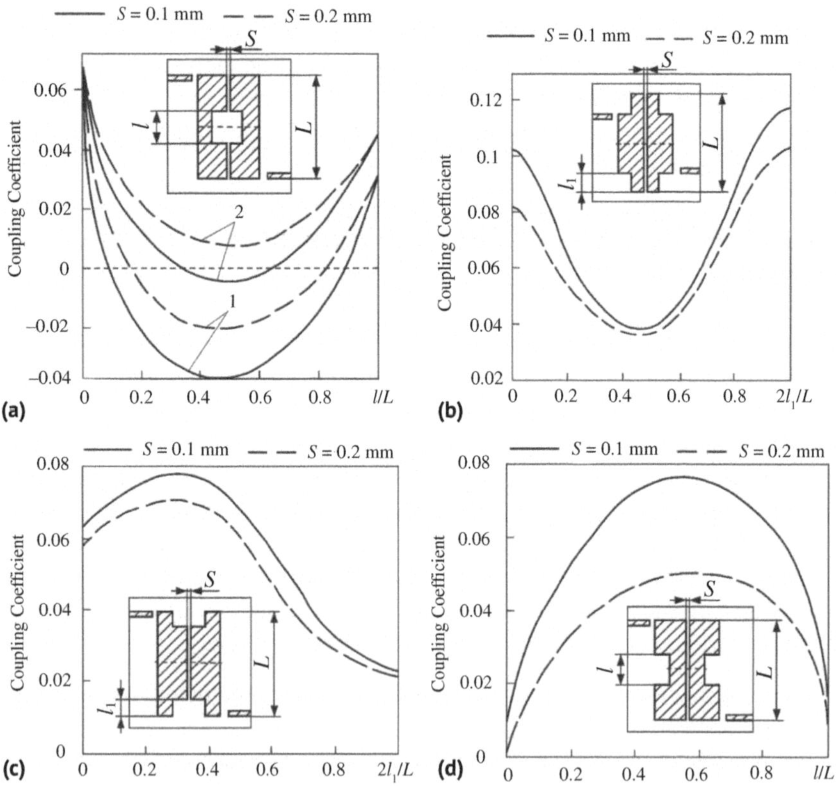

Figure 13.1 Dependences between the coupling coefficients on the shape of microstrip SIRs. (a) Narrower conductors are situated in the middle of the external part of the resonators (*1* for $w_2 = 1$ mm and *2* for $w_2 = 1.5$ mm). (b) Narrower conductors are placed in the internal parts of resonators, beginning from their ends. (c) Narrower conductors are observed in the external parts of resonators, beginning from their ends. (d) Narrower conductors are located in the middle of the internal parts of resonators.

permittivity of the substrate material is $\varepsilon_r = 92$, the resonator length $L = 10$ mm, the largest possible width of central resonator conductors is $w_1 = 2$ mm, and the metallic shield height above a substrate is $H = 10$ mm. In the case of variations in resonator shape according to Figure 13.1a, the dependences were constructed at two values of the width of narrow resonator conductors: $w_2 = 1$ mm (curves *1*) and $w_2 = 1.5$ mm (curves *2*). In all other cases (Figure 13.1b–d), the width of narrow conductors is $w_2 = 1$ mm. In all cases, the resonant frequency of the resonators f_0 varied within the range of 1.5–2.0 GHz.

The length of the narrow part of a resonator, which varied symmetrically about its midline, is designated by l in Figure 13.1a. Coupling coefficient K decreases with increasing l and vanishes at a specific l_0. In this case, the magnetic and electric components of electromagnetic coupling compensate each other. Thus, quantity l_0 indicates the state of "equilibrium" between interactions opposite in sign. This mixed coupling is called resonant coupling [3]. The functions $K = K(l/L)$ (Figure 13.1a) are concave downward, and their minima are observed at $l/L \approx 1/2$. Their curves intersect the abscissa axis (horizontal dashed line) twice. A maximum change in K occurs when the gap between resonators is 0.1 mm and the width of narrow conductors is $w_2 = 1$ mm: $-0.095 \leq K \leq 0.161$.

The length of narrow resonator conductors is designated by l_1 in Figure 13.1b. They are located in the internal part of resonators, beginning with their ends. The values are laid along the abscissa axis. As in the preceding case, the dependencies are concave downward and have minima at $2l_1/L \approx 1/2$. Moreover, there is no crossing between these curves and the abscissa axis. The influence of variations in the shape of SIRs, illustrated in Figure 13.1c, d, does not require additional explanations. The same can be said about the designations used. Here, the dependences of the coupling coefficients K are concave upward.

The discussed stepped-impedance microstrip resonators with open-circuited ends (Figure 13.1) are symmetric about the horizontal axis, i.e., the dashed line passing through their middles. If cutting and shorting of resonators are performed along this axis, new pairs of resonators, which are shorted at one end, will have analogous variations in the coupling coefficient.

The negative coupling can be increased by consolidating the effects intrinsic to the resonators in Figure 13.1a, b. This leads to two microstrip resonators with conductor widths of 1 mm, the shape of which is presented in Figure 13.2a. Such a pair of resonators has the coupling coefficient $K = -0.138$.

Owing to the consolidation of the effects attainable in the resonators in Figure 13.1c, d, a pair of resonators (Figure 13.2b) reaches the increased level of positive coupling ($K = 0.256$). If the resonators in Figure 13.2 are cut and shorted along the dashed line, pairs of resonators with one short-circuited end are formed. They will also have significant coefficients of inductive or capacitive coupling.

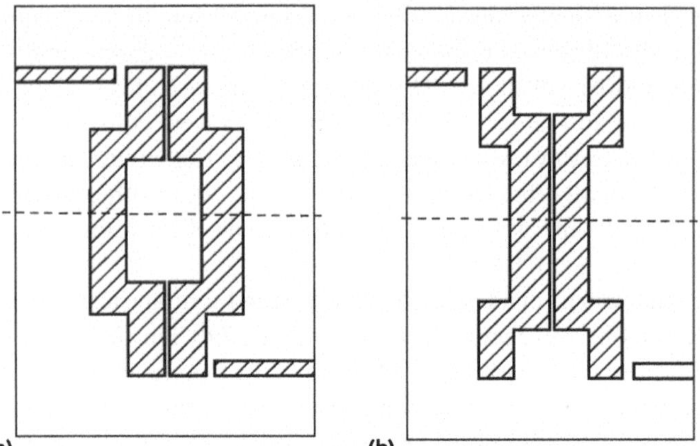

Figure 13.2 Pairs of microstrip SIRs with extreme coupling coefficients. (a) $K = -0.138$. (b) $K = 0.256$.

From the analysis of the coupling coefficients of microstrip SIRs, it has been established that the values of K can vary appreciably in the presence of a fixed gap between them: $-0.138 \leq K \leq 0.256$. This inference is of great practical significance.

Note that the curves in Figure 13.1a, b are similar to the curves for stripline SIRs in Figure 12.5a, b, respectively. The difference is that they are raised by some amount corresponding to the difference in the coupling coefficients of uniform resonators. If a horizontal line corresponding to a given value of K is drawn in Figure 13.1, as in Figure 12.5, it will intersect the curves at several points. This means that a given K can be implemented using SIRs of different shapes, which is ensured by various values of components K_m and K_e, and other positions of the transmission zero (TZ).

Mixed coupling is also present in pairs of microstrip resonators, the pairs of which are shown in Figure 12.8. The mixed coupling between SIRs with a moving coupling strip (Figure 12.8e), which introduces a magnetic component, is considered in Ref. [4] in detail.

13.1.2 Coupling Between Nonadjacent Resonators

The coupling coefficient of nonadjacent resonators R_1 can be changed using a "connection" segment (CS) arranged "near" these resonators (Figure 13.3a, b). This coupling coefficient is also mixed. In computer simulation, it was assumed that its width is 0.2 mm, the gap between these connection segments and resonators is 0.05 mm, extension l over resonators R_1 is a variable quantity ($0 \leq l \leq 2$ mm), and the distance between resonators R_1 is 2.4 mm. The dependences of the coupling coefficient on l (Figure 13.3c) have been constructed in

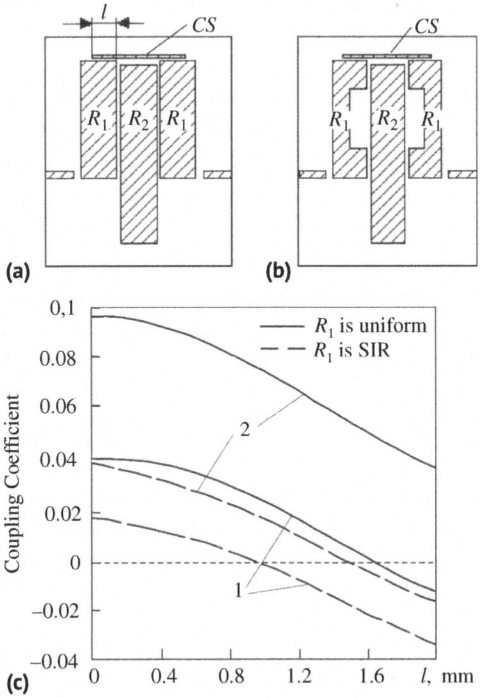

Figure 13.3 Control of electromagnetic interaction between two nonadjacent resonators R_1 using connection segment CS. (a) R_1 is uniform. (b) R_1 is SIR. (c) Dependence of K on the length l; here curves 1 and 2 represent the presence and absence of intermediate resonator R_2, respectively.

the presence and absence of additional central resonator R_2. Its length exceeded the length of R_1. The width and length of resonator R_2 were 2 mm and 15 mm, respectively.

SIRs (Figure 13.3b) have a wide part with the value $w_1 = 2$ mm, a narrow part of the length 5 mm, and a width $w_2 = 1$ mm.

Note the following features typical of the curves $K = K(l)$ presented in Figure 13.3c:

- **i.** They are steadily decreasing.
- **ii.** A microstrip conductor between resonators leads to a decrease in magnetic coupling.
- **iii.** A stepped-impedance irregularity (Figure 13.3b) also decreases the magnetic coupling of resonators.
- **iv.** Using connection segments enables us to vary the coupling coefficient between nonadjacent resonators over a wide range, achieving their positive and negative values.

13.1.3 Shifting Transmission Zero at Constant *K* Value

The influence of the shape of microstrip SIRs and the gap between them on the TZ position is clearly illustrated by the example of six pairs of resonators [5], the topologies of which are shown in the insets to Figure 13.4. All pairs contain quarter-wave SIRs, one end of which is short-circuited. They are implemented on a substrate of TMM-10i (Rogers), with $\varepsilon_r = 9.8$, $\tan\delta = 0.002$, and $h = 1.905$ mm.

The first three pairs of SIRs (Figure 13.4a) are characterized by a positive mixed coupling coefficient $K_{12} = 0.0414$ and a value of $f_0 = 1650$ MHz.

Using the expression for the coupling coefficient (3.10), we determine the even $f_e = 1621$ MHz and odd $f_o = 1689$ MHz modes corresponding to the value $K_{12} = 0.0414$. Figure 13.4a shows the frequency responses of these resonator pairs, which are simulated using the Microwave Office (AWR) software. The wide

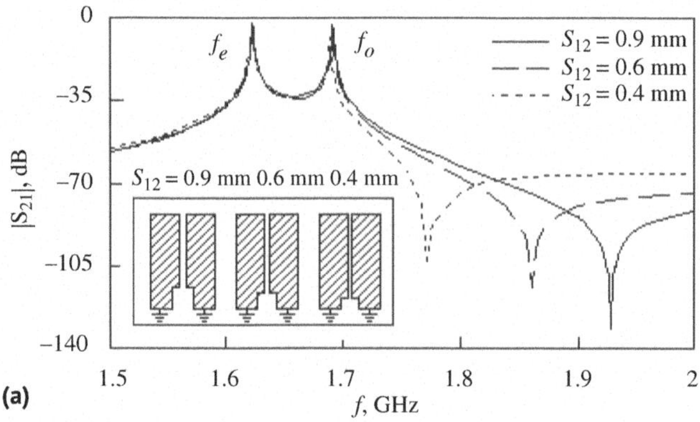

(a)

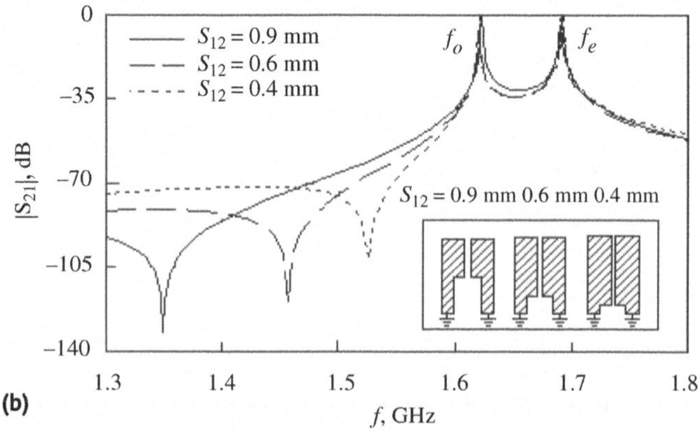

(b)

Figure 13.4 Shifting the TZ at a constant K value. (a) Simulated insertion loss function for positive mixed coupling coefficient $K = \text{const}$. (b) Simulated insertion loss function for negative mixed coupling coefficient $K = \text{const}$.

$w = 3.6$ mm and narrow $w_1 = 3.0$ mm parts of the resonators are fixed. The lengths of the resonators L and their narrow part l_1 are varied so that two peaks on the insertion loss curve coincide with the frequencies of 1.621 and 1.689 GHz. The gap S_{12} of each pair takes the values 0.9, 0.6, and 0.4 mm.

If $S_{12} = 0.4$ mm, then with geometric parameters $L = 15.6$ mm, $l_1 = 3.1$ mm, TZ is provided at the frequency $f_z = 1770$ MHz, which is reflected in Figure 13.4a by the dotted line. With an increase in the gap S_{12}, the length l_1 of the narrow part of the SIR increases, and TZ shifts to the right and moves away from the coupling frequencies f_e and f_o.

Figure 13.4b illustrates the implementation of the negative coefficient of mixed coupling $K_{12} = -0.0414$.

In this case, we have $f_o = 1621$ MHz and $f_e = 1689$ MHz. As before, the value $w = 3.6$ mm remains the same. The width of the narrow part of the SIR has decreased $w_1 = 1.8$ mm. The values w and w_1 are fixed, while the values L, l_1, and S_{12} are varied.

According to (3.10), we have $f_o = 1621$ MHz and $f_e = 1689$ MHz. If $S_{12} = 0.4$ mm, then with geometric parameters $L = 14.5$ mm, $l_1 = 1.9$ mm, TZ is provided at the frequency $f_z = 1530$ MHz, as shown in Figure 13.4b by the dotted line. With an increase in the gap S_{12}, the length l_1 of the narrow part of the SIR increases, and TZ shifts to the left and moves away from the even f_e and odd f_o modes.

The considered procedure assumes adjusting the geometric parameters of each pair of SIRs according to the given values of the three frequencies f_o, f_e, and f_z. This procedure is sometimes called optimization and is performed using an EM simulator.

In a pair of closely spaced $\lambda/4$ resonators, the electric component of electromagnetic coupling is most pronounced near the open end, and the magnetic component of the coupling is most pronounced near the short-circuited end. In the pairs of SIRs in Figure 13.4a, b, a rectangular area with removed metallization is made in the region of the short-circuited ends, which reduces the magnetic component of the coupling. This rectangular area is small in the pairs of SIRs in Figure 13.4a. It somewhat decreases the magnetic component of the coupling, but it remains dominant. In the pairs of SIRs in Figure 13.4b, this rectangular area has an increased size. As a result, it greatly reduces the magnetic component of the coupling, making the electrical component dominant.

It is important to note that reducing the gap between microstrip SIRs allows f_z to be closer to f_0 (Figure 13.4a, b), which is also inherent in stripline SIRs (Figure 12.5a). Reducing the gap increases the components of the mixed coupling K_m and $|K_e|$, and their ratio approaches unity $K_m/|K_e| \to 1$. The degree of convergence of the frequencies f_z and f_0 is limited by the minimum allowable gap S between microstrip SIRs.

13.2 Compact Mixed Coupled Microstrip Triplet BPFs

The use of mixed couplings [6–8], also called frequency-dependent [9–12], in N-order BPFs allowed to increase the number of their TZs by one. This difference is most significant in BPFs with few resonators $N = 2, 3$. For this reason, a substantial number of works [13–24] were devoted to mixed coupled triplet BPFs ($N = 3$). Note that triplet BPFs with simple couplings [25, 26] have only one TZ at real frequencies and two options for its location: to the left or the right of the passband. Triplet BPFs with mixed couplings have two TZs on the complex plane $S = \sigma + j\Omega$ and 10 options for their location (Table 7.1). The main advantage of mixed coupled triplet BPF is the many options for the location of two TZs, making it possible to implement a significant variety of frequency responses.

In the following section, we consider compact microstrip triplet BPFs of practical interest. Compactness means a dense arrangement of filter resonators on a dielectric substrate. In this case, the metallized layer occupies most of the substrate. Some of the compact triplet BPFs are considered in Chapter 7 and are shown in Figures 7.3, 7.4, 7.9, and 7.11. The resonators of such filters are wide enough, which provides them with a relatively high unloaded quality factor Q_u.

13.2.1 Some Variants of Compact Microstrip Implementation of Triplet BPF

Figure 13.5 shows the topologies of some other compact triplet BPFs of the first type. The triplet BPFs in Figure 13.5a were proposed relatively long ago [19] when the term "mixed coupling" was not yet in use. All resonators of this compact filter are quarter-wave. In the triplet BPF in Figure 13.5b [20, 21], the outer resonators are quarter-wave. The middle resonator of this filter is a half-wave with two short-circuited ends, and it does not change the signs of the coupling coefficients to the opposite. Therefore, this filter is a triplet BPF of the first type with mixed cross-coupling K_{13}. In the filters in Figure 13.5, the magnetic component of the cross-coupling K_{m13} is formed by the parasitic interaction between resonators 1 and 3. The electrical component K_{e13} is determined by the capacitive gap between the open ends of resonators 1 and 3. To increase the values of $|K_{e13}|$, a microstrip interdigital capacitor ([1], Figure 4.6a) connected between the open ends of these resonators can be used. Each filter in Figure 13.5 can implement the first seven variants of the arrangement of the two TZs in Table 7.1 and the corresponding frequency responses.

Figure 13.6 shows the topologies of some compact triplet BPFs of the second type. The triplet BPF in Figure 13.6a differs from the triplet BPF in Figure 13.5a in that its central $\lambda/4$ resonator is replaced by a U-shaped $\lambda/2$ resonator with both open ends. The triplet BPF in Figure 13.6b differs from the triplet BPF in Figure 13.5b in that both ends of its central half-wave zig-zag resonator are open.

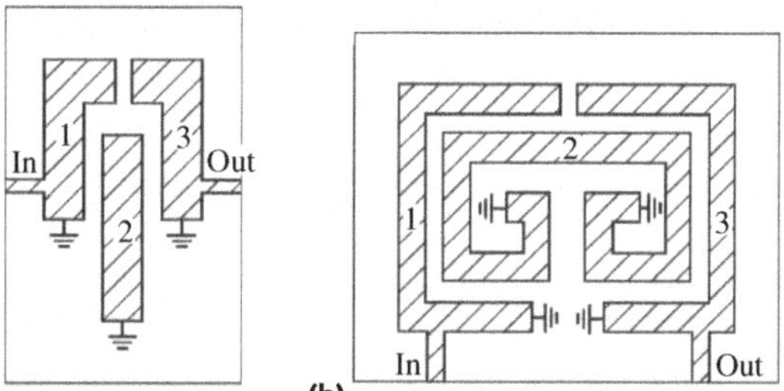

Figure 13.5 Topology of first-type microstrip triplet BPFs with mixed cross-coupling. (a) With quarter-wave central resonator. *Source*: Adapted from [19]. (b) Half-wave zig-zag central resonator has two short-circuited ends [20, 21].

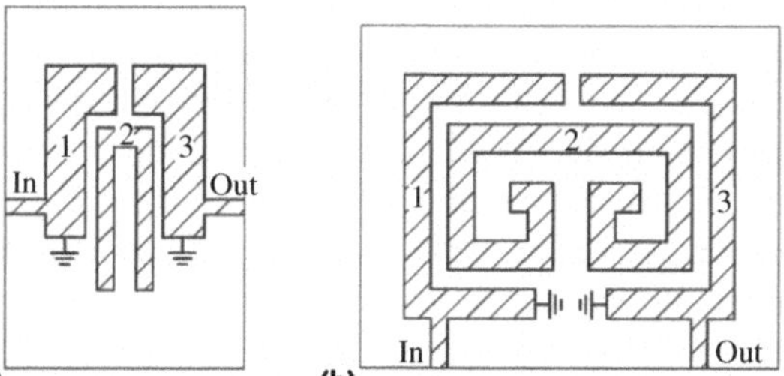

Figure 13.6 Topology of second-type microstrip triplet BPFs with mixed cross-coupling. (a) With half-wave U-shaped central resonator. *Source*: Adapted from [19]. (b) Half-wave zig-zag central resonator has two open-circuited ends [20, 21].

The magnetic K_{m13} and electric K_{e13} components of the mixed cross-coupling K_{13} of these filters are determined by the same factors as those of the filters in Figure 13.5.

Each filter in Figure 13.6 can have three options for arranging the two TZs (Nos. 8–10) from Table 7.1 and the corresponding frequency responses.

13.2.2 Mixed Coupled Triplet BPFs with Closely Placed Resonators 1 and 3

The equalities given in Chapter 7

$$\sigma_{1,2} = \pm\frac{m_{012}}{\sqrt{a_{13}}}, \tag{13.1a}$$

$$j\Omega_{1,2} = \pm j\frac{m_{012}}{\sqrt{a_{13}}}, \tag{13.1b}$$

for the first and second types of triplet BPFs, respectively, with resonant cross-coupling ($K_{13} = 0$), reflect that the convergence degree of two equidistant TZs depends on the slope parameter $a_{13} = K_{m13} = |K_{e13}|$. The larger the parameter a_{13}, the closer these TZs are to each other. The triplet BPFs considered earlier do not have a sufficiently large value of a_{13} since resonators 1 and 3 are at some distance from each other, and the K_{m13} component is comparatively small.

Figures 13.7 and 13.8 show triplet BPFs of the first [22] and second [18] types, respectively, in which resonators 1 and 3 are SIRs located next to each other. This allows increasing the practically feasible value of a_{13}. The filters use the substrate TMM-10i (Rogers), with $\varepsilon_r = 9.8$, $\tan\delta = 0.002$, $h = 1.905$ mm, and they cover a square of 40 mm × 17 mm.

The first-type microstrip triplet BPF implementing position No. 1 from Table 7.1 (Figure 13.7) contains two SIRs 1 and 3 tuned to the resonant frequency $f_0 = 1.75$ GHz. They are characterized by the following geometric parameters: length $L = 14$ mm, the greatest width $w = 3.6$ mm, the smaller width $w_1 = 2.4$ mm, the length of the narrow part $l_1 = 4.7$ mm, and the gap between the resonators $S = 1$ mm. These parameters provide mixed resonator coupling ($K_{13} = 0$) between them with the steepness parameter and components $a_{13} = K_{m13} = |K_{e13}| = 0.09$. The central resonator 2 is a quarter-wave resonator located on the second tier and is oriented antiparallel to resonators 1 and 3. It has a length of $L' = 14.3$ mm and a width of $w' = 6.2$ mm.

This resonator is separated from the SIRs by a gap of $S' = 1$ mm, which provides the filter with the main coupling coefficients $K_{12} = K_{23} = -0.0388$. With the specified parameters, the filter has the frequency responses shown in Figure 13.7b.

The measured data are as follows: $f_0 = 1.75$ GHz, BW = 80 MHz, the insertion loss at center frequency $L_{A0} = 1.5$ dB, and the group delay $\tau = 7.3 \pm 0.1$ ns within the frequency range of 38 MHz.

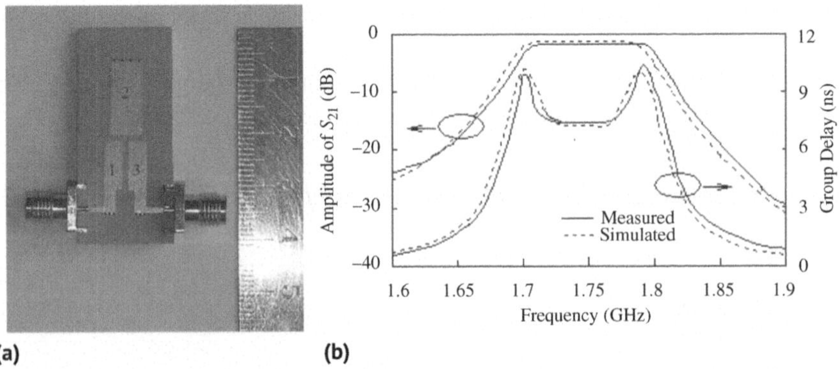

Figure 13.7 First-type microstrip triplet BPF with closely spaced resonators 1 and 3. (a) Photograph of fabricated filter. *Source*: [22] / with permission of IEEE. (b) Frequency responses with resonant cross-coupling.

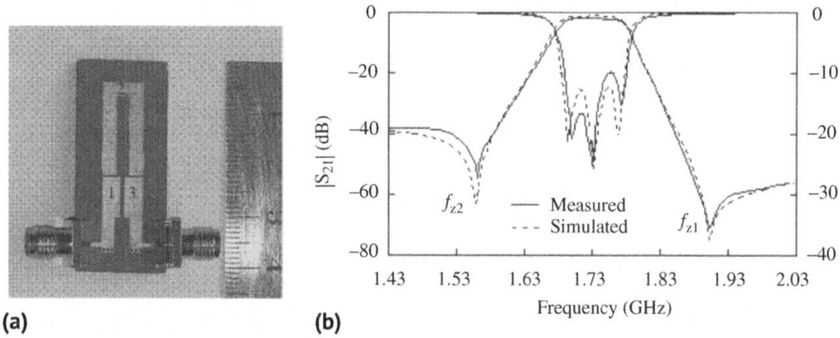

Figure 13.8 Second-type microstrip triplet BPF with closely spaced resonators 1 and 3. (a) Photograph of the fabricated filter. *Source*: [18] / with permission of IEEE. (b) Frequency responses with resonant cross-coupling.

Using (13.1a), it is easy to determine the TZs of this filter $\sigma_{1,2} = \pm 2.83$.

The filter shown in Figure 13.7 is configured to be a delay filter with a constant τ value in a frequency band of 47.5% of the passband width. However, it can arrange two TZs in positions Nos. 2–7 of Table 7.1 and their corresponding frequency responses.

The second-type microstrip triplet BPF (Figure 13.8), implementing position No. 8 (Table 7.1), also contains two SIRs 1 and 3 tuned to the resonant frequency $f_0 = 1.73$ GHz. Their geometric parameters are as follows: length $L = 14$ mm, greatest width $w = 3.6$ mm, least width $w_1 = 2.4$ mm, length of narrow part $l_1 = 6.2$ mm, and gap between resonators $S = 0.9$ mm. These parameters provide mixed resonator coupling ($K_{13} = 0$) between them, characterized by the values $a_{13} = K_{m13} = |K_{e13}| = 0.062$.

The central resonator 2 is a U-shaped half-wave resonator located on the second level, with lengths of 36.5 mm and a width of 2.6 mm. It is separated from the SIRs by a gap $S' = 0.6$ mm, which provides the filter with the main coupling coefficients $K_{12} = K_{23} = -0.046$. The filter has a quasi-elliptic transfer function response with the specified parameters (Figure 13.8b).

The measured data are as follows: $f_0 = 1.73$ GHz; BW = 85 MHz; $L_{A0} = 1.4$ dB; the return loss $L_R > 10$ dB. The TZs f_{z1} and f_{z2} are equidistant to f_0: $|f_{z1,2} - f_0| =$ 170 MHz. They correspond to TZs on the $j\Omega$-axis: $j\Omega_{1,2} = \pm j4$.

The filter under consideration has equidistant TZs at real frequencies, corresponding to position No. 8 (Table 7.1). It can also have the arrangement of two TZs indicated in positions No. 9 and No. 10 (Table 7.1) and the corresponding frequency responses.

When designing microstrip triplet BPFs with resonant cross-coupling ($K_{13} = 0$), a key question arises: How will a pair of SIRs increase the slope parameter $a_{13} = K_{m13} = |K_{e13}|$ of the coupling? The answer to this question is illustrated in Figure 13.9 [22]. The simulation of the frequency response of a two-resonator

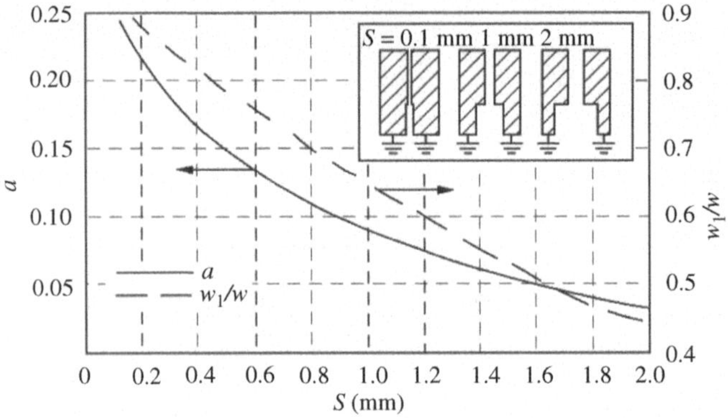

Figure 13.9 Slope parameter a and ratio w_1/w of microstrip SIR for resonant coupling ($K_m = |K_e|$, $K = 0$).

filter, whose SIRs are weakly coupled with loads, was used to build these graphs. The simulation was performed using Microwave Office software (AWR).

In this case, the SIR parameters were fixed: length $L = 14$ mm, maximum width $w = 3.6$ mm, and length of the narrow part $l_1 = 4.7$ mm. The gap size S and the width of the narrow part w_1 of SIR were variable. For each given value of S, w_1 was chosen so that the equality $K_{13} = 0$ was fulfilled.

The increase in $a_{13} = K_{m13} = |K_{e13}|$ can be achieved by decreasing the gap S between the resonators and increasing its narrow part w_1. In the inserted picture in Figure 13.9, the topologies of the pairs of resonators for the gaps with widths of 0.1, 1.0, and 2.0 mm are depicted. If the value of l_1 is changed, then the nature of the dependencies in Figure 13.9 will remain the same.

The main factor limiting the parameter a_{13} is the permissible convergence degree of microstrip SIRs. In practice, it is possible to achieve $a_{13} \approx 0.3$, the substitution of which into (13.1) gives $\sigma_{1,2} = \Omega_{1,2} \approx \pm 1.83 m_{012} \approx \pm 1.83/\sqrt{g_1 g_2}$, where g_1 and g_2 are the parameters of the low-frequency prototype. When using the Chebyshev prototype with a pulsation value of 0.2 dB (Table 2.6), $g_1 = 1.2275$, $g_2 = 1.1525$, and $\sigma_{1,2} = \Omega_{1,2} \approx \pm 1.54$. This value characterizes the potential capabilities of microstrip triplet BPFs with closely placed SIRs regarding the maximum convergence of two TZs.

13.2.3 Miniature Second-type Triplet BPF with All Mixed Couplings on High Dielectric Constant Substrate

The use of mixed couplings between adjacent resonators instead of simple ones leads to a reduction in the transverse dimensions of the triplet BPF due to a significant reduction in the gaps between the resonators.

Significant miniaturization of the triplet BPF with all mixed couplings is achieved using high dielectric constant substrates. One such triplet BPF, which is symmetrical, is shown in Figure 7.9.

Asymmetrical triplet BPFs with all mixed couplings (Figure 7.7b) are not always convenient to use. These BPFs can be symmetrical if the central reflection-type resonator is replaced by a through-type resonator with both open-circuited ends, as shown in Figure 13.10 by a double circle. This filter is named the second-type triplet BPF with all mixed couplings.

Figure 13.11a shows the topology of a microstrip second-type triplet BPF with all mixed couplings [23], which corresponds to the coupling structure in Figure 13.10b. Its central resonator is a U-shaped resonator with both open-circuited ends. This filter uses a substrate with $\varepsilon_r = 92$, $\tan\delta = 0.0004$, and $h = 1$ mm. A gap of $S = 0.2$ mm separates the resonators, and the filter occupies an area of 10.2 mm × 6 mm.

Figure 13.11b and the inset depict this filter's frequency responses and photograph. Its center frequency $f_0 = 2015$ MHz and BW = 100 MHz. TZs f_{z2} and f_{z1} are equidistant concerning f_0: $|f_{z1,2} - f_0| = 285$ MHz; they correspond to TZs at the $j\Omega$-axis: $j\Omega_{1,2} = \pm j5.7$. Other measured data: $L_{A0} = 1.5$ dB; $L_R > 12$ dB. The selectivity of this filter 35 dB ($f_0 \pm 165$ MHz) can be expressed at the normalized frequency $j\Omega$: 35 dB ($\pm j3.3$).

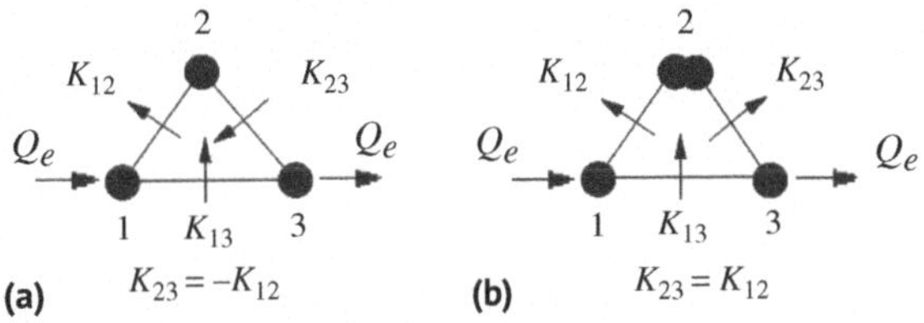

Figure 13.10 Equivalent triplet BPFs with all mixed couplings. (a) Asymmetric. (b) Symmetric.

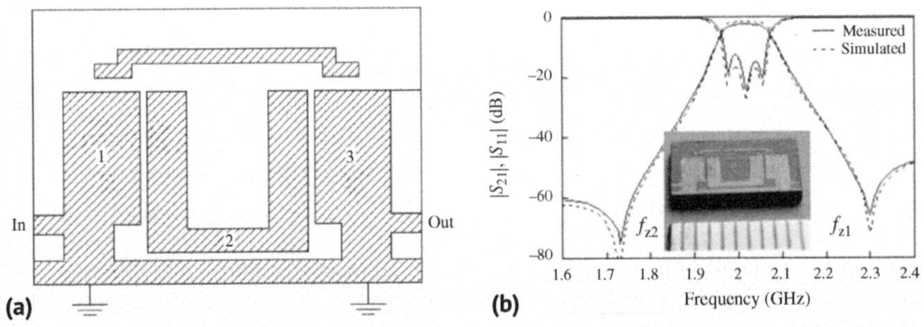

Figure 13.11 Microstrip second-type triplet BPF with all mixed coupling. (a) Topology. (b) Measured and simulated frequency responses. *Source*: [23] / with permission of IEEE.

This filter is quasi-elliptic and corresponds to position No. 8 (Table 7.1). Positions No. 9 and No. 10 can also be implemented by it.

13.2.4 Universal Microstrip Triplet BPF with All Mixed Couplings

Figure 13.12a shows the topology of the universal microstrip triplet BPF with all mixed couplings [24]. It contains three half-wave SIRs and a thin strip connecting the outer resonators via capacitive gaps. This triplet BPF is universal because it allows the realization of 10 different options for arranging two TZs (Table 7.1). Two other options provide this effect for conductively connecting the loads to the end resonators of the filter. When connecting the loads to ports *1–2*, the arrangements of two TZs and the frequency responses of the first-type triplet BPF corresponding to positions Nos. 1–7 (Table 7.1) are realized. When connecting the loads to ports *1′–2′*, the frequency responses of the second-type triplet BPF corresponding to positions Nos. 8–10 in this table are realized. This filter has three signal paths from input to output, which explains many different frequency responses [24].

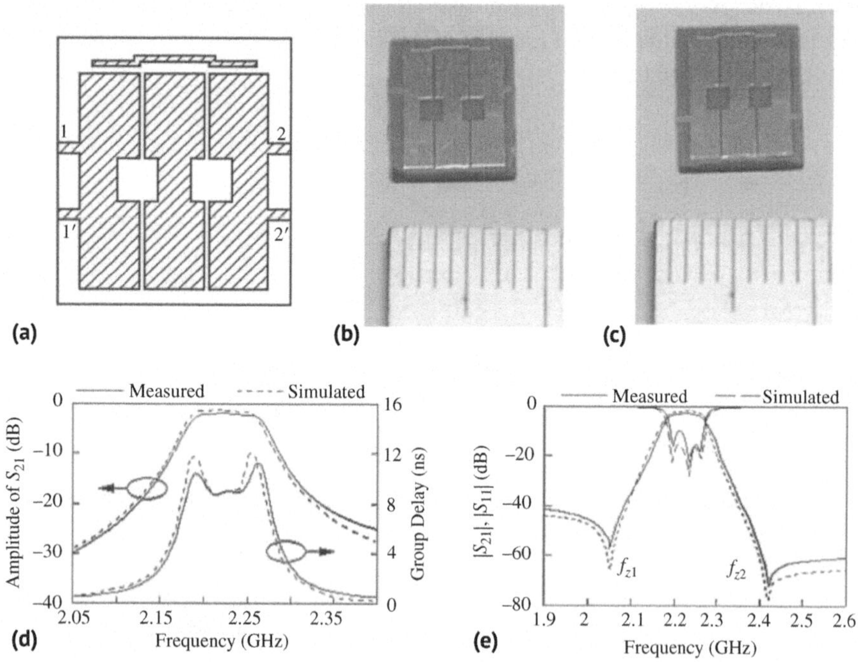

Figure 13.12 Universal microstrip triplet BPF with all mixed couplings and different port positions. (a) Topology. (b) Photograph with port positions 1–2. (c) Photograph with port positions $1' - 2'$. (d) Frequency responses on port positions 1–2. (e) Frequency responses on port positions $1' - 2'$.

Figure 13.12b, c shows the photographs of these microstrip BPFs with different input and output port positions. The filter is implemented on a substrate with the following parameters, with $\varepsilon_r = 92$, tan $\delta = 0.0004$, $h = 1$ mm, and occupies an area of 9 mm × 8 mm. The filter resonators are separated from each other by a gap of $S = 0.2$ mm. Both filters have a center frequency of $f_0 = 2.23$ GHz and a bandwidth of BW = 70 MHz. Figure 13.12d shows the frequency responses of the BPF of the first type (Figure 13.12b) with port positions 1–2. The filter is configured to implement position No. 1 (Table 7.1). This BPF is a delay line filter and provides the group delay $\tau = 8.7 \pm 0.1$ ns within the frequency band of 35 MHz, and its $L_{A0} = 1.8$ dB. When connected to ports 1–2, positions Nos. 2–7 (Table 7.1) can also be provided.

Figure 13.12e shows the frequency responses of the BPF of the second type (Figure 13.12c) with port positions 1′–2′. The filter parameters are chosen to implement position No. 8 (Table 7.1). It is a quasi-elliptic BPF with two TZs, which are equidistant concerning f_0: $|f_{z1,2} - f_0| = 180$ MHz, corresponding to TZs on the $j\Omega$-axis: $j\Omega_{1,2} = \pm j5.14$. The measured parameters are as follows: $f_0 = 2230$ MHz; $L_{A0} = 1.8$ dB; $L_R > 12$ dB. The selectivity of the filter is 30 dB ($f_0 \pm 102$ MHz) or 30 dB ($\pm j2.91$). When connecting to ports 1–2, positions No. 9 and No. 10 (Table 7.1) can also be implemented.

An important parameter of resonators is their unloaded quality factor Q_u. Table 13.1 compares the Q_u values of microstrip uniform half-wave resonators and SIR of half-wave type, whose middle section of length l_1 has a smaller width w_1. The Q_u value of SIRs is slightly smaller than that of $\lambda/2$ resonators. It should be considered that lowering f_0 leads to a decrease in Q_u of microstrip resonators. Therefore, reducing the f_0 of $\lambda/2$ resonator by increasing its length to the resonant frequency of the SIR will decrease its Q_u value.

Rectangular dielectric coaxial resonators DRR030 3 mm × 3 mm with $\varepsilon_r = 92$ are widely used in BPFs [27]. They have a value of $Q_u \approx 250$ at $f_0 = 1500$ MHz, close enough to the Q_u values of SIRs (Table 13.1). Therefore, the insertion losses of BPFs with DRR030 coaxial resonators and microstrip BPFs with SIRs will be approximately the same.

Table 13.1 Comparison of measured unloaded quality factor Q_u of microstrip uniform half-wave resonators and SIRs on a substrate with $\varepsilon_r = 92$, tan $\delta = 0.0004$, $h = 1$ mm.

Resonator	L, mm	w, mm	l_1, mm	w_1, mm	f_0, MHz	Q_u
Uniform	9	3	—	—	1861	285
SIR	9	3	3	1.5	1638	270
Uniform	9	2	—	—	1916	275
SIR	9	2	3	1	1723	255

13.2.5 Duplexer with Universal Microstrip Triplet BPF

In Zakharov and Rozenko [28], a duplexer was considered based on two universal microstrip triplet BPFs with all mixed couplings. The width of the transmission TX (1750...1780 MHz) and receiver RX (1840...1870 MHz) channels is 30 MHz. The frequency interval between the channels is 60 MHz.

Both filters were fabricated on substrates with parameters $\varepsilon_r = 92$, $\tan\delta = 0.0004$, $h = 1$ mm, the area of which was 11 mm × 8 mm. The resonators of the RX filter were separated by a gap of $S = 0.2$ mm, and in the RX filter, a gap of $S = 0.1$ mm was used. The frequency responses of these triplet BPFs are shown in Figure 13.13. To reduce insertion losses, the passbands of these filters were made wider. The vertical dashed lines mark the frequency boundaries of the RX and TX channels.

The aforementioned filters RX and TX, with dimensions of 11 mm × 8 mm, were combined into a duplexer. For this purpose, an additional substrate (Figure 13.14)

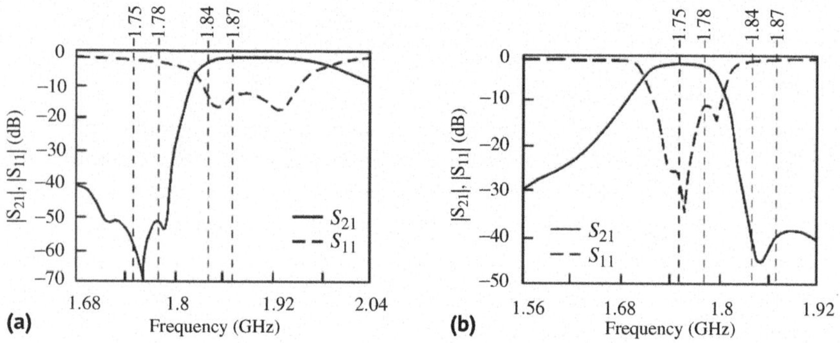

Figure 13.13 Measured frequency responses of universal triplet BPFs for duplexer. (a) RX channel. (b) TX channel.

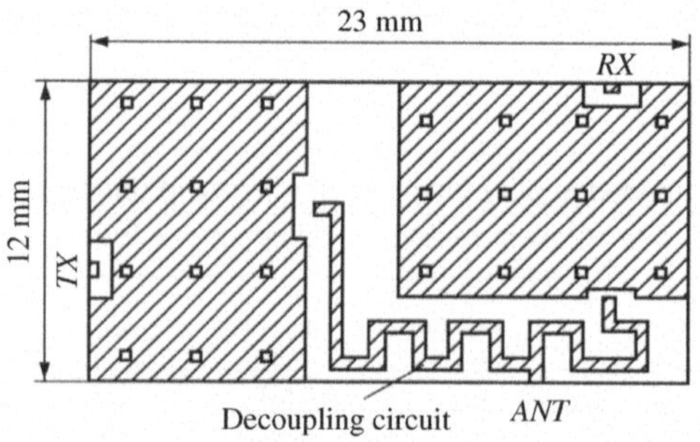

Figure 13.14 Topology of substrate for installing duplexer's bandpass microstrip filters, which also contain circuits for decoupling channels RX and TX.

RT/Duroid 6010.2 LM (Rogers) with the square of 23 mm × 12 mm, and parameters $\varepsilon_r = 10.2$, $\tan\delta = 0.0023$, $h = 0.635$ mm, was used. The two BPFs were located perpendicular to each other on metallized rectangles with metallized holes, which reduced their "parasitic" influence on each other. The decoupling circuit (zig-zag microstrip segment) plays an important role in the duplexer. This circuit is formed by two line sections with L_{RX} and L_{TX} lengths, whose ends are connected to the filters of corresponding channels on one side and to a common point on the other. This common point is connected to the antenna ANT. The objective of the decoupling circuit is to eliminate interferences between BPFs RX and TX.

The photograph of the microstrip duplexer is presented in Figure 13.15a, and its measured insertion loss is shown in Figure 13.15b. Specifications on the duplexer are contained in Table 13.2.

The considered duplexer contains triplet BPFs with SIRs of half-wave type and occupies an area of 23 mm × 12 mm. If triplet BPFs with SIRs of quarter-wave type (Figure 7.9) are used in the duplexer, the area it occupies will decrease to 17 mm × 10 mm [28]. Considering the small size and the figures in Table 13.2, it can be regarded that duplexers on high dielectric constant substrates with mixed coupled triplet BPFs are promising.

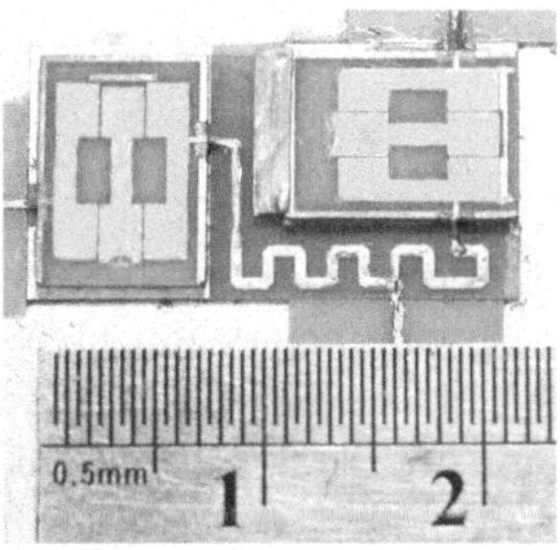

(a)

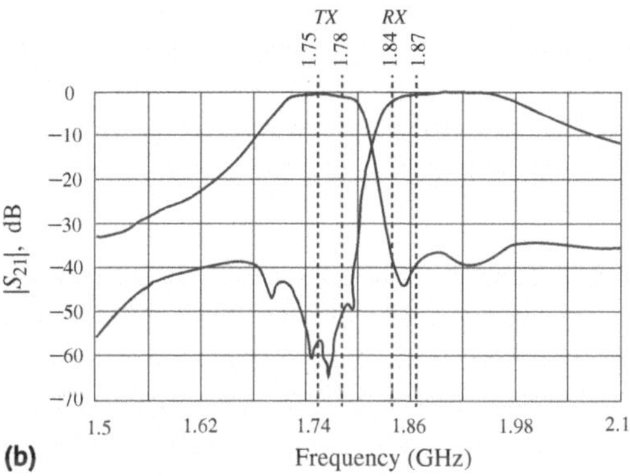

(b)

Figure 13.15 Duplexer with universal triplet BPFs. (a) Photograph. *Source*: [28] / with permission of Springer Nature. (b) Measured insertion loss.

Table 13.2 Specifications on a duplexer with microstrip mixed coupled triplet BPFs.

f_0, MHz		BW, MHz	L_{Amax}, dB	Ripple, dB	VSWR	Attenuation dB (MHz)
RX	1855	$f_0 \pm 15$	3.2	1.0	1.7	52 (1750–1780)
TX	1765	$f_0 \pm 15$	2.7	0.8	1.7	38 (1840–1870)

13.3 Higher-order Mixed Coupled Compact Microstrip BPFs

In microstrip BPFs, contrary to other designs, e.g., stripline designs, a sufficiently strong magnetic interaction exists between nonadjacent resonators. The thicker the dielectric substrate and the closer to the nonadjacent resonators, the stronger this interaction. Because of the cross-couplings, these filters "have to become" nonminimum phase with a steeper right-hand slope. This is confirmed by the frequency responses of microstrip combline, interdigital, and pseudocombline microstrip filters [1]. If we simulate the frequency response of an idealized filter with only mixed coupling and then "connect" weak cross-coupling, then the original response of the filter will be degraded. To reduce this interaction, it is necessary to use a unique arrangement of resonators on the substrate.

13.3.1 BPF with Two Pairs of Antiparallel Resonators

Parasitic coupling between nonadjacent resonators can be reduced in two-level BPF designs.

A fourth-order microstrip BPF with two pairs of antiparallel mixed coupled SIRs (Figure 13.16) was proposed in Ref. [5]. These pairs of resonators are located at different levels. The filter has two mixed and one simple couplings.

Let us design the quasi-elliptic BPF with follows data: $f_0 = 1650$ MHz; BW = 75 MHz (FBW = 0.04545); passband ripple $L_{Ar} = 0.04321$ dB; two TZs $\left|f_{z1,2} - f_0\right| = 120$ MHz $\left(\Omega_{z1,2} = \pm 3.2\right)$.

Stage 1: Using Table 2.6, we can find g parameters: $g_0 = 1, g_1 = 0.9314, g_2 = 1.2920$, $g_3 = 1.5775$, $g_4 = 0.7628$, $g_5 = 1.2210$. Substituting these g parameters and FBW value into expressions (3.8), we obtain $|K_{12}| = |K_{34}| = 0.0414$, $|K_{23}| = 0.0318$, $Q_e = 20.49$. The filter with the specified parameters is implemented on a substrate of TMM-10i (Rogers), with $\varepsilon_r = 9.8$, $\tan\delta = 0.002$, $h = 1.905$ mm. It has a size of 40 mm × 20.4 mm. The topology and photograph of the filter are shown in Figure 13.16a, b, respectively. The frequency responses of the filter are shown in Figure 13.16c. The filter contains two pairs of SIRs for implementing mixed couplings K_{12} and K_{34}. Metallized holes provide a short circuit at the ends of the resonators with a diameter of 0.4 mm. These pairs are placed at different levels. Between resonators 2 and 3, there is the pure electric coupling K_{23}.

Stage 2: Let us assume that $K_{12} > 0$ for the first pair of resonators and $K_{34} < 0$ for the second pair of resonators. Using the specification, we define the components of the mixed couplings in both representation forms. Since $K_{12} = 0.0414$ and FBW = 0.04545, then we obtain $m_{012} = K_{12}/\text{FBW} = 0.9109$. Then we determine the steepness parameter $a_{12} = m_{012}/\Omega_z = 0.9109/3.2 = 0.2846$, where $\Omega_z = 3.2$ is TZ. The linear representation form (7.6) is defined as $m_{12}(\Omega) = 0.9109 - 0.2846\,\Omega$. The same calculation sequence gives

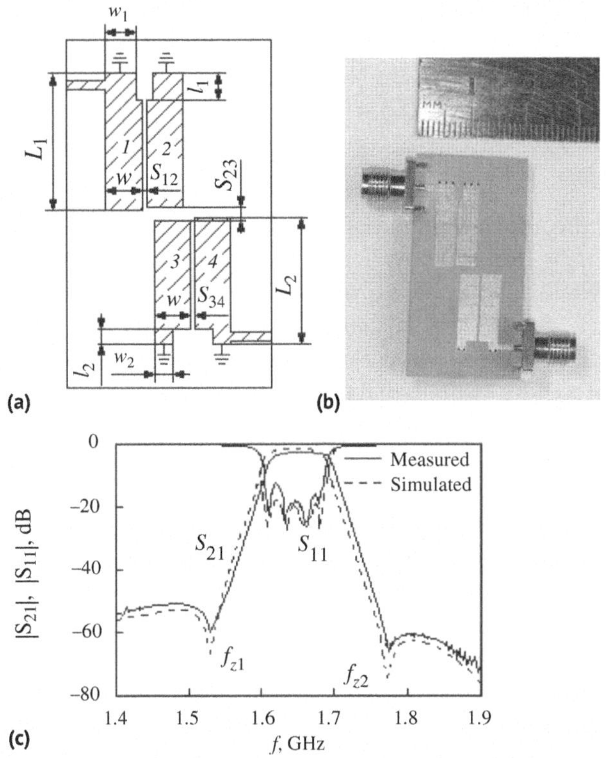

Figure 13.16 Microstrip mixed coupled dual-level BPF with both-sided TZs ($K_{12} > 0$, $K_{34} < 0$). (a) Topology $w = 3.6$ mm, $L_1 = 15.6$ mm, $w_1 = 3$ mm, $l_1 = 3.1$ mm, $S_{12} = 0.4$ mm, $L_2 = 14.5$ mm, $w_2 = 1.8$ mm, $l_2 = 1.9$ mm, $S_{34} = 0.4$ mm, $S_{23} = 1.5$ mm. (b) Photograph of fabricated filter. *Source*: [5] / with permission of IEEE (edited). (c) Measured and simulated frequency responses.

$m_{34}\,(\Omega) = -0.9109 - 0.2846\ \Omega$. The transition from the mixed coupling form (7.6) to (7.2) is carried out by formulas (7.8) and (7.9). For $K_{12} = 0.0414$, we get $K_{m12} = 0.3061$ and $K_{e12} = -0.2647$, and for $K_{34} = -0.0414$, we have $K_{m34} = 0.2647$ and $K_{e34} = -0.3061$.

Stage 3: Let us define the parameters of each pair of SIRs with given mixed couplings. Figure 13.4a illustrates the implementation of $K_{12} = 0.0414$ and $f_0 = 1650$ MHz. Using the expression for the coupling coefficient $K = (f_o^2 - f_e^2) / (f_o^2 + f_e^2)$, we determine the even $f_e = 1621$ MHz and odd $f_o = 1689$ MHz modes corresponding to the value $K_{12} = 0.0414$. Figure 13.4a shows the frequency responses of some pairs of these SIRs, which are simulated using the Microwave Office (AWR) software. The values of w and w_1 of the resonators are fixed. The lengths of the resonators L_1 and their narrow part l_1 are varied so that two peaks on the insertion loss curve coincide with the frequencies of 1.621 and 1.689 GHz. Three pairs of resonators with $K_{12} = 0.0414$ are presented in the inset in Figure 13.4a. The gap S_{12} of each pair takes the values 0.9, 0.6, and 0.4 mm. To have the required mixed coupling ($K_{m12} = 0.3061$, $K_{e12} = -0.2647$ and $m_{12}(\Omega) = 0.9109 - 0.2846\ \Omega$), it is necessary to provide the value $f_z = 1770$ MHz, which is realized at $S_{12} = 0.4$ mm, $L_1 = 15.6$ mm, $l_1 = 3.1$ mm.

Figure 13.4b illustrates the implementation of $K_{34} = -0.0414$. In this case, we have $f_o = 1621$ MHz and $f_e = 1689$ MHz. As before, the values $w = 3.6$ mm and $w_2 = 1.8$ mm are fixed, while L_2, l_2, and S_{34} are varied. The required mixed coupling ($K_{m34} = 0.2647$, $K_{e34} = -0.3061$ and $m_{34}(\Omega) = -0.9109 - 0.2846\ \Omega$) is provided at $f_z = 1530$ MHz. It is realized at $S_{34} = 0.4$ mm, $L_2 = 14.5$ mm, $l_2 =$ 1.9 mm. The considered procedure assumes that the geometric parameters of each pair of SIRs should be adjusted according to the given values of the frequencies f_o, f_e, and f_z. This procedure is sometimes called optimization, performed using an EM simulator [12].

In a pair of closely spaced $\lambda/4$ resonators, the electric component of electromagnetic coupling is most pronounced near the open end, and the magnetic component of the coupling is most pronounced near the short-circuited end. In the pairs of SIRs in Figure 13.4a, b, a rectangular area with removed metallization is made in the region of the short-circuited ends, which reduces the magnetic component of the coupling. This rectangular area is small in the pairs of SIRs in Figure 13.4a. It somewhat decreases the magnetic component of the coupling, but it remains dominant. In the pairs of SIRs in Figure 13.4b, this rectangular area has an increased size. As a result, it greatly reduces the magnetic component of the coupling, making the electrical component dominant.

It is important to note that reducing the gap between microstrip SIRs allows f_z to be closer to f_0 (Figure 13.4a, b). Reducing the gap leads to an increase in the components of the mixed coupling K_m and $|K_e|$ and an increase in the slope parameter a of the linear representation form of the mixed coupling (7.6).

The gap $S_{23} = 1.5$ mm between the open ends of resonators 2 and 3 provides the required coupling coefficient $K_{23} = -0.0318$. By choosing the coordinate for connecting the loads to the end resonators, the value $Q_e = 20.49$ is realized.

The measured frequency responses (Figure 13.16c) of the filter are characterized as follows: $f_0 = 1650$ MHz, BW = 75 MHz, insertion loss $L_{A0} = 1.7$ dB, return loss $L_R > 16$ dB. The TZs ($f_{z1} = 1771$ MHz and $f_{z2} = 1531$ MHz) are almost equidistantly relative to f_0.

Figure 13.17 shows the topology and frequency responses of a microstrip BPF with two pairs of antiparallel resonators, which implement two prescribed left-handed TZs. The two-level arrangement of the resonators in the considered fourth-order microstrip BPF reduces the parasitic interaction between nonadjacent resonators. It allows more accurate reproduction of the required frequency responses.

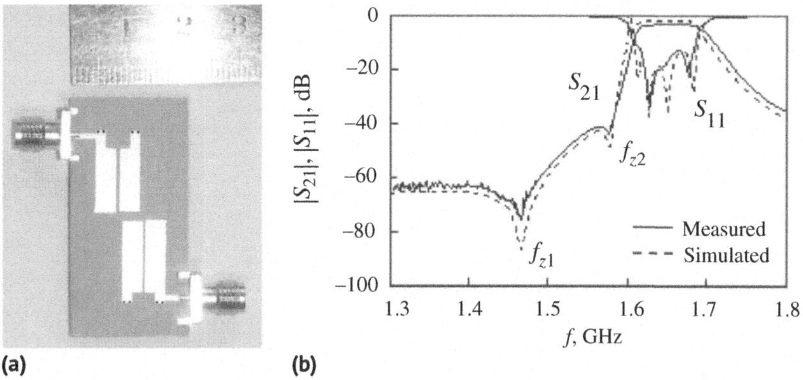

Figure 13.17 Microstrip mixed coupled dual-level BPF with left-sided TZs ($K_{12} < 0$, $K_{34} < 0$): (a) Photograph. (b) Frequency responses.

13.3.2 Microstrip Quadruplet BPF with Mixed Cross-coupling

Microstrip quadruplet BPF (Figure 13.18) was considered in Ref. [29]. A TMM-10i (Rogers) dielectric substrate was used for fabrication, with $\varepsilon_r = 9.8$, $\tan\delta = 0.002$, $h = 1.905$ mm. Let us consider two cases when the mixed cross-coupling is negative, $K_{14} < 0$ and positive $K_{14} > 0$. In both cases, the area occupied by the filter is 38 mm × 17.4 mm.

Figure 13.18a shows the quadruplet BPF topology for the case of $K_{14} < 0$ with dimensions indicated. Figure 13.18b shows a photograph of the filter. The filter contains two SIRs (1 and 4) and two modified quarter-wave resonators (2 and 3), whose resonant frequency is $f_0 = 1.74$ GHz. All resonators are short-circuited at one end using metallized holes with a diameter of 0.6 mm. In the short-circuit region of resonators 1 and 4, rectangles of size 6.2 mm × 1.2 mm are removed from their central conductors, thereby forming SIRs. Resonators 1 and 4 are coupled to resonators 2 and 3 through capacitive gaps. Therefore, the main coupling coefficients $K_{12} = K_{34}$ between the resonators are negative. The main coupling coefficient between resonators 2 and 3 is positive $K_{23} > 0$.

With the parameters shown in Figure 13.18a, the filter under consideration has the main coupling coefficients $K_{12} = K_{34} = -0.0330$, $K_{23} = 0.0319$. Resonators 2 and 3 are modified since the short circuit is implemented nonstandardly (Figure 13.18a). This technique made it possible to obtain a relatively small coupling coefficient $K_{23} = 0.0319$ value when using a thick ($h = 1.905$ mm) substrate and a relatively small gap between the resonators $S_{23} = 2.4$ mm.

The pair of SIRs 1 and 4 with the used parameters is characterized by odd $f_o = 1.709$ GHz and even $f_e = 1.736$ GHz modes of coupled oscillations and also generates TZ at the frequency $f_z = 1.60$ GHz. These values allow us to determine the coefficient of mixed coupling $K_{14} = -0.0157$ (3.10) and its components

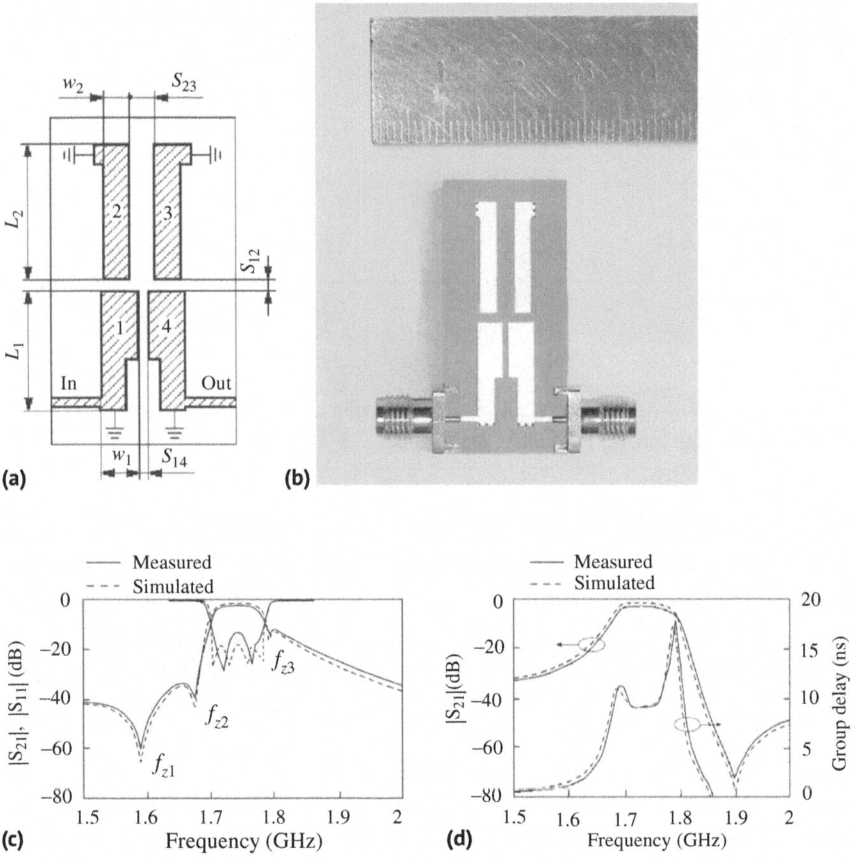

Figure 13.18 Microstrip quadruplet BPF with mixed cross-coupling. (a) Topology: $L_1 = 14$ mm, $w_1 = 3.6$ mm, $L_2 = 15.8$ mm, $w_2 = 2.4$ mm, $S_{12} = 1.3$ mm, $S_{14} = 0.8$ mm, and $S_{23} = 2.4$ mm. (b) Photograph. *Source*: [29] / with permission of Springer Nature (edited). (c) Frequency responses for case $K_{14} < 0$ and $K_{23} > 0$. (d) Frequency responses for case $K_{14} > 0$ and $K_{23} > 0$.

$K_m = 0.0989$, $K_e = -0.1146$ (7.5). The external quality factor of the extreme resonators Q_e is determined by the choice of the coordinate of connection of short sections 1.2 mm wide to resonators 1 and 4. In this case, $Q_e = 26$. The measured and simulated frequency responses of the filter under consideration are shown in Figure 13.18c. Since the main coupling coefficient K_{23} is positive and the cross-mixed coupling coefficient K_{14} is negative, the filter has three TZs at real frequencies, as indicated in Table 7.3 (position No. 1).

The filter has the following measured parameters: center frequency $f_0 =$ 1740 MHz; bandwidth BW = 75 MHz; insertion loss at center frequency

$L_{A0} = 1.5$ dB; return loss $L_R > 13$ dB. Three TZs are located at real frequencies $f_{z1} = 1590$ MHz, $f_{z2} = 1674$ MHz, $f_{z3} = 1788$ MHz. The waveguide design of such a quadruplet BPF was considered in Ref. [30].

The mixed cross-coupling of the filter in Figure 13.18a becomes positive $K_{14} > 0$, if the gap S_{14} between the SIRs is increased by 0.4 mm, from $S_{14} = 0.8$ mm to $S_{14} = 1.2$ mm. The main coupling coefficients of the filter $K_{12} = K_{34}$ and K_{23} remain unchanged. The pair of SIRs 1 and 4 with the gap between them $S_{14} = 1.2$ mm is characterized by even $f_e = 1.719$ GHz and odd $f_o = 1.7223$ coupled oscillation modes and also generates TZ at the frequency $f_z = 1.747$ GHz. These values allow us to determine the mixed coupling coefficient $K_{14} = 0.00157$ (3.10) and its components $K_m = 0.05221$, $K_e = -0.05064$ (7.5). The unloaded quality factor $Q_e = 21$ was used in this filter.

The frequency responses of the filter are shown in Figure 13.18d. Since $K_{23} > 0$ and mixed cross-coupling $K_{14} > 0$, the filter has a constant group delay with increased right-side selectivity, as indicated in Table 7.3 (item No. 7). The filter has the following measured parameters: center frequency $f_0 = 1740$ MHz; passband width BW = 75 MHz; insertion loss at center frequency $L_{A0} = 1.5$ dB, return loss $L_R > 18$ dB. One TZ is located at the real frequency $f_z = 1890$ MHz. The other two TZs are located equidistant on the σ-axis. The filter has the delay time $\tau = 9.2 \pm 0.2$ ns in a frequency band of 38 MHz, which is the half of passband. A similar filter was implemented in Ref. [31] based on substrate integrated waveguide technology.

Note that the microstrip quadruplet BPF in Figure 13.18 always has a positive sign of the main coupling coefficient K_{23}. Therefore, it cannot implement the arrangement of some TZs indicated in Table 7.3.

13.3.3 Microstrip Mixed Coupled BPFs with Coupling Segments

To regulate the electromagnetic interaction between nonadjacent resonators in microstrip mixed coupled BPFs, it is advisable to use "coupling" segments connecting these resonators through capacitive gaps [2]. The most suitable SIRs for use in such BPFs are the half-wave type, which has two open-circuited ends. An example of such a BPF with the number of resonators $N = 4$ is shown in Figure 13.19. The filter topology is shown in Figure 13.19a, where the dimensions are given in millimeters. The filter is fabricated on a dielectric substrate with $\varepsilon_r = 92$, $\tan \delta = 0.0004$, and $h = 1$ mm. The maximum width of each resonator is 2 mm, and the gap between the resonators is 0.2 mm. Two coupling segments connect resonators 1–3 and 2–4, regulating the mixed cross-coupling between these resonators. Resonators 1 and 4 are far from each other, so their parasitic interaction is minimal.

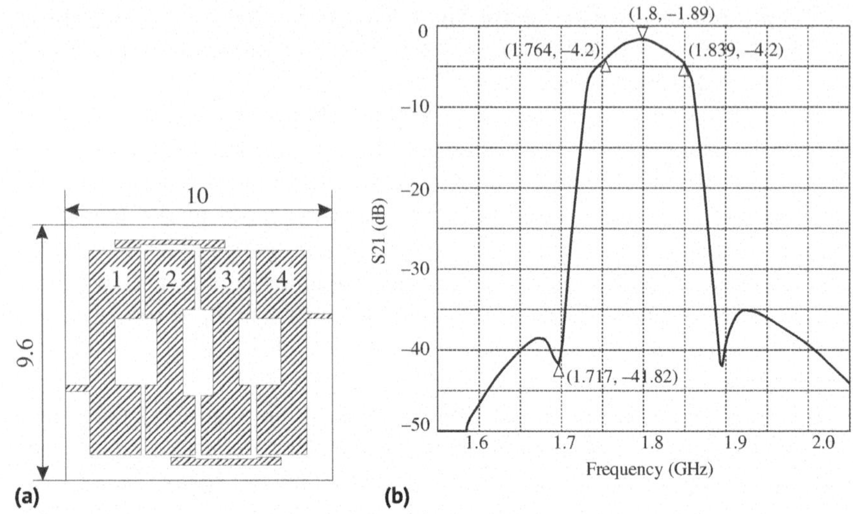

Figure 13.19 Fourth-order microstrip mixed coupled BPFs with coupling segments. (a) Topology. (b) Measured insertion loss.

There are three mixed couplings between the filter's adjacent resonators. The coupling between the two middle resonators is positive and generates the right-hand TZ. The other two couplings are negative, and they form the right-hand TZs. This filter's measured insertion loss function is shown in Figure 13.19b, where the frequencies (GHz) and corresponding insertion losses (dB) are given in parentheses. The filter has the following measured parameters: center frequency $f_0 = 1.8$ GHz; passband width BW = 75 MHz (FBW = 0.042); insertion loss at the center frequency $L_{A0} = 1.9$ dB. Since the two TZs are close to the passband, the filter has a relatively high-frequency selectivity of 35 dB ($f_0 \pm 75$ MHz) or 35 dB ($\pm j2$).

If the number of BPF resonators in Figure 13.19 increases by one, its frequency selectivity will also increase. The topology of the filter with the number of resonators $N = 5$ is shown in Figure 13.20a. The mixed couplings K_{23} and K_{34} of this filter are positive and generate the right-hand TZ. The mixed couplings K_{12} and K_{45} of this filter are negative, and they generate the left-hand TZ. Three coupling segments between resonators 1–3, 2–4, and 3–5 regulate the cross-coupling between these resonators.

This filter's measured insertion loss function is shown in Figure 13.20b. The filter is characterized by the following data: center frequency $f_0 = 1.845$ GHz; bandwidth BW = 85 MHz (FBW = 0.046); insertion loss at the center frequency $L_{A0} = 2.5$ dB. The frequency selectivity of the filter is characterized by the index of 40 dB ($f_0 \pm 75$ MHz) or 40 dB ($\pm j1.76$).

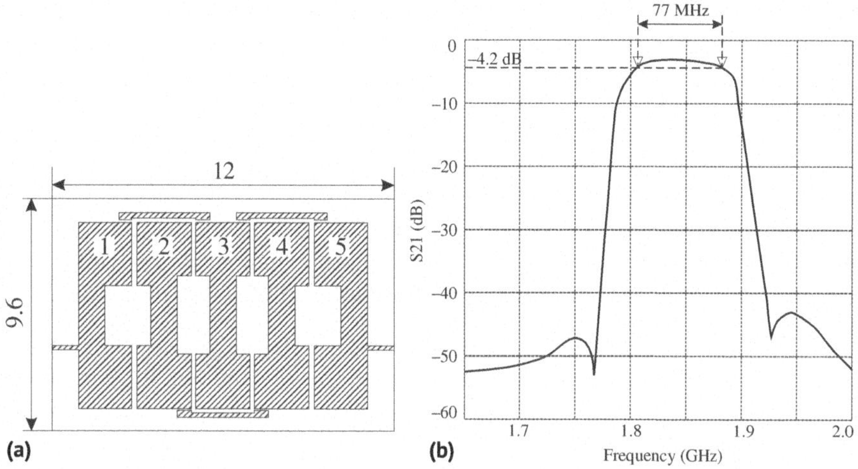

Figure 13.20 Fifth-order microstrip mixed coupled BPFs with coupling segments. (a) Topology. (b) Measured insertion loss.

13.4 Mixed Coupled Stepped-impedance Resonators in Microstrip BPFs on Thick Substrates and Low-mounted Screen

In most microstrip BPFs, thin substrates whose thickness does not exceed $h \approx 1$ mm are used. Due to thin substrates, the unloaded quality factor Q_u of microstrip resonators is relatively low and does not exceed $Q_u \approx 250 - 300$. As a rule, the metal screen was previously installed at a height H, which is eight times the thickness of the substrate $H = 8h$, to eliminate the screen's influence on the filter's frequency responses. As a result, the BPF occupies a significant volume, and the small Q_u value of the resonators makes it unsuitable for the implementation of narrow passbands, the relative width of which (FBW) is FBW = 0.01–0.03. This drawback can be eliminated in three stages.

Stage 1: An increase in the value of Q_u of microstrip resonators can be achieved by increasing the substrate thickness h, making it possible to realize narrow passbands. The most compact are the combline and pseudocombline [1] topologies of BPFs, which use quarter-wave or half-wave resonators parallel to each other without mutual shift. The implementation of narrow bandwidths requires the use of small coupling coefficients K between the resonators, which leads to increased intervals between them. These intervals will increase even further with increasing substrate thickness h, which will significantly increase the lateral dimensions of microstrip BPFs.

Stage 2: The lateral dimensions of narrowband microstrip BPFs can be reduced by using mixed coupling between resonators. To implement mixed couplings, closely spaced microstrip SIRs can be used, which has not been sufficiently studied when thick substrates and a low screen are used.

Stage 3: The advent of EM simulators makes it possible to consider the screen's influence on the frequency responses of microstrip BPFs. This makes it possible to position the screen at a low height, $H = (1 - 3)h$, and reduce the filter volume. Let the screen influence the filter performance, but simulation takes this influence into account. The screen is involved in the formation of frequency responses.

The above dictates the need for a detailed study of the mixed coupling coefficients between SIRs in microstrip combline and pseudocombline BPFs on thick substrates and a low screen. This knowledge will form the basis for the design of microstrip narrowband BPFs.

13.4.1 General Patterns

To provide a relatively small FBW value, it is necessary to use microstrip resonators with an increased unloaded quality factor Q_u. Figure 13.21 shows the dependences of Q_u of half-wave microstrip resonators on their width w and substrate thickness h for two resonant frequencies. (sic)

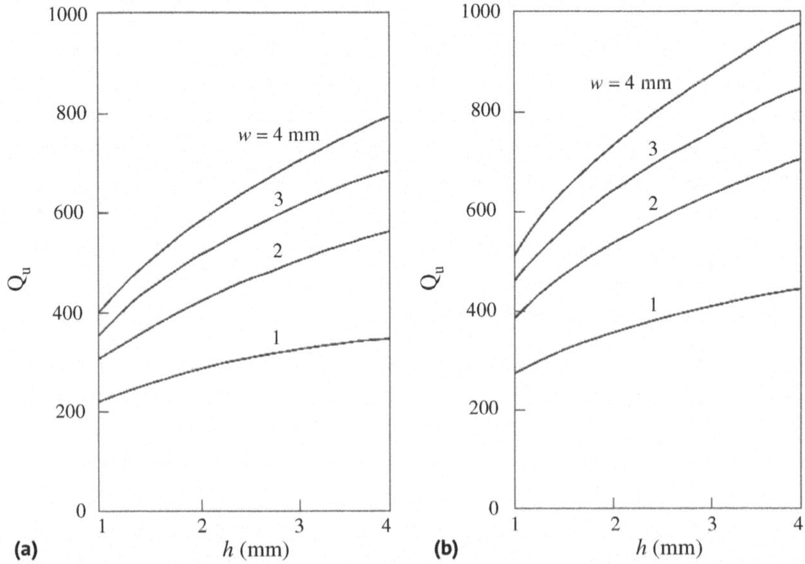

Figure 13.21 Unloaded quality factor Q_u of microstrip half-wave resonators versus their width w and substrate thickness h. (a) $f_0 = 2$ GHz. (b) $f_0 = 3$ GHz.

thickness h, calculated using formulas from Section 4.4.1 for the case $\varepsilon_r = 9.7$. As h and w increase, the value of Q_u increases significantly. So, at $h = 3$ mm and $w = 4$ mm, the value is $Q_u \approx 700$ at the frequency of 2 GHz and $Q_u \approx 850$ at the frequency of 3 GHz. Note that a change in the dielectric constant ε_r has little effect on the value of Q_u. When calculating Q_u values, the roughness of the substrate and losses in the adhesion layer were not considered. The influence of these factors reduces the Q_u values of real microstrip resonators, as a rule, by 10–15%. If the resonator has a short-circuited end, the Q_u values are reduced by another 5–10%.

The dependences in Figure 13.21 show that to increase Q_u, the thicker substrates must be used, but this leads to a negative effect, expressed by a significant increase in the transverse dimensions of combline and pseudocombline BPFs containing parallel $\lambda/4$ or $\lambda/2$ resonators, respectively.

This effect is illustrated in Figure 13.22, which shows the dependence of the coupling coefficient K of two quarter-wave microstrip resonators ($f_0 = 1.75$ GHz) with a width $w = 2$ mm ($\varepsilon_r = 92$) on the size of the gap S between them. The EM simulator of Microwave Office software and the following formula (3.10) were used to build this dependence.

As S increases, the K values decrease monotonically. To realize the value $K = 0.03$, it is necessary to use the following gaps: $S = 3.6$ mm for $h = 1$ mm; $S = 7$ mm for $h = 2$ mm; $S = 10$ mm for $h = 3$ mm; $S = 13$ mm for $h = 4$ mm. To realize the values of $K < 0.03$, it is necessary to use even more significant gaps. When thick substrates are used, the Q_u value increases, but a substantial increase

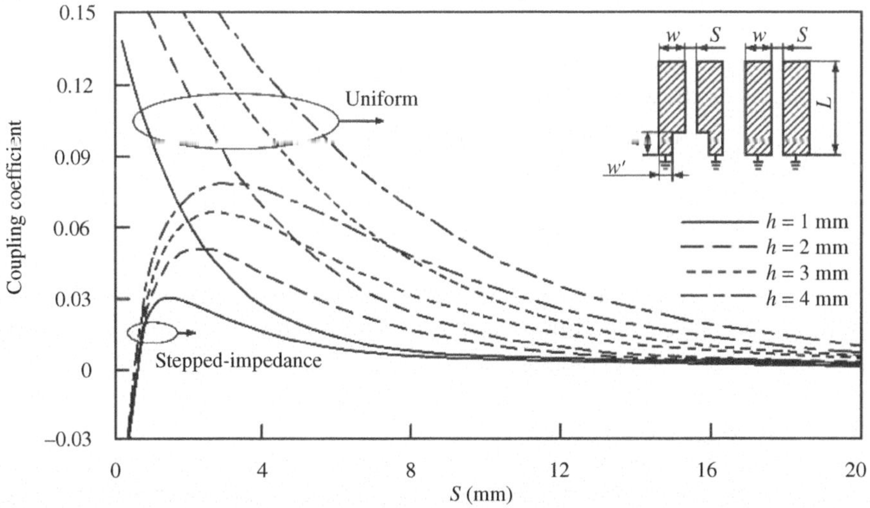

Figure 13.22 Coupling coefficient between microstrip uniform and SIRs, $L = 5$ mm, $w = 2$ mm, $l = 1.5$ mm, $w' = 1$ mm, $\varepsilon_r = 92$.

in the lateral dimensions of these BPFs accompanies this. This increase in size can be avoided by using the SIRs shown in the inset of Figure 13.22. The dependence $K = K(S)$ of the SIRs differs significantly from the similar dependence of uniform resonators. As S decreases, the K value first increases, reaching its maximum value, and then decreases, taking zero ($K = 0$) and negative ($K < 0$) values. Small values of S ($S < 1$ mm) are required to realize small values of K, even when using thick substrates. Moreover, K values can be both positive and negative. The dependences in Figure 13.22 are plotted for a quarter-wave SIR. The same dependencies are also valid for a half-wave SIR, in which the narrower part is located in the middle part of the resonator.

13.4.2 Mixed Couplings Between Microstrip SIRs on Thick Substrate and Low Screen

We will conduct a detailed analysis of mixed coupling between microstrip SIRs with the thick substrate ($h = 2$ mm, $\varepsilon_r = 92$) and low screen ($H = 3$ mm) as follows.

We will consider two variants of changing the shape of SIRs: when changes occur in the vertical direction (Figure 13.23) and when the shape of SIRs changes in the horizontal direction (Figure 13.24).

Figure 13.23 shows the dependence of K on the vertical change in the shape of closely spaced microstrip SIRs, where their geometric dimensions are also indicated. These dependencies are plotted for two different resonator shapes as shown in the insets when l varies starting from the short-circuited end (Figure 13.23a) and from the open end (Figure 13.23b).

As the length l of the narrow part of the resonator in Figure 13.23a, whose width is w', increases, the value of K initially decreases to zero. Then, it becomes negative and takes on the minimum value. After this, the value of K increases again, takes a zero value, and then becomes a positive value. The function $K = K(l/L)$ is convex downward, the minimum of which is provided at the ratio $l/L \approx 1/2$. With a gap between the resonators $S = 0.2$ mm, the K values vary within the $-0.087 \leq K \leq 0.12$ range.

For a pair of resonators (Figure 13.23b), the coupling coefficient K is always positive. The shape of the curve $K = K(l/L)$ is close to bell-shaped; the maximum of the function under consideration is achieved at a value of $l/L \approx 1/2$. With the gaps between the resonators $S = 0.2$ mm and $S = 0.4$ mm, the value of K varies within the range: $0.062 \leq K \leq 0.12$.

In Figure 13.23a, points *1′*, *2′*, *3′*, and *4′* correspond to the coordinate's l/L, providing the same positive values of the coupling coefficients $K = 0.03$. Each curve in Figure 13.23a intersects line $K = 0.03$ at two points. The gap between

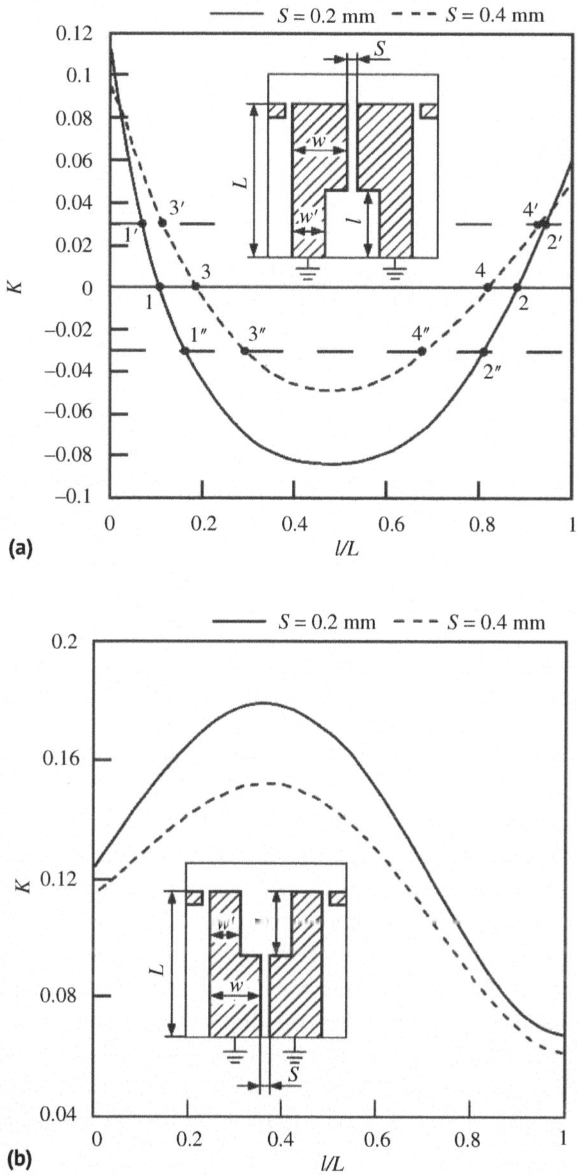

Figure 13.23 Coupling coefficient between microstrip SIRs when their shape changes vertically: $\varepsilon_r = 92$, $h = 2$ mm, $H = 3$ mm, $L = 5$ mm, $w = 2$ mm, $w' = 1.2$ mm, $S = 0.2$ and 0.4 mm, l = var. (a) Short-circuited end of SIR is narrow. (b) Open-circuited end of SIR is narrow.

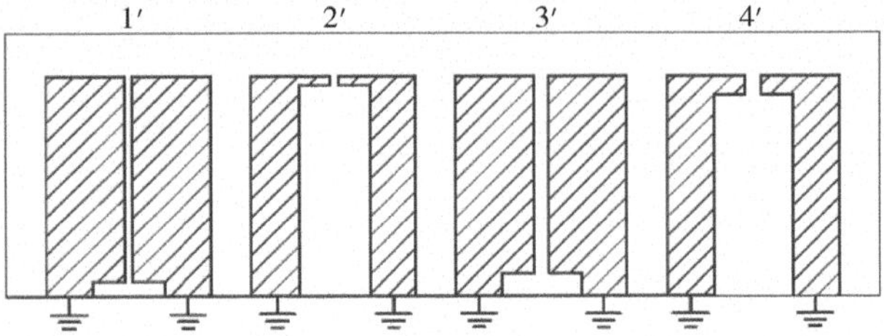

Figure 13.24 Pairs of SIRs with positive mixed coupling coefficient $K = 0.03$ corresponding to points $1', 2', 3', 4'$ in Figure 13.23a.

Table 13.3 Components of mixed coupling $K = 0.03$ at points $1'$, $2'$, $3'$, $4'$ in Figure 13.23a.

No.	$1'$	$2'$	$3'$	$4'$
S, mm	0.2	0.2	0.4	0.4
l/L	0.06	0.95	0.1	0.93
f_z/f_0	1.053	1.156	1.066	1.203
K_m	0.306	0.119	0.250	0.097
K_e	-0.276	-0.089	-0.220	-0.067
a	0.291	0.103	0.235	0.081

the resonators $S = 0.2$ mm corresponds to the points $1'$ and $2'$, and the gap $S = 0.4$ mm corresponds to the points $3'$ and $4'$. Figure 13.24 shows the topologies of pairs of resonators that correspond to the indicated points. These pairs have the same value, $K = 0.03$, but different values TZ (f_z). Using EM simulation and formulas (7.5), the components of the mixed coupling K_m, K_e, and its slope parameter at the indicated points are determined. These values are contained in Table 13.3.

In Figure 13.23a, points $1''$, $2''$, $3''$, and $4''$ correspond to the coordinate's l/L, at which a negative value of the coupling coefficient $K = -0.03$ is ensured. Figure 13.25 shows the topologies of pairs of resonators that correspond to these points. These pairs are characterized by the same value of $K = -0.03$ but differ in the position of the TZ f_z they generated. The mixed coupling components K_m and K_e corresponding to these pairs and the slope parameter a are contained in Table 13.4.

Figures 13.24 and 13.25 and data from Tables 13.3 and 13.4 show that the given value K of the mixed coupling coefficient can have several pairs of SIRs with different gaps S between resonators with various shapes. Pairs of SIRs with the same K value differ in the relative positions of f_z and f_0, reflected in the different components K_m, K_e, and the slope parameter a. The gap between the SIRs and their shape makes changing the relative position of f_z and f_0 for the given K possible.

Table 13.4 Components of mixed coupling $K = -0.03$ at points 1″, 2″, 3″, 4″ in Figure 13.23a.

No.	1″	2″	3″	4″
S, mm	0.2	0.2	0.4	0.4
l/L	0.17	0.81	0.3	0.69
f_z/f_0	0.947	0.898	0.926	0.894
K_m	0.261	0.125	0.180	0.119
K_e	–0.291	–0.155	–0.210	–0.149
a	0.275	0.139	0.194	0.133

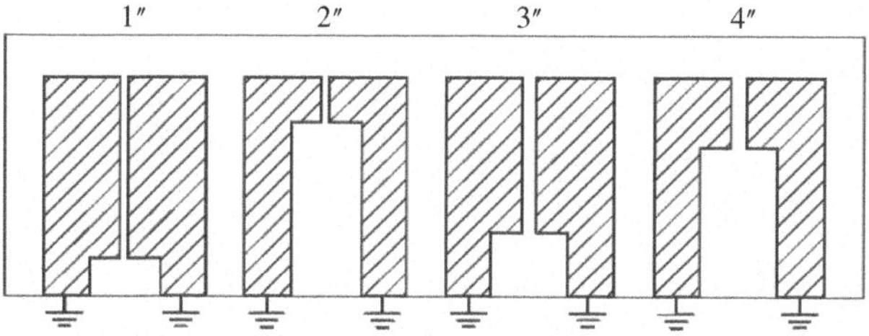

Figure 13.25 Pairs of SIRs with negative mixed coupling coefficient $K = -0.03$ corresponding to points 1″, 2″, 3″, 4″ in Figure 13.23a.

In many cases, achieving the maximum convergence of the frequencies f_z and f_0 at a given value of K is desirable. The performed analysis indicates the following patterns:

1. For a fixed gap S between SIRs, a smaller l/L ratio is accompanied by a smaller frequency interval between f_z and f_0.
2. Reducing the gap S between the SIRs allows for a greater degree of convergence between f_z and f_0.
3. The limitation on the convergence degree of the frequencies f_z and f_0 is determined by the minimum permissible gap S between the SIRs.

Mixed coupling, in which the magnetic and electrical components compensate for each other $K_m = |K_e|$, $K = 0$ $(f_z = f_0)$, is resonant coupling [3]. Its use as a K_{13} cross-coupling in triplet BPFs leads to quasi-elliptic frequency response or flat group delay. In Figure 13.23a, points *1*, *2*, *3*, and *4* correspond to the coordinate's l/L, at which resonant coupling $K = 0$ is ensured. The l/L values at these points are contained in Table 13.5. They occupy an intermediate position between similar values in Tables 13.3 and 13.4. This is also reflected in Figure 13.26, which shows the topologies of pairs of SIRs with resonant coupling.

Table 13.5 Components of mixed resonant coupling $K = 0$ at points *1, 2, 3, 4* in Figure 13.23a.

No.	*1*	*2*	*3*	*4*
S, mm	0.2	0.2	0.4	0.4
l/L	0.11	0.89	0.18	0.83
f_z/f_0	1	1	1	1
$a = K_m = \|K_e\|$	0.283	0.121	0.215	0.107

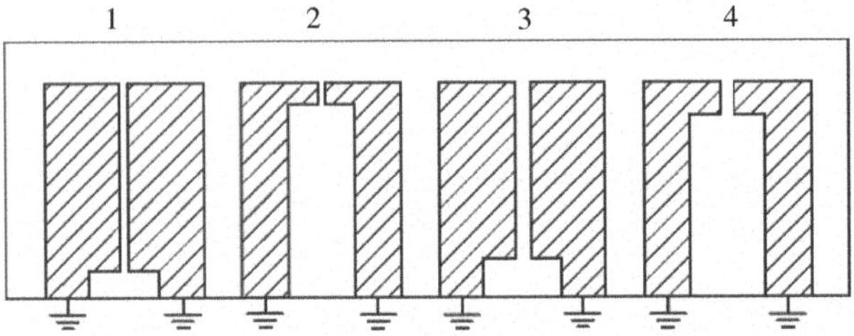

Figure 13.26 Pairs of SIRs with mixed resonant coupling $K = 0$ corresponding to points *1, 2, 3, 4* in Figure 13.23a.

Table 13.5 also contains the values of the slope parameter a. Since with resonant coupling $f_z = f_0$, expressions (7.5) become unsuitable for calculating K_m, K_e, and the slope parameter a. Let us denote the small values of the mixed coupling coefficients K_1 and K_2, which have opposite signs and satisfy the equality $K_2 = -K_1$. Let these couplings be characterized by the corresponding slope parameters a_1 and a_2, which can be determined using (7.5). Then, the parameter a of resonant coupling is determined by the approximate equality

$$a \approx \frac{(a_1 + a_2)}{2}. \tag{13.2}$$

The smaller the values of K_1 and K_2, the more accurately the approximate equality (13.2) is satisfied.

The values of the slope parameter a in Table 13.5 satisfy equality (13.2), in which a_1 and a_2 are taken from Tables 13.3 and 13.4, respectively. The minimum permissible value of S limits the maximum possible value of a.

The aforementioned principles are valid when using substrates with different dielectric constants ε_r and half-wave type SIRs.

Figure 13.27 shows the dependence of K on the horizontal change in shape of closely spaced microstrip SIRs. Their geometric dimensions are also indicated and do not require additional explanation. These dependencies are plotted for two different shapes of SIRs when w' changes near the short-circuited end

(Figure 13.27a) and near the open-circuited end (Figure 13.27b). For the pair of SIRs in Figure 13.27a, the K values vary within the range of $-0.120 \leq K \leq 0.116$, and for the pair of SIRs in Figure 13.27b, the change in K is limited to the interval $-0.067 \leq K \leq 0.02$. Notably, the dependence $K = K(w')$ in both cases is linear, significantly simplifying designing BPFs based on these resonators.

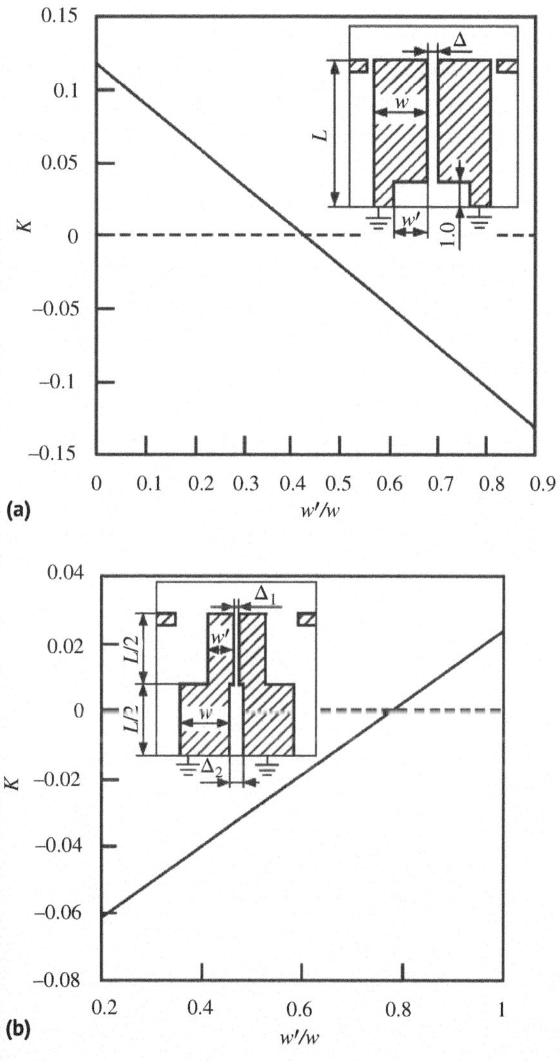

Figure 13.27 Mixed coupling coefficient between microstrip SIRs when their shape changes horizontally: $\varepsilon_r = 92$, $h = 2$ mm, $H = 3$ mm, $L = 5$ mm, $w = 2$ mm, $w' =$ var. (a) Short-circuited end of SIR is narrow, $\Delta = 0.4$ mm. (b) Open-circuited end of SIR is narrow, $\Delta_1 = 0.1$ mm, $\Delta_2 = 0.5$ mm.

The aforementioned considerations indicate that mixed coupling coefficients can be implemented in pairs of microstrip SIRs on a substrate with $h = 2$ and a low-mounted screen ($H = 3$ mm), varying over a wide range.

13.4.3 Influence of Screen Height and Substrate Thickness on Mixed Coupling Coefficients

Figure 13.28 shows the graphs of the influence of the screen height H on the mixed coupling coefficient and its component of the SIRs pair in Figure 13.27a for $w' = 0.6$ mm. With decreasing H, the components of the mixed coupling K_m and $|K_e|$ decrease, with the values of K_m decreasing faster than $|K_e|$. The slope of the curves $K_m\,(H)$ and $|K_e\,(H)|$ is positive. At some value of H_0, both curves intersect, corresponding to the resonant coupling case $|K_e| = K_m$ ($K = 0$). The nature of the change in K_m and $|K_e|$ determines the nature of the change in the mixed coupling coefficient K. The $K\,(H)$ function also has a positive slope. The values of $K < 0$, $K = 0$, and $K > 0$ for $H < H_0$, $H = H_0$, and $H > H_0$, respectively. For arbitrary parameters of the SIRs pair, the curves $K_m\,(H)$ and $|K_e\,(H)|$ have a positive slope but do not always intersect. In these cases, the value of K is either always positive or always negative; it does not take the zero value. Note that the components K_m and $|K_e|$ can significantly exceed the coefficient of the mixed coupling K.

Since the slope parameter of the mixed coupling is the geometric mean of the values of K_m and $|K_e|$, the graph of the dependence $a\,(H)$ is always located between the curves $K_m\,(H)$ and $|K_e\,(H)|$. The function $a\,(H)$ also increases. The

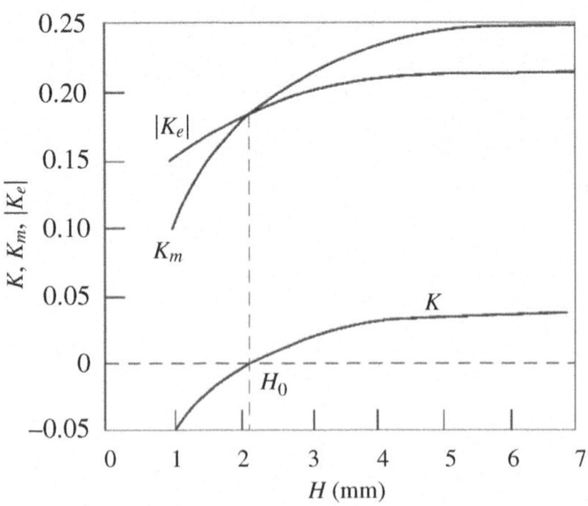

Figure 13.28 Influence of screen height on mixed coupling coefficient and its component of SIRs pair in Figure 13.27a at $w' = 0.6$ mm.

aforementioned regularities are also characteristic of the pair of resonators in Figure 13.27b.

For comparison, Figure 13.29 shows the effect of the screen height on the coupling coefficient of a pair of microstrip $\lambda/4$ resonators, corresponding to the pair of resonators in Figure 13.27a at $w' = 0$. The nature of the change in the $K(H)$ dependence (Figure 13.29) is the same as in Figure 13.28. Only the values of K are always positive. Thus, the screen height H can adjust the mixed coupling between microstrip SIRs, ensuring the value $K = 0$.

An important characteristic of mixed coupling is slope parameter a, which increases the BPF selectivity. To calculate a, using the components K_m and $|K_e|$ is not necessary. If we substitute the components from (7.5) into the expression $a = \sqrt{K_m\,|K_e|}$, it will take the form

$$a = \frac{K}{f_z/f_0 - f_0/f_z}, \tag{13.3}$$

where f_0 is the center frequency between the even f_e and odd f_o modes: $f_0 = (f_o + f_e)/2$.

Calculating a using three characteristic frequencies, f_e, f_o, and f_z, is sufficient. The closer f_z to f_0, the smaller the denominator (13.3) and the larger the parameter a for a given coupling coefficient K. This is the physical meaning of the slope parameter a. The approach of f_z to f_0 is limited by the frequency f_o of the odd oscillation mode. In the case when $f_z \to f_o$, the upper boundary of the values $\{a\}$ follows from (13.3)

$$\text{Sup}\,\{a\} = \frac{2+K}{2+K/2}. \tag{13.4}$$

Equation (13.4) shows that an increase in K is accompanied by a rise in Sup $\{a\}$. Since $\text{Sup}\,\{K\} = 2$, then from (13.4), it follows

$$\text{Sup}\,\{\text{Sup}\,a\} = \frac{4}{3}. \tag{13.5}$$

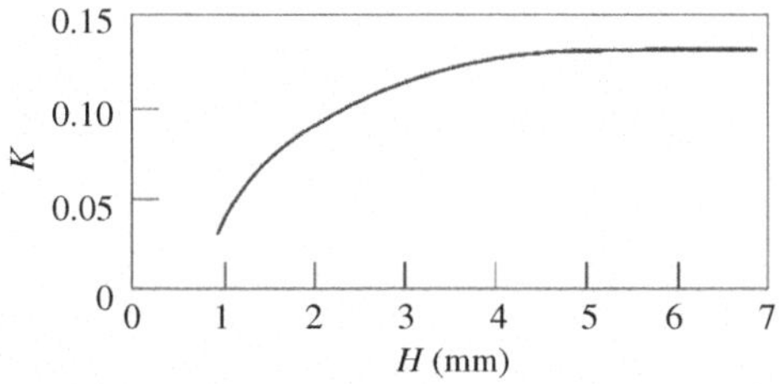

Figure 13.29 Influence of screen height on coupling coefficient between microstrip $\lambda/4$ resonators.

The upper boundaries (13.4) and (13.5) are practically unattainable but represent a specific reference point with scientific significance.

Tables 13.3–13.5 show the a values for pairs of SIRs with gaps $\Delta = 0.2$ mm and $\Delta = 0.4$ mm. Reducing the gap Δ leads to an increase in the parameter a. At $\Delta = 0.2$ mm, the value $a = 0.291$ was obtained (Table 13.3). To further increase a, we use the gap $\Delta = 0.1$ mm and establish the possibility of increasing a using thicker substrates.

Figure 13.30 shows the graphs of the influence of the substrate thickness h on the mixed coupling coefficient K and parameter a of the SIRs pair in Figure 13.27a with $\Delta = 0.1$ mm and $H = 20$ mm. Figure 13.30a corresponds to a positive coupling coefficient K, which is ensured by choosing the width $w' = 0.4$ mm. In Figure 13.30b, the values of K are negative, which is ensured by the width $w' = 0.7$ mm. Note that the change slope in K will be positive at width $w' = 0.5$ mm, but the K values will be negative and positive. The value $K = 0$, corresponding to the resonator coupling, will also be realized. In all cases, the change slope in K and a is positive. The highest slope parameter value $a = 0.428$ is achieved at the substrate thickness $h = 5$ mm (Figure 13.30a). In this figure, a value varied within the $0.340 \le a \le 0.428$ range.

Figure 13.31 shows the influence of the substrate thickness h on K and a of the SIRs pair in Figure 13.27b at $\Delta_1 = 0.1$ mm and $H = 20$ mm. Figure 13.31a corresponds to the gap $\Delta_2 = 0.3$ mm and width $w' = 0.7$ mm, providing positive K values. In Figure 13.31b, the K values are negative, positive, and equal to zero, provided by the gap $\Delta_2 = 0.4$ mm and width $w' = 0.7$ mm. As in the previous case, the dependences $K(h)$ and $a(h)$ have a positive slope. In Figure 13.31a,

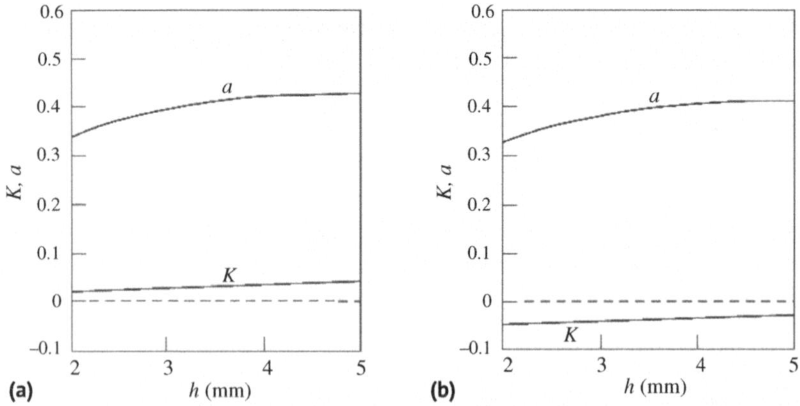

Figure 13.30 Influence of substrate thickness h on mixed coupling coefficient and slope parameter a of SIRs pair in Figure 13.27a at $\Delta = 0.1$ mm and $H = 20$ mm. (a) For case $w' = 0.4$ mm, $K > 0$. (b) For case $w' = 0.7$ mm, $K < 0$.

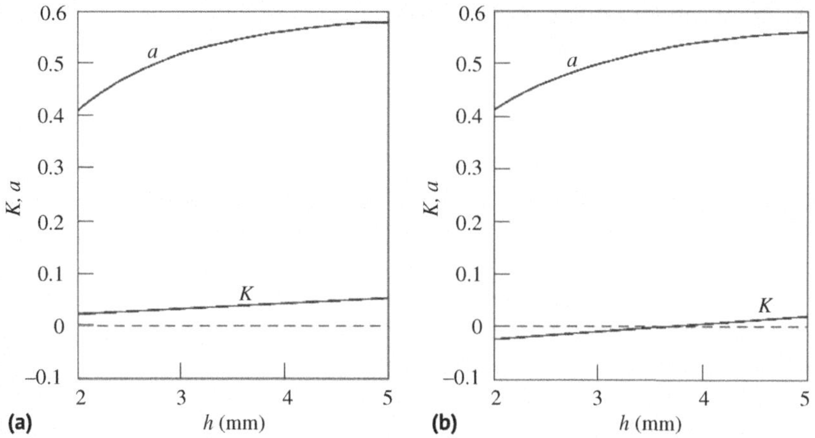

Figure 13.31 Influence of substrate thickness h on mixed coupling coefficient and slope parameter a of SIRs pair in Figure 13.27b at $\Delta_1 = 0.1$ mm and $H = 20$ mm. (a) For case $\Delta_2 = 0.3$ mm and $w' = 0.7$ mm. (b) For case $\Delta_2 = 0.4$ mm and $w' = 0.7$ mm.

the slope parameter a changes in the $0.422 \leq a \leq 0.576$ range. The highest value of $a = 0.576$, corresponding to the coupling coefficient $K = 0.0521$, was achieved at $h = 5$ mm. According to (13.4), the value $K = 0.0521$ corresponds to the upper boundary $\text{Sup}\{a\} = 1.014$, which is 1.76 times greater than the a value obtained at $h = 5$ mm.

This difference can be reduced by increasing the substrate thickness to $h = 10$ mm and taking the gaps between the SIRs in Figure 13.27b to be $\Delta_1 = 0.025$ mm and $\Delta_2 = 0.075$ mm. These parameters will increase the mixed coupling coefficient $K = 0.0717$ and slope parameter $a = 0.689$. In this case, the upper boundary (13.4) $\text{Sup}\{a\} = 1.018$ is only 1.48 times greater than the obtained value of a. There may be other configurations of microstrip resonators with mixed coupling, the steepness parameter a of which is even closer to the boundary value (13.4).

The analysis results in Figures 13.30 and 13.31 indicate that the substrate thickness h can regulate the mixed coupling between microstrip SIRs, including ensuring the value of $K = 0$. An increase in h leads to an increase in the slope parameter a of mixed coupled SIRs. An increase in h is accompanied by the rise in the unloaded quality factor Q_u of microstrip SIRs, while the gap between SIRs remains the same. This triple effect makes using thick substrates in microstrip BPF with SIRs promising.

13.4.4 New Configurations of Microstrip Filters with SIRs

The topology of a pair of microstrip SIRs shown in Figure 13.27b is new. At the same time, it provides a more significant slope parameter a of mixed coupling than the more common pair of SIRs shown in Figure 13.27a. It is advisable to consider BPF based on this pair of SIRs. Figure 13.32 shows simulated frequency responses of second-order microstrip BPFs with the new configuration. We used the EM simulator of Microwave Office software. Both filters use a substrate with a dielectric constant $\varepsilon_r = 92$ and thickness $h = 2$ mm. The height of the screen above the substrate is $H = 3$ mm, and the size of each filter is 6.5 mm × 5.7 mm × 5.0 mm. The insets show the filter topologies that differ from the topology in Figure 13.27b in gaps Δ_1, Δ_2, and width w'.

In the BPF in Figure 13.32a, the gaps are $\Delta_1 = 0.1$ mm, $\Delta_2 = 0.3$ mm, and the width is $w' = 0.6$ mm. This ensures a positive mixed coupling between the resonators $K > 0$ and the right-hand TZ. The filter is matched to the loads by selecting the coordinate of their conductive connection to the resonators. With a weak coupling of the resonators to the loads, we determine three characteristic frequencies $f_e = 2072$ MHz, $f_o = 2108$ MHz, $f_z = 2136$ MHz, which allows us to calculate the mixed coupling coefficient $K = 0.0172$ (3.10) and its slope parameter (13.3): $a = 0.395$. This filter has the following parameters: central frequency $f_0 = 2093$ MHz, bandwidth by level 0.2 dB is $\mathrm{BW}_{0.2\,\mathrm{dB}} = 28$ MHz, return loss in the passband $L_R = 14$ dB, $f_z = 2139$ MHz. The presence of TZ leads to an increase in the filter's right-side selectivity. An attenuation of 20 dB is provided with shifting from the right edge of the passband by 22 MHz.

In the BPF in Figure 13.32b, the gaps are $\Delta_1 = 0.1$ mm, $\Delta_2 = 0.5$, and the width $w' = 0.6$ mm. With these parameters, the mixed coupling between the SIRs is negative $K < 0$, which a left-sided TZ accompanies. With a weak coupling of the resonators with the loads, we determine the characteristic frequencies $f_e = 2070$ MHz, $f_o = 2034$ MHz, and $f_z = 2005$ MHz, which allows us to calculate the mixed coupling coefficient $K = -0.0175$ (3.10) and its slope parameter (13.3): $a = 0.378$. This filter is characterized by the following parameters: $f_0 = 2055$ MHz, $\mathrm{BW}_{0.3\,\mathrm{dB}} = 36$ MHz, $L_R = 12$ dB, $f_z = 1999$ MHz. The presence of TZ leads to an increase in the filter's left-side selectivity. An attenuation of 20 dB is provided with shifting from the left edge of the passband at 24 MHz.

Figure 13.33a shows the topology of a symmetrical two-level microstrip triplet BPF, with the resonators arranged on two levels. The SIRs 1 and 3 are located next to each other. They form a mixed cross-coupling K_{13}. The central resonator 2 of this filter is half-wave. It is connected to resonators 1 and 3 by simple couplings $K_{12} = K_{23}$, which are positive. The filter uses a substrate with a permittivity of $\varepsilon_r = 92$, $h = 2$ mm, and the screen's distance above the substrate is $H = 3$ mm.

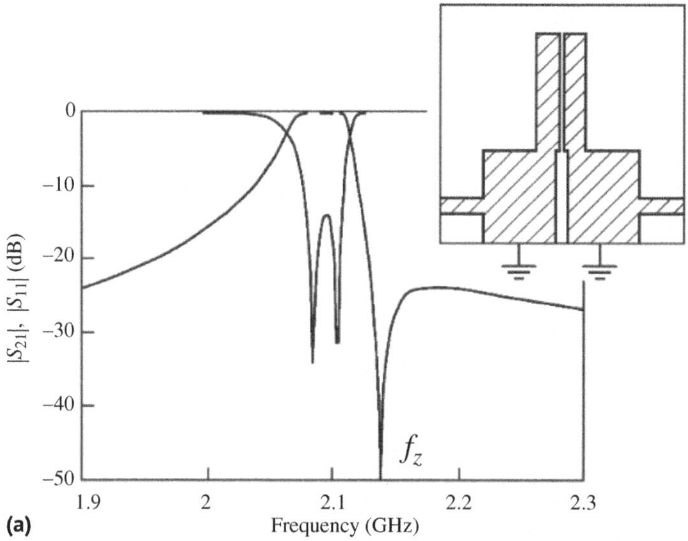

(a)

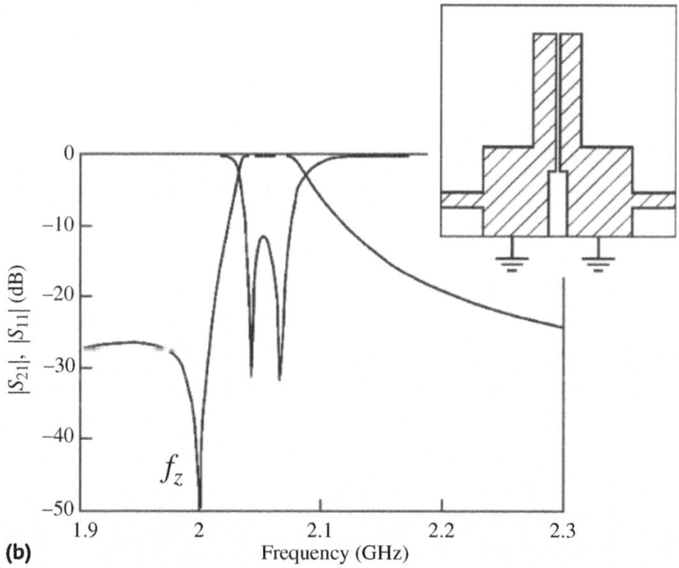

(b)

Figure 13.32 Simulated frequency responses of second-order microstrip BPFs with SIRs shown in Figure 13.27b. (a) Sharp right slope, $K > 0$, $\Delta_1 = 0.1$ mm, $\Delta_2 = 0.3$ mm, $w' = 0.6$ mm. (b) Sharp left slope, $K < 0$, $\Delta_1 = 0.1$ mm, $\Delta_2 = 0.5$ mm, $w' = 0.6$ mm.

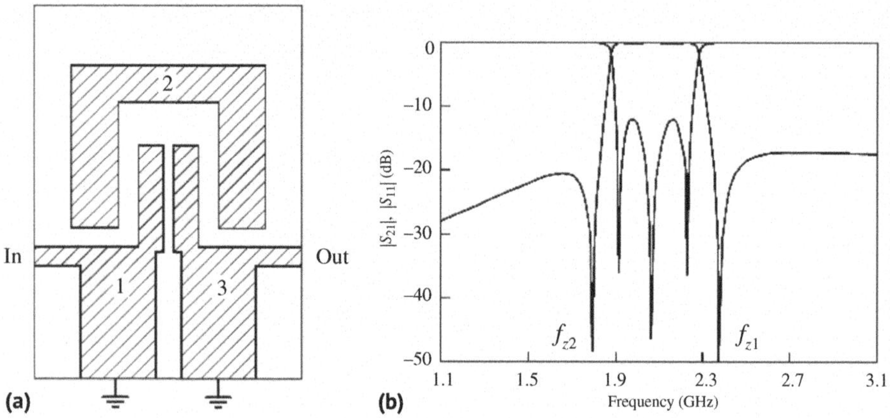

Figure 13.33 Two-level microstrip triplet BPF with closely spaced resonators 1 and 3. (a) Topology. (b) Simulated frequency responses.

The filter size is 8 mm × 6.4 mm × 5 mm. The filter is based on a pair of previously used closely spaced SIRs, the topology of which differs from the topology in Figure 13.27b in the size of the gaps Δ_1, Δ_2 and the width w': $\Delta_1 = 0.2$ mm, $\Delta_2 = 0.6$ mm, $w' = 0.6$ mm. With the used parameters, the mixed cross-coupling K_{13} becomes resonator $K_{13} = 0$ with slope parameter $a = 0.19$. The width of the central half-wave resonator is 1.1 mm. This resonator is coupled to the SIRs by coupling coefficients $K_{12} = K_{23} = 0.1215$, which correspond to coupling frequencies $f_e = 1948$ MHz, $f_o = 2200$ MHz, and normalized value $m_{12} = m_{23} = 0.719$. With resonant cross-coupling, this triplet BPF generates two equidistant TZs at real frequencies, the normalized value of which $j\Omega_{z1,2}$ is determined by the equality (7.46): $j\Omega_{z1,2} = \pm jm_{12}\sqrt{a}$. Substituting m_{12} and a into this equality gives the calculated value: $j\Omega_{z1,2} = \pm j1.65$.

Figure 13.33b shows the simulated frequency responses of this filter. This BPF has the following parameters: $f_0 = 2085$ MHz, $\text{BW}_{0.3\text{ dB}} = 352$ MHz (FBW = 0.169), $L_R = 12$ dB. Its two TZs are located at frequencies $f_{z1,2} = f_0 \pm 294$ MHz, corresponding to the normalized value $j\Omega_{z1,2} = \pm j1.64$, which is very close to the calculated value. The presence of two TZs close to the passband provides the filter with increased selectivity of 20 dB ($f_0 \pm 264$ MHz) or 20 dB ($\pm j1.5$).

If the mixed cross-coupling is not resonant $K_{13} \neq 0$, then two TZs in the triplet BPF will not be equidistant. At $K > 0$, the left-hand TZ will be closer to the passband than the right-hand one. At $K < 0$, the situation will be the opposite. The asymmetric arrangement of two TZs can be used in duplexers. When cascading BPFs of the new configuration (Figures 13.32 and 13.33), a transition to higher-order BPFs is achieved.

13.4.5 Design of Narrowband Quasi-elliptic Microstrip BPF with Low Screen

Let us implement the design of narrowband (FBW ≤ 0.02) microstrip BPF with a low-mounted screen and quasi-elliptic frequency response. The filter parameters will be set as follows: center frequency $f_0 = 1615$ MHz, BW = 30 MHz (FBW = 0.0186), and passband ripple $L_{\mathrm{Ar}} = 0.1$ dB, which corresponds to the return loss $L_R = 16.426$ dB. Table 2.6 shows g parameters: $g_0 = 1, g_1 = g_3 = 1.0315$, $g_2 = 1.1474$. Substituting these parameters and FBW value into formulas (3.8), we determine the coupling coefficients $K_{12} = K_{23} = 0.0171$ and the external quality factor of end resonators $Q_{e1} = Q_{e3} = Q = 55.46$.

To implement the initial data, we use a substrate with a dielectric constant $\varepsilon_r = 36$, the thickness of which is $h = 2$ mm, $\tan\delta = 0.0004$, and the BPF topology (Figure 13.12a) with half-wave SIRs and port positions $1'$–$2'$. The area occupied by the substrate is 23 mm × 15 mm. The screen is installed above the substrate at a height of $H = 3$ mm. This filter is designed for surface mounting, so double-sided photolithography was used in its manufacture, which is reflected in the topologies for the face side (Figure 13.34a) and for the back side (Figure 13.34b). The side ends of the substrate are also metallized.

The filter uses fairly wide resonators $w = 3.4$ mm, the length of which is $L = 17.4$ mm and separated by the gap $S = 0.4$ mm. In the central part, the resonators have a narrower part with a width of $w' = 2.2$ mm and length $l = 2.4$ mm.

If $w' = w = 3.4$ mm, then between closely spaced resonators 1 and 2 (2 and 3), there is a relatively strong magnetic coupling $K_{12} = 0.12$. As w' decreases, the value of K_{12} also decreases. It reaches the required value $K_{12} = 0.0171$ at $w' = 2.2$ mm. In this case, the characteristic frequencies determined by computer simulation take the following values: $f_e = 1601$ MHz, $f_o = 1629$ MHz, $f_z = 1846$ MHz ($f_z/f_0 = 1.143$), $f_0 = 1615$. These values and formulas (7.5) and (13.3) allow us to determine the components of the mixed coupling between adjacent resonators $K_{m12} = 0.0729$, $K_{e12} = -0.0558$ and slope parameter $a_{12} = 0.0637$.

There is a K_{13} mixed cross-coupling between resonators 1 and 3. Its magnetic component representing the parasitic interaction between nonadjacent resonators equals $K_{m13} = 0.063$. For the TZs of this filter to be located equidistant relative to f_0, the mixed cross-coupling K_{13} must be resonant $K_{13} = 0$ ($|K_{e13}| = K_{m13}$). For this purpose, a 0.2 mm wide zig-zag strip is used, which connects resonators 1 and 3 through capacitive gaps S', which introduces the electrical component K_{e13} into K_{13}. Increase in $|K_{e13}|$ occurs as the gap S' decreases and the length l' of this strip above the resonators increases. The required mutual compensation $|K_{e13}| = K_{m13}$ ($K_{13} = 0$) is provided at $S' = 0.2$ mm and $l' = 2.7$ mm.

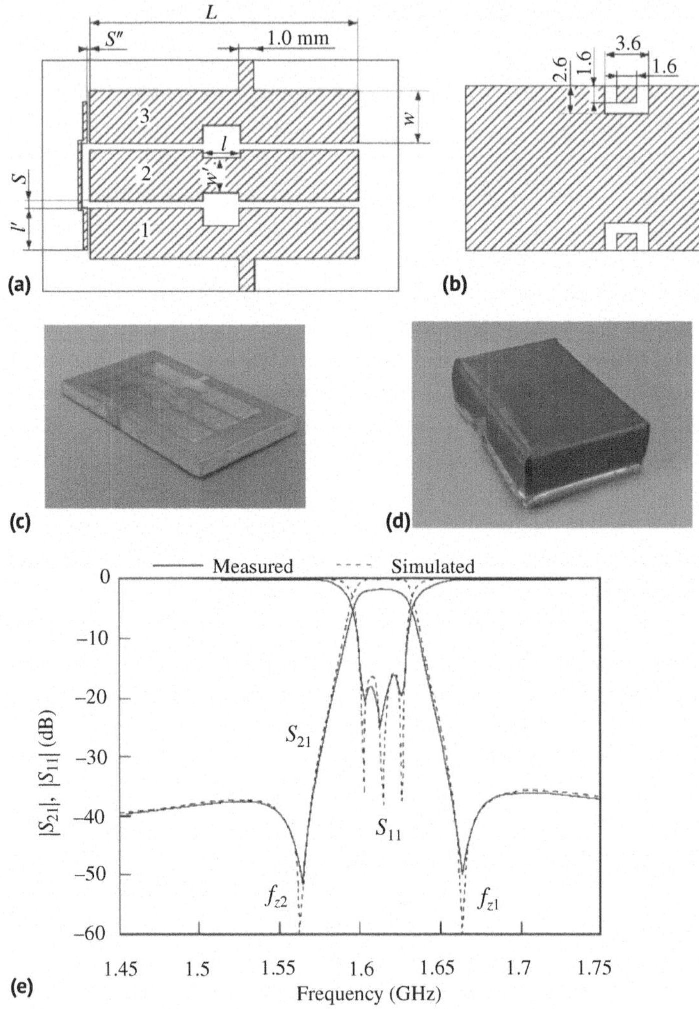

Figure 13.34 Microstrip narrowband triplet BPF with low screen and all mixed couplings ($\varepsilon_r = 36$, $h = 2$ mm, $H = 3$ mm). (a) Topology of face side ($L = 17.4$ mm, $w = 3.4$ mm, $w' = 2.2$ mm, $l = 2.4$ mm, $S = 0.4$ mm, $S' = 0.2$ mm, $l' = 2.7$ mm. (b) Topology of back side. (c) Photograph of BPF without screen. (d) Photograph of BPF with screen. (e) Frequency responses.

The position of the short-tapped segment determines the external Q-factor $Q_e = 55.46$ of end resonators. For this, the EM simulation of group delay of parameter S_{11} for a resonator with one-side load and the expression (3.20) are used.

Figure 13.34c shows the BPF photo without the screen. The metal strips 1 mm wide at the ends of the substrate connect the face and back sides of the filter.

Figure 13.34d shows a photograph of this filter with a screen, and its size is 23 mm × 15 mm × 5 mm. The advantage of using thick substrates is that the screen can be soldered directly to their ends, as shown in Figure 13.34d.

The frequency responses of the BPF are depicted in Figure 13.34e. The measured parameters are as follows: $f_0 = 1615$ MHz, BW = 30 MHz, $L_{A0} = 1.9$ dB, $L_R >$ 16 dB, TZs $f_{z1,2} = f_0 \pm 50$ MHz. TZs close to the passband provide the filter with an increased selectivity of 35 dB ($f_0 \pm 43$ MHz), which can be expressed at the normalized frequency $j\Omega$: 30 dB ($\pm j2.87$). The SIRs used in the filter have an unloaded quality factor $Q_u \approx 400$. If we use a substrate with a thickness of $h = 3$ mm at the same screen height $H = 3$ mm, then the L_{A0} value of the microstrip BPF under consideration will decrease with a slight increase in size (23 mm × 15 mm × 6 mm).

Note that several options exist for placing the screen low above the substrate. Thus, the screen can be soldered not to the ends of the substrate but to a conductive frame located along the perimeter of the face side. The technology widely used in BPFs and duplexers with coaxial ceramic resonators of rectangular cross-section can be used to install the screen. In this case, an additional thin substrate made of a "soft" dielectric is used, to which a thick and rigid substrate with a topological conductive pattern is soldered. Suppose thick and soft laminated substrates are used in the filter. In that case, the number of different shielding options for microstrip BPFs increases due to the possibility of using multiple metallized holes.

Thus, using thick substrates, low-mounted screens, and SIRs allows the creation of compact microstrip narrowband BPFs that have increased selectivity.

References

1 J.-S. Hong, *Microstrip Filters for RF/Microwave Applications*, 2nd ed. New Jersey: Wiley, 2011, https://doi.org/10.1002/9780470937297.

2 A. V. Zakharov, S. A. Rozenko, and N. A. Zakharova, Microstrip bandpass filters on substrates with high permittivities, *J. Commun. Technol. Electron.*, vol. 57, no. 3, pp. 342–351, 2012, https://doi.org/10.1134/S1064226912020143.

3 R. Levy, New cascaded trisections with resonant cross-couplings (CTR Sections) applied to the design of optimal filters, in *IEEE MTT-S International Microwave Symposium Digest*, 2004, vol. 2 pp. 447–450, https://doi.org/10.1109/mwsym.2004.1336007.

4 F. Zhu, W. Hong, J.-X. Chen, and K. Wu, Quarter-wavelength stepped-impedance resonator filter with mixed electric and magnetic coupling, *IEEE Microw. Wirel. Components Lett.*, vol. 24, no. 2, pp. 90–92, 2014, https://doi.org/10.1109/LMWC.2013.2290225.

5 A. Zakharov, S. Rozenko, L. Pinchuk, and S. Litvintsev, Microstrip quazi-elliptic bandpass filter with two pairs of antiparallel mixed-coupled SIRs, *IEEE Microw. Wirel. Components Lett.*, vol. 31, no. 5, pp. 433–436, 2021, https://doi.org/10.1109/LMWC.2021.3065394.

6 K. Ma, J.-G. Ma, K. S. Yeo, and M. A. Do, A compact size coupling controllable filter with separate electric and magnetic coupling paths, *IEEE Trans. Microw. Theory Tech.*, vol. 54, no. 3, pp. 1113–1119, 2006, https://doi.org/10.1109/TMTT.2005.864118.

7 Q.-X. Chu and H. Wang, A compact open-loop filter with mixed electric and magnetic coupling, *IEEE Trans. Microw. Theory Tech.*, vol. 56, no. 2, pp. 431–439, 2008, https://doi.org/10.1109/TMTT.2007.914642.

8 H. Wang and Q.-X. Chu, An inline coaxial quasi-elliptic filter with controllable mixed electric and magnetic coupling, *IEEE Trans. Microw. Theory Tech.*, vol. 57, no. 3, pp. 667–673, 2009, https://doi.org/10.1109/TMTT.2009.2013290.

9 S. Amari, M. Bekheit, and F. Seyfert, Notes on bandpass filters whose inter-resonator coupling coefficients are linear functions of frequency, in *2008 IEEE MTT-S International Microwave Symposium Digest*, 2008, pp. 1207–1210, https://doi.org/10.1109/MWSYM.2008.4633275.

10 L. Szydlowski, A. Lamecki, and M. Mrozowski, Coupled-resonator filters with frequency-dependent couplings: coupling matrix synthesis, *IEEE Microw. Wirel. Components Lett.*, vol. 22, no. 6, pp. 312–314, 2012, https://doi.org/10.1109/LMWC.2012.2197386.

11 S. Tamiazzo and G. Macchiarella, Synthesis of cross-coupled filters with frequency-dependent couplings, *IEEE Trans. Microw. Theory Tech.*, vol. 65, no. 3, pp. 775–782, 2017, https://doi.org/10.1109/TMTT.2016.2633258.

12 L. Balewski, G. Fotyga, M. Mrozowski, M. Mul, P. Sypek, D. Szypulski, and A. Lamecki, Step on it bringing fullwave finite-element microwave filter design up to speed, *IEEE Microw. Mag.*, vol. 21, no. 3, pp. 34–49, 2020, https://doi.org/10.1109/MMM.2019.2958165.

13 W. Shen, L.-S. Wu, X.-W. Sun, W.-Y. Yin, and J.-F. Mao, Novel substrate integrated waveguide filters with mixed cross coupling (MCC), *IEEE Microw. Wirel. Components Lett.*, vol. 19, no. 11, pp. 701–703, 2009, https://doi.org/10.1109/LMWC.2009.2032007.

14 M. Hoft and T. Shimamura, Design of symmetric trisection filters for compact low-temperature Co-fired ceramic realization, *IEEE Trans. Microw. Theory Tech.*, vol. 58, no. 1, pp. 165–175, 2010, https://doi.org/10.1109/TMTT.2009.2035870.

15 C.-L. Hsu, C.-H. Yu, and J.-T. Kuo, Microstrip trisection filters with quasi-elliptic and flat group delay responses, in *2012 4th International High Speed Intelligent Communication Forum*, 2012, pp. 1–2, https://doi.org/10.1109/HSIC.2012.6212988.

16 Z. Q. Xu, P. Wang, J. X. Liao, and Y. Shi, Substrate integrated waveguide filter with mixed coupled modified trisections, *Electron. Lett.*, vol. 49, no. 7, pp. 482–483, 2013, https://doi.org/10.1049/el.2012.3826.

17 A. Zakharov, Transmission zeros of trisection and quadruplet bandpass filters with mixed cross coupling, *IEEE Trans. Microw. Theory Tech.*, vol. 69, no. 1, pp. 89–100, 2021, https://doi.org/10.1109/TMTT.2020.3034663.

18 A. Zakharov, S. Rozenko, and M. Ilchenko, Two types of trisection bandpass filters with mixed cross-coupling, *IEEE Microw. Wirel. Components Lett.*, vol. 28, no. 7, pp. 585–587, 2018, https://doi.org/10.1109/LMWC.2018.2837905.

19 C.-C. Chen, Y.-R. Chen, and C.-Y. Chang, Miniaturized microstrip cross-coupled filters using quarter-wave or quasi-quarter-wave resonators, *IEEE Trans. Microw. Theory Tech.*, vol. 51, no. 1, pp. 120–131, 2003, https://doi.org/10.1109/TMTT.2002.806924.

20 D. Lu, Y. Yang, Y. Wu, and M. Yu, Planar/waveguide quasi-elliptic linear-phase filters using mixed cascaded frequency-dependent triplets with extracted pole sections, *IEEE Trans. Microw. Theory Tech.*, vol. 69, no. 6, pp. 3070–3083, 2021, https://doi.org/10.1109/TMTT.2021.3074602.

21 D. Lu, B. Zhu, and M. Yu, Synthesis approach to planar filter design using new cascade hybrid triplet block with arbitrary-placed TZ pair, *IEEE Trans. Microw. Theory Tech.*, vol. 70, no. 4, pp. 2196–2206, 2022, https://doi.org/10.1109/TMTT.2022.3148361.

22 A. Zakharov and M. Ilchenko, Trisection microstrip delay line filter with mixed cross-coupling, *IEEE Microw. Wirel. Components Lett.*, vol. 27, no. 12, pp. 1083–1085, 2017, https://doi.org/10.1109/LMWC.2017.2759724.

23 A. Zakharov, S. Litvintsev, and M. Ilchenko, Trisection bandpass filters with all mixed couplings, *IEEE Microw. Wirel. Components Lett.*, vol. 29, no. 9, pp. 592–594, 2019, https://doi.org/10.1109/LMWC.2019.2929650.

24 A. Zakharov, S. Rozenko, S. Litvintsev, and M. Ilchenko, Trisection bandpass filter with mixed cross-coupling and different paths for signal propagation, *IEEE Microw. Wirel. Components Lett.*, vol. 30, no. 1, pp. 12–15, 2020, https://doi.org/10.1109/LMWC.2019.2957207.

25 R. M. Kurzrok, General three-resonator filters in waveguide, *IEEE Trans. Microw. Theory Tech.*, vol. 14, no. 1, pp. 46–47, 1966, https://doi.org/10.1109/TMTT.1966.1126154.

26 R. J. Cameron, C. M. Kudsia, and R. R. Mansour, *Microwave Filters for Communication Systems*, 2nd ed. Hoboken, NJ: John Wiley & Sons, Inc., 2018, https://doi.org/10.1002/9781119292371.

27 Microwave and RF Components, TEM coaxial resonators. Data Sheet list, *Web-site*, 2025. https://www.rf-microwave.com/en/tem-coaxial-resonators/359/.

28 A. V. Zakharov and S. A. Rozenko, Duplexer designed on the basis of microstrip filters using high dielectric constant substrates, *J. Commun. Technol. Electron.*, vol. 57, no. 6, pp. 649–655, 2012, https://doi.org/10.1134/S1064226912030187.

29 A. V. Zakharov, S. A. Rozenko, and S. Litvintsev, Fourth-order microstrip band-pass filter with mixed cross-coupling, *J. Commun. Technol. Electron.*, vol. 66, no. 2, pp. 211–219, 2021, https://doi.org/10.1134/S1064226921020157.

30 L. Szydlowski, A. Lamecki, and M. Mrozowski, Coupled-resonator waveguide filter in quadruplet topology with frequency-dependent coupling – a design based on coupling matrix, *IEEE Microw. Wirel. Components Lett.*, vol. 22, no. 11, pp. 553–555, 2012, https://doi.org/10.1109/LMWC.2012.2225604.

31 L. Szydlowski, N. Leszczynska, and M. Mrozowski, A linear phase filter in quadruplet topology with frequency-dependent couplings, *IEEE Microw. Wirel. Components Lett.*, vol. 24, no. 1, pp. 32–34, 2014, https://doi.org/10.1109/LMWC.2013.2288178.

14

Bandpass Filters with Wide Stopband

Bandpass filters (BPFs) with increased stopband, located between the main and first parasitic passbands, are of practical use. There are many different techniques to expand the stopband. Only the most well-known methods are listed in the following sections. One of them is to move the main f_1 and first spurious f_2 resonant frequencies of the resonators away from each other. Combline resonators with a large capacitance value [1] and stepped-impedance resonators (SIRs) [2–8] are used for this purpose. This method is illustrated in Figure 5.18b in the case of using SIRs.

For the same purpose, resonators with the same main resonant frequency f_1 and different higher resonant frequencies can be used in BPFs. In Sagawa et al. [9], $\lambda/4$ resonators and quarter-wave SIRs were used, which alternated with each other. The filters in Ref. [10] contained different half-wave SIRs.

Quite popular is the method that uses discriminating couplings between resonators to suppress spurious passbands [11–14]. This effect is achieved by using open-loop resonators [11], $\lambda/4$ and $\lambda/2$ resonators simultaneously [12], and SIRs only [13, 14]. Figure 12.18 shows the case of expanding the stopband of stripline BPF with SIRs. An extensive rejection band is realized in integrated filters [15–17], where BPFs are combined with low-pass filters. We consider little-known methods for expanding the BPF's stopband as follows.

14.1 Special Connecting Points on Quarter-wave and Half-wave Resonators

One method uses quarter-wave or half-wave resonators and special connection points for coupling elements [18]. Figure 14.1a shows $\lambda/4$ resonator, where the input susceptance $B(\theta)$ is monitored at a point with electric length

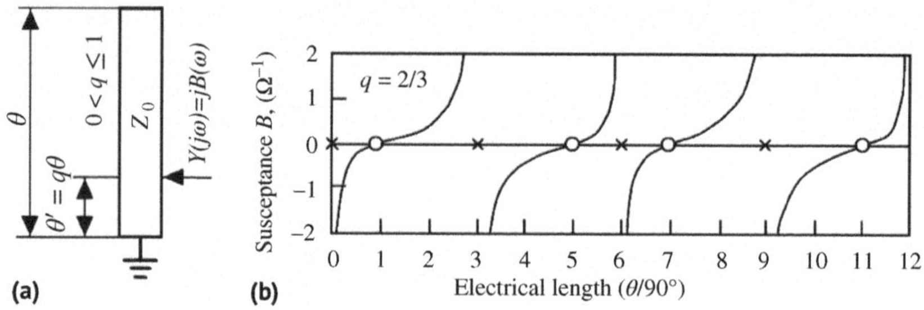

Figure 14.1 Quarter-wave resonator at different connection options. (a) Schematic. (b) Susceptance with $q = 2/3$.

Table 14.1 Dependence of maximum value of R on k for BPF with $\lambda/4$ resonators.

k	1	2	3	4	5	6	7
q	2/3	2/5	2/7	2/11	2/15	2/17	2/21
R	5	7	11	13	17	19	23

$\theta' = q\theta\ (0 \le q \le 1)$ from the short-circuited end. In this case, the input susceptance is

$$B(\theta) = -Z_0^{-1} \frac{\cos\theta}{\sin(q\theta)\cos[(1-q)\theta]}. \tag{14.1}$$

This resonator was partially discussed in the previous sections. It has a set of resonant frequencies f_n, $n = 1, 2, \ldots$, which have corresponding electrical lengths $\theta_n = (2n-1)(\pi/2)$. They are arranged as odd numbers f_1, $f_2 = 3f_1$, $f_3 = 5f_1$, ... From the function (14.1), it follows that along the resonator length, there are the points q, connection to which leads to the disappearance of some resonant frequencies. These points are the intersection points of the trajectories of the electrical lengths (Figure 9.9b) corresponding to the antiresonant frequencies of the quarter-wave resonator. The indicated trajectories are obtained from the condition: the denominator (14.1) equals zero. Thus, at $q = 2/3$, the resonant frequency $f_2 = 3f_1$ is eliminated (Figure 14.1b). To eliminate the resonant frequency $f_3 = 5f_1$, we can use one of two coordinates, $q = 2/5$ or $q = 4/5$. The resonant frequency f_n will be eliminated at n different values of the coordinate q: $2/(2n-1); 4/(2n-1); \ldots; 2(n-1)/(2n-1)$.

The function $B(\theta)$ (14.1) is periodic with period $\theta_T = \pi/(1-q_{\max})$, where $q_{\max}$ is the largest possible q value. Denote by η an odd number, which corresponds to the number of the excluded harmonic ($3f_0, 5f_0, \ldots$). Since $\eta = 1/(1-q_{\max})$, then

$$\theta_T = \eta\pi. \tag{14.2}$$

Thus, from (14.2) it follows that the suppression of the frequency $3f_0$ ($\eta = 3$, $q = 2/3$, $\theta_T = 540°$, $\theta_T/2 = 270°$) entails the suppression of the frequencies $9f_0$, $15f_0$, and $21f_0$, ... (Figure 14.1b). Using the specified values of q leads to the suppression of spurious bandwidths in the combline BPF. We will characterize the expansion of the stopband by the ratio

$$R = \frac{f'_n}{f_1}, \tag{14.3}$$

where f'_n is the closest to f_1 unsuppressed resonant frequency. Using the pattern (14.2), Table 14.1 was compiled, which shows the relationship between the number k of special points used in the combline filter with $\lambda/4$ resonators and the maximum value of R (14.3).

Figure 14.2a shows $\lambda/2$ resonator, where the input susceptance $B(\theta)$ is monitored at a point with an electric length $\theta' = q\theta$ ($0 \le q \le 1$) from the open-circuited end. In this case, the input susceptance is

$$B(\theta) = -Z_0^{-1} \frac{\sin \theta}{\cos (q\theta) \cos [(1-q)\theta]}. \tag{14.4}$$

A partial consideration of this resonator was given earlier. This resonator has a set of resonant frequencies f_n, with corresponding electrical lengths $\theta_n = n\pi$, which are related to each other as natural numbers. From the function (14.4), it follows that along the resonator length, there are points q, to which a connection leads to the disappearance of some resonant frequencies. These points are the intersection points of the trajectories of the electrical lengths (Figure 9.10b) corresponding to the antiresonant frequencies of the half-wave resonator. The indicated trajectories are obtained from the condition that the denominator (14.4) is

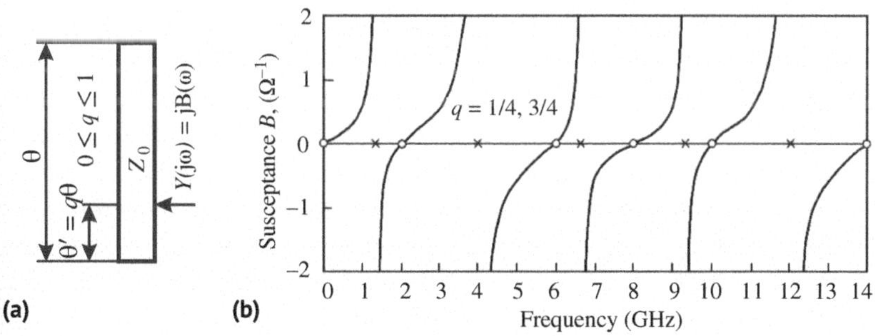

Figure 14.2 Half-wave resonator at different connection options. (a) Schematic. (b) Susceptance with $q = 1/4$ or $3/4$.

Table 14.2 Dependence of maximum value of R on k for BPF with $\lambda/2$ resonators.

k	1	2	3	4	5	6	7
q	1/4	1/6	1/8	1/10	1/12	1/14	1/16
R	3	4	5	7	8	11	13

equal to zero. So, at $q = 1/2$, the fundamental resonant frequency f_1 is eliminated. However, this case is unacceptable to us. To eliminate the next resonant frequency, $f_2 = 2f_1$, two connection coordinates, $q = 1/4$ and $q = 3/4$, can be used (Figure 14.2b). Suppression of $f_3 = 3f_1$ occurs at three values of the coordinate $q = 1/6,\ 3/6,\ 5/6$. In the general case, the resonant frequency with number n ($f_n = nf_1$) will be eliminated for n different values of the coordinate q: $1/2n; 3/2n, \ldots; (2n - 1)/2n$.

Function (14.4) is periodic, as is function (14.1). If the resonant frequency nf_1 is excluded from the frequency spectrum of the $\lambda/2$ segment, then other resonant frequencies spaced $2nf_1$ apart are also excluded. This pattern is illustrated in Figure 14.2b for the case $n = 2$.

Using these patterns, Table 14.2 was compiled, which shows the relationship between the number k of special points used in the pseudocombline BPF [19] with $\lambda/2$ resonators and the maximum value of R (14.3).

Quarter-wave and half-wave resonators with special connection points can be used in a BPF with both simple and mixed couplings.

14.2 Microstrip BPF with $\lambda/4$ Resonators and Special Connecting Points

Let us consider the second-order BPF with $\lambda/4$ resonators and all simple couplings. We utilize the established regularity (Table 14.1) to design a BPF with a wide stopband, characterized by $R = 13$ and the following parameters: $f_0 = 1$ GHz, BW = 50 MHz (FBW = 0.05), and $L_{Ar} = 0.2$ dB. For $R = 13$, it is sufficient to use four ($k = 4$) different q values and the number of resonators $N = 2$ (Figure 14.3a). This filter contains two $\lambda/4$ resonators with $Z_0 = 10\ \Omega$ and three inductors connected to four points $q = 2/3, 2/5, 2/7, 8/11$. From Table 2.6 for $L_{Ar} = 0.2$ dB, we find $g_0 = 1$, $g_1 = 1.0378$, $g_2 = 0.6745$, $g_3 = 1.5386$. Substituting these values and FBW = 0.05 into (3.8), we obtain $K_{12} = 0.0598$ and $Q_{e1} = Q_{e2} = Q_e = 20.756$. The value of K_{12} is implemented by the inductance L_{12}, which is connected to the points with $q = 2/5$ and $q = 2/7$, having the slope parameters (3.15a): $b_1 = 0.2273\ \Omega^{-1}$ and $b_2 = 0.4172\ \Omega^{-1}$, respectively. Substituting K_{12}, b_1, and b_2 in (7.3a), we obtain $L_{12} = 8.642$ nH. To realize the value of $Q_e = 20.756$, we use the inductance

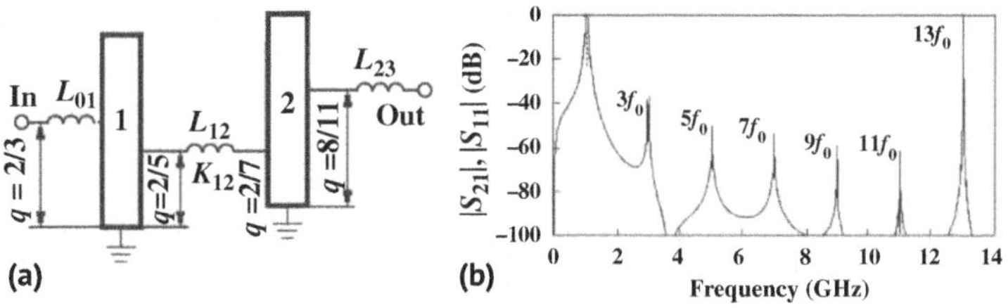

Figure 14.3 Second-order combline BPF with extended stopband. (a) Schematic. (b) Simulated frequency responses.

$L_{01} = 12.75$ nH with $q = 2/3$ and the inductance $L_{23} = 13.53$ nH with $q = 8/11$. The frequency responses of this BPF are shown in Figure 14.3b. The reliability of Table 14.1 is confirmed: $R = 13$ when $k = 4$. In a third-order BPF, we can use six different q points and get $R = 19$.

Let us consider a two-order microstrip BPF with mixed coupling between resonators. Set the parameters of the BPF: $f_0 = 1$ GHz, BW = 30 MHz (FBW = 0.03), $L_{Ar} = 0.2$ dB, $R = 7$, $f_z = 0.85$ GHz, which has an extended stopband ($R = 7$) and increased left-handed selectivity. For $R = 7$, it is sufficient to use two ($k = 2$) different q values and the number of quarter-wave resonators $N = 2$. In this case, the g parameters are the same as in the previous BPF (Figure 14.3a). Substituting the g parameters and value FBW = 0.03 into (3.8), we obtain $|K_{12}| = 0.0358$ and $Q_{e1} = Q_{e2} = Q_e = 34.59$. Since TZ is left-handed with frequency $f_z = 0.85$ GHz, then the coupling coefficient between resonators is negative $K_{12} = -0.0358$. Expressions (7.5) are also applicable to determine the magnetic and electric components of the mixed coupling coefficient [18]. Substituting the values of K_{12}, f_z, and f_0 into these expressions, we obtain $K_{m12} = 0.0932$, $K_{e12} = -0.129$.

The obtained values of K_{12} and Q_e are realized in a microstrip BPF (Figure 14.4a) fabricated on a substrate TMM-10i (Rogers), with $\varepsilon_r = 9.8$, $h = 1.905$ mm, tan $\delta = 0.002$. The filter uses $\lambda/4$ resonators of length $L = 29.2$ mm and width $w = 3$ mm ($Z_0 = 38.4\ \Omega$). The short circuits at the ends of the resonators are formed by metallized holes with a diameter of 0.4 mm.

The filter utilizes two pairs of points with $q = 2/3$ and $q = 2/5$, which should yield a ratio of $R = 7$. Between the points $q = 2/5$ of the resonators, the capacitance C is connected, providing a mixed lumped-distributed coupling. This enables the increase in left-handed selectivity. The magnetic component K_{m12} of the mixed coupling K_{12} is distributed along the length of the resonators. The electric component K_{e12} is composed of two parts, one of which is distributed along the length of the resonators, and a lumped capacitor realizes the other. The required values of $K_{m12} = 0.0932$, $K_{e12} = -0.129$, and $K_{12} = -0.0358$ are provided when the specific frequencies (f_o, f_e, f_z) take the values $f_o = 982$ MHz, $f_e = 1018$ MHz, and

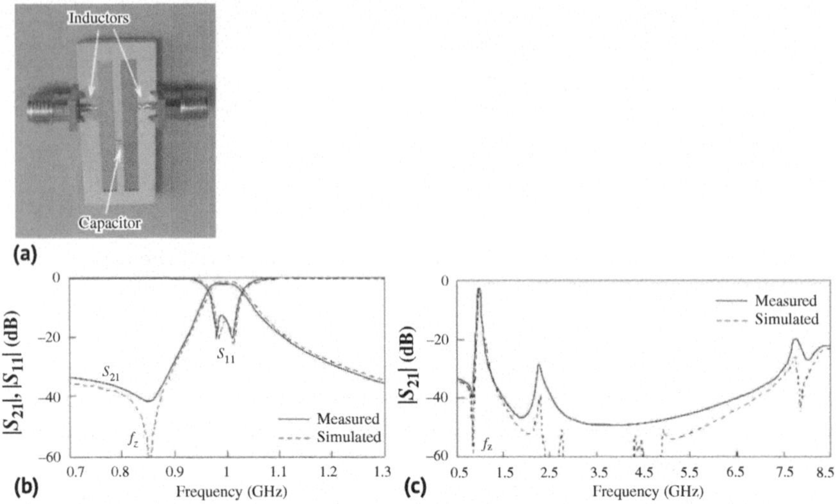

Figure 14.4 Combline second-order mixed coupled microstrip BPF. (a) Photograph. *Source*: [18] / with permission of IEEE. (b) Frequency responses in narrow frequency band. (c) Frequency responses in wide frequency band.

$f_z = 0.85$ MHz. The gap Δ between resonators and the value of lumped capacitance C are varied in such a way that the specific frequencies are realized. To achieve this, we utilize the EM simulator in Microwave Office. This is ensured at $\Delta = 1.6$ mm and $C = 1.02$ pF. The pair of points with $q = 2/3$ is used to connect the resonators with loads by using inductors L.

The value of Q_e can be determined using the EM simulation of the group delay of parameter S_{11} for a resonator with a one-sided load by formula (3.20). The inductance $L = 26.8$ nH provides the required value of $Q_e = 34.59$. This filter utilizes two inductors with an inductance of $L = 27$ nH and a capacitor with a capacitance of $C = 1$ pF. The frequency responses of the filter are shown in Figure 14.4b, c. The measured data are $f_0 = 1$ GHz; BW = 30 MHz; $L_{A0} = 1.5$ dB; $L_R > 13.5$ dB; $f_z = 0.85$ MHz, $R = 7.7$.

Note that this filter can have TZ located to the right of the passband, which will provide it with increased right-hand selectivity.

14.3 BPF with $\lambda/4$ Resonators and SIRs Coupled by Quarter-wave Segments

BPFs contain quarter-wave resonators and SIRs with equal electrical lengths for the component sections, as well as quarter-wave line segments serving as coupling elements. An example of such a filter with three resonators is shown in Figure 14.5.

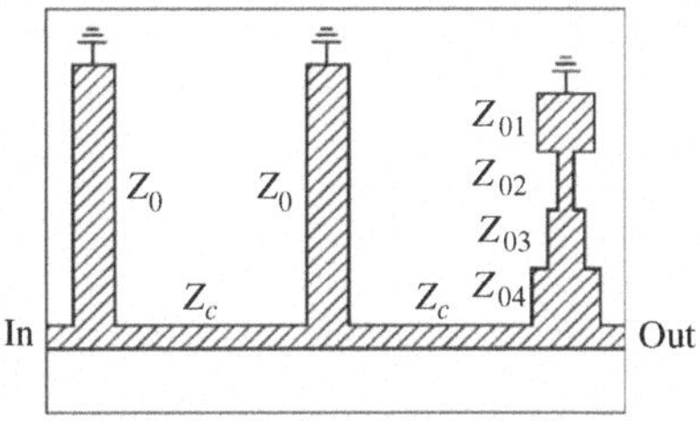

Figure 14.5 BPF with $\lambda/4$ resonators and SIR coupled by quarter-wave segments.

The stopband expansion effect is related to the nature of the mutual arrangement of the resonant and antiresonant frequencies of the $\lambda/4$ resonator and SIR. If the BPF contains only $\lambda/4$ resonators, then its passbands coincide with the resonant frequencies and are related to each other as 1:3:5..., and so on. To suppress parasitic passbands, it is necessary to combine the antiresonant frequencies of the SIR with the corresponding resonant frequencies of the $\lambda/4$ resonator. In addition, to prevent the resonant frequencies of the SIR from leading to parasitic passbands, they must be combined with the antiresonant frequencies of the $\lambda/4$ resonator. In other words, the arrangement of the specified frequencies of the two resonators must be mutually opposite, except for the first resonant frequency, which leads to the main passband.

A natural question arises: How best to determine the parameters of the SIR with N sections to ensure the maximum value of R (14.3), which characterizes the stopband width? By the parameters of the SIR, we mean its length and the values of the characteristic impedances of the component sections, Z_{0i}, $i = 1, 2, \ldots, N$.

The answer to this question is partially illustrated in Figure 14.6, which shows the mutually opposite arrangement of the zeros and poles of the input admittance Y_{in} of $\lambda/4$ resonator and SIR with four sections ($N = 4$). In this figure, the sequential numbering of the critical frequencies is used. The zeros of the Y_{in} function (resonant frequencies) are marked with circles, and the poles of the Y_{in} function (antiresonant frequencies) are marked with crosses. The critical frequencies of $\lambda/4$ resonator are marked with the "°" symbol above.

The length of the SIR is less than the length of the $\lambda/4$ resonator. When analyzing transmission line resonators, it is convenient to use their delay time rather than their geometric length. We denote the delay time of the SIR as t_d and the delay time of the $\lambda/4$ resonator as t_d^0. It is required to determine the value of t_d. We use the properties of the SIR for this, which are discussed in Section 5.3. The critical frequencies of SIRs are periodic. Within each period, they are located symmetrically about the center of the period. The critical frequency ω_N with number N is located in the center of the first period. It can be either zero or a pole. This frequency is determined by the equality (5.21) $\omega_N = N\pi/2t_d$. The highest value of

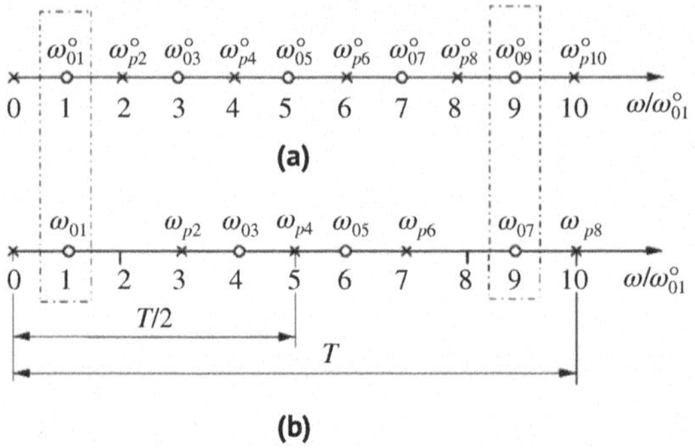

Figure 14.6 Zeros and poles of input admittance of transmission line resonators. (a) $\lambda/4$ resonator. (b) SIR with four sections.

R in the BPF will be realized if the critical frequency of the SIR with number N is combined with the critical frequency of the $\lambda/4$ resonator with number $(N+1)$

$$\omega_N = \omega_{N+1}^0.$$

Since for $\lambda/4$ resonator $\omega_{N+1}^0 = (N+1)\pi/2t_d^0$, then the expression above takes the form $N\pi/2t_d = (N+1)\,\pi/2t_d^0$. From which it follows

$$t_d = \frac{N}{N+1}t_d^0. \tag{14.5}$$

The delay time of the SIR is less than the delay time of the $\lambda/4$ resonator and is determined by the factor $N/\,(N+1)$. As the number of sections N increases, this difference decreases. The electrical length θ_{01}^0 of the $\lambda/4$ resonator at the main resonant frequency ω_{01}^0 is 90°. The values of θ_{01} SIRs will be less than $\theta_{01} = 90°N/\,(N+1)$ and are given in Table 14.3.

With mutually opposite distributions of critical frequencies of the two resonators under consideration, the maximum value of R is determined by the equality

$$R = 2N + 1. \tag{14.6}$$

The validity of (14.6) is illustrated in Figure 14.7 using the SIR with $N = 4$ sections. The resonant frequencies inside the rectangles result in the BPF passbands. In this case, $R = 9$. The values of R (14.6) are also included in Table 14.3.

The characteristic impedances Z_{0i} of SIRs with given critical frequencies can be easily determined using the calculation sequence from Sections 5.3, 5.4. The calculated values of Z_{0i} are shown in Table 14.3 in normalized form. The value Z_{0N} of the section adjacent to the SIR input was used as the normalization. This table also contains the practically important parameter $m = Z_{0\max}/Z_{0\min}$.

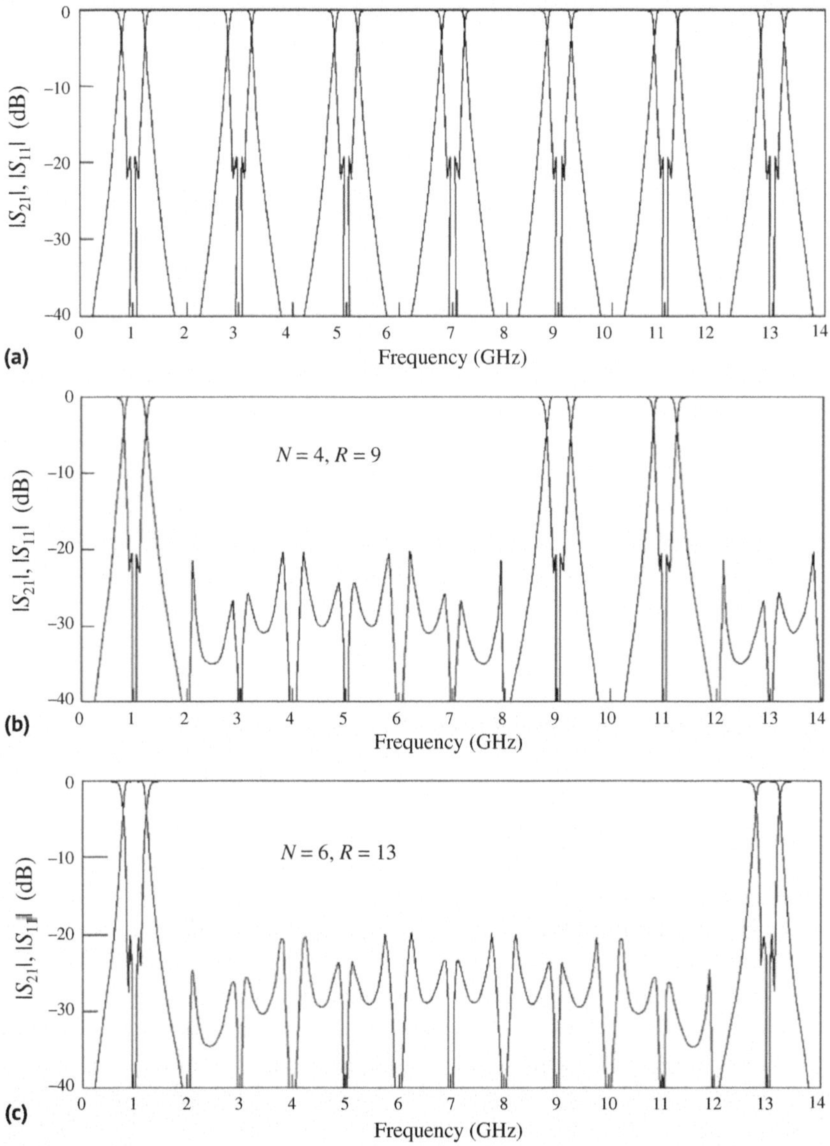

Figure 14.7 Simulated frequency responses of third-order BPF in Figure 14.5. (a) All resonators are quarter-wave. (b) One resonator is SIR with four sections. (c) One resonator is SIR with six sections.

Table 14.3 Characteristics of SIRs for BPF with wide stopband.

N	θ_{01}	R	Z_{0i}			m
3	(3/4)90°	7	$Z_{03} = 1$	$Z_{02} = 2.414$	$Z_{01} = 2.414$	2.414
4	(4/5)90°	9	$Z_{04} = 1$ $Z_{01} = 1.618$	$Z_{03} = 1.894$	$Z_{02} = 3.065$	3.065
5	(5/6)90°	11	$Z_{05} = 1$ $Z_{02} = 2.828$	$Z_{04} = 1.604$ $Z_{01} = 1.097$	$Z_{03} = 2.757$	2.828
6	(6/7)90°	13	$Z_{06} = 1$ $Z_{03} = 3.092$	$Z_{05} = 1.434$ $Z_{02} = 2.312$	$Z_{04} = 2.353$ $Z_{01} = 0.780$	3.964

Simulated frequency responses of the considered third-order BPF with stubs are shown in Figure 14.7. Figure 14.7a corresponds to the case of using all $\lambda/4$ resonators in the filter. The filter, which uses one SIR with four sections, corresponds to Figure 14.7b. If the SIR is composed of six sections, its frequency responses are shown in Figure 14.7c. The linear simulator of Microwave Office (AWR) software was used to obtain the frequency responses. The characteristic impedances of the quarter-wave resonators $Z_0 = 15\ \Omega$ and the coupling sections $Z_c = 50\ \Omega$ were used. The characteristic impedances Z_{0i} of the SIRs corresponded to the values from Table 14.3, multiplied by 10.

14.4 Transmission Line BPF with One Passband

In BPF with quarter-wave coupling segments, it is possible to obtain a single passband by eliminating all spurious passbands. This can be achieved by using simultaneously $\lambda/4$ or $\lambda/2$ resonators, and nonuniform resonators with a smooth change in characteristic impedance (CIR), discussed in Section 5.2. The process of obtaining a single passband in BPF with quarter-wave coupling segments is illustrated in Figures 14.8–14.11.

Let us turn to a distributed circuit with parallel-type resonators, whose topology is shown in Figure 14.8 by a dashed line. It contains uniform transmission line resonators of half-wave and quarter-wave types with input impedances $Z_{\text{in}} = -jZ_0 \cot(\omega t_d)$ and $Z_{\text{in}} = jZ_0 \tan(\omega t_d)$, respectively. The delay time of the resonators and coupling segments is the same and equal to $t_d = l/v$, where l is the length of the specified elements and v is the speed of propagation of the electromagnetic wave.

The critical frequencies of these input impedances are shown in Figure 14.9. The zeros, marked by circles, are antiresonant frequencies, and the poles, marked by crosses, are resonant frequencies.

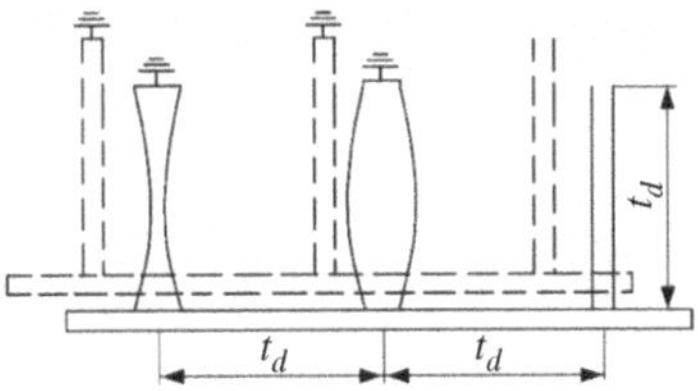

Figure 14.8 Distributed circuit topologies with uniform and nonuniform transmission line resonators, coupled by quarter-wave segments.

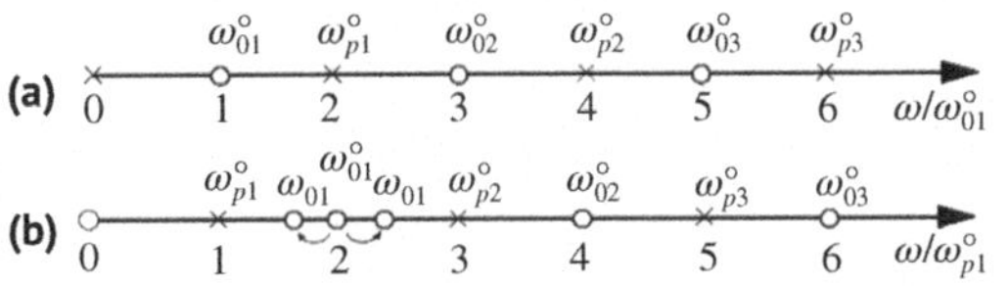

Figure 14.9 Critical frequencies of input impedances of resonators in Figure 14.8. (a) Half-wave resonator. (b) Quarter-wave-type resonators.

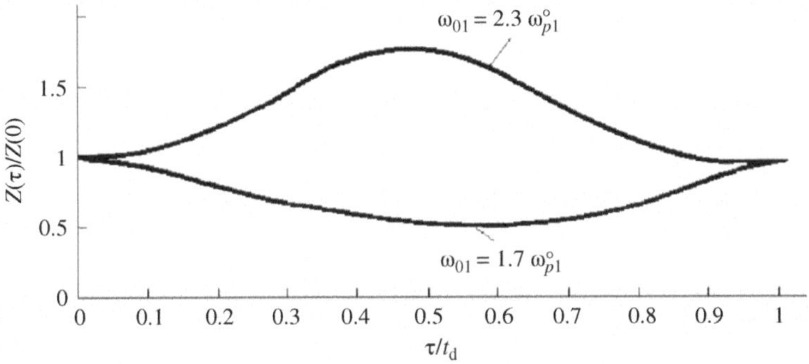

Figure 14.10 Characteristic impedances of nonuniform TL resonators of quarter-wave type.

Since the distribution of the critical frequencies of the two types of resonators is mutually opposite, the circuit in question does not have passbands, and it is all-blocking. The insertion loss function of this circuit is shown in Figure 14.11a. The linear simulator of Microwave Office (AWR) software was used to construct the frequency responses. The characteristic impedances of all resonators $Z_0 = 25\ \Omega$ and coupling sections $Z_c = 90\ \Omega$ were used.

We assume that a shift of one of the zeros ω_{0n}° of the input impedance of a quarter-wave uniform resonator (Figure 14.9b) will result in the appearance of a

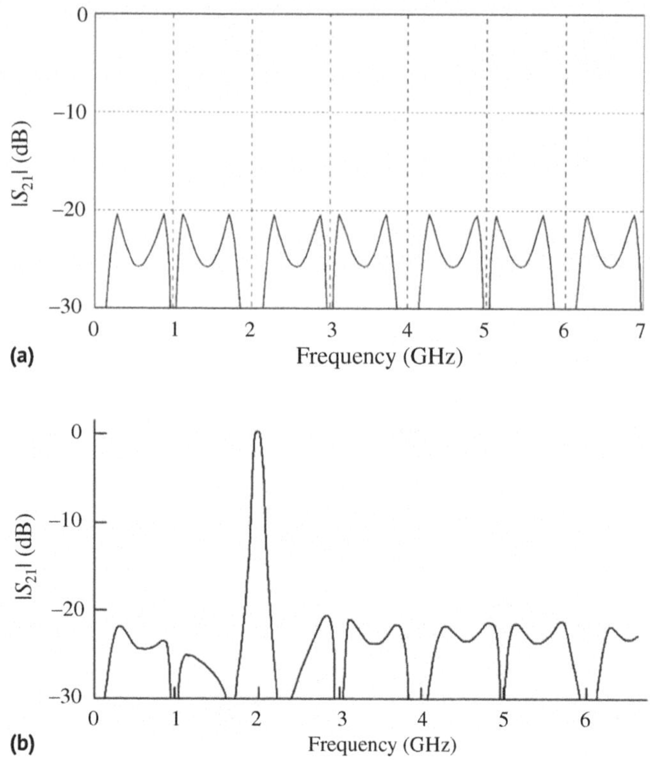

Figure 14.11 Insertion loss curves of distributed circuits in Figure 14.8. (a) All-blocking circuit with $\lambda/4$ and $\lambda/2$ resonators. (b) Single-band filter with one $\lambda/2$ resonator and two CIRs of quarter-wave type.

passband in the distributed circuit under consideration. Replacing ω°_{0n} with ω_{0n} will result in a CIR with input impedance Z_{in} (5.15), which corresponds to the characteristic impedance $Z_0(\tau)$ function (5.16).

The graphs of the function $Z_0(\tau)$ are plotted in Figure 14.10 for various arrangements of the antiresonant frequency ω_{01}: $\omega_{01} = 1.7\omega^{\circ}_{p1}$; $\omega_{01} = 2.3\omega^{\circ}_{p1}$. In the distributed circuit shown in Figure 14.8 (solid line), we will use two different nonuniform quarter-wave CIRs that correspond to two options for placing the antiresonant frequency: $\omega_{01} = 1.7\omega^{\circ}_{p1}$ and $\omega_{01} = 2.3\omega^{\circ}_{p1}$.

Note that Figure 14.9 shows the critical frequencies of the $Z_{\text{in}}(\omega)$ function of the resonators and not the $Y_{\text{in}}(\omega)$ function. This is because, for the subsequent transition to the characteristic impedance function $Z_0(\tau)$ of CIR, the $Z_{\text{in}}(\omega)$ function is used.

To calculate the insertion loss function of the distributed circuit under consideration, the classical expression is used

$$L_{\mathrm{Ar}} = -10 \log \frac{4}{|A + B/R + CR + D|^2} \text{ [dB]}, \tag{14.7}$$

which contains the elements of the transfer matrix $[ABCD]$ of the filter, considered as a two-port network, and the load resistance R. The matrix $[ABCD]$ is obtained by multiplying the transfer matrices of resonators

$$\begin{bmatrix} a_r & b_r \\ c_r & d_r \end{bmatrix} = \begin{bmatrix} 1 & 0 \\ 1/Z_{\mathrm{in}}(s) & 1 \end{bmatrix} \tag{14.8}$$

and coupling segments

$$\begin{bmatrix} a_c & b_c \\ c_c & d_c \end{bmatrix} = \begin{bmatrix} \cosh(st_d) & Z_c \sinh(st_d) \\ Z_c^{-1} \sinh(st_d) & \cosh(st_d) \end{bmatrix} \tag{14.9}$$

In expressions (14.8) and (14.9), $s = j\omega$.

Using expressions (5.15), (14.7)–(14.9), the insertion loss function of the distributed circuit in Figure 14.8 (solid curve) is plotted in Figure 14.11b. The following parameters were used: characteristic impedance of the uniform resonator $Z_0 = 25\ \Omega$; initial value of the characteristic impedance CIRs $Z_0\ (0) = 25\ \Omega$; characteristic impedance of the coupling sections $Z_c = 90\ \Omega$; $R = 50\ \Omega$. All resonators and coupling sections are quarter-wave at a frequency of 1 GHz.

The distributed circuit in question is a BPF with a single passband. The symmetry of the frequency response in Figure 14.11a is achieved because the values of ω_{01} in two CIRs are shifted symmetrically with respect to ω_{01}° in opposite directions.

These frequencies are the transmission zeros of this filter. The closer they are to the passband, the higher the selectivity of the BPF.

There are other possible combinations of uniform and nonuniform resonators that lead to BPFs with a single passband.

References

1 G. L. Matthaei, L. Young, and E. M. T. Jones, *Microwave Filters, Impedance-Matching Networks, and Coupling Structures*. New York: Artech House Books, 1980.

2 M. Makimoto and S. Yamashita, Compact bandpass filters using stepped impedance resonators, *Proc. IEEE*, vol. 67, no. 1, pp. 16–19, 1979, https://doi.org/10.1109/PROC.1979.11196.

3 M. Makimoto and S. Yamashita, Bandpass filters using parallel coupled stripline stepped impedance resonators, *IEEE Trans. Microw. Theory Tech.*, vol. 28, no. 12, pp. 1413–1417, 1980, https://doi.org/10.1109/TMTT.1980.1130258.

4 M. Sagawa, M. Makimoto, and S. Yamashita, Geometrical structures and fundamental characteristics of microwave stepped-impedance resonators, *IEEE Trans. Microw. Theory Tech.*, vol. 45, no. 7, pp. 1078–1085, 1997, https://doi.org/10.1109/22.598444.

5 T. Ishizaki and T. Uwano, A stepped impedance comb-line filter fabricated by using ceramic lamination technique, in *1994 IEEE MTT-S International Microwave Symposium Digest (Cat. No.94CH3389-4)*, 1994, pp. 617–620, https://doi.org/10.1109/MWSYM.1994.335498.

6 J. T. Kuo and E. Shih, Microstrip stepped impedance resonator bandpass filter with an extended optimal rejection bandwidth, *IEEE Trans. Microw. Theory Tech.*, vol. 51, no. 5, pp. 1554–1559, 2003, https://doi.org/10.1109/TMTT.2003.810138.

7 C.-L. Hsu and J.-T. Kuo, A two-stage SIR bandpass filter with an ultra-wide upper rejection band, *IEEE Microw. Wirel. Components Lett.*, vol. 17, no. 1, pp. 34–36, 2007, https://doi.org/10.1109/LMWC.2006.887248.

8 A. Zakharov, Parametric and structural-parametric synthesis of nonuniform transmission line resonators, *IEEE Trans. Circuits Syst. I Regul. Pap.*, vol. 68, no. 3, pp. 1055–1067, 2021, https://doi.org/10.1109/TCSI.2020.3044925.

9 M. Sagawa, M. Makimoto, and S. Yamashita, A design method of bandpass filters using dielectric-filled coaxial resonators (short papers), *IEEE Trans. Microw. Theory Tech.*, vol. 33, no. 2, pp. 152–157, 1985, https://doi.org/10.1109/TMTT.1985.1132971.

10 C.-F. Chen, T.-Y. Huang, and R.-B. Wu, Design of microstrip bandpass filters with multiorder spurious-mode suppression, *IEEE Trans. Microw. Theory Tech.*, vol. 53, no. 12, pp. 3788–3793, 2005, https://doi.org/10.1109/TMTT.2005.859869.

11 X. Y. Zhang and Q. Xue, Harmonic-suppressed bandpass filter based on discriminating coupling, *IEEE Microw. Wirel. Components Lett.*, vol. 19, no. 11, pp. 695–697, 2009, https://doi.org/10.1109/LMWC.2009.2032002.

12 S. C. Lin, Y. S. Lin, and C. H. Chen, Extended-stopband bandpass filter using both half- and quarter-wavelength resonators, *IEEE Microw. Wirel. Components Lett.*, vol. 16, no. 1, pp. 43–45, 2006, https://doi.org/10.1109/LMWC.2005.860014.

13 Y. C. Li, X. Y. Zhang, and Q. Xue, Bandpass filter using discriminating coupling for extended out-of-band suppression, *IEEE Microw. Wirel. Components Lett.*, vol. 20, no. 7, pp. 369–371, 2010, https://doi.org/10.1109/LMWC.2010.2049425.

14 A. Zakharov and M. Ilchenko, Coupling coefficients between resonators in stripline combline and pseudocombline bandpass filters, *IEEE Trans. Microw. Theory Tech.*, vol. 68, no. 7, pp. 2679–2690, 2020, https://doi.org/10.1109/TMTT.2020.2988866.

15 C. Quendo, E. Rius, C. Person, and M. Ney, Integration of optimized low-pass filters in a bandpass filter for out-of-band improvement, *IEEE Trans. Microw. Theory Tech.*, vol. 49, no. 12, pp. 2376–2383, 2001, https://doi.org/10.1109/22.971624.

16 A. Manchec, C. Quendo, E. Rius, C. Person, and J.-F. Favennec, Synthesis of dual behavior resonator (DBR) filters with integrated low-pass structures for spurious responses suppression, *IEEE Microw. Wirel. Components Lett.*, vol. 16, no. 1, pp. 4–6, 2006, https://doi.org/10.1109/LMWC.2005.861361.

17 A. V. Zakharov, Expanding the rejection band of planar bandpass filters, *J. Commun. Technol. Electron.*, vol. 57, no. 1, pp. 95–100, 2012, https://doi.org/10.1134/S1064226911090105.

18 A. Zakharov, S. Rozenko, and S. Litvintsev, Combline filters with increased stopband and one-sided selectivity, *IEEE Microw. Wirel. Technol. Lett.*, vol. 33, no. 4, pp. 407–410, 2023, https://doi.org/10.1109/LMWT.2022.3221269.

19 J.-S. Hong, *Microstrip Filters for RF/Microwave Applications*, 2nd ed. New Jersey: Wiley, 2011, https://doi.org/10.1002/9780470937297.

15

Tunable Resonators

The electrically tunable/reconfigurable filters are used in diplexers [1], duplexers [2], and as standalone devices [3, 4]. Most of these filters contain resonators from segments of transmission lines (TLs) and variable capacitors as tuning elements. Semiconductor varactors [5–7], ferroelectric capacitors [8, 9], and sets of lumped capacitors commutated by microelectromechanical systems (MEMS) switches [10] or *pin* diodes [11] are used as variable capacitors. These resonators have a multitude of resonant frequencies, ω_n, where $n = 1, 2, \ldots$. Of practical interest is tuning at one specific resonance frequency ω_n, within a predetermined frequency band $\omega_n \in [\omega_{n\min}, \omega_{n\max}]$.

15.1 Resonators with Variable Capacitances, Tuning Limitations

The frequency tunability of resonators refers to the ability to change the resonance frequency by adjusting the capacitance. The tunability of the resonator in a frequency band is characterized by the relationship between frequency ratio $K_n = \omega_{n\max}/\omega_{n\min}$ and the relative variation of the capacitance $\chi = C_{\max}/C_{\min}$. Increasing the tunability of resonators means increasing K_n for a specified value of χ, or reducing χ for specified value of K_n. Since the values K_n and χ are not functions of the electrical circuit, the problem of tunability increase in the resonator is solved by varying the resonator parameters, which improves the relationship between K_n and χ.

The explanation of the obtained result is typically not provided. The existing state is due to the absence of a circuit function (similar to the input or transfer function) that would directly express the tunability of the resonators. A circuit function characterizing the tunability and its properties [12] will be discussed as follows. This function is based on elements of the sensitivity theory [13–16], which has proven itself in many applications. Based on this function, theoretical limitations on the tunability of resonators with a single capacitance will be established, and some resonators with enhanced tunability will be synthesized.

15.1.1 Sensitivity Function and Its Properties

Let us consider a tunable resonator (Figure 15.1) consisting of a one-port network without losses, to the terminals of which a variable capacitance C is connected. The one-port network can contain distributed and lumped elements. When the capacitance changes, the resonance frequencies of the resonator, ω_n, $n = 1, 2, \ldots$, also change. They are the roots of the resonance equation

$$\widetilde{B}(\omega) = B(\omega) + \omega C = 0, \tag{15.1}$$

where $B(\omega)$ is the input susceptance of a one-port network, and $\widetilde{B}(\omega)$ is the input susceptance of the resonator.

A positive definite value

$$S_C = -\left.\frac{d\omega/\omega}{dC/C}\right|_{\omega=\omega_n} = -\left.\frac{d\ln\omega}{d\ln C}\right|_{\omega=\omega_n} \tag{15.2}$$

is the sensitivity of the resonance frequencies ω_n to the capacitance C changing. It is defined in the region of negative values of $B(\omega)$, where (15.1) has solutions. Using the resonance equation (15.1), we can express the sensitivity S_C (15.2) through input susceptance $B(\omega)$

$$S_C = \left.\left(1 - \frac{dB}{d\omega}/\frac{B}{\omega}\right)^{-1}\right|_{\omega=\omega_n}, B \leq 0. \tag{15.3}$$

If function $B(\omega)$ is given by its zeros ω_{0i} and poles ω_{pi}, $i = 1, 2, \ldots$

$$B(\omega) = -H\frac{\left(\omega_{01}^2 - \omega^2\right)\left(\omega_{02}^2 - \omega^2\right)\ldots}{\omega\left(\omega_{p1}^2 - \omega^2\right)\left(\omega_{p2}^2 - \omega^2\right)\ldots}, \tag{15.4}$$

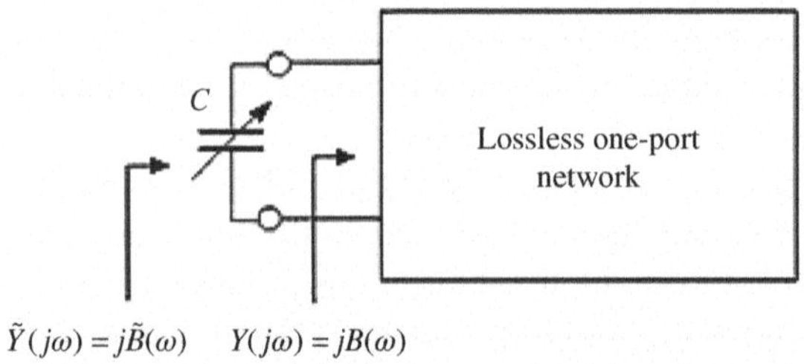

Figure 15.1 Tunable resonator with variable capacitance.

then the sensitivity is expressed through these critical frequencies

$$S_C(\omega) = \frac{1}{2}\left[1 - \sum_i \frac{1}{1-(\omega_{0i}/\omega)^2} + \sum_i \frac{1}{1-(\omega_{pi}/\omega)^2}\right]^{-1}, \tag{15.5}$$

which follows from (15.3) and (15.4). Here, H is a constant value, called the conductivity coefficient, which does not affect the sensitivity S_C. The applicability of sensitivity S_C is due to the fact that S_C is a circuit function, in contrast to the values K and χ. It contains information about the properties of a circuit, and these can be determined by changing the critical frequencies ω_{0i} and ω_{pi}.

A combline resonator is used most widely in varactor-tuned filters (Figure 15.2a). The resonant equation (15.1) for it has the form

$$\omega C = -B(\omega) = Z_0^{-1}\cot\theta, \tag{15.6a}$$

where Z_0 is the characteristic impedance, $\theta = \omega L/v$ is the electric length, L is the length of the segment, and v is the propagation speed of electromagnetic wave.

Substituting the input susceptance $B(\omega)$ from (15.6a) in (15.3) leads to the sensitivity function of this resonator

$$S_C(\theta) = \left(1 + \frac{2\theta}{\sin 2\theta}\right)^{-1}, \tag{15.6b}$$

$$S_C(\theta) \ge 0.$$

The graph of the sensitivity function (15.6b) for the first three oscillations is shown in Figure 15.2b. The abscissa is the electric length normalized to 90°.

A resonator shown in Figure 15.3a is also used as a tunable one. One end of the resonator is open. The resonant equation (15.1) for it is expressed by the formula

$$\omega C = -B(\omega) = -Z_0^{-1}\tan\theta. \tag{15.7a}$$

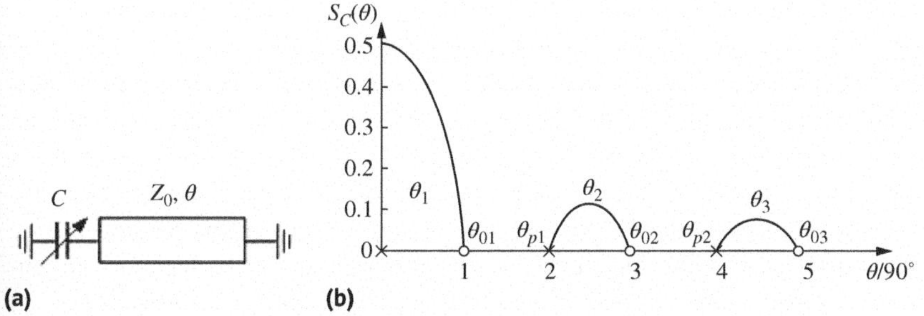

Figure 15.2 Tunable combline resonator. (a) Schematic. (b) Function $S_C(\theta)$.

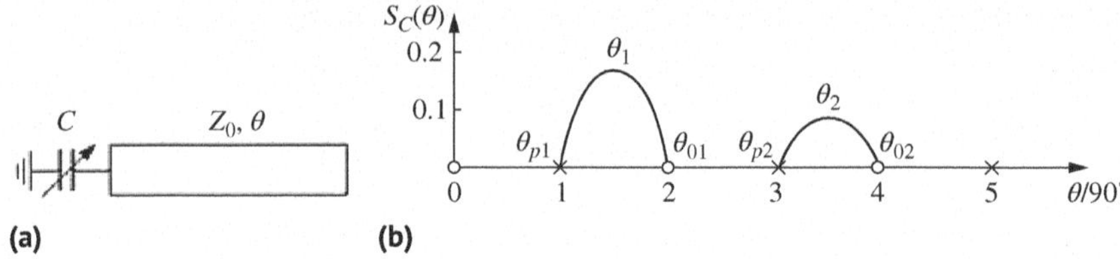

Figure 15.3 Tunable resonator with open end. (a) Schematic. (b) Function $S_C\ (\theta)$.

Substituting the input susceptance $B\ (\omega)$ from (15.7a) into (15.3) gives the sensitivity function of this resonator

$$S_C\ (\theta) = \left(1 - \frac{2\theta}{\sin 2\theta}\right)^{-1}, \tag{15.7b}$$

$$S_C\ (\theta) \geq 0.$$

The graph of the function S_C (15.7b) for the first two oscillations is shown in Figure 15.3b.

The following properties characterize a positive-definite circuit function $S_C\ (\omega)$:

- The coincidence of the zeros of the sensitivity function with the internal critical frequencies of the function $B\ (\omega)$.
- The strict convexity of the function $S_C\ (\omega)$ between its zeros.

The sensitivity S_C characterizes the local tunability of the resonator. At the same time, it connects the values K and χ. If the sensitivity change in the tuning band from $\omega_{n\min}$ to $\omega_{n\max} = K\omega_{n\min}$ is known, then

$$\ln \chi = \int_{\omega_{n\min}}^{K\omega_{n\min}} \frac{d \ln \omega}{S_C\ (\omega)} \tag{15.8}$$

which follows immediately from definition (15.2). Expression (15.8) shows that an increasing S_C in a tuning band leads to a decrease in χ for a defined value of K.

It is widely known that the tunability of a combline resonator at a main resonance frequency ω_1 increases when the electrical length of the resonator decreases. This fact is explained by the function S_C (Figure 15.2b). The sensitivity function increases with decreasing θ up to the value of 1/2. For a resonator with an open end, the function S_C at ω_1 has a maximum (Figure 15.3b). For more effective tuning, this maximum should be situated in the middle of the tuning range.

It is important to note that the sensitivity function S_C (15.5) depends only on the critical frequencies. It does not rely on the level of input susceptance H (15.4) and on capacitance C. This feature is retained when TL resonators are used. In this case, the function S_C is independent of capacitance C and characteristic impedance Z_0. It depends only on the electrical length θ of the TL segment

(15.6), (15.7). Three main values, Z_0, C, and θ, are used for the design of a tunable resonator.

Due to the existing feature of the S_C function, the electrical length θ is determined first. The values of Z_0 and C are determined secondarily. For this, based on the resonance equations of the resonators shown in Figures 15.2a and 15.3a, we write down at the upper frequency of the tuning range $\omega_{1\max}$ as

$$Z_0 C_{\min} = \omega_{1\max}^{-1} \cot(\theta_{1\max}) \text{ and}$$

$$Z_0 C_{\min} = -\omega_{1\max}^{-1} \tan(\theta_{1\max})$$

respectively. The right side of these equalities is a positive number. We denote it by A and write these equations in a single form

$$Z_0 C_{\min} = A.$$

This representation shows that the values of Z_0 and $C_{\min}$ are mathematically equal. Additionally, they are inversely proportional to each other; increasing the value of Z_0 results in a decrease in the value of $C_{\min}$, and vice versa. The choice of Z_0 and $C_{\min}$ values is somewhat arbitrary. This choice is made by the designer, and the main factor in choosing is the equality of the product of Z_0 and $C_{\min}$ values to the number of A.

The electrical length θ is determined first by the condition of obtaining the highest possible sensitivity S_C. This is equivalent to minimizing the relative capacitance change $\chi = C_{\max}/C_{\min}$. Therefore, the starting point of the design is the minimization of the value χ.

15.1.2 Tunability Limitation of Resonators

The use of the sensitivity function $S_C(\omega)$ allows us to determine the tunability limitation of resonators in general terms. We consider some different cases.

15.1.2.1 Resonator Without Losses with One Capacitance

If the sensitivity is constant $[S_C(\omega) = S_0 = \text{const}]$ in the tuning band, then from (15.8) it follows $\chi = K_n{}^{1/S_0}$ or $K_n = \chi^{S_0}$. The last equality makes it evident that the maximum value of the sensitivity function $S_{C\max}$ in the tuning band limits the frequency ratio K_n:

$$K_n \leq \chi^{S_{C\max}} \tag{15.9}$$

The input admittance of a lossless one-port network $B(\omega)$ is a reactance function. The fundamental property of a lossless one-port network is expressed by the inequality [17]

$$\frac{dB}{d\omega} \geq \frac{|B|}{\omega}. \tag{15.10}$$

Inequality (15.10) and definition (15.3) lead to an essential property of the function $S_C(\omega)$ of the resonator without losses – limitation of this function on the entire frequency axis

$$S_C(\omega) \leq \frac{1}{2}. \tag{15.11}$$

In addition, (15.11) and (15.9) lead to a limitation of the frequency ratio

$$K_n \leq \sqrt{\chi}. \tag{15.12}$$

Limitations (15.11) and (15.12) are fulfilled with an equal sign only in the case when a tunable resonator is a lumped *LC* circuit.

15.1.2.2 Resonator Without Losses with Several Capacitances

Initially, we consider a one-port network (Figure 15.1), which contains lumped inductances of constant value and m variable capacitances: $C_1, C_2, ..., C_m$. Without loss of generality, we will consider the lowest resonance frequency ω_1, which depends on these capacitances

$$\omega_1 = \omega_1 (C_1, C_2, ... C_m). \tag{15.13}$$

The function (15.13) is a homogeneous function of the variables C_i, $i = 1, 2, \ldots, m$, of degree $(-1/2)$:

$$\omega_1 (\chi C_1, \chi C_2, \ldots, \chi C_m) = \chi^{-1/2} \omega_1 (C_1, C_2, \ldots, C_m), \tag{15.14}$$

$$\sum_{i=1}^{m} C_i \frac{\partial \omega_1}{\partial C_1} = -\frac{1}{2}\omega_1. \tag{15.15}$$

In particular, we have $\omega_1(C) = \text{const} \cdot C^{-1/2}$ for a single *LC* circuit, where $\text{const} = L^{-1/2}$. In this case, expressions (15.14) and (15.15) become obvious

$$\omega_1(\chi C) = \chi^{-1/2}\omega_1(C) \text{ and}$$

$$C(d\omega_1/dC) = -(1/2)\,\omega_1.$$

Definitions (15.14) and (15.15) of a homogeneous function of degree $(-1/2)$ are equal. Equation (15.14) means that increasing all capacitances by χ times is accompanied by a decrease of ω_1 by $\sqrt{\chi}$ times. Formula (15.15) expresses the Euler theorem on homogeneous functions [18]. After dividing both sides of (15.15) by $(-\omega_1)$, we obtain

$$\sum_{i=1}^{m} S_{Ci} = \frac{1}{2}, \tag{15.16}$$

where S_{Ci} is the sensitivity of resonance frequency ω_1 to the change of capacitance C_i (15.2). Expression (15.16) is known [13, 14] as an invariant of sensitivity ω_1 to the change of all capacitances of a lumped LC circuit. It shows that the sum of all sensitivities S_{Ci} of a lumped circuit without losses is 1/2, and this sum is invariant with respect to the structure of the circuit. Equation (15.16) expresses the mutual dependence of sensitivities, which is one of the fundamental properties of such circuits. An increase in one sensitivity accompanies a decrease in another, but their sum remains constant.

If several inductances are added to the initial circuit, then the character of the function (15.13) does not change. It will still be a homogeneous function of degree (−1/2) and satisfy conditions (15.14)–(15.16). However, if several constant capacitances are added to the initial circuit (let their total number become equal to M), then (15.16) goes to the inequality

$$\sum_{i=1}^{m} S_{Ci} \leq \frac{1}{2}, \tag{15.17}$$

(since the number of variable capacitances is $m < M$ on which the summation is carried out), then the restriction for the frequency ratio will take the form

$$K \leq \sqrt{\chi_{\max}} \tag{15.18}$$

where $\chi_{\max}$ is the largest value among $\chi_1, \chi_2, \ldots, \chi_m$.

A circuit with distributed parameters without losses can be regarded as the limiting case of a lumped LC circuit, when the number of elements tends to infinity. Adding a distributed element (several elements) to an initial lumped circuit is equivalent to connecting an infinite number of lumped constant capacitances and inductances. Therefore, the limitations of tunability (15.17) and (15.18) also apply to lumped-distributed circuits without losses, which are tuned by several capacitors. The tunability limitations of resonators with several variable capacitances (15.17) and (15.18) and with one variable capacitance (15.11) and (15.12) are the same. This means that we cannot significantly increase the tunability by increasing the number of variable capacitances.

In the varactor-tuned bandpass filter [19], microstrip TL resonators were used. For tuning in the frequency range of 225–400 MHz ($K = 1.778$), one varactor was used in each resonator, the capacitance of which changed by $\chi = 3.67$ times. The S_C values were quite large $0.402 \leq S_C \leq 0.470$. The use of additional varactors will slightly reduce the χ value. This suggests that it is not advisable to use additional varactors in cases of high-value S_C.

15.1.2.3 Influence of Dissipative Losses

In the presence of losses, the resonant frequencies of resonators will be determined by two conditions

$$\begin{aligned} \tilde{B}(\omega) &= B(\omega) + \omega C = 0, \\ \left.\frac{d\tilde{B}}{d\omega}\right|_{\omega=\omega_n} &= C + \left.\frac{dB}{d\omega}\right|_{\omega=\omega_n} \geq 0, \end{aligned} \tag{15.19}$$

where functions $\widetilde{B}(\omega)$ and $B(\omega)$ are no longer reactance functions. They are imaginary parts of the corresponding input admittance $\widetilde{Y}(\omega)$ and $Y(\omega)$. The first of conditions (15.19) coincides with the resonance condition for the case without loss (15.1), but its solution yields a large number of zeros on the frequency axis. Some of these zeros are not resonant frequencies, since the function at these frequencies has a negative slope. The second of conditions (15.19) chooses those zeros where function $\widetilde{B}(\omega)$ has a positive slope and they are resonant frequencies. When losses are absent, the second of conditions (15.19) is redundant, since the slope of the input susceptance is always positive.

Since the first of the resonance conditions (15.19) gives $C = -B/\omega$, the second of the conditions (15.19) is equivalent to the inequality

$$\frac{dB}{d\omega} \geq \left.\frac{|B|}{\omega}\right|_{\omega=\omega_n}.$$

This inequality coincides with inequality (15.10) for the input susceptance, which led to the limitation of tunability (15.11) and (15.12). The same inequalities also hold for resonators with losses. Therefore, the idea of increasing the tunability of resonators by artificially introducing losses has no factual basis.

15.1.3 Ways of S_C Increasing

Since $S_C(\omega)$ is a circuit function, a natural question arises. How to increase $S_C(\omega)$ due to a special distribution of critical frequencies ω_{0i} and ω_{pi}?

On the example of the function $S_C(\omega)$ determined by critical frequencies ω_{01}, ω_{p1}, ω_{02}, ω_{p2} (Figure 15.4), we consider the possibility of increasing the sensitivity at the second resonant frequency $\omega_2 \in [\omega_{p1}, \omega_{02}]$. Obviously, the increase in frequency interval $[\omega_{p1}, \omega_{02}]$, i.e., expansion of the resonance region, will lead to a rise in the values of $S_C(\omega)$. Figure 15.4 illustrates the effect of critical frequencies ω_{01} and ω_{p2} located to the left and right of the resonance region. In the case under consideration, the sensitivity function $S_C(\omega)$ contains four critical frequencies ω_{01}, ω_{p1}, ω_{02}, and ω_{p2}.

Curve *1* in Figure 15.4 corresponds to the location of the critical frequencies $\omega_{01} \to 0$, $\omega_{p2} \to \infty$. In this case, the sensitivity takes the smallest values. Curve *2* was obtained at $\omega_{01} \to \omega_{p1}$, $\omega_{p2} \to \infty$ and curve *3* corresponds to $\omega_{p2} \to \omega_{02}$, $\omega_{01} \to 0$. Approaching the frequency interval $[\omega_{p1}, \omega_{02}]$ of only one critical frequency increases $S_C(\omega)$ in the region shifted toward this frequency. Suppose two critical frequencies ω_{01} and ω_{p2} are closing to the interval $[\omega_{p1}, \omega_{02}]$ simultaneously. In that case, the sensitivity $S_C(\omega)$ in this interval will increase substantially (curve *4* in Figure 15.4).

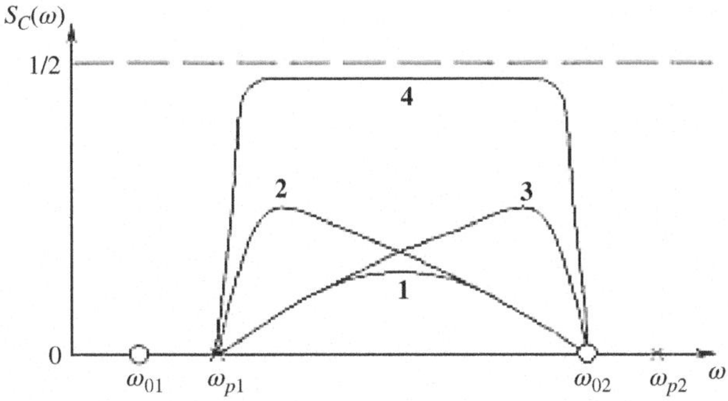

Figure 15.4 Dependence of function $S_C(\omega)$ versus location of critical frequencies: $\omega_{01} \to 0$, $\omega_{p2} \to \infty$ (1); $\omega_{01} \to \omega_{p1}$, $\omega_{p2} \to \infty$ (2); $\omega_{p2} \to \omega_{02}$, $\omega_{01} \to 0$ (3); $\omega_{01} \to \omega_{p1}$, $\omega_{p2} \to \omega_{02}$ (4).

As we can see earlier, the widest bandwidth resonator is a lumped *LC* circuit. In this case, the limitations (15.11) and (15.12) are fulfilled with an equal sign. Then the input susceptance (15.4) is the inductance conductivity

$$B(\omega) = -\frac{H}{\omega}, \tag{15.20a}$$

where $H = 1/L$. According to (15.2), this inductance conductivity corresponds to the sensitivity function equal to 1/2 on the entire frequency axis

$$S_C(\omega) = \frac{1}{2}. \tag{15.20b}$$

The function (15.20a) has no internal critical frequencies. It has a pole at zero frequency and zero at infinity.

The question arises: Could we build a circuit whose input susceptance $B(\omega)$ contains internal critical frequencies (15.4) with sensitivity (15.5) close to 1/2 on the entire frequency axis? Using the analysis results presented in Figure 15.4, we construct the function $B(\omega)$ as follows.

Let all internal poles ω_{pi} of the function differ from zeros ω_{0i} by a small positive value ε: $\omega_{pi} - \omega_{0i} = \varepsilon$. The function has a pole at zero frequency and a zero at infinity. An example of such a function is shown in Figure 15.5a, where the dotted line represents the inductance conductivity (15.20a). At $\varepsilon \to 0$, we obtain the idealized function $B^*(\omega)$, whose module can be represented as

$$|B^*(\omega)| = \frac{H}{\omega} + \sum_i \delta(\omega - \omega_{pi}), \tag{15.21a}$$

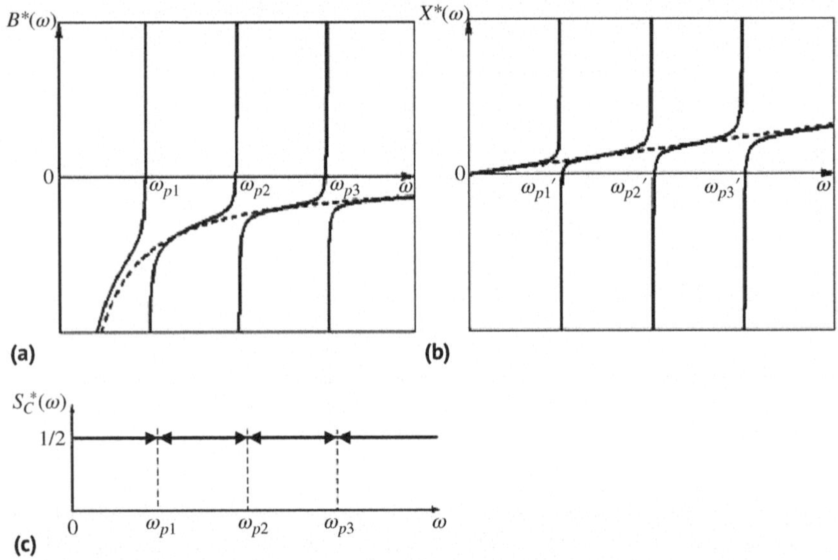

Figure 15.5 Characteristic graphs of circuit functions: (a) Input susceptance $B^*(\omega)$ with pairwise close critical frequencies. (b) Input reactance $X^*(\omega)$ with pairwise close critical frequencies. (c) Idealized sensitivity function $S_C{}^*(\omega)$.

where $\delta(\omega - \omega_{pi})$ is the delta-Dirac function [18]. The idealized function $B^*(\omega)$ differs from the inductance conductivity (15.20a) by the presence of discontinuities of the second kind, where it takes the values $\pm\infty$. The idealized input susceptance (15.21a) corresponds to the idealized sensitivity function

$$S_C{}^*(\omega) = \begin{cases} 1/2, & \omega \neq \omega_{pi}, \\ 0, & \omega = \omega_{pi}. \end{cases} \tag{15.21b}$$

The function (15.21b) is shown in Figure 15.5c. The input reactance $X(\omega) = -1/B(\omega)$ corresponds to the input susceptance $B(\omega)$ with pairwise close critical frequencies. The zeros ω'_{0i} and poles ω'_{pi} of this function are also pairwise close. An example of such a function is presented in Figure 15.5b, where the dotted line shows the inductance reactance $X(\omega) = L\omega$. Extremely close to the critical frequencies ω'_{0i} and ω'_{pi} to each other, we obtain an idealized input reactance $X^*(\omega)$, whose modulus can be represented as

$$|X^*(\omega)| = H\omega + \sum_i \delta\left(\omega - \omega'_{pi}\right).$$

This idealized function differs from the reactance of inductance by the presence of discontinuities of the second kind, where it takes the values of $\pm\infty$. The graphs in Figure 15.5a, b illustrate the input susceptance and input reactance with pairwise close critical frequencies.

Let us give a qualitative interpretation of the result presented in Figure 15.5. Expression (15.5) shows that the pairwise convergence of the zeros and the adjacent right-hand poles of the function $B(\omega)$ brings mutual compensation of their influence on the sensitivity function $S(\omega)$ of the circuit. The influence of the pole and zero located at the origin and at infinity remains uncompensated. As a result, a sensitivity close to inductance is provided. This method of increasing tunability can be described as the mutual compensation of zeros and poles of the input susceptance.

We illustrate this method by the example of a quarter-wave resonator with a length L and a characteristic impedance Z_0 (Figure 15.6a). The variable capacitance C is connected to a point located at distance l from the short-circuited end of the resonator.

At the connection point of the capacitance, the input susceptance is determined by the expression:

$$B(\omega) = Z_0^{-1}\left\{-\cot\left(\frac{\omega l}{\upsilon}\right) + \tan\left[\frac{\omega(L-l)}{\upsilon}\right]\right\}, \tag{15.22}$$

where υ is the electromagnetic wave propagation velocity. The graph of the function (15.22) is shown in Figure 15.6b at $l = L/10$. The smaller the distance l, the closer the zeros ω_{0i} and poles ω_{pi} are to each other. The tunability of the resonator in Figure 15.6a is higher than that of a resonator with a traditional capacitance connection (Figure 15.2a), which is tuning in the region of electrical lengths $\theta = 60 - 75°$ when capacitance C changes 7.61 times.

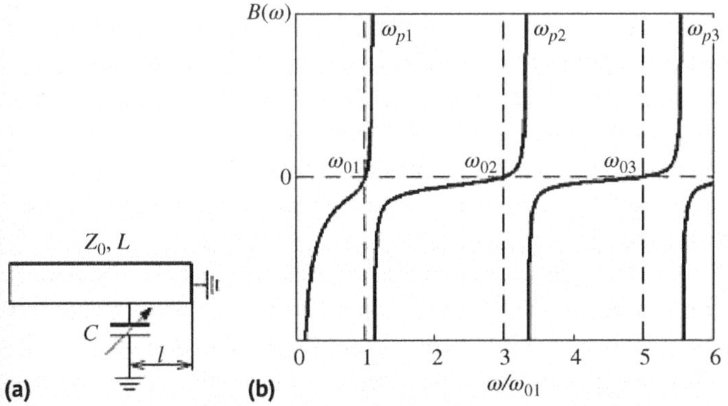

Figure 15.6 Implementation example of the mutual compensation method of zeros and poles of input susceptance: (a) Resonator circuit. (b) Input susceptance at the point of connecting the capacitor to the resonator for $l = L/10$.

Using calculations based on (15.22), it was established that the resonator (Figure 15.6a) at $l = L/10$ is tuned to the same electrical length when the capacitance C changes by a factor of 4.13. However, this significantly increases the minimum value of the capacitance C_{min}. This increase may be useful at very high frequencies, where the minimum value of capacitance C may be minimal, for example, $C_{min} = 0.1$ pF. Moving the connection point of capacitance C toward the short-circuited end of the resonator increases the value of C_{min}.

Each resonant frequency ω_i can change only within its own resonant region, which is limited by zero ω_{0i} and the left adjacent pole $\omega_{p(i-1)}$ of input susceptance $B(\omega)$. The expansion of the resonance region is accompanied by an increase in sensitivity S_C and frequency range at the corresponding resonant frequency.

When designing tunable resonators, it is necessary to use both methods of tunability enhancement. It is an expansion of the resonance region and the use of the technique of mutual compensation of zeros and poles of the input susceptance $B(\omega)$.

15.1.4 Resonators with Improved Tunability

The established features of the location of the input susceptance $Y(j\omega) = jB(\omega)$ critical frequencies were used to create resonators with improved tunability. These resonators are presented in Table 15.1. They are SIRs with an electric length θ consisting of N segments of the same length $\theta' = \theta/N$. Here, θ_{0i} are resonant electrical lengths, and θ_{pi} are antiresonant electrical lengths. The characteristic impedances of composite segments have only two values: the minimum Z_0 and the maximum mZ_0, $m > 1$. The arrow shows the input admittance Y of each resonator, which is a function of Richards' frequency variable S (5.18).

Most of the resonators in Table 15.1 were synthesized in Section 5.6 by the parametric synthesis method and are contained in Tables 5.6 and 5.7: resonator No. 1 from Table 15.1 corresponds to resonator No. 2 from Table 5.6; resonator No. 2 from Table 15.1 corresponds to resonator No. 3 from Table 5.7; resonator No. 3 from Table 15.1 corresponds to resonator No. 5 from Table 5.6. The synthesis of the resonators from Tables 5.6 and 5.7 was carried out using the input impedance $Z = 1/Y$. Therefore, the critical frequencies indicated there are mutually opposite in nature compared to those presented in Table 15.1.

Resonators No. 1 and No. 2 are synthesized for tuning the first (fundamental) resonant frequency ω_1. With the help of resonators No. 3 and No. 4, the second resonance frequency ω_2 is tuned. Table 15.1 also shows the location of the critical frequencies of these resonators, providing improved tunability. They include both the expansion of a resonance region and the closing of neighboring critical frequencies to it.

Table 15.2 presents the input susceptance of the resonators from Table 15.1, along with the ratios of their critical frequencies as a function of m. They can help calculate and evaluate the achieved effect. The value ω_{01}° of resonator No. 1

Table 15.1 Tunable resonators and their crit cal frequencies.

No.	Topology of resonators	Critical frequencies
No. 1 θ_1	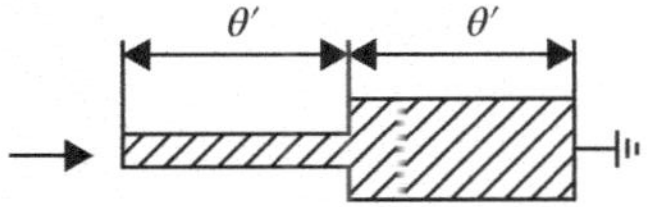	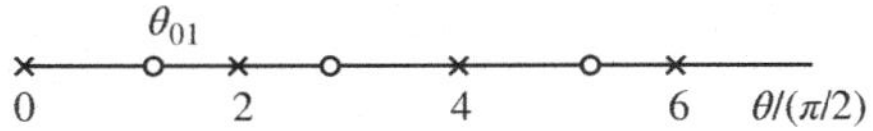
No. 2 θ_1	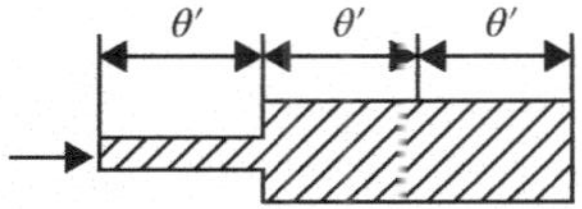	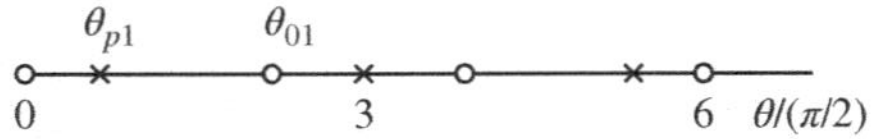
No. 3 θ_2	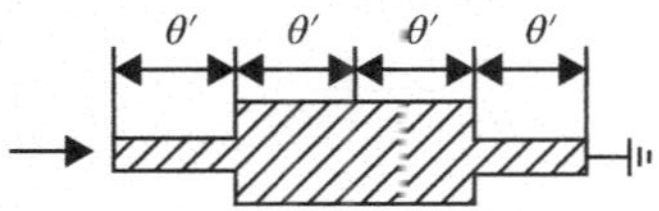	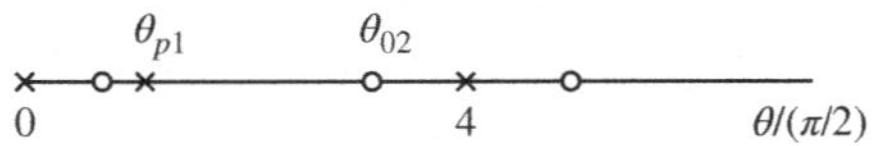
No. 4 θ_2	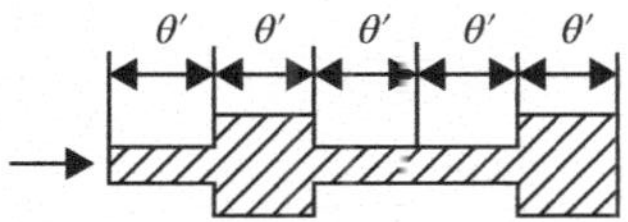	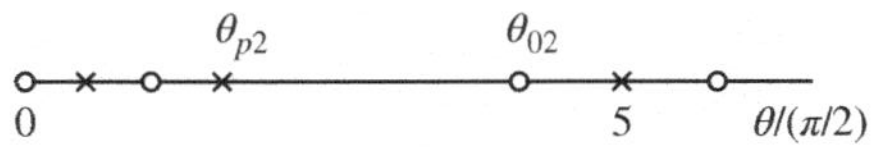

Table 15.2 Input susceptance and ratio of critical frequencies of tunable resonators.

No.	Input susceptance	Ratio of critical frequencies
No. 1 ω_1	$Y^2(S) = Z_0^{-1} \dfrac{1 + m^{-1}S^2}{(1 + 1/m)\,S}$	$\dfrac{\omega_{01}}{\omega_{01}^0} = \dfrac{4\tan^{-1}\sqrt{m}}{\pi}$
No. 2 ω_1	$Y^3(S) = Z_0^{-1} \dfrac{(1 + 2m)\,S + S^3}{1 + (1 + 2m)\,S^2}$	$\dfrac{\omega_{01}}{\omega_{p1}} = \dfrac{\tan^{-1}\sqrt{1 + 2m}}{\tan^{-1}\sqrt{1/(1 + 2m)}}$
No. 3 ω_2	$Y^4(S) = Z_0^{-1} \dfrac{1 + (2 + 2m + 2/m)\,S^2 + S^4}{(2 + 2/m)\,S + (2 + 2m)\,S^3}$	$\dfrac{\omega_{02}}{\omega_{p1}} = \dfrac{\tan^{-1}\sqrt{1 + m + 1/m + \sqrt{(1 + m + 1/m)^2 - 1}}}{\tan^{-1}\sqrt{1/m}}$
No. 4 ω_2	$Y^5(S) = Z_0^{-1} \dfrac{b_1S + b_3S^3 + b_5S^5}{a_0 + a_2S^2 + a_4S^4}$, $a_0 = 1,\ a_2 = 4 + 4m + 2/m,\ a_4 = 1 + 2m + 2m^2$, $b_1 = 3 + 2m,\ b_3 = 2 + 4m + 2m^2 + 2/m,\ b_5 = 1$	$\dfrac{\omega_{02}}{\omega_{p2}} = \dfrac{\tan^{-1}\sqrt{b_3/(2b_5) + \sqrt{(b_3/(2b_5))^2 - b_1/b_5}}}{\tan^{-1}\sqrt{a_2/(2a_4) + \sqrt{(a_2/(2a_4))^2 - a_0/a_4}}}$

is the zero of input susceptance when resonator is uniform ($m = 1$). At this frequency we have $\theta_{01}^0 = 90°$. At $m = 3$, the ratio $\omega_{01}/\omega_{01}^0 = 1.33$, which indicates an extension of the resonant region by 1.33 times. Its right boundary $\theta_{01}^0 = 90°$ is increased up to $\theta_{01} = 120°$. At $m = 5$, the ratio $\omega_{01}/\omega_{01}^0 = 1.46$, which indicates an increase in the right boundary up to $\theta_{01} = 131.8°$. The resonator No. 2 at $m = 1$ is a half-wave resonator with critical frequency ratio $\omega_{01}/\omega_{p1} = 2$. At $m = 3$, we have $\omega_{01}/\omega_{p1} = 3.37$, and for $m = 5$, we obtain $\omega_{01}/\omega_{p1} = 4.36$, which indicates a significant expansion of the resonance region.

We note that the expansion effect of the resonant region of resonator No. 1 is maximal when the electrical lengths of the two segments with high and low characteristic impedance are equal. If this relation is changed, the resonance region becomes narrower. For resonator No. 2, the maximum expansion effect of the resonant region occurs when the relation of the electrical lengths of composite segments is equal to the value shown in Table 15.1.

The substitution of the input susceptance of resonator No. 1 (Table 15.2) into (15.3) leads to its sensitivity function

$$S_C = \left[1 + \theta\left(1 + \Omega^2\right)\left(\frac{1}{\Omega} + \frac{2\Omega}{m - \Omega^2}\right)\right]^{-1}, \tag{15.23}$$

where $\Omega = \tan(\theta/2)$. The function (15.23) is shown in Figure 15.7 for $m = 3$ and $m = 1$ (uniform resonator). For $m = 3$, the function S_C increases due to the expansion of the resonance region. It becomes equal to zero for $\theta_{01} = 120°$. The graphs in Figure 15.7 show that the advantage of stepped-impedance resonator No. 1 increases when tuning in the region with increased electrical lengths. Let us illustrate this for the typical value $\chi = C_{max}/C_{min} = 2.2$ for ferroelectric capacitors.

To characterize the frequency adjustment, in practice, the fractional tuning range (FTR) is often used, expressed in percent:

$$\text{FTR} = 2\frac{\theta_{nmax} - \theta_{nmin}}{\theta_{nmax} + \theta_{nmin}} = 2\frac{K_n - 1}{K_n + 1}$$

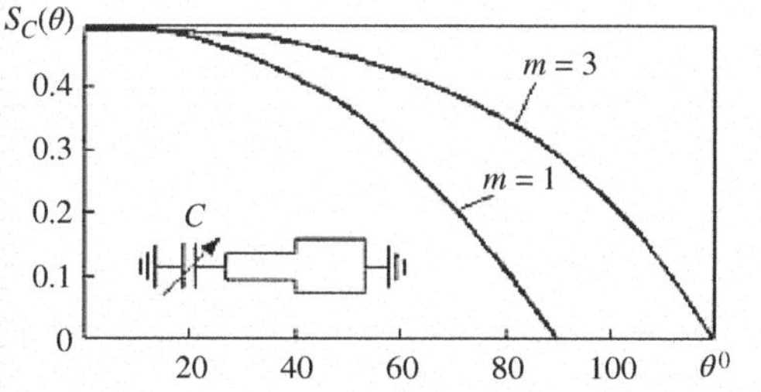

Figure 15.7 Function $S_C(\theta)$ of resonator No. 1.

The FTR values for different electrical lengths of resonator No. 1 are shown in Table 15.3 for $m = 3$ and $m = 1$. These values are determined directly from the resonance equation (15.1). The quantitative evaluation of this advantage is given in Table 15.3 as the FTR ratio.

The substitution of the input susceptance of resonator No. 2 (Table 15.2) into (15.3) gives its sensitivity function

$$S_C = \left[1 - \theta\left(1 + \Omega^2\right)\left(\frac{2a\Omega}{1 - a\Omega^2} + \frac{3\Omega^2 - a}{\Omega^3 - a}\right)\right]^{-1}, \tag{15.24}$$

where $\Omega = \tan(\theta/3)$, $a = 2m + 1$.

The function (15.24) is illustrated in Figure 15.8 for $m = 3$ and $m = 1$, exhibiting a bell-shaped character. In view of this, the resonator has an optimal tuning band, within which the tuning is most effective. At $m = 1$, the maximum of the function $S_{C\max} = 0.178$ is achieved for electric length $\theta = 129°$, which is located in the middle of this band. Using the resonance equation (15.1), we determine that for $\chi = 2.2$, this band is located in the region of electric lengths with $120 \le \theta \le 137.9°$, and its FTR = 13.86%. At $m = 3$, the function S_C increases due to the expansion of the resonance region and the closing of neighboring critical frequencies to it. The maximum of the function $S_{C\max} = 0.29$ is achieved for an electric length $\theta = 118°$. At $\chi = 2.2$, the most effective tuning takes place in the region of electric lengths $110 \le \theta \le 137.7°$, which corresponds to FTR = 22.36%. The tuning range is increased 1.61 times in comparison with a uniform resonator, for which $m = 1$. As the parameter m increases, the sensitivity function S_C and the tuning range of the considered stepped-impedance resonators also increase.

Table 15.3 Tuning bandwidth of resonator No. 1 at $\chi = 2.2$ for $m = 3$ and $m = 1$.

θ ($m = 1$), °	40–54	50–64	60–72.4	70–79.4	80–85.1	—
FTR ($m = 1$), %	29.8	24.6	18.8	12.6	6.2	—
θ ($m = 3$), °	40–56.4	50–68.5	60–79.4	70–89.2	80–97.2	90–104.2
FTR ($m = 3$), %	34	31.2	27.9	24.1	19.4	14.6
FTR ($m = 3$)/FTR ($m = 1$)	1.14	1.27	1.48	1.91	3.13	—

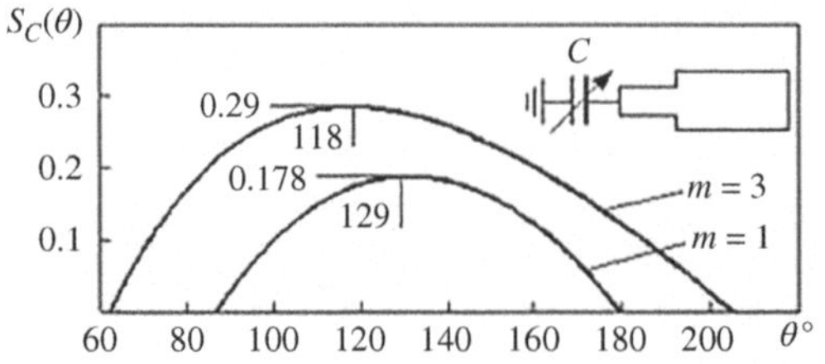

Figure 15.8 Function S_C (θ) of resonator No. 2.

A quarter-wave SIR (No. 1 in Table 15.1) is not only one that has an increased tunability at the main resonant frequency ω_1. Figure 15.9 shows resonators, which are equivalent at a frequency of ω_1 for specific capacitances. Their equivalence is due to the fact that at a frequency of ω_1, their resonance equations coincide. The notation used in Figure 15.9 requires no explanation. Note that at the frequency of the next resonance ω_2, these resonators are not equivalent, since their resonance equations differ from each other. The individual properties determine the practical use of the resonators shown in Figure 15.9. Thus, the resonator (Figure 15.9d) uses twice the value $2C$ of capacitance. Additionally, this resonator has an increased length, which is a beneficial factor at high frequencies.

A half-wave SIR (No. 2 in Table 15.1) is also not the only one with an increased bandwidth at the main resonant frequency of ω_1. Figure 15.10 shows resonators, which are equivalent at a frequency of ω_1 for specific capacitances. The notation in Figure 15.10 requires no explanation. At the same time, these resonators are not equivalent at the next resonant frequency ω_2. Some SIRs with variable capacitances have been considered in Refs. [6, 7], but without specifying the reasons for the increase in tunability.

Resonators No. 3 and No. 4 (Tables 15.1 and 15.2) utilize the second resonant frequency ω_2. This allows us to use the higher electrical lengths of resonators, which is particularly useful for shorter wavelengths, such as millimeters. An analysis of these resonators can be carried out in the sequence discussed earlier.

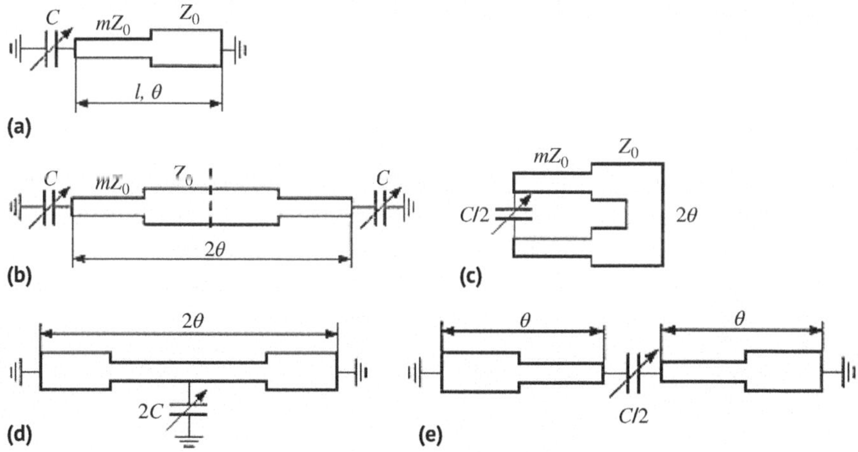

Figure 15.9 Resonators with the same tuning range at the main resonant frequency as quarter-wave SIR: (a) Initial resonator. (b) Symmetric resonator with two capacitors. (c) Loop hairpin resonator. (d) Symmetric resonator with parallel capacitor. (e) Symmetric resonator with series capacitor.

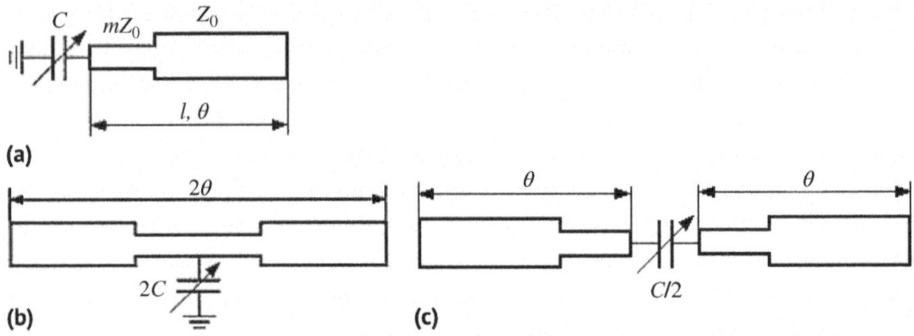

Figure 15.10 Resonators with the same range tuning at the main resonant frequency as the half-wave SIR: (a) Initial resonator. (b) Symmetric resonator with parallel capacitor. (c) Symmetric resonator with series capacitor.

Table 15.1 and Figures 15.9 and 15.10 show SIRs with improved tunability. They implement the established features of the critical frequencies location of function $B(\omega)$. An important design parameter of these resonators is $m = Z_{0\max}/Z_{0\min}$, the value of which is limited.

15.1.5 Intersection of Resonant Regions in Tunable Resonators

Let us obtain the resonance equation for the considered symmetric SIR (Figure 15.11a) with two capacitances. The resonance equation of the lossless resonator, represented as a two-port network with transfer matrix $[ABCD]$, can be written as [20]:

$$C = 0. \tag{15.25}$$

When (15.25) is satisfied, the input impedance of the resonator $Z_{11} = A/C$ and the output impedance $Z_{22} = D/C$ become infinite. The transfer matrix of the symmetrical two-port network $[ABCD]$ can be expressed in terms of the transfer matrix elements A', B', C', and D' of its left half (2.6). In this case, $C = 2C'D'$, and the resonance equation (15.25) is replaced by two equations

$$\begin{aligned} C' &= 0, \\ D' &= 0. \end{aligned} \tag{15.26}$$

The transfer matrix of the left half of the symmetric SIR is denoted by $[a'b'c'd']$. The elements of the matrix A', B', C', D' can be found by multiplying two matrices

$$\begin{bmatrix} A' & B' \\ C' & D' \end{bmatrix} = \begin{bmatrix} 1 & 0 \\ Y_{C1} & 1 \end{bmatrix} \begin{bmatrix} a' & b' \\ c' & d' \end{bmatrix} = \begin{bmatrix} a' & b' \\ Y_{C1}a' + c' & Y_{C1}b' + d' \end{bmatrix},$$

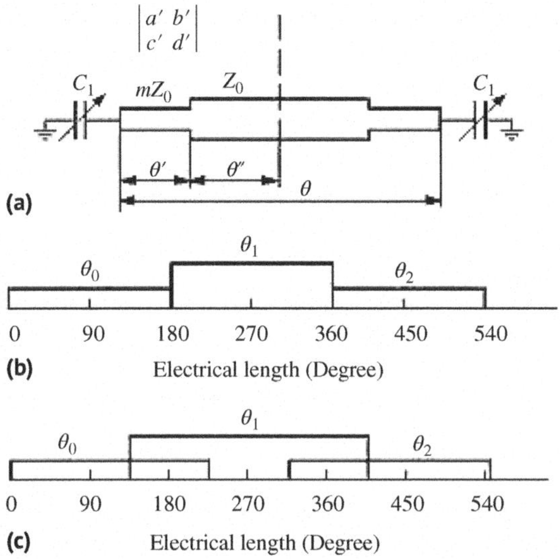

Figure 15.11 Stepped-impedance resonator with two capacitors: (a) Topology. (b) Resonance regions at $m = 1$. (c) Resonance regions at $m = 3.14$ and $\theta''/\theta' = 2.05$.

where $Y_{C1} = j\omega C_1$ is the admittance of the capacitance. Substitution of the obtained elements D' and C' into (15.26) gives us two resonance equations:

$$Y_{C1} = \frac{-d'}{b'}, n \text{ is even,} \tag{15.27a}$$

$$Y_{C1} = \frac{-c'}{a'}, n \text{ is odd.} \tag{15.27b}$$

If a uniform TL segment ($m = 1$) is used in a resonator, then the transmission matrix of half of this segment has the form:

$$\begin{bmatrix} a' & b' \\ c' & d' \end{bmatrix} = \begin{bmatrix} \cos(\theta/2) & jZ_0 \sin(\theta/2) \\ jZ_0^{-1} \sin(\theta/2) & \cos(\theta/2) \end{bmatrix}. \tag{15.28}$$

In this case, the substitution of the elements of the right-hand side of (15.28) into (15.27) gives the resonance equations for this case:

$$\omega C_1 = Z_0^{-1} \cot(\theta/2), n \text{ is even,} \tag{15.29a}$$

$$\omega C_1 = -Z_0^{-1} \tan(\theta/2), n \text{ is odd.} \tag{15.29b}$$

The resonance regions for the first three oscillations corresponding to the resonance equations (15.29) are shown in Figure 15.11b. The width of all resonance regions is the same and equal to π. These regions are closely adjacent without intersecting each other. We note that the properties of this resonator and the location of resonant regions (Figure 15.11b) were derived in Ref. [21].

For $m \neq 1$ the transfer matrix $[a'b'c'd']$ is expressed by the formula

$$\begin{bmatrix} a' & b' \\ c' & d' \end{bmatrix} = \left[\left(1+\Omega_1^2\right)\left(1+\Omega_2^2\right)\right]^{-1/2} \begin{bmatrix} 1 - m\Omega_1\Omega_2 & jZ_0\left(m\Omega_1 + \Omega_2\right) \\ jZ_0^{-1}\left(\Omega_1/m + \Omega_2\right) & 1 - \Omega_1\Omega_2/m \end{bmatrix},$$

where $\Omega_1 = \tan\theta'$, $\Omega_2 = \tan\theta''$.

Substitution of elements of this matrix into (15.27) gives the following resonance equations:

$$\omega C_1 = \frac{1}{Z_0}\frac{1 - \Omega_1\Omega_2/m}{m\Omega_1 + \Omega_2}, \; n \text{ is even,} \tag{15.30a}$$

$$\omega C_1 = -\frac{1}{Z_0}\frac{\Omega_1/m + \Omega_2}{1 - m\Omega_1\Omega_2}, \; n \text{ is odd.} \tag{15.30b}$$

Assuming $m = 3.14$ and $\theta''/\theta' = 2.05$ in (15.30), we determine the boundaries of the variation of the electrical lengths for the first three resonance frequencies: $0 \le \theta_0 \le 227.3°$, $122.3 \le \theta_1 \le 418.9°$, $310.9 \le \theta_2 \le 543.7°$ (Figure 15.11c). In case the parameter m and the ratio θ''/θ' change the resonance, the regions of this resonator also change. In the resonator under consideration (Figure 15.11a), the effect of intersecting the resonance regions occurs.

For example, consider a microstrip SIR with weak coupling to its input and output loads (Figure 15.12a). The length of the resonator is 60 mm; the remaining dimensions are shown in Figure 15.12a in millimeters. For this resonator, a substrate with a dielectric constant $\varepsilon_r = 10.2$ and a thickness of $h = 1.905$ mm is used. The width of the microstrip line in the central part of the resonator is $w = 5.5$ mm, corresponding to the characteristic impedance of $Z_0 = 25.9\ \Omega$. The width of the microstrip line at the ends of the resonator is $w_1 = 0.5$ mm, and its characteristic impedance $mZ_0 = 81.47\ \Omega$, $m = 3.14$, $\theta''/\theta' = 2.05$.

Figure 15.12b, c shows the simulated frequency responses of this resonator for $C_1 = 0.5$ pF and $C_1 = 10$ pF, respectively. The simulation was performed using Microwave Office software (AWR).

The placement of the resonance frequencies f_0, f_1, and f_2 in these figures illustrates the effect of the intersection of the resonance regions.

The change in capacitance C_1 from 0.5 to 10 pF caused a change in the first three oscillations in the frequency ranges $430.7 \le f_0 \le 1064$ MHz, $802.5 \le f_1 \le 1727$ MHz, $1627 \le f_2 \le 2179$ MHz, which intersect each other. If three oscillations

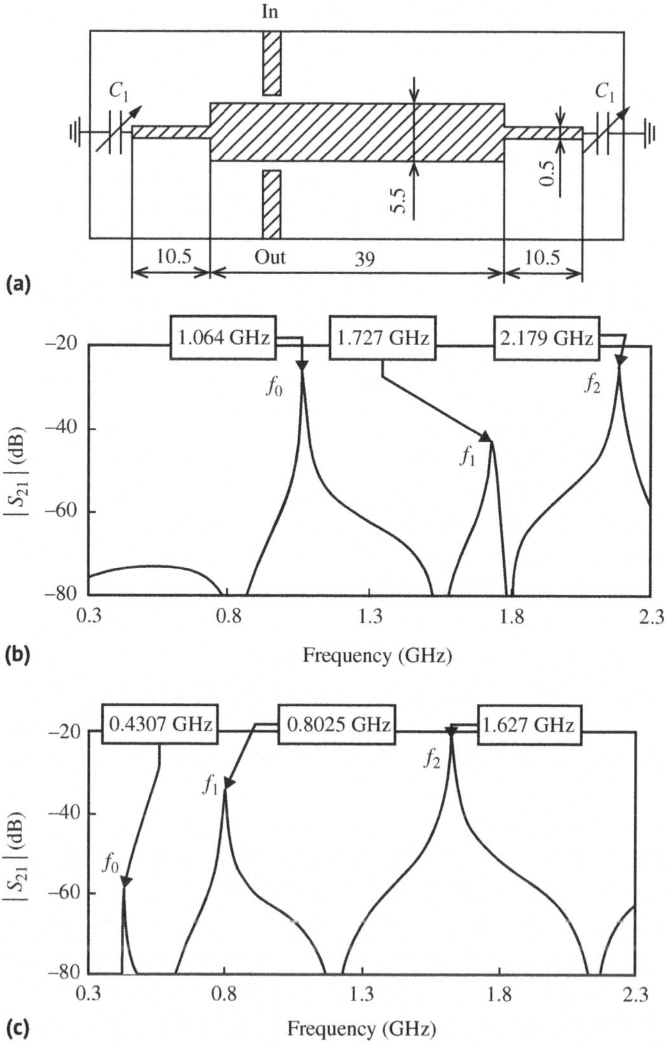

Figure 15.12 Tunable microstrip SIR with two capacitances: (a) Topology. All dimensions are in millimeters. (b) Insertion loss (IL) response at $C_1 = 0.5$ pF. (c) IL response at $C_1 = 10$ pF.

are used simultaneously for tuning, the frequency tuning range will be extensive, from 430.7 to 2179 MHz, which corresponds to a ratio of $f_{max}/f_{min} = 5.06$.

Apparently, to utilize this effect, in addition to continuous changes in capacitance, it is necessary to perform an additional switching operation, similar to that in Ref. [22]. By switching, it is possible to connect loads to different points of the resonator, attenuating two out of three oscillations due to a higher external Q-factor Q_e.

15.1.6 Regularities of Tunable Resonators with Capacitances

Note the general patterns of tunable resonators with capacitances, which become apparent when using the sensitivity function S_C (15.3), (15.5).

1. Tunable at the main resonant frequency, resonators with input susceptance B (15.4), having a pole at zero frequency, have a higher frequency range than resonators with input susceptance having zero at zero frequency. This is explained by the differences in their sensitivity functions $S_C(\theta)$, which are shown in Figures 15.2 and 15.3, respectively.
2. When the electrical length of resonators with input susceptance $B(\omega)$, having a pole at zero frequency (Figure 15.2b), decreases, the sensitivity S_C approaches 1/2, and their frequency range tends to the LC circuit. Frequency range comparison of such resonators can be correctly carried out with the same electrical lengths.
3. The sensitivity function $S_C(\omega)$ of resonators with input susceptance $B(\omega)$, which has zero at zero frequency, has a bell-shaped character at the main resonant frequency (Figure 15.3b). Such resonators are characterized by an "optimal" tuning band, in the middle of which the maximum of the function $S_C(\omega)$ is located. Frequency range comparisons of such resonators can be performed with different electrical lengths relative to each other.
4. If a resonator with a single variable capacitance does not have enough high sensitivity S_C and tunability, then they can be improved by using an additional variable capacitance. For example, a resonator with single-variable capacitance (Figure 15.8) has a maximum sensitivity of $S_{C\max} = 0.29$. Using the second variable capacitance and the transition to the resonator (Figure 15.11a) increases the maximum sensitivity value to 1/2 and improves tunability.
5. If the sensitivity $S_C(\omega)$ of a resonator with single variable capacitance is high enough, then the use of additional variable capacitances cannot lead to any significant increase in tunability. This is indicated by inequalities (15.17) and (15.18).
6. The addition of dissipative losses into the resonator is not a means of increasing its frequency range, since the restrictions (15.11) and (15.12) are valid in this case as well.
7. The established features of the sensitivity function $S_C(\omega)$, presented in Figure 15.4, made it possible to synthesize SIRs with increased tunability (Table 15.1) at the fundamental and subsequent resonant frequencies.

8. The effect of intersection resonant regions occurs in SIR with two variable capacitances (Figure 15.12).

15.2 Resonators with Variable Capacitances and Inductances, Tuning Limitations

The tunability of resonators with one and several variable capacitances is limited by inequalities (15.17) and (15.18). Therefore, it is advisable to study resonators with combined frequency tuning when resonators contain several variable capacitances and inductances.

15.2.1 Sensitivity of Resonant Frequencies to Changes in Inductance

If the capacitance C in Figure 15.1 is replaced by inductance L, the resonance equation (15.1) will change

$$\frac{1}{\omega_n L} - B(\omega_n) = 0, \quad B(\omega) \geq 0. \tag{15.31}$$

A positive definite value

$$S_L = -\left.\frac{d\omega/\omega}{dL/L}\right|_{\omega=\omega_n} = -\left.\frac{d\ln\omega}{d\ln L}\right|_{\omega=\omega_n} \tag{15.32}$$

is the sensitivity of the resonance frequencies ω_n to the change in inductance L. It is defined in the region of positive values of $B(\omega)$, where (15.31) has solutions. Using the resonance equation (15.31), we can express the sensitivity S_L (15.32) through input susceptance $B(\omega)$

$$S_L(\omega) = \left.\left(1 + \frac{dB}{d\omega}\Big/\frac{B}{\omega}\right)^{-1}\right|_{\omega=\omega_n}, \quad B \geq 0. \tag{15.33}$$

The properties of the S_L function are the same to S_C function (15.3). It is bounded on the entire frequency axis $S_L \leq 1/2$. The zeros of this function coincide with the critical frequencies of the $B(\omega)$ function.

A combline resonator in Figure 15.13a, tuned by inductance, has the input susceptance $B(\omega) = -Z_0^{-1}\cot\theta$. Substituting $B(\omega)$ in (15.33), we obtain the sensitivity function

$$S_L(\theta) = \left(1 - \frac{2\theta}{\sin 2\theta}\right)^{-1}. \tag{15.34}$$

The graph of function (15.34) is shown in Figure 15.13b.

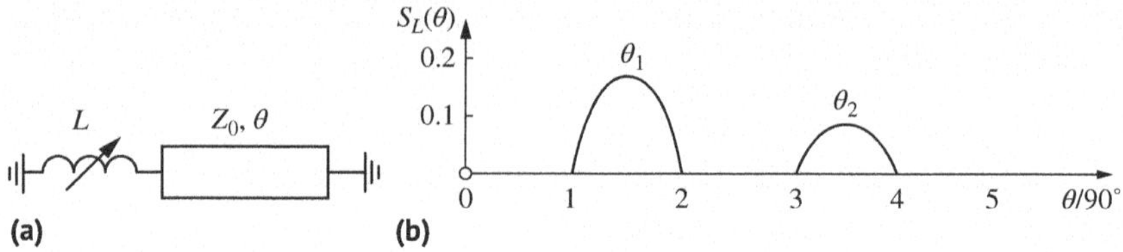

Figure 15.13 Combline resonator with inductance. (a) Schematic. (b) Function $S_L(\theta)$.

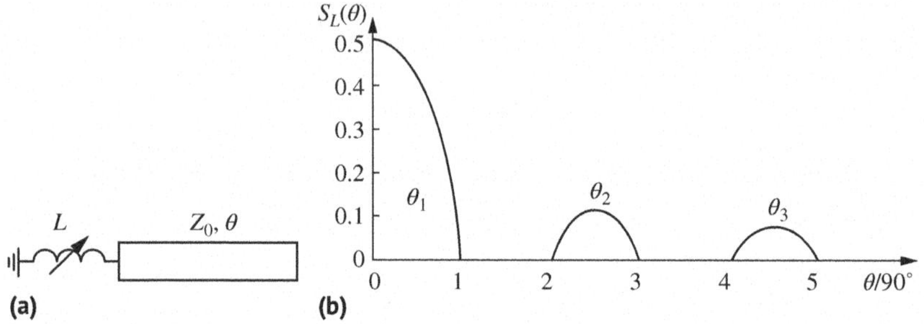

Figure 15.14 Resonator with open ends and inductance. (a) Schematic. (b) Function $S_L(\theta)$.

A resonator with open ends (Figure 15.14a), tuned by inductance, has the input susceptance $B(\omega) = Z_0^{-1} \tan\theta$. Substituting $B(\omega)$ in (15.33), we obtain the sensitivity function

$$S_L(\theta) = \left(1 + \frac{2\theta}{\sin 2\theta}\right)^{-1}. \tag{15.35}$$

The graph of function (15.35) is shown in Figure 15.14b.

15.2.2 Tunability Limitation

Initially, we consider a lossless lumped circuit that contains m variable capacitances C_i $(i = 1, 2, \ldots, m)$ and n variable inductances L_j $(j = 1, 2, \ldots, n)$. The fundamental resonant frequency ω_0 of this circuit is a function of C_i and L_j

$$\omega_0 = \omega_0\left(\{C_i\},\ \{L_j\}\right). \tag{15.36}$$

The function (15.36) is the homogeneous function of variables C_i, L_j of degree (−1):

$$\omega_0\left(\{\chi C_i\}, \{\chi L_j\}\right) = \chi^{-1}\omega_0\left(\{C_i\}, \{L_j\}\right), \tag{15.37}$$

where χ is the positive number, $\chi > 1$. Equality (15.37) shows that an increase in all values of C_i and L_j by χ times will lead to a decrease in ω_0 by χ times.

Using the Euler theorem [18], we represent expression (15.37) as

$$\sum_{i=1}^{m} C_i \frac{\partial \omega_0}{\partial C_i} + \sum_{j=1}^{n} L_j \frac{\partial \omega_0}{\partial L_j} = -\omega_0. \tag{15.38}$$

We divide both parts of (15.38) by $(-\omega_0)$. As a result, we get

$$-\sum_{i=1}^{m} \frac{\partial \omega_0/\omega_0}{\partial C_i/C_i} - \sum_{j=1}^{n} \frac{\partial \omega_0/\omega_0}{\partial L_j/L_j} = 1. \tag{15.39}$$

When we use the sensitivities (15.2) and (15.32), equality (15.39) will be written in the form

$$\sum_{i=1}^{m} S_{Ci} + \sum_{j=1}^{n} S_{Lj} = 1. \tag{15.40}$$

Expression (15.40) is known [14] as an invariant of sensitivity ω_0 to the change of all capacitances and inductances of a lumped *LC* circuit. It shows that the sum of all sensitivities S_{Ci} and S_{Lj} of a lumped lossless circuit is invariant with respect to the structure of the circuit. The invariant of sensitivity (15.40) is the fundamental property of lossless electrical circuits.

The parameter $\chi = C_{i\max}/C_{i\min} = L_{j\max}/L_{j\min}$ in (15.37) represents the relative variation of capacitances and inductances. The ratio of the maximum and minimum values of the resonant frequency ω_0 will be denoted by $K = \omega_{0\max}/\omega_{0\min}$. The value of K characterizes the tuning range. Then, equality (15.37) can be written as

$$K = \chi. \tag{15.41}$$

If we add TL segments to the circuit under consideration, or use different values of χ_i, χ_j, then in the expressions (15.40) and (15.41), the equal sign must be replaced with the inequality sign

$$\sum_{i=1}^{m} S_{Ci} + \sum_{j=1}^{n} S_{Lj} \leq 1, \tag{15.42}$$

$$K \leq \chi_{\max} \tag{15.43}$$

where $\chi_{\max}$ is the largest value among $\chi_1, \chi_2, \ldots, \chi_m$ for capacitances and among $\chi_1, \chi_2, \ldots, \chi_n$ for inductances.

Conditions (15.42) and (15.43) express the tunability limitations of resonators derived from lossless TL sections, which are tunable through several capacitances and inductances.

Similar restrictions for resonators tunable with several capacitances are expressed by inequalities (15.17) and (15.18).

The right-hand side of (15.42) and (15.43) significantly exceeds the right-hand side of (15.17) and (15.18). This testifies to the higher potential capabilities of resonators tunable with multiple capacitances and inductances, indicating that combined frequency tuning is promising.

15.2.3 Uniform TL Resonators with Combined Frequency Tuning

Let us analyze the resonator with combined tuning shown in Figure 15.15a. The notation in these figures does not require explanation. In this case, the resonance equations (15.1) and (15.31) take the form

$$\omega C = -\frac{\omega L Z_0^{-1} \tan\theta - 1}{\omega L + Z_0 \tan\theta}, \tag{15.44a}$$

$$\frac{1}{\omega L} = \frac{\omega C + Z_0^{-1} \tan\theta}{\omega C Z_0 \tan\theta - 1}. \tag{15.44b}$$

Using (15.44) and (15.3), (15.33), it is possible to calculate the sensitivity S_C, S_L, determine their sum $S_C + S_L$, and perform the analysis. However, such an analysis is rather cumbersome and uninformative. Without loss of generality, we can assume that L and C change synchronously and are related by the condition

$$\frac{L}{C} = Z_0^2. \tag{15.45}$$

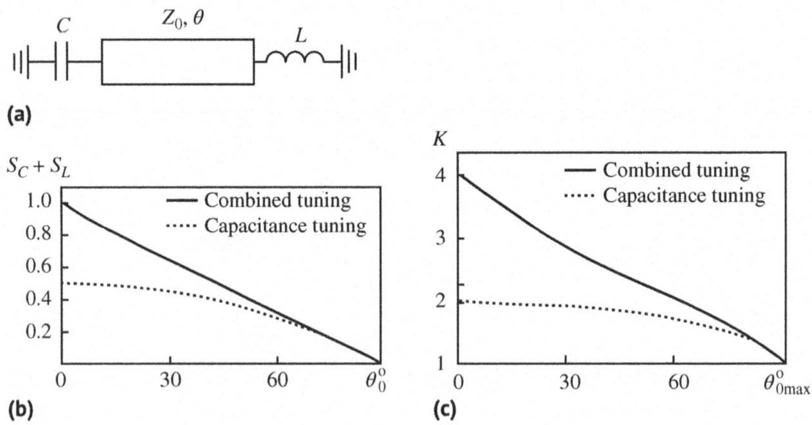

Figure 15.15 Resonator with combined tuning. (a) Schematic. (b) Total sensitivity of fundamental resonance frequency to variations in capacitance and inductance. (c) Ratio K versus maximum electric length θ_{0max} of resonator for $\chi = 4$.

Using the resonance equation for a lossless two-port network (15.25) and excluding the inductance from it ($L = CZ_0^2$), we obtain the simplest resonance equation

$$\omega C = \frac{\pm 1 - \sin\theta}{Z_0 \cos\theta} = -P(\omega). \tag{15.46}$$

The "+" sign in front of 1 refers to ω_n with even n, and the minus sign refers to ω_n with odd n. The function $P(\omega)$ on the right-hand side of (15.46) differs from $B(\omega)$ in (15.1); it is not a susceptance function. However, it can be used to calculate the sum of sensitivities $S_C + S_L$ by substituting it into equation (15.3). As a result, we obtain

$$S_C + S_L = \left(1 \pm \frac{\theta}{\cos\theta}\right)^{-1}. \tag{15.47}$$

The "+/−" signs in expression (15.47) have the same meaning. The plot of total sensitivity (15.47) for the main resonant frequency ω_0 is presented in Figure 15.15b by a solid line.

From the resonance equation (15.46) for the main resonance frequency ω_0, the next expression follows

$$\chi = K\frac{\cos\theta_{0\max}(1 - \sin\theta_{0\min})}{\cos\theta_{0\min}(1 - \sin\theta_{0\max})}. \tag{15.48}$$

Since $\theta_{0\min}$ and $\theta_{0\max}$ are related by the condition $\theta_{0\min} = \theta_{0\max}/K$, the expression (15.48) determines the relationship between three quantities χ, K, and $\theta_{0\max}$. We use the value $\chi = 4$, at which expression (15.48) will determine the dependence of the frequency ratio K on $\theta_{0\max}$.

The solid line in Figure 15.15c represents the function $K = K(\theta_{0\max})$. These plots show that, with a decrease in the electric length of the resonators, their indices of tunability approach their limiting values: $S_C + S_L \to 1$, $K \to \chi$.

The plots in Figure 15.15b, c show that, with an increase in θ_0, the tunability indices of the resonators decrease. The electric length is a factor restraining the tunability. In the same figures, dotted lines indicate the same characteristics for resonators with a single type of tuning element. The comparison of these characteristics demonstrates the advantage of the combined frequency tuning. Microstrip BPF with combined frequency tuning is discussed in Section 16.3.

15.2.4 SIR with Combined Frequency Tuning

Figure 15.16a shows SIR with combined frequency tuning, where parameter m is the ratio $m = Z_{0\max}/Z_{0\min} (m \geq 1)$. To simplify the analysis, we will assume that

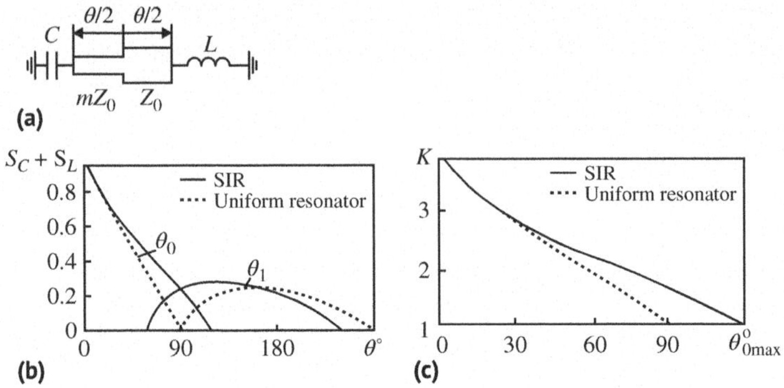

Figure 15.16 SIR combined tuned by L and C. (a) Schematic. (b) The total sensitivity versus θ. (c) Dependence of K on $\theta_{0\max}$ for case $\chi = 4$.

L and C are related by the condition $L/C = mZ_0^2$. This allows us to eliminate the inductance L and represent the resonance equation in the form

$$\omega C = \frac{\pm\sqrt{m}\,[1 + \tan^2(\theta/2)] - (1+m)\,[\tan(\theta/2)]}{mZ_0\,[1 - \tan^2(\theta/2)]}. \tag{15.49}$$

Even resonant frequencies ω_n correspond to "+" in the numerator (15.49), and odd ω_n correspond to the "–" sign.

Substitution of (15.49) into (15.3) gives the result

$$S_C + S_L = \left[1 + \frac{\theta}{\cos^2(\theta/2)}\left(\frac{m\tan(\theta/2)}{m\tan^2(\theta/2) - 1} - \frac{\pm\sqrt{m}\tan(\theta/2) - 2(1+m)}{\pm\sqrt{m}\,[\tan^2(\theta/2) - 1] - (1+m)\tan(\theta/2)}\right)\right]^{-1} \tag{15.50}$$

If we put $m = 3$ in (15.50), then we get the dependences shown in Figure 15.16b by solid lines. The solid curve in Figure 15.16c plots a dependence similar to that shown in Figure 15.15c. Expression (15.49) was used in its construction. Figure 15.16b, c shows the dotted curves for the uniform resonator ($m = 1$).

We can see from Figures 15.15 and 15.16 that the electrical length θ of both types of resonators is a factor limiting the tunability of the resonant frequency ω_0. The SIR has better tunability than the uniform resonator.

The SIRs shown in Figure 15.17 have three tuning elements, and their electrical length is 2θ. These SIRs, with the notation adopted in Figure 15.17, and the resonator in Figure 15.16a, are equivalent at the frequency ω_0. Tunability of these resonators is the same.

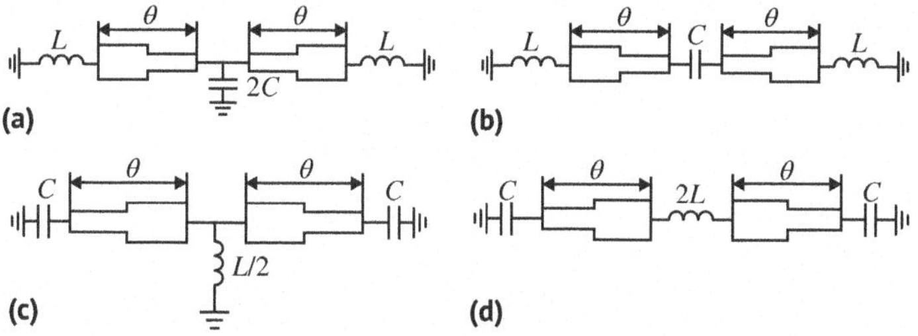

Figure 15.17 Tunable SIRs with three tuning elements. (a) SIR with two inductances and parallel capacitance. (b) SIR with two inductances and series capacitance. (c) SIR with two capacitances and parallel inductance. (d) SIR with two capacitances and series inductance.

15.2.5 Tuning Curves

Figure 15.18a illustrates the topology of a microstrip resonator featuring three tuning elements and exhibiting weak coupling to the input and output loads. This resonator corresponds to the circuit in Figure 15.17a. The dielectric substrate used in the resonator is characterized by $\varepsilon_r = 10.2$ and thickness $h = 0.635$ mm. The width of the two low-impedance sections of the microstrip line is 3 mm ($Z_0 = 17.8\ \Omega$). The width of the central section of the microstrip line is 0.5 mm ($mZ_0 = 53.8\ \Omega$, $m = 3.02$). The length of the resonator is 28 mm, and the lengths of the component sections are given in Figure 15.18a in millimeters.

To simulate the frequency response of the resonator under consideration, we use the Microwave Office (AWR) software. The values of the lumped elements changed by $\chi = 6$ times: $2C = 2 - 12$ pF $(C = 1 - 6$ pF$)$; $L = 1 - 6$ nH.

In Figure 15.18b, the dashed line corresponds to the dependence of the resonator's f_0 on the values of the synchronously varying inductances and capacitance C. This dependence is also called a tuning curve. The variation of the lumped elements within the above-mentioned boundaries ($\chi = 6$) results in a change in frequency from 526 to 1412 MHz, which corresponds to a frequency ratio of $K = 2.68$. The electrical length of this resonator at the frequency of the fundamental resonance changes within the limits $47.4 \leq 2\theta_0 \leq 127.6°$.

There is no need for a smooth variation of the inductance; it may be varied in steps using *pin* diodes or MEMS technology. The solid lines in Figure 15.18b are the resonator's tuning curves corresponding to the fixed inductance values $L = 6$ nH (curve *1*) and $L = 1$ nH (curve *2*). Frequency f_0 varies, respectively, within the limits $526 \leq f_0 \leq 796$ MHz and $784 \leq f_0 \leq 1412$ MHz. A step change in inductance from 6 to 1 nH (single switch) provides the same frequency ratio $K = 2.68$. When combined tuning is used, the value of the inductance can be changed by switching, whereas the capacitance can be smoothly varied, thus implementing the principle of continuous frequency variation.

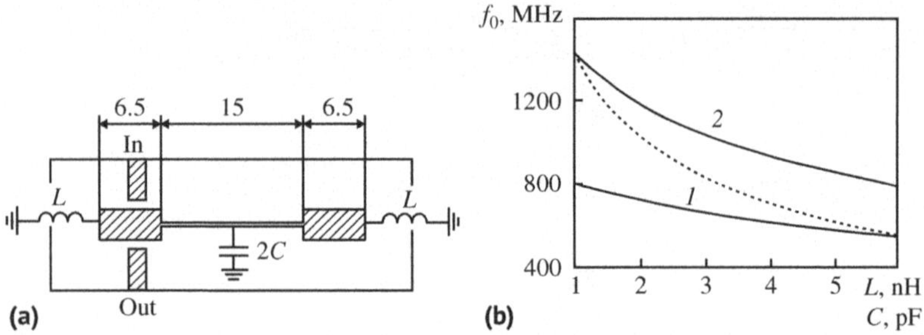

Figure 15.18 Microstrip SIR with combined frequency tuning. (a) Topology. (b) Tuning curves for resonance frequency ω_0.

Table 15.4 Relative variation of tuning elements.

Tuning	Resonator	χ
Combined	SIR	6
	Uniform	7.64
Capacitive	SIR	9.52
	Uniform	12.29

Let us compare the resonator under consideration to other tunable resonators. If the microstrip resonator (Figure 15.18a) is a uniform one with a 0.5 mm wide central conductor, then, as the computer simulations show, for the resonator to be tunable in the same frequency range $526 \leq f_0 \leq 1412$ MHz $(K = 2.68)$, the required relative variation of the capacitance and inductance will be $\chi = 7.64$.

We use two capacitances for reconfiguration, which are connected to the open ends of the SIR shown in Figure 15.11a. The end sections of this resonator are high-impedance ones, whereas the section in the center part is a low-impedance one. Setting the above-mentioned widths of the central conductors of the SIR (0.5 and 3 mm) and the resonator's length $l = 28$ mm, we find by the computer simulation that the relative capacitance variation providing tuning in the same frequency range is $\chi = 9.52$. If the resonator is made uniform, the value of χ increases up to $\chi = 12.29$.

The values of χ for the above-considered cases are presented in Table 15.4. It is clearly seen that, when we change the tuning from capacitive to the combined one and from the uniform resonator to the SIR, the value of χ decreases.

15.3 Hairpin Resonators Tuned by Capacitance

Let us consider two cases: hairpin open-ended [19] and short-ended [23] resonators.

15.3.1 Hairpin Open-ended Resonator

The hairpin open-ended resonator is shown in Figure 15.19a. Its TL segment has a characteristic impedance Z_0 and an electrical length $\theta = \omega L/\upsilon$, where L is the length of the segment and υ is the propagation speed of the electromagnetic wave. The set of electrical lengths θ_n corresponds to the set of resonance frequencies ω_n. In the further analysis of the resonators, we will neglect their dissipation losses. In this case, their frequencies ω_n are defined by a single condition of resonance: $Y = 0$, where Y is the input susceptance of a resonator.

To obtain the equation for the function Y of the resonator shown in Figure 15.19a, let us consider the content of Figure 15.19b, where the conversion of a two-port network into a one-port network using the loop connection of the ports *2–2′* and *1–1′*.

To describe the relation between currents and voltages of the two-port network, let us use the system of *ABCD* parameters [17]:

$$\begin{aligned} U_1 &= AU_2 - BI_2, \\ I_1 &= CU_2 - DI_2. \end{aligned} \tag{15.51a}$$

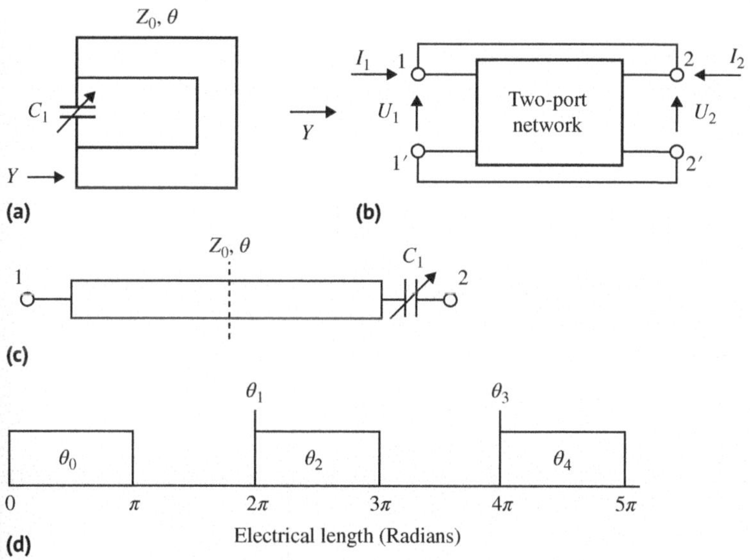

Figure 15.19 Tunable loop-type hairpin resonator. (a) Topology. (b) Loop connection that converts two-port network into one-port network. (c) Straightened shape of hairpin resonator. (d) Distribution of resonant electrical lengths.

In case of the loop connection, the voltages at the input and the output of a two-port network are equal $U_1 = U_2 = U$. Therefore, (15.51a) can be expressed as:

$$U = AU - BI_2,$$
$$I_1 = CU - DI_2. \qquad (15.51b)$$

Solving (15.51b) regarding currents, we obtain:

$$I_1 = U\left[C - D\left(A-1\right)/B\right],$$
$$I_2 = U\left(A-1\right)/B. \qquad (15.51c)$$

The resulting current of a one-port network is equal to the sum of the currents at the input and output of a two-port network. Adding I_1 and I_2 from (15.51c), we can obtain

$$I = I_1 + I_2 = \frac{U\left(A+D-2\right)}{B}.$$

Input admittance of a one-port network, which was made from a two-port network by the loop connection, is defined by the following expression:

$$Y = \frac{I}{U} = \frac{A+D-2}{B}. \qquad (15.52)$$

The considered loop hairpin resonator with a capacitor can be represented in a detailed form (Figure 15.19c). Let us denote the transfer matrix of the half of a TL segment as

$$\begin{bmatrix} a' & b' \\ c' & d' \end{bmatrix} = \begin{bmatrix} \cos\left(\theta/2\right) & jZ_0\sin\left(\theta/2\right) \\ jZ_0^{-1}\sin\left(\theta/2\right) & \cos\left(\theta/2\right) \end{bmatrix}, \qquad (15.53)$$

where $j = \sqrt{-1}$. The transfer $[ABCD]$ matrix of the two-port network shown in Figure 15.19c can be expressed through the parameters a', b', c', d' of the matrix (15.53) and the admittance of a capacitor $Y_{C1} = j\omega C_1$ [17]:

$$\begin{bmatrix} A & B \\ C & D \end{bmatrix} = \begin{bmatrix} 1+2b'c' & 2a'b \\ 2c'd' & 1+2b'c' \end{bmatrix} \times \begin{bmatrix} 1 & 1/Y_{C1} \\ 0 & 1 \end{bmatrix}$$
$$= \begin{bmatrix} 1+2b'c' & \left(1+2b'c'\right)/Y_{C1} + 2a'b' \\ 2c'd' & 2c'd'/Y_{C1} + \left(1+2b'c'\right) \end{bmatrix}. \qquad (15.54)$$

The substitution of the elements of the matrix (15.53) into (15.54) gives the admittance of a loop hairpin resonator (Figure 15.19a):

$$Y = 2jZ_0^{-1}\frac{\sin\left(\theta/2\right)\left[2jZ_0Y_{C1}\sin\left(\theta/2\right) + \cos\left(\theta/2\right)\right]}{1 - 2\sin^2\left(\theta/2\right) + 2jZ_0Y_{C1}\cos\left(\theta/2\right)\sin\left(\theta/2\right)}. \qquad (15.55)$$

Equating the numerators of the admittance (15.55) to zero, we obtain two equations of resonance:

$$\omega_n C_1 = \frac{1}{2} Z_0^{-1} \cot\left(\theta_n/2\right), n \text{ is even}, \tag{15.56a}$$

$$\sin\left(\theta_n/2\right) = 0, \quad \theta_n = (n+1)\,\pi, n \text{ is odd}. \tag{15.56b}$$

The equation of resonance (15.56a) shows that the resonance frequencies ω_n with even numbers n depend on the capacitance C_1. The electrical lengths at these frequencies vary within the limits $n\pi \leq \theta_n \leq (n+1)\,\pi$. The resonance frequencies ω_n with odd numbers n (15.56b) do not depend on the capacitance C_1. The resonance electrical lengths at these frequencies have the constant value $\theta_n = (n+1)\,\pi$, so that these oscillations are divisible by λ. The resonance domains for the first five oscillation modes are shown in Figure 15.19d. These domains cover only half of the axis θ. If the capacitance C_1 increases, the frequencies ω_n for even numbers n approach from the right to the frequencies ω_n for odd numbers n.

Using the resonance equations (15.56), the graph of the ratio f_1/f_0 versus θ_0 for a tunable hairpin resonator is plotted in Figure 15.20. Even and odd frequencies ω_n of a combline resonator can be calculated from the single equation of resonance (15.6), which does not have a 1/2 factor on the right side. The same plot for a combline resonator is depicted in Figure 15.20 by a dashed line. For this curve, the doubled value of the electrical length $2\theta_0$ is marked along the x-axis.

If $C_1 = 0$, the combline resonator has the electric length $\theta_0 = 90°$ and the ratio $f_1/f_0 = 3$. The loop hairpin resonator at $C_1 = 0$ has $\theta_0 = 180°$ and the ratio $f_1/f_0 = 2$.

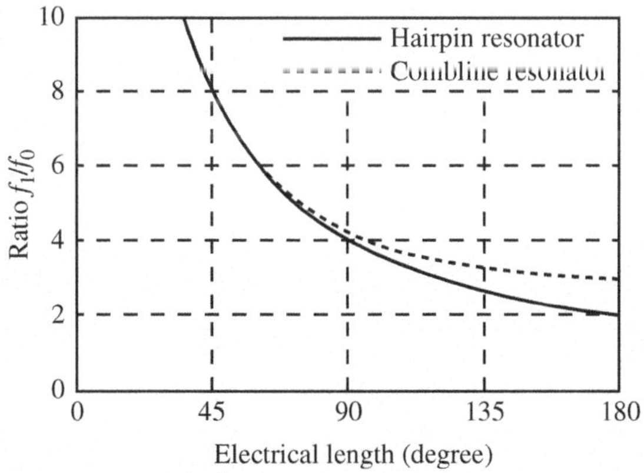

Figure 15.20 Comparison of ratio $R = f_1/f_0$ for hairpin and combline resonators, where f_0 and f_1 are the main and first parasitic resonant frequencies.

With increasing C_1 (decreasing θ_0), both frequencies f_1 and f_0 of the combline resonator decrease, and their ratio f_1/f_0 increases.

In the loop hairpin resonator, the fundamental resonant frequency f_0 also decreases, and the frequency of the first spurious resonance f_1 remains stationary. As a result, the loop hairpin resonator ratio f_1/f_0 increases faster than the combline resonator. A difference in the value of f_1/f_0 essentially decreases as the electrical length decreases. For $\theta_0 = 90°$ (45°), the difference between the values of f_1/f_0 is 7%, and for $\theta_0 = 60°$ (30°), the difference is 2.8%.

Thus, the tunability of hairpin and combline resonators coincides, and the capacitance C_1 of the hairpin resonator used for tuning is two times smaller than that of the combline resonator. The ratio of resonant frequencies f_1/f_0 of the hairpin resonator approaches the ratio f_1/f_0 of the combline resonator as the resonant electrical length θ_0 decreases. These features allow the hairpin and combline resonators to be combined in a single BPF by synchronously tuning it using varactors. Such a filter is discussed in Section 16.1.

15.3.2 Hairpin Short-ended Resonators

Figure 15.21 illustrates symmetrical hairpin short-ended resonators with various capacitance connections to their middle part. The dashed line through the middle of the resonators divides them into two equal upper and lower halves. The transfer matrix of the lower half of the resonators is noted as $[A'B'C'D']$, and the transfer matrix of a distributed circuit of the same half is denoted as $[a'b'c'd']$. Resonance in the symmetrical resonators under consideration occurs when the voltage at the

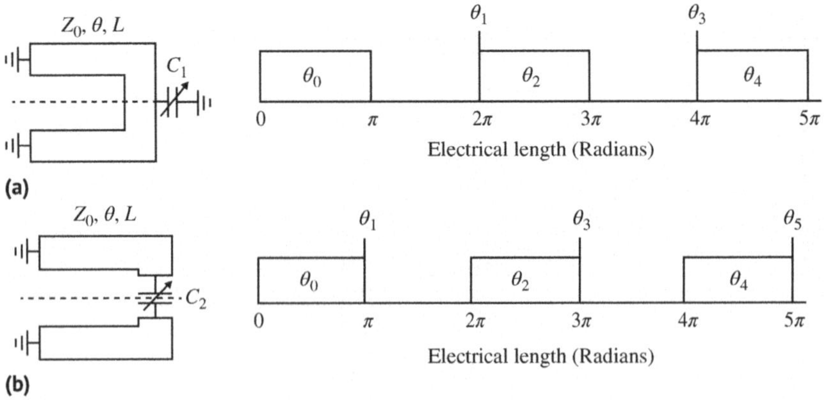

Figure 15.21 Tunable hairpin short-ended resonators and distribution of their resonant electrical lengths. (a) With a parallel-connected capacitor. (b) With a series-connected capacitor.

center of the resonator reaches its maximum or minimum value. Resonance conditions can be expressed in the form:

$$y_{22} = \frac{A'}{B'} = 0,$$
$$\frac{1}{y_{22}} = \frac{B'}{A'} = 0. \tag{15.57}$$

The resonance conditions (15.57) apply to the general case when a distributed part of the resonator contains stubs and a nonuniform TL. In this case, the variables A' and B' in (15.57) can have poles at finite frequencies and some coinciding zeros. In a particular case, when parameters A' and B' do not have the specified poles and coinciding zeros, resonance conditions (15.57) take the form:

$$A' = 0,$$
$$B' = 0. \tag{15.58a}$$

In the resonator in Figure 15.21a, the middle line divides the capacitance C_1 into two equal parts ($C_1/2$), so the matrix $[A'B'C'D']$ takes the form:

$$\begin{bmatrix} A' & B' \\ C' & D' \end{bmatrix} = \begin{bmatrix} a' & b' \\ c' & d' \end{bmatrix} \begin{bmatrix} 1 & 0 \\ Y_{C1}/2 & 1 \end{bmatrix} = \begin{bmatrix} a' + b'Y_{C1}/2 & b' \\ c' + d'Y_{C1}/2 & d' \end{bmatrix},$$

where $Y_{C1} = j\omega C_1$ is the admittance of the capacitance C_1. Substitution of the elements A' and B' into (15.58a) leads to the resonance equations

$$Y_{C1} = -2a'/b', \; n \text{ is even}, \tag{15.58b}$$

$$b' = 0, \; n \text{ is odd}. \tag{15.58c}$$

At the resonator in Figure 15.21b, the capacitance C_2 can be represented by a series connection of two capacitances of the value ($2C_2$); therefore, the $[A'B'C'D']$ matrix can be defined as:

$$\begin{bmatrix} A' & B' \\ C' & D' \end{bmatrix} = \begin{bmatrix} a' & b' \\ c' & d' \end{bmatrix} \begin{bmatrix} 1 & 1/2Y_{C2} \\ 0 & 1 \end{bmatrix} = \begin{bmatrix} a' & b' + a'/2Y_{C2} \\ c' & d' + c'/2Y_{C2} \end{bmatrix},$$

where $Y_{C2} = j\omega C_2$ is the admittance of the capacitance C_2. Substitution of the elements A' and B' into (15.58a) leads to the resonance equations

$$Y_{C2} = -a'/(2b'), \; n \text{ is even}, \tag{15.59a}$$

$$a' = 0, \; n \text{ is odd}. \tag{15.59b}$$

Equations (15.58a) and (15.59) apply to symmetrical resonators, which may contain a nonuniform TL.

If we use the segment of a uniform TL, then the transfer matrix $[a'b'c'd']$ is determined by the expression (15.53). Substituting it into (15.58a) gives the resonance equations for the resonator shown in Figure 15.21a:

$$\omega_n C_1 = 2Z_0^{-1} \cot\left(\frac{\theta}{2}\right), \quad n \text{ is even,} \tag{15.60a}$$

$$\sin\left(\frac{\theta_n}{2}\right) = 0, \quad \theta_n = (n+1)\,\pi, \quad n \text{ is odd.} \tag{15.60b}$$

If matrix elements (15.53) are substituted into (15.59), we obtain the resonance equations for the resonator shown in Figure 15.21b:

$$\omega_n C_2 = \frac{1}{2} Z_0^{-1} \cot\left(\frac{\theta_n}{2}\right), \quad n \text{ is even,} \tag{15.61a}$$

$$\cos\left(\frac{\theta_n}{2}\right) = 0, \quad \theta_n = n\pi, \quad n \text{ is odd.} \tag{15.61b}$$

Odd resonant frequencies of hairpin short-ended resonators are independent of variable capacitance (15.60b, 15.61b). The even resonant frequencies of these resonators depend on the variable capacitance (15.60a, 15.61a). The electrical lengths θ_n at these frequencies vary within the resonant regions $n\pi \le \theta_n \le (n+1)\,\pi$. The resonance domains for the first oscillation modes of the resonators with parallel and series capacitance connections are shown in Figure 15.21a, b, respectively.

15.3.3 Comparison of Tunable Hairpin Resonators

The hairpin open-ended resonators (Figure 15.19) and hairpin short-ended resonators (Figure 15.21) are characterized by the same tunability. This follows from the fact that their resonance equations for the fundamental resonant frequency (15.56a, 15.60a, and 15.61a) coincide, except for the constant factor on the right-hand side. The capacitance of the short-ended hairpin resonator with a parallel-connected capacitor (Figure 15.21) used for tuning is four times greater than that of the other two hairpin resonators. This makes it promising for use at high frequencies, where the capacitance values become extremely small.

An important characteristic of resonators is the f_1/f_0 ratio, on which the width of the filter's rejection band depends. Using the resonance equations, the graph of the ratio f_1/f_0 versus θ_0 for the tunable hairpin resonators is plotted in Figure 15.22. The hairpin short-ended resonator (Figure 15.21a) and loop hairpin resonator (Figure 15.19a) have the same f_1/f_0 ratio, which is plotted by a solid line (Figure 15.22). If $\theta = 90°$, then $f_1/f_0 = 4$, and if $\theta = 40°$, then $f_1/f_0 = 9$. The hairpin short-ended resonator with series capacitance (Figure 15.21b) is characterized by a smaller f_1/f_0 ratio, which is shown by a dashed line (Figure 15.22).

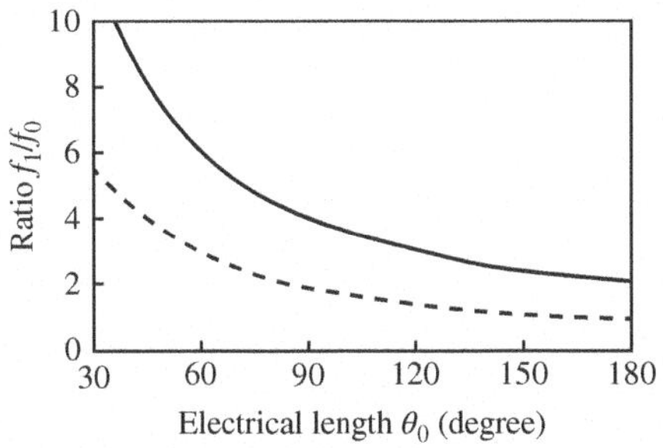

Figure 15.22 Comparison of the ratio f_1/f_0 for different tunable hairpin resonators.

Analysis based on resonance equations showed that in BPFs, it is preferable to use the resonators shown in Figures 15.21a and 15.19a because they have a high f_1/f_0 ratio. The variable capacitance of the first one is four times greater than that of the second one. Therefore, it is advisable to use the hairpin short-ended resonator (Figure 15.21a) at higher frequencies and the loop-type hairpin resonator (Figure 15.19a) at lower frequencies.

15.4 Unloaded Quality Factors of Tunable Resonators

15.4.1 Unloaded Quality Factor Q_u of Resonators with One Capacitor

Let us consider a compound resonator (Figure 15.23) formed by a distributed circuit of lossy TL segments and a capacitor with a finite quality factor Q_C connected in parallel. This resonator is equivalent to a one-port network with an input admittance $\tilde{Y}$. The one-port network includes capacitance C of the capacitor, its equivalent parallel conductance G_L, and a distributed circuit with input admittance Y. We consider the resonator behavior at its fundamental resonance frequency ω_0.

The initial data of the problem are the quality factors of the capacitor Q_C and TL Q_l, which represent the unloaded quality factor of the resonator from a half-wave or quarter-wave segment of this line. The value of Q_l is discussed in Chapter 4.

It is necessary to obtain the formula expressing the unloaded quality Q_u of the compound resonator (Figure 15.23) in terms of Q_C and Q_l [24]. Let us start with the simplest case, for a lossless TL. In this case $Q_l = \infty$.

15.4.1.1 Lossless Transmission Line

The input admittance of the distributed circuit is purely imaginary: $Y = jB$. The resonance condition for a one-port circuit (Figure 15.23) is expressed by equality (15.1).

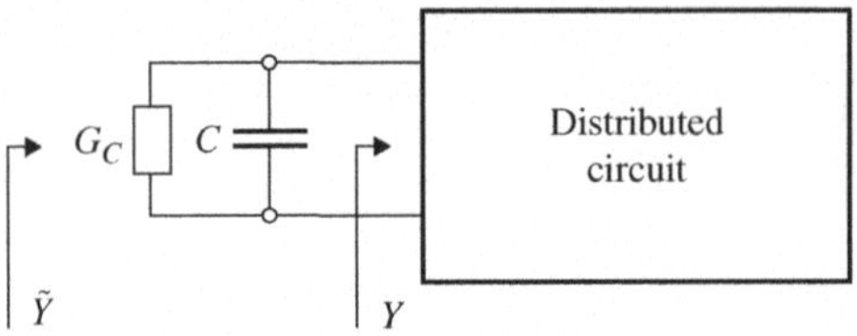

Figure 15.23 Resonator containing distributed lossy circuit and a capacitor with finite quality factor.

The unloaded quality Q_u of the resonator is determined as follows [25]

$$Q_u = \frac{b}{G_C}, \tag{15.62}$$

where b is the susceptance slope parameter (3.2).

Taking into account (15.1) and (3.2), the expression for Q_u of the resonator in question takes the form

$$Q_u = \frac{\omega C}{2G_C}\left(1 - \frac{dB/d\omega}{B/\omega}\right). \tag{15.63}$$

Taking into account expressions for sensitivity (15.3) and capacitor quality $Q_C = \omega C/G_C$ [25], we rewrite the expression (15.63) for the unloaded quality Q_u of the compound resonator in the absence of loss in the distributed circuit. As a result, we obtain

$$Q_u = \frac{\omega C}{2S_C G_C} = \frac{Q_C}{2S_C}. \tag{15.64}$$

Expression (15.64) is accurate because it is obtained without any assumptions. It follows from (15.64) that $Q_u \to \infty$ for $S_C \to 0$, even in the case of small values of Q_C.

15.4.1.2 General Case

If the relationship between Q-factors Q_C and Q_l was subject to a widely used pattern

$$\frac{1}{Q_u} = \frac{1}{Q_C} + \frac{1}{Q_l}, \tag{15.65}$$

then we would have $Q_u = Q_C$ for $Q_l = \infty$. However, this is not the case, since equality (15.64) holds. Consequently, it is necessary to use a form of combining Q_C and Q_l different from (15.65).

As a result, we utilize the pattern of a capacitor's Q factor, obtained by either parallel or series connection of two other capacitors. If two capacitors with capacitances C_1 and C_2 and corresponding loss conductance G_1 and G_2 are connected in parallel, the quality factor Q of the resulting capacitor is determined by the expression [25]

$$\frac{1}{Q} = \frac{G_1 + G_2}{\omega (C_1 + C_2)} = \frac{G_1 C_1}{\omega C_1 (C_1 + C_2)} + \frac{G_2 C_2}{\omega C_2 (C_1 + C_2)}.$$

Taking into account that

$$\frac{1}{Q_1} = \frac{G_1}{\omega C_1} \text{ and } \frac{1}{Q_2} = \frac{G_2}{\omega C_2},$$

we rewrite the previous equality in the form

$$\frac{1}{Q} = \frac{A_1}{Q_1} + \frac{A_2}{Q_2}, \tag{15.66}$$

where $A_1 = C_1/(C_1 + C_2)$ and $A_2 = C_2/(C_1 + C_2)$ are the weighting coefficients of the corresponding capacitors. Formula (15.66) expresses the quality factor Q of the resulting capacitor in terms of quality factors Q_1 and Q_2 of two other capacitors connected in parallel. It can be shown that, in the case of the series connection of two capacitors, the quality of the resulting capacitor is also determined by (15.66). In this case, the weighting coefficients are $A_1 = C_2/(C_1 + C_2), A_2 = C_1/(C_1 + C_2)$. The larger the weighting coefficient, the larger the contribution introduced by this element into the resulting quality factor Q. The quality factor Q of a composite inductor is also subject to pattern (15.66).

The weighting (or influence) coefficients A_1 and A_2 in (15.65) satisfy the condition

$$A_1 + A_2 = 1. \tag{15.67}$$

If the quality factors of two capacitors are equal, $Q_1 = Q_2$, the quality factor of the resulting capacitor is the same in the cases of the parallel and series connection. Equalities (15.66) and (15.67) reflect this regularity. If two inductors are connected, equalities (15.66) and (15.67) are also satisfied.

Let us find the solution to the formulated problem in the form (15.66), which can be written as follows

$$\frac{1}{Q_u} = \frac{A_C}{Q_C} + \frac{A_l}{Q_l}, \tag{15.68}$$

where A_C is the influence coefficient of the capacitor and A_l is the influence coefficient of the TL segment on the unloaded quality Q_u of the resonator. Formulas (15.66) and (15.68) differ from each other by the notation $A_1 = A_C, A_2 = A_l$. If one of the terms in (15.68) is known, the second term can also be found using relationship (15.67), which solves the formulated problem.

We represent the previously set expression (15.64) in the form

$$\frac{1}{Q_u} = \frac{2S_C}{Q_C},$$

from which we obtain the influence coefficient of the capacitor on the quality factor Q_u:

$$A_C = 2S_C. \tag{15.69}$$

Since the influence coefficient A_C is found (15.69), from condition (15.67), we determine the influence coefficient of the TL segment

$$A_l = 1 - 2S_C. \tag{15.70}$$

The sensitivity limitation $0 \le S_C \le 1/2$ leads to a limited influence coefficients:

$$0 \le A_C \le 1,$$
$$0 \le A_l \le 1.$$

Substituting (15.69) and (15.70) into (15.68), we finally obtain

$$\frac{1}{Q_u} = \frac{2S_C}{Q_C} + \frac{1 - 2S_C}{Q_l}. \tag{15.71}$$

If a lossless capacitor is used, then $Q_C = \infty$ and

$$Q_u = \frac{Q_l}{1 - 2S_C}.$$

If $S_C = 0$, then $Q_u = Q_l$, but if $S_C = 1/2$, then $Q_u = \infty$.

The values of the sensitivity in (15.71) are determined by (15.3), where $B(\omega)$ is the input susceptance of the lossless distributed circuit. This circumstance substantially simplifies the Q_u calculation, since it is not necessary to find the total admittance Y of the distributed circuit and separate its imaginary part.

Expression (15.71) can be used to determine the unloaded quality factor Q_C of ferroelectric capacitors, as demonstrated in Section 16.4.

15.4.2 Analysis Q_u of Some Resonators

Let us consider two different resonators with a capacitor. The first of them (Figure 15.24a) is a well-known combline resonator with a capacitor. Let us analyze the unloaded quality Q_u of the resonator for various values of its electric length θ_0 using (15.71). For this purpose, we use the sensitivity function (15.6b). For clarity of presentation, the graph of this function for the fundamental resonant frequency is shown in Figure 15.24b.

We consider two variants of values Q_C and Q_l: $Q_C = 50$, $Q_l = 200$ (curve *1*); $Q_C = 200$, $Q_l = 50$ (curve *2*). The plots of Q_u variation for these two variants are shown in Figure 15.24c. Curve *1* corresponds to the first variant, and curve *2* corresponds to the second one. The resonator unloaded quality Q_u increases with increasing θ_0 if $Q_l > Q_C$ (curve *1*). If $Q_l < Q_C$, the value Q_u decreases with increasing θ_0 (curve *2*).

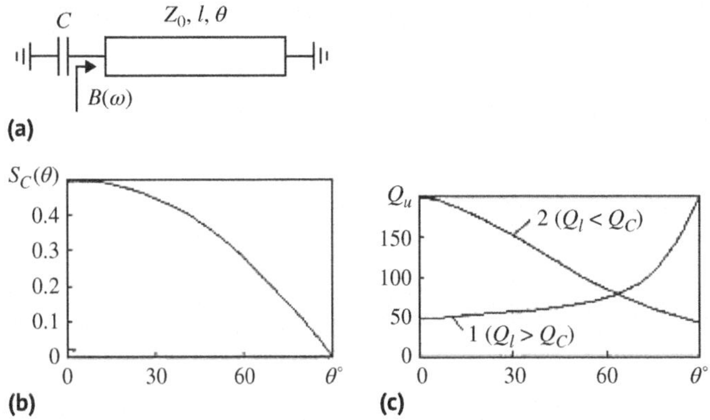

Figure 15.24 Combline resonator. (a) Schematic. (b) Sensitivity function $S_C(\theta)$ of fundamental resonance frequency. (c) Unloaded quality Q_u versus electrical length. Curve *1* corresponds to $Q_l = 200$, $Q_C = 50$, and curve *2* to $Q_l = 50$, $Q_C = 200$.

It is reasonable to compare the character of Q_u variation (Figure 15.24c) and sensitivity S_C (Figure 15.24b). In the first case ($Q_l > Q_C$), variations of Q_u and S_C are opposite to each other. In the second case ($Q_l < Q_C$), the unloaded quality Q_u and the sensitivity S_C decrease as the electric length increases. The unloaded quality Q_u is equal to the capacitor quality factor Q_C if the electric length tends to zero, and it is equal to the line quality factor Q_l for $\theta_0 = 90°$. Therefore, the ratio between Q_C and Q_l determines the increasing or decreasing character of Q_u variation with increasing electric length.

For a small value of θ_0, the capacitor makes a larger contribution to the resonator unloaded quality Q_u, and for large θ_0, it has a larger influence on the TL quality factor Q_l. This behavior is due to the influence coefficients in (15.71) $A_C = 2S$ and $A_l = 1 - 2S$. Suppose $A_C = A_l$, the influences of two elements on Q_u are identical. The equality is achieved if the following condition is satisfied:

$$S_C = \frac{1}{4}, \tag{15.72}$$

which, together with (15.6), yields the value of the resonator electric length $\theta_0 = 65.3°$. For electric lengths $0 < \theta_0 < 65.3°$ ($A_C > A_l$), the capacitor has a larger influence on Q_u, and for electric lengths $65.3° < \theta_0 \le 90°$ ($A_C < A_l$), the TL segment has a larger impact on Q_u.

The second resonator (Figure 15.25a) contains the open-end TL segment. For the analysis, we use (15.71) and the sensitivity function (15.7b). The plot of sensitivity (15.7b) in the region of the electric length $90° \le \theta_0 \le 180°$ at the fundamental resonance frequency is shown in Figure 15.25b. The function S_C is a bell-shaped function.

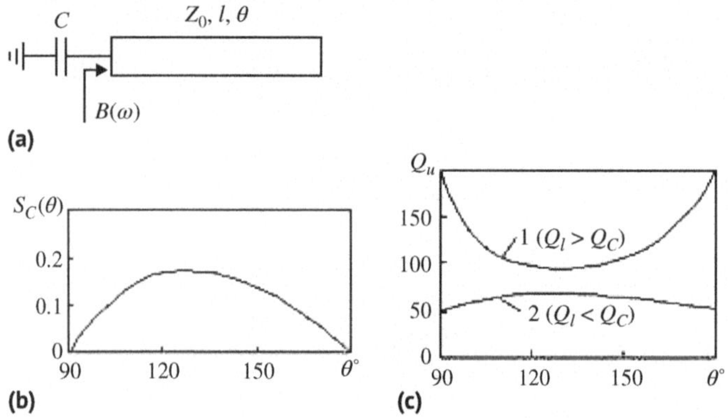

Figure 15.25 TL resonator (Figure 15.21) with capacitor and one open end. (a) Schematic. (b) Sensitivity function S_C (θ) of the fundamental resonance frequency. (c) Unloaded quality Q_u versus its electrical length. Curve *1* corresponds to $Q_l = 200$, $Q_C = 50$, and curve *2* to $Q_l = 50$, $Q_C = 200$.

The open-end resonator exhibits lower tunability than the combline resonator. Its advantage is that it has a larger length at higher frequencies, for example, in the X-range. The variation Q_u of the open-end resonator is shown in Figure 15.25c and exhibits a bell-shaped character. As in the case of the combline resonator, we consider two variants of Q_C and Q_l values: $Q_C = 50$, $Q_l = 200$ (curve *1*); $Q_C = 200$, $Q_l = 50$ (curve *2*). The dependence shown by downward convex curve *1* corresponds to the first case ($Q_l > Q_C$). Upward convex curve *2* is obtained for the second case ($Q_l < Q_C$). In the considered case, as in the previous one, dependence Q_u (θ) is opposite with respect to the sensitivity function S_C (θ) if $Q_l > Q_C$. If $Q_l < Q_C$, the character of variation of these functions coincides.

Since the sensitivity S_C for this resonator is always smaller than 1/4, condition (15.72) cannot be satisfied. As a result, for this resonator, the condition $A_l > A_C$ is always satisfied. Therefore, the influence of the line quality factor Q_l on Q_u is always larger than the influence of Q_C.

The following specific features of unloaded quality Q_u of considered two resonators were established:

- Unloaded quality Q_u of combline resonator for electric length $0 < \theta_0 \leq 90°$ increases with increasing θ_0 if $Q_l > Q_C$.
- Unloaded quality Q_u of combline resonator decreases with increasing θ_0 if $Q_l < Q_C$.
- For combline resonator, the influences of Q_l and Q_C on resonator unloaded quality Q_u are identical for electric length $\theta_{0e} = 65.3°$. For electric length $0 < \theta_0 < 65.3°$, the Q_C has a larger influence on Q_u, and for electric length $65.3° < \theta_0 \leq 90°$, the Q_l has a larger impact on Q_u.

- The Q_u variation of the open-end resonator for admissible electric length $90 \leq \theta_0 \leq 180°$ has a bell-shaped character. The curve $Q_u(\theta)$ is downward convex for $Q_l > Q_C$ and upward convex for $Q_l < Q_C$.
- For the open-end resonator, the influence of Q_l on Q_u is always larger than the influence of Q_C.
- For both resonators, the variation characters of $Q_u(\theta)$ and $S_C(\theta)$ are opposite if $Q_l > Q_C$ and coincide if $Q_l < Q_C$.

In capacitance-tuning TL resonators, the values of Q_C and Q_l are not constant; they change in the tuning range, and there are formulas for calculating them. Although the values of Q_C and Q_l have changed, the sensitivity function S_C (15.3) and the character of the frequency dependence Q_u are constant (Figures 15.24c and 15.25c).

15.4.3 Verification of Q_u Formula

Formula (15.64), which determines the resonator's unloaded quality Q_u in the absence of loss in the TL, was rigorously obtained; therefore, its reliability is beyond doubt. The transition from (15.64) to the general expression for unloaded quality Q_u (15.71) was made under certain assumptions. Therefore, (15.71) requires some additional verification. The values of Q_l, Q_C, and S_C are determined a priori and introduced into the formula to calculate the Q_u value. The verification can be performed using a computer simulation of the IL response L_A of the resonator for weak coupling with input and output loads with the same parameters Q_l, Q_C, and S_C. The simulated IL response L_A is used to determine the unloaded quality Q_u of the resonator. Software such as Microwave Office (AWR) and HFSS (Ansoft) simulate electromagnetic processes at the electromagnetic level, and a high level of simulation accuracy characterizes them. To determine Q_u, expressions (4.3)–(4.5) are used.

Figure 15.26a shows the topology of the combline microstrip resonator in the case of weak coupling with input and output loads. The substrate, with a dielectric constant of 10.2, a thickness of $h = 1.905$ mm, and a height of a metallic screen above the substrate of $H = 11.43$ mm, is used for this resonator. The width of the microstrip line in the resonator is $w = 4$ mm, corresponding to $Z_0 = 31.6\ \Omega$. The resonator length is 28.5 mm. Other dimensions are shown in Figure 15.26a in millimeters. As a result, its resonance frequency is 973 MHz at $C = 0$, which corresponds to electric length $\theta_0 = 90°$.

In the simulation, we use the capacitor with $Q_C = 50$ and the microstrip line with $Q_l = 200$. If the dielectric dissipation $\tan\delta$ of substrate and air are equal, then the quality factor Q_l of microstrip line is expressed by the equality $Q_l = 1/\tan\delta$. To provide the value of $Q_l = 200$, it is necessary to use the dielectric dissipation $\tan\delta = 1/Q_l = 0.005$. The electric length of the resonator θ_0 is controlled by

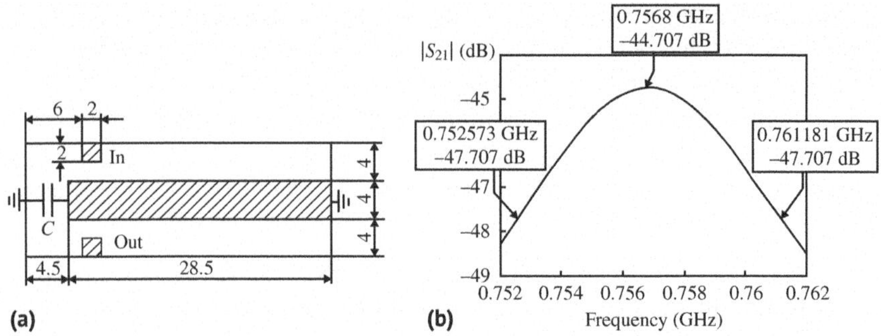

Figure 15.26 Definition of unloaded quality factor Q_u of microstrip combline resonator by EM simulation. (a) Resonator topology in case of weak coupling with loads. (b) Simulated IL of resonator with electrical length of 70°.

capacitance variation. The larger the capacitance C, the smaller θ_0. Thus, the electric length $\theta_0 = 70°$ corresponds to the resonance frequency 756.8 MHz. To obtain these values, a capacitance of $C = 2.391$ pF is required. Figure 15.26b shows the IL response of the resonator, simulated using Microwave Office software. The simulation results shown in Figure 15.26b yield the following values: $L_{A0} = 44.707$ dB; $\text{BW}_{3\,\text{dB}} = 8.608$ MHz. Substituting these values and $f_0 = 756.8$ MHz into (4.3)–(4.5), we obtain $Q_u = 88.43$. These data are presented in Table 15.5, which contains similar values for an electric length of 20 and 90° of the resonator. Table 15.5 also includes the values of sensitivity S_C and unloaded quality Q_u obtained from (15.71). The difference between simulated and calculated values of Q_u did not exceed 0.5%. The comparison confirms the reliability of the general expression (15.71) for unloaded quality Q_u.

15.4.4 Unloaded Quality Q_u of Resonators with Some Capacitors

Consider resonators containing m capacitors with capacitance C_i, $i = 1, 2, \ldots, m$, that are characterized by quality factors Q_{Ci}. Denote the input susceptance B_i at the connection point of C_i. Using equation (15.3), it is possible to calculate the sensitivity S_{Ci} of the resonant frequency to the change in capacitance of this capacitor.

15.4.4.1 Formula for Q_u Calculation

Since the considered resonators are linear circuits, the formula (15.71) for resonators with one capacitor can be extended to resonators with multiple capacitors. It takes the form

$$\frac{1}{Q_u} = 2\sum\nolimits_{i=1}^{m} \frac{S_{Ci}}{Q_{Ci}} + \frac{1 - 2\sum\nolimits_{i=1}^{m} S_{Ci}}{Q_l}. \tag{15.73}$$

Table 15.5 Comparison of resonator quality factor Q_u obtained by means of computer simulation and calculation.

θ_0, °	f_0, MHz	C, pF	$\mathrm{BW}_{3\,\mathrm{dB}}$, MHz	L_{A0}, dB	Q_u, simulated	S_C	Q_u, calculated	Difference, %
90	973	0	4.991	−31.35	200.4	0	200	0.2
70	756.8	2.391	8.608	−44.707	88.43	0.21	88.5	0.08
20	216.2	61.85	4.211	−58.144	51.4	0.479	51.63	0.44

In (15.73), the quantities

$$A_{Ci} = 2S_{Ci},$$
$$A_l = 1 - \sum_{i=1}^{m} 2S_{Ci} \tag{15.74}$$

are, respectively, the influence coefficient of the capacitor with number i on the quality factor Q_u of the resonator and the influence coefficient of the TL on Q_u. The greater the coefficients (15.74), the more substantial the influence of the corresponding elements on Q_u.

Influence coefficients (15.74) have the following restrictions:

$$0 \le \sum_{i=1}^{m} A_{Ci} \le 1; \quad 0 \le A_l \le 1; \quad A_l + \sum_{i=1}^{m} A_{Ci} = 1,$$

which follow from the sensitivity limitation (15.17).

Formula (15.73) enables us to state the following.

Statement: If two resonators contain a section of one TL and different numbers m_1 and m_2 of capacitors with equal Q factor, these resonators have equal unloaded quality $Q_{u1} = Q_{u2}$ at the same frequencies when the sums of sensitivities are equal

$$\sum_{i=1}^{m_1} S_{Ci} = \sum_{j=1}^{m_2} S_{Cj}. \tag{15.75}$$

Proof: Since the Q factors of capacitors are equal, $Q_{Ci} = Q_{Cj} = Q_C$, the substitution of Q_C into (15.73) yields the expressions for the unloaded quality of these resonators Q_{u1} and Q_{u2}:

$$\frac{1}{Q_{u1}} = 2\sum_{i=1}^{m_1} \frac{S_{Ci}}{Q_C} + \frac{1 - 2\sum_{i=1}^{m_1} S_{Ci}}{Q_l},$$
$$\frac{1}{Q_{u2}} = 2\sum_{j=1}^{m_2} \frac{S_{Cj}}{Q_C} + \frac{1 - 2\sum_{j=1}^{m_2} S_{Cj}}{Q_l}.$$

Since condition (15.75) is fulfilled, the right-hand sides of these expressions are identical, which yields the equality $Q_{u1} = Q_{u2}$. The statement is proved.

Expression (15.73) does not contain capacitances C_i but instead takes into account only the degree of their influence on the resonance frequency, as expressed by sensitivity S_{Ci}. The statement proved earlier indicates that the unloaded quality of two resonators can coincide for different numbers of capacitors and with different values of their capacitances.

15.4.4.2 Experimental Verification of Statement

The concept formulated in the statement is essential for practice. Consider it in more detail using experimental verification that enables us to determine the unloaded quality factor Q_u of resonators directly.

Figure 15.27 shows the topologies and photographs of identical hairpin microstrip resonators with two and one capacitors, which exhibit weak coupling with loads. The dimensions are in millimeters. In the microstrip resonators, a substrate made of the RT/Duroid 6010 LM laminate (Rogers) with $\varepsilon_r = 10.2$, $h = 1.905$ mm, and $\tan\delta = 0.002$ was used. The width of the microstrip line of the hairpin resonator is $w = 4.5$ mm ($Z_0 = 29.3\ \Omega$). To apply the control voltage to the varactors of the hairpin resonators, a 10 kΩ chip resistor is connected to the middle of these resonators (Figure 15.27c, d).

The main resonance frequency ω_0 of a resonator with two capacitances (Figure 15.27a) is determined by the resonance equation (15.29a). The substitution of (15.29a) into (15.3) yields the total sensitivity of resonance frequency ω_0 to synchronous variation of two capacitances of the resonator:

$$S_{C1} + S_{C2} = 2S_C = \left(1 + \frac{\theta}{\sin\theta}\right)^{-1}. \tag{15.76}$$

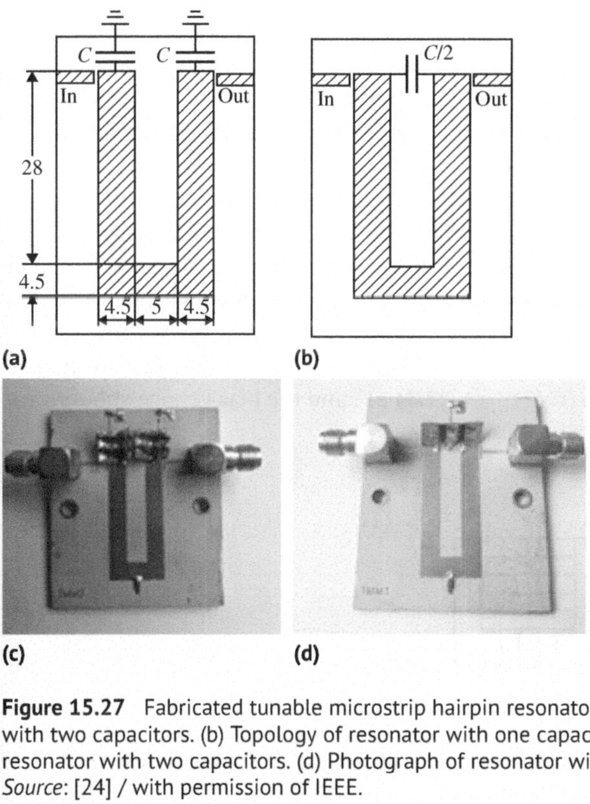

Figure 15.27 Fabricated tunable microstrip hairpin resonators. (a) Topology of resonator with two capacitors. (b) Topology of resonator with one capacitor. (c) Photograph of resonator with two capacitors. (d) Photograph of resonator with one capacitor.
Source: [24] / with permission of IEEE.

The resonator in Figure 15.27b contains only one capacitor, the value of which is $C/2$. It has the same resonance frequency ω_0 as the previous resonator because it is described by the same resonance equation (15.56a) with a multiplier (1/2) in its right part. The sensitivity of ω_0 of this resonator to changes in one capacitance is the same as the sensitivity of the previous resonator to changes in two capacitances (15.76). It follows from the coincidence of the resonance equations.

When all capacitances of the resonators are zero, their resonance frequency is $f_0 = 850$ MHz. This value corresponds to the electric length of the resonators $\theta_0 = 180°$. We will tune these two resonators from $f_{0\min} = 225$ MHz to $f_{0\max} = 400$ MHz, which corresponds to a change in their electrical length from $\theta_{0\min} = 47.64°$ to $\theta_{0\max} = 84.7°$. Substituting these values of θ_0 into (15.29), we determine the boundary values of the tuning capacitance: $C_{\min} = 14.9$ pF, $C_{\max} = 54.68$ pF.

In the loop hairpin resonator (Figure 15.27b), the value of a single capacitance changes within the range from $(C/2)_{\min} = 7.45$ pF to $(C/2)_{\max} = 27.34$ pF. To implement the required values of capacitance $\frac{C}{2} = 7.45 - 27.34$ pF, we use several varactors BB135 (NXP company) with a minimal capacitance $C_{\min} \approx 2$ pF, which are connected in parallel and in series to form a unit. A loop hairpin resonator with the unit consisting of two pairs of oppositely connected varactors is presented in Figure 15.28a. Its capacitance is the same as that of a single varactor.

A controlling voltage is applied to the interior of the unit through the additional metallized plate 3, which is disposed between the arms 1 and 2 of the resonator.

In Figure 15.28b, the structure of the unit for assembling a few metal strips is denoted by numbers 1, 2, and 3. The varactors are soldered to both sides of these strips. Such a unit has small dimensions. For example, a unit with dimensions of 7 mm × 6 mm can be used to place 10 pairs of varactors in SOT23 packages. The varactors are mounted on both sides of the unit. After that, this unit is soldered in a vertical position to the left *1* and the right *2* parts of the loop hairpin resonator, as well as to its additional metal plate *3* (Figure 15.28a).

The unit must contain 15 varactors BB135 to change the $C/2$ capacitance from $(C/2)_{\min} = 7.45$ pF to $(C/2)_{\max} = 27.34$ pF, and the bias voltage should vary

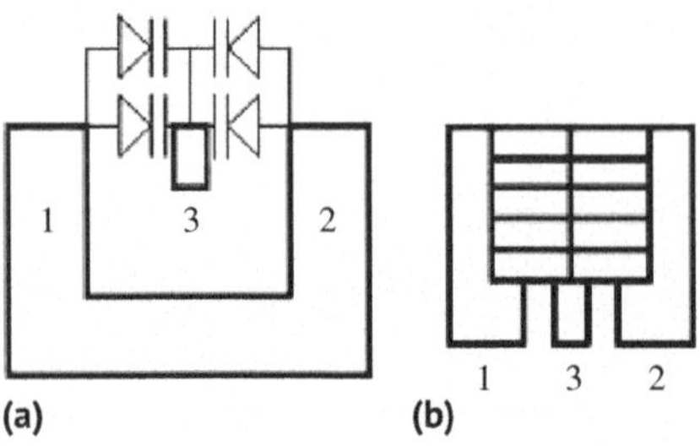

Figure 15.28 Unit with set of varactors used in resonators. (a) Schematic with four varactors. (b) One side of unit with varactors.

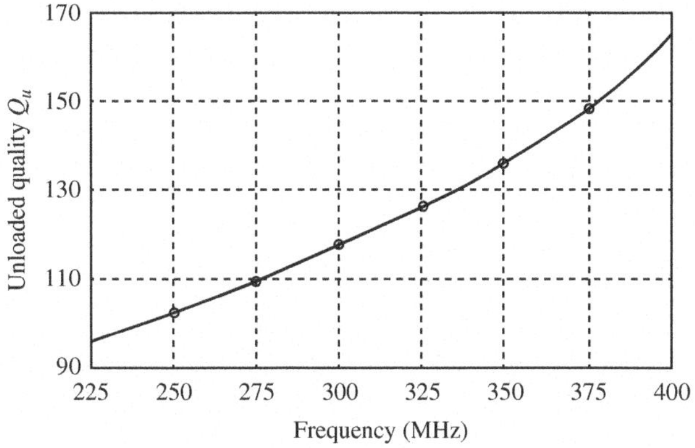

Figure 15.29 Measured unloaded quality factor Q_u of tunable hairpin resonators in Figures 15.27c, d.

from 6 to 27.2 V. The photograph of the loop hairpin resonator with one unit is shown in Figure 15.27d. The hairpin resonator with two capacitors (Figure 15.27a) contains four units, and its photograph is shown in Figure 15.27c.

Figure 15.29 shows the measured unloaded quality Q_u of the two tunable hairpin resonators in the frequency range 225–400 MHz. The Q_u values of both resonators are the same, and they change from $Q_u = 98$ (225 MHz) to $Q_u = 167$ (400 MHz). Resonator in Figure 15.27d contains 15 varactors, and resonator in Figure 15.27c contains 60 varactors. The use of a large number of varactors averages the spread of their parameters, making measurements more accurate.

The transition from the resonator shown in Figure 15.27d to the resonator shown in Figure 15.27c has a significant feature. Resonator in Figure 15.27d contains one capacitor with capacitance $C/2$, and the resonator in Figure 15.27c has two capacitors, each of which has a capacitance C. In this case, the unloaded quality Q_u of both resonators is the same.

The considered measured data confirm the correctness of the aforementioned statement and formula (15.73) for unloaded quality Q_u of a resonator containing several capacitors.

15.4.5 Unloaded Quality Q_u of Resonators with Some Capacitors and Inductors

Let the resonator contain n inductors. Replacing in expression (15.73) the sensitivities S_{Ci} with the sensitivities S_{Lj} (15.32), $j = 1, 2, \dots, n$, we obtain an expression that determines the unloaded quality factor of a resonator with several inductors:

$$\frac{1}{Q_u} = 2\sum_{i=1}^{n} \frac{S_{Lj}}{Q_{Lj}} + \frac{1 - 2\sum_{i=1}^{n} S_{Lj}}{Q_l}. \tag{15.77}$$

Note that the influence coefficient of Q_l on Q_u is equal to $\left(1 - 2\sum S_{Lj}\right)$. This coefficient cannot be negative, since the sum included in it does not exceed 1/2: $\sum S_{Lj} \leq 1/2$.

To obtain an expression for the unloaded quality factor of a resonator with several capacitors and inductors, it is sufficient to combine formulas (15.73) and (15.77):

$$\frac{1}{Q_u} = 2\left(\sum_{i=1}^{m} \frac{S_{Ci}}{Q_{Ci}} + \sum_{j=1}^{n} \frac{S_{Lj}}{Q_{Lj}}\right) + \frac{2}{Q_l}\left(1 - \sum_{i=1}^{m} S_{Ci} - \sum_{j=1}^{n} S_{Lj}\right). \tag{15.78}$$

If the electric circuit contains only one capacitor and one inductor, then sensitivity values $S_C = S_L = 1/2$, and expression (15.78) turns into the well-known formula for the quality factor of a single LC circuit:

$$\frac{1}{Q_u} = \frac{1}{Q_C} + \frac{1}{Q_L}.$$

When connecting multiple capacitors and inductors to TL sections, calculating the sensitivities S_{Ci} and S_{Lj} using formulas (15.3) and (15.33) becomes difficult. This is because the expressions for the reactive conductivities B_i and B_j become quite cumbersome. In these cases, the values of the resonant frequency sensitivities can be determined by computer modeling or experimentally. By setting a small relative change in capacitance $\Delta C/C$ or inductance $\Delta L/L$, we measure the relative change in resonant frequency $\Delta\omega_0/\omega_0$. The ratio of these values, taken with the opposite sign, will give an approximate value of sensitivity.

It is important to note that the formulas for calculating Q_u (15.73), (15.77), and (15.78) do not include the values of the capacitances C_i and inductances L_j, but only take into account the degree of their influence on the resonant frequency, expressed by the sensitivities S_{Ci} and S_{Lj}.

References

1 W. Feng, Y. Zhang, and W. Che, Tunable dual-band filter and diplexer based on folded open loop ring resonators, *IEEE Trans. Circuits Syst. II Express Briefs*, vol. 64, no. 9, pp. 1047–1051, 2017, https://doi.org/10.1109/TCSII.2016.2634555.

2 I. ud Din, J. Wernehag, S. Andersson, and H. Sjoland, Two tunable frequency duplexer architectures for cellular transceivers, *IEEE Trans. Circuits Syst. I Regul. Pap.*, vol. 64, no. 11, pp. 2988–2999, 2017, https://doi.org/10.1109/TCSI.2017.2706969.

3 P. Wong and I. Hunter, Electronically tunable filters, *IEEE Microw. Mag.*, vol. 10, no. 6, pp. 46–54, 2009, https://doi.org/10.1109/MMM.2009.933593.

4 J. Hong, Reconfigurable planar filters, *IEEE Microw. Mag.*, vol. 10, no. 6, pp. 73–83, 2009, https://doi.org/10.1109/MMM.2009.933590.

5 J. X. Chen, Y. Ma, J. Cai, L. H. Zhou, Z. H. Bao, and W. Che, Novel frequency-agile bandpass filter with wide tuning range and spurious suppression, *IEEE Trans. Ind. Electron.*, vol. 62, no. 10, pp. 6428–6435, 2015, https://doi.org/10.1109/TIE.2015.2427122.

6 W. Qin, J. Cai, Y.-L. Li, and J.-X. Chen, Wideband tunable bandpass filter using optimized varactor-loaded SIRs, *IEEE Microw. Wirel. Components Lett.*, vol. 27, no. 9, pp. 812–814, 2017, https://doi.org/10.1109/LMWC.2017.2734848.

7 B. Kapilevich, A varactor-tunable filter with constant bandwidth and loss compensation, *Microw. J.*, no. 4, pp. 106–114, 2007, https://www.microwavejournal.com/articles/4575-a-varactor-tunable-filter-with-constant-bandwidth-and-loss-compensation.

8 J. Sigman, C. D. Nordquist, P. G. Clem, G. M. Kraus, and P. S. Finnegan, Voltage-controlled Ku-band and X-band tunable combline filters using barium-strontium-titanate, *IEEE Microw. Wirel. Components Lett.*, vol. 18, no. 9, pp. 593–595, 2008, https://doi.org/10.1109/LMWC.2008.2002453.

9 J. Nath, D. Ghosh, J.-P. Maria, A. I. Kingon, W. Fathelbab, P. D. Franzon, and M. B. Steer, An electronically tunable microstrip bandpass filter using thin-film Barium-Strontium-Titanate (BST) varactors, *IEEE Trans. Microw. Theory Tech.*, vol. 53, no. 9, pp. 2707–2712, 2005, https://doi.org/10.1109/TMTT.2005.854196.

10 G. M. Rebeiz, *RF MEMS: Theory, Design, and Technology*. Wiley, 2003, https://www.wiley.com/en-us/RF+MEMS%3A+Theory%2C+Design%2C+and+Technology-p-9780471201694.

11 J. F. White, *Microwave Semiconductor Engineering*. Dordrecht: Springer Netherlands, 1982, https://doi.org/10.1007/978-94-011-7065-9.

12 A. Zakharov and M. Ilchenko, Circuit function characterizing tunability of resonators, *IEEE Trans. Circuits Syst. I Regul. Pap.*, vol. 67, no. 1, pp. 98–107, 2020, https://doi.org/10.1109/TCSI.2019.2940066.

13 H. W. Bode, *Network Analysis and Feedback Amplifier Design*. New Jersey: D. Van Nostrand Company, Inc., 1945.

14 K. Géher, *Theory of Network Tolerances*. Budapest: Akadémiai Kiadó, 1971, http://real-eod.mtak.hu/id/eprint/12705.

15 L. Sibul, Sensitivity analysis of linear control systems with random plan parameters, *IEEE Trans. Autom. Control*, vol. 15, no. 4, pp. 459–462, 1970, https://doi.org/10.1109/TAC.1970.1099490.

16 A. G. Yepes, F. D. Freijedo, O. Lopez, and J. Doval-Gandoy, Analysis and design of resonant current controllers for voltage-source converters by means of Nyquist diagrams and sensitivity function, *IEEE Trans. Ind. Electron.*, vol. 58, no. 11, pp. 5231–5250, 2011, https://doi.org/10.1109/TIE.2011.2126535.

17 E. A. Guillemin, *Synthesis of Passive Networks: Theory and Methods Appropriate to the Realization and Approximation Problems*. New York: Wiley, 1957.

18 G. A. Korn and T. M. Korn, *Mathematical Handbook for Scientists and Engineers: Definitions, Theorems, and Formulas for Reference and Review*. New York, NY: McGraw-Hill, 1961.

19 A. Zakharov, S. Rozenko, and M. Ilchenko, Varactor-tuned microstrip bandpass filter with loop hairpin and combline resonators, *IEEE Trans. Circuits Syst. II Express Briefs*, vol. 66, no. 6, pp. 953–957, 2019, https://doi.org/10.1109/TCSII.2018.2873227.

20 A. Zakharov, S. Litvintsev, and M. Ilchenko, Transmission line tunable resonators with intersecting resonance regions, *IEEE Trans. Circuits Syst. II Express Briefs*, vol. 67, no. 4, pp. 660–664, 2020, https://doi.org/10.1109/TCSII.2019.2922429.

21 G. Megla, *Dezimeterwellentechnik: Theorie Und Technik Der Dezimeterschaltungen*, 4th ed. Leipzig: Fachbuchverlag, 1955.

22 C.-F. Chen, G.-Y. Wang, and J.-J. Li, Microstrip switchable and fully tunable bandpass filter with continuous frequency tuning range, *IEEE Microw. Wirel. Components Lett.*, vol. 28, no. 6, pp. 500–502, 2018, https://doi.org/10.1109/LMWC.2018.2831440.

23 A. Zakharov, S. Rozenko, S. Litvintsev, and M. Ilchenko, Hairpin resonators in varactor-tuned microstrip bandpass filters, *IEEE Trans. Circuits Syst. II Express Briefs*, vol. 67, no. 10, pp. 1874–1878, 2020, https://doi.org/10.1109/TCSII.2019.2953247.

24 A. Zakharov and M. Ilchenko, Unloaded quality factor of transmission line resonators with capacitors, *IEEE Trans. Circuits Syst. I Regul. Pap.*, vol. 67, no. 7, pp. 2204–2215, 2020, https://doi.org/10.1109/TCSI.2020.2971112.

25 G. L. Matthaei, L. Young, and E. M. T. Jones, *Microwave Filters, Impedance-Matching Networks, and Coupling Structures*. New York: Artech House Books, 1980.

16

Tunable Bandpass Filters

Electronically tunable microwave bandpass filters (BPFs) find wide practical application and expand the capabilities of existing and future wireless systems. These filters are the subject of numerous publications in scientific and technical literature. In this section, only some of these filters will be considered, which have a microstrip construction and contain hairpin and combline resonators. Semiconductor varactors, ferroelectric capacitors, and switched inductors will be used for their reconfiguration.

16.1 Varactor-tuned Microstrip BPF with Hairpin and Combline Resonators

A microstrip combline resonator is a favorite for creating tunable BPFs [1–6]. Such filters have an asymmetric transfer function, in which the right slope is steeper than the left. This drawback arises due to parasitic coupling between nonadjacent microstrip combline resonators [7, Section 5.2.5]. A more symmetric transfer function is provided by a tunable fourth-order microstrip BPF [8], which contains two end combline resonators and two "inner" loop-type hairpin resonators. Combining hairpin and combline resonators is possible because they have the same tunability (Section 15.3). The capacitance of the hairpin resonator used for tuning is two times smaller than that of the combline resonator.

Consider the specification of the narrowband tunable BPF with Chebyshev response.

Tuning range of f_0	225–400 MHz
Passband fractional bandwidth FBW	0.025
Frequency of spurious resonance f_1	>1350 MHz
Selectivity	40 dB ($f_0 \pm 5\%$)
Return loss	$L_R > 10$ dB
Load impedance	50 Ω

16.1.1 Calculation and Simulation of Tunable BPF

The required attenuation −40 dB should be provided in the frequency band of the filter with a relative width of FBWs = 0.1. Analysis of the attenuation plots of low-frequency prototypes with Chebyshev responses [9] shows that at FBW = 0.025, the required selectivity can be achieved in a fourth-order filter ($N = 4$) with a ripple $L_{Ar} = 0.1$ dB. We derive the g parameters of such a filter from Table 2.6: $g_1 = 1.1088$, $g_2 = 1.3061$, $g_3 = 1.7703$, $g_4 = 0.8180$, and $g_5 = 1.3554$. Substitution of these parameters and the value of FBW = 0.025 into (3.8) gives us the required values of the coupling coefficients between the resonators and the external Q-factors of the end resonators: $K_{12} = K_{34} = 0.0208$, $K_{23} = 0.0164$, $Q_{e1} = Q_{e2} = Q_e = 44.35$.

These parameters were realized at the middle frequency of the tuning range, $f_0 = 300$ MHz, of a fourth-order microstrip filter with two types of resonators, the topology of which is shown in Figure 16.1a. The dimensions are in millimeters. In the microstrip filter, a substrate made of the RT/Duroid 6010 LM laminate (Rogers) with $\varepsilon_r = 10.2$, $h = 1.905$ mm, and $\tan\delta = 0.002$ was used. This filter has an area of 86 mm × 48 mm. The width of the microstrip line of the hairpin resonator is $w = 4.5$ mm ($Z_0 = 29.3\ \Omega$), and the width of the combline resonator is $w = 1.25$ mm ($Z_0 = 58.6\ \Omega$).

In the specification, the location of the spurious resonance frequency is $f_1 > 1350$ MHz. This condition enables us to determine the electrical length of the loop hairpin resonator at the maximum frequency of the tuning range θ_{0max}. The larger f_1, the smaller the value θ_{0max}. From the resonance equations (15.56), we obtain the condition

$$\theta_{0max} < 2\pi / (f_1/f_0)_{min} \, . \tag{16.1}$$

In our case, $(f_1/f_0)_{min} = 1350/400 = 3.375$; $2\pi/(f_1/f_0)_{min} = 106.67°$. To satisfy condition (16.1), we take the value $\theta_{0max} = 84.7°$. The value of θ_{0min} is determined automatically: $\theta_{0min} = \theta_{0max} \times (225/400) = 47.64°$. Knowing θ_{0max} and θ_{0min}, we determine the capacitance values from the resonance equation (15.56a): $C_{1min} = 7.45$ pF and $C_{1max} = 27.34$ pF. The electrical lengths of the combline resonator are half the values: $\theta_{0max} = 42.35°$; $\theta_{0min} = 23.82°$. Other details of designing a tunable filter are well known [8], and we omit them.

The simulated responses of the filter shown in Figure 16.1b confirm that the parasitic bandpass remains fixed when the main bandpass changes. The simulation is performed using the Microwave Office (AWR) program.

The value

$$\chi = \frac{L_{Amin}(f_0 \pm \Delta)}{L_{Amax}(f_0 \pm \Delta)}, \chi \leq 1 \tag{16.2}$$

is the symmetry factor of the frequency response, and Δ is the frequency shift. The closer χ to unity, the more symmetrical the frequency response. The filter

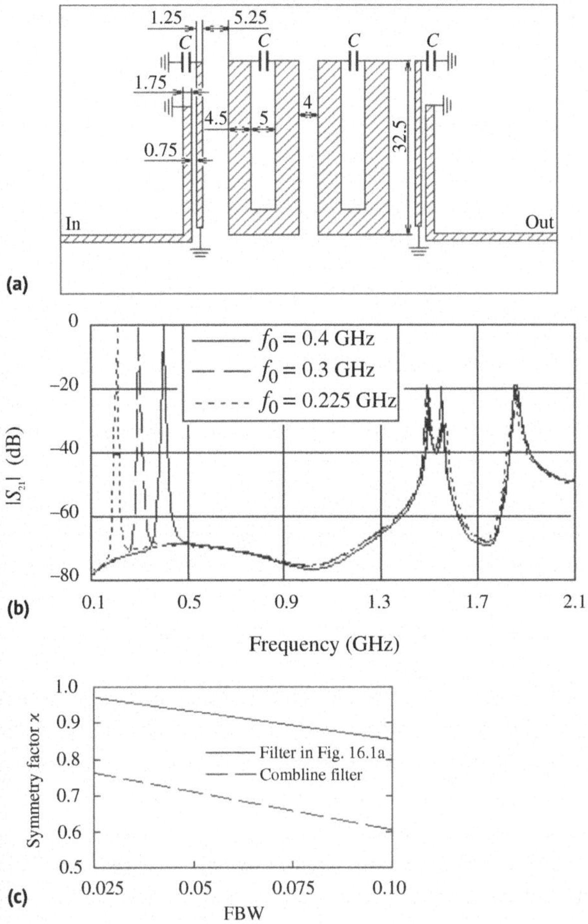

Figure 16.1 Microstrip tunable filter with hairpin and combline resonators. (a) Topology. All dimensions are in millimeters. (b) Simulated performances for three different f_0 values. (c) Symmetry factor χ.

(Figure 16.1a) at the frequency $f_0 = 400$ MHz has the value of $\text{FBW}_{3\text{ dB}} = 0.025$. If $\Delta = 20$ MHz, then $L_{A\min}$ $(f_0 - \Delta)$ =47.8 dB, $L_{A\max}$ $(f_0 + \Delta) = 46.9$ dB and the symmetry factor (16.2): $\chi = L_{A\min}/L_{A\max} = 0.981$. Figure 16.1c shows plots of χ versus $\text{FBW}_{3\text{ dB}}$ for the filter in Fig. 16.1a and for the combline filter. The combline filter with $\text{FBW}_{3\text{ dB}} = 0.025$ at the same frequency $f_0 = 400$ MHz is characterized by a symmetry factor $\chi = 0.77$. The advantage of a filter with combined resonators is $\chi_{\max}/\chi_{\min} = 0.981/0.77 = 1.27$. With $\text{FBW}_{3\text{ dB}} = 0.1$, the combined filter has a value of $\chi = 0.86$, while the combline filter has a value of $\chi = 0.605$. The advantage of the combined filter is $\chi_{\max}/\chi_{\min} = 1.42$. With an increase in FBW, the symmetry factor decreases in both filters, and the advantage of the combined filter increases.

16.1.2 Manufacturing and Measurement

To decrease the losses of the filter within the passband, it is necessary to increase the unloaded Q-factor Q_u of the composite resonator, which consists of a TL segment with the Q-factor Q_l and a varactor with the Q-factor

$$Q_C = \frac{1}{2\pi f R_s C}, \tag{16.3}$$

where R_s is an equivalent series resistance corresponding to losses in the varactor. The expression (16.3) indicates that decreasing the product $R_s C_{\min}$ results in an increase in Q_C at the specified frequency. Decreasing the initial capacitance of the varactor $C_{\min}$ typically results in a decrease in the product $R_s C_{\min}$ and an increase in the Q-factor of the varactor. It can be shown that the series or parallel connection of two capacitors with equal Q-factors (Q_C) leads to a resultant capacitor with the same Q-factor. Therefore, it is advisable to replace one varactor with several connected in parallel, each having a smaller capacitances. This allows for obtaining greater values of tuning capacitance while maintaining the same high Q-factor of the resultant capacitor.

Since varactors are nonlinear electronic components, filters that comprise them exhibit intermodulation distortion of the third-order IP3 (dBm). The higher the IP3, the smaller the nonlinear distortion, and the higher the handling power of a filter. In the case of an opposite connection of varactors, the value of IP3 increases due to partial mutual compensation of nonlinear effects.

In Figure 15.28a, a loop hairpin resonator with a unit consisting of two pairs of oppositely connected varactors is presented. Its capacitance is the same as that of a single varactor. A controlling voltage is applied to the interior of the unit through the additional metallized plate 3, which is disposed between the sides of the resonators 1 and 2. In Figure 15.28b, the structure of the unit for assembling a few metal strips is denoted by numbers 1, 2, and 3. The varactors in SOT23 packages are mounted on both sides of the unit.

The photograph of the filter under consideration without a shield is shown in Figure 16.2a. Its measured parameters are depicted in Figure 16.2b, c. The filter utilizes varactors BB135 (NXP) with a minimal capacitance ($C_{\min} = 2$ pF), a capacitance ratio $C_{\max}/C_{\min} \approx 10$, and a voltage range of 0.5–28 V. Each resonator is connected to units comprising 15 varactors, arranged in parallel and in series.

To apply the controlling voltage to the varactors of the hairpin resonators, a 10 kΩ chip resistor is connected to the middle of these resonators. Measured value $(f_1/f_0)_{\min} = 3.47$ (Figure 16.2c).

The filter specification is given in Table 16.1.

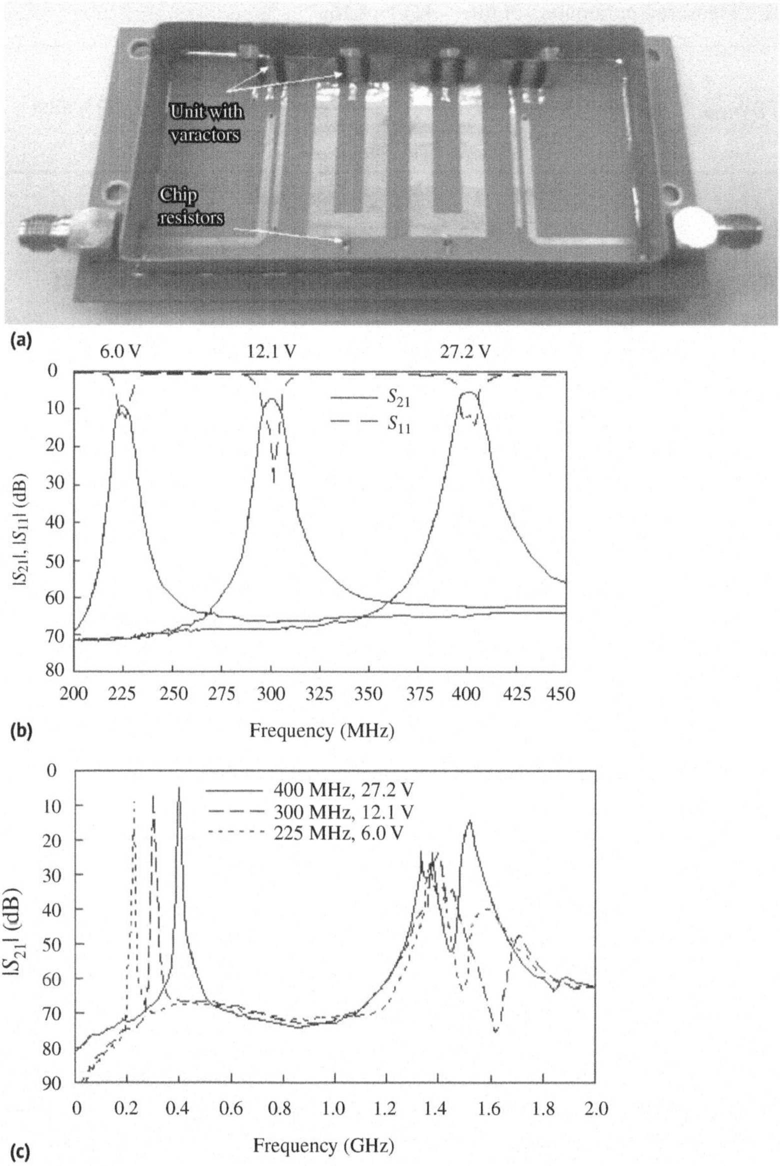

Figure 16.2 Fabricated tunable filter with combline and loop hairpin resonators: (a) Photograph. *Source*: [8] / with permission of IEEE. (b) Measured performance in tuning range. (c) Measured performance in wide frequency band.

Table 16.1 Measured parameters of filter in Figure 16.2.

f_0, MHz	$FBW_{3\,dB}$	L_{A0}, dB	Attenuation, dB		χ	f_1/f_0	IP3, dBm
			$0.95f_0$	$1.05f_0$			
225	0.0262	9.2	46	44	0.956	7.01	16
300	0.0253	7.4	45	43	0.955	4.63	18
400	0.0250	5.6	43	41	0.953	3.47	22

In the filter under consideration, the fractional bandwidth is practically constant within the tuning range (FBW ≈ const). This is because the coupling coefficients between the resonators practically do not change $K_{i,i+1} \approx$ const. There are other tunable filters, in which the bandwidth remains constant, BW ≈ const. One of these filters is proposed in Ref. [2]. It is a combline filter that contains one-end short-circuited stepped-impedance resonators, between which there is a mixed coupling.

The measurement data show that the microstrip tunable filter with combined resonators is characterized by a symmetrical frequency response, a narrow bandwidth $FBW_{3\,dB} \approx 0.025$, a wide tuning range (56%), a wide stopband $[(f_1/f_0)_{min} = 3.47]$, and a relatively high IP3 value for a narrow passband.

16.2 Varactor-tuned Microstrip BPF with Two Operating Regimes

The previous narrowband varactor-tuned fourth-order BPF had a maximum insertion loss of $L_{A0max} = 9.2$ dB in the tuning band. To reduce the value of L_{A0max}, two operating modes can be used in the filter [10], as described as follows.

As an example, we consider the design of the narrowband high-selectivity tunable BPF with an extended rejection band and Chebyshev response with the following parameters:

Tuning range of f_0	225–400 MHz
Passband fractional bandwidth FBW	0.024
Selectivity	40 dB ($f_0 \pm 5\%$)
Attenuation at $3f_0$	≥40 dB
Return loss	$L_R \geq 12$ dB
Load impedance	50 Ω

16.2.1 Calculation and Simulation of Tunable BPF

In the filter, we will use hairpin loop-type resonators (Figure 15.19a). The required selectivity can be achieved in a fourth-order filter with a ripple $L_{Ar} = 0.04321$ dB. In this case, the return loss is $L_R = 20$ dB. We write out from [7] the g parameters of such a filter: $g_0 = 1, g_1 = 0.9314, g_2 = 1.2920, g_3 = 1.5775, g_4 = 0.7628, g_5 = 1.2210$. Using these parameters and the value FBW = 0.024 based on the formulas (3.8), we determine the coupling coefficients between the resonators and the external Q-factors of the end resonators: $K_{12} = K_{34} = 0.0218$, $K_{23} = 0.0168$, $Q_{e1} = Q_{e2} = Q_e = 38.81$.

These parameters were realized at the middle frequency of the tuning range $f_0 = 300$ MHz of a microstrip BPF, the topology of which is shown in Figure 16.3a. In the fabricated microstrip filter (Figure 16.3b), a substrate made of the RT/-Duroid 6010 LM (Rogers) with $\varepsilon_r = 10.2$, $h = 1.905$ mm, and $\tan\delta = 0.002$ was used. The width of the microstrip line of the hairpin resonator is $w = 4.5$ mm ($Z_0 = 29.3\ \Omega$). The dimensions of the filter with shield are 90 mm × 46 mm × 15 mm.

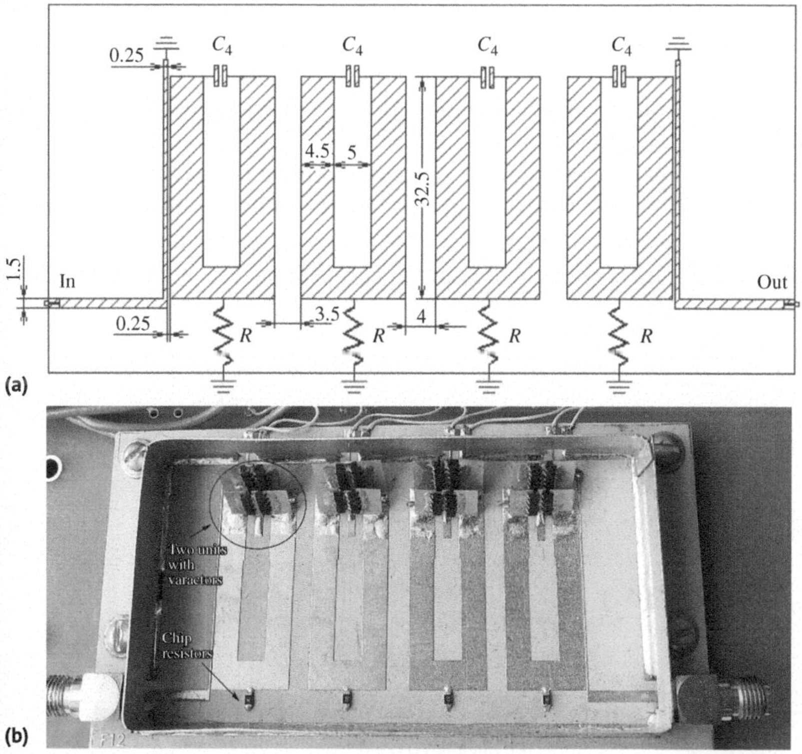

Figure 16.3 Microstrip tunable BPF with hairpin resonators. (a) Topology. All dimensions are in millimeters. (b) Photograph of a fabricated filter without shield. *Source*: [10] / with permission of IEEE.

New in the specification is the attenuation of 40 dB at the frequency $3f_0$. The maximum operating frequency $f_0 = 400$ MHz corresponds to $3f_0 = 1200$ MHz. To have attenuation 40 dB at this frequency, the spurious resonance f_1 of its resonators must satisfy the condition $f_1 > 1250$ MHz. This condition enables us to determine the electrical length θ_{0max} of the hairpin resonator at the maximum frequency f_{0max} of the tuning range. From the resonance equations (15.56), we obtain the condition $\theta_{0max} < 2\pi/(f_1/f_0)_{min}$. In our case, $(f_1/f_0)_{min} = 1250/400 = 3.125$ and $2\pi/(f_1/f_0)_{min} = 115.2°$. To satisfy this condition, we set $\theta_{0max} = 84.7°$. Substituting this value and $f_{0max} = 0.4$ GHz into the resonance equation (15.56a), we determine the capacitance value $C_{4min} = 7.45$ pF. At the minimum frequency $f_{0min} = 0.225$ GHz of the tuning range, we obtain the electrical length $\theta_{0min} = 47.64°$. Substitution of these two values into (15.56a) gives $C_{4max} = 27.34$ pF. If $f_0 = 0.3$ GHz, then $\theta_0 = 63.52°$ and $C_4 = 14.62$ pF.

The required gaps between the resonators, corresponding to the coupling coefficients $K_{12} = K_{34} = 0.0218$ and $K_{23} = 0.0168$, are determined by computer simulation at a frequency of $f_0 = 300$ MHz. The value $K_{12} = 0.0218$ corresponds to a gap of 3.5 mm, and the value $K_{23} = 0.0168$ corresponds to a gap of 4 mm. To connect the end resonators to loads, we use a thin line, 0.25 mm wide, which is short-circuited at one end. The value of the external quality factor $Q_e = 38.81$ is ensured by removing this line by 0.25 mm from the hairpin resonator. To the middle of each resonator is connected a resistor, $R = 10\,\text{k}\Omega$, which supplies voltage to the varactors.

Figure 16.4 shows the variation of coupling coefficients and Q_e in the tuning range. With increasing frequency, the coupling coefficients decrease slightly. Changes do not exceed 3%, so the filter's FBW should remain almost constant. The value of Q_e also decreases, but more significantly (by a factor of 1.33).

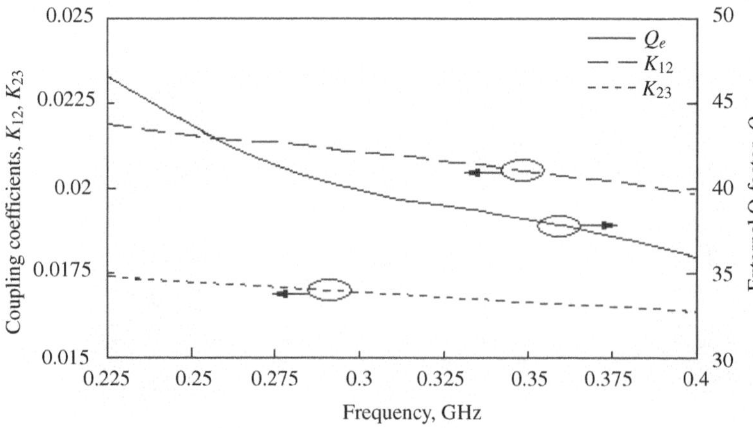

Figure 16.4 Coupling coefficients and external Q-factor of a tunable BPF in Figure 16.3.

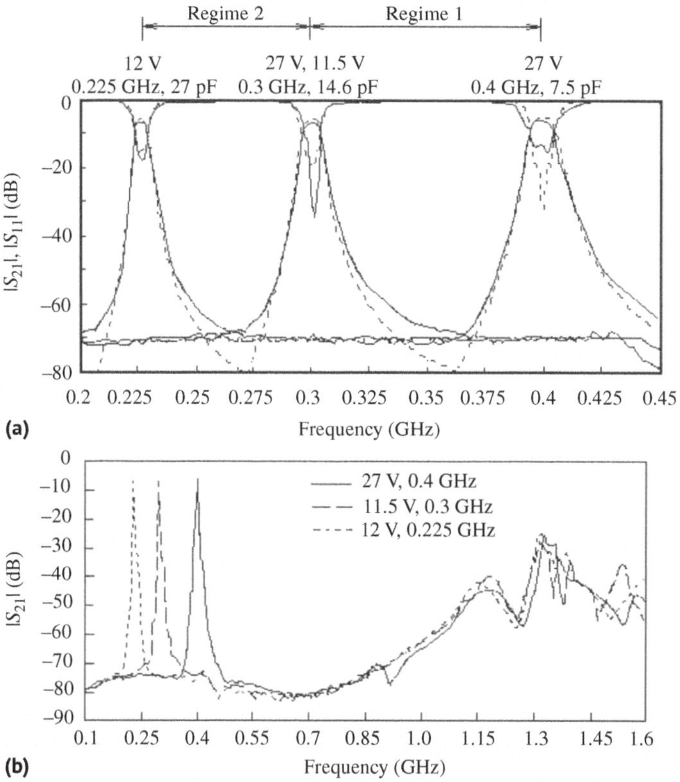

Figure 16.5 Frequency responses of tunable BPF with hairpin resonators. (a) Measured and simulated frequency responses in tuning range. (b) Measured insertion loss in wide frequency band.

In Figure 16.5a, the dashed line shows the simulated frequency responses of the filter in the tuning range.

16.2.2 Manufacturing and Measurement

To decrease the insertion loss of the BPF, it is necessary to increase the unloaded Q-factor Q_u of the composite resonator, which consists of a TL segment with Q-factor Q_l, and a varactor with Q-factor (16.3). This expression indicates that decreasing the product R_sC_{min} increases Q_C. It was shown in Ref. [6] that if one large capacitance varactor is replaced by several small capacitance varactors that are connected in parallel and in series, then an increase in Q_C will occur.

To implement the required capacitance values of $C_4 = 7.5 - 27$ pF, we use several varactors BB135 (NXP company) with a minimal capacitance of $C_{min} \approx 2$ pF

Table 16.2 Measured parameters of tunable filter in Figure 16.3.

Regime	f_0, MHz	U, V	FBW, %	L_{A0}, dB	Attenuation, dB		f_1/f_0
					$0.95f_0$	$1.05f_0$	
Regime 1	400	27	2.45	5.4	51	42	3.6
	350	18.5	2.42	6.4	51.5	42.5	4.1
	300	11.5	2.4	7.5	52	43	4.8
Regime 2	300	27	2.4	5.7	52	43	4.8
	260	18	2.37	6.6	53	44	5.6
	225	12	2.35	7.6	54	45	6.4

and a voltage range of 0.5–28 V. As the voltage on a varactor decreases, its Q_v decreases. To mitigate this effect, we will utilize two units with different numbers of varactors connected in parallel and in series, operating within the high-voltage range of 11.5–27 V. The measured frequency responses of the filter are depicted in Figure 16.5a, b. Measurements were performed using the network analyzer R&S ZNB8 (Rohde & Schwarz). The measured insertion loss L_{A0} at frequency f_0 exceeds the simulated L_{A0} by no more than 0.5 dB. The filter specification is given in Table 16.2. This filter has two operation regimes. In the first regime, the first unit with $N_1 = 15$ varactors is connected to each resonator. It provides filter tuning at the frequency of 400 MHz with a voltage of 27 V. Reducing the voltage increases the capacitance of the varactor and lowers the operating frequency. This occurs until $f_0 = 300$ MHz is reached and the voltage reaches 11.5 V. At this point, the filter switches to the second operational regime, where a second unit with $N_2 = 14$ varactors is connected to each resonator, and the voltage rises to 27 V. The filter is tuned to 300 MHz at a voltage of 27 V. In the second operational regime, the voltage is sequentially lowered to 12 V, at which the lower frequency of 225 MHz within the tuning range is reached.

The measurement data show that a big tuning range, a high selectivity, and a wide rejection band characterize the microstrip tunable filter with hairpin resonators (Figure 16.5). The spurious frequency response peak is located at a frequency of $f_1 = 1345$ MHz, characterized by a minimum value of the ratio $(f_1/f_0)_{\min}$=3.36. Due to this, the filter has attenuation exceeding 41 dB at $3f_0$, which is required in the specification.

In the filter under consideration, two operating regimes and two varactor units were used. This allowed the avoidance of varactor operation in the low-voltage section, where their Q_C factor is relatively low. Compared to the previous filter in Figure 16.2, the minimum voltage on the varactors was increased from 6 to 11.5 V, resulting in a decrease in insertion losses from 9.2 to 7.6 dB.

16.3 Bandpass Filter with Combined Frequency Tuning

Figure 16.6 shows the topology of a second-order microstrip BPF with combined frequency tuning and its photograph [11]. Lumped capacitances C change continuously, and inductance L takes two discrete values L_1 and L_2. For this BPF, an RT/Duroid 6010 LM (Rogers) substrate was chosen, which has the parameters: $\varepsilon_r = 10.2$, $h = 1.905$ mm, $\tan\delta = 0.002$. Microstrip resonators have a width of $w = 3.5$ mm ($Z_0 = 34.4\,\Omega$) and a gap between resonators of $S = 2$ mm. A thin line 0.125 mm wide is used to connect each resonator to the load. This line is 29 mm long and is removed by 0.125 mm from the resonator. The size of the filter with shield is 70 mm × 40 mm × 15 mm.

We will compare the tunability of this filter with the filter in Figure 16.2, which was tuned only by varactors. A correct comparison of the tunability of two resonators can be performed if their electrical lengths are approximately the same. The microstrip BPF in Figure 16.2 was tuned by varactors in the frequency range from $f_{0\min} = 225$ MHz to $f_{0\max} = 400$ MHz (56%) with a change in the varactor capacitance by $\chi = 3.67$ times. The electrical length of the used resonators was changed from $\theta_{0\min} = 23.82°$ to $\theta_{0\max} = 42.35°$.

(a)

(b)

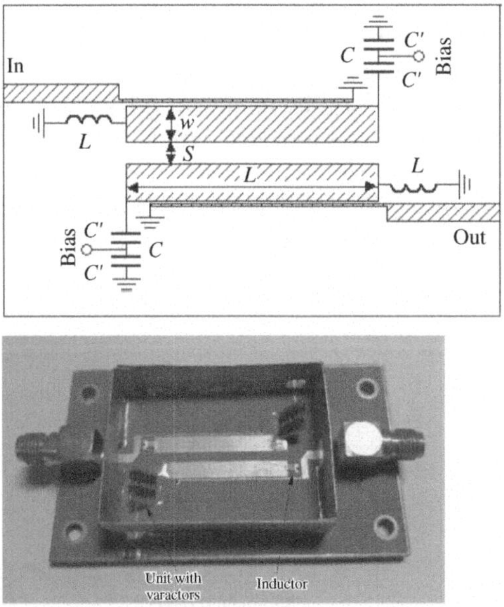

Figure 16.6 Fabricated microstrip tunable BPF. (a) Topology: $L = 22$ mm, $w = 3.5$ mm, $S = 2$ mm. (b) Photograph of BPF without shield. *Source*: [11] / with permission of Radioelectronics and Communications Systems.

To provide the value $\theta_{0min} = 23.82°$ ($f_{0min} = 225$ MHz), the resonator length should be $l = 33$ mm. Assuming the value of the first inductance $L_1 = 12$ nH, from (15.44a), we determine the maximum capacitance $C_{max} = 15$ pF. We will assume that the capacitance C changes by a factor of 3, and then its minimum value is $C_{min} = 5$ pF. This value will lead to an increase in the resonant frequency of the resonator to $f_0 = 310$ MHz ($\theta_0 = 32.82°$). At this intermediate frequency ($f_0 = 310$ MHz), we replace the inductance L_1 with the inductance $L_2 = 3.9$ nH and from (15.44a) determine the capacitance $C = 13.3$ pF. When the capacitance changes to its minimum value, $C_{min} = 5$ pF, the resonant frequency reaches its maximum value, $f_{0max} = 420$ MHz ($\theta_{0max} = 44.46°$). In our case, $C = 3$ pF, $L =$ 3.08 nH, and the frequency changes from 225 to 420 MHz (60.5%). This tuning band is split into two sub-bands: 225–310 MHz at L_1 =12 nH and 310–420 MHz at $L_2 = 3.9$ nH.

Figure 16.7a shows the frequency dependence of the coupling coefficient k_{12}. Figure 16.7b shows the variation of the external quality factor Q_e. Both dependences are built based on EM simulation using the Microwave Office software.

The BPF (Figure 16.6) utilized two inductors with $L_1 = 12$ nH and $L_2 = 3.9$ nH. To implement the capacitance values $C = 5$–15 pF, we use several varactors BB135 (NXP): $C'_{min} \approx 2$ pF; $U = 0.5$–28 V. To increase the varactor's quality factor Q_v we used unit with multiple varactors (Figures 16.2, 16.3), which are connected in parallel and in series, and the high-voltage area: $U = 7$–28 V. Each unit contains five pairs of varactors connected in parallel, so its $C_{min} = 5$ pF is a need for us.

Figure 16.8 shows the frequency responses of the filter, which has two tuning sub-bands. The first sub-band, 225–310 MHz, is provided with $L_1 = 12$ nH and a voltage variation of 7–28 V. The second sub-band, 310–420 MHz, is provided with $L_2 = 3.9$ nH and a voltage variation of 8–28 V. At the border of the sub-bands ($f_0 = 310$ MHz), the filter parameters change abruptly, as reflected in Table 16.3 and Figure 16.7.

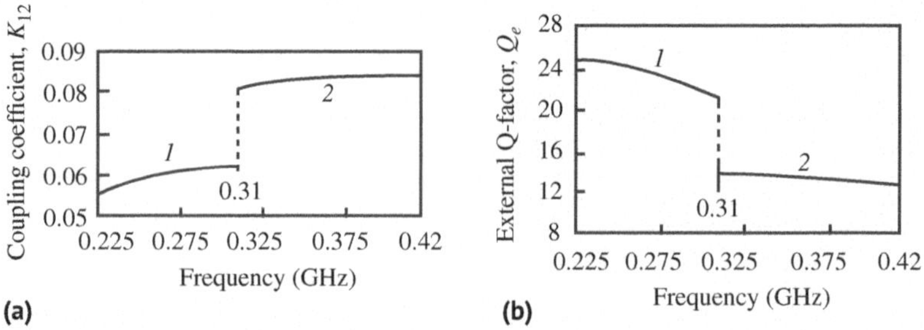

Figure 16.7 (a) Coupling coefficient and (b) external Q-factor of a microstrip tunable BPF in Figure 16.6 for $L = 12$ nH (curve *1*) and $L = 3.9$ nH (curve *2*).

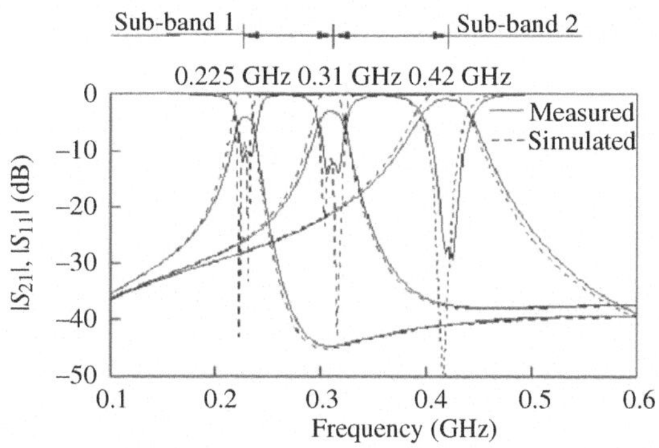

Figure 16.8 Frequency responses of a microstrip second-order BPF with combined frequency tuning.

Table 16.3 Measured parameters of tunable BPF in Figure 16.6.

Band	f_0, MHz	U, V	$\text{FBW}_{3\text{ dB}}$	L_{A0}, dB	L_R, dB
Sub-band 1, $L = 12$ nH	225	7	0.075	4.1	10
	280	15.7	0.078	3.6	11.7
	310	28	0.079	3.2	12.2
Sub-band 2, $L = 3.9$ nH	310	8	0.113	1.8	18
	370	15.7	0.114	1.5	24
	420	28	0.114	1.4	26

For the BPF, we have $\chi_C = 3$, $\chi_L = 3.08$, and the tuning frequency band is 60.5%. In BPF in Figure 16.2 with varactor frequency tuning, a larger change in capacitance $\chi_C = 3.67$ led to a narrower tuning band of 56%.

16.4 Combline Bandpass Filter with Ferroelectric Capacitors

Compared to varactor-tuned BPFs, filters with ferroelectric capacitors are used at higher frequencies, where their quality factor Q_C retains relatively high values. Of practical interest are both the frequency responses of such BPFs and the Q_C values of the ferroelectric capacitors used in them.

16.4.1 Frequency Responses of Bandpass Filters with Ferroelectric Capacitors

Let us consider four fifth-order microstrip combline BPFs (Figure 16.9a) that are tuned by BST (barium–strontium–titanium) capacitors [6]. The filters are fabricated on a Rogers TMM-3 substrate with a dielectric constant $\varepsilon_r = 3.27$ and $\tan\delta = 0.002$. The substrate thickness $h = 0.762$ mm, and the height of the metal shield above the substrate $H = 4$ mm. One end of the combline resonator is short-ended, and the other is loaded with a BST capacitor with capacitance C. The resonance frequencies of combline resonators are determined by (15.7a): $\omega C = -Z_0^{-1} \cot\theta$.

Figure 16.9b shows the photograph of one of the filters without a shield. All filters have the same dimensions 23.6 mm × 19 mm × 5 mm. The width of the central conductor of the microstrip resonators is $w = 1.4$ mm, and their characteristic impedance is $Z_0 = 57.4\ \Omega$. The four filters differ from each other in the length of the resonators: $l_1 = 13$ mm, $l_2 = 10$ mm, $l_3 = 8.4$ mm, and $l_4 = 7$ mm. Metallized holes provide a short circuit at the ends of the resonators, which have a diameter of 0.4 mm.

The voltage applied to the capacitors was varied from 5 to 100 V. Figure 16.10 shows the insertion loss of the filters. Curves *1*, *2*, and *3* correspond to a voltage of 5, 50, and 100 V, respectively.

A change in the bias voltages from 5 to 100 V resulted in tuning of the first filter (Figure 16.2a) in a frequency range from $f_{0\min} = 1778$ MHz to $f_{0\max} = 2320$ MHz, which corresponds to the fractional tuning range $\text{FTR} = 2(f_{0\max} - f_{0\min})/(f_{0\max} + f_{0\min}) = 0.26$. The capacitance of the BST capacitors was varied from 0.76 to 1.6 pF, corresponding to a capacitance ratio of $C_{\max}/C_{\min} = 2.1$. At a voltage of 5 V, the center frequency of the first filter $f_{0\min} = 1784$ MHz, the 3 dB bandwidth $\text{BW}_{3\,\text{dB}} = 218$ MHz, which corresponds to the fractional bandwidth $\text{FBW} = 0.122$, and the minimum passband insertion loss $L_{A0} = 4.26$ dB. At

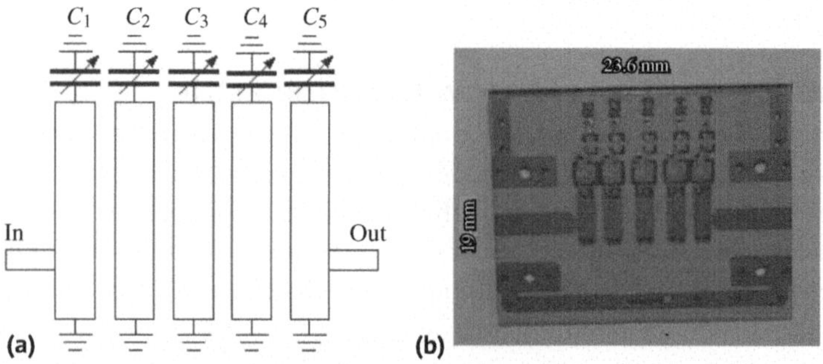

Figure 16.9 Tunable microstrip filter with ferroelectric capacitors. (a) Schematic. (b) Photograph without shield. *Source*: [6] / with permission of IEEE.

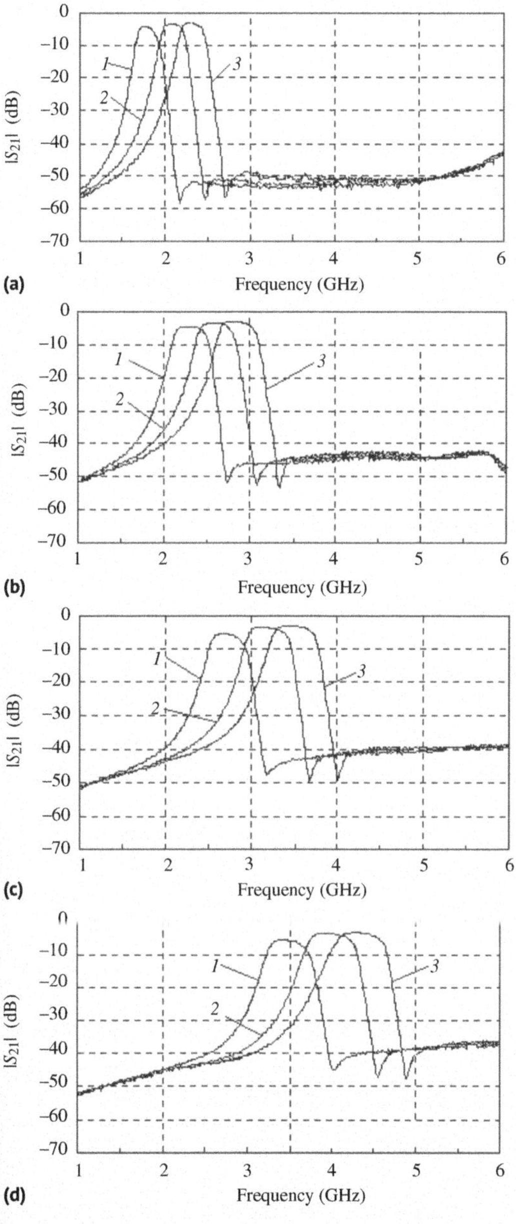

Figure 16.10 Insertion loss of microstrip filters with ferroelectric capacitors, operating in different frequency ranges: (a) 1.78–2.32 GHz, (b) 2.30–2.82 GHz, (c) 2.71–3.49 GHz, (d) 3.47–4.33 GHz.

a voltage of 100 V, the filter's center frequency $f_{0max} = 2320$ MHz, the bandwidth $BW_{3\,dB} = 273$ MHz, which corresponds to FBW = 0.118, and $L_{A0} = 3.24$ dB.

These values, which characterize the first filter, are listed in Table 16.4. The similar values for the other three filters and the values of the return loss L_R and the resonator electrical lengths θ are also included in Table 16.4. This table uses the calculated values of the variable capacitance obtained in the design process.

The simultaneous use of four filters with ferroelectric capacitors enables tuning from 1.78 to 4.33 GHz, corresponding to 83% tuning range.

Next, we will solve the problem of determining the quality factor of the ferroelectric capacitors Q_C from the measured frequency responses of the filters in two stages. At the first stage, we will find the unloaded quality Q_u of the filter's resonators, which include transmission line segments and ferroelectric capacitors. At the second stage, based on the obtained values of Q_u, we will determine the quality factor Q_C of the ferroelectric capacitors.

16.4.2 Unloaded Quality Q_u of Tunable Resonators

In designing BPFs, the Chebyshev, or equal-ripple, attenuation characteristics of lossless filters are most often used [9]. Suppose that an N-order BPF should be designed with ripple L_{Ar}, midband frequency f_0, and bandwidth BW. We note that the L_{Ar} level determines the value of BW. The normalized g parameters of low-pass prototype filters are used for the filter design. The insertion loss of a lossless BPF with an odd number of resonators is zero at the midband frequency: $L_{A0} = 0$. If N is even, then $L_{A0} = L_{Ar}$. In the presence of losses, the increase in attenuation ΔL_{A0} at the filter's center frequency is determined by the S. Cohn formula (4.1). If all filter resonators are identical, then their unloaded quality factor Q_u can be determined from expression (4.7), which we present as

$$Q_u = \frac{4.343 f_0}{\mathrm{BW}\,\Delta L_{A0}} \sum_{i=1}^{n} g_i. \tag{16.4}$$

Table 16.4 Characteristics of microstrip fifth-order BPFs with ferroelectric capacitors.

Characteristics	Filter *1*	Filter *2*	Filter *3*	Filter *4*
$f_{0min} - f_{0max}$, MHz	1784–2320	2306–2851	2710–3490	3477–4333
FTR, %	26	21	25	22
$C_{min} - C_{max}$, pF	0.76–1.6	0.7–1.26	0.54–1.08	0.4–0.76
C_{max}/C_{min}	2.1	1.8	2.0	1.9
FBW_{max}, %	11.8–12.2	14.5–16.1	13.8–14.2	12.7–13.3
L_{A0min}–L_{A0max}, dB	3.24–4.26	3.03–4.26	3.14–5.3	3.15–5.63
L_{Rmin}–L_{Rmax}, dB	12–21	11–13	13–20	12–17
$\theta_{min} - \theta_{max}$, °	44.1–57.4	44–54.4	43.4–55.9	46.5–58

We will use the calculation procedure for Q_u (16.4), discussed in Section 4.3, as applied to filters with ferroelectric capacitors, the insertion loss function of which is shown in Figure 16.10. To this end, we must determine some values in (16.4). The main difficulty in solving the problem is due to the low unloaded quality Q_u. This circumstance results in significant insertion loss variations in the passband filter. It seems impossible to measure directly the lossless value of BW and ripple level L_{Ar}, which corresponds to g_i sum in (16.4).

To solve this problem, it is appropriate to use the filter characteristics that are least affected by the dissipative loss. Among these characteristics are return loss L_R and the width of the "barrier band" Δf_b. The "barrier band" is the passband width measured by a high attenuation level. The measured frequency responses determine the values L_R and Δf_b. The transition from Δf_b to lossless BW is performed with the use of the filter's insertion loss plots [9]. Knowing L_R, we can determine the ripple level L_{Ar} by (2.33).

The aforementioned solution enables us to define the unloaded quality Q_u of the resonators roughly. The Chebyshev characteristics serve as the basis for designing these filters. Let us assume that the average return loss of the filters $L_R = 13.5$ dB (Table 16.4). Using (2.33), we find the ripple level corresponding to this value of return loss $L_{Ar} = 0.2$ dB. Next, we find the sum of g parameters for a fifth-order Chebyshev filter with $L_{Ar} = 0.2$ dB (Table 2.6): $\Sigma g_i = 7.52$.

For the first BPF at the lower center frequency $f_{0\text{min}} = 1784$ MHz (Figure 16.10a, curve *1*), the measured 20 dB "barrier band" width is $\Delta f_b = 400$ MHz. From the insertion loss curve [9], for $L_{Ar} = 0.2$ dB and $N = 5$, we find the ratio $\Delta f_b/\text{BW} = 1.44$. The BW of a lossless filter is determined as BW = 400/1.44 ≈ 278 MHz. The measured insertion loss of the BPF at f_0 is $L_{A0} = \Delta L_{A0} = 4.26$ dB. Substituting the found values $\Sigma g_i = 7.52$, BW = 278 MHz, $\Delta L_{A0} = 4.26$ dB, $f_0 = 1784$ MHz into (16.4), we obtain $Q_u \approx 49.2$.

For the same filter, at the upper center frequency $f_{0\text{max}} = 2320$ MHz (Figure 16.10a, curve *3*), we find the 20 dB "barrier band" width: $\Delta f_b = 450$ MHz. Next, we turn to the lossless BW = 312 MHz. Taking into account the measured insertion loss $\Delta L_{A0} = 3.24$ dB, we find $Q_u \approx 75$ from (16.4). Curve *1* in Figure 16.11 shows the resonator Q_u variation of the first BPF over its tuning range. The values of Q_u vary from 49.2 to 75. The other three curves in Figure 16.11 correspond to the other three BPFs from Table 16.4.

Determining Q_u, we arbitrarily assumed that the average level of the filter return loss $L_R = 13.5$ dB. Let us analyze the potential outcomes of a change at this level. If $L_R = 16$ dB, then $L_{Ar} = 0.1$ dB, $\Sigma g_i = 7.01$, and $\Delta f_b/\text{BW} = 1.52$. If we take $L_R = 9.5$ dB, then $L_{Ar} = 0.5$ dB, $\Sigma g_i = 8.42$, and $\Delta f_b/\text{BW} = 1.35$.

In the above-considered cases, the unloaded quality Q_u calculated using (16.4) will differ from the prior values Q_u corresponding to $L_R = 13.5$ dB. In the first case, they will be underestimated by 2%, and in the second case, they will be overestimated by 5%. In actual practice, this accuracy is acceptable.

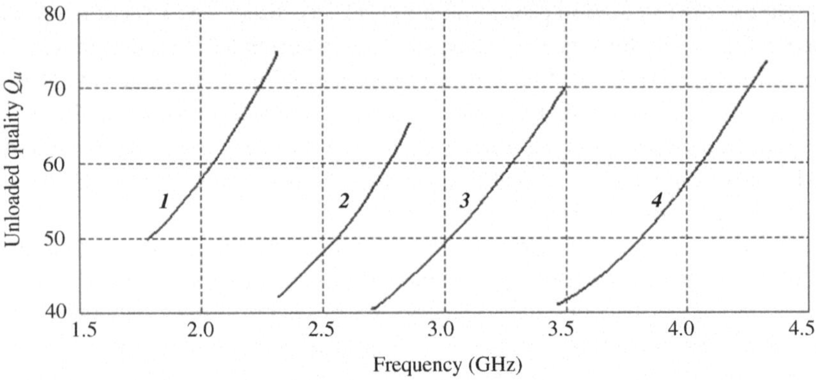

Figure 16.11 Unloaded quality Q_u of a combline resonator with a ferroelectric capacitor over tuning range of each BPF.

16.4.3 Quality Factor Q_C of Ferroelectric Capacitors

Let us consider the second stage of solving the problem of obtaining the quality factor of the ferroelectric capacitors Q_C from the found values of the resonator's unloaded quality Q_u and the calculated quality factor Q_l of the resonators' transmission line. For the microstrip transmission line used in the resonators under consideration, the calculated formulas (4.8), (4.13), and (4.14) give the values of Q_l, which are plotted in Figure 16.12.

Now, we have the values of Q_u (Figure 16.11) and Q_l (Figure 16.12). Based on these values, it is necessary to determine the quality factor Q_C of the ferroelectric capacitors. To solve this problem, we use (15.71), which determines the unloaded quality Q_u of a composite resonator containing a TL segment and a capacitor.

For the first BPF at the lower center frequency $f_{0\min} = 1784$ MHz, the known values are $Q_u = 49.2$, $Q_l = 151$, and $\theta = 44.14°$. At the specified electric length, using (15.6b), we find the sensitivity $S_C = 0.393$. Substitution of these values Q_u, Q_l, and S_C into (15.71) gives the quality factor of the ferroelectric capacitor $Q_C \approx 41.6$.

Computing Q_C, we use the calculated quality factor of the microstrip transmission line $Q_l = 151$. The formulas for calculating Q_l (4.8), (4.13), and (4.14) do not take into account the roughness of the dielectric substrate and the loss in the short-ended circuit of the resonator. We assume that these effects will reduce the estimated Q_l by about 30%, i.e., $Q_l = 106$, and repeat the previous calculations. As a result, we obtain $Q_C \approx 42.9$. Thus, as Q_l is decreased by 30%, the calculated quality factor of the ferroelectric capacitor increases from 41.6 to 42.9, i.e., by 3%. Note that the roughness of the substrate does not affect the quality factor of the capacitor Q_C soldered onto the substrate.

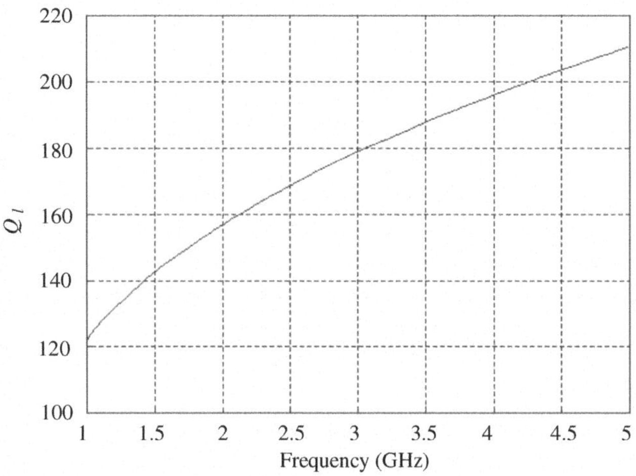

Figure 16.12 Quality factor of a microstrip TL with width $w = 1.4$ mm made on a TMM-3 substrate with thickness $h = 0.761$ mm.

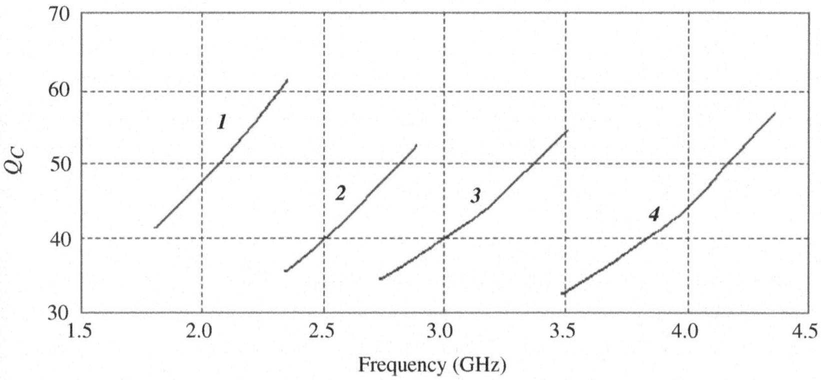

Figure 16.13 Variation in quality factor Q_C of the ferroelectric capacitor over the tuning range of each BPF.

To determine Q_C, we use the values of Q_l reduced by 30%. Determining Q_C at the upper center frequency $f_{0\max} = 2320$ MHz, we use the following values: $Q_u = 75$, $Q_l = 123$, $\theta = 57.4°$, and $S = 0.312$. Substitution of these values into (5.71) gives $Q_C = 55.7$.

The plots of the quality factor of the ferroelectric capacitors Q_C over the same tuning range for each BPF are shown in Figure 16.13. Over the tuning range from 1.78 to 4.33 GHz, the quality factor of the ferroelectric capacitors under consideration takes the values $Q_C = 33.5–60.7$.

References

1 A. R. Brown and G. M. Rebeiz, A varactor-tuned RF filter, *IEEE Trans. Microw. Theory Tech.*, vol. 48, no. 7, pp. 1157–1160, 2000, https://doi.org/10.1109/22.848501.

2 B.-W. Kim and S.-W. Yun, Varactor-tuned combline bandpass filter using step-impedance microstrip lines, *IEEE Trans. Microw. Theory Tech.*, vol. 52, no. 4, pp. 1279–1283, 2004, https://doi.org/10.1109/TMTT.2004.825626.

3 M. Sanchez-Renedo, R. Gomez-Garcia, J. I. Alonso, and C. Briso-Rodriguez, Tunable combline filter with continuous control of center frequency and bandwidth, *IEEE Trans. Microw. Theory Tech.*, vol. 53, no. 1, pp. 191–199, 2005, https://doi.org/10.1109/TMTT.2004.839309.

4 J.-S. Hong, Reconfigurable planar filters, *IEEE Microw. Mag.*, vol. 10, no. 6, pp. 73–83, 2009, https://doi.org/10.1109/MMM.2009.933590.

5 J. Sigman, C. D. Nordquist, P. G. Clem, G. M. Kraus, and P. S. Finnegan, Voltage-controlled Ku-band and X-band tunable combline filters using barium-strontium-titanate, *IEEE Microw. Wirel. Components Lett.*, vol. 18, no. 9, pp. 593–595, 2008, https://doi.org/10.1109/LMWC.2008.2002453.

6 A. Zakharov and M. Ilchenko, Unloaded quality factor of transmission line resonators with capacitors, *IEEE Trans. Circuits Syst. I Regul. Pap.*, vol. 67, no. 7, pp. 2204–2215, 2020, https://doi.org/10.1109/TCSI.2020.2971112.

7 J.-S. Hong, *Microstrip Filters for RF/Microwave Applications*, 2nd ed. New Jersey: Wiley, 2011, https://doi.org/10.1002/9780470937297.

8 A. Zakharov, S. Rozenko, and M. Ilchenko, Varactor-tuned microstrip bandpass filter with loop hairpin and combline resonators, *IEEE Trans. Circuits Syst. II Express Briefs*, vol. 66, no. 6, pp. 953–957, 2019, https://doi.org/10.1109/TCSII.2018.2873227.

9 G. L. Matthaei, L. Young, and E. M. T. Jones, *Microwave Filters, Impedance-Matching Networks, and Coupling Structures*. New York: Artech House Books, 1980.

10 A. Zakharov, S. Rozenko, S. Litvintsev, and M. Ilchenko, Hairpin resonators in varactor-tuned microstrip bandpass filters, *IEEE Trans. Circuits Syst. II Express Briefs*, vol. 67, no. 10, pp. 1874–1878, 2020, https://doi.org/10.1109/TCSII.2019.2953247.

11 S. Litvintsev and A. Zakharov, Combined frequency tuning of resonators from TL segments, *Radioelectron. Commun. Syst.*, vol. 67, no. 10, pp. 538–549, 2024, https://doi.org/10.3103/S0735272723030019.

17

Dual-mode Resonators and Bandpass Filters

A special place among resonators is occupied by dual-mode resonators, since their use leads to an increase in the bandpass filter (BPF) selectivity with a smaller number of resonators. It was first established in Ref. [1] that a microstrip ring resonator is a dual-mode resonator, a finding confirmed in Refs. [2, 3]. Later, other dual-mode resonators and BPFs based on them were proposed, some of which are discussed in this chapter.

17.1 Dual-mode Resonators with Uniform TL Segments

Such resonators are formed from half-wave sections of transmission line (TL) with open- or short-circuited ends (Table 17.1, No. 1) when stubs or lumped elements are connected to them.

17.1.1 Resonators with Open-circuit Ends

Consider the symmetric resonators shown in Figure 17.1. They are half-wave open-circuited TL segments, to the middle of which an open or short-circuited stub is connected. These resonators can be represented by a two-port network with a transfer matrix $[ABCD]$. In the absence of loss, the resonance condition $Y_{\text{in}} = 0$ is expressed by equality (15.25), which it is advisable to rewrite

$$C = 0. \tag{17.1}$$

The transfer matrix of the symmetrical two-port network $[ABCD]$ can be expressed through the transfer matrix elements A', B', C', D' of its left half [4]

$$\begin{bmatrix} A & B \\ C & D \end{bmatrix} = \begin{bmatrix} 1 + 2B'C' & 2A'B' \\ 2C'D' & 1 + 2B'C' \end{bmatrix},$$

Table 17.1 Transmission line dual-mode resonators.

No.	Topology	Resonant frequencies
No. 1	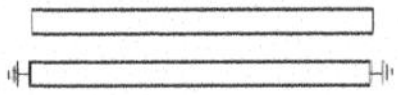	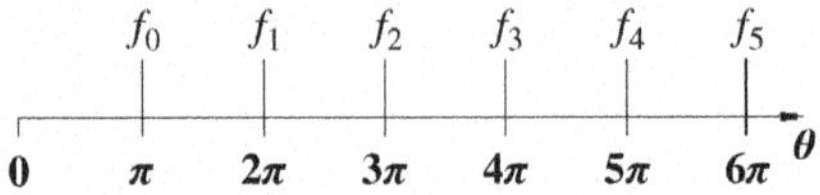
No. 2	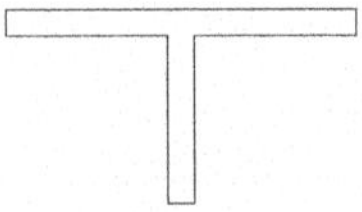	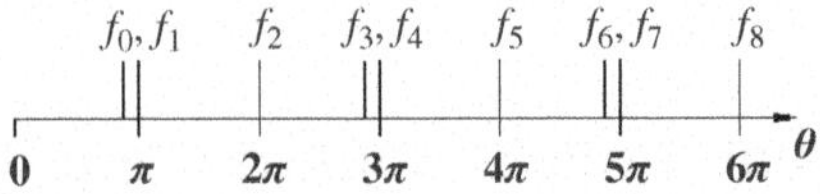
No. 3	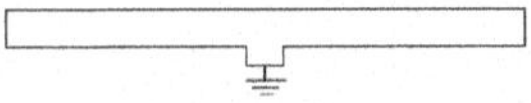	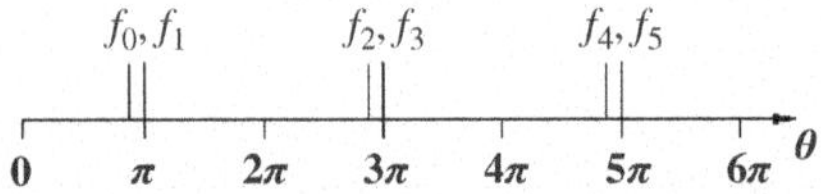
No. 4	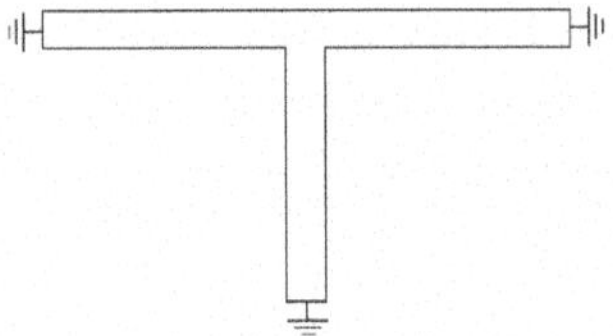	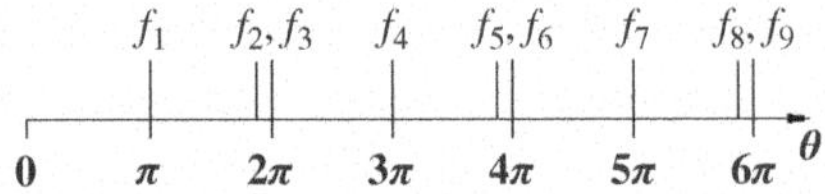
No. 5	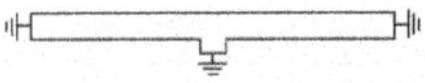	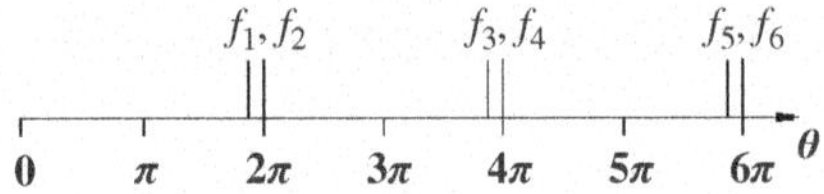
No. 6	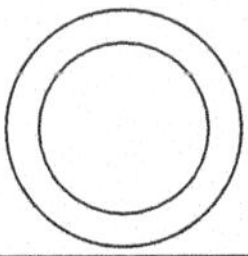	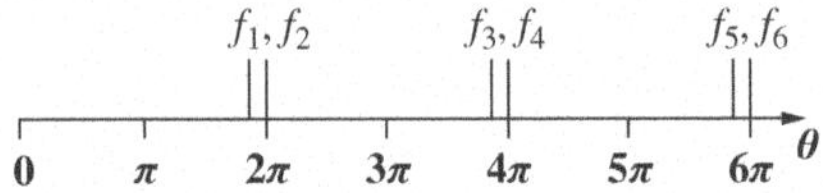

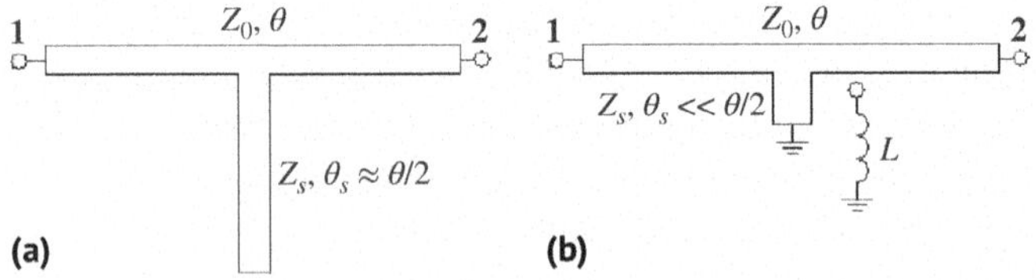

Figure 17.1 Dual-mode resonator with open-circuited ends. (a) With open-circuited stub. (b) With short-circuited stub.

which replaces the resonance condition (17.1) with two conditions

$$
\begin{aligned}
C' &= 0, \\
D' &= 0.
\end{aligned}
\tag{17.2}
$$

Let us compose the resonance equations for the resonators shown in Figure 17.1. We will use the input admittance of the stub at the point of its connection to the TL segment Y_s and the transfer matrix $[abcd]$ of the left half of the TL segment. Obviously, the matrix $[A'B'C'D']$ represents the product of two matrices [5]

$$\begin{bmatrix} A' & B' \\ C' & D' \end{bmatrix} = \begin{bmatrix} a & b \\ c & d \end{bmatrix} \begin{bmatrix} 1 & 0 \\ \frac{Y_s}{2} & 1 \end{bmatrix} = \begin{bmatrix} a + \frac{bY_s}{2} & b \\ c + \frac{dY_s}{2} & d \end{bmatrix}$$

From the last expression, it follows that the resonance equations (17.2) take the form

$$d = 0, \tag{17.3a}$$

$$Y_s = \frac{-2c}{d}. \tag{17.3b}$$

If we use a segment of uniform TL with characteristic impedance Z_0 and electrical length θ, then [4]

$$\begin{bmatrix} a & b \\ c & d \end{bmatrix} = \begin{bmatrix} \cos\left(\frac{\theta}{2}\right) & jZ_0 \sin\left(\frac{\theta}{2}\right) \\ jZ_0^{-1} \sin\left(\frac{\theta}{2}\right) & \cos\left(\frac{\theta}{2}\right) \end{bmatrix}. \tag{17.4}$$

In this case, resonance equations (17.3) are as follows:

$$\cos\left(\frac{\theta}{2}\right) = 0, \tag{17.5a}$$

$$Y_s = -j2Z_0^{-1} \tan\left(\frac{\theta}{2}\right). \tag{17.5b}$$

Equation (17.5a) determines the part of the resonant electrical lengths

$$\theta_n = \pi,\ 3\pi,\ 5\pi, \dots . \tag{17.6}$$

Another part of the resonant electrical lengths is determined by equation (17.5b), and it depends on the stub parameters.

17.1.1.1 Quarter-wave Open-circuited Stub

If an open-circuited stub is used, the resulting dual-mode resonator [6–9] has the T-shape, as shown in Figure 17.1a. The open-circuited stub is characterized by the parameters Z_s and $\theta_s \approx \theta/2$. The input admittance of this stub is $Y_s = jZ_s^{-1} \tan(\theta_s)$, and the resonance equation (17.5b) takes the form

$$Z_s^{-1} \tan(\theta_s) = -2Z_0^{-1} \tan\left(\frac{\theta}{2}\right) \tag{17.7}$$

If the stub is quarter-wave ($\theta_s = \theta/2$), then the solutions of equation (17.7) are the electrical lengths θ_n, for which the function $\tan(\theta/2) = 0$ or $\pm\infty$:

$$\theta_n = \pi,\ 2\pi,\ 3\pi,\ 4\pi,\ 5\pi, \ldots = (n+1)\pi,\ n = 0,\ 1,\ 2,\ \ldots. \tag{17.8}$$

The half of the resonant electrical lengths (17.8) with even numbers n coincides with the previously obtained resonant electrical lengths (17.5). This coincidence characterizes the dual-mode oscillations.

The resonator in question is notable for its dual-mode oscillations, which alternate with single-mode oscillations [10]. Table 17.1 (No. 2) shows the distribution of the resonant electrical lengths of the considered resonator. The continuous numbering of resonant frequencies is used in Table 17.1. Paired lines correspond to dual-mode oscillations.

If the length of the stub is slightly increased to $\theta_s = (\theta/2) + \Delta$, where Δ is a small value, then the part of the resonant electrical lengths associated with dual-mode oscillation will decrease; we denote this by θ''_k. If the value of θ_s is somewhat reduced $\theta_s = (\theta/2) - \Delta$, then the values of θ''_k will increase. They are shown in Figure 17.2a with dotted lines. The remaining resonant electric lengths of the resonator are denoted as θ'_n.

Figure 17.2b shows the insertion loss frequency responses of this resonator for weak coupling with I/O loads. These responses have a double-sided TZ at the frequency f_z. If the stub length is less than $\lambda/4$, f_z is right-handed with respect to dual-mode oscillations f_e and f_o. If the stub length is greater than $\lambda/4$, then f_z is left-handed.

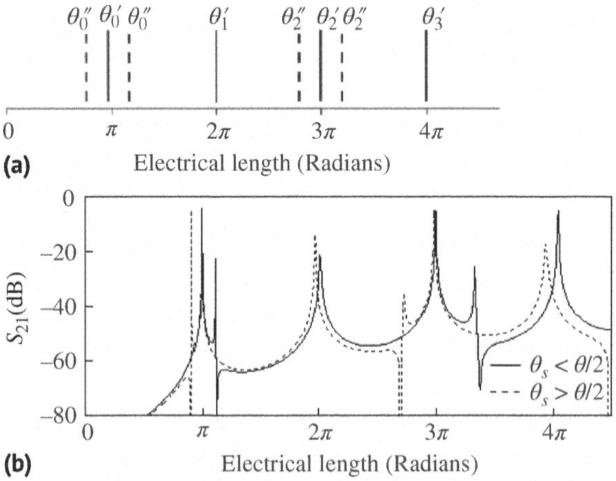

Figure 17.2 Individual characteristics of dual-mode resonator in Figure 17.1a. (a) Resonance electrical lengths distribution. (b) Frequency responses.

As in the case of two resonators with mixed coupling (Chapter 7), the frequency f_z is located closer to the odd mode of oscillations f_o than to the even f_e. This pattern enables us to identify the frequencies of dual-mode oscillations during measurements, i.e., to determine which ones are even and which are odd.

The quarter-wave open stub produces odd modes because f_z is located closer to these frequencies. The half-wave section of the resonator under consideration gives rise to even modes of dual-mode oscillations.

Note that the use of resonance equation (17.2), rather than resonance equations derived from even-odd mode analysis [7–9, 11], allowed us to establish a new property of the dual-mode resonator under consideration. This property consists of the existence of simple oscillations that alternate with dual-mode oscillations.

The resonance equation (17.7) makes it possible to determine the frequency of the odd mode f_o accurately, and then the coupling coefficient between two oscillation modes by formula (3.10), which is advisable to rewrite

$$K = \frac{f_o^2 - f_e^2}{f_o^2 + f_e^2}. \tag{17.9}$$

Since the value of f_o can be either greater or less than f_e, the value of the coupling coefficient between dual-mode oscillations (17.9) can be either positive or negative.

In some applications of the dual-mode resonator in question, its quarter-wave open-circuited stub took a T-shape, which does not change its purpose designation.

17.1.1.2 Short-circuited Stub with Small Length

The resonator in Figure 17.1b contains a short-circuited stub with short length $\theta_s \ll \theta/2$ [12–14]. A metallized hole or an inductance can realize this stub. The part of its resonant electrical lengths θ'_n is still determined by expression (17.6). The other part of the resonant electrical lengths θ''_k is determined by (17.5b). The input admittance of a short-circuited stub is $Y_s = -jZ_s^{-1} \cot(\theta_s)$, which brings the resonance equation (17.5b) to the form

$$-Z_s^{-1} \cot(\theta_s) = -2Z_0^{-1} \tan\left(\frac{\theta}{2}\right). \tag{17.10}$$

The resonant electrical lengths θ''_k (17.10) are shown in Figure 17.3a with dashed vertical lines. They are always located to the left of θ'_n. The greater the stub length and its characteristic impedance Z_s, the further apart the electrical lengths θ''_k and θ'_n are. When replacing the stub with inductance L, the location of the resonant electrical lengths remains.

Figure 17.3b shows the insertion loss frequency responses of this resonator for weak coupling with I/O loads. The resonant frequencies f''_n, caused by the stub, are not located in the vicinity of the transmission zeros, so they are even modes in the dual-mode oscillation. Thus, a short-circuited stub of small length generates

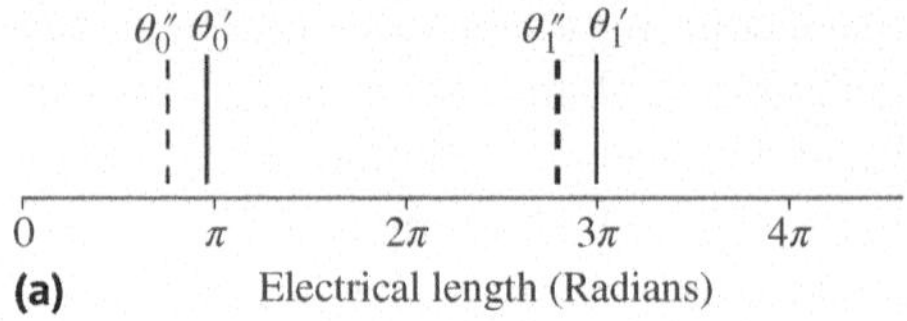

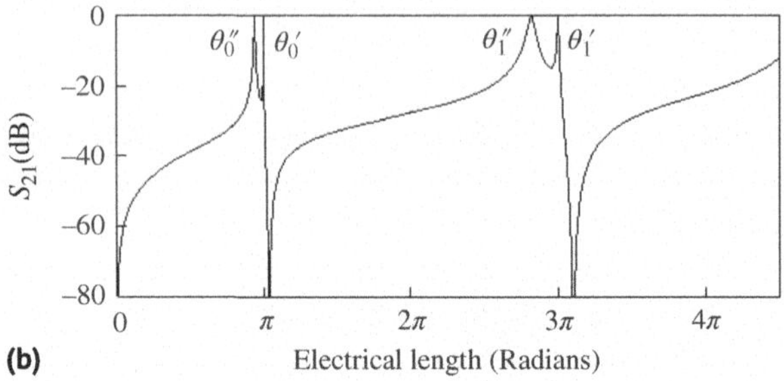

Figure 17.3 Individual characteristics of resonator in Figure 17.1b. (a) Resonance electrical lengths distribution. (b) Frequency response.

even modes, and a half-wave section generates odd modes in the resonator under consideration.

Using (17.10), we can determine the even mode and calculate the coupling coefficient K of dual-mode oscillation (17.9). Since the condition $f_e \leq f_o$ is satisfied, the value of K is always positive. If the length of the stub tends to zero ($\theta_s \to 0$), then the considered resonator degenerates into a resonator short-circuited in the middle part.

17.1.2 Resonators with Short-circuited Ends

Consider symmetric resonators shown in Figure 17.4. They are half-wave short-circuited TL segment, to the middle of which a short-circuited stub of various lengths is connected [10]. Since all three ends of this resonator are short-circuited, its resonant frequencies cannot be determined using resonance equations derived from even-odd mode analysis.

Resonance in the symmetrical resonators under consideration occurs when the voltage at the center of the resonator reaches its maximum or minimum value. This is expressed by the resonance equation (15.58), which are written as

$$A' = 0, \tag{17.11}$$

and

$$B' = 0, \tag{17.12}$$

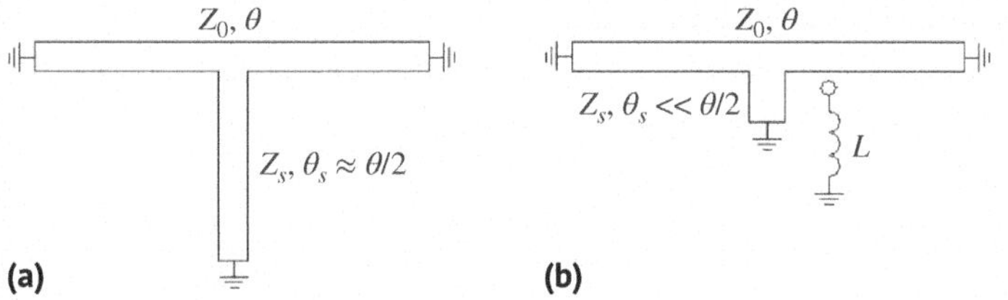

Figure 17.4 Dual-mode resonators with short-circuited ends and short-circuited stub. (a) Stub of quarter-wave length. (b) Stub with small length.

where A' and B' are transfer matrix parameters of the left half of the resonator.

Condition (17.11) is satisfied when the voltage in the middle of the resonator reaches maximum, which corresponds, in particular, to $\lambda/2$-type oscillations. Condition (17.12) is provided at a minimum voltage in the middle of the resonator. It corresponds, in particular, to λ-type oscillations.

Substituting expressions for A' and B' in (17.11) and (17.12), we obtain

$$a + \frac{bY_s}{2} = 0, \tag{17.13a}$$

$$b = 0. \tag{17.13b}$$

Substituting parameters a and b of matrix (17.4) into (17.13) yields the final resonance equations:

$$Y_s = -\frac{2a}{b} = 2jZ_0^{-1} \cot\left(\frac{\theta}{2}\right), \tag{17.14a}$$

$$b = jZ_0^{-1} \sin\left(\frac{\theta}{2}\right) = 0. \tag{17.14b}$$

Equation (17.14b) determines the part of the resonant electrical lengths

$$\theta_n = 2\pi,\ 4\pi,\ 6\pi, \dots . \tag{17.15}$$

Since the input admittance of the stub is $Y_s = -jZ_s^{-1} \cot(\theta_s)$, then equation (17.14a), which determines the other part of the resonant electrical lengths, takes the form

$$-Z_s^{-1} \cot(\theta_s) = -2Z_0^{-1} \cot\left(\frac{\theta}{2}\right). \tag{17.16}$$

The roots of the equation (17.16) depend on the stub parameters.

17.1.2.1 Quarter-wave Short-circuited Stub

A resonator with quarter-wave stub ($\theta_s = \theta/2$) is shown in Figure 17.4a. In this case, the roots of (17.16) are the electrical lengths θ_n, for which the function $\cot(\theta/2) = 0$ or $\pm\infty$:

$$\theta_n = \pi,\ 2\pi,\ 3\pi,\ 4\pi,\ 5\pi, \ldots = (n+1)\ \pi,\ n = 0,\ 1,\ 2,\ \ldots\ . \tag{17.17}$$

The half of the resonant electrical lengths (17.17) with even numbers n coincides with the previously obtained resonant electrical lengths (17.15), which leads to the dual-mode oscillations. In the resonator under consideration, dual-mode oscillations alternate with single oscillations. Table 17.1 (No. 4) shows the distribution of the resonant electrical lengths of the considered resonator. Unlike resonator No. 2 (Table 17.1), dual mode is observed at electrical lengths $n\pi$ with even values of n.

If the length of the stub is slightly increased $\theta_s = (\theta/2) + \Delta$, then the part of the resonant electrical lengths θ_k'' of dual-mode oscillation decreases. Suppose the value of θ_s is somewhat reduced $\theta_s = (\theta/2) - \Delta$, then the values of θ_k'' increase. These values are shown in Figure 17.5a with dotted lines. The remaining resonant electrical lengths of the resonator are denoted as θ_n'.

Figure 17.5b shows the insertion loss frequency responses of this resonator for weak coupling with I/O loads. These responses have a double-sided TZ at the frequency f_z. The quarter-wave open stub produces odd modes because f_z is located

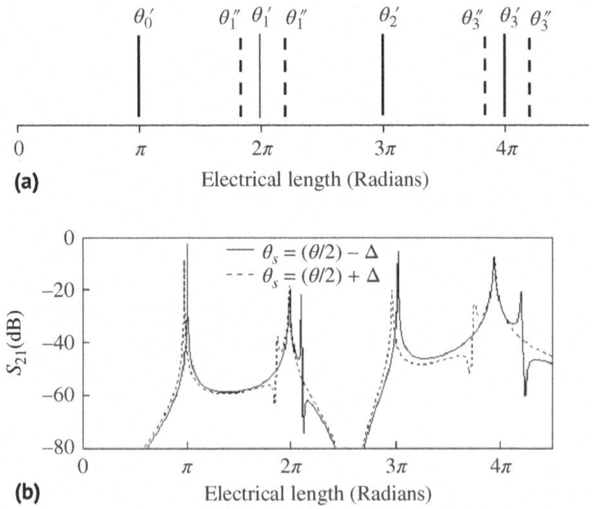

Figure 17.5 Individual characteristics of resonator in Figure 17.4a. (a) Resonance electrical lengths distribution. (b) Frequency responses.

closer to these frequencies. The half-wave section of the resonator under consideration gives rise to even modes of dual-mode oscillations.

The resonator shown in Figure 17.4a, as well as the resonator in Figure 17.1a, in addition to dual-mode oscillations, has simple oscillations. These two oscillation modes alternate with each other. The difference lies in the sequence of these oscillations, which is clearly illustrated in Table 17.1 (positions No. 2 and No. 4).

17.1.2.2 Short-circuited Stub with Small Length

The resonator in Figure 17.4b contains a short-circuit stub of short length $\theta_s \ll \theta/2$. Part of its resonant electrical lengths is still determined by (17.15). The other part of the resonant electrical lengths is determined by equation (17.16) at $\theta_s \ll \theta/2$. The roots of (17.16) are the electrical lengths θ_k'', which are shown in Figure 17.6a with dashed vertical lines. They are always located to the left of θ_n'. The greater the length of the stub and its characteristic impedance Z_s, the further apart the electrical lengths θ_n'' and θ_n' are. The location of the resonant electrical lengths is preserved when replacing the stub with inductance L.

Figure 17.6b shows the insertion loss frequency responses of this resonator for weak coupling with I/O loads.

The resonant frequencies f_n'', caused by the stub, are not located in the vicinity of the transmission zeros, so they are even modes in the dual-mode oscillation. Thus, a short-circuited stub of small length generates even modes in the resonator in Figure 17.4b. Since the condition $f_e \leq f_o$ is fulfilled in this resonator, the value of the coupling coefficient between two modes K (17.9) is always positive.

If the length of the stub in Figure 17.4b tends to zero $\theta_s \to 0$, then the considered resonator degenerates into a resonator, short-circuited in the middle part, in which the frequencies of even and odd modes coincide.

The emergence of dual-mode resonators with all short-circuited ends (Figure 17.4a, b) is due to the use of the resonance equations (17.11)–(17.14). These resonators are practically useful.

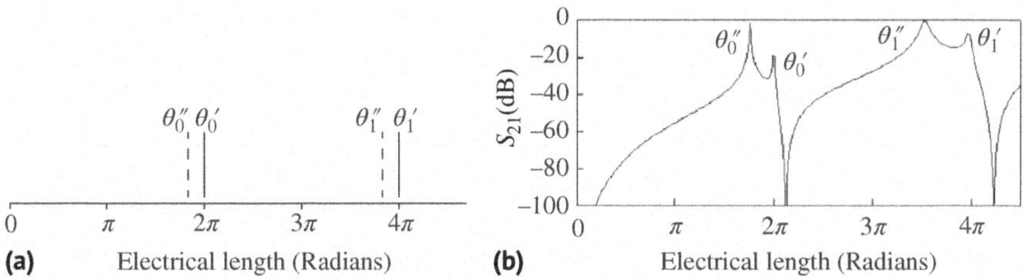

Figure 17.6 Individual characteristics of resonator in Figure 17.4b. (a) Resonance electrical lengths distribution. (b) Frequency responses.

17.1.3 Ring Resonator

The ring resonator No. 6 (Table 17.1) can be considered as the limiting case of the loop resonator with capacitance C_1 (Figure 15.19a), when $C_1 \to \infty$. If the capacitance C_1 increases, the resonant electrical length θ_n, which is an odd number (Figure 15.19d), remains stationary, and θ_n with an even number approaches them from the right. At $C_1 \to \infty$, the resonant electrical length θ_0 vanishes, while the remaining θ_n with even numbers become equal in pairs to the θ_n with odd numbers. This case corresponds to a ring resonator (Table 17.1, No. 6), where two-mode oscillations occur.

To obtain an expression for the input admittance Y of a ring resonator, we use a half-wave section of TL and connect its two open ends. In this case, the function Y is determined by formula (15.52), in which the parameters of the transfer matrix are well known

$$\begin{bmatrix} A & B \\ C & D \end{bmatrix} = \begin{bmatrix} \cos(\theta) & jZ_0 \sin(\theta) \\ jZ_0^{-1} \sin(\theta) & \cos(\theta) \end{bmatrix}. \tag{17.18}$$

Substituting the elements of matrix (17.18) into (15.52) gives

$$Y = \frac{2\cos\theta - 2}{jZ_0 \sin\theta} = 2jZ_0^{-1} \tan\left(\frac{\theta}{2}\right). \tag{17.19}$$

Let us note the characteristic features of dual-mode resonators in Table 17.1:

1. In resonators with a quarter-wave stub (No. 2 and No. 4), alternation of dual-mode and single oscillations is observed.
2. In a resonator with short stub in the middle (No. 3 and No. 5), only dual-mode oscillations take place.
3. In resonators with open-circuited ends (No. 2 and No. 3), dual-mode oscillations correspond to electrical lengths $n\pi$ with odd values of n.
4. In resonators with short-circuited ends (No. 4 and No. 5), dual-mode oscillations correspond to electrical lengths $n\pi$ with even values of n.
5. Quarter-wave open-circuited stub of resonators (No. 2 and No. 4) forms odd modes f_o of dual-mode oscillations, and TZ of these resonators could be left-handed and right-handed.
6. Short-circuited stub with small electric length of resonators (No. 3 and No. 5) forms even modes f_e of dual-mode oscillations, and TZ of these resonators could be right-handed only.
7. The dual-mode oscillations of the ring resonator (No. 6) and resonator (No. 5) are identical.

17.1.4 Microstrip Dual-mode BPFs Short-circuited in the Middle

Let us compare two dual-mode resonators, No. 3 and No. 5 (Table 17.1), which have short stub in the middle. Figure 17.7a, b shows the topology of microstrip dual-mode resonators short-circuited in the middle, which replaces the short stub. The dimensions are in millimeters. In the microstrip resonators, a substrate RT/Duroid 6010LM (Rogers) with $\varepsilon_r = 10.2$, $h = 1.905$ mm, and $\tan\delta = 0.002$ is used. The width of the microstrip line of the hairpin resonator is $w = 4.5$ mm $(Z_0 = 29.3\ \Omega)$.

The resonators No. 3 and No. 5 (Table 17.1) were considered earlier, and they have now adopted a hairpin shape. A short circuit in the middle of the resonators is provided by metallized holes with a diameter of 1 mm. The diameter of metallized holes in the ends of the resonator in Figure 17.7b is 0.6 mm. The load connection coordinate controls the external quality factor Q_e.

Figure 17.7c shows the simulated frequency responses of a dual-mode BPF with two open-circuited ends. The filter is characterized by the following data: center frequency $f_0 = 0.825$ GHz, bandwidth BW = 105 MHz (FBW = 0.127), insertion loss at the center frequency $L_{A0} = 0.6$ dB, and return loss $L_R > 15$ dB.

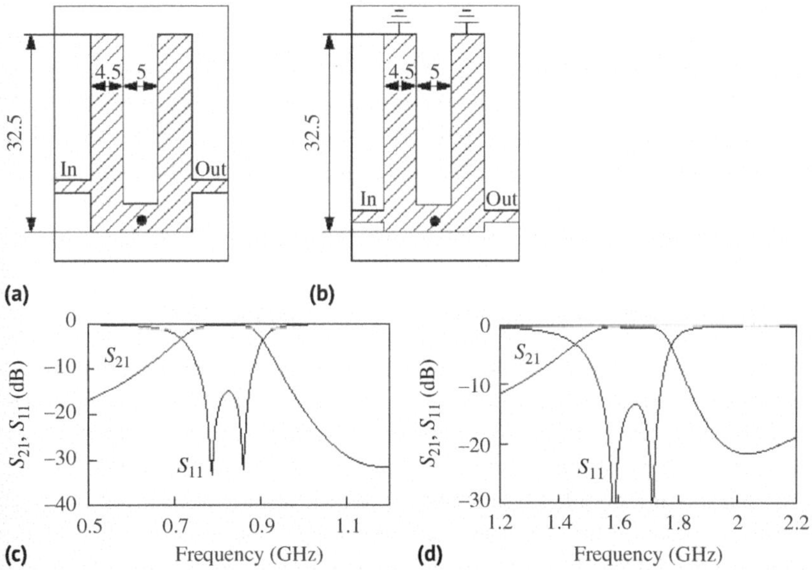

Figure 17.7 Microstrip dual-mode BPFs short-circuited in the middle. (a) Topology of BPF with two open-circuited ends. (b) Topology of BPF with two short-circuited ends. (c) Simulated frequency responses of BPF with two open-circuited ends. (d) Simulated frequency responses of BPF with two short-circuited ends.

Figure 17.7d shows the simulated frequency responses of BPF with two short-circuited ends. The filter has the following parameters: $f_0 = 1.64$ GHz, BW = 200 MHz (FBW = 0.122), $L_{A0} = 0.5$ dB, and $L_R > 13$ dB. The center frequency f_0 of this filter is about twice the center frequency of the BPF with two open ends, which can be useful at higher frequencies. This is consistent with the patterns shown in Table 17.1 (No. 3 and No. 5). The right slope of both filters is steeper than the left due to the right-handed TZ (Figures 17.3b and 17.6b).

17.2 Dual-mode Resonators with SILS

Dual-mode resonators with uniform TL segments (Table 17.1) are characterized by limited functionality due to the equidistant arrangement of resonant frequencies. If stepped-impedance transmission line segments (SILS) with the required arrangement of resonant frequencies are used in dual-mode resonators, this will lead to an expansion of their functionality. Optimal synthesis of SILS can be implemented based on the parametric synthesis method discussed in Section 5.6. The optimal synthesis means that we obtain an exceptional ratio between the resonance frequencies of structural elements and have the minimum value of parameter $m = Z_{0\max}/Z_{0\min}$. It leads to SILS that have the resonant frequencies with the required placement and $m = m_{\min}$. The formulated problem was solved in Ref. [15]. Following is a description of the solutions obtained. Traditionally, SILS synthesis is performed using the input impedance Z [16], the poles of which f_{pn} are resonant frequencies, and the zeros f_{0n} are antiresonant frequencies. For brevity, we will denote the resonant frequencies f_{pn} as f_n. In the following section, we will adhere to the accepted terminology, which should not be misleading.

17.2.1 Dual-mode Resonators with Open-circuited Ends SILS

Dual-mode resonators No. 2 and No. 3 (Table 17.1) contain a uniform half-wave section with open ends, which limits their capabilities. The fundamental dual-mode oscillation of these resonators with a resonant frequency f_1 corresponds to the electrical length of the half-wave resonance $\theta_1 = \pi$. Resonator No. 2 has a simple oscillation at a doubled frequency $2f_1$, which limits the width of the BPF rejection band based on this resonator. Resonator No. 3 has a second dual-mode oscillation at a tripled frequency $3f_1$, which also limits the width of the BPF rejection band. To expand the rejection band, it is necessary to remove the specified frequencies from the fundamental dual-mode oscillation. In addition, the lengths of resonators No. 2 and No. 3 are quite large, and it would be desirable to reduce them.

17.2.1.1 SILS-1 with Open-circuited Ends and Increased Ratio f_2/f_1

The synthesized SILS-1, as shown in Figure 17.8a and Table 5.7 (No. 1), has an increased ratio of the two resonant frequencies f_2/f_1.

A third-order reactance function describes its input impedance

$$Z^3(S) = Z_0 \frac{1 + (1 + m + 1/m)\, S^2}{(2 + 1/m)\, S + mS^3}, \tag{17.20}$$

where $S = j \tan(2\theta/3)$ is frequency variable. And its critical frequencies are shown in Figure 17.8b. The frequency ω_{01}° in this figure is the fundamental antiresonant frequency of the half-wave resonator. It corresponds to the electrical length $\theta_{01}^{0} = \pi/2$. The resonant frequencies of the resonator are determined by the zeros of the denominator (17.20), from which it follows

$$\frac{f_2}{f_1} = \frac{\pi}{\tan^{-1}\sqrt{2/m + 1/m^2}} - 1. \tag{17.21}$$

The graph of the dependence f_2/f_1 versus the parameter m (17.21) is shown in Figure 17.8c. For $m = 5$, equation (17.21) gives $f_2/f_1 = 4.36$, and we have $f_2/f_1 = 2$ for a uniform segment.

The parameter

$$\eta = \frac{f_1}{f_1^0} = \frac{2\theta}{\pi} \tag{17.22}$$

is important. This parameter shows the shift of the resonant frequency f_1 of SILS-1 with respect to the same resonant frequency of a uniform half-wave segment f_1^0. If $\eta < 1$, then SILS-1 is shorter than the half-wave segment when their

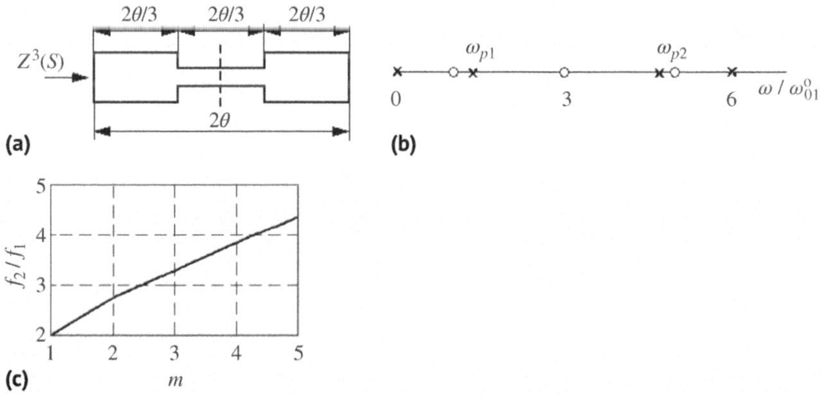

Figure 17.8 SILS-1 with increased ratio f_2/f_1 of resonance frequencies. (a) Topology. (b) Distribution of critical frequencies. (c) Ratio $f_2/f_1\,(m)$.

resonant frequencies coincide. If $\eta > 1$, then SILS-1 has a higher operating frequency than the half-wave segment when their lengths coincide.

From (17.20), we obtain for the SILS-1 under investigation

$$\eta = \frac{3}{\pi} \tan^{-1} \sqrt{\frac{2}{m} + \frac{1}{m^2}}. \tag{17.23}$$

If $m = 5$, then from (17.23) we have the shortening degree of the SILS-1: $\eta = 0.559$.

In Figure 17.8a, the vertical line divides SILS-1 into two equal parts. The transfer matrix of its left half is written as

$$\begin{bmatrix} a & b \\ c & d \end{bmatrix} = \begin{bmatrix} 1 + (1 + 2m) S^2 & Z_0 \left[(2 + m) S + mS^3 \right] \\ Z_0^{-1} \left[\left(2 + \frac{1}{m} \right) S + \left(\frac{1}{m} \right) S^3 \right] & 1 + \left(1 + \frac{2}{m} \right) S^2 \end{bmatrix}, \tag{17.24}$$

where $S = j \tan(\theta/3)$.

17.2.1.2 SILS-2 with Open-circuited Ends and Increased Ratio f_3/f_1

The synthesized SILS-2 has an increased ratio of the two resonant frequencies f_3/f_1. This is illustrated in Figure 17.9a and is also listed in Table 5.7 (position No. 6). A fourth-order reactance function describes its input impedance

$$Z^4(S) = Z_0 \frac{1 + 2(1 + m + 1/m) S^2 + S^4}{2(1 + 1/m) S + 2(1 + m) S^3}, \tag{17.25}$$

where $S = j \tan(\theta/2)$ is the frequency variable. The critical frequencies of the resonator are shown in Figure 17.9b, where the normalizing frequency ω_{01}° has the same meaning as before.

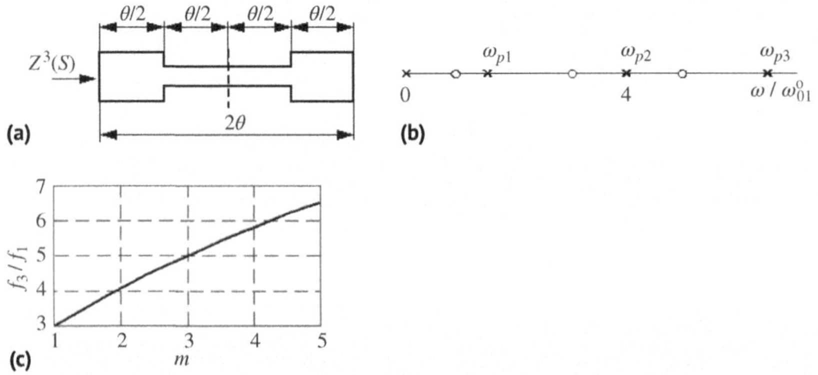

Figure 17.9 SILS-2 with increased ratio f_3/f_1 of resonance frequencies. (a) Topology. (b) Distribution of critical frequencies. (c) Ratio $f_3/f_1\,(m)$.

By equating the denominator (17.25) to zero, we determine the resonant frequencies of the resonators and the relationship of f_3/f_1 versus parameter m

$$\frac{f_3}{f_1} = \frac{\pi}{\tan^{-1}\sqrt{1/m}} - 1. \tag{17.26}$$

The relationship (17.26) is shown in Figure 17.9c. For $m = 5$, expression (17.26) gives $f_3/f_1 = 6.47$, and for a uniform segment, we have $f_3/f_1 = 3$.

It follows from (17.25) that the length shortening parameter η (17.22) has the form

$$\eta = \frac{4}{\pi}\tan^{-1}\sqrt{\frac{1}{m}}. \tag{17.27}$$

If $m = 5$, then from (17.27), we obtain the shortening degree of the SILS-2: $\eta = 0.535$.

In Figure 17.9a, a vertical line divides SILS-2 into two equal parts. The transfer matrix of its left half is written as

$$\begin{bmatrix} a & b \\ c & d \end{bmatrix} = \begin{bmatrix} 1 + \left(\frac{1}{m}\right) S^2 & Z_0\,(1+m)\,S \\ Z_0^{-1}\left(1 + \frac{1}{m}\right) S & 1 + mS^2 \end{bmatrix}, \tag{17.28}$$

where $S = j\tan(\theta/2)$.

17.2.1.3 Dual-mode Resonator with SILS-1

This resonator is shown in Figure 17.10a. Its resonant frequencies are determined by equations (17.3). Taking into account the elements of the transfer matrix (17.24), these equations take the form

$$d = 1 + \left(1 + \frac{2}{m}\right) S^2 = 0, \tag{17.29a}$$

$$Y_s\,(S^*) + \frac{2c}{d} = 0, \tag{17.29b}$$

where

$$\frac{c}{d} = Z_0^{-1}\frac{(2 + 1/m)\,S + (1/m)\,S^3}{1 + (1 + 2/m)\,S^2} \tag{17.30}$$

and $S = j\tan(\theta/3)$. The admittance of the stub $Y_s\,(S^*)$ is expressed by the reactance function $Y_s\,(S^*) = c/d$ (17.30) with frequency variable $S^* = j\tan[(\theta \pm \Delta')/3]$.

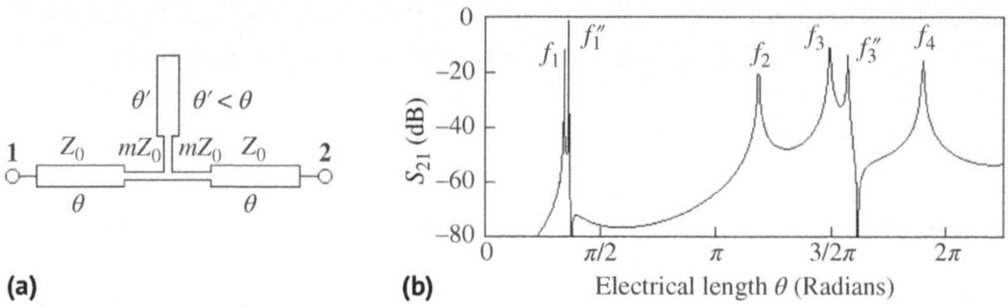

Figure 17.10 Dual-mode resonator with SILS-1 and $\theta' < \theta$. (a) Topology. (b) Resonance frequencies.

Figure 17.10b shows the insertion loss frequency responses of the considered dual-mode resonator for the case $\theta' = \theta - \Delta'$ and weak coupling with I/O loads. When plotting, we assumed $Z_0 = 10\ \Omega$ and $m = 3$.

The resonance equation (17.29a) describes the oscillations with odd numbers $f_1, f_3, \ldots$. The resonance equation (17.29b) describes the oscillations with even numbers $f_2, f_4, \ldots$, as well as right-hand oscillations, with odd numbers f_1'' and f_3'', which form dual-mode oscillations. This resonator features dual-mode and single oscillations that alternate with each other.

We use SILS-1 to distance the simple oscillation f_2 from the lowest dual-mode oscillation. The distancing of these frequencies is described by the ratio $f_2/f_1 = 3.35$ for $m = 3$. This value aligns with the result obtained by equation (17.21). So, for $m = 3$ and $m = 1$, we have $f_1 = 0.69$ GHz and $f_1^0 = 1$ GHz, respectively. As a result, using SILS-1 reduces the main resonance ($\eta = 0.69$), which coincides with the result calculated by (17.23).

17.2.1.4 Dual-mode Resonator with SILS-2

This resonator is shown in Figure 17.11a, along with SILS-2. It contains a short-circuited stub with a small electrical length ($\Delta' \ll \theta$), with input admittance $Y_s = -jZ_0^{-1}\cot\Delta'$. The resonant frequencies of this resonator are determined by (17.3). Taking into account the elements of the transfer matrix (17.28), these equations take the form:

$$d = 1 + mS^2 = 0, \tag{17.31a}$$

$$-Z_0^{-1}\cot\Delta' + \frac{2c}{d} = 0, \tag{17.31b}$$

where

$$\frac{c}{d} = Z_0^{-1}\frac{\left(1 + \frac{1}{m}\right)S}{1 + mS^2}, \tag{17.32}$$

and $S = j\tan\left(\dfrac{\theta}{2}\right)$.

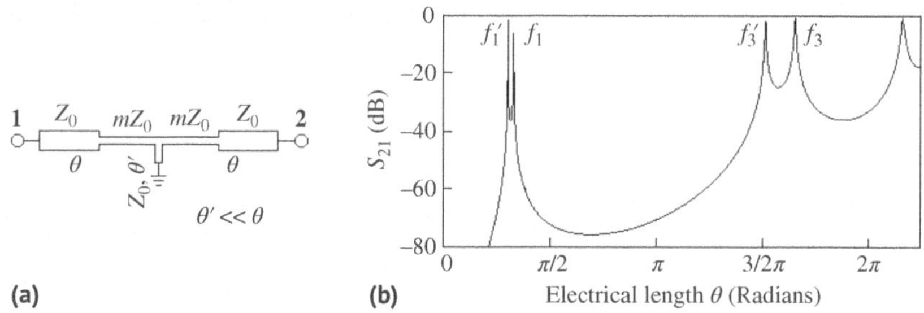

Figure 17.11 Dual-mode resonator with SILS-2 and short-circuited stub of short length. (a) Topology. (b) Resonance frequencies.

The insertion loss of this resonator at $m = 3$ is presented in Figure 17.11b. The resonance equation (17.31a) describes the oscillations with odd numbers $f_1, f_3, \ldots$. The resonance equation (17.31b) describes the left-handed oscillations with odd numbers, which form dual-mode oscillations. This resonator has only dual-mode oscillations.

For $m = 3$, we have $f_1 = 0.667$ GHz and if $m = 1$, then $f_1^0 = 1$ GHz. We use SILS-2 to reduce the main resonance ($\eta = 0.667$), which coincides with the result by (17.27). We use SILS-2 to distance both dual-mode oscillations from each other. The ratio f_3/f_1 at $m = 5$ is equal to $f_3/f_1 = 6.47$ (17.26). If in SILS-2 the characteristic impedances Z_0 and mZ_0 are swapped, then in (17.26) we need to replace m by $1/m$. Then we obtain $f_3/f_1 = 1.73$ at $m = 5$. The passband with frequency f_3 has significantly approached the bandwidth with frequency f_1. In this filter, the f_3/f_1 ratio is varied over a wide range $1.73 \leq f_3/f_1 \leq 6.47$. This allows using this filter as a dual-band filter with tunable positions of the first and second passbands.

17.2.2 Dual-mode Resonators with Short-circuited Ends SILS

Dual-mode resonators No. 4 and No. 5 (Table 17.1), also shown in Figure 17.4, contain a uniform half-wave section with short-circuited ends, which limits their capabilities. For the resonator in Figure 17.4a, it is necessary to increase the ratio of resonant frequencies $f_3/f_1 = 3$, and for the resonator in Figure 17.4b, it is necessary to increase the ratio $f_4/f_2 = 2$. These shortcomings are due to a horizontal uniform TL segment with an electrical length 2θ, and it must be replaced.

To solve the created problems, we need to synthesize a SILS that has resonant frequencies at a specific placement. The main difficulty is that both ends of the

SILS are short-circuited, and it does not have an input similar to SILS with two open ends or SILS with one open and one short-circuited end. The design of SILS is carried out based on the input impedance $Z(s)$, which is not applicable in this case. What to do in this case?

We will solve this problem using certain statements from the theory of nonuniform TLs [17]. Let us take a segment of an inhomogeneous TL with a characteristic impedance $Z_0(s)$ that is open at the ends. It is characterized by a set of resonant frequencies $\{f_{pi}\}$. If we change the impedance $Z_0(s)$ according to the formula $Z_0^*(x) = Z^2/Z_0(x)$ and simultaneously change the boundary conditions at the ends to short-circuited ones, then the set of resonant frequencies $\{f_{pi}\}$ will not change. This approach allows us to synthesize SILS with two open ends and then go to the variant with two short-circuited ends by exchanging Z_0 and mZ_0.

17.2.2.1 SILS-3 with Short-circuited Ends and Increased Ratio f_3/f_1

For a dual-mode resonator (Figure 17.4a), it is required to increase the resonant frequency ratio f_3/f_1. This problem has already been solved when using SILS-2 with two open ends (Figure 17.9a). A line segment with electrical length 2θ is divided into four equal parts, the characteristic impedances of which are given by $Z_{01} = Z_{04} = Z_{0\min} = Z_0$ and $Z_{02} = Z_{03} = Z_{0\max} = mZ_0$. We will use this solution for a line segment that is short-circuited at both ends. Following the pattern [17], let us reverse the indicated characteristic impedances with respect to the value of mZ_0^2. As a result, we get the desired solution

$$\begin{aligned} Z_{01} &= Z_{04} = Z_{0\max} = mZ_0, \\ Z_{02} &= Z_{03} = Z_{0\min} = Z_0, \end{aligned} \tag{17.33}$$

which is shown in Figure 17.12a and which corresponds to the distribution of resonant frequencies in Figure 17.12b.

The dependence of the ratio f_3/f_1 on parameter m of the synthesized SILS-3 (17.33) with two short-circuited ends is determined by (17.26). This dependence is shown in Figure 17.9c.

In Figure 17.12a, the vertical dashed line divides SILS-3 into two equal parts. The transfer matrix of its left half has the form:

$$\begin{bmatrix} a & b \\ c & d \end{bmatrix} = \begin{bmatrix} 1 + mS^2 & Z_0 (1+m) S \\ Z_0^{-1} \left(1 + \dfrac{1}{m}\right) S & 1 + \left(\dfrac{1}{m}\right) S^2 \end{bmatrix}, \tag{17.34}$$

where $S = j \tan(\theta/2)$.

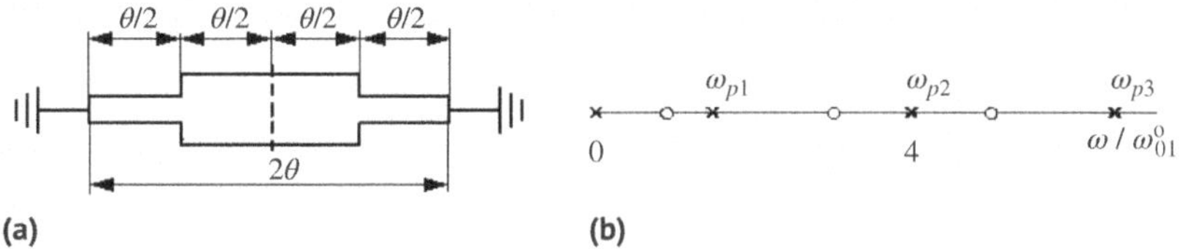

Figure 17.12 SILS-3 with increased ratio f_3/f_1 of resonance frequencies. (a) Topology. (b) Distribution of resonance frequencies.

17.2.2.2 SILS-4 with Short-circuited Ends and Increased Ratio f_4/f_2

For a dual-mode resonator (Figure 17.4b), it is required to increase the resonant frequency ratio f_4/f_2 (Table 17.1). To solve it, at the first step, we use a SILS with two open ends. The reactance function $Z^N(S)$ should be used with the minimum order of $N_{\min} = 3$, and SILS will have characteristic impedances

$$\begin{aligned} Z_{01} &= Z_{03} = Z_{0\max} = mZ_0, \\ Z_{02} &= Z_{0\min} = Z_0. \end{aligned} \tag{17.35}$$

For SILS (17.35) with two open ends, the input impedance is written as

$$Z^3(S) = Z_0 \frac{1 + (1 + m + 1/m)\, S^2}{(1 + 2/m)\, S + (1/m^2)\, S^3}. \tag{17.36}$$

From (17.36) the ratio of resonance frequencies f_4/f_2 follows as the function of parameter m

$$\frac{f_4}{f_2} = \frac{\pi + \tan^{-1}\sqrt{m\,(m+2)}}{\pi - \tan^{-1}\sqrt{m\,(m+2)}}. \tag{17.37}$$

Let us reverse the indicated characteristic impedances (17.35) of SILS with two open ends with respect to the value of mZ_0^2. As a result, we obtain the desired solution for the SILS with two short-circuited ends

$$\begin{aligned} Z_{01} &= Z_{03} = Z_{0\min} = Z_0, \\ Z_{02} &= Z_{0\max} = mZ_0. \end{aligned} \tag{17.38}$$

which is shown in Figure 17.13a and corresponds to the distribution of resonant frequencies (Figure 17.13b).

The resonance frequency ratio f_4/f_2 of SILS-4 (17.38) is determined by (17.37), and plotted in Figure 17.13c. If $m = 5$, then from (17.37) we obtain $f_4/f_2 = 2.61$. The uniform resonator has a ratio of $f_4/f_2 = 2$.

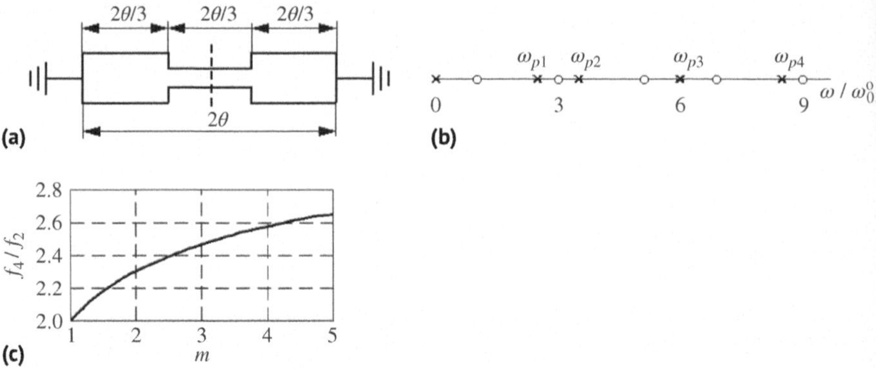

Figure 17.13 SILS-4 with increased ratio f_4/f_2 of resonance frequencies. (a) Topology. (b) Distribution of resonance frequencies. (c) Ratio $f_4/f_2\,(m)$.

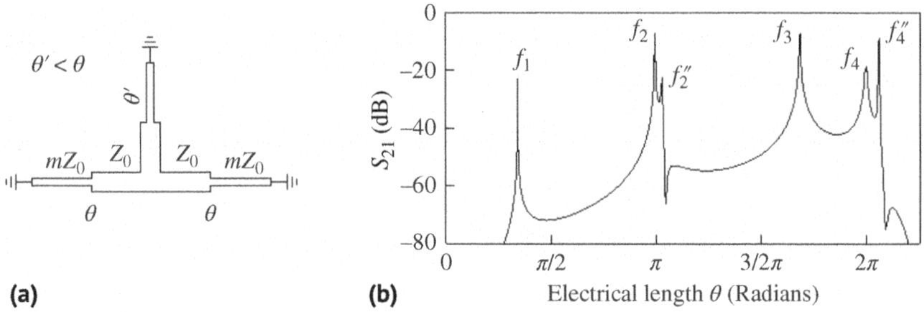

Figure 17.14 Dual-mode resonator with SILS-3 and $\theta' < \theta$. (a) Topology. (b) Resonance frequencies.

17.2.2.3 Dual-mode Resonator with SILS-3

This resonator is shown in Figure 17.14a. Its resonant frequencies are determined by equations (17.13). Taking into account the elements of the transfer matrix (17.34), these equations take the form

$$b = Z_0\,(1 + m)\,S = 0, \tag{17.39a}$$

$$Y_s\,(S^*) + \frac{2a}{b} = 0, \tag{17.39b}$$

where

$$\frac{a}{b} = Z_0^{-1}\frac{1\;+\;mS^2}{(1 + m)\,S}, \tag{17.40}$$

and $S = j\tan(\theta/2)$. The admittance of the stub $Y_s(S^*)$ is expressed by the reactance function $Y_s(S^*) = a/b$ (17.40) with frequency variable $S^* = j\tan[(\theta \pm \Delta')/2]$.

The insertion loss of the considered dual-mode resonator for reduced electrical length $\theta' = \theta - \Delta'$ of the stub is presented in Figure 17.14b. When plotting, we assumed $Z_0 = 10\ \Omega$ and $m = 3$. The resonant electrical lengths with even numbers $\theta_n = \pi,\ 2\pi,\ 3\pi, \ldots$ are determined by (17.39a). These resonant electrical lengths correspond to the even resonant frequencies $f_2, f_4, \ldots$, which do not depend on the parameter m.

The resonance equation (17.39b) describes simple oscillations with odd numbers $f_1, f_3, \ldots$, as well as right-hand oscillations f_2'', f_4'', which form dual-mode oscillations. This resonator features dual-mode and single oscillations that alternate with each other. If the stub has an increased length $\theta' = \theta + \Delta'$, then resonant frequencies $f_2', f_4', \ldots$ will be left-handed.

Using SILS-3 allows us to distance simple oscillation with resonant frequencies f_1 and f_3 from dual-mode oscillation with frequency f_2. The distancing of these frequencies from each other is determined by f_3/f_1 (17.26). For $m = 5$ we have the value $f_3/f_1 = 6.47$, and if $m = 1$, then $f_3/f_1 = 3$.

17.2.2.4 Dual-mode Resonator with SILS-4

This resonator is shown in Figure 17.15a. It contains a short-circuited stub with a small electrical length $\Delta' \ll \theta$, which has an input admittance $Y_s = -j(mZ_0)^{-1}\cot\Delta'$. Its resonant frequencies are determined by equations (17.13). Taking into account the elements of its transfer matrix (17.24), these equations take the form

$$b = Z_0\left[(2+m)S + mS^3\right] = 0, \tag{17.41a}$$

$$-(mZ_0)^{-1}\cot\Delta' + \frac{2a}{b} = 0, \tag{17.41b}$$

where $a/b = Z_0^{-1}\dfrac{1+(1+2m)S^2}{(2+m)S + mS^3}$ and $S = j\tan\left(\dfrac{\theta}{3}\right)$.

The insertion loss of this resonator at $m = 3$ and weak coupling with I/O loads is presented in Figure 17.15b. Equation (17.41a) defines the resonant electrical lengths with even numbers, which correspond to the resonant frequencies $f_2, f_4, \ldots$. These frequencies depend on parameter m of SILS-4. The resonance equation (17.41b) generates left-handed oscillations $f_2'', f_4'', \ldots$, adjacent to oscillations defined by equation (17.41a), which leads to dual-mode oscillations. This resonator has only dual-mode oscillations.

The resonant electrical length of its lowest dual-mode oscillation is

$$\theta_2 = \frac{3}{2}\left[\pi - \tan^{-1}\sqrt{m(m+2)}\right].$$

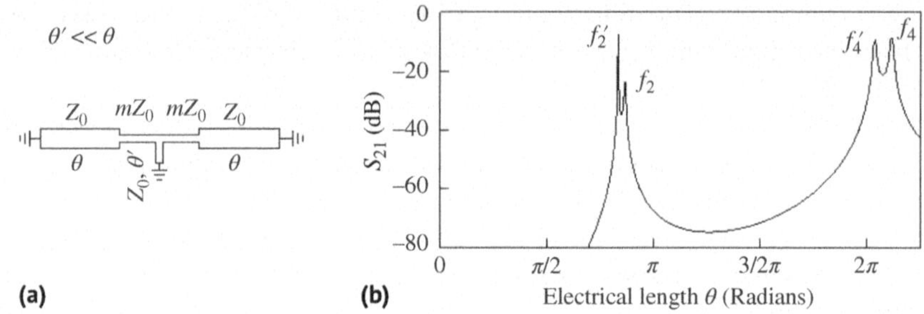

Figure 17.15 Dual-mode resonator with SILS-4 and short-circuited stub of short length. (a) Topology. (b) Resonance frequencies.

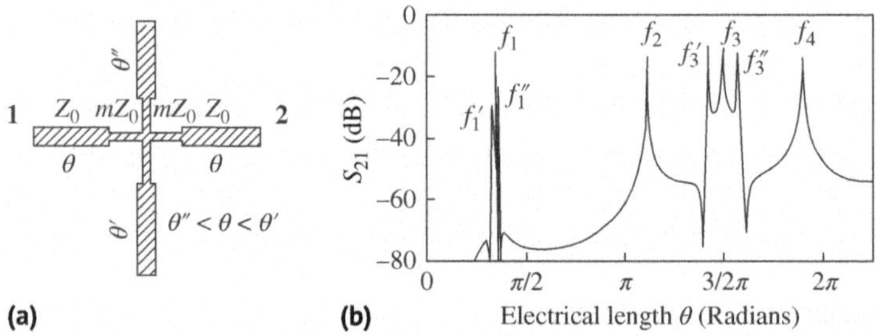

Figure 17.16 Triple-mode resonator with SILS-1. (a) Topology. (b) Resonance frequencies (increased ratio f_2/f_1, reduced f_1).

The parameter η (17.22) in this case is written as

$$\eta = \frac{3}{\pi}\left[\pi - \tan^{-1}\sqrt{m\,(m+2)}\right]. \tag{17.42}$$

From (17.42) we obtain $\eta = 1.66$ for $m = 5$. Thus, the lowest dual-mode oscillation of the SILS (Figure 17.15a) exceeds the fundamental oscillation of a uniform half-wave resonator by a factor of 1.66.

We use the SILS-4 to distance dual-mode oscillations f_2 and f_4 from each other. For $m = 5$, the f_4/f_2 ratio is $f_4/f_2 = 2.61$ (17.37).

If characteristic impedances Z_0 and mZ_0 are interchanged in SILS-4, then in (17.37) we need to replace m by $1/m$. Then we obtain $f_4/f_2 = 1.46$ and $\eta = 2.44$ at $m = 5$. In the discussed filter, the ratio f_4/f_2 has an extensive range, $1.46 \leq f_4/f_2 \leq 2.61$. As a result, we can use it as a dual-band filter with tunable placement of two passbands.

The synthesized SILS-1...SILS-4 can also be used in resonators with triple-mode oscillation [18, 19]. The triple-mode resonators with SILS and their resonant frequencies are illustrated in Figures 17.16–17.19, which are self-explanatory.

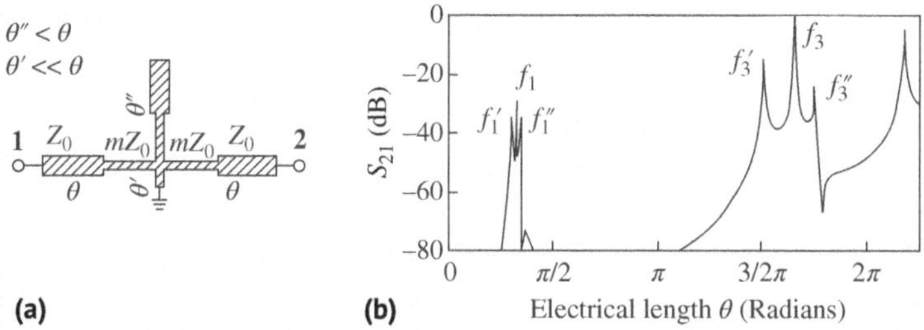

Figure 17.17 Triple-mode resonators with SILS-2 and short-circuited stub of short length. (a) Topology. (b) Resonance frequencies (increased ratio f_3/f_1, reduced f_1).

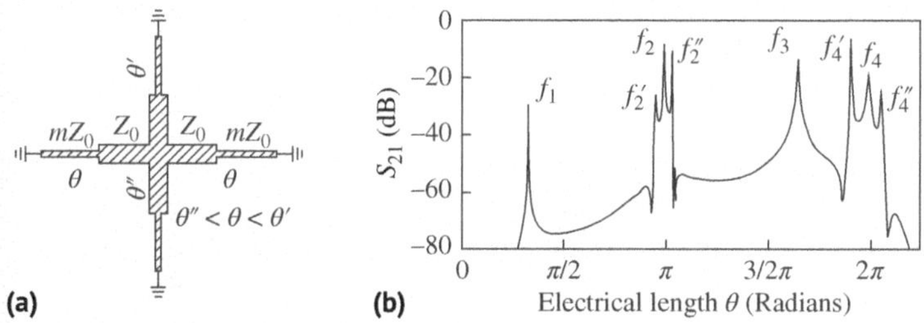

Figure 17.18 Triple-mode resonator with SILS-3. (a) Topology. (b) Resonance frequencies (increased ratio f_3/f_1).

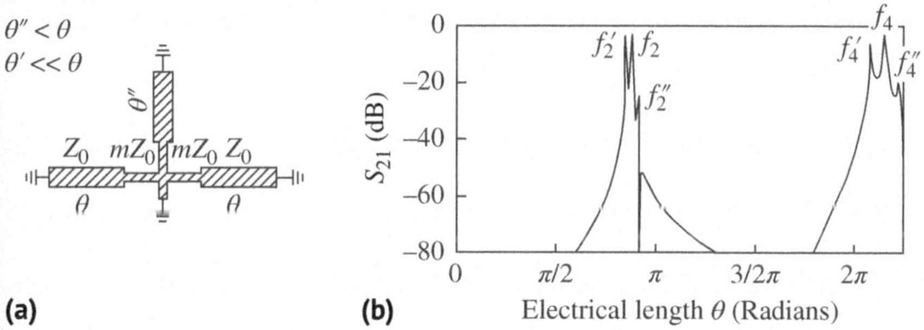

Figure 17.19 Triple-mode resonators with SILS-4 and short-circuited stub of short length. (a) Topology. (b) Resonance frequencies (increased ratio f_4/f_2, increased f_2).

The resonators in Figures 17.16 and 17.18 have triple-mode and simple oscillations, which alternate with each other. All oscillations of the resonators in Figures 17.17 and 17.19 are triple-mode.

17.2.3 Microstrip Dual-mode BPFs with SILS

We consider two dual-mode filters with SILS (Figures 17.20 and 17.21). These filters are implemented on a substrate of TMM-10i (Rogers), with $\varepsilon_r = 9.8$, $\tan\delta = 0.002$, and $h = 1.905$ mm. Both filters contain structural elements with a maximum and minimum width of 3.2 mm ($Z_{0\min} = 37.1\ \Omega$) and 0.8 mm ($Z_{0\max} = 70.8\ \Omega$), respectively. Then we obtain $m = 1.91$. The coupling coefficient K is obtained by using EM simulation in the Microwave Office and using (17.9).

17.2.3.1 Filter with SILS-1

The filter's topology is depicted in Figure 17.20a. All dimensions are in millimeters, and the filter measures 34 mm × 25 mm. The photograph and measured and simulated frequency responses are shown in Figure 17.20b, c, respectively.

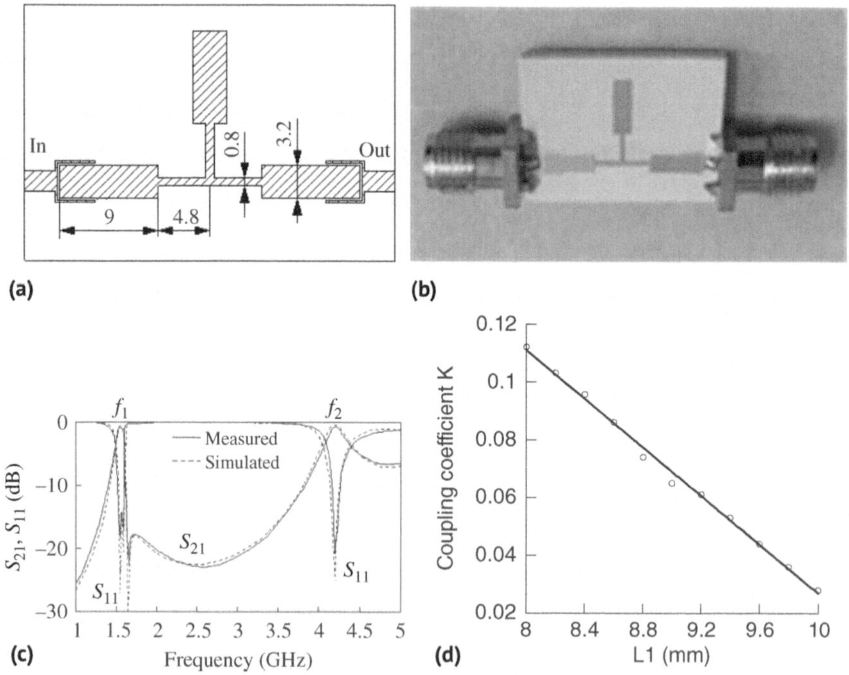

Figure 17.20 Microstrip dual-mode resonator with SILS-1. (a) Topology. (b) Photograph of fabricated filter. *Source*: [15] / with permission of IEEE. (c) Measured and simulated frequency responses. (d) Simulated coupling coefficient.

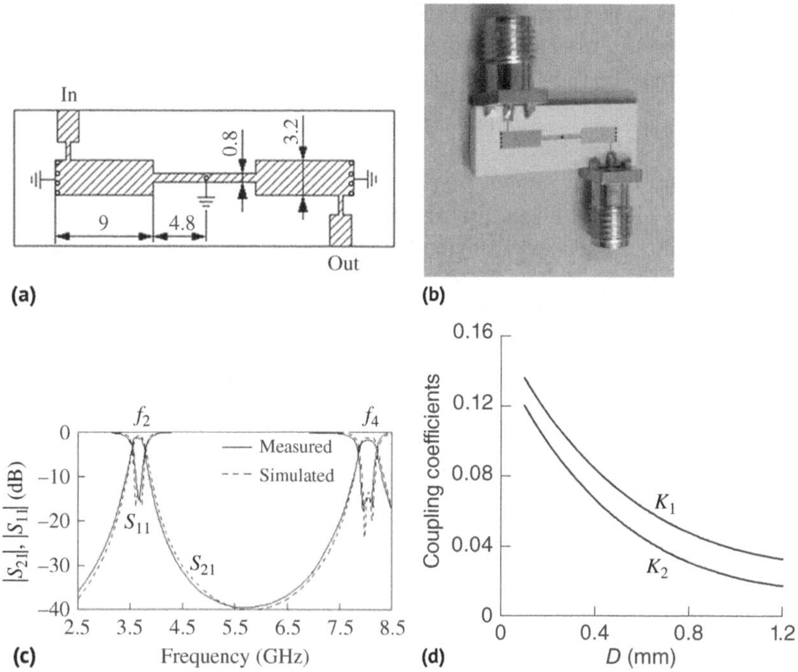

Figure 17.21 Microstrip dual-mode resonators with SILS-4. (a) Topology. (b) Photograph of fabricated filter. *Source*: [15] / with permission of IEEE. (c) Measured and simulated frequency responses. (d) Simulated coupling coefficient.

Then, for the given parameters, we obtain the coupling coefficient $K = 0.065$. Figure 17.20d shows the simulated coupling coefficient versus the length L_1 of the low-impedance part of the stub.

The measured parameters are as follows: center frequency $f_1 = 1575$ MHz; BW = 80 MHz (FBW = 0.051); minimum $L_{A0} = 0.7$ dB, $L_R > 14$ dB. We use the shorter stub in the filter, having electrical length θ' ($\theta' < \theta$), which forms the right-hand transmission zero $f_z = 1647$ MHz. If we use a longer stub ($\theta' > \theta$), then transmission zero f_z will shift to the left of the bandpass.

We have the closest parasitic bandpass at $f_2 = 4.2$ GHz due to the simple oscillation. From measurements, we can observe that both dual-mode and simple oscillations coexist in the filter prototype simultaneously. This filter has an extended stopband compared to the BPF on uniform segments due to an increased $f_2/f_1 = 2.67$ ratio. Equation (17.21) yields the same value at $m = 1.91$.

The horizontal segment of the resonator has length $L = 27.6$ mm (SILS-1). At the same time, the uniform TL with the same width of 3.2 mm ($Z_0 = 37.1\ \Omega$) has a length of $\lambda_0/2 = 36$ mm for the horizontal segment. As we can see, the length of SILS-1 turned out to be shorter by $\eta = 0.77$ $(m = 1.91)$. The capacitances at the input and output of the filter contributed to this shortening. Using (17.23), we obtain $\eta = 0.81$ at $m = 1.91$.

17.2.3.2 Filter with SILS-4

The topology of the filter is shown in Figure 17.21a. It occupies a 38.2 mm × 12.8 mm substrate. The metallized holes, with diameters of 0.6 mm and 0.4 mm, are made in the central part and at the ends of the filter, respectively. The photograph and measured and simulated frequency responses of the prototype filter are presented in Figure 17.21b, c, respectively. This filter can be used as a dual-band filter. With the specified parameters, we obtain the coupling coefficients $K_1 = 0.048$ and $K_2 = 0.031$ in the first and second passbands. Figure 17.21d shows the simulated coupling coefficient versus the diameter D of the metallized hole.

The measured parameters in the first passband are center frequency $f_1 = 3.64$ GHz; fractional bandwidth by level 3 dB is $\text{FBW}_{3\,\text{dB}} = 0.07$, minimum $L_{A0} = 0.8$ dB, and $L_R > 13$ dB. The measured data in the second passband are as follows: center frequency $f_2 = 8.06$ GHz, $\text{FBW}_{3\,\text{dB}} = 0.045$, minimum $L_{A0} = 1.2$ dB, and $L_R = 16$ dB.

The measured ratio $f_2/f_1 = 2.21$ for center frequencies of two passbands. At the same time, the same calculated value using (17.37) is equal to $f_2/f_1 = 2.26$ at $m = 1.91$. The difference between these values is 2%. This two-band filter is distinguished by the fact that we can define the relative position of the two passbands $(1.46 \leq f_2/f_1 \leq 2.61)$ by changing m. Additionally, the operating frequency of the proposed filter has been shifted to higher frequencies. Its central frequency $f_1 = 3.64$ GHz of the lower passband is increased by 1.83 times (17.42) compared to the half-wave resonator.

17.2.4 Properties of Dual-mode Resonators with SILS

The properties of dual-mode resonators with stepped-impedance TL segments are systematized in Table 17.2. The table uses continuous numbering of resonant frequencies.

Table 17.2 Properties of dual-mode resonators with SILS.

SILS	Single mode	Dual-band	Ratio f_2/f_1 or f_3/f_1	η
SILS-1	Yes	No	$2 < f_2/f_1 \leq 4.36$	$0.558 \leq \eta \leq 1$
SILS-2	No	Yes	$1.73 \leq f_2/f_1 \leq 6.47$	$0.535 \leq \eta \leq 1$
SILS-3	Yes	No	$3 \leq f_3/f_1 \leq 6.47$	—
SILS-4	No	Yes	$1.46 \leq f_2/f_1 \leq 2.61$	$1.66 \leq \eta \leq 2.44$

The dual-mode resonator with SILS-1 has an extended stopband, as the f_2/f_1 ratio can reach a value of $f_2/f_1 = 4.36$. This resonator is much shorter than the half-wave one. Its shortening factor η can get the value $\eta = 0.558$.

The dual-mode resonator with SILS-2 has a wide rejection band, since the f_2/f_1 ratio can reach $f_2/f_1 = 6.47$. This resonator has the maximum shortening $\eta = 0.535$. This resonator is very promising for use in dual-band BPF, as it allows for varying the relative position of two passbands within wide limits, $1.73 < f_2/f_1 < 6.47$.

The dual-mode resonator with SILS-3 is characterized by the fact that its simple oscillations with frequencies f_1, f_3, ..., and dual-mode oscillations with frequencies f_2, f_4, ..., alternate with each other. The SILS-3 allows us to distance parasitic simple oscillations with frequencies f_1 and f_3 from the main dual-mode oscillation with frequency f_2. The ratio f_3/f_1 can reach the value $f_3/f_1 = 6.47$.

The dual-mode resonator with SILS-4 has an extended stopband. Its f_2/f_1 ratio can reach the value $f_2/f_1 = 2.61$. This resonator differs in that its main resonant frequency is located much higher than that of the half-wave resonator ($1.66 \leq \eta \leq 2.44$). It allows us to increase the operating frequencies. This resonator is promising for use in a dual-band BPF. We can change the relative position of its passbands over a wide range ($1.46 \leq f_2/f_1 \leq 2.61$).

These change limits in values correspond to a moderate value of the stepped-impedance planar TL segments parameter $m = 5$. In fact, for planar segments, the parameter m can reach the value $m \approx 9.2$ (Table 5.3). For coaxial structures, the parameter m is even greater than 30. Therefore, the limits of the change in parameters in Table 17.2 can be significantly expanded, which indicates the broad functional capabilities of the stepped-impedance dual-mode resonator.

The aforementioned analysis showed that there are resonators in which only two-mode oscillations occur, but there are also resonators in which two-mode oscillations alternate with simple oscillations. The same gradation occurs in triple-mode resonators (Figures 17.16–17.19). The analysis of symmetric resonators based on even and odd waves is not omnipotent. It does not apply to resonators with all short-circuited ends.

17.3 Dual-mode Resonators with Two Transmission Zeros

Some dual-mode resonators have two transmission zeros (TZs), which makes their frequency responses more varied. These include loop-type hairpin resonators with open or short-circuited ends. We will consider them in turn.

17.3.1 Loop-type Hairpin Resonator with Open-circuited Ends

This resonator [20–23] is a hairpin resonator with two open ends, between which a capacitor C is connected, which makes the resonator a loop type. The schematic drawings of this resonator, which has several modifications, are shown in

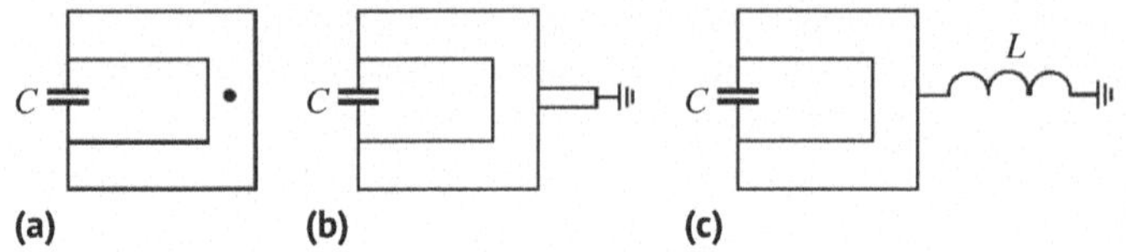

Figure 17.22 Dual-mode loop-type hairpin resonator with open-circuited ends. (a) With metallized hole. (b) With short-circuited stub of short length. (c) With inductance.

Figure 17.22. One of them (Figure 17.22a) has a metallized hole in the middle part [20]. Another variation [21, 22] contains a short-circuited stub of short length in the middle part (Figure 17.22b). The third modification [23] features an inductance in the middle part (Figure 17.22c). The determination of the resonant frequencies of these resonators is discussed in detail in the cited literature.

Figure 17.23 shows the topology and frequency responses of a microstrip BPF based on a dual-mode resonator with a short-circuited stub of short length [21]. The filter was designed on a substrate with a relative dielectric constant of 2.2 and a thickness of 0.508 mm. The filter size amounted to 15.8 mm × 16.5 mm. The center frequency of the BPF is $f_0 = 1.35$ GHz, and the fractional bandwidth is around FBW = 0.05.

Note that the equidistant arrangement of two TZs relative to f_0 is quite problematic in a BPF based on a loop-type hairpin resonator with open-circuited ends (Figure 7.22). This is illustrated in Figure 17.23b, which serves as a limitation when implementing a quasi-elliptic transfer function.

17.3.2 Loop-type Hairpin Resonator with Short-circuited Ends

The considered resonator is shown in Figure 17.24a. It represents a half-wave TL segment with both short-circuited ends. The resonator is characterized by characteristic impedance Z_0 and electrical length $\theta = \omega L/v$, where L is the length of the segment, and v is the propagation speed of the electromagnetic wave. Both arms of the resonator are connected by a capacitance C located at a distance $q\theta$ $(0 < q < 1/2)$ from the short-circuited ends, which makes the resonator a loop type. This resonator has many resonant frequencies ω_n $(n = 0,\ 1,\ \ldots)$ and corresponding resonant electrical lengths θ_n, defined by the resonance condition: $Y = 0$. We will monitor the input admittance Y of the resonator at a distance $q\theta$ from the short-circuited end.

To determine the Y function, let us imagine the resonator in a straightened shape (Figure 17.24b), which is a two-port network with a transfer matrix $[ABCD]$. The transformation of the two-port network in Figure 17.24b into a loop-type resonator in Figure 17.24a occurs when ports *2* and *1* are connected. In this case, the function Y is determined by the expression (15.52): $Y = (A + D - 2)/B$.

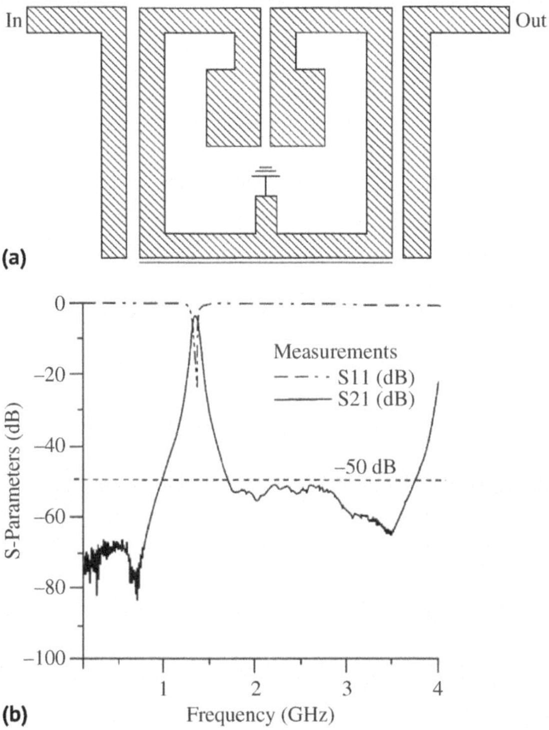

Figure 17.23 Microstrip BPF based on dual-mode resonator with short-circuited stub of short length [21]. (a) Topology. (b) Frequency responses.

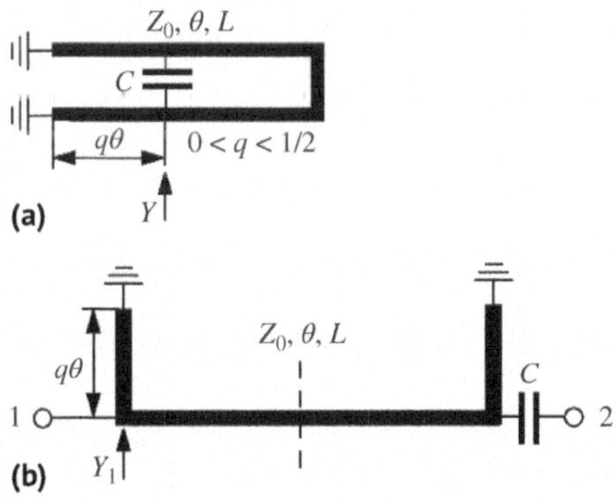

Figure 17.24 Dual-mode loop-type hairpin resonator with both short-circuited ends. (a) Topology. (b) Straightened shape of hairpin resonator.

17.3.2.1 Resonant Frequencies

The condition $Y = 0$ leads to the resonance equation

$$A + D - 2 = 0. \tag{17.43}$$

We will divide the process of obtaining the resonance equation for the proposed resonator into two stages.

Stage 1: Let us denote the transfer matrix $[A'B'C'D']$ of the distributed part of the two-port network in Figure 17.24b, and $Y_C = j\omega C$ is the conductivity of the capacitance. With these notations

$$\begin{bmatrix} A & B \\ C & D \end{bmatrix} = \begin{bmatrix} A' & B' \\ C' & D' \end{bmatrix} \times \begin{bmatrix} 1 & \dfrac{1}{Y_C} \\ 0 & 1 \end{bmatrix}. \tag{17.44}$$

Since the distributed part in Figure 17.24b is symmetric, its matrix $[A'B'C'D']$ is expressed through the matrix $[a'b'c'd']$ of its left half [4]

$$\begin{bmatrix} A' & B' \\ C' & D' \end{bmatrix} = \begin{bmatrix} 1 + 2b'c' & 2a'b' \\ 2c'd' & 1 + 2b'c' \end{bmatrix}. \tag{17.45}$$

Substituting (17.45) into (17.44) gives

$$\begin{bmatrix} A & B \\ C & D \end{bmatrix} = \begin{bmatrix} 1 + 2b'c' & \dfrac{(1 + 2b'c')}{Y_{C1}} + 2a'b' \\ 2c'd' & \dfrac{2c'd'}{Y_{C1}} + (1 + 2b'c') \end{bmatrix} \tag{17.46}$$

Taking into account (17.46), the resonant equation (17.43) splits into two equations

$$c' = 0. \tag{17.47a}$$

$$Y_C = -\frac{d'}{2b'}. \tag{17.47b}$$

Stage 2: Let us define the elements of the matrix $[a'b'c'd']$, which is equal to the product of two matrices

$$\begin{bmatrix} a' & b' \\ c' & d' \end{bmatrix} = \begin{bmatrix} 1 & 0 \\ Y_1 & 1 \end{bmatrix} \times \begin{bmatrix} \cos(0.5 - q)\theta & jZ_0 \sin(0.5 - q)\theta \\ jZ_0 \sin(0.5 - q)\theta & \cos(0.5 - q)\theta \end{bmatrix},$$

where $Y_1 = -jZ_0^{-1}\cot(q\theta)$ is the input admittance of the leftmost segment (Figure 17.24b). As a result of matrix multiplication, we obtain:

$$a' = \cos\left[(0.5 - q)\,\theta\right], b' = jZ_0 \sin\left[(0.5 - q)\,\theta\right],$$
$$c' = -jZ_0^{-1}\left[\cot(q\theta)\cos\left[(0.5 - q)\,\theta\right] - \sin\left[(0.5 - q)\,\theta\right]\right],$$
$$d' = \cot(q\theta)\sin\left[(0.5 - q)\right] + \cos\left[(0.5 - q)\,\theta\right].$$

Substituting these elements into (17.47) leads to the final resonance equations

$$\cos\left(\frac{\theta}{2}\right) = 0, \quad \theta_n = (n + 1)\,\pi, \quad n \text{ is even,} \tag{17.48a}$$

$$\omega C = 0.5Z_0^{-1}\frac{\sin(\theta/2)}{\sin(q\theta)\sin\left[(0.5 - q)\,\theta\right]}, \quad n \text{ is odd.} \tag{17.48b}$$

Equation (17.48a) determines even resonant frequencies with electrical lengths $\theta_0 = \pi$, $\theta_2 = 3\pi$, $\theta_4 = 5\pi$, ..., which are stationary and do not depend on capacitance C. Equation (17.48b) defines odd resonant frequencies dependent on capacitance C, which at $C = 0$ correspond to electrical lengths $\theta_n = (n + 1)\,\pi$, $\theta_1 = 2\pi$, $\theta_3 = 4\pi, \theta_5 = 6\pi, \ldots$. When $C = \infty$, then $\theta_n = (n - 1)\,\pi, \theta_1 = 0, \theta_3 = 2\pi, \theta_5 = 4\pi, \ldots$. The distribution of the resonant regions of the resonator under consideration is shown in Figure 17.25a. These regions completely overlap the θ axis, tightly adjacent to each other.

For comparison, Figure 17.25b illustrates the distribution of resonant regions in the traditional hairpin resonator with capacitance C. The odd resonant frequencies ω_i do not depend on the capacitance C. The variations in θ_n of even resonances cover half of the θ axis.

In the future, we will be interested in cases where the even ω_0 (ω_e) and odd ω_1 (ω_o) oscillation modes of the resonator under consideration are close. In this case, the oscillations become dual-mode. Using the resonance equation (17.48), we can determine the values of ω_e and ω_o, and then calculate the coupling coefficient K using formula (17.9).

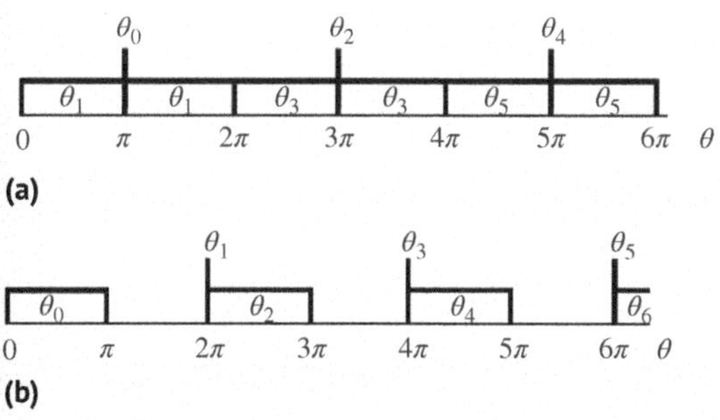

Figure 17.25 Distribution of resonant electrical lengths. (a) For dual-mode resonator in Figure 17.24a. (b) Traditional hairpin resonator in Figure 15.19a with uniform TL.

In addition to dual-mode oscillations ω_e and ω_o, the resonator has two TZs at real frequency. We denote them TZ1 (f_{z1}) and TZ2 (f_{z2}). The TZ1 features dual-mode oscillation, similar to other dual-mode resonators.

The TZ2 corresponds to the resonator's antiresonant frequency, which is closest to f_e and f_o. Let us consider these two TZs in turn. The value of TZ1 is not determined analytically. We will use the linear simulator of the Microwave Office software and the schematic (Figure 17.26a) of the proposed dual-mode hairpin resonator. Small values of capacitance $C_1 = 0.2$ pF provide weak coupling of the resonator with the loads.

If $q = 0.25$ and $C = 2.93$ pF, then from (17.48) follows $f_e = 0.98$ GHz, $f_o = 1.018$ GHz, corresponding to $K = 0.038$. Using the simulated insertion loss curve for the case (Figure 17.26b), we determine the position of TZ1: $f_{z1} = 1.1$ GHz. These values are presented in the second row of Table 17.3. This table also contains similar values for $q = 0.3$, $q = 0.4$, and C, corresponding to the coupling coefficient $K = 0.038$. The data in Table 17.3 show that for case $K = \text{const}$ $(K > 0)$, the TZ1 value is regulated and right-sided.

If $q = 0.25$ and $C = 3.5$ pF, then from (17.48) follows that $f_e = 1.02$ GHz and $f_o = 0.978$ GHz, which corresponds to $K = -0.042$. Using the simulated insertion loss curve for the case (Figure 17.26c), we determine the position of TZ1:

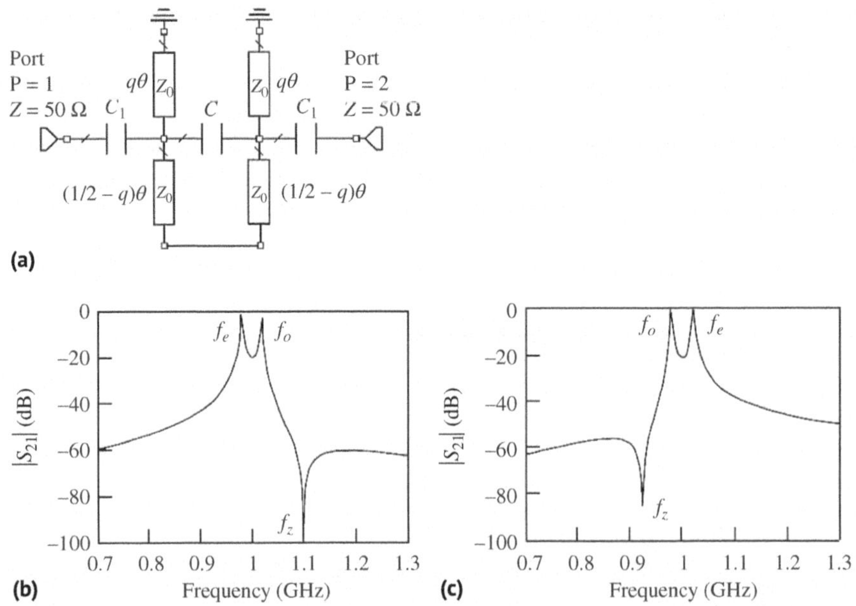

Figure 17.26 Location of TZ1 in hairpin dual-mode resonator in Figure 17.24. (a) Schematic of resonator: $Z_0 = 50\ \Omega$, $C_1 = 0.2$ pF, $\theta = \pi$ at $f = 1$ GHz, q and C are var. (b) Insertion loss for case $K > 0$ $(q = 1/4,\ C = 2.93\ \text{pF})$. (c) Insertion loss for case $K < 0$ $(q = 1/4,\ C = 3.5\ \text{pF})$.

Table 17.3 Adjustment of FZ1 for case K = const ($K > 0$).

q	C, pF	f_e, GHz	f_o, GHz	K	f_{z1}, GHz
0.25	2.93	0.980	1.018	0.038	1.10
0.3	3.14	0.974	1.012	0.038	1.048
0.4	5.27	0.966	1.004	0.038	1.014

Table 17.4 Adjustment of FZ1 for case K = const ($K < 0$).

q	C, pF	f_e, GHz	f_o, GHz	K	f_{z1}, GHz
0.25	3.5	1.02	0.978	−0.042	0.924
0.3	3.68	1.02	0.978	−0.042	0.950
0.4	6.14	1.008	0.966	−0.042	0.954

$f_{z1} = 0.924$ GHz. These values are presented in the second row of Table 17.4. This table also contains similar values for $q = 0.3$, $q = 0.4$, and C, corresponding to the coupling coefficient $K = -0.042$. The data in Table 17.4 show that for case $K = \text{const}\ (K < 0)$, the TZ1 value is regulated and left-sided.

Let us note the general regularities displayed in Tables 17.3 and 17.4:

1. For a given q, the coupling coefficient of two modes K can be either positive or negative, and the value of K is regulated by the capacitance C.
2. The TZ1 is always located on the side of the odd mode f_o.
3. The greater q and C, the closer f_{z1} is located to f_e, f_o, both for $K > 0$ and $K < 0$.

17.3.2.2 Antiresonance Frequencies

It is known that antiresonant frequencies ω_{pi}, which are the poles of input admittance $Y(j\omega_{pi}) = \pm j\infty$ of the resonator, generate TZs in the BPF, located at the same frequencies. From (15.52) it follows that the ω_{pi} of loop resonators is determined by the equation $B = 0$. From (17.46), it follows that for the resonator under consideration, this equation has the form

$$B = \frac{1 + 2b'c'}{Y_C} + 2a'b' = 0$$

After substituting elements a', b', and c' into the equation, we obtain

$$\omega_p C = Z_0^{-1}\left[\cot\left(q\theta_p\right) - \tan\left[(0.5 - q)\,\theta_p\right] + \csc\left[(1 - 2q)\,\theta_p\right]\right], \tag{17.49}$$

where electrical length θ_p corresponds to ω_p.

We denote the antiresonant frequency closest to the two oscillation modes f_e and f_o as f_p. From (17.49), the frequency f_p is always located to the right of the pair f_e and f_o ($f_p > f_e, f_o$). The TZ2 (f_{z2}) of the resonator under consideration coincides with its antiresonant frequency f_p ($f_{z2} = f_p$), determined by equation (17.49).

Various options for the location of the four characteristic frequencies of the resonator under consideration are shown in Figure 17.27. If the coupling coefficient K between f_e and f_o is positive, then the two TZs f_{z1} and f_{z2} are always right-sided. Moreover, their position is adjustable with a fixed K value. This is illustrated in (Figure 17.27a) for $K = 0.05$. If the coupling coefficient K is negative, then two TZs are both-sided: f_{z1} is left-sided, and f_{z2} is right-sided. Their position is adjustable with a fixed K value, including the case of an equidistant arrangement. This is illustrated in (Figure 17.27b) for $K = -0.08$.

If we extend the frequency band in Figure 17.26b, c, TZ2 will be visible in these frequency responses as well.

Thus, the dual-mode resonator has extended functionality compared to the dual-mode resonators (Figure 17.22).

17.3.3 Microstrip Quasi-elliptic Dual-mode BPF

The dual-mode resonator in question can be used as a second-order BPF with two adjustable TZs. Let us set the following BPF parameters: central frequency $f_0 = 1$ GHz, bandwidth BW = 40 MHz (FBW = 0.04), pulsation $L_{Ar} = 0.2$ dB (return loss $L_R = 13.467$ dB), TZs f_{z1} and f_{z2} should be equidistant relative to $f_0 = 1$ GHz.

Let's write down the g parameters of the low-frequency prototype (Table 2.6) for $N = 2$ and $L_{Ar} = 0.2$ dB: $g_0 = 1$, $g_1 = 1.0378$, $g_2 = 0.6746$, $g_3 = 1.5386$. Using g

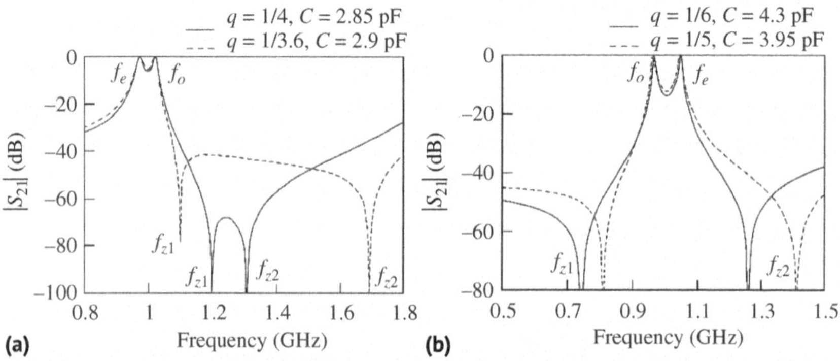

Figure 17.27 Characteristic frequencies of the dual-mode resonator in Figure 17.24. (a) For positive $K = 0.05$ ($f_e < f_o$). (b) For negative $K = -0.08$ ($f_e > f_o$).

parameters, FBW = 0.04, and formulas (3.8), we determine the coupling coefficient between the even and odd modes $K = 0.047$ and the external quality factor $Q_e = 25.9$.

To implement the BPF, a Rogers TMM-10i substrate was used, with $\varepsilon_r = 9.8$, $h = 1.016$ mm, and $\tan\delta = 0.002$. The width of the microstrip line of the hairpin resonator is $w = 1.8$ mm ($Z_0 = 36\ \Omega$), and the length $L = 54.8$ mm. Metallized holes were used to provide a short circuit at the ends of the resonator, which had a diameter of 0.6 mm. Distance between the two arms of hairpin resonator $\Delta = 2.4$ mm. The filter occupies an area of 35 mm × 16 mm. Its photograph is shown in Figure 17.28a.

To ensure the coupling coefficient $K = 0.047$, it is necessary to arrange the modes of oscillations at the frequencies: $f_o = 0.977$ GHz and $f_e = 1.023$ GHz. In this case, ensuring an equidistant location of f_{z1} and f_{z2} relative to f_0 is necessary. This is achieved by appropriately selecting the values of q and C for use. The solution is the values $q = 0.075$ and $C = 6.7$ pF. In this case, the EM simulator of Microwave Office software has two TZs located equidistantly relative to f_0 at the following frequencies: $f_{z1} = (f_0 - 140\ \text{MHz})$ and $f_{z2} = (f_0 + 140\ \text{MHz})$.

Quarter-wave line segments are used to match the filter with the loads. The degree of matching depends on the characteristic impedance Z_c of these segments.

The required return loss $L_R = 13.467$ dB is provided at the value of $Z_c = 72.5\ \Omega$, which is realized with a width of segments $w_c = 0.4$ mm and length $L_c = 30$ mm. In manufacturing the filter, a capacitor with a size code of 0402 and a capacitance of 6.8 pF was used, which differs from the calculated value of 6.7 pF.

Figure 17.28b shows the frequency responses of the filter. The measured dates of the filter are as follows: $f_0 = 1$ GHz, BW = 40 MHz, the insertion loss $L_{A0} = 0.8$ dB, the return loss in bandwidth $L_R > 13$ dB, the TZs are equidistant and located at the

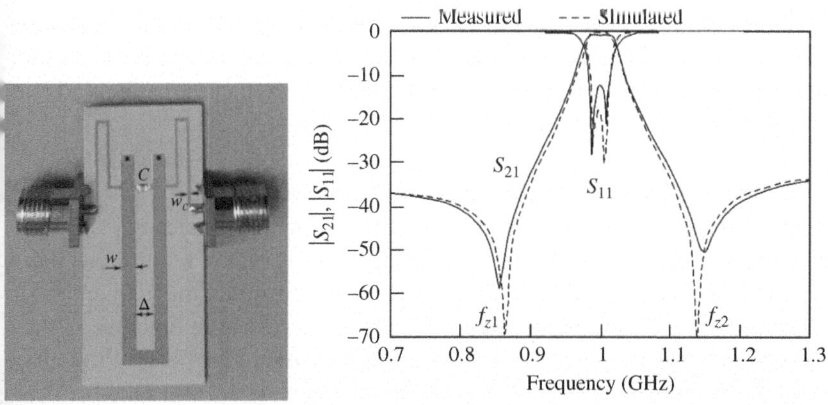

(a) (b)

Figure 17.28 Microstrip dual-mode BPF. (a) Photograph. (b) Frequency responses.

frequencies: $f_{z1} = (f_o - 140\,\text{MHz})$, $f_{z2} = (f_o + 140\,\text{MHz})$. The selectivity of this filter is 40 dB ($f_0 \pm 120$ MHz), which can be expressed in terms of the normalized frequency $j\Omega$ as 40 dB ($\pm j6.0$).

The resonator in Figure 17.24a has two TZs, which can be right-handed and provide a dual-mode BPF with improved right-hand selectivity, which is essential when using a BPF in a duplexer. The two TZs can also be located on different sides of f_e and f_o, including equidistant, making it possible to implement a quasi-elliptic dual-mode BPF.

17.4 Meander Dual-mode Resonators and Bandpass Filters

Two dimensions characterize the dual-mode resonators discussed earlier, since they contain stubs or U-shaped TL segments. Dual-mode patch resonators are also two-dimensional [24, 25]. This complicates their implementation when using TL designs such as metallic or dielectric coaxial. In the following section, we consider TL dual-mode resonators, which are characterized by one dimension. These resonators are SIRs, the characteristic impedance of which takes only two alternating values. Therefore, they look like a meander.

17.4.1 Meander Dual-mode Resonators with Open-circuited Ends

Dual-mode TL resonators are symmetrical, in which the resonant frequencies of two oscillations are close to each other, and these oscillations are coupled to each other.

To synthesize resonators with two closely spaced resonant frequencies, we will use SIRs containing N composite segments with different characteristic impedances Z_{0i} ($i = 1, 2, \ldots, N$) and the same electrical length $\theta_s = \theta/N$ (Figure 5.5a), where θ is the electrical length of the entire resonator. The numbering of the component segments is carried out from the far end of the resonator. The input admittance function of such an SIR at the Richards' frequency variable S (5.18) is expressed by the reactance function $Y_{\text{in}}^N(S)$ of the degree N.

17.4.1.1 Meander Dual-mode Resonators with Three Sections

We synthesize SIR with two closely spaced resonant frequencies ω_{01} and ω_{02}, which correspond to electrical lengths θ_{01} and θ_{02}. Moreover, we require that these frequencies are located equidistantly relative to the antiresonant frequency ω_{p2} (θ_{p2}) (Figure 17.29a). Resonant frequencies are the zeros of the function $Y_{\text{in}}^N(S)$. They are marked with circles. Antiresonant frequencies are the poles of the function $Y_{\text{in}}^N(S)$. They are depicted as crosses. These frequencies are also called critical.

According to the parametric synthesis method we use (Chapter 5), the minimum order of the reactance function $Y_{\text{in}}^N(S)$ ensuring the location of the critical

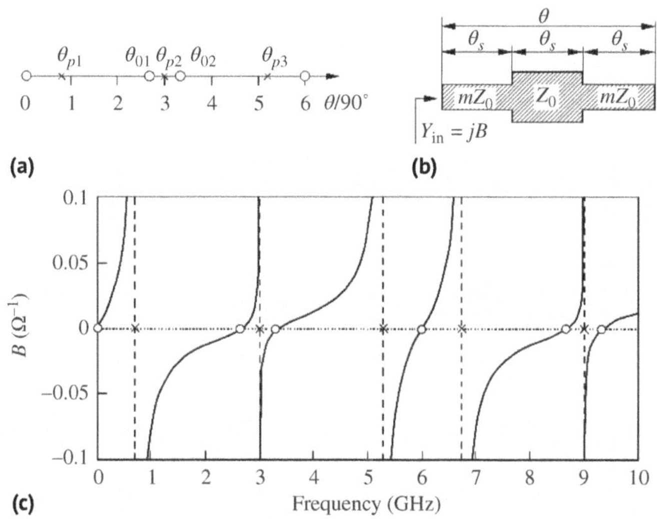

Figure 17.29 Meander dual-mode resonators with three sections and both open-circuited ends. (a) Critical electrical length. (b) Topology. (c) Input admittance for case $Z_0 = 10\,\Omega$, $m = 5$, $f_{p2} = 3$ GHz.

frequencies ω_{01}, ω_{02}, and ω_{p2} is equal to three. Let us write this function in general form

$$Y^3(S) = \frac{b_1^{(3)}S + b_3^{(3)}S^3}{a_0^{(3)} + a_2^{(3)}S^2}, \tag{17.50}$$

where $S = j\Omega = j\tan\theta_s = j\tan(\theta/3)$.

The coefficients of the numerator and denominator polynomials (17.50) are determined by the equalities from Table 5.5a.

$$b_1^{(3)} = \frac{1}{Z_{01}} + \frac{1}{Z_{02}} + \frac{1}{Z_{03}},\quad b_3^{(3)} = \frac{Z_{02}}{Z_{01}Z_{03}} \tag{17.51}$$

$$a_0^{(3)} = 1,\quad a_2^{(3)} = \frac{Z_{02}}{Z_{01}} + \frac{Z_{03}}{Z_{01}} + \frac{Z_{03}}{Z_{02}}, \tag{17.52}$$

At $S = \pm j\infty$, the input admittance (17.50) is infinite, which is ensured when the electrical length of the resonator is $\theta_{p2} = 3\pi/2$, which corresponds to a pole of this function. The zero ω_{01} of the function (17.50) is located at the frequency $\Omega_{01}^2 = b_1^{(3)}/b_3^{(3)}$. Its increase leads to a decrease in the ratio ω_{02}/ω_{01}. Using the equalities (17.51), we determine

$$\Omega_{01}^2 = \frac{b_1^{(3)}}{b_3^{(3)}} = \frac{Z_{01} + Z_{03}}{Z_{02}} + \frac{Z_{01}Z_{03}}{Z_{02}^2}$$

From the last equality, it follows that the maximum value of Ω_{01}^2 is achieved when the conditions are met

$$Z_{01} = Z_{03} = Z_{0\max} = mZ_0, \quad Z_{02} = Z_{0\min} = Z_0. \tag{17.53}$$

The parametric synthesis is completed. The synthesized SIR is presented in Figure 17.29b, which is a meander. Taking into account (17.51) and (17.52), its input admittance (17.50) takes the form:

$$Y^3(S) = Z_0^{-1} \frac{(1 + 2/m) S + (1/m^2) S^3}{1 + (1 + m + 1/m) S^2}. \tag{17.54}$$

Equating the numerator (17.54) to zero, we determine the electrical length θ_{01} corresponding to the resonant frequency ω_{01}

$$\theta_{01} = 3 \tan^{-1} \sqrt{(m + 2)\, m}. \tag{17.55}$$

Since function (17.54) is periodic with period $\theta = 3\pi$, then the electrical length θ_{02}, corresponding to the resonant frequency ω_{02}, is determined by the equality

$$\theta_{02} = 3\pi - \theta_{01} = 3\pi - 3 \tan^{-1} \sqrt{(m + 2)\, m}. \tag{17.56}$$

The electrical lengths θ_{01} (17.55) and θ_{02} (17.56) are located equidistantly with respect to the electrical length at antiresonance $\theta_{p2} = 3\pi/2$, which does not depend on the parameter m of the resonator. Using (17.55) and (17.56), we obtain the ratio of the resonant frequencies ω_{02}/ω_{01} of the considered meander resonator

$$\frac{\omega_{02}}{\omega_{01}} = \frac{\theta_{02}}{\theta_{01}} = \frac{\pi}{\tan^{-1} \sqrt{(m + 2)\, m}} - 1. \tag{17.57}$$

The dependence (17.57) is displayed in Table 17.5. The resonance frequencies ω_{02} and ω_{01} approach each other significantly as the parameter m increases. At $m = 30$, the degree of convergence is 4%. Note that in coaxial SIRs, implementing the parameter $m = 30$ does not cause difficulties. In microstrip SIRs, the parameter $m = 30$ can be achieved by using combined substrates with different dielectric constants, for example, $\varepsilon_r = 92$ and $\varepsilon_r = 2.2$.

The synthesized SIR (Figure 17.29b) is dual-mode; its resonance frequencies ω_{01} and ω_{02} can be located close to each other, and their mutual arrangement is regulated by the parameter m (17.57). The input admittance of the resonator for the

Table 17.5 Ratio ω_{02}/ω_{01} versus m (17.57).

m	1	5	10	20	30
$\frac{\omega_{02}}{\omega_{01}}$	2	1.24	1.12	1.06	1.04

case $Z_0 = 10\ \Omega$, $m = 5$, $f_{p2} = 3$ GHz is shown in Figure 17.29c. This function was created using linear simulator of Microwave Office software. In this resonator, dual-mode and simple oscillations alternate with each other.

17.4.1.2 Meander Dual-mode Resonator with Five Sections

We synthesize the second SIR with two close to each other resonant frequencies ω_{02} and ω_{03}, which correspond to electrical lengths θ_{02} and θ_{03}. Moreover, we will require that these frequencies be located equidistantly relative to the antiresonant frequency ω_{p3} (θ_{p3}), as shown in Figure 17.30a. Such an arrangement of ω_{02} and ω_{03} allows the resonator to operate at higher frequencies with the same length.

The minimum order of the reactance function $Y_{\text{in}}^{N}(S)$, which ensures the equidistant location of the resonant frequencies ω_{02} and ω_{03} relative to the antiresonant frequency ω_{p3} (Figure 17.30a), is five. Let us write the function $Y^5(S)$ in general form

$$Y^5(S) = \frac{b_1^{(5)}S + b_3^{(5)}S^3 + b_5^{(5)}S^5}{a_0^{(5)} + a_2^{(5)}S^2 + a_4^{(5)}S^4}, \tag{17.58}$$

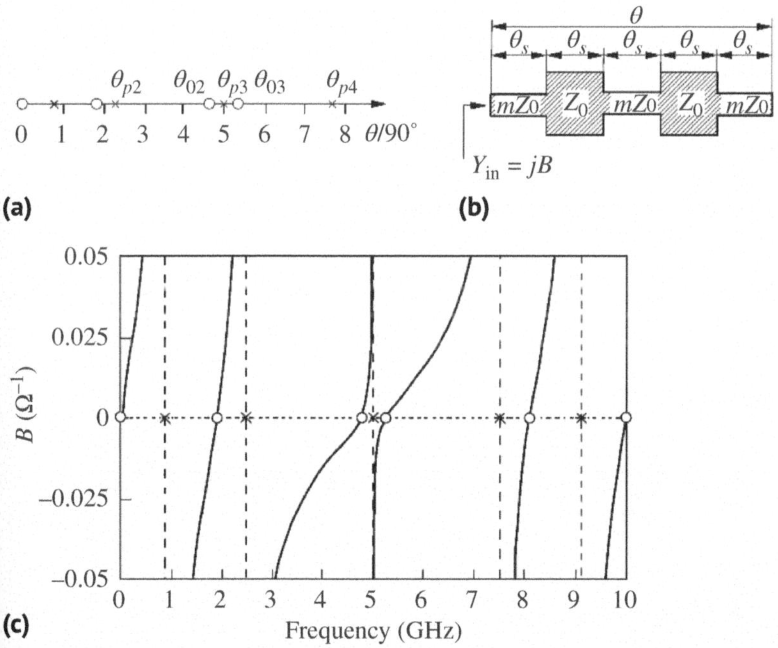

Figure 17.30 Meander dual-mode resonator with five sections and both open-circuited ends. (a) Critical electrical length. (b) Topology. (c) Input admittance for case $Z_0 = 10\ \Omega$, $m = 3$, $f_{p3} = 5$ GHz.

where

$$S = j\Omega = j\tan\left(\frac{\theta}{5}\right), \tag{17.59}$$

$$b_5^{(5)} = \frac{Z_{02}Z_{04}}{Z_{01}Z_{03}Z_{05}}. \tag{17.60}$$

The remaining coefficients of the function (17.58) are determined from Table 5.5b.

At $S = \pm j\infty$, the input admittance (17.58) is infinity. It is ensured when the electrical length of the resonator is $\theta_{p3} = 5\pi/2$, which corresponds to the third antiresonant frequency ω_{p3}. From (17.58), it follows that the leading root Ω_{02}^2 of the algebraic equation of the second degree determines the zero ω_{02} of the function $Y^5(S)$

$$\Omega_{02}^2 = \frac{b_3^{(5)}}{2b_5^{(5)}} + \sqrt{\left(\frac{b_3^{(5)}}{2b_5^{(5)}}\right)^2 - \frac{b_1^{(5)}}{b_5^{(5)}}}. \tag{17.61}$$

The maximum value of Ω_{02}^2 is ensured at the minimum value of the coefficient (17.60), which leads to the final solution:

$$\begin{aligned} &Z_{02} = Z_{04} = Z_{0\min} = Z_0 \\ &Z_{01} = Z_{03} = Z_{05} = Z_{0\max} = mZ_0 \end{aligned} \tag{17.62}$$

The parametric synthesis is completed. The synthesized SIR is presented in Figure 17.30b, which is a meander. With these values of characteristic impedances, the input admittance coefficients (17.58) are written as (Table 5.5b):

$$b_1^{(5)} = Z_0^{-1}\left(2 + \frac{3}{m}\right), \quad b_3^{(5)} = Z_0^{-1}\left(2 + m + \frac{3}{m} + \frac{4}{m^2}\right), \quad b_5^{(5)} = Z_0^{-1}\left(\frac{1}{m^3}\right),$$

$$a_0^{(5)} = 1, \quad a_2^{(5)} = 4 + 3m + \frac{3}{m}, \quad a_4^{(5)} = 1 + m + \frac{1}{m} + m^2 + \frac{1}{m^2}.$$

The input admittance (17.58) takes the form:

$$Y^5(S) = Z_0^{-1}\frac{(2 + 3/m)S + (2 + m + 3/m + 4/m^2)S^3 + (1/m^3)S^5}{1 + (4 + 3m + 3/m)S^2 + (1 + m + 1/m + m^2 + 1/m^2)S^4}. \tag{17.63}$$

The resonant frequencies of this resonator on the Ω-axis are the roots of the quadratic equation

$$\Omega_0^4 - (4m + 3m^2 + 2m^3 + m^4)\,\Omega_0^2 + m^2(3 + 2m) = 0. \tag{17.64}$$

The leading root of equation (17.64) determines the electrical length of the synthesized resonator θ_{02} at the resonant frequency ω_{02}

$$\theta_{02}(m) = 5\tan^{-1}\left[\frac{A}{2} + \sqrt{\left(\frac{A}{2}\right)^2 - m^2(3+2m)}\right]^{1/2}, \quad (17.65)$$

where $A = 4m + 3m^2 + 2m^3 + m^4$.

Since function (17.65) is periodic with period $\theta = 5\pi$, then the electrical length θ_{03}, corresponding to the resonant frequency ω_{03}, is determined by the equality

$$\theta_{03}(m) = 5\pi - \theta_{02}(m). \quad (17.66)$$

The electrical lengths θ_{02} (17.65) and θ_{03} (17.66) are located equidistantly with respect to the electrical length at antiresonance $\theta_{p3} = 5\pi/2$, which does not depend on the parameter m of the resonator. Using (17.65) and (17.66), we obtain the ratio

$$\frac{\omega_{03}}{\omega_{02}} = \frac{\theta_{03}}{\theta_{02}} = \frac{\pi}{\tan^{-1}\left[\frac{A}{2} + \sqrt{\left(\frac{A}{2}\right)^2 - m^2(3+2m)}\right]^{1/2}} - 1, \quad (17.67)$$

where A is the same as in (17.65).

The dependence (17.67) is displayed in Table 17.6. The resonance frequencies θ_{02} and θ_{03} approach each other significantly as the parameter m increases. At $m = 5$, the degree of convergence is 4%.

The synthesized SIR (Figure 17.30b) is dual-mode, its resonant frequencies ω_{02} and ω_{03} can be located close to each other, and their mutual arrangement is regulated by the parameter m (17.67). The input admittance of the resonator for $Z_0 = 10\ \Omega$, $m = 3$, and $f_{p3} = 5$ GHz is shown in Figure 17.30c. This resonator has dual-mode and simple oscillations.

17.4.2 Meander Dual-mode Resonators with Short-circuited Ends

Meander dual-mode resonators can also have two short-circuited ends. Such resonators are synthesized using the method of structural-parametric synthesis (Chapter 5), the results of which are presented as follows.

Table 17.6 Ratio ω_{p3}/ω_{p2} versus m (17.67).

m	1	2	3	4	5
ω_{p3}/ω_{p2}	1.5	1.19	1.10	1.06	1.04

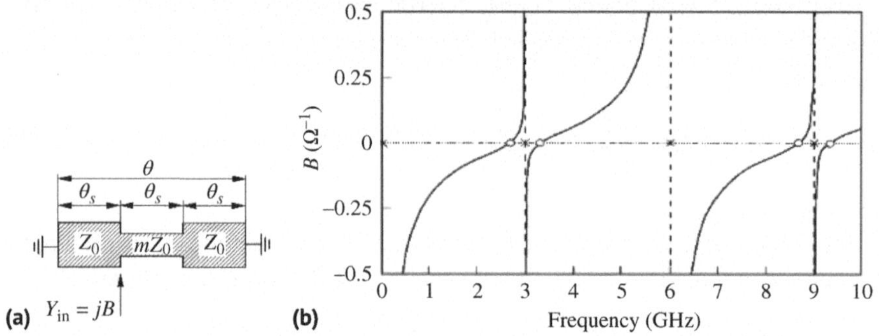

Figure 17.31 Meander dual-mode resonator with three sections and both short-circuited ends. (a) Topology. (b) Input admittance for case $Z_0 = 10\ \Omega$, $m = 5$, $f_{p1} = 3$ GHz.

17.4.2.1 Meander Dual-mode Resonators with Short-circuited Ends and Three Sections

Figure 17.31a shows a synthesized meander dual-mode resonator with three sections, which ensures mutual convergence of the resonant frequencies ω_{01} and ω_{02}, located equidistantly with respect to the antiresonant frequency ω_{p1}. The resonator input is located on the side, as shown in this figure. In this case, the input admittance is expressed by a second-order reactance function

$$Y^2(S) = Z_0^{-1} \frac{(2+m) + (1/m)\,S^2}{(1+m)\,S}, \tag{17.68}$$

where $S = j\Omega = j\tan\theta_s = j\tan(\theta/3)$.

The resonant frequencies ω_{01} and ω_{02} of this resonator, which form dual-mode oscillation, coincide with similar resonant frequencies of dual-mode resonators with both open-circuited ends (Figure 17.29a). This follows from a comparison of the numerators of the input admittances of these resonators, as shown in (17.54) and (17.68). This coincidence enables us to extend the previously obtained expressions for the resonant frequencies (17.55)–(17.57) and Table 17.5 to the dual-mode resonator with both short-circuited ends (Figure 17.31a). The input admittance of the resonator for the case $Z_0 = 10\ \Omega$, $m = 5$, $f_{p1} = 3$ GHz is shown in Figure 17.31b. This resonator contains only dual-mode oscillations.

17.4.2.2 Meander Dual-mode Resonators with Short-circuited Ends and Five Sections

Figure 17.32a shows a synthesized meander dual-mode resonator with five sections, which provides mutual convergence of the resonant frequencies ω_{02} and ω_{03},

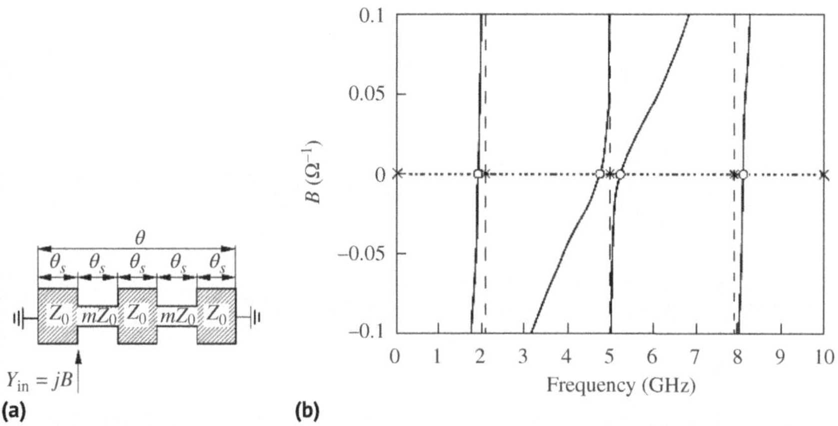

Figure 17.32 Meander dual-mode resonator with five sections and both short-circuited ends. (a) Topology. (b) Input admittance for case $Z_0 = 10\ \Omega$, $m = 3$, $f_{p3} = 5$ GHz.

located equidistantly with respect to the antiresonant frequency ω_{p2}. The resonator input is located on the side, as shown in this figure. The fourth-order reactance function expresses the input admittance of this resonator

$$Y^4(S) = Z_0^{-1} \frac{(3+2m) + (3+2m+4/m+m^2)\, S^2 + (1/m^2)\, S^4}{2(1+m)\, S + (1+m+1/m+m^2)\, S^3}, \qquad (17.69)$$

where $S = j \tan \theta_s = j \tan(\theta/5)$.

The resonant frequencies ω_{02} and ω_{03} of this resonator, which form dual-mode oscillations, coincide with similar resonant frequencies of dual-mode resonators with both open-circuited ends (Figure 17.30a). This follows from a comparison of the numerators of the input admittances of these resonators (17.63) and (17.69). This coincidence allows us to extend the previously obtained expressions for the resonant frequencies (17.65)–(17.67) and Table 17.6 to the dual-mode resonator with both short-circuited ends (Figure 17.32a). The input admittance of the resonator for the case $Z_0 = 10\ \Omega$, $m = 3$, $f_{p2} = 5$ GHz is shown in Figure 17.32b. This resonator contains both dual-mode and simple oscillations.

The information about the synthesized meander dual-mode resonators is presented in Table 17.7, facilitating a more visual comparison. The synthesized resonators have interesting properties.

Resonators No. 1 and No. 2 have both open-circuited ends, while resonators No. 3 and No. 4 have these ends short-circuited. Resonators No. 1 and No. 3 have three sections. They bring the first ω_{01} and second ω_{02} resonant frequencies closer together, thus forming a dual-mode pair of oscillations. The dual-mode oscillations of both resonators coincide, despite having different boundary conditions at

Table 17.7 Meander dual-mode resonators without stubs.

No.	Schematic	Input admittance graph	Input admittance function
No. 1	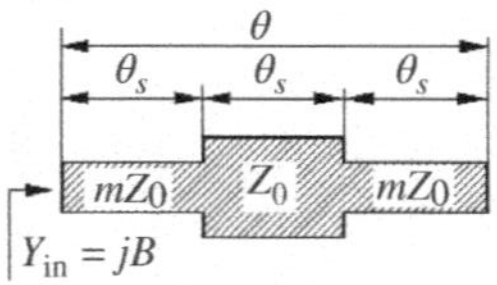	Figure 17.29c	$Y_{\text{in}}^{(3)}(S) = Z_0^{-1} \dfrac{(1+2/m)S + (1/m^2)S^3}{1 + (1+m+1/m)S^2}$
No. 2	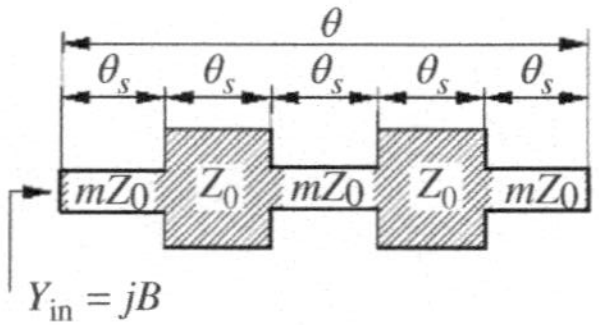	Figure 17.30c	$Y_{\text{in}}^{(5)}(S) = Z_0^{-1} \dfrac{b_1^{(5)}S + b_3^{(5)}S^3 + b_5^{(5)}S^5}{a_0^{(5)} + a_2^{(5)}S^2 + a_4^{(5)}S^4}$ $b_1^{(5)} = 2 + \dfrac{3}{m}$ $b_3^{(5)} = 2 + m + \dfrac{3}{m} + \dfrac{4}{m^2}$ $b_5^{(5)} = 1/m^3, \quad a_0^{(5)} = 1$ $a_2^{(5)} = 4 + 3m + \dfrac{3}{m}$ $a_4^{(5)} = \dfrac{1+m+1}{m+m^2+1/m^2}$

No. 3	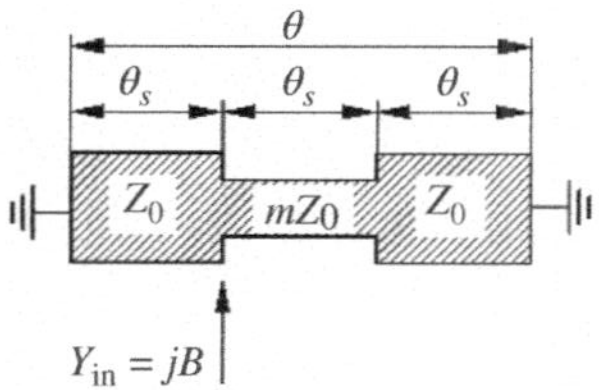	Figure 17.31b	$Y_{\text{in}}^{(2)}(S) = Z_0^{-1}\dfrac{(2+m)+(1/m)S^2}{(1+m)S}$
No. 4	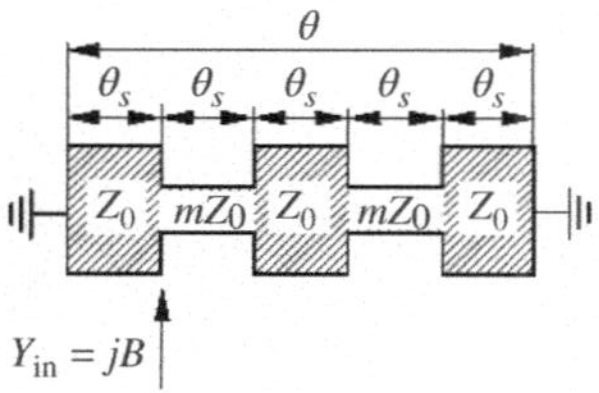	Figure 17.32b	$Y_{\text{in}}^{(4)}(S) = Z_0^{-1}\dfrac{b_0^{(4)} + b_2^{(4)}S^2 + b_4^{(4)}S^4}{a_1^{(4)}S + a_3^{(4)}S^3}$ $b_0^{(4)} = 3 + 2m$ $b_2^{(4)} = \dfrac{3+2m+4}{m+m^2}$ $b_4^{(4)} = 1/m^2, \quad a_1^{(4)} = 2(1+m)$ $a_3^{(4)} = \dfrac{1+m+1}{m+m^2}$

the ends and mutually opposite characteristic impedance functions. It is also noteworthy that their input admittances are described by reactance functions of different orders (third and second). Resonator No. 1 alternates between dual-mode and simple oscillations. Resonator No. 3 has only dual-mode oscillations, and the density of its critical frequencies on the frequency axis is lower than that of resonator No. 1.

Resonators No. 2 and No. 4 have five sections. They bring the second ω_{02} and third ω_{03} resonant frequencies closer together, resulting in the formation of dual-mode oscillations. These resonators are characterized by different boundary conditions at the ends, a mutually opposite nature of the change in the characteristic impedance along the length, and other orders of the reactance function of the input admittance (fifth and fourth). At the same time, the frequencies of their dual-mode oscillations coincide. Along with dual-mode oscillations, these resonators have simple oscillations. The density of the critical frequencies on the frequency axis of resonator No. 4 is lower than resonator No. 2.

17.4.3 Meander Dual-mode Bandpass Filters

Two coupled electromagnetic oscillations are characterized by the coupling coefficient (17.9). A reasonably accurate approximate equality can replace this equality

$$K \approx \frac{f_o - f_e}{(f_o + f_e)/2}, \tag{17.70}$$

where f_e and f_o are the frequencies of the even and odd modes, respectively (they are also called the coupling frequencies).

We will use expression (17.70) in the following.

17.4.3.1 Coupling Coefficients in Dual-mode Resonators

For meander dual-mode resonators with three sections, the coupling frequencies are the resonant frequencies f_{01} and f_{02}, which correspond to the resonant electrical lengths θ_{01} (17.55) and θ_{02} (17.56). Since $(\theta_{02} + \theta_{01})/2 = \theta_{p1}$ or θ_{p2}, then expression (17.70) is written as

$$K \approx \frac{\theta_{02} - \theta_{01}}{\theta_{p1,2}}. \tag{17.71}$$

Taking into account the value $\theta_{p1,2} = 3\pi/2$ and equations (17.55), (17.56), and (17.71), we write the final expression for the coupling coefficient

$$K \approx 2\left[1 - \frac{2}{\pi}\tan^{-1}\sqrt{(m+2)\,m}\right]. \tag{17.72}$$

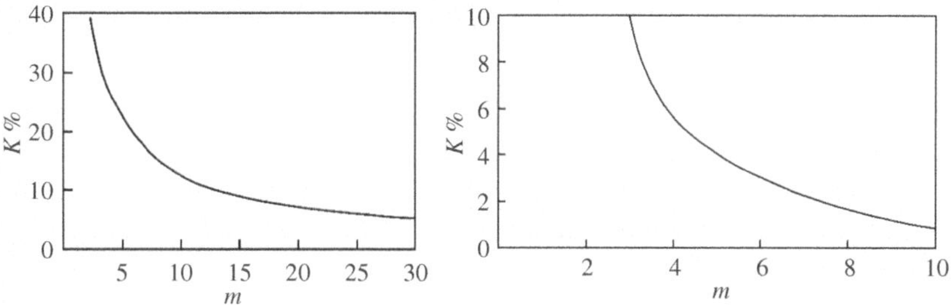

Figure 17.33 Dependence of coupling coefficient K of dual-mode oscillations on parameter m of meander resonators. (a) Resonators with three sections. (b) Resonators with five sections.

The dependence of K (17.72) on the parameter m of meander dual-mode resonators with three sections is shown in Figure 17.33a. The larger the parameter m, the smaller the coupling coefficient. At $m = 30$, the value of $K = 0.05$ (5%).

For meander dual-mode resonators with five sections, the coupling frequencies are the resonant frequencies f_{02} and f_{03}, which correspond to the resonant electrical lengths θ_{02} (17.65) and θ_{03} (17.66). Since $(\theta_{03} + \theta_{02})/2 = \theta_{p2}$ or θ_{p3}, then expression (17.70) is written as

$$K \approx \frac{\theta_{03} - \theta_{02}}{\theta_{p2,3}}. \tag{17.73}$$

Taking into account the value $\theta_{p2,3} = 5\pi/2$ and equations (17.65), (17.66), and (17.73), we write the final expression for the coupling coefficient

$$K \approx 2\left(1 - \frac{2}{\pi}\tan^{-1}\left[\frac{A}{2} + \sqrt{\left(\frac{A}{2}\right)^2 - m^2(3 + 2m)}\right]^{1/2}\right), \tag{17.74}$$

where A has the same meaning as in (17.65).

The dependence (17.74) is shown in Figure 17.33b. Just as in Figure 17.33a, an increase in the parameter m leads to a decrease in the values of K. Moreover, the degree of reduction in K is more significant. Thus, at $m = 5$, the coupling coefficient in Figure 17.33b is $K = 0.0432$, and in Figure 17.33a, the value is $K = 0.207$. A dual-mode BPF with three sections is preferable for implementing wider passbands. It is advisable to use a dual-mode BPF with five sections to ensure narrow passbands at higher frequencies.

17.4.3.2 Design of Meander Dual-mode BPFs

The design of the meander dual-mode BPFs is based on the G. Matthaei's method, discussed in Chapter 3 and using g parameters of a low-frequency prototype. We will design BPFs based on each of the two synthesized resonators No. 3 and No. 2 (Table 17.7).

Meander BPF with three sections: Let us design a meander dual-mode BPF with the following parameters: center frequency $f_0 = 3$ GHz, bandwidth BW = 300 MHz, fractional bandwidth FBW = 0.1, passband ripple $L_{Ar} = 0.2$ dB, which corresponds to the return loss $L_R = 13.467$ dB. From Table 2.6, we find g parameters: $g_0 = 1$, $g_1 = 1.0378$, $g_2 = 0.6745$, $g_3 = 1.5386$.

Substituting these values and FBW = 0.1 into formulas (3.8), we determine the coupling coefficient between even f_e and odd f_o modes $K = 0.1195$ and the external quality factor $Q_{e1} = Q_{e2} = Q_e = 10.378$. The value of Z_0 will be set equal to 10 Ω. To determine the parameter m, providing the value $K = 0.1195$, we use formula (17.72). As a result, we obtain $m = 9.7$.

Figure 17.34a shows the BPF circuit based on a dual-mode resonator with both short-circuited ends (No. 3, Table 17.7). The filter is coupled to the loads using J inverters, which contain capacitors; their transverse branches with negative capacitors are not shown. In the synthesized resonator, the pole ω_{p1} of the input admittance coincides with the central frequency of the passband f_0 and is characterized by the electrical length $\theta_{p1} = 3\pi/2$. Therefore, the electrical lengths of each of the three sections is $\theta_s = \theta_{p1}/3 = 90°$ at the frequency of 3 GHz.

Figure 17.34b shows the frequency responses of the filter, simulated by a linear simulator. The value of the inverter capacitance, $C = 0.685$ pF, was determined through simulation, which provided the required value of return loss $L_R = 13.467$ dB.

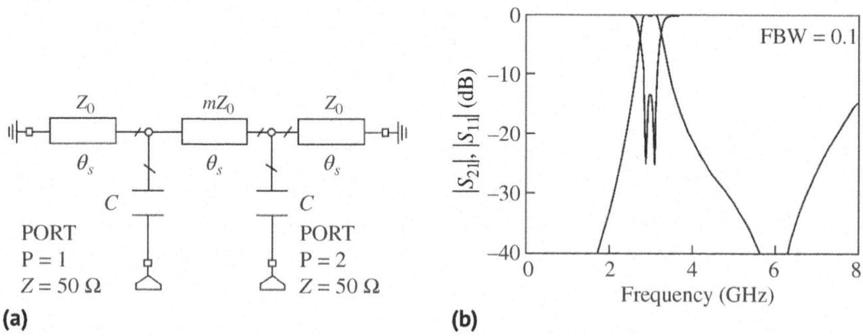

Figure 17.34 Meander dual-mode BPF with three sections and both short-circuited ends. (a) Schematic: $Z_0 = 10$ Ω, $m = 9.7$, $\theta_s = 90°$ at $f_0 = 3$ GHz, $C = 0.685$ pF. (b) Frequency responses.

The insertion loss function of the dual-mode BPF under consideration is not symmetrical; its left slope is steeper than the right one. Left-side selectivity is characterized by values of 20 dB (f_0 – 590 MHz) or 20 dB (–j3.93). Values of 20 dB (f_0 + 900 MHz) or 20 dB (+j6.0) characterize right-side selectivity.

Meander BPFs with five sections: Let us design a meander dual-mode BPF with the following parameters: center frequency $f_0 = 5$ GHz, bandwidth BW = 200 MHz, fractional bandwidth FBW = 0.04, and passband ripple $L_{\text{Ar}} =$ 0.2 dB ($L_R = 13.467$ dB). In this case, the g parameters remain the same. Substituting FBW = 0.04 and g parameters into formulas (3.8), we determine the coupling coefficient $K = 0.0478$ and the external quality factor $Q_{e1} = Q_{e2} = Q_e =$ 25.945. In the filter, we will use dual-mode resonators No. 2 with five sections and both open-circuited ends (Table 17.7). The value of Z_0 will be set equal to 10 Ω. To determine the parameter m that provides the value $K = 0.0478$, we use formula (17.74). As a result, we obtain $m = 4.6$.

Figure 17.35a shows the BPF circuit based on a dual-mode resonator with five sections. The filter is connected to the loads via J inverters containing capacitors; their transverse branches, with negative capacitances, are not shown.

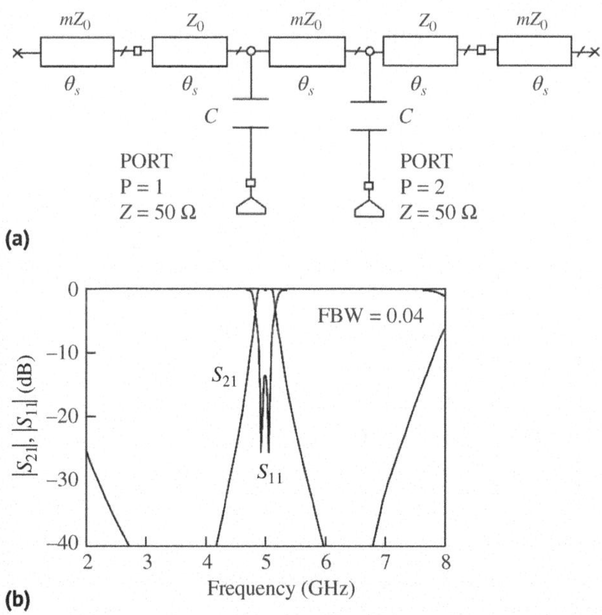

Figure 17.35 Meander dual-mode BPF with five sections and both open-circuited ends. (a) Schematic: $Z_0 = 10$ Ω, $m = 4.6$, $\theta_s = 90°$ at $f_0 = 5$ GHz, $C = 0.596$ pF. (b) Frequency responses.

In the synthesized resonators, the pole ω_{p3} of the input admittance coincides with the central frequency of the passband f_0 and is characterized by the electrical length $\theta_{p3} = 5\pi/2$. Therefore, the electrical lengths of each of the five sections is $\theta_s = \theta_{p3}/5 = 90°$ at the frequency of 5 GHz.

Figure 17.35b shows the frequency responses of the filter, simulated by a linear simulator. The value of the inverter capacitance, $C = 0.596$ pF, is determined by simulation, providing the required degree of matching characterized by a return loss of $L_R = 13.467$ dB.

The insertion loss function of the dual-mode BPF under consideration is symmetrical. The filter frequency selectivity is characterized by the index 20 dB ($f_0 \pm 470$ MHz) or 20 dB ($\pm j4.7$).

The design process of synthesized meander dual-mode BPFs is quite simple due to the established relationships (17.72) and (17.74).

17.5 Reconfigurable Loop-type Dual-mode Resonators and Filters

There are a significant number of different tunable dual-mode resonators and filters [26–30]. Below, we consider only a few of them, which have a loop-type structure. The basis for analyzing and designing such resonators and filters is the resonance equations.

17.5.1 Dual-mode Tunable Loop-type Resonator with Stub

Consider the tunable resonator [31] shown in Figure 17.36a, which features two variable capacitances C_1 and C_2. Capacitance C_1 connects the two ends of the TL segment, making this resonator loop-type. Capacitance C_2 is connected to the open end of the stub. Figure 17.36b depicts the straightened shape of this resonator, which represents a two-port network. The *ABCD* matrix of this two-port network can be expressed as the product of four matrices:

$$\begin{bmatrix} A & B \\ C & D \end{bmatrix} = \begin{bmatrix} a & b \\ c & d \end{bmatrix} \begin{bmatrix} 1 & 0 \\ Y_s & 1 \end{bmatrix} \begin{bmatrix} d & b \\ c & a \end{bmatrix} \begin{bmatrix} 1 & \dfrac{1}{Y_{C1}} \\ 0 & 1 \end{bmatrix}, \tag{17.75}$$

where *a*, *b*, *c*, *d* parameters are the elements of the transmission matrix of the half TL segment (17.4), Y_s is the input admittance of the stub at the point of connection to the TL segment, Y_{C1} is the admittance of capacitance C_1.

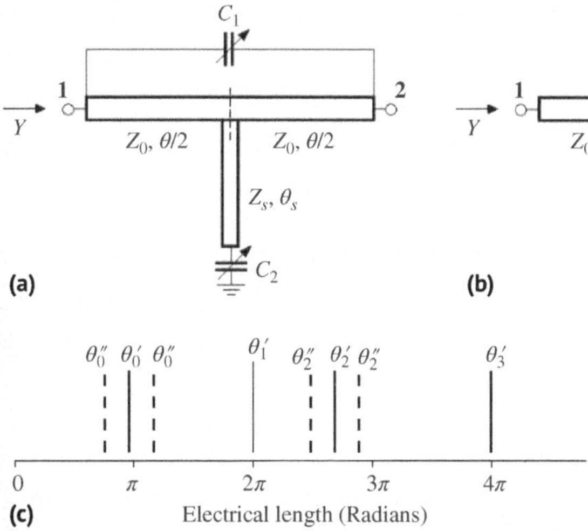

Figure 17.36 Tunable loop-type dual-mode resonator. (a) Schematic. (b) Straightened shape of the resonator. (c) Distribution of resonance electrical lengths.

Multiplying the matrices on the right-hand side of (17.75) gives the elements A and D:

$$A = bdY_s + ad + bc;$$
$$D = d^2\frac{Y_s}{Y_{C1}} + 2dc\frac{1}{Y_{C1}} + bdY_s + ad + bc.$$

Substituting these elements into (17.43), we obtain the resonance equation of the loop-type resonator (Figure 17.36a):

$$2dc\frac{1}{Y_{C1}} + 2bdY_s + d^2\frac{Y_s}{Y_{C1}} + 4bc = 0. \tag{17.76}$$

To obtain equation (17.76), we take into account the reciprocal condition $ad - bc = 1$. Two types of oscillations characterize the considered resonator. The resonant frequencies of the first-type oscillations are denoted by f_n', and resonance frequencies of the second-type are denoted by f_k''. The resonance frequencies of the first-type oscillations f_n' are independent of the stub. In this case, we can set in (17.76) $Y_s = 0$, which will bring this equation to the form

$$c\left(2b + \frac{d}{Y_{C1}}\right) = 0, \quad \text{for } f_n'. \tag{17.77a}$$

The resonance frequencies of the second-type oscillations f''_k are caused by the stub and are independent of the capacitance C_1. In this case, we can set in (17.76) $Y_{C1} = 0$, which leads to the equation for these resonance frequencies

$$Y_s = \frac{-2c}{d}, \quad \text{for } f''_k. \tag{17.77b}$$

Without loss of generality, we assume that the electric length of stub $\theta_s = \theta/2$. In this case, the transfer matrix $[a_s b_s c_s d_s]$ of the stub is

$$\begin{bmatrix} a_s & b_s \\ c_s & d_s \end{bmatrix} = \begin{bmatrix} a & mb \\ c/m & d \end{bmatrix}, \tag{17.78}$$

where a, b, c, and d are the elements of matrix (17.4), and m is the ratio of characteristic impedances, $m = Z_s/Z_0$.

The input admittance of the stub Y_s is obtained as a result of the transformation of the conductivity $Y_{C2} = j\omega C_2$ of the capacitance C_2 by the stub with the transfer matrix (17.78)

$$Y_s = \frac{aY_{C2} + c/m}{mbY_{C2} + d}. \tag{17.79}$$

The substitution of (17.79) into the left-hand side of the resonance equation (17.77b) leads to the form

$$\frac{aY_{C2} + c/m}{mbY_{C2} + d} = -\frac{2c}{d}, \quad \text{for } f''_k. \tag{17.80}$$

The resonance equations (17.77a) and (17.80) can be written in a slightly different form:

$$Y_{C1} = \frac{-d}{2b}, \quad \text{for } f'_n, \quad n \text{ is even}, \tag{17.81a}$$

$$c = 0, \quad \text{for } f'_n, \quad n \text{ is odd}, \tag{17.81b}$$

$$Y_{C2} = -\frac{2 + 1/m}{a/c + 2mb/d}, \quad \text{for } f''_k. \tag{17.81c}$$

Substituting in (17.81), $Y_{C1} = j\omega C_1$, $Y_{C2} = j\omega C_2$, and elements a, b, c, d of the transfer matrix (17.4), we obtain the final resonance equations for the resonance frequencies f'_n, f''_k and the corresponding resonance electrical lengths:

$$\omega'_n C_1 = \frac{1}{2} Z_0^{-1} \cot\left(\frac{\theta'_n}{2}\right), \quad \text{for } f'_n, \quad n \text{ is even}, \tag{17.82a}$$

$$\sin\left(\frac{\theta'_n}{2}\right) = 0, \quad \theta'_n = (n+1)\pi \quad \text{for } f'_n, \quad n \text{ is odd}, \tag{17.82b}$$

$$\omega_k'' C_2 = \frac{Z_0^{-1}(2 + 1/m)}{2m \tan(\theta_k''/2) - \cot(\theta_k''/2)}, \quad \text{for } f_k''. \tag{17.82c}$$

The distribution of the resonant electrical lengths of the resonator is shown in Figure 17.36c. They correspond to the resonant frequencies f_n' and f_k''. The resonant frequencies f_n' (even n) depend on the capacitance C_1 (17.82a). They are even modes in dual-mode oscillations. The resonant frequencies f_k'' depend on the capacitance C_2 of the stub (17.82c). They are odd modes in dual-mode oscillations. Depending on the ratio between capacitances C_1 and C_2, the odd modes f_k'' can be either larger or smaller than the even modes. The resonant frequencies f_n' (odd n) represent single oscillations, and they are independent of C_1 and C_2 (17.82b). The considered loop-type resonator with a stub exhibits dual-mode and simple oscillations that alternate with each other.

The tunability of even and odd modes in dual-mode oscillations is different. Let us illustrate these differences using the resonance frequency sensitivity S_C, as shown in (15.2) and (15.3).

The even modes f_n' of the resonator are determined by expression (17.82a), the right side of which, taken with the opposite sign, determines the susceptance $B(\omega)$:

$$B(\omega) = -Z_0^{-1} \cot\left(\frac{\theta}{2}\right). \tag{17.83}$$

Substituting (17.83) into (15.3), we determine the sensitivity of f_n' (even n) to the capacitance C_1 change

$$S_{C1} = \left(1 + \frac{\theta}{\sin\theta}\right)^{-1}. \tag{17.84}$$

The sensitivity plot (17.84) for the lowest even mode f_e' is shown in Figure 17.37 by a solid line. As the electrical length θ decreases (C_1 increases), the sensitivity S_{C1} monotonously increases. For $\theta \to 0$, we have the value $S_{C1} \to 1/2$.

The odd modes f_k'' are defined by the expression (17.82c). In this case, the susceptance

$$B(\omega) = \frac{Z_0^{-1}(2 + 1/m)}{\cot(\theta/2) - 2m\tan(\theta/2)}. \tag{17.85}$$

Substitution (17.85) in (15.3) determines the sensitivity of the odd modes f_k'' to the capacitance C_2 change

$$S_{C2} = \left[1 - \frac{\theta}{2}\,\frac{\sin^{-2}(\theta/2) + 2m\cos^{-2}(\theta/2)}{\cot(\theta/2) - 2m\tan(\theta/2)}\right]. \tag{17.86}$$

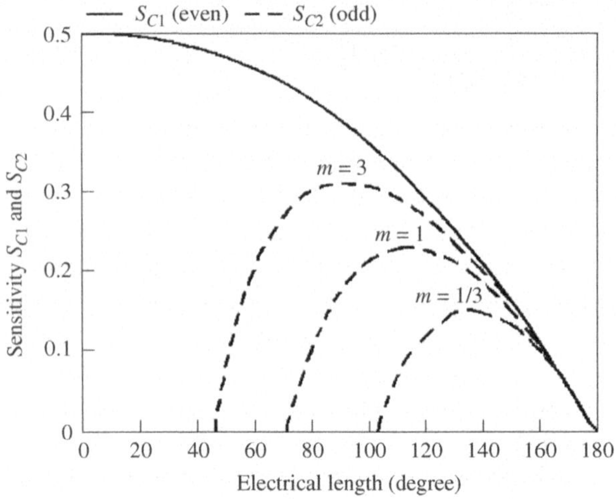

Figure 17.37 Functions $S_C(\theta)$ of loop-type dual-mode resonator in Figure 17.36a.

The sensitivity plots S_{C2} (17.86) for the lowest odd mode f_o oscillations are shown in Figure 17.37 by dotted lines for different values of $m = Z_s/Z_0$. These dependencies are bell-shaped. For more effective tuning of the odd mode, this maximum should be situated in the middle of the tuning range. An increase in m leads to an increase in sensitivity. This is explained by the expansion of the resonance region, which is bounded from below by the pole of susceptance $B(\omega)$ (17.85).

Equating the denominator (17.85) to zero, we determine the electrical length $\theta_p = 2\tan^{-1}\left(1/\sqrt{2m}\right)$ of this pole. The last equality yields the values θ_p: 101.5° for $m = 1/3$; 70.53° for $m = 1$, and 44.42° for $m = 3$. These values are shown in Figure 17.37. The odd modes of dual-mode oscillation cannot exist to the left of these values. The sensitivity analysis performed (Figure 17.37) indicates that it is advisable to use a stub with an increased Z_s value, as this leads to an increase in both sensitivity and tunability of the resonator.

To achieve a synchronous change in the even and odd modes of dual-mode oscillations, it is necessary to make a specific ratio between capacities C_1 and C_2

$$\frac{C_2}{C_1} = \frac{2 + 1/m}{m - 0.5\cot^2(\theta/2)}, \tag{17.87}$$

which follows from (17.82a) and (17.82c).

Expression (17.87) shows that $C_2 > C_1$ and change in the capacitance C_2 of the stub always is greater than the change in the capacitance C_1: $C_{2\max}/C_{2\min} > C_{1\max}/C_{1\min}$ for tuning dual-mode oscillations in a given frequency band. These differences decrease with increasing m.

A similar dual-mode loop resonator was considered in Ref. [28]. Its resonance frequencies were determined using two resonance equations derived from even-odd mode analysis. A third part of all resonant frequencies was skipped. It was proved earlier that resonator (Figure 17.36a) has not only dual-mode oscillations but also single oscillations, and these oscillations alternate with each other. The resonance frequencies of these resonators are described with three resonance equations (17.82a), (17.82b), (17.82c) – two equations for even and odd modes of dual-mode oscillations and one equation for single oscillations.

As a result of the analysis carried out based on the resonance equation for a loop-type resonator (17.43), it was possible to establish new properties of the tunable loop dual-mode resonator with stub (Figure 17.36a). The frequencies of simple oscillations are fixed; they do not depend on the capacitances C_1 and C_2. The resonance region of the odd f_o mode is narrower than even f_e mode, which reduces the tunability of these filters. An increase in the characteristic impedance of the stub $Z_s = mZ_0$ increases the tunability of dual-mode oscillations by expanding the resonant region of the odd mode.

17.5.2 Microstrip Varactor-tuned Dual-mode Filter with Stub

Let us examine the experimental microstrip filter, whose topology is illustrated in Figure 17.38a. In the filter, a substrate made of the RT/Duroid 6010LM (Rogers) with $\varepsilon_r = 10.2$, $h = 1.905$ mm, and $\tan\delta = 0.002$ was used. The width of microstrip line of hairpin resonator $w_1 = 4.5$ mm ($Z_0 = 29.3\ \Omega$), the width of stub $w_2 = 1$ mm ($Z_s = mZ_0 = 63.74\ \Omega$, $m = 2.18$). The area occupied by the filter is 53 mm × 31 mm.

The tuning capacitors C_1 and C_2 are implemented using a back-to-back diode configuration, which makes it easier to control biasing in the common node. The resistor $R = 10$ kΩ is used to supply voltage to the varactors. The filter used a varactor NXP BB135 with capacitance $C \approx 2$–20 pF, series resistance $R_s \approx 0.75\ \Omega$, and voltage $U = 0.5$–28 V.

The photograph of the filter is shown in Figure 17.38b. The filter utilizes two units (Figure 15.28) that are mounted vertically, combining two pairs of varactors in a back-to-back configuration.

We omit the details of the filter design and dwell only on its most essential properties.

Figure 17.38c shows the measured and simulated frequency responses in the tuning range, and Figure 17.38d shows the measured insertion loss in a wide frequency band. These frequency responses correspond to the case when the frequency of the odd mode is less than the frequency of the even mode $f_o < f_e$, which generates the left-handed transmission zero. If the condition $f_o > f_e$ is satisfied, then the frequency responses of the filter will have a right-handed transmission zero.

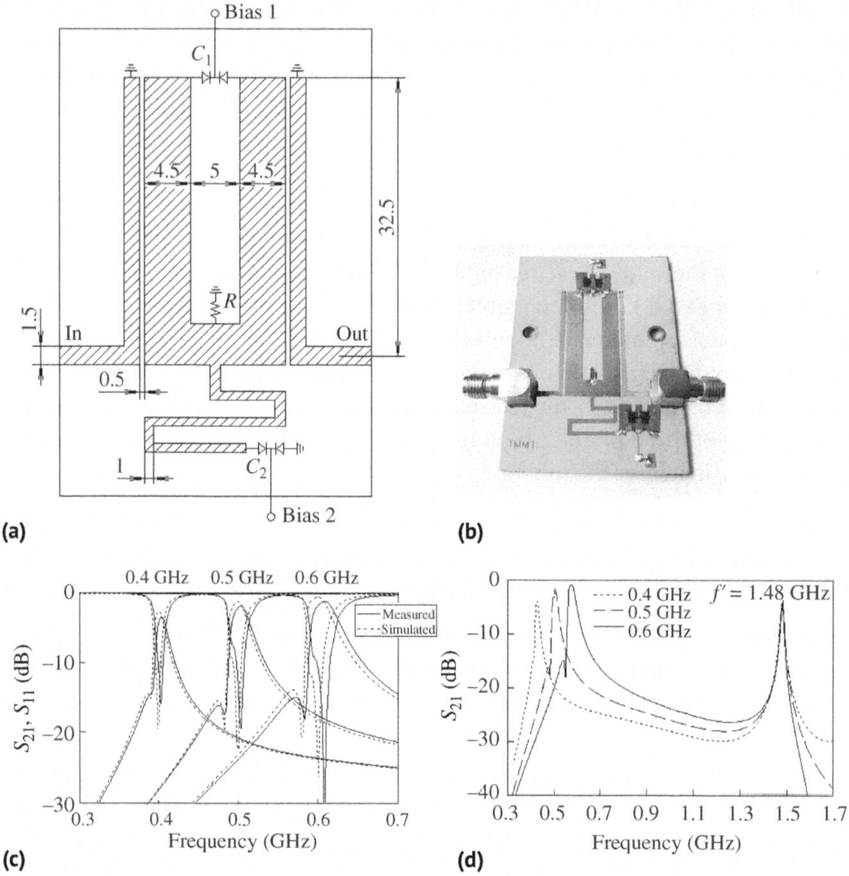

Figure 17.38 Microstrip varactor-tuned dual-mode filter with stub. (a) Topology. All dimensions are in millimeters. (b) Photograph of fabricated filter. *Source*: [31] / with permission of Springer Nature. (c) Measured and simulated frequency responses in tuning range. (d) Measured insertion loss in wide frequency band. Three frequencies are used: $f_0 = 0.4$ GHz: $C_1 = 8$ pF ($U_1 = 5.2$ V), $C_2 = 13.5$ pF ($U_2 = 2.1$ V); $f_0 = 0.5$ GHz: $C_1 = 4.4$ pF ($U_1 = 10.5$ V), $C_2 = 6.5$ pF ($U_2 = 7.1$ V); $f_0 = 0.6$ GHz: $C_1 = 2.36$ pF ($U_1 = 22.3$ V), $C_2 = 3.4$ pF ($U_2 = 12.5$ V).

The tuning range of the filter 0.4–0.6 GHz (40%) is provided when capacitor C_1 changes from 2.36 pF to 8 pF ($C_{1\max}/C_{1\min} \approx 3.4$) and capacitor C_2 changes from 3.4 pF to 13.5 pF ($C_{2\max}/C_{2\min} \approx 4$). As predicted by expression (17.87): $(C_{2\max}/C_{2\min}) > (C_{1\max}/C_{1\min})$. The filter has the following measured data: fractional bandwidth FBW by 3 dB level was changed within FBW = 0.035–0.05, insertion loss at the center frequency $L_{A0} \leq 3.6$ dB; return loss $L_R \geq 12$ dB.

Along with dual-mode oscillations, the experimental filter exhibits single oscillations at a frequency of $f_1' = 1.48$ GHz (Figure 17.38d). The resonance frequency of a single oscillation $f_1' = 1.48$ GHz is independent of C_1 and C_2, and fixed.

17.5.3 Reconfigurable Dual-mode Stepped-impedance Hairpin Resonator

Figure 17.39a shows a schematic of the stepped-impedance hairpin resonator with capacitor C_1 [32], and Figure 17.39b gives a straightened shape of the resonator. The capacitor C_1 connects both ends of the TL segment, resulting in a loop structure for the resonator. The segment of the symmetric stepped-impedance TL has the characteristic impedance Z_0 in its middle part. At both ends, it has areas with a higher characteristic impedance mZ_0 ($m \geq 1$). Its electrical length is $\theta = 2\,(\theta' + \theta'')$ and electrical lengths θ' and θ'' are shown in Figure 17.39b.

The transfer matrix of the left half of the stepped-impedance segment is denoted by $[a'b'c'd']$; then the transfer matrix $[ABCD]$ of the two-port circuit in Figure 17.39b can be represented in the form [4]

$$\begin{bmatrix} A & B \\ C & D \end{bmatrix} = \begin{bmatrix} (1 + 2b'c') & \dfrac{1 + 2b'c'}{Y_C} + 2a'b' \\ (2c'd') & \dfrac{2c'd'}{Y_C} + (1 + 2b'c') \end{bmatrix} \tag{17.88}$$

where $Y_C = j\omega C_1$ is the admittance of the capacitor. The two-port network in Figure 17.39b is converted by means of a loop connection of the output to the input into the one-port network (Figure 17.39a).

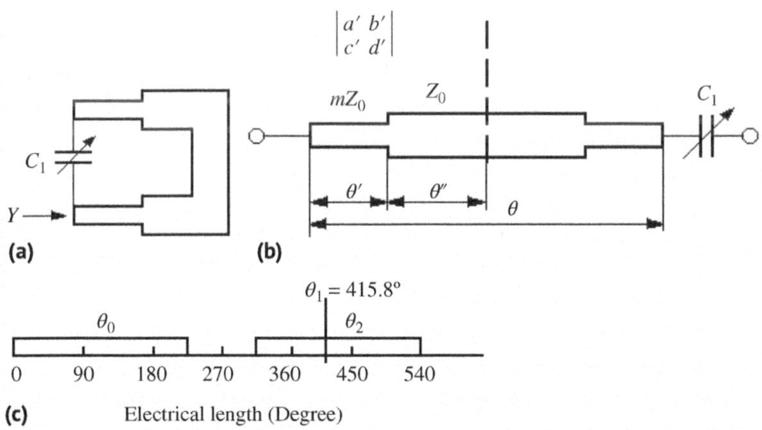

Figure 17.39 Reconfigurable dual-mode stepped-impedance hairpin resonator. (a) Schematic. (b) Straightened shape of the resonator. (c) Resonance regions at $m = 3$ and $\theta''/\theta' = 2$.

The substitution of the elements of matrix (17.88) into (15.52) gives the input admittance of the loop-type resonator

$$Y = \frac{2c'\,(2b' + d'/Y_C)}{(1 + 2b'c')\,/Y_C + 2a'b'}. \tag{17.89}$$

Equating the numerator of the input admittance (17.89) to zero, we obtain two resonance equations:

$$Y_C = -\frac{d'}{2b'}, \quad n \text{ is even,} \tag{17.90a}$$

$$c' = 0, \quad n \text{ is odd.} \tag{17.90b}$$

Equation (17.90b) shows that the odd resonance frequencies do not depend on the capacitance C_1 and are fixed.

The transfer matrix of the left half of the stepped-impedance segment in Figure 17.39b is

$$\begin{vmatrix} a' & b' \\ c' & d' \end{vmatrix} = \left[\left(1 + \Omega_1^2\right)\left(1 + \Omega_2^2\right)\right]^{-1/2} \begin{vmatrix} 1 - m\Omega_1\Omega_2 & jZ_0\,(m\Omega_1 + \Omega_2) \\ jZ_0^{-1}\,(\Omega_1/m + \Omega_2) & 1 - \Omega_1\Omega_2/m \end{vmatrix}, \tag{17.91}$$

where $\Omega_1 = \tan\theta'$, $\Omega_2 = \tan\theta''$.

Substituting the elements of the matrix (17.91) into expressions (17.90), we obtain the resonance equations for the loop-type resonator under consideration:

$$\omega C_1 = \frac{1}{2Z_0}\frac{1 - \Omega_1\Omega_2/m}{m\Omega_1 + \Omega_2}, \quad n \text{ is even,} \tag{17.92a}$$

$$\Omega_1/m + \Omega_2 = 0, \quad n \text{ is odd.} \tag{17.92b}$$

Assuming $m = 3$ and $\theta''/\theta' = 2$ in the resonance equation (17.92), we determine the boundaries of the variation of the electrical lengths for the first three resonant frequencies, when $C_1 \in [0, \infty)$: $0 \le \theta_0 \le 226.6°$, $\theta_1 = 415.8° = \text{const}$, $313.4 \le \theta_2 \le 540°$. They are shown in Figure 17.39c. The resonance region θ_1 is represented by a single value that is located in the middle part of the resonance region θ_2. We have established that in the resonator under consideration (Figure 17.39a), the effect of intersecting the resonance regions takes place.

At some value of $C_1 = C_{10}$, the even mode θ_2, defined by (17.90a), is close to the odd mode θ_1, which is stationary and defined by (17.90b). As a result of the convergence, a dual-mode oscillation is formed, which is accompanied by a transmission zero located on the side of the odd mode. If $C_1 < C_{10}$, then $f_o < f_e$, and the TZ is left-handed. If $C_1 > C_{10}$, then $f_o > f_e$, and the TZ is right-handed.

17.5.4 Microstrip Reconfigurable Dual-mode Filter with Stepped-impedance Hairpin Resonator

The reconfigurable microstrip dual-mode filter with a single capacitor is presented in Figure 17.40a. Figure 17.40b demonstrates a photograph of the fabricated filter. The filter utilizes a substrate of RT/Duroid 6010LM (Rogers) with the following parameters: $\varepsilon_r = 10.2$, $h = 1.905$ mm, and $\tan\delta = 0.002$. The width of the microstrip line in the central part of the resonator is $w = 4.5$ mm, corresponding to a characteristic impedance $Z_0 = 29.5\ \Omega$. The width of the microstrip line at the ends of the resonator is $w_1 = 1$ mm and its characteristic impedance $mZ_0 = 64\ \Omega$, $m = 2.17$, $\theta''/\theta' = 1.82$.

Some series- and parallel-connected varactors (Figure 15.28) are used to implement capacitance C_1. A resistor with a high value, $R = 10$ kΩ, connected to the middle of the loop hairpin resonator, supplies voltage to the varactors.

The simulated frequency responses of this resonator for $C_1 = 1.05$ pF and $C_1 = 12$ pF are shown in Figure 17.40c, d by dashed lines. Under the adopted parameters, this resonator is a dual-mode filter with an attenuation pole f_z adjoining the odd oscillation mode f_o ($f_o = f_1$). The even mode of oscillation f_e corresponds to the resonance frequency f_2. The resonant frequencies f_1 and f_2 are determined with and without resistor $R = 10$ kΩ. In both cases, the values of resonant frequencies coincided. At $C_1 = 1.05$ pF, we have the resonance frequencies $f_1 = 1.723$ GHz and $f_2 = 1.818$ GHz, and attenuation pole f_z is located to the left of the passband. At $C_1 = 12$ pF, the resonance frequencies are equal to $f_1 = 1.723$ GHz, $f_2 = 1.608$ GHz, and the attenuation pole f_z is located to the right of the passband.

The filter utilizes varactors NXP BB135 with a minimal capacitance $C_{min} \approx 2$ pF, $C_{max}/C_{min} \approx 10$, and a control voltage range of 0.5–28 V. These varactors are combined in a series and parallel connection to form a special unit. To apply the control voltage to the unit of the hairpin resonators, a 10 kΩ SMD resistor is connected to the middle of this resonator.

The measured frequency responses are shown in Figure 17.40c, d by solid lines. To implement the capacitance $C_1 = 1.05$ pF, one pair of varactors and a voltage of 27 V were used. The measured parameters are as follows: center frequency $f_0 = 1.77$ GHz, insertion loss $L_{A0} = 1.4$ dB, BW = 120 MHz, return loss $L_R > 15$ dB, attenuation pole $f_z = 1.67$ GHz. To realize the capacitance $C_1 = 12$ pF, eight pairs of varactors and a voltage of 16 V were used. The measured parameters are as follows: $f_0 = 1.66$ GHz, $L_{A0} = 1.5$ dB, BW = 165 MHz, $L_R > 11$ dB, $f_z = 1.754$ GHz. The performed measurements confirmed the existence of dual-mode oscillations in the loop hairpin stepped-impedance resonator with one capacitor.

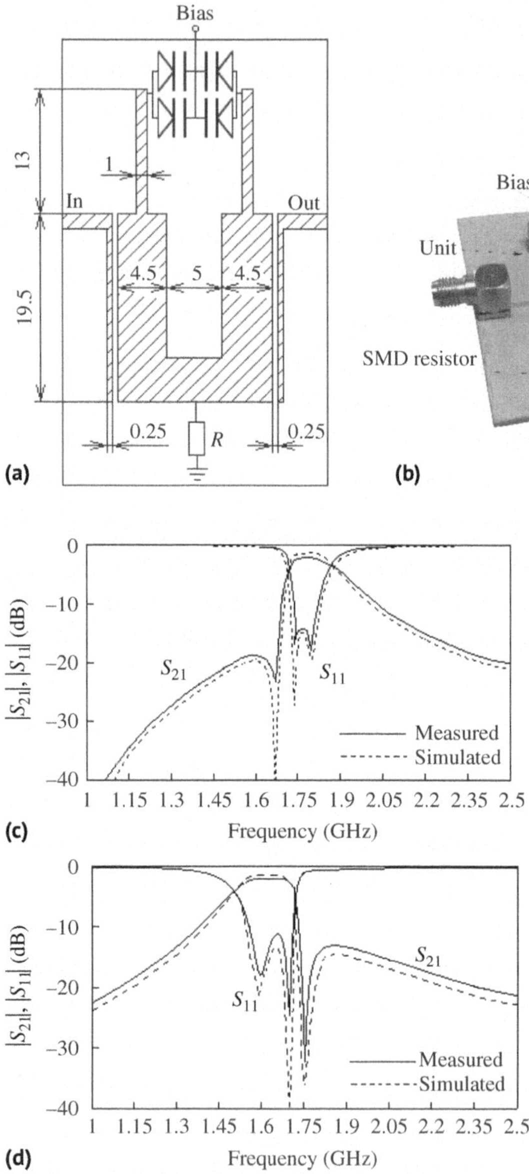

Figure 17.40 Microstrip reconfigurable dual-mode filter with stepped-impedance hairpin resonator. (a) Topology. All dimensions are in millimeters. (b) Photograph of fabricated filter. *Source*: [32] / with permission of IEEE. (c) Measured and simulated frequency responses at $C = 1.05$ pF, $f_1 = 1.723$ GHz, $f_2 = 1.818$ GHz. (d) Measured and simulated frequency responses at $C = 12$ pF, $f_1 = 1.723$ GHz, $f_2 = 1.608$ GHz.

References

1 I. Wolff, Microstrip bandpass filter using degenerate modes of a microstrip ring resonator, *Electron. Lett.*, vol. 8, no. 12, p. 302, 1972, https://doi.org/10.1049/el:19720223.

2 M. Makimoto and M. Sagawa, Varactor tuned bandpass filters using microstrip-line ring resonators, in *MTT-S International Microwave Symposium Digest*, 1986, vol. 86, pp. 411–414, https://doi.org/10.1109/MWSYM.1986.1132206.

3 H. Yabuki, M. Sagawa, M. Matsuo, and M. Makimoto, Stripline dual-mode ring resonators and their application to microwave devices, *IEEE Trans. Microw. Theory Tech.*, vol. 44, no. 5, pp. 723–729, 1996, https://doi.org/10.1109/22.493926.

4 G. L. Matthaei, L. Young, and E. M. T. Jones, *Microwave Filters, Impedance-Matching Networks, and Coupling Structures*. New York: Artech House Books, 1980.

5 E. A. Guillemin, *Synthesis of Passive Networks: Theory and Methods Appropriate to the Realization and Approximation Problems*. New York: Wiley, 1957.

6 M.-S. Chung, I.-S. Kim, and S.-W. Yun, Varactor-tuned hairpin bandpass filter with an attenuation pole, in *2005 Asia-Pacific Microwave Conference Proceedings*, 2005, vol. 4, pp. 1–4, https://doi.org/10.1109/APMC.2005.1606748.

7 X. Y. Zhang, J.-X. Chen, Q. Xue, and S.-M. Li, Dual-band bandpass filters using stub-loaded resonators, *IEEE Microw. Wirel. Components Lett.*, vol. 17, no. 8, pp. 583–585, 2007, https://doi.org/10.1109/LMWC.2007.901768.

8 W. Feng and W. Che, Novel wideband differential bandpass filters based on T-shaped structure, *IEEE Trans. Microw. Theory Tech.*, vol. 60, no. 6, pp. 1560–1568, 2012, https://doi.org/10.1109/TMTT.2012.2188538.

9 H.-J. Tsai, B.-C. Huang, N.-W. Chen, and S.-K. Jeng, A reconfigurable bandpass filter based on a varactor-perturbed, T-shaped dual-mode resonator, *IEEE Microw. Wirel. Components Lett.*, vol. 24, no. 5, pp. 297–299, 2014, https://doi.org/10.1109/LMWC.2014.2306893.

10 S. Litvintsev and A. V. Zakharov, Analysis of dual-mode resonators from transmission line segments, *Radioelectron. Commun. Syst.*, vol. 65, no. 4, pp. 186–199, 2022, https://doi.org/10.3103/S0735272722040033.

11 J.-S. Hong, *Microstrip Filters for RF/Microwave Applications*, 2nd ed. New Jersey: Wiley, 2011, https://doi.org/10.1002/9780470937297.

12 C.-H. Wang, Y.-S. Lin, and C. H. Chen, Novel inductance-incorporated microstrip coupled-line bandpass filters with two attenuation poles, in *2004 IEEE MTT-S International Microwave Symposium Digest (IEEE Cat. No.04CH37535)*, 2004, pp. 1979–1982, https://doi.org/10.1109/MWSYM.2004.1338999.

13 Y.-S. Lin, C.-H. Wang, C.-H. Wu, and C. H. Chen, Novel compact parallel-coupled microstrip bandpass filters with lumped-element K-inverters, *IEEE Trans. Microw. Theory Tech.*, vol. 53, no. 7, pp. 2324–2328, 2005, https://doi.org/10.1109/TMTT.2005.850445.

14 C.-H. Wu, Y.-S. Lin, C.-H. Wang, and C. H. Chen, Novel microstrip coupled-line bandpass filters with shortened coupled sections for stopband extension, *IEEE Trans. Microw. Theory Tech.*, vol. 54, no. 2, pp. 540–546, 2006, https://doi.org/10.1109/TMTT.2005.862710.

15 A. Zakharov and S. Litvintsev, Expanding functionality of dual-mode resonators and filters using nonuniform transmission line structural elements, *IEEE Trans. Circuits Syst. I Regul. Pap.*, vol. 69, no. 8, pp. 3124–3135, 2022, https://doi.org/10.1109/TCSI.2022.3169472.

16 A. Zakharov, Parametric and structural-parametric synthesis of nonuniform transmission line resonators, *IEEE Trans. Circuits Syst. I Regul. Pap.*, vol. 68, no. 3, pp. 1055–1067, 2021, https://doi.org/10.1109/TCSI.2020.3044925.

17 O. N. Litvinenko and V. I. Soshnikov, *Theory of Nonuniform Lines and Their Application in Radio Engineering* [in Russian]. Moscow: Sov. Radio, 1964.

18 I. S. Krishna and S. Mukherjee, Triple-mode substrate integrated coaxial resonator based bandpass filter featuring flexible transmission zeros and adjustable bandwidth, *IEEE Trans. Circuits Syst. II Express Briefs*, vol. 68, no. 4, pp. 1223–1227, 2021, https://doi.org/10.1109/TCSII.2020.3035717.

19 Z.-C. Guo, L. Zhu, and S.-W. Wong, A quantitative approach for direct synthesis of bandpass filters composed of transversal resonators, *IEEE Trans. Circuits Syst. II Express Briefs*, vol. 66, no. 4, pp. 577–581, 2019, https://doi.org/10.1109/TCSII.2018.2866101.

20 R.-J. Mao, X.-H. Tang, and F. Xiao, Novel compact quarter-wavelength resonator filter using lumped coupling elements, *IEEE Microw. Wirel. Components Lett.*, vol. 17, no. 2, pp. 112–114, 2007, https://doi.org/10.1109/LMWC.2006.890332.

21 L. Athukorala and D. Budimir, Compact dual-mode open loop microstrip resonators and filters, *IEEE Microw. Wirel. Components Lett.*, vol. 19, no. 11, pp. 698–700, 2009, https://doi.org/10.1109/LMWC.2009.2032003.

22 S. Zhang, L. Zhu, and R. Weerasekera, Synthesis of inline mixed coupled quasi-elliptic bandpass filters based on $\lambda/4$ resonators, *IEEE Trans. Microw. Theory Tech.*, vol. 63, no. 10, pp. 3487–3493, 2015, https://doi.org/10.1109/TMTT.2015.2467380.

23 A. V. Zakharov, M. Y. Ilchenko, and L. S. Pinchuk, Microwave resonators with dual-mode oscillations, *Radioelectron. Commun.*, vol. 55, no. 11, pp. 496–505, 2012, https://doi.org/10.3103/S0735272712110040.

24 K. N. Mohan and Y. K. Choukiker, Higher order bandpass single and dual band frequency selective surfaces with aperture coupled patch resonators, *IEEE Access*, vol. 12, pp. 25254–25264, 2024, https://doi.org/10.1109/ACCESS.2024.3366452.

25 C. Fan, S. Ma, K. Gong, J. Sun, H. Qian, Q. Liu, and Y. Liu, Highly selective microstrip bandpass filters using dual-mode patch resonator loaded with short via, *AEU - Int. J. Electron. Commun.*, vol. 185, p. 155457, 2024, https://doi.org/10.1016/j.aeue.2024.155457.

26 W. Tang and J.-S. Hong, Varactor-tuned dual-mode bandpass filters, *IEEE Trans. Microw. Theory Tech.*, vol. 58, no. 8, pp. 2213–2219, 2010, https://doi.org/10.1109/TMTT.2010.2052958.

27 Y. Wan, J. Zhou, Y. Rao, J. Xie, Q. Li, and X. Luo, Independently tunable compact dual-band bandpass filter with high selectivity and wide stopband using multilayer folded dual-mode SIDGS resonator, in *2023 IEEE/MTT-S International Microwave Symposium - IMS 2023*, 2023, pp. 827–830, https://doi.org/10.1109/IMS37964.2023.10187986.

28 D. Lu, X. Tang, N. S. Barker, M. Li, and T. Yan, Synthesis-applied highly selective tunable dual-mode BPF with element-variable coupling matrix, *IEEE Trans. Microw. Theory Tech.*, vol. 66, no. 4, pp. 1804–1816, 2018, https://doi.org/10.1109/TMTT.2017.2783376.

29 G. Chaudhary and Y. Jeong, Arbitrary termination complex impedances tunable bandpass filter using single and dual-mode resonators, *IEEE Access*, vol. 11, pp. 63013–63026, 2023, https://doi.org/10.1109/ACCESS.2023.3287551.

30 G. Chaudhary and Y. Jeong, Quasi-elliptic higher-order tunable bandpass filter with constant absolute bandwidth using synchronously tuned dual-mode resonators, *IEEE Access*, vol. 11, pp. 95051–95059, 2023, https://doi.org/10.1109/ACCESS.2023.3310934.

31 S. Litvintsev, A. Zakharov, S. Rozenko, and M. Kozachuk, Dual-mode tunable loop resonator with stub, *Radioelectron. Commun. Syst.*, vol. 65, no. 9, pp. 475–484, 2022, https://doi.org/10.3103/S0735272722110048.

32 A. Zakharov, S. Litvintsev, and M. Ilchenko, Transmission line tunable resonators with intersecting resonance regions, *IEEE Trans. Circuits Syst. II Express Briefs*, vol. 67, no. 4, pp. 660–664, 2020, https://doi.org/10.1109/TCSII.2019.2922429.

Index

d

e

f

g

u

v

Wiley The manufacturer's authorized representative according to the EU General Product Safety Regulation is Wiley-VCH GmbH, Boschstr. 12, 69469 Weinheim, Germany, e-mail: Product_Safety@wiley.com.

Printed and bound by CPI Group (UK) Ltd, Croydon, CR0 4YY

30/07/2026

02181221-0001